Springer Series in Computational Mathematics

Volume 58

This is basically a numerical analysis series in which high-level monographs are published.

We develop this series aiming at having more publications in it which are closer to applications. There are several volumes in the series which are linked to some mathematical software.

Josef Dick · Peter Kritzer ·
Friedrich Pillichshammer

Lattice Rules

Numerical Integration, Approximation, and Discrepancy

Springer

Josef Dick
School of Mathematics and Statistics
The University of New South Wales
Sydney, NSW, Australia

Peter Kritzer
RICAM
Austrian Academy of Sciences
Linz, Oberösterreich, Austria

Friedrich Pillichshammer
Johannes Kepler University Linz
Linz, Oberösterreich, Austria

ISSN 0179-3632 ISSN 2198-3712 (electronic)
Springer Series in Computational Mathematics
ISBN 978-3-031-09950-2 ISBN 978-3-031-09951-9 (eBook)
https://doi.org/10.1007/978-3-031-09951-9

Mathematics Subject Classification: 11K36, 11K38, 11K45, 65C05, 65D30, 65D15, 65Y20

This Springer imprint is published by the registered company Springer Nature Switzerland AG
The registered company address is: Gewerbestrasse 11, 6330 Cham, Switzerland

Für

Rafael und Anna
Simon
Hanna, Lea und Johannes

Preface

Lattice rules are particular instances of quasi-Monte Carlo rules for numerical integration of functions over the d-dimensional unit cube $[0, 1]^d$, where the emphasis lies on high dimensions d in modern practical applications. A special case of the concept of lattice rules, the method of good lattice points, was introduced by N.M. Korobov in the year 1959 and a few years later independently by E. Hlawka. The Soviet school around Korobov quickly developed a quite satisfactory theory of the method of good lattice points, which is summarized in a first book on lattice rules published by Korobov himself in 1963. The first steps towards general lattice rules were taken by K.K. Frolov in 1977, but it was I.H. Sloan, from the 1980s onwards, who developed a systematic approach to the subject in cooperation with various co-authors. During these years also an Austrian group around H. Niederreiter, a former student of Hlawka, contributed important results in the context of the method of good lattice points. As a particular example of another researcher in this group, we would like to mention our mentor G. Larcher.

During the 1980s lattice rules became a booming field of research with many exciting results and applications. These developments are very well summarized in two seminal books by H. Niederreiter, entitled "Random Number Generation and Quasi-Monte Carlo Methods" (1992), and by I.H. Sloan and S. Joe, entitled "Lattice Methods for Multiple Integration" (1994), respectively. From the middle of the 1990s onwards, these two books significantly contributed to the popularity of lattice rules and initiated further research on the topic, which has persisted until today. The present book is, in particular, devoted to these more recent developments, comprising, for example, lattice rule integration in weighted reproducing kernel Hilbert spaces, the fast component-by-component construction of lattice rules, questions related to tractability, integration of not necessarily periodic functions, shifted and/or folded lattice rules, and lattice rules for function approximation.

The primal aim of this book is to provide an introduction to the topic with detailed explanations of the basic concepts. Most parts of the book should be accessible to students in mathematics or computer science with basic knowledge of number theory, algebra, and calculus. For readers with no previous knowledge of the topic we suggest starting with Chapter 1, which provides a first overview and a smooth introduction

to the general field, including information on the classical theory of lattice rules and the basics of modern quasi-Monte Carlo error analysis using elegant methods in the context of reproducing kernel Hilbert spaces. Next, it is suggested to continue with Chapters 2 and 3 dealing with lattice rule integration of smooth periodic functions and with the component-by-component construction of lattice rules, respectively. In the latter chapter we recommend studying first the error analysis for a prime number of elements and for product weights, and putting aside the more general cases (i.e., an arbitrary number of nodes, and weights other than of product form). Likewise, also the more involved and subtly modified construction schemes in Chapter 4 should be skipped on a first reading. Parts of Chapter 5, which is concerned with the discrepancy of lattice point sets, and an outlook on Chapter 7, dealing with lattice rule integration for smooth but not necessarily periodic functions, complete a first introduction to the topic. This program, possibly with some restrictions, could also serve as a rough plan for an introductory university course on quasi-Monte Carlo integration and lattice rules.

The book also contains more advanced material, and this corresponds to its second aim, which is to serve as a reference book for practitioners applying lattice rules to real-world problems and for researchers working in numerical analysis and scientific computing. Of particular interest for practitioners might be Chapter 4, in which several refined ways of constructing lattice rules in an efficient way are presented, Chapter 6 dealing with lattice rules which are extensible in the number of integration nodes, Chapter 12 which is concerned with stability of lattice rules, i.e., rules leading to good results simultaneously for different choices of parameters, and Chapter 16, which addresses the fast quasi-Monte Carlo matrix-vector multiplication. The latter is a very useful technique when the main computational cost of evaluating the integrand at a given point is the multiplication of the point with a matrix. It can successfully be applied to many practical problems as, e.g., partial differential equations with random coefficients, which is an important field where quasi-Monte Carlo methods are applied today; we shall give a short outline of this subject in Appendix A. Numerical experiments demonstrating the success of the construction methods presented in this book are provided in Appendix B.

Two very important topics, both from the theoretical as well as the practical point of view, are presented in Chapter 7, which deals with lattice rule integration of not necessarily periodic functions, and in Chapter 8 on lattice rule integration over more general domains than the unit cube. When applying lattice rule integration, finite smoothness of the integrands is usually assumed. In Chapter 9 the integration problem for functions that may even be analytic is discussed. In this case one can obtain exponential convergence of the worst-case error of suitable rules. Although not being lattice rules in the classical sense, rules of a very similar structure are presented in Chapter 10. The rules considered there are based on the so-called Korobov p-sets, and their errors may have a very favorable dependence on the dimension of the integration domain. In particular, they allow us to obtain polynomial tractability results for the integration problem in a certain unweighted subclass of Hölder-continuous Fourier series. In this book the error analysis for lattice rules is usually studied with respect to the worst-case setting. In Chapter 11, however, we

take a different perspective and study the error in the so-called randomized setting. Chapters 13–15 are devoted to the application of lattice rules to the approximation of functions in the L_2- and the L_∞-case, respectively. The emphasis of Chapter 15 is on multiple rank-1 lattice point sets, that are, in a nutshell, multi-set unions of usual rank-1 lattice point sets which allow us to obtain an improved convergence rate in the L_∞-approximation problem.

In our presentation we also always take care of the dependence of the involved errors on the dimension of the underlying integration or approximation problem. This point of view was often neglected in the classical theory of lattice rules but became an utmost important aspect in the modern theory and is intimately related to the applicability of lattice rules to real-world problems of very high dimensionality, where by "high" we mean being in the hundreds or even in the thousands.

We hope that this book will turn out to be useful for teaching, self-study, and as a reference, and that it will encourage many people to study lattice rules or to apply them to real-word problems.

Acknowledgements

Appendix B, containing many numerical experiments supporting the theoretical results in this book, was written by Adrian Ebert, who is a true expert on implementing numerical methods based on lattice rules. We thank Adrian for his valuable contribution, which is an important supplement to the theoretical results presented in the body of the book, and which impressively demonstrates the success of lattice rules.

Furthermore, we would like to thank Erich Novak for valuable comments.

We also would like to acknowledge the support of the University of New South Wales, the Austrian Academy of Sciences, and the Johannes Kepler University Linz.

Sydney and Linz,
March 2022

Josef Dick
Peter Kritzer
Friedrich Pillichshammer

Contents

List of Symbols

$\mathbb{N}$	set of natural numbers $\mathbb{N} := \{1, 2, 3, \ldots\}$
$\mathbb{Z}$	set of integers
$\mathbb{Z}_b$	set of b-adic numbers for $b \geq 2$
$\mathbb{R}$	set of real numbers
$\mathbb{C}$	set of complex numbers
$\mathtt{i}$	$\sqrt{-1}$
e	Euler's number
d	dimension
λ_d	d-dimensional Lebesgue measure
φ	Euler's totient function
ζ	Riemann zeta-function, $\zeta(\alpha) := \sum_{h=1}^{\infty} h^{-\alpha}$ for $\alpha > 1$
Γ	Gamma function, $\Gamma(x) := \int_0^{\infty} t^{x-1}\mathrm{e}^{-t}\,\mathrm{d}t$ for $x > 0$
$\mathbb{E}$	expected value
Var	variance
$\gcd(a, b)$	greatest common divisor of a and b
$\lVert \cdot \rVert_p$	ℓ_p-norm in $\mathbb{R}^d$ for $p \in [1, \infty]$
$\lVert \cdot \rVert_{L_p}$	L_p-seminorm for $p \in [1, \infty]$
$\{\cdot\}$	fractional part, i.e., $\{x\} := x - \lfloor x \rfloor$ for real x
$\lceil \cdot \rceil$	ceiling function
$\lfloor \cdot \rfloor$	floor function
$\lVert \cdot \rVert$	distance to the nearest integer function, $\lVert x \rVert := \min_{m \in \mathbb{Z}} \lvert x - m \rvert$
log	natural logarithm
$\log_b$	logarithm to the base b
$[d]$	the set $\{1, 2, \ldots, d\}$
$\boldsymbol{x}, \boldsymbol{k}, \ldots$	vectors of length d are denoted in bold font and their components in normal font with indices, e.g., $\boldsymbol{x} = (x_1, \ldots, x_d)$ or $\boldsymbol{k} = (k_1, \ldots, k_d)$
$\mathbf{0}$	zero vector (if length is clear from the context)
$\mathbf{0}_d$	zero vector of length d
$\mathbf{1}$	all-1 vector (if length is clear from the context)

$\mathbf{1}_d$	all-1 vector of length d
U_m	identity matrix of size $m \times m$
$\|A\|$	number of elements of a finite set A
χ_A	indicator function of A; $\chi_A(x) = 1$ if $x \in A$ and 0 if $x \notin A$
O	"big O"-notation; for functions $f, g : D \subseteq \mathbb{R} \to \mathbb{R}$, $g \geq 0$, we write $f(x) = O(g(x))$ if there exists a constant $C > 0$ such that $\|f(x)\| \leq Cg(x)$ for all $x \in D$.
$\lesssim$	for functions $f, g : D \subseteq \mathbb{R} \to \mathbb{R}$, $g \geq 0$, we write $f(x) \lesssim g(x)$ if there exists a constant $C > 0$ such that $\|f(x)\| \leq Cg(x)$ for all $x \in D$.
$\asymp$	$f(x) \asymp g(x)$ means that $f(x) \lesssim g(x)$ and $g(x) \lesssim f(x)$
$\ll$	$N \ll M$ is used on an informal level whenever N is "much smaller" than M
$r_1(h)$	$r_1(h) := \max(1, \|h\|)$ for $h \in \mathbb{Z}$
$r_1(\boldsymbol{h})$	$r_1(\boldsymbol{h}) := \prod_{j=1}^d r_1(h_j)$ for $\boldsymbol{h} = (h_1, \ldots, h_d) \in \mathbb{Z}^d$
$\Delta_{\mathcal{P}}$	local discrepancy function of a point set $\mathcal{P}$
D_N^*	star-discrepancy
$D_{N,\boldsymbol{\gamma}}^*$	$\boldsymbol{\gamma}$-weighted star-discrepancy
D_N	extreme discrepancy
$L_{p,N}$	L_p-discrepancy
$L_{p,N,\mathbf{1}}$	combined L_p-discrepancy
$L_{p,N,\boldsymbol{\gamma}}$	$\boldsymbol{\gamma}$-weighted L_p-discrepancy
$\mathrm{err}_{N,d}(f, \mathcal{P})$	error functional
$e(N, d)$	N-th minimal error in dimension d
$e(0, d)$	initial error in dimension d
$N(\varepsilon, d)$	information complexity
I, I_d	integration functional; usually $I(f) = I_d(f) = \int_{[0,1]^d} f(\boldsymbol{x})\,\mathrm{d}\boldsymbol{x}$
$\mathcal{H}(K)$	reproducing kernel Hilbert space with reproducing kernel K
$\mathcal{L}^\perp$	dual of an integration lattice $\mathcal{L}$
$C(M)$	$(-M/2, M/2] \cap \mathbb{Z}$ for integer $M \geq 2$
$C_d(M)$	$(C(M))^d$ for $d \in \mathbb{N}$
$C_d^*(M)$	$C_d(M) \setminus \{\mathbf{0}\}$
$G_d(N)$	$\{0, 1, \ldots, N-1\}^d$
$G_d^\varphi(N)$	$\{g \in G_1(N) \setminus \{0\} \ : \ \gcd(g, N) = 1\}^d$
MC	Monte Carlo
QMC	quasi-Monte Carlo
IBC	Information-Based Complexity
CBC	component-by-component
DBD	digit-by-digit
SCS	successive coordinate search
Λ^{all}	the information class of continuous linear functionals
Λ^{std}	the information class of function evaluations

Chapter 1
Introduction

The problem of numerical integration occurs in many practical applications. These range from computational mathematics, finance, statistics, and computer graphics to life sciences, to name just a few areas where integrals or expected values have to be computed. In most cases this cannot be done analytically, and one has to resort to numerical methods. For integrands depending only on one or maybe very few variables there are classical integration methods which can be applied very successfully. However, in most modern applications the integrands involved depend on a huge number d of variables. An ad hoc approach in such cases is to use tensor products of univariate integration rules. However, the number of function evaluations required in such algorithms grows exponentially with the number of integration variables, which means that tensor product rules are doomed to fail if d is only moderately large. This constraint led to the development of probabilistic methods, which were first applied by E. Fermi in the 1930s, and by J. von Neumann and S. Ulam in a secret research project at Los Alamos National Laboratory in the 1940s, and which can be easily applied to very high-dimensional problems. These methods were named Monte Carlo methods, a code word created in allusion to Ulam's uncle who was a gambler in the casino in Monte Carlo, according to a popular anecdote. Deterministic versions of Monte Carlo methods are quasi-Monte Carlo methods, and one of the most popular instances of these are lattice rules. The development of lattice rules from the 1950s onwards was promoted by a Soviet school and an Austrian school, led by N.M. Korobov and E. Hlawka, respectively.

In this introductory chapter we describe these developments with an emphasis on lattice rules, which are the primary object of this book.

1.1 Monte Carlo and Quasi-Monte Carlo Integration

Monte Carlo and quasi-Monte Carlo rules are generally applicable methods for the numerical integration of multivariate functions, even in high dimensions, with a wide range of possible applications, as for example in finance, physics, and computer

J. Dick et al., *Lattice Rules*, Springer Series in Computational Mathematics 58,
https://doi.org/10.1007/978-3-031-09951-9_1

science. We shall usually normalize the integration domain to be the d-dimensional unit cube $[0,1]^d$ for $d \in \mathbb{N}$ (an exception to this setting will be considered in Chapter 8). For a Lebesgue-integrable function f on $[0,1]^d$, Monte Carlo methods use the estimate

$$\int_{[0,1]^d} f(\boldsymbol{x})\,\mathrm{d}\boldsymbol{x} \approx \frac{1}{N}\sum_{k=0}^{N-1} f(\boldsymbol{x}_k), \tag{1.1}$$

where $\boldsymbol{x}_0, \boldsymbol{x}_1, \ldots, \boldsymbol{x}_{N-1}$ are independent and uniformly distributed random samples from $[0,1]^d$.

The right-hand side in (1.1) is called a *Monte Carlo (MC) rule (or algorithm)* applied to f. Let us denote it by $M_{N,d}(f)$. The MC rule $M_{N,d}(f)$ is an unbiased estimator of the integral $\int_{[0,1]^d} f(\boldsymbol{x})\,\mathrm{d}\boldsymbol{x}$, since, using linearity of expectation, we have

$$\mathbb{E}[M_{N,d}(f)] = \frac{1}{N}\sum_{k=0}^{N-1} \mathbb{E}[f] = \mathbb{E}[f] = \int_{[0,1]^d} f(\boldsymbol{x})\,\mathrm{d}\boldsymbol{x},$$

where we write $\mathbb{E}[f]$ to denote the expected value of $f(X)$, with X being uniformly distributed on $[0,1]^d$.

If $f \in L_2([0,1]^d)$, then the expected absolute error in (1.1) is of order $O(N^{-1/2})$. To be more precise, let

$$\mathrm{Var}[f] := \int_{[0,1]^d} \left(f(\boldsymbol{x}) - \int_{[0,1]^d} f(\boldsymbol{y})\,\mathrm{d}\boldsymbol{y}\right)^2 \mathrm{d}\boldsymbol{x}$$

be the *variance* of the function f. Then, from the Bienaymé formula in conjunction with the independence of $\boldsymbol{x}_0, \boldsymbol{x}_1, \ldots, \boldsymbol{x}_{N-1}$, we directly obtain the following result, which is a fundamental property of the MC method.

Theorem 1.1 *Let $f \in L_2([0,1]^d)$. Then for any $d, N \in \mathbb{N}$ we have*

$$\mathrm{Var}[M_{N,d}(f)] = \frac{\mathrm{Var}[f]}{N}.$$

From Jensen's inequality in probability theory and from Theorem 1.1 we obtain for the expected absolute error,

$$\begin{aligned}
\mathbb{E}\left[\left|\int_{[0,1]^d} f(\boldsymbol{x})\,\mathrm{d}\boldsymbol{x} - M_{N,d}(f)\right|\right] &\le \left(\mathbb{E}\left[\left|\int_{[0,1]^d} f(\boldsymbol{x})\,\mathrm{d}\boldsymbol{x} - M_{N,d}(f)\right|^2\right]\right)^{1/2} \\
&= \sqrt{\mathrm{Var}[M_{N,d}(f)]} \\
&= \frac{\sigma[f]}{\sqrt{N}},
\end{aligned}$$

where $\sigma[f] := \sqrt{\mathrm{Var}[f]}$ is the *standard deviation* of the function f. Hence the expected absolute error of $M_{N,d}(f)$ is bounded by $\sigma[f]/\sqrt{N}$. Note that $\sigma[f]$ is independent of N and therefore the convergence rate is independent of the dimension d (however, the standard deviation $\sigma[f]$ is in general not independent of d, an issue closely related to so-called variance-reduction techniques, see, e.g., [78, 178]).

An advantage of the MC method is that it is very general and easy to implement, although the generation of good "random" samples is in general an involved topic that is beyond the scope of this book. Keeping in mind that only square integrability of the integrand f is demanded, the expected convergence rate is of order $O(N^{-1/2})$. One may hope, though, that smoother integrands may yield an improved convergence rate. This, however, is not the case, for crude MC rules. More smoothness of the integrand does in general not imply a better convergence rate of the expected integration error for crude MC. Nevertheless, the MC convergence rate $O(N^{-1/2})$ will serve as a benchmark throughout the whole book.

If the integrand is smooth, a possible way to take advantage of this property is to switch from MC to *quasi-Monte Carlo (QMC) methods*, which are, in a nutshell, deterministic versions of MC methods. QMC methods for an integral over $[0,1]^d$ again use the approximation (1.1), but in this case $\boldsymbol{x}_0, \boldsymbol{x}_1, \ldots, \boldsymbol{x}_{N-1} \in [0,1]^d$ are deterministic points that are chosen to obtain convergence rates of the deterministic integration error which are better than $O(N^{-1/2})$. The integration rule (1.1) is, in this context, called a QMC rule and will be denoted by $Q_{N,d}$, i.e.,

$$Q_{N,d}(f) := \frac{1}{N} \sum_{k=0}^{N-1} f(\boldsymbol{x}_k).$$

To get an idea about the possible error convergence rate for deterministic QMC rules, consider for instance the following. Assume that the integrand f is continuous and in $L_2([0,1]^d)$. We can then interpret the expected value of the MC error as the average over all possible choices of N quadrature points. It is clear that there always exist particular examples of N-element point sets $\mathcal{P} = \{\boldsymbol{x}_0, \boldsymbol{x}_1, \ldots, \boldsymbol{x}_{N-1}\}$ which yield an error bound that is at least as good as average (by "point set" we mean a multi-set, i.e., points are allowed to occur multiple times). Hence we know that there exist deterministic point sets $\mathcal{P}$, for which the integration error is at most $\sigma[f]/\sqrt{N}$. For integrands with more smoothness, we would like to improve this rate of convergence, and we would like to have one quadrature rule which guarantees this improved rate of convergence for an entire function class.

There are three principal approaches: (i) via Halton sequences and their variants; (ii) via the theory of (t, m, s)-nets; (iii) via the theory of lattice rules. This book is devoted to the third approach, the theory of lattice rules. Previous introductions to this theory were given in the book [199] by Niederreiter and the book [230] by Sloan and Joe. The aim of the present book is to discuss the topic from the current point of view, with a particular focus on developments from recent years.

1.2 Lattice Rules

A lattice rule is a generalization of the classical rectangle rule

$$\int_0^1 f(x)\,\mathrm{d}x \approx \frac{1}{N}\sum_{k=0}^{N-1} f\left(\frac{k}{N}\right), \tag{1.2}$$

which, in the case of an integrand f with $f(0) = f(1)$, coincides with the N-point trapezoidal rule for the interval $[0, 1]$.

We can generalize (1.2) for instance by using the points $\{kg/N\}$ for $k \in \{0, 1, \ldots, N-1\}$, where $g \in \{0, 1, \ldots, N-1\}$ and by $\{z\} = z - \lfloor z \rfloor \in [0, 1)$ we mean the fractional part of a real number z. In the case when N is a prime number this yields exactly the same quadrature rule as (1.2), only the ordering of the points is different (which obviously does not change the value of the sum). However, this approach leads us to a method to generalize a rectangle rule to higher dimensions. For each coordinate $j \in [d]$, where here and throughout the whole book we denote by $[d]$ the set $\{1, 2, \ldots, d\}$ of the first d positive integers, choose a number $g_j \in \{0, 1, \ldots, N-1\}$, or, more generally, $g_j \in \mathbb{Z}$. Then for $f : [0, 1]^d \to \mathbb{R}$ we define the quadrature rule

$$\frac{1}{N}\sum_{k=0}^{N-1} f\left(\left\{\frac{k(g_1, g_2, \ldots, g_d)}{N}\right\}\right). \tag{1.3}$$

Here, the fractional part of a vector is understood component-wise. It turns out that the quadrature points defined in this way are part of a lattice in $\mathbb{R}^d$, hence rules of this form are called *lattice rules* (or more precisely, *rank-1 lattice rules*, the reason for this nomenclature will become more clear later). The numerical integration schemes in (1.3) were historically the first lattice rules and they are collectively known as the *method of good lattice points*. For given $N \in \mathbb{N}$ and given vector $\boldsymbol{g} = (g_1, g_2, \ldots, g_d) \in \mathbb{Z}^d$ we denote the collection of points used in (1.3) by $\mathcal{P}(\boldsymbol{g}, N)$, i.e.,

$$\mathcal{P}(\boldsymbol{g}, N) := \left\{\left\{\frac{k}{N}\boldsymbol{g}\right\} \;:\; k \in \{0, 1, \ldots, N-1\}\right\}, \tag{1.4}$$

which is again to be understood as a multi-set, i.e., points are allowed to occur multiple times. In this context, we call $\boldsymbol{g}$ the *generating vector* or sometimes also *lattice point*, and N the *modulus* of the lattice rule, respectively. Figure 1.1 shows an example of quadrature points $\mathcal{P}(\boldsymbol{g}, N)$ used in the rule (1.3).

Lattice rules can also be introduced from a group theoretic and a geometric perspective. We start with the former point of view, and in the remainder of this section and the next section we discuss algebraic properties of lattices and the underlying point set of lattice rules. Readers less interested in those algebraic properties may skip this part.

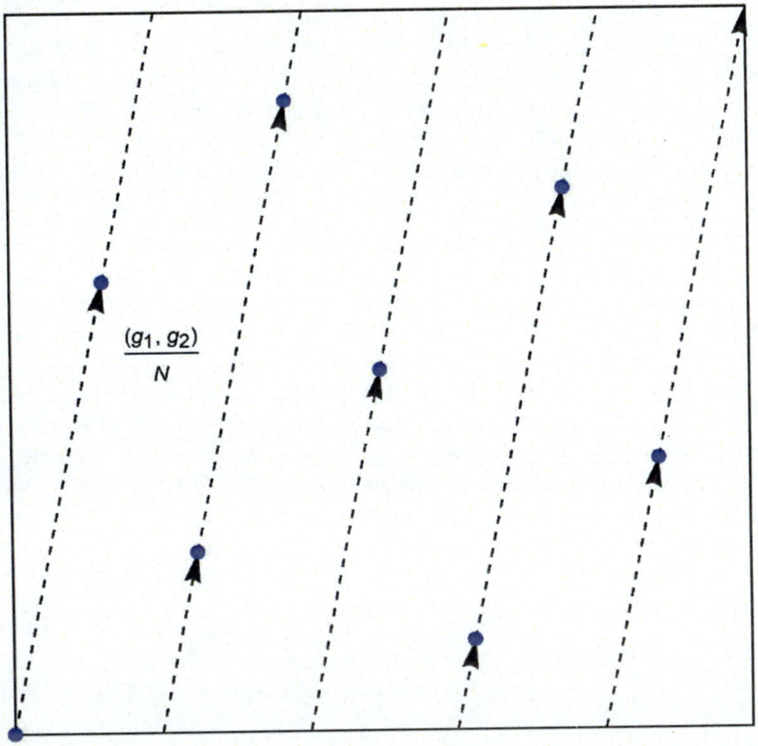

Fig. 1.1: A quadrature point set $\mathcal{P}(\boldsymbol{g}, N)$ as in (1.3) consisting of $N = 8$ elements in the unit square generated by the lattice point $(g_1, g_2) \in \{1, 2, \ldots, N-1\}^2$. The whole point set is obtained by successive addition modulo one of the vector $(g_1, g_2)/N$.

For a given dimension $d \in \mathbb{N}$, the Euclidean space $\mathbb{R}^d$ is an abelian group under addition which has $\mathbb{Z}^d$ as a subgroup. Thus, we can form the factor group $\mathbb{R}^d/\mathbb{Z}^d$, also called the d-dimensional *torus group*. Now we consider, for the moment, the nodes $0, 1/N, \ldots, (N-1)/N$ in (1.2) and their corresponding cosets $0 + \mathbb{Z}$, $1/N + \mathbb{Z}, \ldots, (N-1)/N + \mathbb{Z}$ in the one-dimensional torus group $\mathbb{R}/\mathbb{Z}$. Clearly, these cosets form a finite subgroup of $\mathbb{R}/\mathbb{Z}$; in fact, it is the cyclic group generated by $1/N + \mathbb{Z}$. The generalization to higher dimensions is now obvious. For an arbitrary d, let $\mathcal{L}/\mathbb{Z}^d$ be any finite subgroup of $\mathbb{R}^d/\mathbb{Z}^d$ and let $\boldsymbol{y}_k + \mathbb{Z}^d$ with $\boldsymbol{y}_k \in [0,1)^d$ for $k \in \{0, 1, \ldots, N-1\}$ be the distinct cosets making up the group $\mathcal{L}/\mathbb{Z}^d$. The point set consisting of the points $\boldsymbol{y}_0, \boldsymbol{y}_1, \ldots, \boldsymbol{y}_{N-1}$ is called a *lattice point set*, denoted by $\mathcal{P}(\mathcal{L})$, and the corresponding QMC approximation

$$\int_{[0,1]^d} f(\boldsymbol{x}) \,\mathrm{d}\boldsymbol{x} \approx \frac{1}{N} \sum_{k=0}^{N-1} f(\boldsymbol{y}_k) \tag{1.5}$$

is called a *lattice rule*.

Why do we speak of a "lattice rule" and not, for instance, of a "finite group rule"? The reason is a nice geometric interpretation of lattice rules that we already alluded to above. A d-dimensional *lattice* is defined to be a discrete subgroup of $\mathbb{R}^d$ that is not contained in any proper linear subspace of $\mathbb{R}^d$. Equivalently, a d-dimensional lattice is obtained by taking a basis $\boldsymbol{b}_1, \boldsymbol{b}_2, \ldots, \boldsymbol{b}_d$ of the vector space $\mathbb{R}^d$ and forming the set

$$\mathcal{L} = \left\{ \sum_{j=1}^{d} k_j \boldsymbol{b}_j \;:\; k_j \in \mathbb{Z} \text{ for } j \in [d] \right\}$$

of all linear combinations of $\boldsymbol{b}_1, \boldsymbol{b}_2, \ldots, \boldsymbol{b}_d$ with integer coefficients. The set $\{\boldsymbol{b}_1, \boldsymbol{b}_2, \ldots, \boldsymbol{b}_d\}$ is called a *lattice basis* for $\mathcal{L}$ and the $d \times d$ matrix B with row vectors $\boldsymbol{b}_1, \boldsymbol{b}_2, \ldots, \boldsymbol{b}_d$ is called a *generator matrix*. Note that a lattice has more than one generator matrix, since any matrix of the form UB with U being a $d \times d$ unimodular matrix with integer entries is again a generator matrix of the same lattice.

The lattices corresponding to lattice rules are required to have an additional property stated in the following definition.

Definition 1.2 A d-dimensional lattice is called a d-dimensional *integration lattice* if it contains $\mathbb{Z}^d$ as a subset.

If we take a d-dimensional integration lattice $\mathcal{L}$ as the starting point, then the intersection $\mathcal{L} \cap [0, 1)^d$ is a finite set since $\mathcal{L}$ is discrete, and this finite set of points in $[0, 1)^d$ forms a lattice point set $\mathcal{P}(\mathcal{L}) = \mathcal{L} \cap [0, 1)^d$ (sometimes also called the *abscissa set* of the lattice rule). Furthermore, all lattice point sets can be obtained in this way.

An interesting special case arises when the finite subgroup $\mathcal{L}/\mathbb{Z}^d$ of $\mathbb{R}^d/\mathbb{Z}^d$ is cyclic. Let N be the order of the group $\mathcal{L}/\mathbb{Z}^d$ and let $\boldsymbol{y} + \mathbb{Z}^d$ be a generator of $\mathcal{L}/\mathbb{Z}^d$. Then $N\boldsymbol{y} \in \mathbb{Z}^d$, and so $\boldsymbol{y} = (1/N)\boldsymbol{g}$ for some $\boldsymbol{g} \in \mathbb{Z}^d$. The lattice rule (1.5) then attains the form

$$\int_{[0,1]^d} f(\boldsymbol{x})\, \mathrm{d}\boldsymbol{x} \approx \frac{1}{N} \sum_{k=0}^{N-1} f\left(\left\{\frac{k}{N}\boldsymbol{g}\right\}\right), \tag{1.6}$$

where the curly brackets $\{\cdot\}$ again denote fractional parts, i.e., the points $(k/N)\boldsymbol{g}$, $k \in \{0, 1, \ldots, N-1\}$, are considered modulo 1 in each coordinate. This is exactly what we have already seen in (1.3). We recall that the set of quadrature points is denoted by $\mathcal{P}(\boldsymbol{g}, N)$ in this case, see (1.4).

The dual lattice

An important concept for the analysis of lattice rules is that of the dual lattice.

Definition 1.3 The *dual lattice* $\mathcal{L}^\perp$ of the d-dimensional integration lattice $\mathcal{L}$ is defined by

$$\mathcal{L}^\perp := \left\{ \boldsymbol{h} \in \mathbb{Z}^d \ : \ \boldsymbol{h} \cdot \boldsymbol{y} \in \mathbb{Z} \ \text{ for all } \boldsymbol{y} \in \mathcal{L} \right\},$$

where $\cdot$ denotes the standard inner product on the $\mathbb{R}^d$.

It is easy to see that the dual lattice of a d-dimensional integration lattice is always a subgroup of $\mathbb{Z}^d$.

In case of a lattice rule of the form (1.6) the dual lattice $\mathcal{L}^\perp$ reduces to

$$\mathcal{L}^\perp = \{ \boldsymbol{h} \in \mathbb{Z}^d \ : \ \boldsymbol{h} \cdot \boldsymbol{g} \equiv 0 \pmod{N} \}, \tag{1.7}$$

since $\boldsymbol{h} \cdot \boldsymbol{y} \in \mathbb{Z}$ for all $\boldsymbol{y} \in \mathcal{L}$ is equivalent to the number-theoretic property $\boldsymbol{h} \cdot \boldsymbol{g} \equiv 0 \pmod{N}$. In order to emphasize the dependence of the dual lattice on $\boldsymbol{g}$ and on N we will sometimes also write $\mathcal{L}^\perp(\boldsymbol{g}, N)$.

The determinant of a lattice

We now introduce an important invariant of a lattice. To this end we again turn our attention to the generator matrices B of a lattice. Recall that these are not uniquely determined; however, for any lattice $\mathcal{L}$, the absolute determinant $|\det(B)|$ of the generator matrices of $\mathcal{L}$ is invariant, since unimodular matrices with integer entries have determinant ± 1. This number is called the *determinant of a lattice*, which is denoted by $\det(\mathcal{L})$. Geometrically, $\det(\mathcal{L})$ is the volume of the parallelepiped spanned by the vectors in a lattice basis of $\mathcal{L}$,

$$\mathcal{U} := \{ t_1 \boldsymbol{b}_1 + t_2 \boldsymbol{b}_2 + \cdots + t_d \boldsymbol{b}_d \ : \ t_i \in [0, 1) \ \text{ for all } \ i \in [d] \},$$

called the *unit cell* or the *fundamental parallelepiped* of the lattice. This can be interpreted in the sense that the average number of lattice points per unit volume is the reciprocal of the determinant.

The following result for integration lattices is due to Sloan and Kachoyan [231].

Theorem 1.4 *If $\mathcal{L}$ is an integration lattice yielding an N-element lattice point set, then* $\det(\mathcal{L}) = 1/N$. *If B is a generator matrix of $\mathcal{L}$, then $(B^\top)^{-1}$ is a generator matrix of the dual lattice $\mathcal{L}^\perp$ and* $\det(\mathcal{L}^\perp) = N$.

Proof Since $|\mathcal{L} \cap [0, 1)^d| = N$, for every $k \in \mathbb{N}$ the d-dimensional cube $A := [0, k)^d$ contains exactly $k^d N$ elements of $\mathcal{L}$, i.e.,

$$|\mathcal{L} \cap A| = k^d N. \tag{1.8}$$

Let now

$$A^\circ := \bigcup_{\substack{\boldsymbol{x} \in \mathcal{L} \\ (\boldsymbol{x} + \mathcal{U}) \subseteq A}} (\boldsymbol{x} + \mathcal{U})$$

be the union of all shifted unit cells $x + \mathcal{U}$ with $x \in \mathcal{L}$ that are fully contained in A, and let

$$\overline{A} := \bigcup_{\substack{x \in \mathcal{L} \\ (x+\mathcal{U}) \cap A \neq \emptyset}} (x + \mathcal{U})$$

be the union of all shifted unit cells $x + \mathcal{U}$ with $x \in \mathcal{L}$ that have nonempty intersection with A. It is clear that

$$A^\circ \subseteq A \subseteq \overline{A}.$$

Furthermore, for the volume $\lambda_d(\overline{A})$ of $\overline{A}$ we have

$$\lambda_d(\overline{A}) = \lambda_d(\mathcal{U})\,|\{x \in \mathcal{L} \,:\, (x + \mathcal{U}) \cap A \neq \emptyset\}| = \det(\mathcal{L})\,|\mathcal{L} \cap \overline{A}|,$$

and in the same way we obtain

$$\lambda_d(A^\circ) = \det(\mathcal{L})\,|\mathcal{L} \cap A^\circ|.$$

This yields

$$\det(\mathcal{L})\,|\mathcal{L} \cap A| - k^d \leq \det(\mathcal{L})\,|\mathcal{L} \cap \overline{A}| - \lambda_d(\overline{A}) + \lambda_d(\overline{A} \setminus A) = \lambda_d(\overline{A} \setminus A),$$

and

$$\det(\mathcal{L})\,|\mathcal{L} \cap A| - k^d \geq \det(\mathcal{L})\,|\mathcal{L} \cap A^\circ| - \lambda_d(A^\circ) - \lambda_d(A \setminus A^\circ) = -\lambda_d(A \setminus A^\circ).$$

In summary,

$$-\lambda_d(A \setminus A^\circ) \leq \det(\mathcal{L})\,|\mathcal{L} \cap A| - k^d \leq \lambda_d(\overline{A} \setminus A).$$

Let δ be the diameter of the unit cell $\mathcal{U}$. Note that δ is finite and does not depend on k. We then have, for sufficiently large k,

$$\overline{A} \setminus A \subseteq [-\delta, k + \delta]^d \setminus [0, k]^d \quad \text{and} \quad A \setminus A^\circ \subseteq [0, k]^d \setminus [\delta, k - \delta]^d,$$

and hence

$$\lambda_d(\overline{A} \setminus A) \leq (k + 2\delta)^d - k^d \quad \text{and} \quad \lambda_d(A \setminus A^\circ) \leq k^d - (k - 2\delta)^d.$$

These estimates together with (1.8) imply

$$-(k^d - (k - 2\delta)^d) \leq \det(\mathcal{L}) k^d N - k^d \leq (k + 2\delta)^d - k^d,$$

which implies

$$-\left(1 - \left(1 - \frac{2\delta}{k}\right)^d\right) \leq \det(\mathcal{L})N - 1 \leq \left(1 + \frac{2\delta}{k}\right)^d - 1.$$

Letting k tend to infinity yields $\det(\mathcal{L}) = 1/N$, as desired.

If $\boldsymbol{b}_1, \boldsymbol{b}_2, \ldots, \boldsymbol{b}_d$ are the row vectors of B and $\boldsymbol{a}_1, \boldsymbol{a}_2, \ldots, \boldsymbol{a}_d$ are the row vectors of $(B^\top)^{-1}$, then $\boldsymbol{a}_i \cdot \boldsymbol{b}_j = \delta_{i,j}$ for $i, j \in [d]$, where $\delta_{i,j} = 1$ if $i = j$ and 0 otherwise. This implies that $\boldsymbol{a}_1, \boldsymbol{a}_2, \ldots, \boldsymbol{a}_d \in \mathcal{L}^\perp$ by definition of $\mathcal{L}^\perp$. If $\boldsymbol{h}$ is an arbitrary element of $\mathcal{L}^\perp$, then we have

$$\boldsymbol{h} = \boldsymbol{h} B^\top (B^\top)^{-1} = \sum_{j=1}^{d} (\boldsymbol{h} \cdot \boldsymbol{b}_j) \boldsymbol{a}_j.$$

Since $\boldsymbol{h} \cdot \boldsymbol{b}_j \in \mathbb{Z}$ for all $j \in [d]$ it follows that $\boldsymbol{h}$ is a $\mathbb{Z}$-linear combination of $\boldsymbol{a}_1, \boldsymbol{a}_2, \ldots, \boldsymbol{a}_d$. Thus, these vectors form a basis of $\mathcal{L}^\perp$, and $(B^\top)^{-1}$ is a generator matrix of $\mathcal{L}^\perp$. Furthermore,

$$\det(\mathcal{L}^\perp) = |\det((B^\top)^{-1})| = \frac{1}{|\det(B)|} = \frac{1}{\det(\mathcal{L})} = N. \qquad \square$$

1.3 The Structure of Lattice Rules

In Section 1.2, we have defined lattice rules as QMC rules using the points of an integration lattice in $[0, 1)^d$ as the integration nodes. An obvious question is how such lattice rules can be represented in general. From (1.6), it can be seen that at least some lattice rules for approximating the integral of a function f defined on $[0, 1]^d$ by N lattice points can be written as

$$Q_{N,d}(f) = \frac{1}{N} \sum_{k=0}^{N-1} f\left(\left\{\frac{k}{N} \boldsymbol{g}\right\}\right).$$

One may observe that the above representation is not unique. Indeed, choosing an integer $m \geq 2$ and replacing N and $\boldsymbol{g}$ by mN and $m\boldsymbol{g}$, respectively, yields an integration rule with mN points, each of them occurring with multiplicity m, such that the newly obtained rule is effectively equivalent to the one we started with. Furthermore, it is easy to construct examples where for the same N two distinct generating vectors $\boldsymbol{g}_1$ and $\boldsymbol{g}_2$ yield rules with identical sets of integration nodes. Another question is of course whether there may be other lattice rules that can be represented in a different way than the one above. Hence, it is natural to ask for a canonical way of denoting lattice rules that makes it easier to structure and classify these, and to detect equivalences. Regarding this problem, it were Sloan and Lyness who used group theory to provide a systematic way of representing lattice rules. We state a crucial theorem from [233] here.

Theorem 1.5 *Let $Q_{N,d}$ be a d-dimensional lattice rule with $N \geq 2$ points. Then there exist uniquely determined integers*

- *r, with $r \in [d]$, and*
- *$n_1, \ldots, n_r > 1$ with $n_{k+1} | n_k$ for $1 \leq k \leq r - 1$, and $N = n_1 \cdots n_r$,*

such that $Q_{N,d}$ applied to a function f can be represented as

$$Q_{N,d}(f) = \frac{1}{N}\sum_{k_r=0}^{n_r-1}\cdots\sum_{k_1=0}^{n_1-1} f\left(\left\{\frac{k_1}{n_1}\boldsymbol{g}_1 + \cdots + \frac{k_r}{n_r}\boldsymbol{g}_r\right\}\right),$$

where $\boldsymbol{g}_1, \ldots, \boldsymbol{g}_r$ are linearly independent (over $\mathbb{Q}$) integer vectors in $\mathbb{Z}^d$.

The proof of Theorem 1.5 is based on the group structure of lattice points, and can be found in [233], see also [199, Theorem 5.28], [204, Section 4.3.2], or [230].

Remark 1.6 The integer r in Theorem 1.5 is called the *rank* of the lattice rule $Q_{N,d}$, and $n_1, \ldots, n_r$ are the *invariants*. In this terminology, the lattice rule in (1.3) based on a lattice point set $\mathcal{P}(\boldsymbol{g}, N)$ is, as already indicated, a *rank-1 lattice rule.*

Theorem 1.5 implies that the rank and the invariants of any given lattice rule are determined uniquely. What about the generating vectors $\boldsymbol{g}_1, \ldots, \boldsymbol{g}_r$ of a rank-r lattice rule? In general, these cannot be determined uniquely. However, Sloan and Lyness [234] identified a class of lattice rules for which even the generating vectors can, in a certain sense, be identified unambiguously. This class is known as projection-regular lattice rules, which shall be described briefly here. Given a d-dimensional lattice rule $Q_{N,d}$ and $s \in [d]$, we speak of the s-dimensional principal projection of $Q_{N,d}$ when we consider the s-dimensional lattice rule obtained by omitting the last $d - s$ components of the integration nodes.

Note now that we can modify the representation of lattice rules outlined in Theorem 1.5 to a so-called extended canonical form by setting $n_{r+1} = \cdots = n_d = 1$, and by choosing arbitrary integer vectors $\boldsymbol{g}_{r+1}, \ldots, \boldsymbol{g}_d$. Then we can represent a lattice rule $Q_{N,d}$ as

$$Q_{N,d}(f) = \frac{1}{N}\sum_{k_d=0}^{n_d-1}\cdots\sum_{k_1=0}^{n_1-1} f\left(\left\{\frac{k_1}{n_1}\boldsymbol{g}_1 + \cdots + \frac{k_d}{n_d}\boldsymbol{g}_d\right\}\right),$$

where we now trivially have $N = n_1 \cdots n_d$. Furthermore the rank r is in this case the maximal index r such that $n_r > 1$. Using the latter representation of lattice rules, projection regularity is defined as follows.

Definition 1.7 Let $Q_{N,d}$ be a d-dimensional lattice rule with invariants $n_1, \ldots, n_d$ in its extended canonical form. The rule $Q_{N,d}$ is called *projection regular* if for any choice of $s \in [d]$ the s-dimensional principal projection of $Q_{N,d}$ has $n_1 n_2 \cdots n_{\min(r,s)}$ points.

The benefit of projection-regular lattice rules is that, by demanding some structure of the generating vectors, their choice is unique. Indeed, note that, given a lattice rule $Q_{N,d}$ in its extended canonical form, we can define a $d \times d$ matrix Z with the generating vectors $\boldsymbol{g}_1, \ldots, \boldsymbol{g}_d$ as its rows, i.e., $Z = (\boldsymbol{g}_1, \ldots, \boldsymbol{g}_d)^\top$. The matrix Z corresponding to a lattice rule is simply called Z-matrix. We say that Z is unit upper triangular if Z is upper triangular and has only 1s as the entries on the main diagonal. Using this terminology, we can state the following theorem, which is the main result of [234].

Theorem 1.8 *A lattice rule $Q_{N,d}$ is projection-regular if and only if the rule $Q_{N,d}$ can be represented in an extended canonical form such that the corresponding Z-matrix is unit upper triangular.*

A question that remains is whether a projection-regular lattice rule can be represented in two different ways in an extended canonical form with unit upper triangular Z-matrices? The answer to this question is "no", so it makes sense to speak of *the* standard form of a projection-regular lattice with unit upper triangular Z-matrix. For further details, we refer to the monograph [230] and the original paper [234].

Due to their simple structure, rank-1 lattice rules have many convenient properties and have frequently been studied in many papers. Nevertheless, also lattice rules of higher or even maximal rank are worth being considered, since in some instances they may perform better than rules of rank 1 with respect to certain criteria, such as the quantity P_α used in the classical literature on lattice rules (see Section 1.4).

Copy rules

A prominent way of obtaining higher-rank lattice rules are copy rules. The principle idea of copy rules is elegant and simple. Given a lattice rule $Q_{N,d}$ based on an integration lattice $\mathcal{L}$, we obtain an integration rule consisting of m^d scaled "copies" of $Q_{N,d}$ by considering the lattice rule based on the integration lattice $m^{-1}\mathcal{L}$ for some positive integer m. In other words, a copy rule is obtained by scaling the original rule $Q_{N,d}$ and copying it to each of the cubes obtained by partitioning $[0,1]^d$ into cubes of volume m^{-d}. Using the canonical representation form, let us suppose we start with a rank-1 rule $Q_{N,d}$ with generating vector $\boldsymbol{g}$, i.e.,

$$Q_{N,d}(f) = \frac{1}{N}\sum_{k=0}^{N-1} f\left(\left\{\frac{k}{N}\boldsymbol{g}\right\}\right).$$

The m^d copy rule $Q_{m,N,d}$ is then given by

$$Q_{m,N,d}(f) := \frac{1}{m^d N}\sum_{k_d=0}^{m-1}\cdots\sum_{k_1=0}^{m-1}\sum_{k=0}^{N-1} f\left(\left\{\frac{k}{mN}\boldsymbol{g} + \frac{(k_1,\ldots,k_d)}{m}\right\}\right).$$

It was shown by Sloan and Lyness in [233] that an m^d copy rule has rank d; to be more precise, a d-dimensional lattice rule is of rank d if and only if it is an m^d copy rule obtained from a rule of lower rank.

Embedded lattice rules

The rank plays a role in yet another variant of lattice rules, namely so-called embedded (sometimes also imbedded) lattice rules. The crucial feature of embedded lattice rules is that, at least up to a certain point, these can be extended by adding further points to the rule without having to discard previous ones. This property can be a desirable advantage in practical implementations. In this context one then more precisely speaks of sequences of embedded integration rules, a concept that not only occurs with lattice rules but also other types of quadratures. Sequences of embedded lattice rules, as they shall be presented here, were originally described by Sloan and Joe [122, 230]; they have the additional property that the more points we add, the higher the rank of the lattice rule gets. Using a representation form similar to the canonical representation form introduced above, a sequence of embedded lattice rules is defined as follows. For a fixed positive integer m that is relatively prime to N, let for $r \in [d]$ the rule $Q_{r,m,N,d}$ be defined by

$$Q_{r,m,N,d}(f) := \frac{1}{m^r N} \sum_{k_r=0}^{m-1} \cdots \sum_{k_1=0}^{m-1} \sum_{k=0}^{N-1} f\left(\left\{\frac{k}{N}\boldsymbol{g} + \frac{(k_1,\ldots,k_r,0,\ldots,0)}{m}\right\}\right),$$

where $\boldsymbol{g}$ is a generating vector with components that are relatively prime to m. The $Q_{r,m,N,d}$ have the property that the integration nodes in $Q_{r,m,N,d}$ also occur in $Q_{r+1,m,N,d}$ for $r \in \{0,1,\ldots,d-1\}$, and that $Q_{r,m,N,d}$ is a lattice rule with $m^r N$ points and rank r for $r \in \{0,1,\ldots,d-1\}$, the only exception being the rank of $Q_{0,m,N,d}$, which is 1. Hence we see that $Q_{d,m,N,d}$, which is the rule based on the largest number of integration nodes and, thus, intuitively the "most precise", has maximal rank. From the representation of the $Q_{r,m,N,d}$ we see much similarity to copy rules, and indeed, $Q_{d,m,N,d}$ is nothing but an m^d copy rule obtained from $Q_{0,m,N,d}$. References to further results related to this subject can be found in the "Notes and Remarks" Section at the end of the present chapter. We shall return to embedded lattice rules in Chapter 6.

1.4 Lattice Rules for Numerical Integration—the Classical Theory

Lattice rules are perfectly configured for the numerical integration of smooth functions that are one-periodic in each variable, as is very well known from the classical theory.

Suppose that a one-periodic function f can be represented by an absolutely convergent Fourier series

$$f(\boldsymbol{x}) = \sum_{\boldsymbol{h}\in\mathbb{Z}^d} \widehat{f}(\boldsymbol{h})\, \mathrm{e}^{2\pi \mathrm{i} \boldsymbol{h}\cdot\boldsymbol{x}} \quad \text{for } \boldsymbol{x} \in \mathbb{R}^d, \tag{1.9}$$

where the Fourier coefficients are given by

$$\widehat{f}(\boldsymbol{h}) = \int_{[0,1]^d} f(\boldsymbol{x})\, \mathrm{e}^{-2\pi \mathrm{i}\boldsymbol{h}\cdot\boldsymbol{x}}\, \mathrm{d}\boldsymbol{x} \quad \text{for } \boldsymbol{h} \in \mathbb{Z}^d. \tag{1.10}$$

Obviously, the Fourier coefficient $\widehat{f}(\mathbf{0})$ is the exact value of the integral $\int_{[0,1]^d} f(\boldsymbol{x})\,\mathrm{d}\boldsymbol{x}$. For a d-dimensional lattice rule based on the N-element lattice point set $\mathcal{P}(\mathcal{L}) = \{\boldsymbol{x}_0, \boldsymbol{x}_1, \ldots, \boldsymbol{x}_{N-1}\}$, where $\mathcal{L}$ is an integration lattice, we therefore obtain

$$\begin{aligned} \frac{1}{N}\sum_{k=0}^{N-1} f(\boldsymbol{x}_k) - \int_{[0,1]^d} f(\boldsymbol{x})\,\mathrm{d}\boldsymbol{x} &= \frac{1}{N}\sum_{k=0}^{N-1}\sum_{\boldsymbol{h}\in\mathbb{Z}^d} \widehat{f}(\boldsymbol{h})\,\mathrm{e}^{2\pi\mathrm{i}\boldsymbol{h}\cdot\boldsymbol{x}_k} - \widehat{f}(\mathbf{0}) \\ &= \frac{1}{N}\sum_{\boldsymbol{h}\in\mathbb{Z}^d\setminus\{\mathbf{0}\}} \widehat{f}(\boldsymbol{h}) \sum_{k=0}^{N-1} \mathrm{e}^{2\pi\mathrm{i}\boldsymbol{h}\cdot\boldsymbol{x}_k}. \end{aligned} \tag{1.11}$$

It is a fundamental property of lattices that the exponential sum that appears in (1.11) can have only the two values N or 0, depending on whether $\boldsymbol{h}$ belongs to the dual lattice $\mathcal{L}^\perp$ or not. This property, stated in the following lemma, highlights the importance of the concept of the dual lattice.

Lemma 1.9 *Let $\mathcal{L}$ be an integration lattice and let $\mathcal{P}(\mathcal{L}) = \{\boldsymbol{x}_0, \boldsymbol{x}_1, \ldots, \boldsymbol{x}_{N-1}\}$ be the corresponding lattice point set. Let $\boldsymbol{h} \in \mathbb{Z}^d$, then we have*

$$\sum_{k=0}^{N-1} \mathrm{e}^{2\pi\mathrm{i}\boldsymbol{h}\cdot\boldsymbol{x}_k} = \begin{cases} N & \text{if } \boldsymbol{h} \in \mathcal{L}^\perp, \\ 0 & \text{if } \boldsymbol{h} \notin \mathcal{L}^\perp. \end{cases}$$

Remark 1.10 In the light of (1.7) it is clear that Lemma 1.9 can be stated for rank-1 lattice rules as

$$\sum_{k=0}^{N-1} \mathrm{e}^{2\pi\mathrm{i}\boldsymbol{h}\cdot\boldsymbol{x}_k} = \begin{cases} N & \text{if } \boldsymbol{h}\cdot\boldsymbol{g} \equiv 0 \pmod{N}, \\ 0 & \text{if } \boldsymbol{h}\cdot\boldsymbol{g} \not\equiv 0 \pmod{N}. \end{cases}$$

In this case we can prove this result directly using a geometric sum, since for $\boldsymbol{h}\cdot\boldsymbol{g} \not\equiv 0 \pmod{N}$ we have

$$\sum_{k=0}^{N-1} \mathrm{e}^{2\pi\mathrm{i}\boldsymbol{h}\cdot\boldsymbol{x}_k} = \sum_{k=0}^{N-1} \mathrm{e}^{2\pi\mathrm{i}k\boldsymbol{h}\cdot\boldsymbol{g}/N} = \frac{\mathrm{e}^{2\pi\mathrm{i}N\boldsymbol{h}\cdot\boldsymbol{g}/N} - 1}{\mathrm{e}^{2\pi\mathrm{i}\boldsymbol{h}\cdot\boldsymbol{g}/N} - 1} = 0,$$

and for $\boldsymbol{h}\cdot\boldsymbol{g} \equiv 0 \pmod{N}$, say $\boldsymbol{h}\cdot\boldsymbol{g} = c\,N$ for some $c \in \mathbb{Z}$, we have

$$\sum_{k=0}^{N-1} \mathrm{e}^{2\pi\mathrm{i}\boldsymbol{h}\cdot\boldsymbol{x}_k} = \sum_{k=0}^{N-1} \mathrm{e}^{2\pi\mathrm{i}kc} = \sum_{k=0}^{N-1} 1 = N.$$

We now give the proof of the fundamental result in Lemma 1.9 in general form. Before we do so, we need to make a short detour to the theory of characters of finite abelian groups. Let $(G, \circ)$ be a finite abelian group. A *character* of G is a group homomorphism $\chi : G \to \mathbb{C}^\times$, where $\mathbb{C}^\times$ denotes the multiplicative group of complex numbers. That is, for all $x, y \in G$ we have $\chi(x \circ y) = \chi(x)\chi(y)$. This already implies $\chi(1_G) = 1$, where 1_G is the identity in G. Every finite abelian group of order N has exactly N distinct characters denoted by $\chi_0, \chi_1, \ldots, \chi_{N-1}$, where the character $\chi_0 \equiv 1$, which is 1 for all $x \in G$, is called the *trivial character* or the *principal character*. The set $\widehat{G}$ of all characters of G forms an abelian group under the multiplication $(\chi\psi)(x) = \chi(x)\psi(x)$ for all $x \in G$, for $\chi, \psi \in \widehat{G}$.

Characters have the following important property which can be exploited in many applications, e.g., in the proof of Lemma 1.9.

Lemma 1.11 (Character properties) *Let χ be a character of a finite abelian group $(G, \circ)$. Then we have*

$$\sum_{x \in G} \chi(x) = \begin{cases} |G| & \text{if } \chi \text{ is the trivial character,} \\ 0 & \text{otherwise.} \end{cases}$$

Let $x \in G$. Then we have

$$\sum_{\chi \in \widehat{G}} \chi(x) = \begin{cases} |\widehat{G}| & \text{if } x = 1_G, \\ 0 & \text{otherwise.} \end{cases}$$

Proof We just prove the first identity, the second follows by a similar reasoning. The result is clear when χ is the trivial character. Otherwise there exists some $a \in G$ for which we have $\chi(a) \neq 1$. Then we have

$$\chi(a) \sum_{x \in G} \chi(x) = \sum_{x \in G} \chi(a \circ x) = \sum_{x \in G} \chi(x),$$

since as x runs through all elements of G so does $a \circ x$. Hence we have

$$(\chi(a) - 1) \sum_{x \in G} \chi(x) = 0,$$

and the result follows since $\chi(a) \neq 1$. □

Now we can use Lemma 1.11 to prove Lemma 1.9.

Proof of Lemma 1.9 Let $\boldsymbol{h} \in \mathbb{Z}^d$. Then

$$\chi_{\boldsymbol{h}}(\boldsymbol{x} + \mathbb{Z}^d) := \mathrm{e}^{2\pi \mathrm{i} \boldsymbol{h} \cdot \boldsymbol{x}} \quad \text{for } \boldsymbol{x} \in \mathcal{L}$$

is a well-defined character of the additive group $\mathcal{L}/\mathbb{Z}^d$. This character is trivial if and only if $\boldsymbol{h} \in \mathcal{L}^\perp$, where $\mathcal{L}^\perp$ is the dual lattice as defined in Definition 1.3. Recall that $\mathcal{P}(\mathcal{L}) = \{\boldsymbol{x}_0, \boldsymbol{x}_1, \ldots, \boldsymbol{x}_{N-1}\}$. Now we have

$$\sum_{k=0}^{N-1} \mathrm{e}^{2\pi \mathrm{i} \boldsymbol{h} \cdot \boldsymbol{x}_k} = \sum_{x \in \mathcal{L}/\mathbb{Z}^d} \chi_{\boldsymbol{h}}(x),$$

and therefore the result follows directly from Lemma 1.11. □

Returning to (1.11) and applying Lemma 1.9, we obtain the following result.

Proposition 1.12 *For a lattice rule with integration lattice $\mathcal{L}$ we have*

$$\frac{1}{N}\sum_{k=0}^{N-1} f(\boldsymbol{x}_k) - \int_{[0,1]^d} f(\boldsymbol{x})\,\mathrm{d}\boldsymbol{x} = \sum_{\boldsymbol{h}\in\mathcal{L}^\perp\setminus\{\boldsymbol{0}\}} \widehat{f}(\boldsymbol{h}),$$

where $\mathcal{P}(\mathcal{L}) = \{\boldsymbol{x}_0, \boldsymbol{x}_1, \ldots, \boldsymbol{x}_{N-1}\}$, and where $\mathcal{L}^\perp$ is the dual lattice.

We now classify the functions f with absolutely convergent Fourier series according to the rate of convergence with which the Fourier coefficients $\widehat{f}(\boldsymbol{h})$ tend to zero as $\|\boldsymbol{h}\|_\infty \to \infty$. The rate of decay of the Fourier coefficients can be related to the smoothness of the function. As mentioned at the beginning of this chapter, our goal is to construct QMC rules which can take advantage of smoothness. Considering classes of periodic integrands whose Fourier coefficients decay with a certain rate is one approach to try to find QMC rules (in particular, lattice rules) that yield a convergence rate of the integration error beyond $O(N^{-1/2})$.

For $h \in \mathbb{Z}$ we define $r_1(h) := \max(1, |h|)$, and for $\boldsymbol{h} = (h_1, \ldots, h_d) \in \mathbb{Z}^d$ we define

$$r_1(\boldsymbol{h}) := \prod_{j=1}^{d} r_1(h_j). \tag{1.12}$$

Definition 1.13 Let $\alpha > 1$ and $C > 0$. Then $\mathcal{E}_\alpha^d(C)$ is defined to be the class of all continuous one-periodic functions f on $\mathbb{R}^d$ with

$$|\widehat{f}(\boldsymbol{h})| \le \frac{C}{(r_1(\boldsymbol{h}))^\alpha} \quad \text{for all nonzero } \boldsymbol{h} \in \mathbb{Z}^d.$$

Furthermore, let

$$\mathcal{E}_\alpha^d := \bigcup_{C>0} \mathcal{E}_\alpha^d(C),$$

i.e., the class of all functions f with $f \in \mathcal{E}_\alpha^d(C)$ for some $C > 0$. Occasionally, the function classes $\mathcal{E}_\alpha^d(C)$ or $\mathcal{E}_\alpha^d$ are called *Korobov classes*.

It is known (see [140]) that $f \in \mathcal{E}_\alpha^d$ implies that all mixed partial derivatives of f of order less than $\alpha - 1$ in each variable exist. On the other hand, the existence and continuity of all mixed partial derivatives

$$\frac{\partial^{r_1+\cdots+r_d} f}{\partial x_1^{r_1}\cdots\partial x_d^{r_d}} \quad \text{with } 0 \le r_i \le \alpha \text{ for } i \in [d]$$

of a function f of order up to α in each variable imply that $f \in \mathcal{E}_\alpha^d$ (see again [140] or [199]).

We now motivate the definition of r_1 in (1.12) geometrically. From Proposition 1.12 we see that the integration error of a lattice rule is the sum over all Fourier coefficients with frequencies in the dual lattice, except for the origin which corresponds to the $\mathbf{0}$-th Fourier coefficient (and which equals the integral that we want to approximate).

In Definition 1.13 we define a function class of integrands, where we use $r_1(\boldsymbol{h})$ to put a condition on the decay rate of the Fourier coefficients $\widehat{f}(\boldsymbol{h})$ for $\boldsymbol{h} \in \mathbb{Z}^d$. Our goal is to have a large class of functions for which the integration error is small.

Consider now the function class $\mathcal{E}_\alpha^d(1)$. The frequencies $\boldsymbol{h}$ for which the corresponding Fourier coefficients of functions in $\mathcal{E}_\alpha^d(1)$ are allowed to be large, say $\widehat{f}(\boldsymbol{h}) > c^{-1}$ for some real number $c > 0$, are in the set

$$\Gamma_c := \{\boldsymbol{h} \in \mathbb{R}^d \ : \ (r_1(\boldsymbol{h}))^{-1} > c^{-1}\} = \{\boldsymbol{h} \in \mathbb{R}^d \ : \ r_1(\boldsymbol{h}) < c\},$$

by extending the definition of r_1 from $\mathbb{Z}^d$ to $\mathbb{R}^d$ in the natural way. For a given, generic lattice rule with dual lattice $\mathcal{L}^\perp$, choose $c > 0$ as large as possible such that $\Gamma_c \cap \mathcal{L}^\perp = \{\mathbf{0}\}$, i.e., such that for all frequencies $\boldsymbol{h} \in \Gamma_c \cap \mathbb{Z}^d$ the terms $\widehat{f}(\boldsymbol{h})\, \mathrm{e}^{2\pi \mathrm{i} \boldsymbol{h}\cdot\boldsymbol{x}}$ are integrated exactly by our lattice rule, except for $\boldsymbol{h} = \mathbf{0}$, as the $\mathbf{0}$-th Fourier coefficient is just the integral we would like to approximate. In order to make the function class large, we want to make also Γ_c "large". Note that the choice of r_1 influences the shape of the set Γ_c, whereas c only expands or shrinks it. Hence we focus on the shape, i.e., on r_1.

The function r_1 could be defined in various ways. For instance, for $r_1(\boldsymbol{h}) := \max(|h_1|, \ldots, |h_d|)$ the set Γ_c is a cube, for $r_1(\boldsymbol{h}) := (h_1^2 + \cdots + h_d^2)^{1/2}$ the set Γ_c is a ball, or for $r_1(\boldsymbol{h}) := \prod_{j=1}^d \max(1, |h_j|)$ the set Γ_c is a hyperbolic cross. Out of these, it turns out that in dimensions $d > 1$ the hyperbolic cross is the largest set, see Figure 1.2. Indeed, the choice of r_1 in (1.12) will yield a convergence rate of the integration error in dimension d which is close to the convergence rate in dimension 1, see (1.13). This is not true for other choices of r_1; e.g., for Γ_c being a cube, the convergence rate in dimension d is only of order $N^{-\alpha/d}$.

Hyperbolic crosses play an essential role in many high-dimensional approximation problems. The monograph [33] is devoted to hyperbolic cross approximation.

The quality criterion P_α

Now we define a first important quality criterion for lattice rules, namely a quantity usually referred to as P_α.

Definition 1.14 For a real number $\alpha > 1$ and an integration lattice $\mathcal{L}$ let

$$P_\alpha(\mathcal{L}) := \sum_{\boldsymbol{h} \in \mathcal{L}^\perp \setminus \{\mathbf{0}\}} \frac{1}{(r_1(\boldsymbol{h}))^\alpha}.$$

The following theorem is due to Sloan and Kachoyan [231].

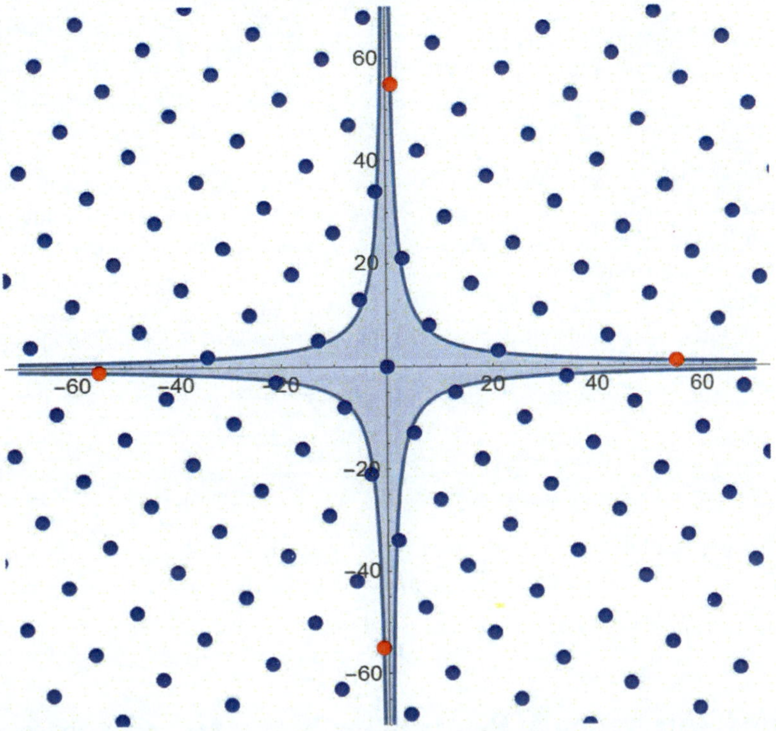

Fig. 1.2: The dual lattice $\mathcal{L}^\perp(\boldsymbol{g}, N)$ of the two-dimensional lattice point set with $\boldsymbol{g} = (1, 89)$ and $N = 144$ with a hyperbolic cross. All the points of the dual lattice, except for the origin, lie outside the interior of the hyperbolic cross. The red points indicate those elements of $\mathcal{L}^\perp(\boldsymbol{g}, N) \setminus \{\boldsymbol{0}\}$ minimizing r_1; here $\min_{\boldsymbol{h} \in \mathcal{L}^\perp(\boldsymbol{g},N)\setminus\{\boldsymbol{0}\}} r_1(\boldsymbol{h}) = 55$, and this value is attained for the elements $(1, 55)$, $(-1, -55)$, $(55, 1)$, $(-55, -1)$ of $\mathcal{L}^\perp(\boldsymbol{g}, N)$.

Theorem 1.15 *For any real number $\alpha > 1$ and $C > 0$, and for a d-dimensional lattice rule based on $\mathcal{P}(\mathcal{L}) = \{\boldsymbol{x}_0, \boldsymbol{x}_1, \ldots, \boldsymbol{x}_{N-1}\}$ we have*

$$\max_{f \in \mathcal{E}_\alpha^d(C)} \left| \frac{1}{N} \sum_{k=0}^{N-1} f(\boldsymbol{x}_k) - \int_{[0,1]^d} f(\boldsymbol{x}) \,\mathrm{d}\boldsymbol{x} \right| = C P_\alpha(\mathcal{L}).$$

Proof It follows from Proposition 1.12 and the definition of $\mathcal{E}_\alpha^d(C)$ that the result holds if we weaken equality to "$\le$". To prove equality we choose a function f_0 such that

$$f_0(\boldsymbol{x}) = C \sum_{\boldsymbol{h} \in \mathbb{Z}^d} \frac{1}{(r_1(\boldsymbol{h}))^\alpha} \mathrm{e}^{2\pi \mathrm{i} \boldsymbol{h} \cdot \boldsymbol{x}} \quad \text{for } \boldsymbol{x} \in \mathbb{R}^d.$$

Then $f_0 \in \mathcal{E}_\alpha^d(C)$ and

$$\frac{1}{N}\sum_{k=0}^{N-1} f_0(\boldsymbol{x}_k) - \int_{[0,1]^d} f_0(\boldsymbol{x})\,\mathrm{d}\boldsymbol{x} = CP_\alpha(\mathcal{L}).$$

This completes the proof. □

Theorem 1.15 shows that, for given $\alpha > 1$ and $N \in \mathbb{N}$, a lattice rule should be chosen such that $P_\alpha(\mathcal{L})$ is small. There are several results available that show that such lattice rules indeed exist, see [113, 199, 230]. For the moment we only present an existence result by Disney and Sloan [60] for rank-1 lattice rules, and refer to that paper for a proof. In the following we write $P_\alpha(\boldsymbol{g}, N)$ for $P_\alpha(\mathcal{L})$ when $\mathcal{L}$ corresponds to a rank-1 lattice rule with lattice point set $\mathcal{P}(\mathcal{L}) = \mathcal{P}(\boldsymbol{g}, N)$.

Theorem 1.16 *For every $d \geq 3$ and every $\alpha > 1$ we have*

$$\min_{\boldsymbol{g}\in\mathbb{Z}^d} P_\alpha(\boldsymbol{g}, N) \leq \left(\left(\frac{2\mathrm{e}}{d}\right)^{\alpha d} + o(1)\right)\frac{(\log N)^{\alpha d}}{N^\alpha} \quad \textit{as } N \to \infty. \tag{1.13}$$

This result implies that with the method of good lattice points one can obtain a convergence rate $O(N^{-\alpha})$ for $f \in \mathcal{E}_\alpha^d$, up to logarithmic factors. We will return to this issue in Chapter 2.

1.5 QMC Integration in Reproducing Kernel Hilbert Spaces

In the current literature, QMC rules are mostly studied in the context of reproducing kernel Hilbert spaces of functions, as this setting makes it possible to tailor the QMC rules (or, respectively, their node sets) to the properties of the function class under consideration, and to represent the integration error in a very convenient way, as we shall see below.

We first give the definition of a reproducing kernel Hilbert space of functions defined on a general domain $D \subseteq \mathbb{R}^d$. In this section, however, the reader may think of D as being equal to the unit cube $[0, 1]^d$ in most cases.

Definition 1.17 A Hilbert space $\mathcal{H}$ of functions $f : D \to \mathbb{C}$ with inner product $\langle\cdot,\cdot\rangle$ is said to be a *reproducing kernel Hilbert space* if there exists a function $K : D \times D \to \mathbb{C}$, called the *reproducing kernel* of $\mathcal{H}$, with the following properties.

1. For all $\boldsymbol{y} \in D$ we have $K(\cdot, \boldsymbol{y}) \in \mathcal{H}$;
2. for all $\boldsymbol{y} \in D$ and for all $f \in \mathcal{H}$ we have $f(\boldsymbol{y}) = \langle f, K(\cdot, \boldsymbol{y})\rangle$.

The second property of a reproducing kernel is called the *reproducing property*, which means that the evaluation of function values can be expressed or "reproduced" as the inner product of the function with the kernel.

Any kernel that satisfies Properties 1 and 2 in Definition 1.17 is symmetric, uniquely defined, and positive semi-definite. To be more precise, the following proposition holds.

Proposition 1.18 *Let $K : D \times D \to \mathbb{C}$ be the reproducing kernel of a Hilbert space $\mathcal{H}$. Then K satisfies the following properties.*

3. *Conjugate symmetry: for all $\boldsymbol{x}, \boldsymbol{y} \in D$ we have $K(\boldsymbol{x}, \boldsymbol{y}) = \overline{K(\boldsymbol{y}, \boldsymbol{x})}$.*
4. *Uniqueness: for any function $\widetilde{K} : D \times D \to \mathbb{C}$ satisfying the properties in Definition 1.17 we have $\widetilde{K} = K$.*
5. *Positive semi-definiteness: for any choice of $a_1, a_2, \ldots, a_n \in \mathbb{C}$ and any choice of $\boldsymbol{x}_1, \boldsymbol{x}_2, \ldots, \boldsymbol{x}_n \in D$ we have*

$$\sum_{i=1}^{n} \sum_{j=1}^{n} \overline{a}_i a_j K(\boldsymbol{x}_i, \boldsymbol{x}_j) \geq 0.$$

Proof For showing Property 3, using the conjugate symmetry of the inner product and the fact that $K(\cdot, \boldsymbol{y}) \in \mathcal{H}$ for every $\boldsymbol{y} \in [0, 1]^d$ we get, for every $\boldsymbol{x} \in D$,

$$K(\boldsymbol{x}, \boldsymbol{y}) = \langle K(\cdot, \boldsymbol{y}), K(\cdot, \boldsymbol{x}) \rangle = \overline{\langle K(\cdot, \boldsymbol{x}), K(\cdot, \boldsymbol{y}) \rangle} = \overline{K(\boldsymbol{y}, \boldsymbol{x})}.$$

Regarding Property 4, we have, for every $\boldsymbol{x}, \boldsymbol{y} \in D$,

$$\widetilde{K}(\boldsymbol{x}, \boldsymbol{y}) = \langle \widetilde{K}(\cdot, \boldsymbol{y}), K(\cdot, \boldsymbol{x}) \rangle = \overline{\langle K(\cdot, \boldsymbol{x}), \widetilde{K}(\cdot, \boldsymbol{y}) \rangle} = \overline{K(\boldsymbol{y}, \boldsymbol{x})} = K(\boldsymbol{x}, \boldsymbol{y}).$$

Finally, for Point 5, note that for every choice of $a_1, a_2, \ldots, a_n \in \mathbb{C}$ and of $\boldsymbol{x}_1, \boldsymbol{x}_2, \ldots, \boldsymbol{x}_n \in D$, we have

$$\begin{aligned}
\sum_{i=1}^{N} \sum_{j=1}^{N} \overline{a}_i a_j K(\boldsymbol{x}_i, \boldsymbol{x}_j) &= \sum_{i=1}^{N} \sum_{j=1}^{N} \overline{a}_i a_j \langle K(\cdot, \boldsymbol{x}_j), K(\cdot, \boldsymbol{x}_i) \rangle \\
&= \left\langle \sum_{j=1}^{N} a_j K(\cdot, \boldsymbol{x}_j), \sum_{i=1}^{N} a_i K(\cdot, \boldsymbol{x}_i) \right\rangle \\
&= \left\| \sum_{i=1}^{N} a_i K(\cdot, \boldsymbol{x}_i) \right\|^2 \geq 0.
\end{aligned}$$

□

On the other hand, it can be shown that a function $K : D \times D \to \mathbb{C}$ satisfying Properties 3 and 5 in Proposition 1.18 uniquely determines a reproducing kernel Hilbert space of functions with kernel K. In this context we will sometimes write $\mathcal{H}(K)$ instead of $\mathcal{H}$ in order to stress that K is the reproducing kernel of $\mathcal{H}$.

Let us now give some examples of reproducing kernel Hilbert spaces.

Example 1.19 Let $\mathcal{H}_r = \{a_0 + a_1 x + \cdots + a_r x^r \ : \ a_0, a_1, \ldots, a_r \in \mathbb{R}\}$ be the space of all polynomials over $\mathbb{R}$ defined on the real line of degree at most r. Define an inner product on $\mathcal{H}_r$ in the following way. For $p(x) = a_0 + a_1 x + \cdots + a_r x^r$ and $q(x) = b_0 + b_1 x + \cdots + b_r x^r$ in $\mathcal{H}_r$ let

$$\langle p, q \rangle = a_0 b_0 + \cdots + a_r b_r.$$

The reproducing kernel of this space is given by

$$K(x,y) = 1 + xy + x^2y^2 + \cdots + x^r y^r \quad \text{for} \quad x, y \in \mathbb{R},$$

because

1. for all $y \in \mathbb{R}$ we obviously have $K(\cdot, y) \in \mathcal{H}_r$, and
2. $\langle p, K(\cdot, y)\rangle = a_0 + a_1 y + \cdots + a_r y^r = p(y)$ holds for all $p \in \mathcal{H}_r$ and all $y \in \mathbb{R}$.

Example 1.20 Let $\mathcal{H}_r = \{a_0 + a_1 \mathrm{e}^{2\pi \mathrm{i} x} + \cdots + a_r \mathrm{e}^{2\pi \mathrm{i} r x} : a_0, a_1, \ldots, a_r \in \mathbb{C}\}$ be the space of all trigonometric polynomials on $[0,1]^d$ of degree at most r. Define an inner product on $\mathcal{H}_r$ in the following way. For $f, g \in \mathcal{H}_r$ with $f(x) = a_0 + a_1 \mathrm{e}^{2\pi \mathrm{i} x} + \cdots + a_r \mathrm{e}^{2\pi \mathrm{i} r x}$ and $g(x) = b_0 + b_1 \mathrm{e}^{2\pi \mathrm{i} x} + \cdots + b_r \mathrm{e}^{2\pi \mathrm{i} r x}$, let

$$\langle f, g\rangle = a_0 \overline{b}_0 + \cdots + a_r \overline{b}_r.$$

The reproducing kernel of this space is

$$K(x,y) = 1 + \mathrm{e}^{2\pi \mathrm{i}(x-y)} + \cdots + \mathrm{e}^{2\pi \mathrm{i} r(x-y)} \quad \text{for} \quad x, y \in [0,1]^d,$$

because

1. for all $y \in [0,1]^d$ we obviously have $K(\cdot, y) \in \mathcal{H}_r$, and
2. $\langle p, K(\cdot, y)\rangle = a_0 + a_1 \mathrm{e}^{2\pi \mathrm{i} y} + \cdots + a_r \mathrm{e}^{2\pi \mathrm{i} r y} = p(y)$ for every $p \in \mathcal{H}_r$ and every $y \in [0,1]^d$.

The following example is of great importance in the context of Koksma–Hlawka type inequalities, which will be introduced in the next section.

Example 1.21 For absolutely continuous functions $f, g : [0,1] \to \mathbb{R}$ define an inner product by

$$\langle f, g\rangle_1 := f(1)g(1) + \int_0^1 f'(x) g'(x) \,\mathrm{d}x \tag{1.14}$$

and the corresponding norm $\|f\|_{1,2} := \sqrt{\langle f, f\rangle_1}$. We have $\|f\|_{1,2} < \infty$ whenever the first derivative f' of f is in $L_2([0,1])$. Based on this norm we define a Hilbert space $\mathcal{H}_1$ by

$$\mathcal{H}_1 = \{f : [0,1] \to \mathbb{R} : f \text{ absolutely continuous and } \|f\|_{1,2} < \infty\}.$$

It is well known that a function $f : [0,1] \to \mathbb{R}$ is *absolutely continuous* if and only if there exists a Lebesgue integrable function $g : [0,1] \to \mathbb{R}$ such that f can be written in the form

$$f(x) = f(0) + \int_0^x g(t) \,\mathrm{d}t = f(1) - \int_x^1 g(t) \,\mathrm{d}t \quad \text{for all} \quad x \in [0,1].$$

In this case f is almost everywhere differentiable and $g = f'$ almost everywhere.

We now show that the function K_1 defined by

$$K_1(x, y) = 1 + \min(1 - x, 1 - y) \tag{1.15}$$

is the reproducing kernel of $\mathcal{H}_1$. For fixed $y \in [0, 1]$ we have

$$K_1(x, y) = \begin{cases} 2 - x & \text{if } x > y, \\ 2 - y & \text{if } x \le y, \end{cases}$$

and

$$\frac{\partial K_1(x, y)}{\partial x} = \begin{cases} -1 & \text{if } x > y, \\ 0 & \text{if } x \le y. \end{cases}$$

Thus with $g : [0, 1] \to \mathbb{R}$ defined by

$$g(x) := \begin{cases} -1 & \text{if } x > y, \\ 0 & \text{if } x \le y, \end{cases}$$

we obtain

$$K_1(1, y) - \int_x^1 g(t)\,\mathrm{d}t = 1 + \int_{\max(x,y)}^1 \mathrm{d}t = 1 + \min(1 - x, 1 - y) = K_1(x, y),$$

and it is checked analogously that $K_1(0, y) + \int_0^x g(t)\,\mathrm{d}t = K_1(x, y)$. Therefore the function $K_1(\cdot, y)$ is absolutely continuous. In particular, $K_1(\cdot, y)$ is also integrable. Furthermore, we have

$$\begin{aligned} \langle K_1(\cdot, y), K_1(\cdot, y)\rangle_1 &= (K_1(1, y))^2 + \int_0^1 \left(\frac{\partial K_1(x, y)}{\partial x}\right)^2 \mathrm{d}x \\ &= 1 + \int_y^1 \mathrm{d}x = 2 - y < \infty, \end{aligned}$$

and hence $\|K_1(\cdot, y)\|_{1,2} < \infty$. This implies that $K_1(\cdot, y) \in \mathcal{H}_1$. It remains to check the reproducing property of K_1. We have

$$\begin{aligned} \langle f, K_1(\cdot, y)\rangle_1 &= f(1)K_1(1, y) + \int_0^1 f'(x)\frac{\partial K_1(x, y)}{\partial x}\,\mathrm{d}x \\ &= f(1) - \int_y^1 f'(x)\,\mathrm{d}x = f(y). \end{aligned}$$

Altogether, we have shown that K_1 is the reproducing kernel of the space $\mathcal{H}_1$.

In this book, we shall frequently study QMC integration over $[0, 1]^d$ of functions from a given Hilbert space $\mathcal{H}$. For the application of QMC rules we require that point evaluation of functions in $\mathcal{H}$, the integrands, makes sense. In particular, we demand that for every $\boldsymbol{y} \in [0,1]^d$ the evaluation functional $T_{\boldsymbol{y}} : \mathcal{H} \to \mathbb{C}$ that evaluates f at $\boldsymbol{y}$ is continuous. It turns out, however, that this requirement is exactly equivalent to the existence of a reproducing kernel of $\mathcal{H}$.

Theorem 1.22 *Let $\mathcal{H}$ be a Hilbert space of functions on $[0,1]^d$ with inner product $\langle\cdot,\cdot\rangle$. Then $\mathcal{H}$ is a reproducing kernel Hilbert space if and only if the evaluation functionals*

$$T_{\boldsymbol{y}}(f) = f(\boldsymbol{y}) \quad \textit{for} \quad f \in \mathcal{H},\ \boldsymbol{y} \in [0,1]^d$$

are continuous.

Proof If the evaluation functionals are continuous, then the representation theorem of Fréchet–Riesz guarantees, for every $\boldsymbol{y} \in [0,1]^d$, the existence of a uniquely determined function $k_{\boldsymbol{y}} \in \mathcal{H}$ with the property that

$$T_{\boldsymbol{y}}(f) = \langle f, k_{\boldsymbol{y}}\rangle \quad \text{for all} \quad f \in \mathcal{H}.$$

If we now define $K(\boldsymbol{x},\boldsymbol{y}) = k_{\boldsymbol{y}}(\boldsymbol{x})$ for $\boldsymbol{x},\boldsymbol{y} \in [0,1]^d$, then K satisfies the two properties in Definition 1.17, and thus $\mathcal{H}$ is a reproducing kernel Hilbert space with reproducing kernel K.

Conversely, assume that K is a reproducing kernel for $\mathcal{H}$. Let $\boldsymbol{y} \in [0,1]^d$. Using the Cauchy–Schwarz inequality, we get for every $f \in \mathcal{H}$ that

$$|T_{\boldsymbol{y}}(f)| = |f(\boldsymbol{y})| = |\langle f, K(\cdot,\boldsymbol{y})\rangle| \le \|f\|\,\|K(\cdot,\boldsymbol{y})\|.$$

From the reproducing property we obtain $\|K(\cdot,\boldsymbol{y})\|^2 = \langle K(\cdot,\boldsymbol{y}), K(\cdot,\boldsymbol{y})\rangle = K(\boldsymbol{y},\boldsymbol{y})$, such that $|T_{\boldsymbol{y}}(f)| \le C$ for every f with $\|f\| \le 1$, where $C := \sqrt{K(\boldsymbol{y},\boldsymbol{y})}$. This means that $T_{\boldsymbol{y}}$ is continuous. □

From now on we will always assume that the functions in $\mathcal{H}(K)$ are absolutely integrable so that $\mathcal{H}(K)$ is embedded in $L_1([0,1]^d)$. For our analysis, we also need to consider the integration functional, which shall be denoted by I, or I_d if we would like to emphasize dependence on d. Indeed, we require that the integration functional

$$I(f) := \int_{[0,1]^d} f(\boldsymbol{x})\,\mathrm{d}\boldsymbol{x} \quad \text{for} \quad f \in \mathcal{H}(K)$$

is continuous. This is guaranteed if the embedding operator $J_{\mathcal{H}} : \mathcal{H}(K) \to L_1([0,1]^d)$ is continuous. The following result is due to Hinrichs [108].

Lemma 1.23 *Let $\mathcal{H}(K)$ be a reproducing kernel Hilbert space of absolutely integrable functions on $[0,1]^d$. Then the embedding operator $J_{\mathcal{H}} : \mathcal{H}(K) \to L_1([0,1]^d)$ is continuous. In particular, the integration functional is continuous and we have*

$$\int_{[0,1]^d} |f(\boldsymbol{x})|\,\mathrm{d}\boldsymbol{x} \le \|J_{\mathcal{H}}\|\,\|f\|_{\mathcal{H}(K)} \quad \textit{for all } f \in \mathcal{H}(K),$$

where $\|J_{\mathcal{H}}\| < \infty$ is the operator norm of $J_{\mathcal{H}}$.

Proof The result is a consequence of the closed graph theorem which states that a linear operator $T : X \to Y$ is continuous if and only if its graph

$$\{(x, y) \in X \times Y \ : \ Tx = y\}$$

is closed in $X \times Y$. In the present case we have to show that the graph

$$\{(f, g) \in \mathcal{H}(K) \times L_1([0, 1]^d) \ : \ f = g\} \tag{1.16}$$

of the embedding operator $J_{\mathcal{H}} : \mathcal{H}(K) \to L_1([0, 1]^d)$ is closed.

Let now $(f_n)_{n\geq 1}$ be a sequence of functions from $\mathcal{H}(K)$ which converges to f in $\mathcal{H}(K)$ and to g in $L_1([0, 1]^d)$. For $\boldsymbol{x} \in [0, 1]^d$ we have

$$\begin{aligned}|f(\boldsymbol{x}) - f_n(\boldsymbol{x})| &= |\langle f, K(\cdot, x)\rangle - \langle f_n, K(\cdot, x)\rangle| \\ &= |\langle f - f_n, K(\cdot, x)\rangle| \leq \|f - f_n\| \, \|K(\cdot, x)\|,\end{aligned}$$

and hence f is also the pointwise limit of the sequence $(f_n)_{n\geq 1}$. On the other hand, convergence in $L_1([0, 1]^d)$ implies convergence in measure with respect to the measure $\mathrm{d}x$. Convergence of $(f_n)_{n\geq 1}$ to g in measure now implies that a subsequence of $(f_n)_{n\geq 1}$ converges to g almost everywhere with respect to $\mathrm{d}x$. Now f and g are equal almost everywhere with respect to $\mathrm{d}x$, so they are equal in $L_1([0, 1]^d)$. Thus, the graph (1.16) is closed, and therefore $J_{\mathcal{H}} : \mathcal{H}(K) \to L_1([0, 1]^d)$ is continuous. □

Remark 1.24 In general, integrability of the kernel function K is not sufficient to guarantee integrability of all elements of $\mathcal{H}(K)$. An example where $K(\cdot, \boldsymbol{y})$ is integrable for every $\boldsymbol{y}$, but $\mathcal{H}(K)$ contains functions that are not integrable can be found in [212, Notes on Section 23.4]. However, we remark that a sufficient condition for the continuity of the integration operator I is the condition

$$\int_{[0,1]^d} \sqrt{K(\boldsymbol{x}, \boldsymbol{x})} \, \mathrm{d}\boldsymbol{x} < \infty, \tag{1.17}$$

since then for all $f \in \mathcal{H}(K)$ with $\|f\| \leq 1$ we have

$$|I(f)| = \left|\int_{[0,1]^d} T_{\boldsymbol{x}}(f) \, \mathrm{d}\boldsymbol{x}\right| \leq \int_{[0,1]^d} |T_{\boldsymbol{x}}(f)| \, \mathrm{d}\boldsymbol{x} \leq \int_{[0,1]^d} \sqrt{K(\boldsymbol{x}, \boldsymbol{x})} \, \mathrm{d}\boldsymbol{x} < \infty.$$

Continuity of the integration functional is important since then one is allowed to interchange inner product and integral, which is made more precise in the following lemma.

Lemma 1.25 *Let $\mathcal{H}(K)$ be a reproducing kernel Hilbert space of absolutely integrable functions on $[0, 1]^d$ with inner product $\langle \cdot, \cdot \rangle$. Then we have*

$$\int_{[0,1]^d} \langle f, K(\cdot, \boldsymbol{y})\rangle \, \mathrm{d}\boldsymbol{y} = \left\langle f, \int_{[0,1]^d} K(\cdot, \boldsymbol{y}) \, \mathrm{d}\boldsymbol{y} \right\rangle.$$

Proof According to Lemma 1.23 the integration functional $I(f)$ is a continuous linear functional on $\mathcal{H}(K)$. Consequently, the representation theorem of Fréchet–Riesz guarantees the existence of a unique function $R \in \mathcal{H}(K)$, such that

$$\int_{[0,1]^d} f(\boldsymbol{y})\,\mathrm{d}\boldsymbol{y} = I(f) = \langle f, R\rangle \quad \text{for all } f \in \mathcal{H}(K).$$

Since $R \in \mathcal{H}(K)$, the evaluation of R at some $\boldsymbol{x}$ can be expressed in terms of the inner product with the kernel function $K(\cdot, \boldsymbol{x})$ and hence

$$R(\boldsymbol{x}) = \langle R, K(\cdot,\boldsymbol{x})\rangle = \overline{\langle K(\cdot,\boldsymbol{x}), R\rangle} = \overline{\int_{[0,1]^d} K(\boldsymbol{y},\boldsymbol{x})\,\mathrm{d}\boldsymbol{y}} = \int_{[0,1]^d} K(\boldsymbol{x},\boldsymbol{y})\,\mathrm{d}\boldsymbol{y},$$

where we used that K is conjugate symmetric. Thus we have

$$\int_{[0,1]^d} \langle f, K(\cdot,\boldsymbol{y})\rangle\,\mathrm{d}\boldsymbol{y} = \int_{[0,1]^d} f(\boldsymbol{y})\,\mathrm{d}\boldsymbol{y} = \langle f, R\rangle = \left\langle f, \int_{[0,1]^d} K(\cdot,\boldsymbol{y})\,\mathrm{d}\boldsymbol{y}\right\rangle. \quad \square$$

QMC integration in reproducing kernel Hilbert spaces

We now study QMC integration for elements of a reproducing kernel Hilbert space $\mathcal{H}(K)$ of absolutely integrable functions in the so-called *worst-case setting*. For an integrable function f defined on $[0,1]^d$ and a point set $\mathcal{P} = \{\boldsymbol{x}_0, \boldsymbol{x}_1, \ldots, \boldsymbol{x}_{N-1}\}$ in $[0,1)^d$ we use the notation

$$\mathrm{err}_{N,d}(f,\mathcal{P}) := \int_{[0,1]^d} f(\boldsymbol{x})\,\mathrm{d}\boldsymbol{x} - \frac{1}{N}\sum_{k=0}^{N-1} f(\boldsymbol{x}_k)$$

for the integration error. We remind the reader that we call

$$Q_{N,d}(f) := \frac{1}{N}\sum_{k=0}^{N-1} f(\boldsymbol{x}_k) \tag{1.18}$$

a quasi-Monte Carlo (QMC) rule (based on $\mathcal{P}$) and $\mathrm{err}_{N,d}(\cdot,\mathcal{P}) : \mathcal{H} \to \mathbb{C}$ the *error functional*.

Definition 1.26 The *worst-case error* of a QMC rule based on an N-element point set $\mathcal{P}$ in $[0,1)^d$ in a Hilbert space $\mathcal{H}$ of absolutely integrable functions on $[0,1]^d$ is defined as

$$\mathrm{err}_{N,d}(\mathcal{H},\mathcal{P}) := \sup_{\substack{f\in\mathcal{H}\\ \|f\|\le 1}} |\mathrm{err}_{N,d}(f,\mathcal{P})|.$$

If $\mathcal{H} = \mathcal{H}(K)$ is a reproducing kernel Hilbert space, the worst-case error can be expressed by means of the kernel function K, as we shall show in the next theorem.

Theorem 1.27 *Let $\mathcal{H}(K)$ be a reproducing kernel Hilbert space of absolutely integrable functions on $[0,1]^d$ and let $\mathcal{P} = \{\boldsymbol{x}_0, \boldsymbol{x}_1, \ldots, \boldsymbol{x}_{N-1}\}$ be an N-element point set in $[0,1)^d$. Then we have*

$$[\mathrm{err}_{N,d}(\mathcal{H}(K), \mathcal{P})]^2 =$$

$$\int_{[0,1]^d}\int_{[0,1]^d} K(\boldsymbol{x},\boldsymbol{y})\,\mathrm{d}\boldsymbol{x}\,\mathrm{d}\boldsymbol{y} - \frac{2}{N}\sum_{k=0}^{N-1}\int_{[0,1]^d} K(\boldsymbol{x}_k,\boldsymbol{y})\,\mathrm{d}\boldsymbol{y} + \frac{1}{N^2}\sum_{k,\ell=0}^{N-1} K(\boldsymbol{x}_k,\boldsymbol{x}_\ell).$$

Proof We have

$$Q_{N,d}(f) = \frac{1}{N}\sum_{k=0}^{N-1}\langle f, K(\cdot,\boldsymbol{x}_k)\rangle = \left\langle f, \frac{1}{N}\sum_{k=0}^{N-1} K(\cdot,\boldsymbol{x}_k)\right\rangle$$

and, according to Lemma 1.25,

$$I(f) = \int_{[0,1]^d}\langle f, K(\cdot,\boldsymbol{y})\rangle\,\mathrm{d}\boldsymbol{y} = \left\langle f, \int_{[0,1]^d} K(\cdot,\boldsymbol{y})\,\mathrm{d}\boldsymbol{y}\right\rangle.$$

Therefore the integration error of the QMC rule $Q_{N,d}$ applied to $f \in \mathcal{H}(K)$ can be expressed as an inner product,

$$\mathrm{err}_{N,d}(f,\mathcal{P}) = \langle f, h\rangle\,, \tag{1.19}$$

where

$$h(\boldsymbol{x}) := \int_{[0,1]^d} K(\boldsymbol{x},\boldsymbol{y})\,\mathrm{d}\boldsymbol{y} - \frac{1}{N}\sum_{k=0}^{N-1} K(\boldsymbol{x},\boldsymbol{x}_k) \quad \text{for} \quad \boldsymbol{x} \in [0,1]^d.$$

This function is frequently called the *representer of the integration error*. Taking the absolute value and applying the Cauchy–Schwarz inequality leads to

$$|\mathrm{err}_{N,d}(f,\mathcal{P})| \le \|f\|\,\|h\|.$$

Moreover it follows from (1.19) that among all functions in the unit ball of $\mathcal{H}(K)$ the normalized representer $h/\|h\|$ is the hardest to integrate. Therefore, the worst-case error can be written as $\mathrm{err}_{N,d}(\mathcal{H}(K),\mathcal{P}) = \|h\|$. Hence, for the squared worst-case error we obtain

$$[\mathrm{err}_{N,d}(\mathcal{H}(K),\mathcal{P})]^2 = \langle h, h\rangle$$

$$= \left\langle \int_{[0,1]^d} K(\cdot,\boldsymbol{y})\,\mathrm{d}\boldsymbol{y} - \frac{1}{N}\sum_{k=0}^{N-1} K(\cdot,\boldsymbol{x}_k), \int_{[0,1]^d} K(\cdot,\boldsymbol{y})\,\mathrm{d}\boldsymbol{y} - \frac{1}{N}\sum_{k=0}^{N-1} K(\cdot,\boldsymbol{x}_k)\right\rangle$$

$$= \int_{[0,1]^d} \int_{[0,1]^d} K(\boldsymbol{x}, \boldsymbol{y}) \, \mathrm{d}\boldsymbol{x} \, \mathrm{d}\boldsymbol{y} - \frac{2}{N} \sum_{k=0}^{N-1} \int_{[0,1]^d} K(\boldsymbol{x}_k, \boldsymbol{y}) \, \mathrm{d}\boldsymbol{y} + \frac{1}{N^2} \sum_{k,\ell=0}^{N-1} K(\boldsymbol{x}_k, \boldsymbol{x}_\ell).$$

□

Remark 1.28 Instead of QMC rules one might also study more general *linear rules* of the form

$$A_{N,d}(f) := \sum_{k=0}^{N-1} w_k f(\boldsymbol{x}_k),$$

where again $\mathcal{P} = \{\boldsymbol{x}_0, \boldsymbol{x}_1, \ldots, \boldsymbol{x}_{N-1}\}$ is a point set in $[0,1)^d$ and where $\boldsymbol{w} = (w_0, w_1, \ldots, w_{N-1}) \in \mathbb{C}^N$ are so-called *integration weights*, not to be confused with the concept of weights for different coordinates, which shall be introduced later in this book. If $\boldsymbol{w} = (1/N, 1/N, \ldots, 1/N)$, we are back to the QMC case. We write

$$\mathrm{err}_{N,d}(f, \mathcal{P}, \boldsymbol{w}) := \int_{[0,1]^d} f(\boldsymbol{x}) \, \mathrm{d}\boldsymbol{x} - A_{N,d}(f)$$

for the error functional and

$$\mathrm{err}_{N,d}(\mathcal{H}, \mathcal{P}, \boldsymbol{w}) := \sup_{\substack{f \in \mathcal{H} \\ \|f\| \le 1}} |\mathrm{err}_{N,d}(f, \mathcal{P}, \boldsymbol{w})|$$

for the worst-case error of the linear rule $A_{N,d}$. Then one can show by the same methods as above that

$$[\mathrm{err}_{N,d}(\mathcal{H}(K), \mathcal{P}, \boldsymbol{w})]^2 =$$

$$\int_{[0,1]^d} \int_{[0,1]^d} K(\boldsymbol{x}, \boldsymbol{y}) \, \mathrm{d}\boldsymbol{x} \, \mathrm{d}\boldsymbol{y} - 2 \, \mathrm{Re} \left(\sum_{k=0}^{N-1} w_k \int_{[0,1]^d} K(\boldsymbol{y}, \boldsymbol{x}_k) \, \mathrm{d}\boldsymbol{y} \right)$$

$$+ \sum_{k,\ell=0}^{N-1} \overline{w_k} w_\ell K(\boldsymbol{x}_k, \boldsymbol{x}_\ell).$$

1.6 Discrepancy and Koksma–Hlawka Type Inequalities

The notion of discrepancy of a point set has its roots in the field of uniform distribution modulo one, going back to a seminal paper of Weyl [264] published in the year 1916. From this paper, the following classical result is known. Let $(\boldsymbol{x}_k)_{k \ge 0}$ be an infinite sequence in $[0,1)^d$, and consider a sequence of QMC rules based on the first $N \ge 1$ points of $(\boldsymbol{x}_k)_{k \ge 0}$ applied to a continuous function f on $[0,1]^d$. Then the approximate values returned by the QMC rules converge to the integral of the function for $N \to \infty$, i.e.,

$$\frac{1}{N}\sum_{k=0}^{N-1} f(\boldsymbol{x}_k) \to \int_{[0,1]^d} f(\boldsymbol{x})\,\mathrm{d}\boldsymbol{x} \quad \text{for } N \to \infty,$$

if and only if the sequence $(\boldsymbol{x}_k)_{k\geq 0}$ in $[0,1)^d$ is uniformly distributed modulo one. The rate at which this convergence takes place can be described in terms of discrepancy, which is a quantitative measure for the uniformity of a point set.

Definition 1.29 For a finite point set $\mathcal{P} = \{\boldsymbol{x}_0, \boldsymbol{x}_1, \ldots, \boldsymbol{x}_{N-1}\}$ in $[0,1)^d$ the *local discrepancy function* $\Delta_{\mathcal{P}} : [0,1]^d \to \mathbb{R}$ is defined as

$$\Delta_{\mathcal{P}}(\boldsymbol{t}) := \frac{|\{k \in \{0,1,\ldots,N-1\} \,:\, \boldsymbol{x}_k \in [\boldsymbol{0},\boldsymbol{t})\}|}{N} - \lambda_d([\boldsymbol{0},\boldsymbol{t})),$$

where, for $\boldsymbol{t} = (t_1,\ldots,t_d) \in [0,1]^d$, we write $[\boldsymbol{0},\boldsymbol{t}) = [0,t_1) \times \cdots \times [0,t_d)$, and where λ_d denotes the d-dimensional Lebesgue measure.

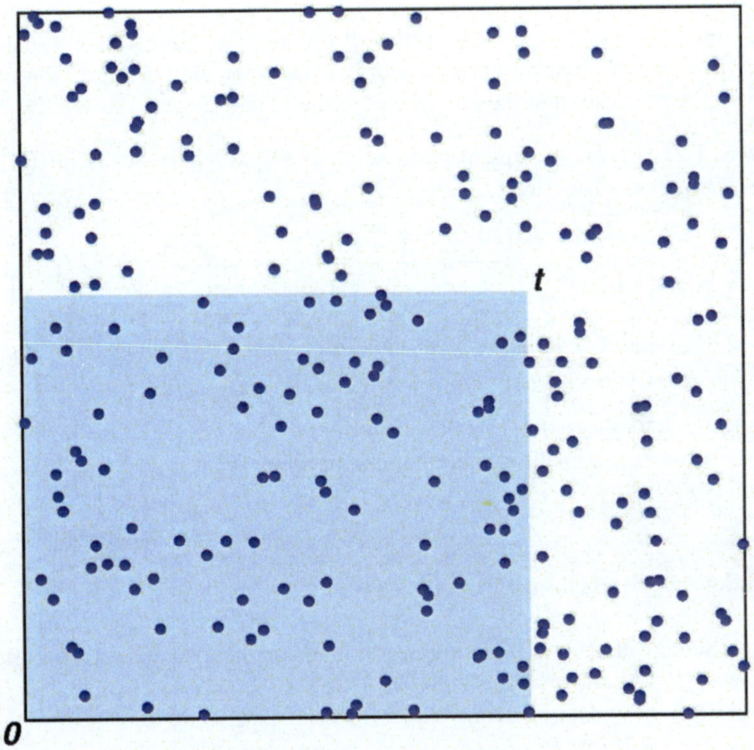

Fig. 1.3: The local discrepancy $\Delta_{\mathcal{P}}(\boldsymbol{t})$ is the difference of the relative number of points in $[\boldsymbol{0},\boldsymbol{t})$ and the volume of this interval (shaded area).

The local discrepancy measures the difference between the relative number of points of the point set $\mathcal{P}$ that belong to the axes-parallel box $[\mathbf{0}, \boldsymbol{t})$, that is anchored in the origin and has its opposite corner in $\boldsymbol{t}$, and the volume of this box (see Figure 1.3). Taking a suitable norm of the local discrepancy function yields a "global" measure for the irregularity of distribution of $\mathcal{P}$ that is called *discrepancy*. Most prominent are the L_p-norms leading to the notions of L_p-discrepancies.

Definition 1.30 For an N-element point set $\mathcal{P}$ in $[0,1)^d$ and for $p \geq 1$ the L_p-*discrepancy* is defined as the L_p-norm of the corresponding local discrepancy function, i.e., if $p \in [1, \infty)$ we define

$$L_{p,N}(\mathcal{P}) := \left(\int_{[0,1]^d} |\Delta_{\mathcal{P}}(\boldsymbol{t})|^p \, \mathrm{d}\boldsymbol{t} \right)^{1/p}.$$

If $p = \infty$ we also speak of the *star-discrepancy*, which is denoted by

$$D_N^*(\mathcal{P}) = L_{\infty,N}(\mathcal{P}) := \sup_{\boldsymbol{t} \in [0,1]^d} |\Delta_{\mathcal{P}}(\boldsymbol{t})|.$$

Sometimes a slightly stricter notion of discrepancy is used, which is called the *extreme discrepancy*. Instead of only intervals anchored in the origin, the extreme discrepancy uses all subintervals of the unit cube as test sets.

Definition 1.31 For an N-element point set $\mathcal{P} = \{\boldsymbol{x}_0, \boldsymbol{x}_1, \ldots, \boldsymbol{x}_{N-1}\}$ in $[0,1)^d$ the *extreme discrepancy* is defined as

$$D_N(\mathcal{P}) := \sup_{\boldsymbol{x} \leq \boldsymbol{y}} \left| \frac{|\{k \in \{0,1,\ldots,N-1\} \,:\, \boldsymbol{x}_k \in [\boldsymbol{x}, \boldsymbol{y})\}|}{N} - \lambda_d([\boldsymbol{x}, \boldsymbol{y})) \right|,$$

where the supremum is extended over all $\boldsymbol{x}, \boldsymbol{y} \in [0,1]^d$ for which $x_j \leq y_j$ for all coordinates $j \in [d]$.

It is clear that for every point set $\mathcal{P}$ we have $D_N^*(\mathcal{P}) \leq D_N(\mathcal{P})$. On the other hand, using the inclusion-exclusion principle it can be shown that $D_N(\mathcal{P}) \leq 2^d D_N^*(\mathcal{P})$.

General lower bounds on the discrepancy

We collect some well-known general lower bounds on the discrepancy of N-element point sets in $[0,1)^d$. For every $p \in (1, \infty]$ and $d \in \mathbb{N}$ there exists a $c_{p,d} > 0$ such that for every N-element point set $\mathcal{P}$ in $[0,1)^d$ with $N \geq 2$ we have

$$L_{p,N}(\mathcal{P}) \geq c_{p,d} \frac{(\log N)^{(d-1)/2}}{N} \quad \text{and} \quad D_N^*(\mathcal{P}) \geq c_{\infty,d} \frac{(\log N)^{(d-1)/2+\eta_d}}{N} \tag{1.20}$$

for some $\eta_d \in (0, 1/2)$. The first result in (1.20) for $p \geq 2$ is a celebrated result by Roth [222] that was extended later by Schmidt [226] to the case $p \in (1, 2)$. The general lower bound for the star-discrepancy is an important result of Bilyk, Lacey, and Vagharshakyan [14]. In the two-dimensional case, the results in (1.20) can be made tighter. There exist positive constants $c_{1,2}$ and $c_{\infty,2}$ such that for every N-element point set $\mathcal{P}$ in $[0, 1)^2$ with $N \geq 2$ we have

$$L_{1,N}(\mathcal{P}) \geq c_{1,2} \frac{\sqrt{\log N}}{N} \quad \text{and} \quad D_N^*(\mathcal{P}) \geq c_{\infty,2} \frac{\log N}{N}. \tag{1.21}$$

The result for the L_1-discrepancy in (1.21) was shown by Halász [91] and the result for the star-discrepancy by Schmidt [224].

We will see later that the notions of the L_p-discrepancies are intimately related to the worst-case errors of QMC rules in certain function spaces. This leads us to error estimates which are known as Koksma–Hlawka type inequalities. Koksma–Hlawka type inequalities are the fundamental error estimates for QMC rules. These inequalities separate the influence of the integrand on the integration error from that of the underlying integration nodes. The classical Koksma–Hlawka inequality based on the star-discrepancy was proven by Koksma [134] for dimension $d = 1$ and later generalized to arbitrary dimension $d \in \mathbb{N}$ by Hlawka [112]. The inequality in its classical form is a very general error estimate valid for all functions of finite variation in the sense of Hardy and Krause, which reduces to the total variation in dimension $d = 1$. We will present the inequality at the end of this section. The deduction of this general version can be found in [155, Chapter 2, Section 5]. Here we choose a slightly more specific but very elegant approach by means of reproducing kernel Hilbert spaces.

The anchored Sobolev space of smoothness one

We consider the reproducing kernel $K_d : [0, 1]^d \times [0, 1]^d \to \mathbb{R}$ given by

$$K_d(\boldsymbol{x}, \boldsymbol{y}) := \prod_{j=1}^{d} K_1(x_j, y_j) = \prod_{j=1}^{d} (1 + \min(1 - x_j, 1 - y_j)), \tag{1.22}$$

where $\boldsymbol{x} = (x_1, \ldots, x_d) \in [0, 1]^d$ and $\boldsymbol{y} = (y_1, \ldots, y_d) \in [0, 1]^d$, and where K_1 is defined as in Example 1.21 by $K_1(x, y) = 1 + \min(1 - x, 1 - y)$ for $x, y \in [0, 1]$. The corresponding inner product is given by

$$\langle f, g \rangle_d := \sum_{\mathfrak{u} \subseteq [d]} \int_{[0,1]^{|\mathfrak{u}|}} f^{(\mathfrak{u})}(\boldsymbol{x}_{\mathfrak{u}}, \mathbf{1}) g^{(\mathfrak{u})}(\boldsymbol{x}_{\mathfrak{u}}, \mathbf{1}) \, \mathrm{d}\boldsymbol{x}_{\mathfrak{u}}. \tag{1.23}$$

Here, for $\mathfrak{u} \subseteq [d]$ and $\boldsymbol{x} = (x_1, \ldots, x_d) \in [0, 1]^d$, we put $(\boldsymbol{x}_{\mathfrak{u}}, \mathbf{1}) = (z_1, \ldots, z_d)$, where

$$z_j = \begin{cases} x_j & \text{if } j \in \mathfrak{u}, \\ 1 & \text{if } j \notin \mathfrak{u}, \end{cases}$$

for $j \in [d]$. Furthermore,

$$f^{(\mathfrak{u})} = \frac{\partial^{|\mathfrak{u}|}}{\partial \boldsymbol{x}_{\mathfrak{u}}} f = \prod_{j \in \mathfrak{u}} \frac{\partial}{\partial x_j} f \quad \text{with } f^{(\emptyset)} = f$$

denotes the mixed first partial derivative of f with respect to those components of $\boldsymbol{x}$ whose index is in $\mathfrak{u}$. We note that, with some abuse of notation, we just write $f(\boldsymbol{x}_{\mathfrak{u}}, \mathbf{1})$ instead of $f((\boldsymbol{x}_{\mathfrak{u}}, \mathbf{1}))$, and likewise for partial derivatives.

Let now $\mathcal{H}_d$ be the reproducing kernel Hilbert space with reproducing kernel K_d and norm $\|\cdot\|_{d,2} := \langle \cdot, \cdot \rangle_d^{1/2}$. We call $\mathcal{H}_d = \mathcal{H}_d(K_d)$ the *anchored Sobolev space of smoothness one with anchor in* $\mathbf{1}$.

This space contains all functions f defined on $[0, 1]^d$ whose mixed partial derivatives $f^{(\mathfrak{u})}$ up to order one in each variable belong to $L_2([0, 1]^d)$ and that can be represented in the form

$$\begin{aligned} f(\boldsymbol{x}) &= \langle f, K_d(\cdot, \boldsymbol{x}) \rangle_d \\ &= \sum_{\mathfrak{u} \subseteq [d]} (-1)^{|\mathfrak{u}|} \int_{[0,1]^{|\mathfrak{u}|}} f^{(\mathfrak{u})}(\boldsymbol{y}_{\mathfrak{u}}, \mathbf{1})\, \chi_{[\mathbf{0}, (\boldsymbol{y}_{\mathfrak{u}}, \mathbf{1}))}(\boldsymbol{x})\, \mathrm{d}\boldsymbol{y}_{\mathfrak{u}} \quad \text{for} \quad \boldsymbol{x} \in [0, 1]^d, \end{aligned}$$

where χ_A denotes the indicator function of a set A.

In Example 1.21 we have already considered the univariate case $\mathcal{H}_1$ and learned that it consists of all absolutely continuous functions f with square integrable first derivative. For $d > 1$ one can show that $\mathcal{H}_d$ is the d-fold tensor product of $\mathcal{H}_1$, i.e.,

$$\mathcal{H}_d := \underbrace{\mathcal{H}_1 \otimes \cdots \otimes \mathcal{H}_1}_{d\text{-fold}} = \overline{\operatorname{span}\left\{ \boldsymbol{x} \mapsto \prod_{j=1}^{d} f_j(x_j) \ :\ f_j \in \mathcal{H}_1 \right\}},$$

where $\boldsymbol{x} = (x_1, \ldots, x_d)$, and where by $\overline{A}$ we denote the closure of a set A. Here the closure is taken with respect to the norm $\|\cdot\|_{d,2}$.

We have

$$\int_{[0,1]^d} \sqrt{K_d(\boldsymbol{y}, \boldsymbol{y})}\, \mathrm{d}\boldsymbol{y} = \left(\int_0^1 \sqrt{2 - y}\, \mathrm{d}y \right)^d = \left(\frac{2(\sqrt{8} - 1)}{3} \right)^d < \infty,$$

and hence the integration operator I is continuous (cf. Remark 1.24). Thus it follows from (1.19) that

$$\mathrm{err}_{N,d}(f, \mathcal{P}) = \langle f, h \rangle_d$$

for all $f \in \mathcal{H}_d$ and all N-element point sets $\mathcal{P}$ in $[0, 1)^d$, where h is the representer of the integration error in $\mathcal{H}_d$. Moreover,

$$
\begin{aligned}
h(\boldsymbol{x}) &= \int_{[0,1]^d} K_d(\boldsymbol{x},\boldsymbol{y})\,\mathrm{d}\boldsymbol{y} - \frac{1}{N}\sum_{k=0}^{N-1} K_d(\boldsymbol{x},\boldsymbol{x}_k) \\
&= \prod_{j=1}^{d}\left(\int_0^1 (1+\min(1-x_j,1-y_j))\,\mathrm{d}y_j\right) \\
&\quad - \frac{1}{N}\sum_{k=0}^{N-1}\prod_{j=1}^{d}(1+\min(1-x_j,1-x_{k,j})) \\
&= \prod_{j=1}^{d}\frac{3-x_j^2}{2} - \frac{1}{N}\sum_{k=0}^{N-1}\prod_{j=1}^{d}(1+\min(1-x_j,1-x_{k,j})),
\end{aligned}
$$

where we write $\boldsymbol{x}_k = (x_{k,1},\ldots,x_{k,d})$. For $\mathfrak{u} \subseteq [d]$ we have

$$
\begin{aligned}
h^{(\mathfrak{u})}(\boldsymbol{x}_{\mathfrak{u}},\boldsymbol{1}) &= (-1)^{|\mathfrak{u}|}\left(\prod_{j\in\mathfrak{u}} x_j - \frac{1}{N}\sum_{k=0}^{N-1}\prod_{j\in\mathfrak{u}}\chi_{[0,x_j)}(x_{k,j})\right) \\
&= (-1)^{|\mathfrak{u}|}\left(\prod_{j\in\mathfrak{u}} x_j - \frac{1}{N}\sum_{k=0}^{N-1}\chi_{[\boldsymbol{0},(\boldsymbol{x}_{\mathfrak{u}},\boldsymbol{1}))}(\boldsymbol{x}_k)\right) \\
&= (-1)^{|\mathfrak{u}|+1}\Delta_{\mathcal{P}}(\boldsymbol{x}_{\mathfrak{u}},\boldsymbol{1}),
\end{aligned}
$$

where $\Delta_{\mathcal{P}}$ denotes the local discrepancy of the N-element point set $\mathcal{P}$ as given in Definition 1.29. Note that $\Delta_{\mathcal{P}}(\boldsymbol{x}_{\emptyset},\boldsymbol{1}) = \Delta_{\mathcal{P}}(\boldsymbol{1}) = 0$.

From these considerations we obtain a formula for the integration error in $\mathcal{H}_d$ which is known as *Hlawka's identity* or *Zaremba's identity*.

Theorem 1.32 (Hlawka's identity, Zaremba's identity) *For $f \in \mathcal{H}_d$ and an N-point set $\mathcal{P}$ in $[0,1)^d$ we have*

$$
\mathrm{err}_{N,d}(f,\mathcal{P}) = \sum_{\emptyset\neq\mathfrak{u}\subseteq[d]}(-1)^{|\mathfrak{u}|}\int_{[0,1]^{|\mathfrak{u}|}} f^{(\mathfrak{u})}(\boldsymbol{x}_{\mathfrak{u}},\boldsymbol{1})\Delta_{\mathcal{P}}(\boldsymbol{x}_{\mathfrak{u}},\boldsymbol{1})\,\mathrm{d}\boldsymbol{x}_{\mathfrak{u}}.
$$

Now, we define a norm, depending on a parameter $q \in [1,\infty]$, by

$$
\|f\|_{d,q}^{*} := \left(\int_{[0,1]^d}\left|\frac{\partial^d f}{\partial \boldsymbol{x}}(\boldsymbol{x})\right|^q \mathrm{d}\boldsymbol{x}\right)^{1/q}
$$

if $q \in [1,\infty)$ and

$$
\|f\|_{d,\infty}^{*} := \sup_{\boldsymbol{x}\in[0,1]^d}\left|\frac{\partial^d f}{\partial \boldsymbol{x}}(\boldsymbol{x})\right|
$$

if $q = \infty$.

Furthermore, we put

$$\|f\|_{d,q} := \left(\sum_{\mathfrak{u} \subseteq [d]} \int_{[0,1]^{|\mathfrak{u}|}} |f^{(\mathfrak{u})}(\boldsymbol{x}_{\mathfrak{u}}, \boldsymbol{1})|^q \, \mathrm{d}\boldsymbol{x}_{\mathfrak{u}} \right)^{1/q} \tag{1.24}$$

for $q \in [1, \infty)$ and

$$\|f\|_{d,\infty} := \sup_{\mathfrak{u} \subseteq [d]} \sup_{\boldsymbol{x} \in [0,1]^d} |f^{(\mathfrak{u})}(\boldsymbol{x}_{\mathfrak{u}}, \boldsymbol{1})| \tag{1.25}$$

for $q = \infty$. In these definitions the term for $\mathfrak{u} = \emptyset$ corresponds to $|f(\boldsymbol{1})|$.

First we consider the space

$$\mathcal{F}_{d,q}^* := \{f \in \mathcal{H}_d : f(\boldsymbol{x}) = 0 \text{ if } x_j = 1 \text{ for some } j \in [d], \text{ and } \|f\|_{d,q}^* < \infty\}.$$

Since we have boundary conditions for functions in $\mathcal{F}_{d,q}^*$, Hlawka's and Zaremba's identity reduces to

$$\mathrm{err}_{N,d}(f, \mathcal{P}) = (-1)^d \int_{[0,1]^d} \frac{\partial^d f}{\partial \boldsymbol{x}}(\boldsymbol{x}) \, \Delta_{\mathcal{P}}(\boldsymbol{x}) \, \mathrm{d}\boldsymbol{x} \quad \text{for } f \in \mathcal{F}_{d,q}^*.$$

Applying the absolute value and Hölder's inequality, we obtain for all $p, q \in [1, \infty]$ with $1/p + 1/q = 1$ that

$$|\mathrm{err}_{N,d}(f, \mathcal{P})| \le \|f\|_{d,q}^* L_{p,N}(\mathcal{P}). \tag{1.26}$$

This is our first variant of a Koksma–Hlawka type inequality, which we can also reformulate as an equality by means of the worst-case error in the sense that the worst-case error of QMC integration in $\mathcal{F}_{d,q}^*$ is exactly the L_p-discrepancy of the underlying point set. For $p, q \in (1, \infty)$ this follows from the fact that Hölder's inequality is sharp for particular choices of functions. The result even holds for $p = \infty$, which follows from [199, Theorem 2.12].

Theorem 1.33 *Let $\mathcal{P}$ be an N-element point set in $[0, 1)^d$. Let $p, q \in [1, \infty]$ be such that $1/p + 1/q = 1$. Then we have*

$$\mathrm{err}_{N,d}(\mathcal{F}_{d,q}^*, \mathcal{P}) = L_{p,N}(\mathcal{P}).$$

Next, we would like to avoid the boundary conditions in the definition of $\mathcal{F}_{d,q}^*$, i.e., we consider the function class

$$\mathcal{F}_{d,q} := \{f \in \mathcal{H}_d : \|f\|_{d,q} < \infty\}, \tag{1.27}$$

where the norm is given by (1.24) or (1.25), respectively. In this context we need to introduce a further notion of discrepancy, namely the so-called combined L_p-discrepancy.

Definition 1.34 For an N-element point set $\mathcal{P}$ in $[0,1)^d$ and for $p \geq 1$ the *combined L_p-discrepancy* is given by

$$L_{p,N,\mathbf{1}}(\mathcal{P}) := \left(\sum_{\emptyset \neq \mathfrak{u} \subseteq [d]} \int_{[0,1]^{|\mathfrak{u}|}} |\Delta_{\mathcal{P}}(\boldsymbol{t}_{\mathfrak{u}}, \mathbf{1})|^p \, \mathrm{d}\boldsymbol{t}_{\mathfrak{u}} \right)^{1/p},$$

with the usual adaptions if $p = \infty$.

Note that $L_{\infty,N,\mathbf{1}}(\mathcal{P}) = L_{\infty,N}(\mathcal{P}) = D_N^*(\mathcal{P})$. The meaning of the index $\mathbf{1}$ in the notation will become clear later when we discuss weighted L_p-discrepancy (see Sections 5.3 and 7.1).

We continue with the error analysis of QMC rules in $\mathcal{F}_{d,q}$. Applying again Hölder's inequality to Hlawka's and Zaremba's identity, now for integrals and sums, we conclude the following.

Theorem 1.35 *Let $\mathcal{P}$ be an N-element point set in $[0,1)^d$, and let $p, q \in [1,\infty]$ be such that $1/p + 1/q = 1$. Then we have*

$$\mathrm{err}_{N,d}(\mathcal{F}_{d,q}, \mathcal{P}) = L_{p,N,\mathbf{1}}(\mathcal{P}).$$

As a corollary we state a "weak" form of the classical Koksma–Hlawka inequality that is obtained by choosing $q = 1$ (and hence $p = \infty$).

Corollary 1.36 (Koksma–Hlawka inequality) *Let $\mathcal{P}$ be an N-element point set in $[0,1)^d$. Then, for all functions f defined on $[0,1]^d$ with $\|f\|_{d,1} < \infty$, we have*

$$|\mathrm{err}_{N,d}(f, \mathcal{P})| \leq \|f\|_{d,1} D_N^*(\mathcal{P}).$$

Remark 1.37 The Koksma–Hlawka inequality in its original, more general form states that

$$|\mathrm{err}_{N,d}(f, \mathcal{P})| \leq V(f) D_N^*(\mathcal{P}),$$

where $V(f)$ denotes the variation of f in the sense of Hardy and Krause. We do not give the definition of the latter notion here (see, e.g., [155, 199]), but we would like to mention that if all partial mixed derivatives of f up to order one in each variable are continuous on $[0,1]^d$, then

$$V(f) = \sum_{\emptyset \neq \mathfrak{u} \subseteq [d]} \int_{[0,1]^{|\mathfrak{u}|}} |f^{(\mathfrak{u})}(\boldsymbol{x}_{\mathfrak{u}}, \mathbf{1})| \, \mathrm{d}\boldsymbol{x}_{\mathfrak{u}} = \|f\|_{d,1} - |f(\mathbf{1})|.$$

We have now seen that the discrepancy is an important figure of merit for the nodes of QMC rules. The discrepancy of lattice point sets will be discussed in Chapter 5, where we will also study further notions of discrepancy, namely the so-called weighted discrepancy and the isotropic discrepancy. The weighted discrepancy is very important in the context of the next section, which deals with the dependence of the error bounds on the dimension.

1.7 The Curse of Dimensionality

The classical worst-case error analysis of integration rules is focused on the decay rate of the errors when the number of nodes used tends to infinity. The dimension d is assumed to be fixed. In many cases one has excellent asymptotic results which, however, may be delusive. We explain this by means of two examples.

Example 1.38 Consider QMC integration in the class $\mathcal{F}_{d,1}$. Then, according to Theorem 1.35, the worst-case error is exactly the star-discrepancy of the underlying integration nodes. This means that we require point sets in $[0,1)^d$ with very low star-discrepancy. The best constructions known so far yield point sets $\mathcal{P}$ with star-discrepancy of order

$$D_N^*(\mathcal{P}) = O\left(\frac{(\log N)^{d-1}}{N}\right). \tag{1.28}$$

Here, also the implicit constant may depend on d in a very adverse form. Examples of such constructions are Hammersley point sets and digital nets in the sense of Niederreiter. But we also know from (1.20) that this order of magnitude cannot be improved substantially for the star-discrepancy. This means that the order of magnitude is essentially best possible for growing N and fixed d. However, the function $N \mapsto (\log N)^{d-1}/N$ is increasing for $N \leq \mathrm{e}^{d-1}$, and only for larger N it decreases to zero with the asymptotic rate of almost $1/N$. This behavior is illustrated in Figure 1.4.

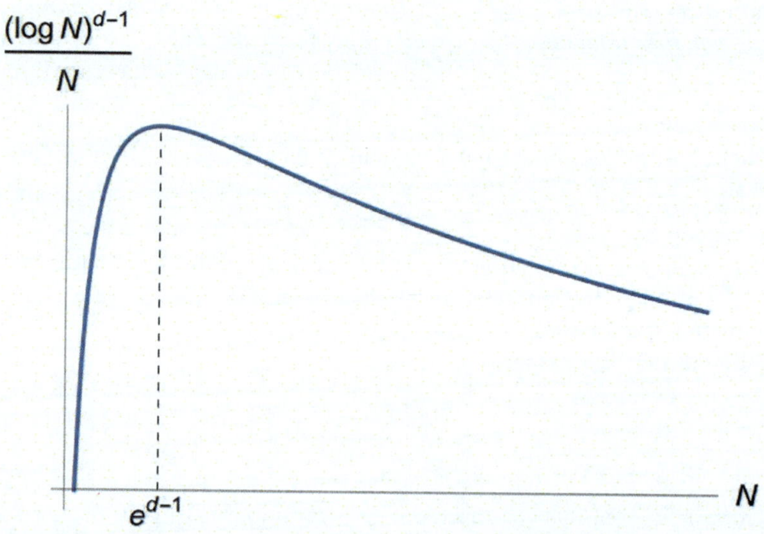

Fig. 1.4: The function $N \mapsto (\log N)^{d-1}/N$.

I.e., in order to have meaningful error bounds for QMC rules one requires point sets in $[0, 1)^d$ with at least e^{d-1} elements or even more, but e^{d-1} is huge, even for moderate choices of d. Furthermore, if we assume, for example, $d = 200$, which may happen in applications from financial mathematics, then $\mathrm{e}^{d-1} \approx 2.65829 \times 10^{86}$, a number exceeding the estimated number of atoms in our known, observable universe (which is estimated to be between 10^{78} and 10^{82}). This means that, according to the classical theory, QMC methods cannot be expected to work for very high dimensions. Nevertheless, and surprisingly, we know from practical applications that QMC rules often do very well and even work much better than we have any right to expect.

Example 1.39 For the numerical integration of functions $f : [0, 1] \to \mathbb{R}$ there are many classical quadrature rules available. Examples are the rectangle or midpoint rule, the trapezoidal rule, Simpson's rule, or the Gauss rule, which all have the general form

$$T_m(f) := \sum_{k=0}^{m} q_k f(x_k) \tag{1.29}$$

with nodes $x_0, x_1, \ldots, x_m \in [0, 1]$ and with integration weights $q_0, q_1, \ldots, q_m \in \mathbb{R}$. As an example consider the trapezoidal rule for which $q_0 = q_m = 1/(2m)$, $q_k = 1/m$ for $k \in \{1, 2, \ldots, m-1\}$, and $x_k = k/m$ for $k \in \{0, 1, \ldots, m\}$. If $f \in C^2([0, 1])$ the error of the trapezoidal rule is well known to be of order $O(1/m^2)$.

In the multidimensional case, i.e., for the integration of functions over the domain $[0, 1]^d$, the classical methods use d-fold Cartesian products of one-dimensional quadrature rules that are then often referred to as *product rules*. This means that one applies a one-dimensional quadrature rule of the form (1.29) to each component, which results in a product rule of the form

$$\sum_{k_1=0}^{m} \cdots \sum_{k_d=0}^{m} q_{k_1} \cdots q_{k_d} f(x_{k_1}, \ldots, x_{k_d}).$$

When rewriting this rule as

$$\sum_{k=0}^{M} w_k f(\boldsymbol{x}_k), \tag{1.30}$$

the set of quadrature points $\{\boldsymbol{x}_0, \boldsymbol{x}_1, \ldots, \boldsymbol{x}_M\}$ is exactly the d-fold product of the collection of the one-dimensional quadrature points $\{x_0, x_1, \ldots, x_m\}$. Note that the product rule (1.30) employs $N = M + 1 = (m + 1)^d$ integration nodes, a number growing exponentially with the dimension d. For example, if $d = 30$, a product rule based on a one-dimensional rule with only two points already involves $N = 2^{30} \approx 10^9$ nodes. Again, if $d = 200$, then a product rule based on a 2-element one-dimensional rule already requires $2^{200} = 1.60694 \times 10^{60}$ nodes. Needless to say, it is infeasible to compute that many function evaluations.

The error analysis of product rules is based on the underlying one-dimensional quadrature rules. For instance, the error of the product rule based on one-dimensional trapezoidal rules applied to functions on $[0, 1]^d$ is of order $O(1/m^2)$, provided that all partial derivatives of order two in each variable are continuous on $[0, 1]^d$. Here,

$m+1$ is the number of nodes used in the underlying one-dimensional rule. This looks quite promising at first sight, but in terms of the actual number N of integration nodes, this error is of order $O(N^{-2/d})$. For large dimensions, which may be in the hundreds or thousands for practical problems, such an error convergence is less than satisfying.

Both of these examples suggest that the classical error analysis yields bounds that are not useful for high-dimensional integration problems. Although we have excellent or even best possible asymptotic error rates in many cases, we do in general not know how long we have to wait to see these excellent rates, especially when the dimension d is large. This problem is the core of the field of *Information-Based Complexity (IBC)*, where also the dependence of the error bounds on the dimension d is crucial.

In IBC the so-called information complexity is studied rather than actual error bounds. Roughly speaking, the information complexity is the minimal amount of information (e.g., in the case of numerical integration, the number of function evaluations) that is necessary to solve a problem within a prescribed level of precision. In order to give the exact definition we need some notation.

Consider the integration problem for functions defined on $[0,1]^d$ from a normed space $(\mathcal{F}_d, \|\cdot\|)$ in the worst-case setting, i.e., we study the worst-case error of integration. It is well known that it is sufficient to study linear algorithms of the form

$$A_{N,d}(f) = \sum_{k=0}^{N-1} w_k f(\boldsymbol{x}_k) \tag{1.31}$$

based on a point set $\mathcal{P} = \{\boldsymbol{x}_0, \boldsymbol{x}_1, \ldots, \boldsymbol{x}_{N-1}\}$ in $[0,1)^d$ and integration weights $\boldsymbol{w} = (w_0, w_1, \ldots, w_{N-1}) \in \mathbb{C}^N$, which we already encountered in Remark 1.28. (A proof of this fact is beyond the scope of the present book. Instead, we refer to the paper [8] by Bakhvalov, see also [210, Section 4.2].)

The worst-case error of the algorithm $A_{N,d}$ is defined as the supremal integration error over the unit ball of $\mathcal{F}_d$, i.e.,

$$\mathrm{err}_{N,d}(\mathcal{F}_d, \mathcal{P}, \boldsymbol{w}) := \sup_{\substack{f \in \mathcal{F}_d \\ \|f\| \le 1}} |I(f) - A_{N,d}(f)|,$$

where I again denotes the integration operator.

Now we can define the N-th minimal error.

Definition 1.40 For $d, N \in \mathbb{N}$, the *N-th minimal worst-case error in dimension d* for the integration problem $I : \mathcal{F}_d \to \mathbb{R}$ is defined as

$$e(N,d) := \inf_{\mathcal{P}, \boldsymbol{w}} |\mathrm{err}_{N,d}(\mathcal{F}_d, \mathcal{P}, \boldsymbol{w})|,$$

where the infimum is extended over all N-element point sets $\mathcal{P}$ in $[0,1)^d$ and over all $\boldsymbol{w} \in \mathbb{C}^N$, i.e., over all linear algorithms of the form (1.31).

For $N = 0$ the *initial error* is defined as

$$e(0,d) := \sup_{\substack{f \in \mathcal{F}_d \\ \|f\| \leq 1}} |I(f)|.$$

The initial error is the norm of the integral operator, and it is used as a reference value for the integration problem in the space $\mathcal{F}_d$.

Remark 1.41 If $\mathcal{F}_d = \mathcal{H}(\mathcal{K})$ is a reproducing kernel Hilbert space with kernel K, then

$$e(0,d) = \left(\int_{[0,1]^d} \int_{[0,1]^d} K(\boldsymbol{x}, \boldsymbol{y}) \, \mathrm{d}\boldsymbol{x} \, \mathrm{d}\boldsymbol{y} \right)^{1/2}.$$

This follows from

$$\begin{aligned} \left| \int_{[0,1]^d} f(\boldsymbol{x}) \, \mathrm{d}\boldsymbol{x} \right| &= \left| \int_{[0,1]^d} \langle f, K(\cdot, \boldsymbol{x}) \rangle \, \mathrm{d}\boldsymbol{x} \right| \\ &= \left| \left\langle f, \int_{[0,1]^d} K(\cdot, \boldsymbol{x}) \, \mathrm{d}\boldsymbol{x} \right\rangle \right| \\ &\leq \|f\| \left| \left\langle \int_{[0,1]^d} K(\cdot, \boldsymbol{x}) \, \mathrm{d}\boldsymbol{x}, \int_{[0,1]^d} K(\cdot, \boldsymbol{y}) \, \mathrm{d}\boldsymbol{y} \right\rangle \right|^{1/2} \\ &= \|f\| \left(\int_{[0,1]^d} \int_{[0,1]^d} K(\boldsymbol{x}, \boldsymbol{y}) \, \mathrm{d}\boldsymbol{x} \, \mathrm{d}\boldsymbol{y} \right)^{1/2}, \end{aligned}$$

where we used the Cauchy–Schwarz inequality. Hence

$$e(0,d) \leq \left(\int_{[0,1]^d} \int_{[0,1]^d} K(\boldsymbol{x}, \boldsymbol{y}) \, \mathrm{d}\boldsymbol{x} \, \mathrm{d}\boldsymbol{y} \right)^{1/2}.$$

Equality follows from the fact that the Cauchy–Schwarz inequality is sharp for the particular choice

$$f = \left(\int_{[0,1]^d} K(\cdot, \boldsymbol{x}) \, \mathrm{d}\boldsymbol{x} \right) / \left\| \int_{[0,1]^d} K(\cdot, \boldsymbol{x}) \, \mathrm{d}\boldsymbol{x} \right\|.$$

Next, we introduce the information complexity, which is the minimal number of function evaluations that is necessary in order to be able to reduce the initial error by a given factor ε, the so-called error demand.

Definition 1.42 For $d \in \mathbb{N}$ and $\varepsilon \in (0,1)$ the *information complexity* is defined as

$$N(\varepsilon, d) := \min\{N \in \mathbb{N} \,:\, e(N,d) \leq \varepsilon \, e(0,d)\}.$$

In IBC one is interested in the growth rate of the information complexity when the error demand ε tends to zero and when the dimension d tends to infinity. This is the idea underlying the concept of *tractability*. There are several notions of tractability which classify the growth rate of the information complexity. The essence of all these notions is that the information complexity, i.e., the minimal number of function evaluations needed to solve the d-dimensional integration problem to within an error demand ε, must not be exponential in ε^{-1} or d. If the latter occurs (i.e., if we have exponential dependence on d), then the problem is said to be suffering from the curse of dimensionality, a phrase that was coined already in 1957 by Bellman [13].

In order for the following definitions to make sense, we need to consider a whole sequence of integration problems, one for each $d \in \mathbb{N}$.

Definition 1.43 The sequence of integration problems $(I : \mathcal{F}_d \to \mathbb{R})_{d\geq 1}$ is said to suffer from the *curse of dimensionality*, if there exist numbers $C > 0$, $\tau > 0$, and $\varepsilon_0 \in (0, 1)$ such that

$$N(\varepsilon, d) \geq C(1+\tau)^d \quad \text{for all } \varepsilon \in (0, \varepsilon_0) \text{ and for infinitely many } d \in \mathbb{N}.$$

Problems which do not suffer from the curse of dimensionality are usually classified by the help of several notions of tractability, some of which we introduce in the next definition.

Definition 1.44 The sequence of integration problems $(I : \mathcal{F}_d \to \mathbb{R})_{d\geq 1}$ is said to be

1. *weakly tractable*, if

$$\lim_{d+\varepsilon^{-1}\to\infty} \frac{\log N(\varepsilon, d)}{d+\varepsilon^{-1}} = 0;$$

2. *polynomially tractable*, if there exist numbers $C, \sigma > 0$ and $\tau \geq 0$ such that

$$N(\varepsilon, d) \leq C d^{\tau} \varepsilon^{-\sigma} \quad \text{for all } \varepsilon \in (0, 1) \text{ and all } d \in \mathbb{N}; \tag{1.32}$$

3. *strongly polynomially tractable*, if there exist numbers $C > 0$ and $\sigma > 0$ such that

$$N(\varepsilon, d) \leq C \varepsilon^{-\sigma} \quad \text{for all } \varepsilon \in (0, 1) \text{ and all } d \in \mathbb{N}. \tag{1.33}$$

The infimum of all $\sigma > 0$ such that a bound of the form (1.33) holds is called the *ε-exponent* of strong polynomial tractability.

We remark that in IBC sometimes much finer notions of tractability are considered, as for example quasi-polynomial tractability, (s, t)-weak tractability, and uniform weak tractability. For this book the notions of weak, polynomial, and strong polynomial tractability suffice. Furthermore, in this book we only consider integration problems, and, in later chapters, function approximation, but in IBC much more general problems are studied.

Example 1.45 Consider again the integration problem for functions from $\mathcal{F}_{d,1}$. The worst-case error of a QMC rule is exactly the star-discrepancy of the underlying point set $\mathcal{P}$. The question is whether this integration problem suffers from the curse of dimensionality or not, and if not, to which tractability class the problem belongs. While the known asymptotic bounds on the star-discrepancy (1.28) are of no help in answering this question, there is a seminal paper by Heinrich, Novak, Wasilkowski, and Woźniakowski [96], who were the first to study the star-discrepancy from the viewpoint of IBC. They proved that there exists an absolute constant $C > 0$ such that for every $d, N \in \mathbb{N}$ there exists an N-element point set $\mathcal{P}$ in $[0, 1)^d$ satisfying

$$D_N^*(\mathcal{P}) \le C\sqrt{\frac{d}{N}}, \tag{1.34}$$

and hence

$$e(N, d) \le C\sqrt{\frac{d}{N}} \quad \text{for all } N, d \in \mathbb{N}.$$

This implies that

$$N(\varepsilon, d) \le C^2 d\varepsilon^{-2},$$

which in turn means that the integration problem in $\mathcal{F}_{d,1}$ is polynomially tractable. In the present context $N(\varepsilon, d)$ is sometimes also referred to as the *inverse of the star-discrepancy*. The currently smallest known admissible value of C is 2.5287 as shown by Gnewuch and Hebbinghaus [79]. We remark that the bound (1.34) is a pure existence result. So far, no explicit construction of a point set with star-discrepancy of at most $C\sqrt{d/N}$ for some positive constant C is known. For further information on this question we refer to the "Notes and Remarks" Section at the end of the present chapter.

Furthermore, we remark that the dependence on the dimension d in the previous result is known to be best possible. Indeed, Hinrichs [107] proved that there exist numbers $c > 0$ and $\varepsilon_0 \in (0, 1)$ such that

$$N(\varepsilon, d) \ge cd\varepsilon^{-1} \quad \text{for all } d \in \mathbb{N} \text{ and } \varepsilon \in (0, \varepsilon_0).$$

Thus, the exact dependence of $N(\varepsilon, d)$ on the dimension d is linear, whereas the exact dependence on ε^{-1} is still an open question.

Strong polynomial tractability of the related concept of weighted star-discrepancy is discussed in Section 5.4 of the present book.

Example 1.46 Consider integration in $\mathcal{E}_\alpha^d(1)$, i.e., integration of functions f satisfying

$$|\widehat{f}(\boldsymbol{h})| \le \frac{1}{(r_1(\boldsymbol{h}))^\alpha} \quad \text{for all } \boldsymbol{h} \in \mathbb{Z}^d \setminus \{\boldsymbol{0}\},$$

where r_1 is given by (1.12), using arbitrary linear algorithms $A_{N,d}$ with integration nodes $\mathcal{P} = \{\boldsymbol{x}_0, \boldsymbol{x}_1, \dots, \boldsymbol{x}_{N-1}\}$ and integration weights $\boldsymbol{w} = (w_0, w_1, \dots, w_{N-1}) \in \mathbb{C}^N$, as introduced in Remark 1.28. Let

$$P_\alpha(\mathcal{P}, \boldsymbol{w}) := \sup_{f \in \mathcal{E}_\alpha^d(1)} \left| \int_{[0,1]^d} f(\boldsymbol{x}) \, \mathrm{d}\boldsymbol{x} - A_{N,d}(f) \right|$$

be the worst-case integration error of the linear algorithm $A_{N,d}$ for the class $\mathcal{E}_\alpha^d(1)$. Define

$$e(N, d) := \inf\{P_\alpha(\mathcal{P}, \boldsymbol{w}) \ : \ \boldsymbol{w} \in \mathbb{C}^N \ \text{ and } \mathcal{P} \subseteq [0,1)^d \text{ with } |\mathcal{P}| = N\}.$$

The following result of Sloan and Woźniakowski [238, Theorem 1] implies that integration in the Korobov class $\mathcal{E}_\alpha^d(1)$ suffers from the curse of dimensionality.

Theorem 1.47 *Consider integration in the Korobov class $\mathcal{E}_\alpha^d(1)$. If $N < 2^d$, then $e(N, d) = 1$.*

This result means that the number of function evaluations required to obtain a worst-case error less than one is exponential in the dimension, and this implies the curse of dimensionality.

Proof of Theorem 1.47 Taking $\boldsymbol{w} = \boldsymbol{0}$ we get

$$P_\alpha(\mathcal{P}, \boldsymbol{0}) = \sup\left\{|\widehat{f}(\boldsymbol{0})| \ : \ f \in \mathcal{E}_\alpha^d(1)\right\} = 1$$

and so $e(N, d) \leq 1$. To show equality, let $B_d := \{0, 1\}^d$. Take an arbitrary linear algorithm $A_{N,d}(f) = \sum_{k=0}^{N-1} w_k f(\boldsymbol{x}_k)$. Define

$$g(\boldsymbol{x}) = \sum_{\boldsymbol{h} \in B_d} b_{\boldsymbol{h}} \, \mathrm{e}^{2\pi \mathrm{i} \boldsymbol{h} \cdot \boldsymbol{x}} \quad \text{for all} \quad \boldsymbol{x} \in [0,1]^d$$

and choose $b_{\boldsymbol{h}}$ for $\boldsymbol{h} \in B_d$ such that $g(\boldsymbol{x}_k) = 0$ for all $k \in \{0, 1, \ldots, N-1\}$. Since we have N homogeneous linear equations and $|B_d| = 2^d > N$ unknowns $b_{\boldsymbol{h}}$, there exists a nonzero vector of such $b_{\boldsymbol{h}}$, and we can normalize the $b_{\boldsymbol{h}}$ by assuming that

$$\max_{\boldsymbol{h} \in B_d} |b_{\boldsymbol{h}}| = b_{\boldsymbol{h}^*} = 1 \quad \text{for some} \quad \boldsymbol{h}^* \in B_d.$$

Define

$$f(\boldsymbol{x}) = \mathrm{e}^{-2\pi \mathrm{i} \boldsymbol{h}^* \cdot \boldsymbol{x}} g(\boldsymbol{x}) = \sum_{\boldsymbol{h} \in B_d} b_{\boldsymbol{h}} \mathrm{e}^{2\pi \mathrm{i} (\boldsymbol{h} - \boldsymbol{h}^*) \cdot \boldsymbol{x}} \quad \text{for all} \quad \boldsymbol{x} \in [0,1]^d.$$

For $\boldsymbol{h}, \boldsymbol{h}^* \in B_d$ we have

$$h_j - h_j^* \in \{-1, 0, 1\} \quad \text{for all} \ \ j \in [d].$$

Hence f is a trigonometric polynomial (see (1.40) for the formal definition of a trigonometric polynomial) of degree at most 1 in each component, and furthermore,

$$r_1(\boldsymbol{h} - \boldsymbol{h}^*) = 1 \quad \text{for} \ \ \boldsymbol{h}, \boldsymbol{h}^* \in B_d.$$

Since $|b_{\boldsymbol{h}}| \leq 1$ for all $\boldsymbol{h} \in B_d$ it follows that $|\widehat{f}(\boldsymbol{h})| \leq 1$ for all $\boldsymbol{h} \in \mathbb{Z}^d$ and therefore $f \in \mathcal{E}_\alpha^d(1)$.

Note that $f(\boldsymbol{x}_k) = 0$ for all $k \in \{0, 1, \ldots, N-1\}$, and this implies that $A_{N,d}(f) = 0$. Consequently,

$$P_\alpha(\mathcal{P}, \boldsymbol{w}) \geq \left| \int_{[0,1]^d} f(\boldsymbol{x}) \,\mathrm{d}\boldsymbol{x} - A_{N,d}(f) \right| = \left| \int_{[0,1]^d} f(\boldsymbol{x}) \,\mathrm{d}\boldsymbol{x} \right| = |b_{\boldsymbol{h}^*}| = 1.$$

Since this holds for all $\boldsymbol{w} \in \mathbb{C}^N$ and $\mathcal{P} = \{\boldsymbol{x}_0, \boldsymbol{x}_1, \ldots, \boldsymbol{x}_{N-1}\}$, we conclude that $e(N, d) \geq 1$. Together with the upper bound we obtain $e(N, d) = 1$, as claimed. □

In this book we will put a focus on the question for which problems the curse of dimensionality can be broken with the help of lattice rules and, if so, which notion of tractability can be obtained then.

1.8 Further Quality Criteria for Lattice Rules

In this final section of our introductory chapter, let us return to lattice rules. So far we have already encountered the most important quality criteria for lattice rules, namely the quantity P_α, the discrepancies of the corresponding lattice point sets, and the worst-case errors of lattice rules with respect to normed spaces of integrands. In this section we briefly discuss further important quality criteria for lattice rules.

The criterion R

The first criterion is a quantity usually referred to as R in the literature, and it is related to P_α as well as to the extreme discrepancy of lattice point sets. Before we state its exact definition, we need to introduce some notation.

For $M \in \mathbb{N}$, $M \geq 2$, let $C(M) := (-M/2, M/2] \cap \mathbb{Z}$, and let

$$C_d(M) := \underbrace{C(M) \times \cdots \times C(M)}_{d \text{ times}}$$

be the d-fold Cartesian product of $C(M)$. Furthermore, we write $C_d^*(M) := C_d(M) \setminus \{\boldsymbol{0}\}$.

Definition 1.48 For a d-dimensional integration lattice $\mathcal{L}$ with $\det(\mathcal{L}^\perp) = N$, with $d, N \geq 2$, we put

$$R(\mathcal{L}) := \sum_{\boldsymbol{h} \in C_d^*(N) \cap \mathcal{L}^\perp} \frac{1}{r_1(\boldsymbol{h})}, \tag{1.35}$$

where r_1 is defined in (1.12). When $\mathcal{L}$ is a rank-1 lattice with lattice point set $\mathcal{P}(\boldsymbol{g}, N)$, we write $R(\boldsymbol{g}, N)$ instead of $R(\mathcal{L})$.

Note that the range of summation in the definition of R in (1.35) is finite. It is easily seen that $|C_d^*(N) \cap \mathcal{L}^\perp| = N^{d-1} - 1$ (see [199, Remark 5.25]).

As already mentioned, the quantity R is related to P_α. The following two results are classical (see [199, Theorem 5.26] and [199, Theorem 5.5]) and show that P_α can be upper-bounded in terms of R. Here and in the following, for $\alpha > 1$, let $\zeta(\alpha) := \sum_{h=1}^{\infty} h^{-\alpha}$ be the Riemann zeta function.

Theorem 1.49 *Let $d, N \in \mathbb{N}$, $d, N \geq 2$, and let $\mathcal{L}$ be a d-dimensional integration lattice with* $\det(\mathcal{L}^\perp) = N$. *Then, for any real $\alpha > 1$, we have*

$$P_\alpha(\mathcal{L}) < (1+\zeta(\alpha)d)(R(\mathcal{L}))^\alpha + \left(1 + \frac{2\zeta(\alpha)}{N^\alpha}\right)^d - 1 + \frac{(2^\alpha\zeta(\alpha))^d}{N^{\alpha d - d + 1}}$$
$$+ \sum_{j=1}^{d-2} \binom{d}{j} (2^\alpha \zeta(\alpha))^{d-j} (1 + 2\zeta(\alpha))^j \frac{1}{N^{(\alpha-1)(d-j)}}.$$

In particular, if either $d = 2$ or $\alpha \geq 2$,

$$P_\alpha(\mathcal{L}) < (1+\zeta(\alpha)d)(R(\mathcal{L}))^\alpha + O\left(\frac{1}{N^\alpha}\right).$$

For rank-1 lattice rules we have the following slight improvement according to Niederreiter [200].

Theorem 1.50 *Let $d, N \in \mathbb{N}$, $d, N \geq 2$, and let $\boldsymbol{g} = (g_1, \ldots, g_d) \in \mathbb{Z}^d$ with* $\gcd(g_j, N) = 1$ *for all $j \in [d]$. Then, for any real $\alpha > 1$, we have*

$$P_\alpha(\boldsymbol{g}, N) < (R(\boldsymbol{g}, N))^\alpha + \left(1 + \frac{2\zeta(\alpha)}{N^\alpha}\right)^d - 1$$
$$+ \frac{1}{N}\left(1 + 2\zeta(\alpha) + \frac{2^\alpha\zeta(\alpha)}{N^{\alpha-1}}\right)^d - \frac{1}{N}(1+2\zeta(\alpha))^d.$$

In particular,

$$P_\alpha(\boldsymbol{g}, N) \leq (R(\boldsymbol{g}, N))^\alpha + O\left(\frac{1}{N^\alpha}\right).$$

The advantage of the quality criterion R is that it is independent of the smoothness parameter α. Hence integration lattices with small R are universal in the sense that a low value of R yields also low values of P_α simultaneously for all $\alpha > 1$. The relation of R to the extreme discrepancy will be discussed in detail in Section 5.1, and the construction of integration lattices with a low value of R in Sections 3.6 and 5.2, respectively. There we will construct, for prime numbers N, generating vectors $\boldsymbol{g} \in \mathbb{Z}^d$ satisfying

$$R(\boldsymbol{g}, N) = O\left(\frac{(\log N)^d}{N}\right).$$

For the error P_α this implies

$$P_\alpha(\boldsymbol{g}, N) = O\left(\frac{(\log N)^{\alpha d}}{N^\alpha}\right)$$

for all $\alpha > 1$.

A weighted version of the quality criterion R will be introduced in Section 3.6 (see Definition 3.12).

The Zaremba index ρ

Another quality measure is the Zaremba index, sometimes also referred to as the figure of merit, which is based on the idea that P_α and R will be small if the nonzero elements of $\mathcal{L}^\perp$ are in a certain sense far away from the origin.

Definition 1.51 For a d-dimensional integration lattice $\mathcal{L}$ with $\det(\mathcal{L}^\perp) = N$ with $d, N \geq 2$, the *Zaremba index* or *figure of merit* is defined as

$$\rho(\mathcal{L}) := \min_{\boldsymbol{h} \in \mathcal{L}^\perp \setminus \{\boldsymbol{0}\}} r_1(\boldsymbol{h}).$$

If $\mathcal{L}$ is a rank-1 lattice with lattice point set $\mathcal{P}(\boldsymbol{g}, N)$, then we write $\rho(\boldsymbol{g}, N)$ instead of $\rho(\mathcal{L})$.

Using the geometric illustration in Figure 1.2, the Zaremba index is the value of $r_1(\boldsymbol{h}^*)$ which is attained by points $\boldsymbol{h}^*$ in the dual lattice $\mathcal{L}^\perp$ which also lie on the boundary of the hyperbolic cross (marked by red dots in Figure 1.2).

The relation of the Zaremba index to numerical integration is well known (see, e.g., [199, 230]). The following theorem illustrates how the quality measures P_α and R can be bounded in terms of the Zaremba index.

Theorem 1.52 *Let $d, N \in \mathbb{N}$, $d, N \geq 2$, and let $\mathcal{L}$ be a d-dimensional integration lattice $\mathcal{L}$ with* $\det(\mathcal{L}^\perp) = N$. *Then we have*

$$\frac{2}{(\rho(\mathcal{L}))^\alpha} \leq P_\alpha(\mathcal{L}) \leq c(d, \alpha) \frac{(1 + \log \rho(\mathcal{L}))^{d-1}}{(\rho(\mathcal{L}))^\alpha} \quad \textit{for all } \alpha > 1, \tag{1.36}$$

where the quantity $c(d, \alpha)$ depends only on d and α, and

$$\frac{1}{\rho(\mathcal{L})} \leq R(\mathcal{L}) \leq \left(\frac{2}{\log 2}\right)^{d-1} \frac{2(\log N)^d + 3(\log N)^{d-1}}{\rho(\mathcal{L})}. \tag{1.37}$$

The upper bound in (1.36) was first shown by Zaremba [268], and the lower bound is trivial. The upper bound in (1.37) goes back to Niederreiter [193], where the presented form is taken from [199, Theorem 5.35]. The lower bound follows from the observation that $1 \leq \rho(\mathcal{L}) \leq N/2$ (see [199, Remark 5.33]). If the rank r of $\mathcal{L}$ satisfies $r \geq 2$, then we even have the improved estimate $1 \leq \rho(\mathcal{L}) \leq n_1$, where n_1 is the first invariant of $\mathcal{L}$ as given in Theorem 1.5 (see [199, Lemma 5.32]).

Theorem 1.52 suggests that the Zaremba index has to be large in order to obtain small values of P_α and R, respectively. In this context Zaremba [269] established the following result for rank-1 lattice rules.

Theorem 1.53 *For every dimension $d \geq 2$ and every sufficiently large integer N, there exists a lattice point $\boldsymbol{g} \in C_d(N)$, with first component $g_1 = 1$, such that*

$$\rho(\boldsymbol{g}, N) > \frac{(d-1)!N}{(2\log N)^{d-1}}.$$

In the two-dimensional case there is an important connection to continued fraction expansions. Let $g \in \{1, 2, \ldots, N-1\}$ with $\gcd(g, N) = 1$. Let $a_1, a_2, \ldots, a_l$ be the partial quotients in the continued fraction expansion of g/N, i.e.,

$$\frac{g}{N} = \cfrac{1}{a_1 + \cfrac{1}{a_2 + \cfrac{1}{\ddots + \cfrac{1}{a_l}}}}, \tag{1.38}$$

and let

$$K\left(\frac{g}{N}\right) := \max_{1 \leq j \leq l} a_j. \tag{1.39}$$

It was shown by Zaremba [266] (see also [199, Theorem 5.17]) that the Zaremba index $\rho(\boldsymbol{g}, N)$ for $\boldsymbol{g} = (1, g)$ can be bounded in terms of $K(g/N)$, namely

$$\frac{N}{K(g/N) + 2} \leq \rho(\boldsymbol{g}, N) \leq \frac{N}{K(g/N)}.$$

Example 1.54 (Fibonacci lattices) Let $(F_n)_{n \geq 1}$ be the sequence of Fibonacci numbers defined recursively by $F_1 = F_2 = 1$ and $F_n = F_{n-1} + F_{n-2}$ for $n \geq 3$. Now let $N = F_n$ for some $n \geq 3$ and let $\boldsymbol{g} = (1, F_{n-1})$. The two-dimensional lattice point set $\mathcal{P}(\boldsymbol{g}, N)$ is then called *Fibonacci lattice point set*. A Fibonacci lattice point set with $N = F_9 = 34$ points and generating vector $\boldsymbol{g} = (1, F_8) = (1, 21)$ is illustrated in Figure 1.5.

Since all partial quotients of F_{n-1}/F_n are equal to 1, we have $K(F_{n-1}/F_n) = 1$ and hence

$$\frac{N}{3} \leq \rho(\boldsymbol{g}, F_n) \leq N.$$

According to a formula of Zaremba [266] we even know that $\rho(\boldsymbol{g}, F_n) = F_{n-2}$. Furthermore, Theorem 1.52 implies

$$P_\alpha(\boldsymbol{g}, N) = O\left(\frac{\log N}{N^\alpha}\right).$$

Remark 1.55 (Zaremba's conjecture) In [268, p. 76] Zaremba conjectured that there exists a number A such that for every denominator $N \geq 2$ there exists a numerator g relatively prime to N such that $K(g/N) \leq A$. This conjecture is known

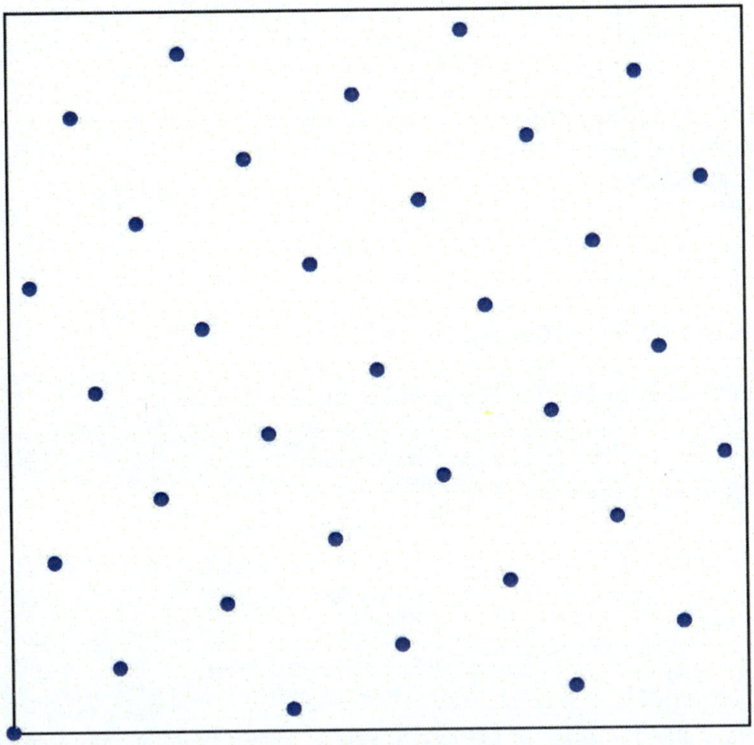

Fig. 1.5: The Fibonacci lattice point set $\mathcal{P}((1,21),34)$.

as *Zaremba's conjecture*. In fact, Zaremba conjectured that A can be chosen equal to 5. Obviously, the conjecture is true when N is a Fibonacci number. Niederreiter [196] proved that Zaremba's conjecture holds true if N is of the form 2^m or 3^m, in which case one may even choose $A = 3$, or of the form 5^m, where A can be chosen equal to 4. Niederreiter's proof provides an effective construction of g with the desired properties.

In [17] Bourgain and Kontorovich proved that Zaremba's conjecture holds true for $A = 50$ for a set of denominators N of density 1, i.e.,

$$\lim_{m\to\infty} \frac{|\{N \in \{2,3,\ldots,m\} \,:\, \exists g \text{ such that } \gcd(g,N)=1 \text{ and } K(g,N) \leq 50\}|}{m} = 1.$$

Returning to the d-dimensional case, it is an open question how to find, for given d and N, lattice points $\boldsymbol{g} \in \mathbb{Z}^d$ for which the lower bound in Theorem 1.53 is attained. Korobov [139] suggested considering lattice point sets with generating vectors $\boldsymbol{g} = (1, g, \ldots, g^{d-1}) \in \mathbb{Z}^d$ with $g \in \{1, 2, \ldots, N-1\}$ such that $\gcd(g, N) = 1$ (generating vectors of this specific form are called *Korobov type generating vectors*, see also Section 3.2). The size of the search space for lattice points of this form reduces to $\varphi(N)$. At least for prime powers N and in dimension $d = 3$, there is an existence result of Larcher and Niederreiter [174] that we state now.

Proposition 1.56 *There exists an absolute constant $C > 0$ such that for every prime power N there exists an integer g (which is coprime with N) such that for $\boldsymbol{g} = (1, g, g^2)$ we have*

$$\rho(\boldsymbol{g}, N) \geq \frac{CN}{(\log N)^2}.$$

The enhanced trigonometric degree

A *trigonometric polynomial* of degree k is a function f of the form

$$f(\boldsymbol{x}) = \sum_{\substack{\boldsymbol{h} \in \mathbb{Z}^d \\ \|\boldsymbol{h}\|_1 \leq k}} a_{\boldsymbol{h}} \, \mathrm{e}^{2\pi \mathrm{i} \boldsymbol{h} \cdot \boldsymbol{x}}, \tag{1.40}$$

where the coefficients $a_{\boldsymbol{h}}$ are complex numbers and where $\|\cdot\|_1$ denotes the ℓ_1-norm of a vector. It is clear that the coefficients $a_{\boldsymbol{h}}$ are exactly the Fourier coefficients of f, i.e.,

$$a_{\boldsymbol{h}} = \widehat{f}(\boldsymbol{h}) = \int_{[0,1]^d} f(\boldsymbol{x}) \, \mathrm{e}^{-2\pi \mathrm{i} \boldsymbol{h} \cdot \boldsymbol{x}} \, \mathrm{d}\boldsymbol{x},$$

and, in particular, $a_{\boldsymbol{0}} = \int_{[0,1]^d} f(\boldsymbol{x}) \, \mathrm{d}\boldsymbol{x}$. Likewise, one may define a trigonometric polynomial in an equivalent way as a function having only a finite number of nonzero Fourier coefficients. It is of degree k, or equivalently of enhanced degree $\kappa = k + 1$, if

$$\widehat{f}(\boldsymbol{h}) = 0 \quad \text{whenever } \|\boldsymbol{h}\|_1 \geq \kappa, \tag{1.41}$$

and it is said to be of strict enhanced degree κ if it is not of enhanced degree $\kappa - 1$.

Definition 1.57 An integration rule based on a lattice $\mathcal{L}$ is said to have *enhanced trigonometric degree* $\kappa(\mathcal{L}) = \kappa$, if it integrates all trigonometric polynomials of enhanced degree κ exactly. We say that it is of strict enhanced degree κ if it is not of enhanced degree $\kappa + 1$. If $\mathcal{L}$ is a rank-1 lattice with lattice point set $\mathcal{P}(\boldsymbol{g}, N)$, then we will write $\kappa(\boldsymbol{g}, N)$ instead of $\kappa(\mathcal{L})$.

For the moment, let κ be an arbitrary positive integer. For a lattice rule with corresponding integration lattice $\mathcal{L}$ we get from Proposition 1.12 that

$$\frac{1}{N} \sum_{k=0}^{N-1} f(\boldsymbol{x}_k) - \int_{[0,1]^d} f(\boldsymbol{x}) \, \mathrm{d}\boldsymbol{x} = \sum_{\substack{\boldsymbol{h} \in \mathcal{L}^\perp \\ 1 \leq \|\boldsymbol{h}\|_1 < \kappa}} \widehat{f}(\boldsymbol{h}) + \sum_{\substack{\boldsymbol{h} \in \mathcal{L}^\perp \\ \|\boldsymbol{h}\|_1 \geq \kappa}} \widehat{f}(\boldsymbol{h}),$$

where $\mathcal{P}(\mathcal{L}) = \{\boldsymbol{x}_0, \boldsymbol{x}_1, \ldots, \boldsymbol{x}_{N-1}\}$, and where $\mathcal{L}^\perp$ is the dual lattice. If f is a trigonometric polynomial of enhanced degree κ, then (1.41) implies that the second sum on the right-hand side of the above equation is zero. If in addition $\mathcal{L}^\perp$ contains no elements with $1 \leq \|\boldsymbol{h}\|_1 < \kappa$, then the entire right-hand side of the above equation vanishes. This means that the lattice rule corresponding to $\mathcal{L}$ integrates the trigonometric polynomial f exactly, which in turn implies that the lattice rule is of enhanced degree κ. From these considerations we find that the enhanced degree of a lattice rule depends only on the location of the elements of the dual lattice $\mathcal{L}^\perp$ with respect to the ℓ_1-norm. To be more precise, we have the following theorem.

Theorem 1.58 *The enhanced trigonometric degree of a lattice $\mathcal{L}$ is*

$$\kappa(\mathcal{L}) = \min_{\boldsymbol{h} \in \mathcal{L}^\perp \setminus \{\boldsymbol{0}\}} \|\boldsymbol{h}\|_1.$$

Additionally, we have the following upper bound on the enhanced trigonometric degree of a lattice.

Theorem 1.59 *The enhanced trigonometric degree $\kappa = \kappa(\mathcal{L})$ of a d-dimensional lattice $\mathcal{L}$ with $\det(\mathcal{L}^\perp) = N$, where $N \geq 2$, satisfies $\kappa(\mathcal{L}) \leq (d!N)^{1/d}$.*

The proof of Theorem 1.59 is based on Minkowski's fundamental theorem, which we state below. For a proof of Theorem 1.60 we refer, e.g., to [93, Theorem 447].

Theorem 1.60 (Minkowski's fundamental theorem) *Let $\mathcal{L}$ be a lattice in $\mathbb{R}^d$ with determinant $\det(\mathcal{L})$. Then any convex set in $\mathbb{R}^d$ which is symmetric with respect to the origin and with volume greater than $2^d \det(\mathcal{L})$ contains a nonzero lattice point of $\mathcal{L}$.*

We now give the proof of Theorem 1.59.

Proof of Theorem 1.59 Let $\mathcal{L}$ be an integration lattice with $\det(\mathcal{L}^\perp) = N$. Consider the centered ℓ_1-ball

$$C_\rho^d := \{\boldsymbol{x} \in \mathbb{R}^d \ : \ \|\boldsymbol{x}\|_1 \leq \rho\},$$

for some $\rho > 0$. Then C_ρ^d is symmetric with respect to the origin and the volume of C_ρ^d is

$$\lambda_d(C_\rho^d) = \frac{2^d \rho^d}{d!}.$$

Hence, by Minkowski's fundamental theorem applied to $\mathcal{L}^\perp$, we have that if

$$\frac{2^d \rho^d}{d!} \geq 2^d \det(\mathcal{L}^\perp) = 2^d N,$$

i.e., if $\rho \geq (d!N)^{1/d}$, then C_ρ^d contains a nonzero point of $\mathcal{L}^\perp$. In other words, $\mathcal{L}^\perp$ contains a nonzero lattice point belonging to $C^d_{(d!N)^{1/d}}$, and therefore we have $\kappa(\mathcal{L}) \leq (d!N)^{1/d}$. □

Next, we show that the upper bound in Theorem 1.59 is essentially best possible. We consider rank-1 lattice rules and show, for prime numbers N, the existence of generators $\boldsymbol{g}$ with large enhanced trigonometric degree. The following proposition is [49, Lemma 2].

Proposition 1.61 *For any prime number N, there exists a $\boldsymbol{g} \in \{0, 1, \ldots, N-1\}^d$ such that*

$$\kappa(\boldsymbol{g}, N) \geq \left\lceil \frac{(d!\,N)^{1/d}}{2} \right\rceil - d.$$

Proof For a given $\boldsymbol{h} = (h_1, \ldots, h_d) \in \mathbb{Z}^d \setminus \{\boldsymbol{0}\}$ with $|h_j| < N$ for all $j \in [d]$, there are N^{d-1} choices of $\boldsymbol{g} \in \{0, 1, \ldots, N-1\}^d$ such that $\boldsymbol{h} \cdot \boldsymbol{g} \equiv 0 \pmod N$. Furthermore

$$|\{\boldsymbol{h} \in \mathbb{Z}^d \ : \ \|\boldsymbol{h}\|_1 = \ell\}| \leq 2^d \binom{\ell + d - 1}{d-1}.$$

Let $\kappa < N$ be a given positive integer (note that we always have $\kappa(\boldsymbol{g}, N) < N$). Then

$$|\{\boldsymbol{h} \in \mathbb{Z}^d \ : \ \|\boldsymbol{h}\|_1 \leq \kappa\}| \leq 2^d \sum_{\ell=0}^{\kappa} \binom{\ell + d - 1}{d-1} = 2^d \binom{\kappa + d}{d}.$$

For every $\boldsymbol{h}$ in the set $\{\boldsymbol{h} \in \mathbb{Z}^d \ : \ \|\boldsymbol{h}\|_1 \leq \kappa\}$ there are N^{d-1} choices $\boldsymbol{g} \in \{0, 1, \ldots, N-1\}^d$ such that $\boldsymbol{h} \cdot \boldsymbol{g} \equiv 0 \pmod N$. For every such $\boldsymbol{g}$ we have $\kappa(\boldsymbol{g}, N) \leq \kappa$. Therefore

$$|\{\boldsymbol{g} \in \{0, 1, \ldots, N-1\}^d \ : \ \kappa(\boldsymbol{g}, N) \leq \kappa\}| \leq N^{d-1} 2^d \binom{\kappa + d}{d}.$$

Note that the total number of possible generators $\boldsymbol{g} \in \{0, 1, \ldots, N-1\}^d$ is N^d. Thus if

$$N^{d-1} 2^d \binom{\kappa + d}{d} < N^d, \tag{1.42}$$

then there exists a $\boldsymbol{g} \in \{0, 1, \ldots, N-1\}^d$ such that $\kappa(\boldsymbol{g}, N) > \kappa$. We estimate

$$2^d \binom{\kappa + d}{d} \leq \frac{2^d (\kappa + d)^d}{d!}.$$

Thus (1.42) is satisfied if $2^d (\kappa + d)^d / d! < N$, i.e., for $\kappa = \lceil 2^{-1} (d!\,N)^{1/d} \rceil - d - 1$. □

Proposition 1.61 is a pure existence result. However, it is possible to give—in some sense—a construction of generating vectors $\boldsymbol{g}$ satisfying

$$\kappa(\boldsymbol{g}, N) \geq c(d) N^{1/d}$$

for a positive but weaker $c(d)$ as compared to Proposition 1.61. To this end, consider an algebraic number field F of degree $d+1$, and let $1, \delta_1, \ldots, \delta_d$ be algebraic integers forming a basis of F. Let $\delta_j^{(1)}, \ldots, \delta_j^{(d)}$ be the conjugates of δ_j for all $j \in [d]$.

For $N \in \mathbb{N}$, let $g_j = g_j(N)$ be the nearest integer to $\delta_j N$ (in case there are two integers with this property, one is free to choose either of them). Then by Dirichlet's theorem (see, e.g., [20, p. 23]), for any $M \in \mathbb{N}$ there exists an $N \in [M]$ such that

$$\max_{j \in [d]} \left| \delta_j - \frac{g_j(N)}{N} \right| \leq \frac{1}{N M^{1/d}}.$$

Let now $\boldsymbol{h} = (h_1, \ldots, h_d) \in \mathbb{Z}^d \setminus \{\boldsymbol{0}\}$ be such that $h_1 g_1 + \cdots + h_d g_d \equiv 0 \pmod{N}$. For $j \in [d]$ define

$$x_j = h_1 \delta_1^{(j)} + \cdots + h_d \delta_d^{(j)} - \frac{h_1 g_1 + \cdots + h_d g_d}{N},$$

and

$$x_{d+1} = h_1 \delta_1 + \cdots + h_d \delta_d - \frac{h_1 g_1 + \cdots + h_d g_d}{N}.$$

Then we have

$$\begin{aligned} |x_{d+1}| &= \left| h_1 \delta_1 + \cdots + h_d \delta_d - h_1 \frac{g_1}{N} - \cdots - h_d \frac{g_d}{N} \right| \\ &\leq \sum_{j=1}^{d} |h_j| \left| \delta_j - \frac{g_j}{N} \right| \leq \frac{\|\boldsymbol{h}\|_1}{N M^{1/d}}, \end{aligned}$$

and for $j \in [d]$ we have

$$\begin{aligned} |x_j| &\leq |x_j - x_{d+1}| + |x_{d+1}| \\ &\leq \sum_{i=1}^{d} |h_i| \, |\delta_i^{(j)} - \delta_i| + \frac{\|\boldsymbol{h}\|_1}{N M^{1/d}} \\ &\leq \|\boldsymbol{h}\|_1 \left(\max_{i \in [d]} |\delta_i^{(j)} - \delta_i| + \frac{1}{N M^{1/d}} \right). \end{aligned}$$

By the definition of the x_j, the product $x_1 \cdots x_{d+1}$ is a nonzero integer, and therefore

$$1 \leq |x_1 \cdots x_{d+1}| \leq \frac{\|\boldsymbol{h}\|_1^{d+1}}{N M^{1/d}} \prod_{j=1}^{d} \left(\max_{i \in [d]} |\delta_i^{(j)} - \delta_i| + \frac{1}{N M^{1/d}} \right).$$

Let $\beta = \prod_{j=1}^{d} \left(\max_{i \in [d]} |\delta_i^{(j)} - \delta_i| + 1 \right)$. Then

$$\|\boldsymbol{h}\|_1 \geq \frac{N^{1/(d+1)} M^{1/(d(d+1))}}{\beta^{1/(d+1)}},$$

and therefore for $\boldsymbol{g}(N) = (g_1(N), \ldots, g_d(N))$ we have

$$\kappa(\boldsymbol{g}(N), N) \geq \frac{N^{1/(d+1)} M^{1/(d(d+1))}}{\beta^{1/(d+1)}}. \tag{1.43}$$

Using the fact that $N \in [M]$, (1.43) implies that

$$\kappa(\boldsymbol{g}(N), N) \geq \frac{N^{1/d}}{\beta^{1/(d+1)}}.$$

We summarize the obtained result in the following proposition.

Proposition 1.62 *With the notation from above, there are infinitely many* $N \in \mathbb{N}$ *such that*

$$\kappa(\boldsymbol{g}(N), N) \geq \frac{N^{1/d}}{\beta^{1/(d+1)}}.$$

Example 1.63 If we take, for example, $\delta_j = 2^{j/(d+1)}$ for $j \in [d]$, then

$$\max_{i \in [d]} |\delta_i^{(j)} - \delta_i| \leq 4$$

for all $j \in [d]$, i.e., $\beta \leq 5^d$. Hence

$$\kappa(\boldsymbol{g}(N), N) \geq \frac{N^{1/d}}{5} \quad \text{for infinitely many } N \in \mathbb{N}.$$

Note that for given N the generating vectors $\boldsymbol{g} = \boldsymbol{g}(N)$ can be computed easily. However, we still need to find suitable values of N. To this end, we can use Inequality (1.43), which is useful to obtain an upper bound on the number of points needed to increase the enhanced trigonometric degree by one.

Assume that for some $N^* > 1$ we have found a generating vector $\boldsymbol{g}^* = \boldsymbol{g}(N^*)$ with enhanced trigonometric degree $\kappa^* = \kappa(\boldsymbol{g}^*, N^*)$ such that $\kappa(\boldsymbol{g}(N), N) < \kappa^*$ for all $1 \leq N < N^*$. Then, from (1.43), we know that there exists an N with $N^* < N \leq M^*$ and a vector $\boldsymbol{g}(N)$ with enhanced trigonometric degree $\kappa(\boldsymbol{g}(N), N) = \kappa^* + 1$, where

$$M^* = \left\lfloor \left(\frac{\beta(\kappa^*)^{d+1}}{N^* + 1} \right)^d \right\rfloor + 1.$$

Thus $M^* - N^*$ is the size of the maximal gap towards the next increase of the enhanced trigonometric degree. In particular, this shows that there exists a sequence of integers $N_1, N_2, \ldots$ with $1 \leq N_1 < N_2 < N_3 < \cdots$ such that

$$\kappa(\boldsymbol{g}(N_1), N_1) < \kappa(\boldsymbol{g}(N_2), N_2) < \kappa(\boldsymbol{g}(N_3), N_3) < \cdots,$$

and

$$\kappa(\boldsymbol{g}(N_k), N_k) \geq \frac{N_k^{1/d}}{\beta^{1/(d+1)}} \quad \text{for } k \in \mathbb{N}.$$

The spectral test

The spectral test is a quantity to assess the coarseness of integration lattices $\mathcal{L}$ in $\mathbb{R}^d$.

Definition 1.64 The *spectral test* of an integration lattice $\mathcal{L}$ is defined as

$$\sigma(\mathcal{L}) := \frac{1}{\min\{\|\boldsymbol{h}\|_2 : \boldsymbol{h} \in \mathcal{L}^\perp \setminus \{\boldsymbol{0}\}\}},$$

where $\|\cdot\|_2$ denotes the ℓ_2-norm in $\mathbb{R}^d$. If $\mathcal{L}$ is a rank-1 lattice with lattice point set $\mathcal{P}(\boldsymbol{g}, N)$, then we will write $\sigma(\boldsymbol{g}, N)$ instead of $\sigma(\mathcal{L})$.

The spectral test has the following geometric interpretation: $\sigma(\mathcal{L})$ is the maximal distance between two adjacent hyperplanes, taken over all families of parallel hyperplanes that cover the lattice $\mathcal{L}$.

The following theorem states a general lower bound on the spectral test of lattice point sets.

Theorem 1.65 *Let $\mathcal{P}(\mathcal{L})$ be an N-element lattice point set in $[0,1)^d$. Then we have*

$$\sigma(\mathcal{L}) \geq \frac{\sqrt{\pi}}{2}\left(\Gamma\left(\frac{d}{2}+1\right)\right)^{-1/d} \frac{1}{N^{1/d}},$$

where Γ denotes the Gamma function $\Gamma(x) := \int_0^\infty t^{x-1} \mathrm{e}^{-t}\, \mathrm{d}t$ for $x > 0$.

Proof The proof is almost the same as the proof of Theorem 1.59. Let $\mathcal{L}$ be the integration lattice yielding the N-element lattice point set and let $\mathcal{L}^\perp$ be the corresponding dual lattice. According to Theorem 1.4 we have $\det(\mathcal{L}^\perp) = N$. Now consider the centered ℓ_2-ball

$$C_r^d = \{\boldsymbol{x} \in \mathbb{R}^d \, : \, x_1^2 + \cdots + x_d^2 \leq r^2\}$$

of radius $r > 0$. Then C_r^d is symmetric with respect to the origin and the volume of C_r^d is

$$\lambda_d(C_r^d) = r^d \frac{\pi^{d/2}}{\Gamma(d/2+1)}.$$

Consequently, by Minkowski's fundamental theorem (see Theorem 1.60) applied to $\mathcal{L}^\perp$, we have that if

$$r^d \frac{\pi^{d/2}}{\Gamma(d/2+1)} \geq 2^d \det(\mathcal{L}^\perp) = 2^d N,$$

i.e., if $r \geq (2/\sqrt{\pi})(\Gamma(d/2+1))^{1/d} N^{1/d} =: \widetilde{r}(d,N)$, then C_r^d contains a nonzero point of $\mathcal{L}^\perp$. In other words, $\mathcal{L}^\perp$ contains a nonzero lattice point which belongs to $C_{\widetilde{r}(d,N)}^d$, and therefore we have $(\sigma(\mathcal{L}))^{-1} \leq \widetilde{r}(d,N)$. Taking the reciprocal values we obtain the desired result. □

The order of magnitude $N^{-1/d}$ is best possible for the spectral test of integration lattices in dimension d. This order can even be attained by using rank-1 lattice point sets.

Proposition 1.66 *For every dimension d there exists a positive number C_d depending only on d with the following property. For every prime number N there exists a lattice point $\boldsymbol{g} \in \{0, 1, \ldots, N-1\}^d$ such that*

$$\sigma(\boldsymbol{g}, N) \leq \frac{C_d}{N^{1/d}}.$$

Proof The result can be shown by the same methods that we already used in the proof of Proposition 1.61 with the ℓ_1-norm replaced by the ℓ_2-norm. Likewise, one may use Proposition 1.61 directly together with the fact that the ℓ_2- and the ℓ_1-norm in $\mathbb{R}^d$ are equivalent. □

Lattices with a low value of the spectral test can be shown to have good isotropic discrepancy. Such point sets may be employed for the numerical integration of functions over the sphere $\mathbb{S}^2$ in $\mathbb{R}^3$. We will discuss this issue further in Section 5.6.

Notes and Remarks

More detailed introductions to Monte Carlo methods can be found in the books by Lemieux [178] and by Müller-Gronbach, Novak, and Ritter [189]. An extensive treatment of applications of MC to problems in finance is given in the book [78] by Glasserman. QMC in the context of finance, in particular techniques to value European and Asian options, are discussed in the paper [177] by L'Ecuyer. See also the survey article [173] by Larcher and Leobacher and the references therein. Thorough surveys of the application of QMC methods in computer graphics can be found in [132, 133] by Keller.

The Halton sequence was first introduced by Halton in [92]. The currently best estimates for the discrepancy of the Halton sequence are based on the method of Atanassov from [5]. For further information see [74, Section 4.2]. The concept of (t, m, s)-nets and digital nets was introduced by Niederreiter in [197]; extensive surveys can be found in [199, Chapter 4] and [180, Chapter 5], and in the book [52].

The method of good lattice points, a special instance of lattice rules, was introduced by Korobov [138] in 1959 and a few years later independently by Hlawka [113]. The Soviet school quickly developed a quite satisfactory theory of the method of good lattice points (see, e.g., Bakhvalov [6]), which is summarized in the book of Korobov [140] from 1963. Further expository texts on early results are the book [115] by Hua and Wang and the survey articles by Zaremba [268] and Niederreiter [195] (with an update in [198]). During the 1980s lattice rules became a booming field of research with many exciting results and applications. These developments are very well summarized in two seminal books by Sloan and Joe [230] and by Niederreiter [199], respectively.

Further details on copy rules can be found in [61] and [230]. The papers [122, 124], and again the book [230], contain further remarks, results, and numerical experiments related to embedded lattice rules. We remark that, though sometimes using slightly different techniques, the question of how to extend the number of points of lattice rules has been addressed in numerous papers from theoretical and practical viewpoints, see, for example, [27, 58, 103, 104, 105, 159, 201]. We will return to this issue in Chapter 6.

The study of numerical integration in reproducing kernel Hilbert spaces of functions goes back to Hickernell in [98, 99] and to Sloan and Woźniakowski in [239]. General information on reproducing kernel Hilbert spaces can be found in [4] by Aronszajn.

A comprehensive introduction to uniform distribution theory (modulo one), which lays the basis for QMC integration, is given in the book [155] by Kuipers and Niederreiter. Roth's paper [222], where the fundamental lower bound on L_2-discrepancy is shown, can nowadays be seen as the starting point of modern discrepancy theory. Detailed information on this fascinating topic, including the proof of Roth's result, can be found in the books by Beck and Chen [12], by Drmota and Tichy [66], by Kuipers and Niederreiter [155], and by Matoušek [188]. A concise proof of Roth's lower bound (1.20) by means of Haar function analysis can be found in [41]. Books presenting constructions of low-discrepancy point sets are [52, 180, 199]. Survey articles about the connection between discrepancy and quasi-Monte Carlo integration are, for example, [53, 195].

The classical reference for Information-Based Complexity is the book [251] by Traub, Wasilkowski, and Woźniakowski. On a rather informal level is the collection [208] containing five essays on the complexity of continuous problems. For a comprehensive survey of tractability theory we refer to the trilogy [210, 211, 212] by Novak and Woźniakowski. A survey of the complexity of numerical integration is [207] by Novak, and another survey, at a very introductory level, regarding the curse of dimensionality is [206].

The result (1.34) in Example 1.45 is only an existence result. Until now no explicit constructions of N-element point sets $\mathcal{P}$ in $[0, 1)^d$ for which $D_N^*(\mathcal{P})$ satisfies (1.34) are known. A first constructive approach was given by Doerr, Gnewuch, and Srivastav [64], which was further improved by Doerr and Gnewuch [62], Doerr, Gnewuch, and Wahlström [65], and Gnewuch, Wahlström, and Winzen [83]. There, a deterministic algorithm is presented that constructs point sets $\mathcal{P}$ in $[0, 1)^d$ satisfying

$$D_N^*(\mathcal{P}) \ll \sqrt{\frac{d}{N}}\,\sqrt{\log(N+1)}$$

in a runtime of $O(d\,\log(d\,N)\,(\sigma N)^d)$ operations, where

$$\sigma = \sigma(d) = O((\log d)^2/(d \log\log d)) \to 0$$

as d tends to infinity, and where the implied constants in the O-notation are independent of d and N. However, this is by far too expensive to obtain point sets for high-dimensional applications. A slight improvement on the runtime is presented in [63], but this improvement has to be paid for by a worse dependence on the dimension of the bound on the star-discrepancy.

Detailed information on classical figures of merit for lattice rules can be found in the books by Niederreiter [199] and Sloan and Joe [230]. The Zaremba index ρ for rank-2 lattices is studied in [51]. For a bound on the worst-case error for numerical integration in function spaces of dominating mixed smoothness on the unit cube with homogeneous boundary condition in terms of the Zaremba index see [57]. Details on the enhanced degree of lattice rules are given in the survey article [185] by Lyness. The construction of lattice point sets with large enhanced trigonometric degree preceding Proposition 1.62 is taken from [49]. In this paper, also a close relation between the enhanced trigonometric degree of lattice rules and the integration error for infinitely smooth periodic integrands is established (see [49, Lemma 1]). Further information on the spectral test can be found in [97] by Hellekalek, and in [123] by Joe and Sloan. An overview of quality and construction criteria for lattice rules can also be found in [32].

Fibonacci lattice point sets are well studied in the literature and have excellent properties with respect to various aspects. See, for example, [15, 16, 19, 110, 154, 248].

Chapter 2
Integration of Smooth Periodic Functions

We have already seen in Section 1.4 that lattice rules are perfectly configured for the numerical integration of smooth functions that are one-periodic in each variable. Nowadays it is most convenient to examine lattice rules in the context of reproducing kernel Hilbert spaces of smooth and one-periodic functions and to analyze their worst-case errors in this setting. In this chapter we will introduce Korobov spaces, which can be seen as some of the most classical examples of function classes to which lattice rules for numerical integration have been applied.

As the name indicates, Korobov spaces go back to theory developed by Korobov in the 1960s, and since the late 1990s it is common to study these spaces in their weighted version, where the term "weighted" refers to a way of reflecting possibly different influence of single variables or groups of variables on the integration problem. We shall outline both the unweighted and the weighted setting here.

2.1 Korobov Spaces

Korobov spaces are sometimes also referred to as periodic Sobolev spaces of dominating mixed smoothness in the literature. However, we will use the former name here, as this is the most commonly used in the context of lattice rules.

For the introduction of Korobov spaces, let us first consider the univariate case.

The univariate case

The Korobov space we are now going to define depends on a real parameter $\alpha > 1/2$, which is related to the decay of the Fourier coefficients of the functions in the space and is therefore often referred to as the smoothness parameter.

We define the Hilbert space $\mathcal{H}_{\mathrm{kor},\alpha}$ of absolutely integrable functions over the unit interval $[0, 1]$ that consists of all one-periodic functions f with absolutely convergent Fourier series

J. Dick et al., *Lattice Rules*, Springer Series in Computational Mathematics 58,
https://doi.org/10.1007/978-3-031-09951-9_2

$$f(x) = \sum_{h\in\mathbb{Z}} \widehat{f}(h)\, \mathrm{e}^{2\pi \mathrm{i} h x}$$

with Fourier coefficients

$$\widehat{f}(h) = \int_0^1 f(x)\, \mathrm{e}^{-2\pi \mathrm{i} h x}\, \mathrm{d}x$$

and with finite norm $\|f\|_{\mathrm{kor},\alpha} = \langle f, f\rangle_{\mathrm{kor},\alpha}^{1/2}$, where the inner product is defined by

$$\langle f, g\rangle_{\mathrm{kor},\alpha} := \sum_{h\in\mathbb{Z}} r_{2\alpha}(h) \widehat{f}(h)\overline{\widehat{g}(h)},$$

with

$$r_\tau(h) := \begin{cases} 1 & \text{if } h = 0, \\ |h|^\tau & \text{if } h \neq 0 \end{cases}$$

for real $\tau > 0$. Note that in accordance with the definition of r_1 in Section 1.4 we can also write $r_\tau(h) := \max(1, |h|^\tau)$.

Definition 2.1 The Hilbert space $\mathcal{H}_{\mathrm{kor},\alpha}$ of Fourier series on $[0, 1]$ given by

$$\mathcal{H}_{\mathrm{kor},\alpha} := \left\{ f \,:\, f(x) = \sum_{h\in\mathbb{Z}} \widehat{f}(h)\, \mathrm{e}^{2\pi \mathrm{i} h x} \quad \text{for } x \in [0, 1], \text{ and } \|f\|_{\mathrm{kor},\alpha} < \infty \right\}$$

is called the (univariate) *Korobov space of smoothness* α.

We now show that $\mathcal{H}_{\mathrm{kor},\alpha}$ is indeed a reproducing kernel Hilbert space, with kernel

$$K_{\mathrm{kor},\alpha}(x, y) = \sum_{h\in\mathbb{Z}} \frac{1}{r_{2\alpha}(h)}\, \mathrm{e}^{2\pi \mathrm{i} h(x-y)} \quad \text{for } x, y \in [0, 1]. \tag{2.1}$$

First, note that, due to $\alpha > 1/2$, the series in (2.1) is absolutely convergent, so $K_{\mathrm{kor},\alpha}(\cdot, y)$ is absolutely integrable and

$$\begin{aligned} \widehat{K}_{\mathrm{kor},\alpha}(\cdot, y)(h) &= \int_0^1 K_\alpha(x, y)\, \mathrm{e}^{-2\pi \mathrm{i} h x}\, \mathrm{d}x \\ &= \sum_{k\in\mathbb{Z}} \frac{1}{r_{2\alpha}(h)}\, \mathrm{e}^{-2\pi \mathrm{i} k y} \int_0^1 \mathrm{e}^{2\pi \mathrm{i}(k-h)x}\, \mathrm{d}x \\ &= \frac{1}{r_{2\alpha}(h)}\, \mathrm{e}^{-2\pi \mathrm{i} h y}. \end{aligned}$$

Thus

$$\begin{aligned} \|K_{\mathrm{kor},\alpha}(\cdot, y)\|_\alpha^2 &= \langle K_{\mathrm{kor},\alpha}(\cdot, y), K_{\mathrm{kor},\alpha}(\cdot, y)\rangle_{\mathrm{kor},\alpha} \\ &= \sum_{h\in\mathbb{Z}} r_{2\alpha}(h)\, \frac{1}{r_{2\alpha}^2(h)} \\ &= 1 + 2\zeta(2\alpha). \end{aligned}$$

For $\alpha > 1/2$ we have $\zeta(2\alpha) < \infty$, and therefore we obtain $K_{\mathrm{kor},\alpha}(\cdot, y) \in \mathcal{H}_\alpha$. Furthermore, for every $y \in [0, 1]$, we have

$$\langle f, K_{\mathrm{kor},\alpha}(\cdot, y)\rangle_{\mathrm{kor},\alpha} = \sum_{h\in\mathbb{Z}} r_{2\alpha}(h)\widehat{f}(h)\frac{\mathrm{e}^{2\pi \mathrm{i} hy}}{r_{2\alpha}(h)} = f(y),$$

which shows that also the reproducing property of $K_{\mathrm{kor},\alpha}$ is satisfied. Hence, in summary, $K_{\mathrm{kor},\alpha}$ is indeed the reproducing kernel of $\mathcal{H}_{\mathrm{kor},\alpha}$, i.e., $\mathcal{H}_{\mathrm{kor},\alpha} = \mathcal{H}(K_{\mathrm{kor},\alpha})$.

Note that $K_{\mathrm{kor},\alpha}$ is actually real-valued since $r_{2\alpha}(h) = r_{2\alpha}(-h)$ for $h \in \mathbb{Z} \setminus \{0\}$ and so

$$\begin{aligned} K_{\mathrm{kor},\alpha}(x, y) &= 1 + \sum_{h=1}^{\infty} \frac{1}{r_{2\alpha}(h)} \left(\mathrm{e}^{2\pi \mathrm{i} h(x-y)} + \mathrm{e}^{-2\pi \mathrm{i} h(x-y)}\right) \\ &= 1 + 2\sum_{h=1}^{\infty} \frac{1}{h^{2\alpha}} \cos(2\pi h(x-y)) \in \mathbb{R}. \end{aligned} \tag{2.2}$$

If $\alpha \geq 1$ is an integer, then it is directly related to the number of existing square integrable derivatives of f. This justifies, once again, why we refer to α as the smoothness parameter.

Proposition 2.2 *If $\alpha \in \mathbb{N}$, then $\mathcal{H}_{\mathrm{kor},\alpha}$ consists of all one-periodic functions whose derivatives up to order $\alpha - 1$ are absolutely continuous, and whose α-th derivative belongs to $L_2([0, 1])$. In particular, for $f \in \mathcal{H}_{\mathrm{kor},\alpha}$, we have*

$$\|f\|^2_{\mathrm{kor},\alpha} = \left(\int_0^1 f(t)\,\mathrm{d}t\right)^2 + \frac{1}{(2\pi)^{2\alpha}} \int_0^1 (f^{(\alpha)}(t))^2\,\mathrm{d}t. \tag{2.3}$$

Proof For $f \in \mathcal{H}_{\mathrm{kor},\alpha}$ we have

$$\|f\|^2_{\mathrm{kor},\alpha} = \sum_{h\in\mathbb{Z}} r_{2\alpha}(h)|\widehat{f}(h)|^2 = |\widehat{f}(0)|^2 + \sum_{h\in\mathbb{Z}\setminus\{0\}} |h|^{2\alpha}|\widehat{f}(h)|^2 < \infty.$$

On the other hand, we know that

$$f(x) = \sum_{h\in\mathbb{Z}} \widehat{f}(h)\,\mathrm{e}^{2\pi \mathrm{i} hx}.$$

Now, for $k \in \{1, 2, \ldots, \alpha\}$, we have

$$f^{(k)}(x) = (2\pi \mathrm{i})^k \sum_{h\in\mathbb{Z}\setminus\{0\}} h^k \widehat{f}(h)\,\mathrm{e}^{2\pi \mathrm{i} hx}.$$

If $k \leq \alpha - 1$, then, using first the triangle inequality and then the Cauchy–Schwarz inequality, we obtain

$$|f^{(k)}(x)| \leq (2\pi)^k \sum_{h\in\mathbb{Z}\setminus\{0\}} |\widehat{f}(h)|(r_{2\alpha}(h))^{1/2}|h|^k(r_{2\alpha}(h))^{-1/2}$$

$$\leq (2\pi)^k \left(\sum_{h \in \mathbb{Z}\setminus\{0\}} |\widehat{f}(h)|^2 r_{2\alpha}(h) \right)^{1/2} \left(\sum_{h \in \mathbb{Z}\setminus\{0\}} |h|^{2k} (r_{2\alpha}(h))^{-1} \right)^{1/2}$$
$$\leq (2\pi)^k \|f\|_{\mathrm{kor},\alpha} (2\,\zeta(2(\alpha-k)))^{1/2} < \infty.$$

Therefore the series in the representation of $f^{(k)}(x)$ is pointwise convergent if $k \leq \alpha - 1$. For $k = \alpha$ we have

$$\int_0^1 |f^{(\alpha)}(x)|^2 \,\mathrm{d}x = (2\pi)^{2\alpha} \sum_{h \in \mathbb{Z}\setminus\{0\}} h^{2\alpha} |\widehat{f}(h)|^2 \leq \|f\|_{\mathrm{kor},\alpha}^2.$$

As a consequence we have

$$\int_0^1 |f^{(\alpha)}(x)|^2 \,\mathrm{d}x < \infty,$$

and hence $f^{(\alpha)} \in L_2([0,1])$.

We have

$$\begin{aligned} |f^{(k)}(x) - f^{(k)}(y)| &\leq (2\pi)^k \sum_{h \in \mathbb{Z}\setminus\{0\}} |h^k| \cdot |\widehat{f}(h)| \cdot |\mathrm{e}^{2\pi \mathrm{i} h x} - \mathrm{e}^{2\pi \mathrm{i} h y}| \\ &= (2\pi)^k \sum_{h \in \mathbb{Z}\setminus\{0\}} |h^k| \cdot |\widehat{f}(h)| \cdot |2 \sin(\pi h (x-y))| \\ &\leq (2\pi)^k \sum_{h \in \mathbb{Z}\setminus\{0\}} |h^k| \cdot |\widehat{f}(h)| \cdot |2\pi h (x-y)| \\ &\leq (2\pi)^{k+1} |x-y| \sum_{h \in \mathbb{Z}\setminus\{0\}} |h|^{k+1} \cdot |\widehat{f}(h)|. \end{aligned}$$

If $k < \alpha - 1$, then one can see in the same way as above that the last series converges. Thus, for $k < \alpha - 1$ the derivatives $f^{(k)}$ are Lipschitz continuous and therefore absolutely continuous.

Now we consider the case $k = \alpha - 1$. For $x \in [0,1]$ we have by Carleson's theorem, which implies that the Fourier series of $f^{(\alpha)}$ converges pointwise almost everywhere, and the dominated convergence theorem that

$$\begin{aligned} \int_0^x f^{(\alpha)}(t) \,\mathrm{d}t &= (2\pi \mathrm{i})^\alpha \sum_{h \in \mathbb{Z}\setminus\{0\}} h^\alpha \widehat{f}(h) \int_0^x \mathrm{e}^{2\pi \mathrm{i} h t} \,\mathrm{d}t \\ &= (2\pi \mathrm{i})^{\alpha-1} \sum_{h \in \mathbb{Z}\setminus\{0\}} h^{\alpha-1} \widehat{f}(h) (\mathrm{e}^{2\pi \mathrm{i} h x} - 1) \\ &= f^{(\alpha-1)}(x) - f^{(\alpha-1)}(0). \end{aligned}$$

This implies

$$f^{(\alpha-1)}(x) = f^{(\alpha-1)}(0) + \int_0^x f^{(\alpha)}(t) \,\mathrm{d}t,$$

and therefore $f^{(\alpha-1)}$ is absolutely continuous as well.

It remains to be shown that (2.3) holds. Since $f^{(\alpha)} \in L_2([0,1])$, we obtain by applying integration by parts α times, and using the one-periodicity of f, that

$$\widehat{f}(h) = \frac{1}{(2\pi\mathtt{i}h)^\alpha}\,\widehat{f^{(\alpha)}}(h) \quad \text{for } h \neq 0.$$

We obtain

$$\begin{aligned}
\|f\|_{\mathrm{kor},\alpha}^2 &= \sum_{h=-\infty}^{\infty} r_{2\alpha}(h)|\widehat{f}(h)|^2 \\
&= |\widehat{f}(0)|^2 + \sum_{h\in\mathbb{Z}\setminus\{0\}} |h|^{2\alpha}\cdot|\widehat{f}(h)|^2 \\
&= \left(\int_0^1 f(t)\,\mathrm{d}t\right)^2 + \frac{1}{(2\pi)^{2\alpha}} \sum_{h\in\mathbb{Z}\setminus\{0\}} |\widehat{f^{(\alpha)}}(h)|^2 \\
&= \left(\int_0^1 f(t)\,\mathrm{d}t\right)^2 + \frac{1}{(2\pi)^{2\alpha}} \int_0^1 (f^{(\alpha)}(t))^2\,\mathrm{d}t,
\end{aligned}$$

where in the last step we used Parseval's identity and the fact that $\widehat{f^{(\alpha)}}(0) = 0$ due to the periodicity of f and its derivatives. □

The multivariate case

Now we turn to the d-variate case, where $d \in \mathbb{N}$. In the following, we write, for $\boldsymbol{h} = (h_1, \ldots, h_d) \in \mathbb{Z}^d$ and $\tau > 0$,

$$r_\tau(\boldsymbol{h}) := \prod_{j=1}^{d} r_\tau(h_j).$$

For $\alpha > 1/2$ we define the Hilbert space $\mathcal{H}_{\mathrm{kor},d,\alpha}$ as the space of all one-periodic functions f with absolutely convergent Fourier series as in (1.9), with Fourier coefficients as in (1.10), and with finite norm $\|f\|_{\mathrm{kor},d,\alpha} = \langle f, f\rangle_{\mathrm{kor},d,\alpha}^{1/2}$, where the inner product is given by

$$\langle f, g\rangle_{\mathrm{kor},d,\alpha} := \sum_{\boldsymbol{h}\in\mathbb{Z}^d} r_{2\alpha}(\boldsymbol{h})\widehat{f}(\boldsymbol{h})\overline{\widehat{g}(\boldsymbol{h})}.$$

Definition 2.3 The Hilbert space $\mathcal{H}_{\mathrm{kor},d,\alpha}$ given by

$$\mathcal{H}_{\mathrm{kor},d,\alpha} = \left\{ f \,:\, f(\boldsymbol{x}) = \sum_{\boldsymbol{h}\in\mathbb{Z}^d} \widehat{f}(\boldsymbol{h})\,\mathrm{e}^{2\pi\mathtt{i}\boldsymbol{h}\cdot\boldsymbol{x}} \text{ for } \boldsymbol{x} \in [0,1]^d, \text{ and } \|f\|_{\mathrm{kor},d,\alpha} < \infty \right\}$$

is called the *Korobov space of smoothness* α.

Again $\mathcal{H}_{\mathrm{kor},d,\alpha}$ is a reproducing kernel Hilbert space, with the kernel given by

$$K_{\mathrm{kor},d,\alpha}(\boldsymbol{x},\boldsymbol{y}) = \prod_{j=1}^{d}\left(\sum_{h_j\in\mathbb{Z}} \frac{\mathrm{e}^{2\pi \mathrm{i} h_j(x_j-y_j)}}{r_{2\alpha}(h_j)}\right) = \sum_{\boldsymbol{h}\in\mathbb{Z}^d}\prod_{j=1}^{d} \frac{\mathrm{e}^{2\pi \mathrm{i} h_j(x_j-y_j)}}{r_{2\alpha}(h_j)} = \sum_{\boldsymbol{h}\in\mathbb{Z}^d} \frac{1}{r_{2\alpha}(\boldsymbol{h})}\,\mathrm{e}^{2\pi \mathrm{i}\boldsymbol{h}\cdot(\boldsymbol{x}-\boldsymbol{y})}, \tag{2.4}$$

where $\boldsymbol{x} = (x_1,\ldots,x_d)$ and $\boldsymbol{y} = (y_1,\ldots,y_d)$, both in $[0,1]^d$. The reproducing property of $K_{\mathrm{kor},d,\alpha}$ can be shown analogously to the univariate case by using (2.4).

Note that $\mathcal{H}_{\mathrm{kor},d,\alpha}$ is the d-fold tensor product of d instances of the univariate space $\mathcal{H}_{\mathrm{kor},\alpha}$, i.e.,

$$\mathcal{H}_{\mathrm{kor},d,\alpha} := \underbrace{\mathcal{H}_{\mathrm{kor},\alpha}\otimes\cdots\otimes\mathcal{H}_{\mathrm{kor},\alpha}}_{d\text{-fold}} = \overline{\mathrm{span}\left\{\boldsymbol{x}\mapsto\prod_{j=1}^{d} f_j(x_j) \;:\; f_j\in\mathcal{H}_{\mathrm{kor},\alpha}\right\}},$$

where the closure is taken with respect to the norm induced by the inner product. Furthermore, the kernel of $\mathcal{H}_{\mathrm{kor},d,\alpha}$ is the d-fold product of the kernel of $\mathcal{H}_{\mathrm{kor},\alpha}$, i.e.,

$$K_{\mathrm{kor},d,\alpha}(\boldsymbol{x},\boldsymbol{y}) = \prod_{j=1}^{d} K_{\mathrm{kor},\alpha}(x_j,y_j).$$

In order to study the smoothness of functions from $\mathcal{H}_{\mathrm{kor},d,\alpha}$ it is convenient to introduce the following operators which are in essence as in [210, Appendix A.1]. For $\alpha\in\mathbb{N}$ and $\emptyset\neq\mathfrak{u}\subseteq[d]$ define the differential operator

$$D_{\mathfrak{u},\alpha}f := \frac{\partial^{\alpha|\mathfrak{u}|}}{\prod_{j\in\mathfrak{u}}\partial x_j^{\alpha}}\, f,$$

and for $\mathfrak{u}=\emptyset$ set $D_{\emptyset,\alpha}f := f$. Furthermore, for $\emptyset\neq\mathfrak{u}\subseteq[d]$ we need the integration operator $I_{\mathfrak{u}}$, given by

$$(I_{\mathfrak{u}}f)(\boldsymbol{x}) := \int_{[0,1]^{|\mathfrak{u}|}} f(\boldsymbol{x})\,\mathrm{d}\boldsymbol{x}_{\mathfrak{u}} \quad \text{for } \boldsymbol{x}\in[0,1]^d.$$

Note that integration in $I_{\mathfrak{u}}$ is with respect to those variables whose indices belong to $\mathfrak{u}$, while the variables with indices not in $\mathfrak{u}$ remain intact. For $\mathfrak{u}=\emptyset$ put $I_{\emptyset}f := f$. Next, define an operator $V_{\mathfrak{u},\alpha}$, where

$$(V_{\mathfrak{u},\alpha}f)(\boldsymbol{x}) := (I_{[d]\setminus\mathfrak{u}}D_{\mathfrak{u},\alpha}f)(\boldsymbol{x}) \quad \text{for } \boldsymbol{x}\in[0,1]^d. \tag{2.5}$$

With this definition we have

$$\left(V_{\emptyset,\alpha}f\right)(\boldsymbol{x}) = \int_{[0,1]^d} f(\boldsymbol{x})\,\mathrm{d}\boldsymbol{x} \quad \text{and} \quad \left(V_{[d],\alpha}f\right)(\boldsymbol{x}) = f^{(\alpha,\alpha,\ldots,\alpha)}(\boldsymbol{x}).$$

Now we can show how the Korobov space norm involves the mixed partial derivatives of f.

Proposition 2.4 *If $\alpha \in \mathbb{N}$, then for $f \in \mathcal{H}_{\mathrm{kor},d,\alpha}$ we have*

$$\|f\|^2_{\mathrm{kor},d,\alpha} = \sum_{\mathfrak{u}\subseteq[d]} \frac{1}{(2\pi)^{2\alpha|\mathfrak{u}|}} \int_{[0,1]^d} |\left(V_{\mathfrak{u},\alpha}f\right)(\boldsymbol{x})|^2\,\mathrm{d}\boldsymbol{x}.$$

Proof We use the Fourier series expansion

$$f(\boldsymbol{x}) = \sum_{\boldsymbol{h}\in\mathbb{Z}^d} \widehat{f}(\boldsymbol{h})\,\mathrm{e}^{2\pi\mathrm{i}\boldsymbol{h}\cdot\boldsymbol{x}}$$

and obtain

$$\left(D_{\mathfrak{u},\alpha}f\right)(\boldsymbol{x}) = \sum_{\boldsymbol{h}\in\mathbb{Z}^d} \widehat{f}(\boldsymbol{h})(2\pi\mathrm{i})^{\alpha|\mathfrak{u}|}\left(\prod_{j\in\mathfrak{u}} h_j^\alpha\right)\mathrm{e}^{2\pi\mathrm{i}\boldsymbol{h}\cdot\boldsymbol{x}}.$$

If a component h_j is zero for some $j \in \mathfrak{u}$, the corresponding term in the sum above vanishes. Hence it suffices to sum only over those terms $\boldsymbol{h}$ for which $\mathfrak{u} \subseteq \mathfrak{u}(\boldsymbol{h})$, where for $\boldsymbol{h} \in \mathbb{Z}^d$ we put

$$\mathfrak{u}(\boldsymbol{h}) := \{j \in [d] \;:\; h_j \neq 0\}.$$

The previous expression for $D_{\mathfrak{u},\alpha}f$ becomes

$$\left(D_{\mathfrak{u},\alpha}f\right)(\boldsymbol{x}) = \sum_{\substack{\boldsymbol{h}\in\mathbb{Z}^d\\ \mathfrak{u}\subseteq\mathfrak{u}(\boldsymbol{h})}} \widehat{f}(\boldsymbol{h})(2\pi\mathrm{i})^{\alpha|\mathfrak{u}|}\left(\prod_{j\in\mathfrak{u}} h_j^\alpha\right)\mathrm{e}^{2\pi\mathrm{i}\boldsymbol{h}\cdot\boldsymbol{x}}.$$

Next, we apply the operator $V_{\mathfrak{u},\alpha}$ to f. We have

$$\begin{aligned}
\left(V_{\mathfrak{u},\alpha}f\right)(\boldsymbol{x}) &= \sum_{\substack{\boldsymbol{h}\in\mathbb{Z}^d\\ \mathfrak{u}\subseteq\mathfrak{u}(\boldsymbol{h})}} \widehat{f}(\boldsymbol{h})(2\pi\mathrm{i})^{\alpha|\mathfrak{u}|}\left(\prod_{j\in\mathfrak{u}} h_j^\alpha\right)\int_{[0,1]^{d-|\mathfrak{u}|}} \mathrm{e}^{2\pi\mathrm{i}\boldsymbol{h}\cdot\boldsymbol{x}}\,\mathrm{d}\boldsymbol{x}_{[d]\setminus\mathfrak{u}} \\
&= \sum_{\substack{\boldsymbol{h}\in\mathbb{Z}^d\\ \mathfrak{u}\subseteq\mathfrak{u}(\boldsymbol{h})\\ h_j=0\ \forall j\notin\mathfrak{u}}} \widehat{f}(\boldsymbol{h})(2\pi\mathrm{i})^{\alpha|\mathfrak{u}|}\left(\prod_{j\in\mathfrak{u}} h_j^\alpha\right)\mathrm{e}^{2\pi\mathrm{i}\sum_{j\in\mathfrak{u}} h_j x_j} \\
&= \sum_{\boldsymbol{h}\in\mathbb{Z}_{\mathfrak{u}}} \widehat{f}(\boldsymbol{h})(2\pi\mathrm{i})^{\alpha|\mathfrak{u}|}\left(\prod_{j\in\mathfrak{u}} h_j^\alpha\right)\mathrm{e}^{2\pi\mathrm{i}\sum_{j\in\mathfrak{u}} h_j x_j},
\end{aligned}$$

where $\mathbb{Z}_{\mathfrak{u}} := \{\boldsymbol{h} \in \mathbb{Z}^d : \mathfrak{u}(\boldsymbol{h}) = \mathfrak{u}\}$. This implies

$$\begin{aligned}\int_{[0,1]^d} |(V_{\mathfrak{u},\alpha} f)(\boldsymbol{x})|^2 \,\mathrm{d}\boldsymbol{x} &= \int_{[0,1]^d} (V_{\mathfrak{u},\alpha} f)(\boldsymbol{x}) \overline{(V_{\mathfrak{u},\alpha} f)(\boldsymbol{x})} \,\mathrm{d}\boldsymbol{x} \\ &= (2\pi)^{2\alpha|\mathfrak{u}|} \sum_{\boldsymbol{h} \in \mathbb{Z}_{\mathfrak{u}}} |\widehat{f}(\boldsymbol{h})|^2 \prod_{j \in \mathfrak{u}} h_j^{2\alpha}.\end{aligned}$$

Now we have

$$\begin{aligned}\|f\|_{\mathrm{kor},d,\alpha}^2 &= \sum_{\boldsymbol{h} \in \mathbb{Z}^d} \left(\prod_{j \in \mathfrak{u}(\boldsymbol{h})} |h_j|^{2\alpha} \right) |\widehat{f}(\boldsymbol{h})|^2 \\ &= \sum_{\mathfrak{u} \subseteq [d]} \sum_{\boldsymbol{h} \in \mathbb{Z}_{\mathfrak{u}}} \left(\prod_{j \in \mathfrak{u}} |h_j|^{2\alpha} \right) |\widehat{f}(\boldsymbol{h})|^2 \\ &= \sum_{\mathfrak{u} \subseteq [d]} \frac{1}{(2\pi)^{2\alpha|\mathfrak{u}|}} \int_{[0,1]^d} |(V_{\mathfrak{u},\alpha} f)(\boldsymbol{x})|^2 \,\mathrm{d}\boldsymbol{x},\end{aligned}$$

as desired. □

Remark 2.5 Note that in the univariate case the formula for the norm in Proposition 2.4 coincides with Formula (2.3) in Proposition 2.2. As a further example we highlight the bivariate case $d = 2$, where we obtain

$$\begin{aligned}\|f\|_{\mathrm{kor},2,\alpha}^2 = & \left| \int_0^1 \int_0^1 f(x_1, x_2) \,\mathrm{d}x_1 \,\mathrm{d}x_2 \right|^2 \\ & + \frac{1}{(2\pi)^{2\alpha}} \int_0^1 \left| \int_0^1 \frac{\partial^\alpha f(x_1, x_2)}{\partial x_1^\alpha} \,\mathrm{d}x_2 \right|^2 \mathrm{d}x_1 \\ & + \frac{1}{(2\pi)^{2\alpha}} \int_0^1 \left| \int_0^1 \frac{\partial^\alpha f(x_1, x_2)}{\partial x_2^\alpha} \,\mathrm{d}x_1 \right|^2 \mathrm{d}x_2 \\ & + \frac{1}{(2\pi)^{4\alpha}} \int_0^1 \int_0^1 \left| \frac{\partial^{2\alpha} f(x_1, x_2)}{\partial x_1^\alpha \partial x_2^\alpha} \right|^2 \mathrm{d}x_1 \,\mathrm{d}x_2.\end{aligned}$$

2.2 Integration in Korobov Spaces

Let us now study integration in $\mathcal{H}_{\mathrm{kor},d,\alpha}$ for $\alpha > 1/2$ by means of lattice rules. The following formula for the worst-case error is fundamental.

Theorem 2.6 *Let $\mathcal{L}$ be an integration lattice with $\det(\mathcal{L}^\perp) = N$. The worst-case error of the corresponding lattice rule based on the lattice point set $\mathcal{P}(\mathcal{L})$ in the Korobov space $\mathcal{H}_{\mathrm{kor},d,\alpha,\gamma}$ with $\alpha > 1/2$ is given by*

$$[\mathrm{err}_{N,d}(\mathcal{H}_{\mathrm{kor},d,\alpha},\mathcal{P}(\mathcal{L}))]^2 = \sum_{\boldsymbol{h}\in\mathcal{L}^\perp\setminus\{\boldsymbol{0}\}} \frac{1}{r_{2\alpha}(\boldsymbol{h})}. \tag{2.6}$$

Remark 2.7 Note that the squared worst-case error in (2.6) is exactly the quantity $P_{2\alpha}$ in Definition 1.14.

Proof The proof of this theorem is based on Theorem 1.27. Since $\alpha > 1/2$ we have

$$K_{\mathrm{kor},d,\alpha}(\boldsymbol{y},\boldsymbol{y}) = (1+2\zeta(2\alpha))^d < \infty.$$

Hence Condition (1.17) is satisfied and we can apply the formula for the squared worst-case error in Theorem 1.27. We have

$$\begin{aligned}\int_{[0,1]^d}\int_{[0,1]^d} K_{\mathrm{kor},d,\alpha}(\boldsymbol{x},\boldsymbol{y})\,\mathrm{d}\boldsymbol{x}\,\mathrm{d}\boldsymbol{y} &= \sum_{\boldsymbol{h}\in\mathbb{Z}^d}\frac{1}{r_{2\alpha}(\boldsymbol{h})}\int_{[0,1]^d}\int_{[0,1]^d} \mathrm{e}^{2\pi\mathrm{i}\boldsymbol{h}\cdot(\boldsymbol{x}-\boldsymbol{y})}\,\mathrm{d}\boldsymbol{x}\,\mathrm{d}\boldsymbol{y}\\ &= \frac{1}{r_{2\alpha}(\boldsymbol{0})}\\ &= 1.\end{aligned}$$

In the same way one can show that, for any fixed $\boldsymbol{x}_k \in [0,1)^d$,

$$\int_{[0,1]^d} K_{\mathrm{kor},d,\alpha}(\boldsymbol{x}_k,\boldsymbol{y})\,\mathrm{d}\boldsymbol{y} = 1.$$

For an arbitrary N-element point set $\mathcal{P} = \{\boldsymbol{x}_0,\boldsymbol{x}_1,\ldots,\boldsymbol{x}_{N-1}\}$ in $[0,1)^d$ we now obtain from Theorem 1.27 that

$$\begin{aligned}[\mathrm{err}_{N,d}(\mathcal{H}_{\mathrm{kor},d,\alpha},\mathcal{P})]^2 &= -1 + \frac{1}{N^2}\sum_{k,\ell=0}^{N-1}\sum_{\boldsymbol{h}\in\mathbb{Z}^d}\frac{1}{r_{2\alpha}(\boldsymbol{h})}\,\mathrm{e}^{2\pi\mathrm{i}\boldsymbol{h}\cdot(\boldsymbol{x}_k-\boldsymbol{x}_\ell)}\\ &= \sum_{\boldsymbol{h}\in\mathbb{Z}^d\setminus\{\boldsymbol{0}\}}\frac{1}{r_{2\alpha}(\boldsymbol{h})}\left|\frac{1}{N}\sum_{k=0}^{N-1}\mathrm{e}^{2\pi\mathrm{i}\boldsymbol{h}\cdot\boldsymbol{x}_k}\right|^2.\end{aligned} \tag{2.7}$$

Let now $\mathcal{P} = \mathcal{P}(\mathcal{L})$. Then the above formula combined with Lemma 1.9 yields

$$[\mathrm{err}_{N,d}(\mathcal{H}_{\mathrm{kor},d,\alpha},\mathcal{P}(\mathcal{L}))]^2 = \sum_{\boldsymbol{h}\in\mathcal{L}^\perp\setminus\{\boldsymbol{0}\}} \frac{1}{r_{2\alpha}(\boldsymbol{h})}.$$ □

From Theorem 2.6 we deduce a result for the special case where the integration lattice is a rank-1 lattice. For a rank-1 lattice with generating vector $\boldsymbol{g}\in\mathbb{Z}^d$, we will in the following frequently write $\mathrm{err}_{N,d,\alpha}(\boldsymbol{g})$ instead of $\mathrm{err}_{N,d}(\mathcal{H}_{\mathrm{kor},d,\alpha},\mathcal{P}(\boldsymbol{g},N))$, i.e.,

$$\mathrm{err}_{N,d,\alpha}(\boldsymbol{g}) := \mathrm{err}_{N,d}(\mathcal{H}_{\mathrm{kor},d,\alpha},\mathcal{P}(\boldsymbol{g},N)).$$

Applying (1.7) to Formula (2.6) in Theorem 2.6 gives the following corollary.

Corollary 2.8 *For any N-point rank-1 lattice rule with generating vector $\boldsymbol{g} \in \mathbb{Z}^d$ we have*

$$[\mathrm{err}_{N,d,\alpha}(\boldsymbol{g})]^2 = \sum_{\substack{\boldsymbol{h} \in \mathbb{Z}^d \setminus \{\boldsymbol{0}\} \\ \boldsymbol{h} \cdot \boldsymbol{g} \equiv 0 \pmod{N}}} \frac{1}{r_{2\alpha}(\boldsymbol{h})}.$$

Remark 2.9 For $\alpha \in \mathbb{N}$ the Bernoulli polynomial $B_{2\alpha}$ of degree 2α has the Fourier expansion

$$B_{2\alpha}(x) = \frac{(-1)^{\alpha+1}(2\alpha)!}{(2\pi)^{2\alpha}} \sum_{h \in \mathbb{Z} \setminus \{0\}} \frac{\mathrm{e}^{2\pi \mathrm{i} h x}}{|h|^{2\alpha}} \quad \text{for all } x \in [0,1). \tag{2.8}$$

Using Corollary 2.8 we therefore obtain for any generating vector $\boldsymbol{g} = (g_1, \ldots, g_d)$ in $\mathbb{Z}^d$,

$$\begin{aligned}
[\mathrm{err}_{N,d,\alpha}(\boldsymbol{g})]^2 &= -1 + {}^{*}\!\!\sum_{\boldsymbol{h} \in \mathbb{Z}^d} \frac{1}{r_{2\alpha}(\boldsymbol{h})} \frac{1}{N} \sum_{k=0}^{N-1} \mathrm{e}^{2\pi \mathrm{i} (\boldsymbol{h} \cdot \boldsymbol{g})/N\, k} \\
&= -1 + \frac{1}{N} \sum_{k=0}^{N-1} \prod_{j=1}^{d} \left(\sum_{h \in \mathbb{Z}} \frac{1}{r_{2\alpha}(h)} \mathrm{e}^{2\pi \mathrm{i} \{k g_j / N\} h} \right) \\
&= -1 + \frac{1}{N} \sum_{k=0}^{N-1} \prod_{j=1}^{d} \left(1 + \frac{(-1)^{\alpha+1}(2\pi)^{2\alpha}}{(2\alpha)!} B_{2\alpha}\left(\left\{ \frac{k g_j}{N} \right\} \right) \right),
\end{aligned}$$

such that $[\mathrm{err}_{N,d,\alpha}(\boldsymbol{g})]^2$ can be computed in $O(d\,N)$ operations.

2.3 Error Bounds for the Unweighted Case

In this section we provide lower and upper bounds on the worst-case error of integration in the Korobov space $\mathcal{H}_{\mathrm{kor},d,\alpha}$.

Lower bounds

According to (2.7) the squared worst-case error of a QMC rule based on an arbitrary N-element point set $\mathcal{P}$ in $[0,1)^d$ can be written as

$$[\mathrm{err}_{N,d}(\mathcal{H}_{\mathrm{kor},d,\alpha}, \mathcal{P})]^2 = \sum_{\boldsymbol{h} \in \mathbb{Z}^d \setminus \{\boldsymbol{0}\}} \frac{1}{r_{2\alpha}(\boldsymbol{h})} \left| \frac{1}{N} \sum_{k=0}^{N-1} \mathrm{e}^{2\pi \mathrm{i} \boldsymbol{h} \cdot \boldsymbol{x}_k} \right|^2 .$$

For general linear rules $A_{N,d}$ as defined in Remark 1.28, this formula can easily be generalized to

$$[\mathrm{err}_{N,d}(\mathcal{H}_{\mathrm{kor},d,\alpha},\mathcal{P},\boldsymbol{w})]^2 = |1-\beta|^2 + \sum_{\boldsymbol{h}\in\mathbb{Z}^d\setminus\{\boldsymbol{0}\}} \frac{1}{r_{2\alpha}(\boldsymbol{h})} \left| \sum_{k=0}^{N-1} w_k \mathrm{e}^{2\pi \mathrm{i}\boldsymbol{h}\cdot\boldsymbol{x}_k} \right|^2, \quad (2.9)$$

where $\beta := \sum_{k=0}^{N-1} w_k$.

We now present a general lower bound on the worst-case error of linear rules $A_{N,d}$ for numerical integration in the Korobov space, which is due to Bakhvalov [6, 9].

Theorem 2.10 *For every $d \in \mathbb{N}$ and every $\alpha > 1/2$ there exists a positive quantity $C(d,\alpha)$ with the following property. For every $N \in \mathbb{N}$, every N-element point set $\mathcal{P}$ in $[0,1)^d$, and every choice of integration weights $\boldsymbol{w} \in \mathbb{C}^N$ we have*

$$\mathrm{err}_{N,d}(\mathcal{H}_{\mathrm{kor},d,\alpha},\mathcal{P},\boldsymbol{w}) \ge C(d,\alpha)\frac{(\log N)^{(d-1)/2}}{N^\alpha}.$$

Proof The principle idea of the proof is to use the formula (2.9) and to obtain a suitable lower bound on the terms $1/r_{2\alpha}(\boldsymbol{h})$. This will be done by cleverly choosing a sequence of functions $f_{\boldsymbol{m}}$ for $\boldsymbol{m} \in \mathbb{N}_0^d$ and by bounding $1/r_{2\alpha}(\boldsymbol{h})$ by the $\boldsymbol{h}$-th Fourier coefficients of some of the $f_{\boldsymbol{m}}$. The choice of the functions $f_{\boldsymbol{m}}$ will depend on the smoothness parameter α.

Indeed, we choose a function $f:\mathbb{R}\to\mathbb{R}$ as an infinitely many times differentiable function such that $f(x) > 0$ for $x \in (0,1)$ and $f^{(r)}(x) = 0$ for $x \in \mathbb{R}\setminus(0,1)$ for all r with $0 \le r \le b := \lceil \alpha \rceil + 1$, namely

$$f(x) = \begin{cases} x^{b+1}(1-x)^{b+1} & \text{for } x \in (0,1), \\ 0 & \text{otherwise.} \end{cases} \quad (2.10)$$

For $m \in \mathbb{N}_0$ let $f_m(x) = f(2^{m+2}x)$, and for $\boldsymbol{m} = (m_1,\dots,m_d) \in \mathbb{N}_0^d$ define

$$f_{\boldsymbol{m}}(\boldsymbol{x}) = \prod_{j=1}^d f_{m_j}(x_j),$$

where $\boldsymbol{x} = (x_1,\dots,x_d)$. We obtain

$$\widehat{f_{\boldsymbol{m}}}(\boldsymbol{0}) = \prod_{j=1}^d \int_0^1 f(2^{m_j+2}x)\,\mathrm{d}x = \prod_{j=1}^d \frac{1}{2^{m_j+2}} \int_0^1 f(y)\,\mathrm{d}y = \frac{1}{2^{\|\boldsymbol{m}\|_1+2d}}(I(f))^d,$$

where $I(f) = \int_0^1 f(y)\,\mathrm{d}y$. As we chose f according to (2.10) we obtain

$$I(f) = B(b+2,b+2) = \frac{((b+1)!)^2}{(2b+3)!},$$

with B denoting the beta function.

Let $t \in \mathbb{N}_0$ be such that

$$2N \le 2^t < 4N,$$

let

$$F(\boldsymbol{y}) = \sum_{k=0}^{N-1} w_k f_{\boldsymbol{m}}(\boldsymbol{x}_k - \boldsymbol{y}),$$

and let

$$B_{\boldsymbol{m}} = \left\{\boldsymbol{y} \in [0,1]^d \ : \ F(\boldsymbol{y}) = 0\right\}.$$

Note that the support of $f_{\boldsymbol{m}}(\boldsymbol{x}_k - \boldsymbol{y})$ as a function of $\boldsymbol{y}$ is contained in the interval $\prod_{j=1}^d (x_{k,j} - 2^{-m_j-2}, x_{k,j})$, where $x_{k,j}$ is the j-th component of $\boldsymbol{x}_k$, and hence the support of F is contained in $\bigcup_{k=0}^{N-1} \prod_{j=1}^d (x_{k,j} - 2^{-m_j-2}, x_{k,j})$. Therefore the volume of the support of F is at most $N 2^{-\|\boldsymbol{m}\|_1}$. Thus for all $\boldsymbol{m}$ such that $\|\boldsymbol{m}\|_1 = t$ we have

$$\lambda_d(B_{\boldsymbol{m}}) \geq 1 - \frac{N}{2^{\|\boldsymbol{m}\|_1}} = 1 - \frac{N}{2^t} > \frac{1}{4}.$$

For the linear rule $A_{N,d}(f) = \sum_{k=0}^{N-1} w_k f(\boldsymbol{x}_k)$ based on $\mathcal{P}$ and $\boldsymbol{w}$ we have

$$A_{N,d}(f_{\boldsymbol{m}}(\cdot - \boldsymbol{y})) - \widehat{f_{\boldsymbol{m}}}(\boldsymbol{0})\beta = \sum_{\boldsymbol{h} \in \mathbb{Z}^d \setminus \{\boldsymbol{0}\}} \left(\widehat{f_{\boldsymbol{m}}}(\boldsymbol{h}) \sum_{k=0}^{N-1} w_k \mathrm{e}^{2\pi \mathrm{i} \boldsymbol{h} \cdot \boldsymbol{x}_k} \right) \mathrm{e}^{-2\pi \mathrm{i} \boldsymbol{h} \cdot \boldsymbol{y}},$$

where we recall that $\beta = \sum_{k=0}^{N-1} w_k$, and for $\boldsymbol{y} \in B_{\boldsymbol{m}}$ we have

$$A_{N,d}(f_{\boldsymbol{m}}(\cdot - \boldsymbol{y})) = F(\boldsymbol{y}) = 0.$$

Therefore,

$$\begin{aligned}
\lambda_d(B_{\boldsymbol{m}}) |\widehat{f_{\boldsymbol{m}}}(\boldsymbol{0})|^2 |\beta|^2 &= \int_{B_{\boldsymbol{m}}} |A_{N,d}(f_{\boldsymbol{m}}(\cdot - \boldsymbol{y})) - \widehat{f_{\boldsymbol{m}}}(\boldsymbol{0})\beta|^2 \, \mathrm{d}\boldsymbol{y} \\
&\leq \int_{[0,1]^d} |A_{N,d}(f_{\boldsymbol{m}}(\cdot - \boldsymbol{y})) - \widehat{f_{\boldsymbol{m}}}(\boldsymbol{0})\beta|^2 \, \mathrm{d}\boldsymbol{y} \\
&= \sum_{\boldsymbol{h} \in \mathbb{Z}^d \setminus \{\boldsymbol{0}\}} |\widehat{f_{\boldsymbol{m}}}(\boldsymbol{h})|^2 \left| \sum_{k=0}^{N-1} w_k \mathrm{e}^{2\pi \mathrm{i} \boldsymbol{h} \cdot \boldsymbol{x}_k} \right|^2,
\end{aligned}$$

where in the last step we used Parseval's identity.

We have

$$\begin{aligned}
\widehat{f_m}(h) &= \int_0^1 f(2^{m+2}x) \, \mathrm{e}^{-2\pi \mathrm{i} h x} \, \mathrm{d}x \\
&= \frac{1}{2^{m+2}} \int_0^1 f(y) \, \mathrm{e}^{-2\pi \mathrm{i} h 2^{-m-2} y} \, \mathrm{d}y \\
&= \frac{1}{2^{m+2}} \widehat{f}\left(\frac{h}{2^{m+2}}\right).
\end{aligned}$$

Since f is infinitely many times differentiable, integration by parts shows that for any $m \in \mathbb{N}_0$ we have

$$|\widehat{f}_m(h)| = \frac{1}{2^{m+2}}\left|\widehat{f}\left(\frac{h}{2^{m+2}}\right)\right| \le C_b \frac{1}{2^{m+2}} \min\left(1, \frac{2^{b(m+2)}}{h^b}\right),$$

where the factor $C_b > 0$ depends only on b and f. Then, for $\boldsymbol{m}$ with $\|\boldsymbol{m}\|_1 = t$, we have

$$\begin{aligned}|\widehat{f}_{\boldsymbol{m}}(\boldsymbol{h})| &\le C(b,d) \prod_{j=1}^{d} \frac{1}{2^{m_j}} \min\left(1, \frac{2^{bm_j}}{r_b(h_j)}\right)\\ &= C(b,d) 2^{(\alpha-1)t} \prod_{j=1}^{d} \frac{1}{2^{\alpha m_j}} \min\left(1, \frac{2^{bm_j}}{r_b(h_j)}\right),\end{aligned}$$

with $C(b,d) := 2^{2(b-1)} C_b^d$. Taking the square and summing over all choices of $\boldsymbol{m}$ with $\|\boldsymbol{m}\|_1 = t$ on both sides of the latter inequality, we obtain

$$\begin{aligned}\sum_{\substack{\boldsymbol{m}\in\mathbb{N}_0^d\\ \|\boldsymbol{m}\|_1=t}} |\widehat{f}_{\boldsymbol{m}}(\boldsymbol{h})|^2 &\le 2^{2(\alpha-1)t} (C(b,d))^2 \sum_{\substack{\boldsymbol{m}\in\mathbb{N}_0^d\\ \|\boldsymbol{m}\|_1=t}} \prod_{j=1}^{d} \frac{1}{2^{2\alpha m_j}} \min\left(1, \frac{2^{2bm_j}}{r_{2b}(h_j)}\right)\\ &\le 2^{2(\alpha-1)t} (C(b,d))^2 \prod_{j=1}^{d} \left(\sum_{m=0}^{\infty} \frac{1}{2^{2\alpha m}} \min\left(1, \frac{2^{2bm}}{r_{2b}(h_j)}\right)\right).\end{aligned}$$

The sum in the latter expression can now be bounded by

$$\begin{aligned}&\sum_{m=0}^{\infty} \frac{1}{2^{2\alpha m}} \min\left(1, \frac{2^{2bm}}{r_{2b}(h_j)}\right)\\ &= \sum_{0\le m\le(\log_2 r_{2b}(h_j))/(2b)} \frac{2^{2(b-\alpha)m}}{r_{2b}(h_j)} + \sum_{m>(\log_2 r_{2b}(h_j))/(2b)} \frac{1}{2^{2\alpha m}}\\ &\le \frac{r_{2(b-\alpha)}(h_j) 2^{2(b-\alpha)} - 1}{2^{2(b-\alpha)} - 1} \frac{1}{r_{2b}(h_j)} + \frac{2^{2\alpha}}{2^{2\alpha} - 1} \frac{1}{r_{2\alpha}(h_j)}\\ &\le \frac{1}{r_{2\alpha}(h_j)}\left(1 + \frac{2^{2\alpha}}{2^{2\alpha}-1}\right)\\ &\le \frac{3}{r_{2\alpha}(h_j)}.\end{aligned}$$

Thus, with $C_1(b,d) := (3^d (C(b,d))^2)^{-1}$ we have

$$C_1(b,d) \frac{1}{2^{2(\alpha-1)t}} \sum_{\substack{\boldsymbol{m}\in\mathbb{N}_0^d\\ \|\boldsymbol{m}\|_1=t}} |\widehat{f}_{\boldsymbol{m}}(\boldsymbol{h})|^2 \le \frac{1}{r_{2\alpha}(\boldsymbol{h})}.$$

Using the formula (2.9) for the worst-case error of the algorithm $A_{N,d}$ we obtain

$$
\begin{aligned}
[\mathrm{err}_{N,d}(\mathcal{H}_{\mathrm{kor},d,\alpha},\mathcal{P},\boldsymbol{w})]^2 &= |1-\beta|^2 + \sum_{\boldsymbol{h}\in\mathbb{Z}^d\setminus\{\boldsymbol{0}\}} \frac{1}{r_{2\alpha}(\boldsymbol{h})}\left|\sum_{k=0}^{N-1} w_k \mathrm{e}^{2\pi\mathrm{i}\boldsymbol{h}\cdot\boldsymbol{x}_k}\right|^2 \\
&\geq |1-\beta|^2 + C_1(b,d)\frac{1}{2^{2(\alpha-1)t}} \sum_{\substack{\boldsymbol{m}\in\mathbb{N}_0^d\\ \|\boldsymbol{m}\|_1=t}} \sum_{\boldsymbol{h}\in\mathbb{Z}^d\setminus\{\boldsymbol{0}\}} |\widehat{f}_{\boldsymbol{m}}(\boldsymbol{h})|^2 \left|\sum_{k=0}^{N-1} w_k \mathrm{e}^{2\pi\mathrm{i}\boldsymbol{h}\cdot\boldsymbol{x}_k}\right|^2 \\
&\geq |1-\beta|^2 + C_1(b,d)\frac{1}{2^{2(\alpha-1)t}} \sum_{\substack{\boldsymbol{m}\in\mathbb{N}_0^d\\ \|\boldsymbol{m}\|_1=t}} \lambda_d(B_{\boldsymbol{m}})|\widehat{f}_{\boldsymbol{m}}(\boldsymbol{0})|^2|\beta|^2 \\
&\geq |1-\beta|^2 + C_2(b,d)|\beta|^2\frac{2^{2t}}{N^{2\alpha}} \sum_{\substack{\boldsymbol{m}\in\mathbb{N}_0^d\\ \|\boldsymbol{m}\|_1=t}} \frac{(I(f))^{2d}}{2^{2t+4d}} \\
&\geq |1-\beta|^2 + C_3(b,d)|\beta|^2\frac{1}{N^{2\alpha}}\binom{t+d-1}{d-1},
\end{aligned}
$$

with suitably chosen positive quantities $C_2(b,d)$ and $C_3(b,d)$. Since

$$
\binom{t+d-1}{d-1} \geq \frac{(t+1)^{d-1}}{(d-1)!},
$$

and since $t \geq \log_2 N$, we can therefore find a $C_4(b,d) > 0$ such that

$$
[\mathrm{err}_{N,d}(\mathcal{H}_{\mathrm{kor},d,\alpha},\mathcal{P},\boldsymbol{w})]^2 \geq |1-\beta|^2 + C_4(b,d)|\beta|^2\frac{(\log N)^{d-1}}{N^{2\alpha}}.
$$

We set $A = C_4(b,d)(\log N)^{d-1}/N^{2\alpha}$. Then the lower bound in the previous inequality can be written as $|1-\beta|^2 + A|\beta|^2$, and by minimizing the latter term with respect to β shows that it is bounded below by $A/(1+A)$. Consequently,

$$
[\mathrm{err}_{N,d}(\mathcal{H}_{\mathrm{kor},d,\alpha},\mathcal{P},\boldsymbol{w})]^2 \geq |1-\beta|^2 + A|\beta|^2 \geq \frac{A}{1+A} \geq C_5(b,d)\frac{(\log N)^{d-1}}{N^{2\alpha}},
$$

with a positive quantity $C_5(b,d)$ depending only on b and d. This implies the desired result. □

Upper bounds

We now study lattice rules for integration in the Korobov space $\mathcal{H}_{\mathrm{kor},d,\alpha}$, whose worst-case error satisfies in some sense optimal upper bounds when compared to Bakhvalov's lower bound in Theorem 2.10. It suffices to restrict our attention to the rank-1 lattice rule case here.

We first show a result which relates the quantity $R(\boldsymbol{g}, N)$, as defined in (1.35), to the worst-case error of a lattice rule in the Korobov space $\mathcal{H}_{\mathrm{kor},d,\alpha}$.

Theorem 2.11 *Let $\alpha > 1/2$ and let $N > 1$ be an integer. Then for all $\boldsymbol{g} \in \mathbb{Z}^d$ we have that*

$$[\mathrm{err}_{N,d,\alpha}(\boldsymbol{g})]^2 \leq (1 + 2\zeta(2\alpha))^d \left(\frac{1}{N^{2\alpha}} + (R(\boldsymbol{g}, N))^{2\alpha}\right).$$

Proof From Corollary 2.8 we have

$$[\mathrm{err}_{N,d,\alpha}(\boldsymbol{g})]^2 = \sum_{\substack{\boldsymbol{h} \in \mathbb{Z}^d \setminus \{\boldsymbol{0}\} \\ \boldsymbol{h} \cdot \boldsymbol{g} \equiv 0 \pmod{N}}} \frac{1}{r_{2\alpha}(\boldsymbol{h})} =: \Sigma_1 + \Sigma_2,$$

where Σ_1 is the sum over all $\boldsymbol{h} \in \mathbb{Z}^d \setminus \{\boldsymbol{0}\}$ which are a multiple of N, i.e., $\boldsymbol{h} = N\boldsymbol{k}$ with $\boldsymbol{k} \in \mathbb{Z}^d \setminus \{\boldsymbol{0}\}$. In this case we obviously have $\boldsymbol{h} \cdot \boldsymbol{g} \equiv 0 \pmod{N}$. In Σ_2 we sum over the remaining $\boldsymbol{h}$ in $\mathbb{Z}^d \setminus \{\boldsymbol{0}\}$ satisfying $\boldsymbol{h} \cdot \boldsymbol{g} \equiv 0 \pmod{N}$.

For $\boldsymbol{h} = N\boldsymbol{k}$ with $\boldsymbol{k} \in \mathbb{Z}^d \setminus \{\boldsymbol{0}\}$ we have

$$\frac{1}{r_{2\alpha}(\boldsymbol{h})} = \frac{1}{r_{2\alpha}(N\boldsymbol{k})} \leq \frac{1}{N^{2\alpha}} \frac{1}{r_{2\alpha}(\boldsymbol{k})},$$

and hence we obtain

$$\Sigma_1 \leq \frac{1}{N^{2\alpha}} \sum_{\boldsymbol{k} \in \mathbb{Z}^d} \frac{1}{r_{2\alpha}(\boldsymbol{k})} = \frac{1}{N^{2\alpha}} \left(\sum_{k \in \mathbb{Z}} \frac{1}{r_{2\alpha}(k)}\right)^d = \frac{1}{N^{2\alpha}} (1 + 2\zeta(2\alpha))^d.$$

Each of the remaining $\boldsymbol{h}$ occurring in Σ_2 can be uniquely represented in the form

$$\boldsymbol{h} = \boldsymbol{h}^* + N\boldsymbol{k},$$

where $\boldsymbol{k} \in \mathbb{Z}^d$ and $\boldsymbol{h}^* = (h_1^*, \ldots, h_d^*) \in C_d^*(N)$ with $\boldsymbol{h}^* \cdot \boldsymbol{g} \equiv 0 \pmod{N}$, and where $C_d^*(N)$ is defined at the beginning of Section 1.8. Thus,

$$\Sigma_2 = \sum_{\boldsymbol{k} \in \mathbb{Z}^d} \sum_{\substack{\boldsymbol{h}^* \in C_d^*(N) \\ \boldsymbol{h}^* \cdot \boldsymbol{g} \equiv 0 \pmod{N}}} \frac{1}{r_{2\alpha}(\boldsymbol{h}^* + N\boldsymbol{k})}.$$

Next, we show that

$$r_1(\boldsymbol{h}^* + N\boldsymbol{k}) \geq r_1(\boldsymbol{h}^*) r_1(\boldsymbol{k}) \tag{2.11}$$

for $\boldsymbol{h}^*$ and $\boldsymbol{k}$ as above. To this end it suffices to show that

$$\max(1, |h_j^* + Nk_j|) \geq \max(1, |h_j^*|) \max(1, |k_j|) \tag{2.12}$$

holds true for all $j \in [d]$. If $h_j^* = 0$ or $k_j = 0$, then (2.12) is obviously satisfied. If $h_j^* \neq 0$ and $k_j \neq 0$, then

$$|h_j^* + Nk_j| \geq N|k_j| - |h_j^*| \geq N|k_j| - \frac{N}{2} = \frac{N}{2}(2|k_j| - 1) \geq |h_j^*| \cdot |k_j|,$$

and (2.12) follows as well, which implies that (2.11) is shown.

From (2.11) we obtain

$$r_{2\alpha}(\boldsymbol{h}^* + N\boldsymbol{k}) = (r_1(\boldsymbol{h}^* + N\boldsymbol{k}))^{2\alpha} \geq (r_1(\boldsymbol{h}^*))^{2\alpha}(r_1(\boldsymbol{k}))^{2\alpha}.$$

Using this estimate we find that

$$\begin{aligned}\Sigma_2 &\leq \sum_{\boldsymbol{k}\in\mathbb{Z}^d} \frac{1}{(r_1(\boldsymbol{k}))^{2\alpha}} \sum_{\substack{\boldsymbol{h}^*\in C_d^*(N)\\ \boldsymbol{h}^*\cdot\boldsymbol{g}\equiv 0 \pmod N}} \frac{1}{(r_1(\boldsymbol{h}^*))^{2\alpha}} \\ &= (1 + 2\zeta(2\alpha))^d \sum_{\substack{\boldsymbol{h}^*\in C_d^*(N)\\ \boldsymbol{h}^*\cdot\boldsymbol{g}\equiv 0 \pmod N}} \frac{1}{(r_1(\boldsymbol{h}^*))^{2\alpha}} \\ &\leq (1 + 2\zeta(2\alpha))^d (R(\boldsymbol{g}, N))^{2\alpha}.\end{aligned}$$

The result now follows by adding the estimates for Σ_1 and Σ_2. □

Remark 2.12 We have already pointed out that the quantity $R(\boldsymbol{g}, N)$ does not depend on the smoothness parameter α. Hence, if we can construct, for given integer $N > 1$, a generating vector (or, using a different term, a lattice point) $\boldsymbol{g} \in \mathbb{Z}^d$ with reasonably small $R(\boldsymbol{g}, N)$, then the worst-case error of the lattice rule based on $\mathcal{P}(\boldsymbol{g}, N)$ is small, simultaneously for all smoothness parameters $\alpha > 1/2$. This is an important feature especially in practice, when one might not know the actual smoothness of an integrand f.

It remains to show the existence of lattice rules with a small value of R, which will be done by means of an averaging argument in the following lemma. In this lemma, and also in the following, we shall write

$$G_d(N) := \{0, 1, \ldots, N - 1\}^d$$

for d and $N \in \mathbb{N}$.

Lemma 2.13 *Let N be a prime number. Then,*

$$\frac{1}{N^d} \sum_{\boldsymbol{g}\in G_d(N)} R(\boldsymbol{g}, N) \leq 2^d \frac{(\log N + 1)^d}{N}. \tag{2.13}$$

In particular, there exists at least one $\boldsymbol{g} \in G_d(N)$ such that

$$R(\boldsymbol{g}, N) \leq 2^d \frac{(\log N + 1)^d}{N}.$$

Proof Averaging the values of $R(\boldsymbol{g}, N)$ over all $\boldsymbol{g} \in G_d(N)$ we obtain

$$\frac{1}{N^d}\sum_{\boldsymbol{g}\in G_d(N)} R(\boldsymbol{g},N) = \frac{1}{N^d}\sum_{\boldsymbol{g}\in G_d(N)} \sum_{\substack{\boldsymbol{h}\in C_d^*(N)\\ \boldsymbol{h}\cdot\boldsymbol{g}\equiv 0 \pmod N}} \frac{1}{r_1(\boldsymbol{h})}$$

$$= \sum_{\boldsymbol{h}\in C_d^*(N)} \frac{1}{r_1(\boldsymbol{h})}\frac{1}{N^d} \sum_{\substack{\boldsymbol{g}\in G_d(N)\\ \boldsymbol{h}\cdot\boldsymbol{g}\equiv 0 \pmod N}} 1,$$

where we just changed the order of summation. The inner sum in the latter expression counts the number of elements $\boldsymbol{g} \in G_d(N)$ such that $\boldsymbol{h}\cdot\boldsymbol{g} \equiv 0 \pmod N$. Let $\boldsymbol{h} = (h_1,\dots,h_d) \in \mathbb{Z}^d \setminus \{\boldsymbol{0}\}$, so there exists an index $j \in [d]$ such that $h_j \neq 0$. Assume first that $h_1 \neq 0$. Then, for arbitrary $g_2,\dots,g_d \in G_1(N)$, the condition $\boldsymbol{h}\cdot\boldsymbol{g} \equiv 0 \pmod N$ is equivalent to

$$h_1 g_1 \equiv -(h_2 g_2 + \cdots + h_d g_d) \pmod N.$$

Since N is a prime number, there exists exactly one $g_1 \in G_1(N)$ satisfying this congruence. However, for choosing $g_2,\dots,g_d$ we have N^{d-1} possibilities altogether. Thus we have

$$|\{\boldsymbol{g} \in G_d(N) \ :\ \boldsymbol{h}\cdot\boldsymbol{g} \equiv 0 \pmod N\}| = N^{d-1}$$

in the case where $h_1 \neq 0$.

The same argument applies if $h_j \neq 0$ for any other $j \in [d]$. So,

$$\frac{1}{N^d}\sum_{\substack{\boldsymbol{g}\in G_d(N)\\ \boldsymbol{h}\cdot\boldsymbol{g}\equiv 0 \pmod N}} 1 = \frac{N^{d-1}}{N^d} = \frac{1}{N}.$$

This implies

$$\frac{1}{N^d}\sum_{\boldsymbol{g}\in G_d(N)} R(\boldsymbol{g},N) = \frac{1}{N}\sum_{\boldsymbol{h}\in C_d^*(N)} \frac{1}{r_1(\boldsymbol{h})} = \frac{1}{N}\left(-1 + (1+S_N)^d\right), \tag{2.14}$$

where we put

$$S_N := \sum_{h\in C_1^*(N)} \frac{1}{|h|}. \tag{2.15}$$

Using a straightforward estimate for the initial segment of the harmonic series we have

$$S_N \le 2\sum_{h=1}^{\lfloor N/2\rfloor} \frac{1}{h} \le 2\left(1 + \int_1^{N/2} \frac{\mathrm{d}t}{t}\right) = 2(\log N + 1 - \log 2). \tag{2.16}$$

Hence

$$\frac{1}{N^d}\sum_{\boldsymbol{g}\in G_d(N)} R(\boldsymbol{g},N) \le \frac{1}{N}\left(1 + 2(\log N + 1 - \log 2)\right)^d \le 2^d\,\frac{(\log N + 1)^d}{N},$$

which shows (2.13). Since there always exists at least one element $\boldsymbol{g} \in G_d(N)$ that is at least as good as average, also the existence result follows. □

Remark 2.14 To obtain the existence result in Lemma 2.13 we used the principle that for any set of real numbers $a_0, a_1, \ldots, a_{N-1} \geq 0$ we have that there exists an a_{k^*}, $k^* \in \{0, 1, \ldots, N-1\}$, which is at least as small as the average, i.e.

$$a_{k^*} \leq \frac{1}{N} \sum_{k=0}^{N-1} a_k.$$

To obtain a more general result, we could also proceed in the following way. Let $\lambda \in (0, 1]$. Then again there exists an $a_{k^*}^\lambda$, $k^* \in \{0, 1, \ldots, N-1\}$ (which may depend on λ), that is at least as small as the average of $a_0^\lambda, a_1^\lambda, \ldots, a_{N-1}^\lambda$, i.e.,

$$a_{k^*}^\lambda \leq \frac{1}{N} \sum_{k=0}^{N-1} a_k^\lambda \quad \text{or, equivalently,} \quad a_{k^*} \leq \frac{1}{N^{1/\lambda}} \left(\sum_{k=0}^{N-1} a_k^\lambda \right)^{1/\lambda}.$$

In particular,

$$\min_{0 \leq k < N} a_k \leq \inf_{0 < \lambda \leq 1} \frac{1}{N^{1/\lambda}} \left(\sum_{k=0}^{N-1} a_k^\lambda \right)^{1/\lambda}.$$

These observations are sometimes referred to as the *standard averaging argument* in this book.

Lemma 2.13 is only an existence result. In its proof we computed the average of R over all lattice points in the set $G_d(N)$ which is a finite set of cardinality N^d. The effective search for good generating vectors $\boldsymbol{g}$ with respect to R will be discussed in Section 3.6 (see Theorem 3.15).

Combining Theorem 2.11 with Lemma 2.13, we obtain the following result.

Theorem 2.15 *Let $d \in \mathbb{N}$ and let N be a prime number. Then there exists a lattice point $\boldsymbol{g} \in G_d(N)$ such that*

$$\mathrm{err}_{N,d,\alpha}(\boldsymbol{g}) \leq \frac{1}{N^\alpha} (1 + 2\zeta(2\alpha))^{d/2} \left(1 + (2(\log N + 1))^{2\alpha d}\right)^{1/2}$$

simultaneously for all $\alpha > 1/2$.

Theorem 2.15 shows the existence of rank-1 lattice rules for which the convergence rate of the worst-case error in the Korobov space $\mathcal{H}_{\mathrm{kor},d,\alpha}$ is of order

$$O_{d,\alpha}\left(\frac{(\log N)^{\alpha d}}{N^\alpha} \right),$$

where by $O_{d,\alpha}$ we indicate that the implied factors may depend on d and α. Note that this bound reflects the smoothness α of the problem. Higher smoothness α leads to improved convergence rates of the worst-case integration error. Furthermore, in comparison to Bakhvalov's lower bound in Theorem 2.10, this convergence rate is—up to logarithmic factors—best possible. However, it remains open how long we have to wait to observe this almost optimal asymptotic rate, especially when the dimension d is large. In practical applications d can be huge, e.g., in the hundreds or thousands for problems in mathematical finance or uncertainty quantification. The dependence of the worst-case error on the dimension d is the subject of a large number of papers and books dealing with the concept of tractability. Naturally, we would like to see that numerical integration in the Korobov space is tractable in some form as introduced in Section 1.7. This, however, is not possible, at least for the unweighted case that is considered in the present section. For technical reasons we show this negative result only for QMC rules here. The general case is treated in the book [211] by Novak and Woźniakowski.

The curse of dimensionality

We will show that if we are only allowed to use QMC rules, then the integration problem in the unweighted Korobov space $\mathcal{H}_{\mathrm{kor},d,\alpha}$ suffers from the curse of dimensionality. To this end, we define a slightly modified reproducing kernel Hilbert space $\widetilde{\mathcal{H}}_{d,\alpha}$ of one-periodic functions with absolutely convergent Fourier series by means of a function $\widetilde{r}_\tau : \mathbb{Z} \to \mathbb{R}$, with

$$\widetilde{r}_\tau(h) := \begin{cases} 1 & \text{if } h = 0, \\ 2\zeta(2\alpha)\,|h|^\tau & \text{if } h \neq 0, \end{cases}$$

for $\tau > 0$, and we put $\widetilde{r}_\tau(\boldsymbol{h}) := \prod_{j=1}^d \widetilde{r}_\tau(h_j)$ for $\boldsymbol{h} \in \mathbb{Z}^d$. In this way, we define an inner product in $\widetilde{\mathcal{H}}_{d,\alpha}$ by

$$\langle f, g\rangle_{d,\alpha,\widetilde{r}} := \sum_{\boldsymbol{h} \in \mathbb{Z}^d} \widetilde{r}_{2\alpha}(\boldsymbol{h}) \widehat{f}(\boldsymbol{h}) \overline{\widehat{g}(\boldsymbol{h})}.$$

It can be checked analogously to what is done in Section 2.1 that $\widetilde{\mathcal{H}}_{d,\alpha}$ with the corresponding norm $\|\cdot\|_{d,\alpha,\widetilde{r}}$ is a reproducing kernel Hilbert space with kernel

$$\widetilde{K}_{d,\alpha}(\boldsymbol{x}, \boldsymbol{y}) = \prod_{j=1}^d \left(\sum_{h_j \in \mathbb{Z}} \frac{\mathrm{e}^{2\pi \mathrm{i} h_j (x_j - y_j)}}{\widetilde{r}_{2\alpha}(h_j)} \right) = \sum_{\boldsymbol{h} \in \mathbb{Z}^d} \frac{1}{\widetilde{r}_{2\alpha}(\boldsymbol{h})} \mathrm{e}^{2\pi \mathrm{i} \boldsymbol{h} \cdot (\boldsymbol{x} - \boldsymbol{y})}.$$

Since $2\zeta(2\alpha) \geq 1$ we obtain $\|f\|_{d,\alpha,\widetilde{r}} \geq \|f\|_{\mathrm{kor},d,\alpha}$ and hence

$$\{f \in \widetilde{\mathcal{H}}_{d,\alpha} \;:\; \|f\|_{d,\alpha,\widetilde{r}} \leq 1\} \subseteq \{f \in \mathcal{H}_{\mathrm{kor},d,\alpha} \;:\; \|f\|_{\mathrm{kor},d,\alpha} \leq 1\}.$$

This implies that integration in $\widetilde{\mathcal{H}}_{d,\alpha}$ is no harder than integration in $\mathcal{H}_{\mathrm{kor},d,\alpha}$, i.e., $\mathrm{err}_{N,d}(\widetilde{\mathcal{H}}_{d,\alpha}, \mathcal{P}) \leq \mathrm{err}_{N,d}(\mathcal{H}_{\mathrm{kor},d,\alpha}, \mathcal{P})$ for any N-element point set $\mathcal{P}$ in $[0,1)^d$. Furthermore, according to the definition of $\widetilde{r}_\tau$, we have that $\widetilde{K}_{d,\alpha}$ is nonnegative.

Due to the product structure of $\widetilde{K}_{d,\alpha}$, it suffices to show this for the univariate case. We have from (2.2) that

$$\widetilde{K}_{1,\alpha}(x,y) = 1 + \frac{2}{2\zeta(2\alpha)} \sum_{h=1}^{\infty} \frac{1}{h^{2\alpha}} \cos(2\pi \mathtt{i} h(x-y)) \geq 1 - \frac{1}{\zeta(2\alpha)} \sum_{h=1}^{\infty} \frac{1}{h^{2\alpha}} = 0.$$

Now we bound $\mathrm{err}_{N,d}(\mathcal{H}_{\mathrm{kor},d,\alpha}, \mathcal{P})$ from below. We have

$$[\mathrm{err}_{N,d}(\mathcal{H}_{\mathrm{kor},d,\alpha}, \mathcal{P})]^2 \geq [\mathrm{err}_{N,d}(\widetilde{\mathcal{H}}_{d,\alpha}, \mathcal{P})]^2 = -1 + \frac{1}{N^2} \sum_{k,\ell=0}^{N-1} \widetilde{K}_{d,\alpha}(\boldsymbol{x}_k, \boldsymbol{x}_\ell),$$

where we used Theorem 1.27. Since $\widetilde{K}_{d,\alpha}$ is nonnegative, we can bound the latter term from below by omitting the nondiagonal terms in the double sum. In this way we find that

$$\begin{aligned} [\mathrm{err}_{N,d}(\mathcal{H}_{\mathrm{kor},d,\alpha}, \mathcal{P})]^2 &\geq -1 + \frac{1}{N^2} \sum_{k=0}^{N-1} \widetilde{K}_{d,\alpha}(\boldsymbol{x}_k, \boldsymbol{x}_k) \\ &= -1 + \frac{1}{N^2} \sum_{k=0}^{N-1} \left(1 + \frac{2\zeta(2\alpha)}{2\zeta(2\alpha)}\right)^d \\ &= -1 + \frac{2^d}{N}. \end{aligned}$$

Consequently,

$$e_{\mathrm{QMC}}(N,d) := \inf_{\mathcal{P}} |\mathrm{err}_{N,d}(\mathcal{H}_{\mathrm{kor},d,\alpha}, \mathcal{P})| \geq \left(\max\left(0, -1 + \frac{2^d}{N}\right)\right)^{1/2},$$

where the infimum is extended over all N-element point sets $\mathcal{P}$ in $[0,1)^d$.

According to Remark 1.41, the initial error in the unweighted Korobov space equals one, since

$$\int_{[0,1]^d} \int_{[0,1]^d} K_{\mathrm{kor},d,\alpha}(\boldsymbol{x}, \boldsymbol{y}) \, \mathrm{d}\boldsymbol{x} \, \mathrm{d}\boldsymbol{y} = 1.$$

Therefore,

$$\begin{aligned} N_{\mathrm{QMC}}(\varepsilon, d) &:= \min\{N \in \mathbb{N} \,:\, e_{\mathrm{QMC}}(N,d) \leq \varepsilon\} \\ &\geq \min\left\{N \in \mathbb{N} \,:\, -1 + \frac{2^d}{N} \leq \varepsilon^2\right\} \\ &= \left\lceil \frac{2^d}{\varepsilon^2 + 1} \right\rceil. \end{aligned}$$

This shows that the information complexity grows exponentially with the dimension d when we consider only QMC rules. Consequently, QMC integration in the (unweighted) Korobov space suffers from the curse of dimensionality.

We remark that this argument can easily be extended to linear algorithms with nonnegative integration weights $\boldsymbol{w} \in [0, \infty)^N$.

2.4 Weighted Korobov Spaces

The reason for the curse of dimensionality in many problems lies in the fact that for standard spaces all variables and all groups of variables are equally important. As a way out, Sloan and Woźniakowski [239] suggested considering weighted spaces, in which the relative importance of variables and groups of variables is modeled by corresponding *weights*. The motivation for introducing weights is that in many practical multivariate problems it can be observed that not all variables or groups of variables have the same influence. An extreme example is the numerical integration of a function $f : [0, 1]^d \to \mathbb{R}$ for $d \in \mathbb{N}$ that is given by $f(x_1, \ldots, x_d) = x_1$. Quite obviously, even though the problem is nominally d-variate, it boils down to a much simpler univariate problem. In functions occurring in applications, a similar phenomenon, though usually to a lesser extent than in the extreme example, can occur. This may, e.g., be caused by a function depending on a number of cash flows, some of which lie in the (distant) future that need to be discounted accordingly. Observations like these led Sloan and Woźniakowski to introducing additional parameters, so-called weights, in the definition of the norm of Hilbert spaces studied in numerical integration problems. We remark that this notion of weights is not to be confused with the notion of integration weights w_k in linear integration rules $A_{n,d}(f) = \sum_{k=0}^{N-1} w_k f(\boldsymbol{x}_k)$.

From the strictly mathematical point of view the introduction of weights may lead to multivariate problems in which the curse of dimensionality can be overcome, and one can have tractability (see Section 2.6 below), provided that suitable conditions on the weights hold. This effect also corresponds to intuition; if a problem depends on many variables, of which only some have significant influence, it is natural to expect that the problem will be easier to solve than one where all variables have the same influence.

As a first example of a weighted function space let us now formally introduce the weighted Korobov space. To this end, as outlined above, we first define additional parameters, which we call weights. For a d-variate problem, in the most general setting, we assign one weight $\gamma_{\mathfrak{u}}$ to each possible group of variables with indices in one of the subsets $\mathfrak{u}$ of $[d]$. Indeed, let

$$\boldsymbol{\gamma} = \{\gamma_{\mathfrak{u}} : \mathfrak{u} \subseteq [d]\}$$

be a set of nonnegative integers, where we use the convention that $\gamma_\emptyset = 1$. For short, we shall also write $\boldsymbol{\gamma} = \{\gamma_{\mathfrak{u}}\}_{\mathfrak{u} \subseteq [d]}$ in the following.

For $\alpha > 1/2$ we define the Hilbert space $\mathcal{H}_{\mathrm{kor},d,\alpha,\boldsymbol{\gamma}}$ as the space of all one-periodic functions f with absolutely convergent Fourier series as in (1.9), with Fourier coefficients as in (1.10), and with finite norm $\|f\|_{\mathrm{kor},d,\alpha,\boldsymbol{\gamma}} := \langle f, f \rangle_{\mathrm{kor},d,\alpha,\boldsymbol{\gamma}}^{1/2}$, where the inner product is given by

$$\langle f, g \rangle_{\mathrm{kor},d,\alpha,\boldsymbol{\gamma}} := \sum_{\boldsymbol{h} \in \mathbb{Z}^d} \left(\frac{1}{\gamma_{\mathfrak{u}(\boldsymbol{h})}} \prod_{j \in \mathfrak{u}(\boldsymbol{h})} |h_j|^{2\alpha} \right) \widehat{f}(\boldsymbol{h}) \overline{\widehat{g}(\boldsymbol{h})},$$

where for $\boldsymbol{h} \in \mathbb{Z}^d$ we put

$$\mathfrak{u}(\boldsymbol{h}) := \{ j \in [d] \ : \ h_j \neq 0 \},$$

and where we define the empty product to be one if $\boldsymbol{h} = \boldsymbol{0}$. Using this notation, we also write, for $\boldsymbol{h} \in \mathbb{Z}^d$ and $\tau > 0$,

$$r_{\tau,\boldsymbol{\gamma}}(\boldsymbol{h}) := \frac{1}{\gamma_{\mathfrak{u}(\boldsymbol{h})}} \prod_{j \in \mathfrak{u}(\boldsymbol{h})} |h_j|^{\tau}. \tag{2.17}$$

If $\gamma_{\mathfrak{u}(\boldsymbol{h})} = 0$, we formally set $r_{\tau,\boldsymbol{\gamma}}(\boldsymbol{h}) := \infty$. Consequently,

$$\langle f, g \rangle_{\mathrm{kor},d,\alpha,\boldsymbol{\gamma}} = \sum_{\boldsymbol{h} \in \mathbb{Z}^d} r_{2\alpha,\boldsymbol{\gamma}}(\boldsymbol{h}) \, \widehat{f}(\boldsymbol{h}) \overline{\widehat{g}(\boldsymbol{h})}, \tag{2.18}$$

which is the weighted analogue of the inner product in the unweighted space $\mathcal{H}_{\mathrm{kor},d,\alpha}$. If for some $\boldsymbol{h} \in \mathbb{Z}^d \setminus \{\boldsymbol{0}\}$ in (2.18) we have $r_{2\alpha,\boldsymbol{\gamma}}(\boldsymbol{h}) = \infty$, then we assume that $\widehat{f}(\boldsymbol{h}) = 0$ for all f in the space, and interpret $\infty \cdot 0$ as equal to 0, so that such terms do not contribute to the sum.

If all weights are positive, we can alternatively rearrange the sums and express the inner product in $\mathcal{H}_{\mathrm{kor},d,\alpha,\boldsymbol{\gamma}}$ as

$$\langle f, g \rangle_{\mathrm{kor},d,\alpha,\boldsymbol{\gamma}} = \sum_{\mathfrak{u} \subseteq [d]} \frac{1}{\gamma_{\mathfrak{u}}} \sum_{\boldsymbol{h}_{\mathfrak{u}} \in (\mathbb{Z} \setminus \{0\})^{|\mathfrak{u}|}} r_{2\alpha}(\boldsymbol{h}_{\mathfrak{u}}) \, \widehat{f}(\boldsymbol{h}_{\mathfrak{u}}, \boldsymbol{0}) \, \overline{\widehat{g}(\boldsymbol{h}_{\mathfrak{u}}, \boldsymbol{0})},$$

where, for $\boldsymbol{h} \in \mathbb{Z}^d$, $(\boldsymbol{h}_{\mathfrak{u}}, \boldsymbol{0})$ denotes the vector $(\ell_1, \ldots, \ell_d)$ with $\ell_j = h_j$ if $j \in \mathfrak{u}$ and $\ell_j = 0$ otherwise, for $j \in [d]$.

Definition 2.16 The Hilbert space $\mathcal{H}_{\mathrm{kor},d,\alpha,\boldsymbol{\gamma}}$ is called the *weighted Korobov space of smoothness* α with weights $\boldsymbol{\gamma}$.

Analogously to the unweighted case, $\mathcal{H}_{\mathrm{kor},d,\alpha,\boldsymbol{\gamma}}$ is a reproducing kernel Hilbert space with kernel

$$
\begin{aligned}
K_{\mathrm{kor},d,\alpha,\boldsymbol{\gamma}}(\boldsymbol{x},\boldsymbol{y}) &= \sum_{\mathfrak{u}\subseteq[d]} \gamma_{\mathfrak{u}} \prod_{j\in\mathfrak{u}} \left(\sum_{h_j\in\mathbb{Z}\setminus\{0\}} \frac{\mathrm{e}^{2\pi\mathrm{i}h_j(x_j-y_j)}}{|h_j|^{2\alpha}} \right) \\
&= \sum_{\boldsymbol{h}\in\mathbb{Z}^d} \gamma_{\mathfrak{u}(\boldsymbol{h})} \prod_{j\in\mathfrak{u}(\boldsymbol{h})} \frac{\mathrm{e}^{2\pi\mathrm{i}h_j(x_j-y_j)}}{|h_j|^{2\alpha}} \\
&= \sum_{\boldsymbol{h}\in\mathbb{Z}^d} \left(\gamma_{\mathfrak{u}(\boldsymbol{h})} \prod_{j\in\mathfrak{u}(\boldsymbol{h})} \frac{1}{|h_j|^{2\alpha}} \right) \mathrm{e}^{2\pi\mathrm{i}\boldsymbol{h}\cdot(\boldsymbol{x}-\boldsymbol{y})} \\
&= \sum_{\boldsymbol{h}\in\mathbb{Z}^d} \frac{1}{r_{2\alpha,\boldsymbol{\gamma}}(\boldsymbol{h})} \mathrm{e}^{2\pi\mathrm{i}\boldsymbol{h}\cdot(\boldsymbol{x}-\boldsymbol{y})}. \qquad (2.19)
\end{aligned}
$$

The reproducing property of $K_{\mathrm{kor},d,\alpha,\boldsymbol{\gamma}}$ can be shown analogously to the unweighted case by using (2.19).

In analogy to Proposition 2.4 one can show how the weighted Korobov space norm involves the mixed partial derivatives of f.

Proposition 2.17 *If $\alpha \in \mathbb{N}$, then we have, for $f \in \mathcal{H}_{\mathrm{kor},d,\alpha,\boldsymbol{\gamma}}$,*

$$
\|f\|^2_{\mathrm{kor},d,\alpha,\boldsymbol{\gamma}} = \sum_{\mathfrak{u}\subseteq[d]} \frac{1}{\gamma_{\mathfrak{u}}} \frac{1}{(2\pi)^{2\alpha|\mathfrak{u}|}} \int_{[0,1]^d} \left| \left(V_{\mathfrak{u},\alpha} f\right)(\boldsymbol{x}) \right|^2 \mathrm{d}\boldsymbol{x},
$$

where $V_{\mathfrak{u},\alpha} f$ is defined as in (2.5).

Example 2.18 In the univariate case with weight γ, Proposition 2.4 yields

$$
\|f\|^2_{\mathrm{kor},1,\alpha,\gamma} = \left(\int_0^1 f(t)\,\mathrm{d}t \right)^2 + \frac{1}{\gamma} \frac{1}{(2\pi)^{2\alpha}} \int_0^1 (f^{(\alpha)}(t))^2\,\mathrm{d}t. \qquad (2.20)
$$

This formula should be compared to the formula for the norm in the univariate unweighted Korobov space given in (2.3).

Product weights

A particularly important type of weights are so-called *product weights* of the form

$$
\gamma_{\mathfrak{u}} = \prod_{j\in\mathfrak{u}} \gamma_j \quad \text{for } \mathfrak{u} \subseteq [d], \qquad (2.21)
$$

where $(\gamma_j)_{j\ge 1}$ is a sequence of positive reals. For $\mathfrak{u} = \emptyset$, we define the empty product to be one, so $\gamma_\emptyset = 1$. We can therefore consider γ_j to represent the influence of the j-th variable in a problem like numerical integration defined on the Hilbert space. For simplicity, we will assume that the variables are ordered according to their influence in the case of product weights, i.e., one usually assumes that

$$
\gamma_1 \ge \gamma_2 \ge \cdots > 0.
$$

This assumption is justified by practical examples. If not explicitly stated otherwise, we will tacitly assume this ordering of product weights throughout the rest of this book. Historically, product weights were the first weights considered in the context of weighted QMC integration, introduced by Sloan and Woźniakowski in [239]. For the case of product weights, we slightly abuse notation, and write $\boldsymbol{\gamma}$ not only to represent the set

$$\boldsymbol{\gamma} = \left\{ \gamma_{\mathfrak{u}} = \prod_{j \in \mathfrak{u}} \gamma_j \ : \ \mathfrak{u} \subseteq [d] \right\},$$

but also the sequence of the weights, i.e., we shall write $\boldsymbol{\gamma} = (\gamma_j)_{j \geq 1}$ (this is justified as the sequence $(\gamma_j)_{j \geq 1}$ determines the collection of the $\gamma_{\mathfrak{u}}$ and vice versa).

In the case of product weights we can write, for $\tau > 0$ and $\boldsymbol{h} \in \mathbb{Z}^d$,

$$r_{\tau,\boldsymbol{\gamma}}(\boldsymbol{h}) = \prod_{j=1}^{d} r_{\tau,\gamma_j}(h_j),$$

where

$$r_{\tau,\gamma_j}(h_j) := \begin{cases} 1 & \text{if } h_j = 0, \\ \gamma_j^{-1} \left| h_j \right|^{\tau} & \text{if } h_j \neq 0. \end{cases} \tag{2.22}$$

In the following, we will frequently write, for $\gamma > 0$, $\mathcal{H}_{\mathrm{kor},\alpha,\gamma} := \mathcal{H}_{\mathrm{kor},1,\alpha,\gamma}$ and $K_{\mathrm{kor},\alpha,\gamma} := K_{\mathrm{kor},1,\alpha,\gamma}$. For product weights $\boldsymbol{\gamma} = (\gamma_j)_{j \geq 1}$ the Korobov space $\mathcal{H}_{\mathrm{kor},d,\alpha,\boldsymbol{\gamma}}$ is the tensor product of the spaces $\mathcal{H}_{\mathrm{kor},\alpha,\gamma_1}, \mathcal{H}_{\mathrm{kor},\alpha,\gamma_2}, \ldots, \mathcal{H}_{\mathrm{kor},\alpha,\gamma_d}$ of univariate functions, i.e.,

$$\begin{aligned} \mathcal{H}_{\mathrm{kor},d,\alpha,\boldsymbol{\gamma}} &= \mathcal{H}_{\mathrm{kor},\alpha,\gamma_1} \otimes \mathcal{H}_{\mathrm{kor},\alpha,\gamma_2} \otimes \cdots \otimes \mathcal{H}_{\mathrm{kor},\alpha,\gamma_d} \\ &= \overline{\mathrm{span} \left\{ \boldsymbol{x} \mapsto \prod_{j=1}^{d} f_j(x_j) \ : \ f_j \in \mathcal{H}_{\mathrm{kor},\alpha,\gamma_j} \right\}}, \end{aligned}$$

where $\boldsymbol{x} = (x_1, \ldots, x_d)$, and where the closure is taken with respect to the norm induced by the inner product. Furthermore,

$$K_{\mathrm{kor},d,\alpha,\boldsymbol{\gamma}}(\boldsymbol{x}, \boldsymbol{y}) = \prod_{j=1}^{d} K_{\mathrm{kor},\alpha,\gamma_j}(x_j, y_j),$$

where $\boldsymbol{x} = (x_1, \ldots, x_d)$ and $\boldsymbol{y} = (y_1, \ldots, y_d)$.

2.5 Integration in Weighted Korobov Spaces

Let us now consider integration in $\mathcal{H}_{\mathrm{kor},d,\alpha,\gamma}$ for $\alpha > 1/2$ by means of lattice rules. The following formula for the worst-case error is of great importance, and will be referred to many times throughout this book, which is why we include the full proof, though it is analogous to the proof of Theorem 2.6.

Theorem 2.19 *Let $\mathcal{L}$ be an integration lattice with* $\det(\mathcal{L}^\perp) = N$. *Let $\alpha > 1/2$. The worst-case error of the corresponding lattice rule based on the lattice point set $\mathcal{P}(\mathcal{L})$ in the weighted Korobov space $\mathcal{H}_{\mathrm{kor},d,\alpha,\gamma}$ is given by*

$$[\mathrm{err}_{N,d}(\mathcal{H}_{\mathrm{kor},d,\alpha,\gamma}, \mathcal{P}(\mathcal{L}))]^2 = \sum_{\boldsymbol{h} \in \mathcal{L}^\perp \setminus \{\boldsymbol{0}\}} \frac{1}{r_{2\alpha,\gamma}(\boldsymbol{h})}. \tag{2.23}$$

Proof The proof of this theorem is based on Theorem 1.27. We have

$$K_{\mathrm{kor},d,\alpha,\gamma}(\boldsymbol{y},\boldsymbol{y}) = \sum_{\mathfrak{u} \subseteq [d]} \gamma_{\mathfrak{u}} (1 + 2\zeta(2\alpha))^{|\mathfrak{u}|} < \infty,$$

because $\alpha > 1/2$. Hence Condition (1.17) is satisfied and we can apply the formula for the worst-case error in Theorem 1.27, which yields

$$\begin{aligned}
&\int_{[0,1]^d} \int_{[0,1]^d} K_{\mathrm{kor},d,\alpha,\gamma}(\boldsymbol{x},\boldsymbol{y}) \,\mathrm{d}\boldsymbol{x} \,\mathrm{d}\boldsymbol{y} \\
&= \sum_{\boldsymbol{h} \in \mathbb{Z}^d} \left(\gamma_{\mathfrak{u}(\boldsymbol{h})} \prod_{j \in \mathfrak{u}(\boldsymbol{h})} \frac{1}{|h_j|^{2\alpha}} \right) \int_{[0,1]^d} \int_{[0,1]^d} \mathrm{e}^{2\pi \mathrm{i} \boldsymbol{h} \cdot (\boldsymbol{x}-\boldsymbol{y})} \,\mathrm{d}\boldsymbol{x} \,\mathrm{d}\boldsymbol{y} \\
&= \gamma_{\mathfrak{u}(\boldsymbol{0})} \prod_{j \in \mathfrak{u}(\boldsymbol{0})} \frac{1}{|h_j|^{2\alpha}} \\
&= 1,
\end{aligned}$$

since $\mathfrak{u}(\boldsymbol{0}) = \emptyset$. In the same way one can show that for any fixed $\boldsymbol{x}_k \in [0,1)^d$ we have

$$\int_{[0,1]^d} K_{\mathrm{kor},d,\alpha,\gamma}(\boldsymbol{x}_k,\boldsymbol{y}) \,\mathrm{d}\boldsymbol{y} = 1.$$

For an arbitrary N-element point set $\mathcal{P} = \{\boldsymbol{x}_0, \boldsymbol{x}_1, \ldots, \boldsymbol{x}_{N-1}\}$ in $[0,1)^d$ we now obtain from Theorem 1.27 and the kernel representation (2.19) that

$$\begin{aligned}
[\mathrm{err}_{N,d}(\mathcal{H}_{\mathrm{kor},d,\alpha,\gamma}, \mathcal{P})]^2 &= -1 + \frac{1}{N^2} \sum_{k,\ell=0}^{N-1} \sum_{\boldsymbol{h} \in \mathbb{Z}^d} \frac{1}{r_{2\alpha,\gamma}(\boldsymbol{h})} \mathrm{e}^{2\pi \mathrm{i} \boldsymbol{h} \cdot (\boldsymbol{x}_k - \boldsymbol{x}_\ell)} \\
&= \sum_{\boldsymbol{h} \in \mathbb{Z}^d \setminus \{\boldsymbol{0}\}} \frac{1}{r_{2\alpha,\gamma}(\boldsymbol{h})} \left| \frac{1}{N} \sum_{k=0}^{N-1} \mathrm{e}^{2\pi \mathrm{i} \boldsymbol{h} \cdot \boldsymbol{x}_k} \right|^2.
\end{aligned} \tag{2.24}$$

Let now $\mathcal{P} = \mathcal{P}(\mathcal{L})$. Then the above formula combined with Lemma 1.9 yields

$$[\mathrm{err}_{N,d}(\mathcal{H}_{\mathrm{kor},d,\alpha,\boldsymbol{\gamma}}, \mathcal{P}(\mathcal{L}))]^2 = \sum_{\boldsymbol{h} \in \mathcal{L}^\perp \setminus \{\boldsymbol{0}\}} \frac{1}{r_{2\alpha,\boldsymbol{\gamma}}(\boldsymbol{h})}. \qquad \square$$

Remark 2.20 Note that in the unweighted case, i.e., $\gamma_{\mathfrak{u}} = 1$ for all $\mathfrak{u} \subseteq [d]$, the squared worst-case error in (2.23) coincides with the quantity $P_{2\alpha}$ in Definition 1.14.

From Theorem 2.19 we deduce a result for the special case when the integration lattice is a rank-1 lattice. In this case, for a generating vector $\boldsymbol{g} \in \mathbb{Z}^d$, we will frequently write $\mathrm{err}_{N,d,\alpha,\boldsymbol{\gamma}}(\boldsymbol{g})$ instead of $\mathrm{err}_{N,d}(\mathcal{H}_{\mathrm{kor},d,\alpha,\boldsymbol{\gamma}}, \mathcal{P}(\boldsymbol{g}, N))$ in the following, i.e.,

$$\mathrm{err}_{N,d,\alpha,\boldsymbol{\gamma}}(\boldsymbol{g}) := \mathrm{err}_{N,d}(\mathcal{H}_{\mathrm{kor},d,\alpha,\boldsymbol{\gamma}}, \mathcal{P}(\boldsymbol{g}, N)).$$

Applying (1.7) to Formula (2.23) in Theorem 2.19 yields the following corollary.

Corollary 2.21 *For any N-point rank-1 lattice rule with generating vector $\boldsymbol{g} \in \mathbb{Z}^d$ we have*

$$[\mathrm{err}_{N,d,\alpha,\boldsymbol{\gamma}}(\boldsymbol{g})]^2 = \sum_{\substack{\boldsymbol{h} \in \mathbb{Z}^d \setminus \{\boldsymbol{0}\} \\ \boldsymbol{h} \cdot \boldsymbol{g} \equiv 0 \ (\mathrm{mod}\ N)}} \frac{1}{r_{2\alpha,\boldsymbol{\gamma}}(\boldsymbol{h})}.$$

Remark 2.22 Ordering the summation according to the nonzero components of $\boldsymbol{h} \in \mathbb{Z}^d$, the formula in Corollary 2.21 can also be written in the form

$$[\mathrm{err}_{N,d,\alpha,\boldsymbol{\gamma}}(\boldsymbol{g})]^2 = \sum_{\emptyset \neq \mathfrak{u} \subseteq [d]} \gamma_{\mathfrak{u}} \sum_{\substack{\boldsymbol{h}_{\mathfrak{u}} \in (\mathbb{Z} \setminus \{0\})^{|\mathfrak{u}|} \\ \boldsymbol{h}_{\mathfrak{u}} \cdot \boldsymbol{g}_{\mathfrak{u}} \equiv 0 \ (\mathrm{mod}\ N)}} \prod_{j \in \mathfrak{u}} \frac{1}{|h_j|^{2\alpha}}.$$

Remark 2.23 As pointed out in Remark 2.9, if $\alpha \in \mathbb{N}$, the Bernoulli polynomial $B_{2\alpha}$ of degree 2α has the Fourier expansion

$$B_{2\alpha}(x) = \frac{(-1)^{\alpha+1}(2\alpha)!}{(2\pi)^{2\alpha}} \sum_{h \in \mathbb{Z} \setminus \{0\}} \frac{\mathrm{e}^{2\pi \mathrm{i} h x}}{|h|^{2\alpha}} \quad \text{for all } x \in [0,1).$$

Using Corollary 2.21, we therefore obtain in the case of product weights that

$$\begin{aligned}
[\mathrm{err}_{N,d,\alpha,\boldsymbol{\gamma}}(\boldsymbol{g})]^2 &= -1 + \sum_{\boldsymbol{h} \in \mathbb{Z}^d} \frac{1}{r_{2\alpha,\boldsymbol{\gamma}}(\boldsymbol{h})} \frac{1}{N} \sum_{k=0}^{N-1} \mathrm{e}^{2\pi \mathrm{i} (\boldsymbol{h} \cdot \boldsymbol{g}/N) k} \\
&= -1 + \frac{1}{N} \sum_{k=0}^{N-1} \prod_{j=1}^{d} \left(\sum_{h \in \mathbb{Z}} \frac{1}{r_{2\alpha,\gamma_j}(h)} \mathrm{e}^{2\pi \mathrm{i} \{k g_j / N\} h} \right) \\
&= -1 + \frac{1}{N} \sum_{k=0}^{N-1} \prod_{j=1}^{d} \left(1 + \gamma_j \frac{(-1)^{\alpha+1}(2\pi)^{2\alpha}}{(2\alpha)!} B_{2\alpha}\left(\left\{ \frac{k g_j}{N} \right\} \right) \right),
\end{aligned}$$

such that $[\mathrm{err}_{N,d,\alpha,\boldsymbol{\gamma}}(\boldsymbol{g})]^2$ can be computed in $O(d\,N)$ operations.

2.6 Tractability

As mentioned above, suitable conditions on the weights can help in overcoming the curse of dimensionality of an integration problem and yield *tractability*. In this section, we study tractability for the weighted Korobov space $\mathcal{H}_{\mathrm{kor},d,\alpha,\boldsymbol{\gamma}}$. The following existence result was shown in [59].

Theorem 2.24 *For any prime number N, any dimension d, and any $\tau \in [1/2, \alpha)$ the following statements hold.*

1. *There exists a generating vector $\boldsymbol{g} \in G_d(N)$ such that*

$$\mathrm{err}_{N,d,\alpha,\boldsymbol{\gamma}}(\boldsymbol{g}) \leq \frac{2^\tau}{N^\tau} \left(\sum_{\emptyset \neq \mathfrak{u} \subseteq [d]} \gamma_{\mathfrak{u}}^{1/(2\tau)} \left(2\zeta\left(\frac{\alpha}{\tau}\right)\right)^{|\mathfrak{u}|} \right)^{\tau}.$$

2. *In the case of product weights the bound in Item 1 can be simplified to*

$$\mathrm{err}_{N,d,\alpha,\boldsymbol{\gamma}}(\boldsymbol{g}) \leq \frac{2^\tau}{N^\tau} \left(-1 + \prod_{j=1}^{d} \left(1 + 2\gamma_j^{1/(2\tau)} \zeta\left(\frac{\alpha}{\tau}\right)\right) \right)^{\tau}.$$

3. *For any real number $c \geq 1$ and*

$$\mathcal{A}_c(\tau) := \left\{ \boldsymbol{g} \in G_d(N) \; : \; \mathrm{err}_{N,d,\alpha,\boldsymbol{\gamma}}(\boldsymbol{g}) \leq \frac{(2c^2)^\tau}{N^\tau} \left(\sum_{\emptyset \neq \mathfrak{u} \subseteq [d]} \gamma_{\mathfrak{u}}^{1/(2\tau)} \left(2\zeta\left(\frac{\alpha}{\tau}\right)\right)^{|\mathfrak{u}|} \right)^{\tau} \right\}$$

it holds that

$$|\mathcal{A}_c(\tau)| \geq N^d \left(1 - \frac{1}{c^2}\right).$$

We first need an inequality which is sometimes referred to as Jensen's inequality, and will be used in the proof of Theorem 2.24 as well as in many other instances in this book.

Lemma 2.25 (Jensen's inequality) *For any $\lambda \in (0, 1]$ and nonnegative reals a_k we have*

$$\left(\sum_k a_k \right)^{\lambda} \leq \sum_k a_k^{\lambda}.$$

Proof We have $0 \leq a_j / (\sum_k a_k) \leq 1$ and thus, since $\lambda \in (0, 1]$,

$$\frac{a_j}{\sum_k a_k} \leq \left(\frac{a_j}{\sum_k a_k} \right)^{\lambda}.$$

Summation over all j implies

$$1 = \frac{\sum_j a_j}{\sum_k a_k} \le \frac{\sum_j a_j^\lambda}{(\sum_k a_k)^\lambda},$$

which finally yields the desired inequality. □

Using Lemma 2.25, we are ready to give the proof of Theorem 2.24.

Proof of Theorem 2.24 We first prove Item 1. Let $\mathrm{av}(N,d)$ denote the average of the squared worst-case error, where the average is taken over all $\boldsymbol{g} \in G_d(N)$, i.e.,

$$\mathrm{av}(N,d) = \frac{1}{N^d} \sum_{\boldsymbol{g} \in G_d(N)} [\mathrm{err}_{N,d,\alpha,\boldsymbol{\gamma}}(\boldsymbol{g})]^2.$$

Then, using Corollary 2.21, we obtain

$$\begin{aligned}
\mathrm{av}(N,d) &= \frac{1}{N^d} \sum_{\boldsymbol{g} \in G_d(N)} \sum_{\substack{\boldsymbol{h} \in \mathbb{Z}^d \setminus \{\boldsymbol{0}\} \\ \boldsymbol{h} \cdot \boldsymbol{g} \equiv 0 \pmod N}} \frac{1}{r_{2\alpha,\boldsymbol{\gamma}}(\boldsymbol{h})} \\
&= \frac{1}{N^d} \sum_{\boldsymbol{h} \in \mathbb{Z}^d \setminus \{\boldsymbol{0}\}} \frac{1}{r_{2\alpha,\boldsymbol{\gamma}}(\boldsymbol{h})} |\{\boldsymbol{g} \in G_d(N) \;:\; \boldsymbol{h} \cdot \boldsymbol{g} \equiv 0 \pmod N\}|.
\end{aligned}$$

If $\boldsymbol{h} \equiv \boldsymbol{0} \pmod N$, then

$$|\{\boldsymbol{g} \in G_d(N) \;:\; \boldsymbol{h} \cdot \boldsymbol{g} \equiv 0 \pmod N\}| = N^d.$$

Otherwise, if $\boldsymbol{h} \not\equiv \boldsymbol{0} \pmod N$, we find as in the proof of Lemma 2.13 that

$$|\{\boldsymbol{g} \in G_d(N) \;:\; \boldsymbol{h} \cdot \boldsymbol{g} \equiv 0 \pmod N\}| = N^{d-1}.$$

Hence it follows that

$$\begin{aligned}
\mathrm{av}(N,d) &= \sum_{\substack{\boldsymbol{h} \in \mathbb{Z}^d \setminus \{\boldsymbol{0}\} \\ \boldsymbol{h} \equiv \boldsymbol{0} \pmod N}} \frac{1}{r_{2\alpha,\boldsymbol{\gamma}}(\boldsymbol{h})} + \frac{1}{N} \sum_{\substack{\boldsymbol{h} \in \mathbb{Z}^d \setminus \{\boldsymbol{0}\} \\ \boldsymbol{h} \not\equiv \boldsymbol{0} \pmod N}} \frac{1}{r_{2\alpha,\boldsymbol{\gamma}}(\boldsymbol{h})} \\
&= \sum_{\boldsymbol{h} \in \mathbb{Z}^d \setminus \{\boldsymbol{0}\}} \frac{1}{r_{2\alpha,\boldsymbol{\gamma}}(N\boldsymbol{h})} + \frac{1}{N} \sum_{\substack{\boldsymbol{h} \in \mathbb{Z}^d \setminus \{\boldsymbol{0}\} \\ \boldsymbol{h} \not\equiv \boldsymbol{0} \pmod N}} \frac{1}{r_{2\alpha,\boldsymbol{\gamma}}(\boldsymbol{h})} \\
&\le \frac{2}{N} \sum_{\boldsymbol{h} \in \mathbb{Z}^d \setminus \{\boldsymbol{0}\}} \frac{1}{r_{2\alpha,\boldsymbol{\gamma}}(\boldsymbol{h})} \\
&= \frac{2}{N} \sum_{\emptyset \ne \mathfrak{u} \subseteq [d]} \gamma_{\mathfrak{u}} \prod_{j \in \mathfrak{u}} \left(\sum_{h_j \in \mathbb{Z} \setminus \{0\}} \frac{1}{|h_j|^{2\alpha}} \right) \\
&= \frac{2}{N} \sum_{\emptyset \ne \mathfrak{u} \subseteq [d]} \gamma_{\mathfrak{u}} (2\zeta(2\alpha))^{|\mathfrak{u}|}.
\end{aligned}$$

Since there is always at least one element which is at least as good as average (see Remark 2.14), we find that there exists at least one particular $\boldsymbol{g} \in G_d(N)$ for which the squared worst-case error satisfies the bound

$$[\mathrm{err}_{N,d,\alpha,\boldsymbol{\gamma}}(\boldsymbol{g})]^2 \leq \frac{2}{N} \sum_{\emptyset \neq \mathfrak{u} \subseteq [d]} \gamma_{\mathfrak{u}} (2\zeta(2\alpha))^{|\mathfrak{u}|}. \tag{2.25}$$

Next, we use Jensen's inequality to improve the convergence rate in (2.25). When applied to the formula for the squared worst-case error, Lemma 2.25 yields

$$\begin{aligned} [\mathrm{err}_{N,d,\alpha,\boldsymbol{\gamma}}(\boldsymbol{g})]^{2\lambda} &= \left(\sum_{\substack{\boldsymbol{h} \in \mathbb{Z}^d \setminus \{\boldsymbol{0}\} \\ \boldsymbol{h} \cdot \boldsymbol{g} \equiv 0 \pmod{N}}} \gamma_{\mathfrak{u}(\boldsymbol{h})} \prod_{j \in \mathfrak{u}(\boldsymbol{h})} \frac{1}{|h_j|^{2\alpha}} \right)^{\lambda} \\ &\leq \sum_{\substack{\boldsymbol{h} \in \mathbb{Z}^d \setminus \{\boldsymbol{0}\} \\ \boldsymbol{h} \cdot \boldsymbol{g} \equiv 0 \pmod{N}}} \left(\gamma_{\mathfrak{u}(\boldsymbol{h})}^{\lambda} \prod_{j \in \mathfrak{u}(\boldsymbol{h})} \frac{1}{|h_j|^{2\alpha\lambda}} \right) \\ &= [\mathrm{err}_{N,d,\alpha\lambda,\boldsymbol{\gamma}^{\lambda}}(\boldsymbol{g})]^2, \end{aligned} \tag{2.26}$$

where we have to restrict λ to the interval $(1/(2\alpha), 1]$ in order to guarantee convergence of the series involved. Here $\boldsymbol{\gamma}^{\lambda}$ denotes the set of weights of the form $\gamma_{\mathfrak{u}}^{\lambda}$ for $\gamma_{\mathfrak{u}}$ in $\boldsymbol{\gamma}$, i.e.,

$$\boldsymbol{\gamma}^{\lambda} = \{\gamma_{\mathfrak{u}}^{\lambda} \,:\, \mathfrak{u} \subseteq [d]\} \quad \text{for} \quad \boldsymbol{\gamma} = \{\gamma_{\mathfrak{u}} \,:\, \mathfrak{u} \subseteq [d]\}.$$

Applying the existence result (2.25) to $[\mathrm{err}_{N,d,\alpha\lambda,\boldsymbol{\gamma}^{\lambda}}(\boldsymbol{g})]^2$ we finally obtain the existence of $\boldsymbol{g} \in G_d(N)$ such that

$$[\mathrm{err}_{N,d,\alpha,\boldsymbol{\gamma}}(\boldsymbol{g})]^{2\lambda} \leq \frac{2}{N} \sum_{\emptyset \neq \mathfrak{u} \subseteq [d]} \gamma_{\mathfrak{u}}^{\lambda} (2\zeta(2\alpha\lambda))^{|\mathfrak{u}|}.$$

Note that $\zeta(2\alpha\lambda) < \infty$ since $\lambda \in (1/(2\alpha), 1]$. Now the desired result follows by substituting τ for $1/(2\lambda)$. This substitution also implies that $\tau \in [1/2, \alpha)$.

Regarding Item 2, if the weights are of product form, $\gamma_{\mathfrak{u}} = \prod_{j \in \mathfrak{u}} \gamma_j$ for $\mathfrak{u} \subseteq [d]$, then

$$\begin{aligned} \sum_{\emptyset \neq \mathfrak{u} \subseteq [d]} \gamma_{\mathfrak{u}}^{1/(2\tau)} \left(2\zeta\left(\frac{\alpha}{\tau}\right)\right)^{|\mathfrak{u}|} &= \sum_{\emptyset \neq \mathfrak{u} \subseteq [d]} \prod_{j \in \mathfrak{u}} \left(\gamma_j^{1/(2\tau)} 2\zeta\left(\frac{\alpha}{\tau}\right)\right) \\ &= -1 + \prod_{j=1}^{d} \left(1 + \gamma_j^{1/(2\tau)} 2\zeta\left(\frac{\alpha}{\tau}\right)\right). \end{aligned}$$

For Item 3, let, for $\lambda \in (1/(2\alpha), 1]$ and $c \geq 1$,

$$\mathcal{B} := \left\{ \boldsymbol{g} \in G_d(N) \ : \ [\mathrm{err}_{N,d,\alpha\lambda,\boldsymbol{\gamma}^\lambda}(\boldsymbol{g})]^2 \le \frac{2c^2}{N} \sum_{\emptyset \ne \mathfrak{u} \subseteq [d]} \gamma_{\mathfrak{u}}^\lambda (2\zeta(2\alpha\lambda))^{|\mathfrak{u}|} \right\}.$$

From the above averaging argument we obtain

$$\begin{aligned} \frac{2}{N} \sum_{\emptyset \ne \mathfrak{u} \subseteq [d]} \gamma_{\mathfrak{u}}^\lambda (2\zeta(2\alpha\lambda))^{|\mathfrak{u}|} &= \frac{1}{N^d} \sum_{\boldsymbol{g} \in G_d(N)} [\mathrm{err}_{N,d,\alpha\lambda,\boldsymbol{\gamma}^\lambda}(\boldsymbol{g})]^2 \\ &> \frac{|\mathcal{B}^c|}{N^d} \frac{2c^2}{N} \sum_{\emptyset \ne \mathfrak{u} \subseteq [d]} \gamma_{\mathfrak{u}}^\lambda (2\zeta(2\alpha\lambda))^{|\mathfrak{u}|}, \end{aligned}$$

where $\mathcal{B}^c := G_d(N) \setminus \mathcal{B}$ is the complement of the set $\mathcal{B}$. Hence

$$|\mathcal{B}^c| < \frac{N^d}{c^2} \quad \text{and therefore} \quad |\mathcal{B}| \ge N^d \left(1 - \frac{1}{c^2}\right).$$

From (2.26) we find that $\mathcal{B} \subseteq \mathcal{A}_c(1/(2\lambda))$. Writing again τ for $1/(2\lambda)$ yields the desired result. □

Remark 2.26 Since τ can be chosen arbitrarily close to α we obtain a convergence rate of $O(N^{-\alpha+\delta})$ for every $\delta > 0$, which is in accordance with Theorem 2.15. The result in Theorem 2.24 can even be made explicit using a component-by-component algorithm, which we will present in Section 3.3.

Theorem 2.24 provides upper bounds on the N-th minimal worst-case error in the weighted Korobov space that we can use to establish conditions on the weights for which the integration problem in the Korobov space is tractable in some form. As in Section 2.3 we consider only QMC rules; the general case is treated in [211, Theorem 16.16]. We define the N-th minimal worst-case error of QMC integration,

$$e_{\mathrm{QMC}}(N,d) := \inf_{\mathcal{P}} \left|\mathrm{err}_{N,d}(\mathcal{H}_{\mathrm{kor},d,\alpha,\boldsymbol{\gamma}}, \mathcal{P})\right|,$$

where the infimum is extended over all N-element point sets $\mathcal{P}$ in $[0,1)^d$. Obviously,

$$e(N,d) \le e_{\mathrm{QMC}}(N,d)$$

for all $d, N \in \mathbb{N}$, since in the definition of the N-th minimal error $e(N,d)$, the infimum is extended over all linear rules, which comprises the class of QMC rules. From this observation we obtain

$$N_{\mathrm{QMC}}(\varepsilon, d) := \min\{N \in \mathbb{N} \ : \ e_{\mathrm{QMC}}(N,d) \le \varepsilon\, e(0,d)\} \ge N(\varepsilon, d).$$

Note again that according to Remark 1.41 the initial error $e(0,d)$ in the weighted Korobov space equals one for all $d \in \mathbb{N}$.

Let us study tractability with respect to $N_{\mathrm{QMC}}(\varepsilon, d)$ instead of the general information complexity $N(\varepsilon, d)$. We give necessary and sufficient conditions on the weights $\boldsymbol{\gamma}$ under which integration in the $\boldsymbol{\gamma}$-weighted Korobov space is tractable for

QMC rules. Obviously, sufficient conditions on the weights for tractability for QMC rules are then also sufficient conditions for the respective notion of tractability for the general class of linear algorithms. However, necessary conditions in the QMC case do not imply necessary conditions for the general case where arbitrary linear algorithms are allowed.

The case of general weights

The formulation of the following corollary and its proof is based on [211, Theorem 16.4].

Corollary 2.27 *Consider multivariate integration in the weighted Korobov space $\mathcal{H}_{\mathrm{kor},d,\alpha,\gamma}$ with $\alpha > 1/2$. For $\tau \in [1/2, \alpha)$ and $q \geq 0$ define*

$$B_{\tau,q} := \sup_{d\in\mathbb{N}} \left(\frac{1}{d^q} \sum_{\emptyset\neq\mathfrak{u}\subseteq[d]} \gamma_{\mathfrak{u}}^{1/(2\tau)} \left(2\zeta\left(\frac{\alpha}{\tau}\right)\right)^{|\mathfrak{u}|} \right).$$

Then the following statements hold true.

1. *If $B_{\tau,q} < \infty$ for some $\tau \in [1/2, \alpha)$ and $q \geq 0$, then the integration problem is polynomially tractable.*
2. *If $B_{\tau,0} < \infty$ for some $\tau \in [1/2, \alpha)$, then the integration problem is strongly polynomially tractable with an ε-exponent of at most $1/\tau$. If $B_{\tau,0} < \infty$ holds for all $\tau \in [1/2, \alpha)$, then the ε-exponent is $1/\alpha$, which is optimal.*
3. *Assume that the weights $\boldsymbol{\gamma}$ are chosen such that $K_{\mathrm{kor},d,\alpha,\gamma}$ is nonnegative. Then integration is strongly polynomially tractable for QMC rules if and only if $B_{1/2,0} < \infty$, and it is polynomially tractable for QMC rules if and only if $B_{1/2,q} < \infty$ for some number q.*
4. *If*

$$\lim_{d\to\infty} \frac{1}{d} \log\left(\sum_{\mathfrak{u}\subseteq[d]} \gamma_{\mathfrak{u}} (2\zeta(2\alpha))^{|\mathfrak{u}|} \right) = 0, \tag{2.27}$$

then integration in $\mathcal{H}_{\mathrm{kor},d,\alpha,\gamma}$ is weakly tractable. If the weights $\boldsymbol{\gamma}$ are chosen such that $K_{\mathrm{kor},d,\alpha,\gamma}$ is nonnegative, then (2.27) is also necessary for weak tractability for QMC rules.

The positive results on tractability stated in Items 1–4 can be obtained by means of rank-1 lattice rules.

Proof We show the four items in the corollary in sequence.

Regarding Item 1, assume that $B := B_{\tau,q}$ is finite for some $\tau \in [1/2, \alpha)$ and $q \geq 0$. Then we have for all $d \in \mathbb{N}$ that

$$\sum_{\emptyset\neq\mathfrak{u}\subseteq[d]} \gamma_{\mathfrak{u}}^{1/(2\tau)} \left(2\zeta\left(\frac{\alpha}{\tau}\right)\right)^{|\mathfrak{u}|} \leq B\, d^q.$$

Moreover, for every prime number N, Theorem 2.24 implies the existence of a generating vector $\boldsymbol{g} \in G_d(N)$ such that

$$\mathrm{err}_{N,d,\alpha,\gamma}(\boldsymbol{g}) \le \frac{(2B)^\tau d^{q\tau}}{N^\tau}.$$

For $\varepsilon > 0$, let N be the smallest prime number that is greater than or equal to $\lceil 2Bd^q\varepsilon^{-1/\tau}\rceil =: M$. Then we have $e(N,d) \le \varepsilon$ and hence

$$N_{\mathrm{QMC}}(\varepsilon, d) \le N < 2M = 2\lceil 2Bd^q\varepsilon^{-1/\tau}\rceil,$$

where we used Bertrand's postulate which states that $M \le N < 2M$. Therefore, we have polynomial tractability.

Regarding Item 2, if $B_{\tau,0} < \infty$ for some $\tau \in [1/2, \alpha)$, we get, from the previous considerations, strong polynomial tractability with an ε-exponent of at most $1/\tau$. If, however, $B_{\tau,0} < \infty$ for all $\tau \in [1/2, \alpha)$, then the ε-exponent is at most

$$\inf\left\{\frac{1}{\tau} \,:\, \tau \in \left[\frac{1}{2}, \alpha\right)\right\} = \frac{1}{\alpha}.$$

From Theorem 2.10 we know that $N^{-\alpha}$ is the best possible convergence rate of numerical integration in the (weighted) Korobov space by linear rules $A_{N,d}$, even in the case $d = 1$. Thus, an ε-exponent of $1/\alpha$ is optimal.

As regards Item 3, according to Theorem 1.27 the squared worst-case error of an arbitrary QMC rule based on a point set $\mathcal{P} = \{\boldsymbol{x}_0, \boldsymbol{x}_1, \ldots, \boldsymbol{x}_{N-1}\}$ in $[0,1)^d$ is

$$[\mathrm{err}_{N,d}(\mathcal{H}_{\mathrm{kor},d,\alpha,\gamma}, \mathcal{P})]^2 = -1 + \frac{1}{N^2}\sum_{k,\ell=0}^{N-1} K_{\mathrm{kor},d,\alpha,\gamma}(\boldsymbol{x}_k, \boldsymbol{x}_\ell).$$

Since $K_{\mathrm{kor},d,\alpha,\gamma}$ is assumed to be nonnegative we can bound the latter double sum from below by omitting the nondiagonal terms. In this way we obtain

$$[\mathrm{err}_{N,d}(\mathcal{H}_{\mathrm{kor},d,\alpha,\gamma}, \mathcal{P})]^2 \ge -1 + \frac{1}{N^2}\sum_{k=0}^{N-1} K_{\mathrm{kor},d,\alpha,\gamma}(\boldsymbol{x}_k, \boldsymbol{x}_k).$$

We have

$$\begin{aligned} K_{\mathrm{kor},d,\alpha,\gamma}(\boldsymbol{x}_k, \boldsymbol{x}_k) &= K_{\mathrm{kor},d,\alpha,\gamma}(\mathbf{0}, \mathbf{0}) \\ &= \sum_{\mathfrak{u}\subseteq[d]} \gamma_{\mathfrak{u}}(1 + 2\zeta(2\alpha))^{|\mathfrak{u}|} \ge \sum_{\emptyset\neq\mathfrak{u}\subseteq[d]} \gamma_{\mathfrak{u}}(2\zeta(2\alpha))^{|\mathfrak{u}|}, \end{aligned}$$

and so

$$[\mathrm{err}^2_{N,d}(\mathcal{H}_{\mathrm{kor},d,\alpha,\gamma}, \mathcal{P})]^2 \ge -1 + \frac{1}{N}\sum_{\emptyset\neq\mathfrak{u}\subseteq[d]} \gamma_{\mathfrak{u}}(2\zeta(2\alpha))^{|\mathfrak{u}|}. \tag{2.28}$$

Assume now that we have (strong) polynomial tractability for QMC algorithms. This means that there exists an N-element point set $\mathcal{P}$ in $[0,1)^d$ such that

$$\mathrm{err}_{N,d}(\mathcal{H}_{\mathrm{kor},d,\alpha,\boldsymbol{\gamma}},\mathcal{P}) \le \varepsilon \quad \text{for } N = N_{\mathrm{QMC}}(\varepsilon,d) \le C\varepsilon^{-p} d^q$$

for some nonnegative numbers C, p, and q, where $q = 0$ in the case of strong polynomial tractability. Fix ε to some value, say $\varepsilon = 1/2$. Then the lower bound (2.28) for $N = N_{\mathrm{QMC}}(1/2,d)$ yields

$$\sum_{\emptyset \ne \mathfrak{u} \subseteq [d]} \gamma_{\mathfrak{u}} (2\zeta(2\alpha))^{|\mathfrak{u}|} \le \left(1 + [\mathrm{err}_{N,d}(\mathcal{H}_{\mathrm{kor},d,\alpha,\boldsymbol{\gamma}},\mathcal{P})]^2\right) N \le \left(1 + \frac{1}{4}\right) C\, 2^p d^q,$$

such that we obtain

$$\sup_{d\in\mathbb{N}} \frac{1}{d^q} \sum_{\emptyset \ne \mathfrak{u} \subseteq [d]} \gamma_{\mathfrak{u}} (2\zeta(2\alpha))^{|\mathfrak{u}|} \le \frac{5C}{4} 2^p < \infty.$$

Hence we have $B_{1/2,q} < \infty$ as desired.

Finally, for Item 4, we know from Theorem 2.24 that we have, for prime numbers N and any $\tau \in [1/2, \alpha)$,

$$\mathrm{err}_{N,d,\alpha,\boldsymbol{\gamma}}(\boldsymbol{g}) \le \frac{C(\tau,d)}{N^\tau}$$

for some $\boldsymbol{g} \in G_d(N)$, where

$$C(\tau,d) := \left(2 \sum_{\emptyset \ne \mathfrak{u} \subseteq [d]} \gamma_{\mathfrak{u}}^{1/(2\tau)} \left(2\zeta\left(\frac{\alpha}{\tau}\right)\right)^{|\mathfrak{u}|}\right)^{\tau}.$$

We choose $\tau = 1/2$ and use the abbreviation $C(d) := C(1/2,d)$. Let $\varepsilon > 0$ and let N be the smallest prime number greater than or equal to $\lceil (C(d)\varepsilon^{-1})^2 \rceil =: M$. Then we have $e_{\mathrm{QMC}}(N,d) \le \varepsilon$ and therefore $N_{\mathrm{QMC}}(\varepsilon,d) \le N \le 2M$, where

$$M = \lceil (C(d)\varepsilon^{-1})^2 \rceil \le (C(d)\varepsilon^{-1})^2 + 1 \le \varepsilon^{-2}(1 + (C(d))^2).$$

Consequently,

$$\frac{\log N_{\mathrm{QMC}}(\varepsilon,d)}{\varepsilon^{-1}+d} \le \frac{\log 2 + 2\log\varepsilon^{-1} + \log(1+(C(d))^2)}{\varepsilon^{-1}+d}.$$

Letting $\varepsilon^{-1} + d$ tend to infinity we therefore get

$$\lim_{\varepsilon^{-1}+d\to\infty} \frac{\log N_{\mathrm{QMC}}(\varepsilon,d)}{\varepsilon^{-1}+d} \le \lim_{\varepsilon^{-1}+d\to\infty} \frac{\log(1+(C(d))^2)}{\varepsilon^{-1}+d}.$$

The limit on the right-hand side of the latter inequality is zero, since

$$1 + (C(d))^2 \le 2 \sum_{\mathfrak{u}\subseteq[d]} \gamma_{\mathfrak{u}} (2\zeta(2\alpha))^{|\mathfrak{u}|},$$

and hence

$$\lim_{d\to\infty} \frac{\log(1 + (C(d))^2)}{d} = 0$$

when Condition (2.27) is satisfied.

Now assume that the kernel $K_{\mathrm{kor},d,\alpha,\boldsymbol{\gamma}}$ is nonnegative and that we have weak tractability for QMC rules. This means that there exists a QMC rule with a worst-case error of at most ε that is based on $N_{\mathrm{QMC}}(\varepsilon, d)$ points in $[0, 1)^d$, and that

$$\lim_{\varepsilon^{-1}+d\to\infty} \frac{\log N_{\mathrm{QMC}}(\varepsilon, d)}{\varepsilon^{-1} + d} = 0.$$

From (2.28) we then obtain

$$\sum_{\emptyset\neq\mathfrak{u}\subseteq[d]} \gamma_{\mathfrak{u}} (2\zeta(2\alpha))^{|\mathfrak{u}|} \le (1 + \varepsilon^2) N_{\mathrm{QMC}}(\varepsilon, d).$$

Fix $\varepsilon > 0$, take the logarithm, divide both sides of the latter inequality by d, and let d tend to infinity. This yields

$$\lim_{d\to\infty} \frac{1}{d} \log\left(\sum_{\mathfrak{u}\subseteq[d]} \gamma_{\mathfrak{u}} (2\zeta(2\alpha))^{|\mathfrak{u}|} \right) = 0,$$

which is Condition (2.27). □

The case of product weights

We close this section with an application of Corollary 2.27 to product weights of the form (2.21).

Corollary 2.28 *Let $\boldsymbol{\gamma} = (\gamma_j)_{j\ge1}$ be a sequence of positive weights. Then the following statements hold true for QMC integration in the product-weighted Korobov space $\mathcal{H}_{\mathrm{kor},d,\alpha,\boldsymbol{\gamma}}$.*

1. *The problem suffers from the curse of dimensionality if* $\lim_{j\to\infty} \gamma_j > 0$.
2. *We have strong polynomial tractability if and only if* $\sum_{j=1}^{\infty} \gamma_j < \infty$. *If*

$$\sum_{j=1}^{\infty} \gamma_j^{1/(2\tau)} < \infty \qquad (2.29)$$

for some $\tau \in [1/2, \alpha)$, then the ε-exponent is at most $1/\tau$. If (2.29) *holds for all $\tau \in [1/2, \alpha)$, then the ε-exponent is $1/\alpha$, which is optimal.*

3. *We have polynomial tractability if and only if*

$$\limsup_{d\to\infty} \frac{\sum_{j=1}^{d} \gamma_j}{\log d} < \infty.$$

4. *We have weak tractability if and only if*

$$\lim_{d\to\infty} \frac{\sum_{j=1}^{d} \gamma_j}{d} = 0.$$

Proof To show Item 1, let $\gamma_j' = \min(\gamma_j, 1/(2\zeta(2\alpha)))$ for $j \in \mathbb{N}$, let $\gamma_{\mathfrak{u}}' = \prod_{j\in\mathfrak{u}} \gamma_j'$ for $\mathfrak{u} \subseteq [d]$, and let $\boldsymbol{\gamma}' = \{\gamma_{\mathfrak{u}}' : \mathfrak{u} \subseteq [d]\}$. From the fact that $\gamma_j' \le \gamma_j$ for all $j \in \mathbb{N}$ we obtain

$$\{f \in \mathcal{H}_{\mathrm{kor},d,\alpha,\boldsymbol{\gamma}'} : \|f\|_{\mathrm{kor},d,\alpha,\boldsymbol{\gamma}'} \le 1\} \subseteq \{f \in \mathcal{H}_{\mathrm{kor},d,\alpha,\boldsymbol{\gamma}} : \|f\|_{\mathrm{kor},d,\alpha,\boldsymbol{\gamma}} \le 1\},$$

which means that integration in $\mathcal{H}_{\mathrm{kor},d,\alpha,\boldsymbol{\gamma}'}$ is no harder than integration in $\mathcal{H}_{\mathrm{kor},d,\alpha,\boldsymbol{\gamma}}$, and so

$$\mathrm{err}_{N,d}(\mathcal{H}_{\mathrm{kor},d,\alpha,\boldsymbol{\gamma}'}, \mathcal{P}) \le \mathrm{err}_{N,d}(\mathcal{H}_{\mathrm{kor},d,\alpha,\boldsymbol{\gamma}}, \mathcal{P}).$$

According to the definition of $\boldsymbol{\gamma}'$ we find that the kernel $K_{\mathrm{kor},d,\alpha,\boldsymbol{\gamma}'}$ is nonnegative. Therefore, for every point set $\mathcal{P} = \{\boldsymbol{x}_0, \boldsymbol{x}_1, \ldots, \boldsymbol{x}_{N-1}\}$ in $[0,1)^d$, we have

$$\begin{aligned}
[\mathrm{err}_{N,d}(\mathcal{H}_{\mathrm{kor},d,\alpha,\boldsymbol{\gamma}}, \mathcal{P})]^2 &\ge [\mathrm{err}_{N,d}(\mathcal{H}_{\mathrm{kor},d,\alpha,\boldsymbol{\gamma}'}, \mathcal{P})]^2 \\
&= -1 + \frac{1}{N^2} \sum_{k,\ell=0}^{N-1} K_{\mathrm{kor},d,\alpha,\boldsymbol{\gamma}'}(\boldsymbol{x}_k, \boldsymbol{x}_\ell) \\
&\ge -1 + \frac{1}{N^2} \sum_{k=0}^{N-1} K_{\mathrm{kor},d,\alpha,\boldsymbol{\gamma}'}(\boldsymbol{x}_k, \boldsymbol{x}_k) \\
&= -1 + \frac{1}{N^2} \sum_{k=0}^{N-1} \prod_{j=1}^{d} \left(1 + 2\gamma_j'\zeta(2\alpha)\right) \\
&= -1 + \frac{1}{N} \prod_{j=1}^{d} \left(1 + 2\gamma_j'\zeta(2\alpha)\right).
\end{aligned} \tag{2.30}$$

If $\gamma_j \ge \gamma_*$ for some positive γ_* for all $j \in \mathbb{N}$, then $2\gamma_j'\zeta(2\alpha) \ge \min(1, 2\gamma_*\zeta(2\alpha)) =: c_*$ uniformly in j, and thus

$$[\mathrm{err}_{N,d}(\mathcal{H}_{\mathrm{kor},d,\alpha,\boldsymbol{\gamma}}, \mathcal{P})]^2 \ge -1 + \frac{(1+c_*)^d}{N}.$$

This implies

$$e_{\mathrm{QMC}}(N,d) \ge \left(\max\left(0, -1 + \frac{(1+c_*)^d}{N}\right)\right)^{1/2},$$

and furthermore

$$N_{\mathrm{QMC}}(\varepsilon,d) \geq \min\left\{N \in \mathbb{N} \,:\, -1+\frac{(1+c_*)^d}{N} \leq \varepsilon^2\right\} = \left\lceil \frac{(1+c_*)^d}{\varepsilon^2+1} \right\rceil.$$

This shows that the information complexity grows exponentially in d, i.e., integration in $\mathcal{H}_{\mathrm{kor},d,\alpha,\gamma}$ suffers from the curse of dimensionality and is intractable.

Regarding Item 2, the necessary and sufficient condition $B_{1/2,0} < \infty$ for strong polynomial tractability in Corollary 2.27 is, for product weights, equivalent to

$$\prod_{j=1}^{\infty} (1+2\gamma_j \zeta(2\alpha)) < \infty.$$

Assume that we have strong polynomial tractability of QMC integration. Then

$$2\zeta(2\alpha)\sum_{j=1}^{\infty} \gamma_j \leq \prod_{j=1}^{\infty} (1+2\gamma_j\zeta(2\alpha)) < \infty.$$

This implies the necessity of the condition $\sum_{j=1}^{\infty} \gamma_j < \infty$ for strong polynomial tractability of QMC integration.

Regarding the converse direction, we have

$$\prod_{j=1}^{d} (1+2\gamma_j\zeta(2\alpha)) = \exp\left(\sum_{j=1}^{d} \log(1+2\gamma_j\zeta(2\alpha))\right) \leq \exp\left(2\zeta(2\alpha)\sum_{j=1}^{d} \gamma_j\right),$$

where we have used the estimate $\log(1+x) \leq x$ for $x \geq 0$. If now $\sum_{j=1}^{\infty} \gamma_j < \infty$, then the right-hand side of the latter inequality is uniformly bounded in d, and so is the left-hand side. This implies strong polynomial tractability of QMC integration.

It remains to study the ε-exponent. Since, for product weights,

$$\begin{aligned}\sum_{\emptyset \neq \mathfrak{u} \subseteq [d]} \gamma_{\mathfrak{u}}^{1/(2\tau)} \left(2\zeta\left(\frac{\alpha}{\tau}\right)\right)^{|\mathfrak{u}|} &= -1 + \prod_{j=1}^{d} \left(1+2\gamma_j^{1/(2\tau)}\zeta\left(\frac{\alpha}{\tau}\right)\right) \\ &\leq \exp\left(2\zeta\left(\frac{\alpha}{\tau}\right)\sum_{j=1}^{d} \gamma_j^{1/(2\tau)}\right), \end{aligned} \tag{2.31}$$

we find that (2.29) implies $B_{\tau,0} < \infty$, and hence the statements on the ε-exponent follow from Corollary 2.27.

For Item 3, the necessary and sufficient condition $B_{1/2,q} < \infty$ for polynomial tractability from Corollary 2.27 is, for product weights, equivalent to

$$\sup_{d\in\mathbb{N}} \frac{1}{d^q} \prod_{j=1}^{d} (1+2\gamma_j\zeta(2\alpha)) < \infty. \tag{2.32}$$

Assume first that we have $A := \limsup_{d\to\infty} \sum_{j=1}^d \gamma_j/\log d < \infty$. Then for every $\delta > 0$ there exists a positive d_δ such that for all $d \geq d_\delta$ it is true that

$$\frac{\sum_{j=1}^d \gamma_j}{\log d} < A + \delta.$$

Now we have

$$\begin{aligned}\prod_{j=1}^d \big(1 + 2\gamma_j\zeta(2\alpha)\big) &\leq \exp\left(2\zeta(2\alpha)\sum_{j=1}^d \gamma_j\right)\\ &= d^{2\zeta(2\alpha)(\sum_{j=1}^d \gamma_j)/\log d}\\ &\leq d^{2\zeta(2\alpha)(A+\delta)}\end{aligned}$$

for $d \geq d_\delta$. Therefore, there exists a positive number c_δ such that

$$\prod_{j=1}^d \big(1 + 2\gamma_j\zeta(2\alpha)\big) \leq c_\delta\, d^{2\zeta(2\alpha)(A+\delta)} \quad \text{for all } d \in \mathbb{N}. \tag{2.33}$$

This implies that Condition (2.32) is satisfied for every $q > 2A\zeta(2\alpha)$, and this in turn implies polynomial tractability.

Assume now that we have polynomial tractability. Then there exists a $q > 0$ such that (2.32) is satisfied. Since we have polynomial tractability, we also have $\lim_{j\to\infty} \gamma_j = 0$, and in particular $\sup_j \gamma_j < \infty$. Note that there exists a $c > 0$ such that

$$\log(1 + x) \geq cx \quad \text{for all } x \in [0, 2\zeta(2\alpha)\sup_j \gamma_j].$$

Thus we find that

$$\log\left(\prod_{j=1}^d (1 + 2\gamma_j\zeta(2\alpha))\right) = \sum_{j=1}^d \log(1 + 2\gamma_j\zeta(2\alpha)) \geq 2c\zeta(2\alpha)\sum_{j=1}^d \gamma_j$$

and therefore

$$\prod_{j=1}^d (1 + 2\gamma_j\zeta(2\alpha)) \geq \exp\left(2c\zeta(2\alpha)\sum_{j=1}^d \gamma_j\right) = d^{(2c\zeta(2\alpha)/(\log d))\sum_{j=1}^d \gamma_j}.$$

Consequently, with the existence of the number $q > 0$ from above, we obtain

$$\begin{aligned} d^{(2c\zeta(2\alpha)/(\log d))\sum_{j=1}^d \gamma_j - q} &\leq \frac{1}{d^q}\prod_{j=1}^d (1 + 2\gamma_j\zeta(2\alpha))\\ &\leq \sup_{d\in\mathbb{N}} \frac{1}{d^q}\prod_{j=1}^d (1 + 2\gamma_j\zeta(2\alpha)) < \infty.\end{aligned}$$

This implies that, for sufficiently large d, we must have

$$\frac{\sum_{j=1}^{d} \gamma_j}{\log d} < \frac{q}{2c\zeta(2\alpha)},$$

and hence

$$\limsup_{d\to\infty} \frac{\sum_{j=1}^{d} \gamma_j}{\log d} < \infty.$$

Finally, we show Item 4. For product weights Condition (2.27) in Corollary 2.27 can be reformulated as

$$\lim_{d\to\infty} \frac{1}{d} \sum_{j=1}^{d} \log\left(1 + 2\zeta(2\alpha)\gamma_j\right) = 0.$$

This, in turn, is equivalent to

$$\lim_{d\to\infty} \frac{\sum_{j=1}^{d} \gamma_j}{d} = 0,$$

which follows easily from the fact that $\lim_{x\to 0} \log(1+x)/x = 1$ and thus $x/2 \le \log(1+x) \le x$ for x close enough to 0. □

Notes and Remarks

Korobov spaces can be seen as L_2-versions of the function class $\mathcal{E}_\alpha^d(C)$ from Definition 1.13. The use of the smoothness parameter α is not unified in the literature. Some authors, e.g., in [240], describe the smoothness by a parameter that differs from α, as used here and in many other texts, by a factor of two. The reason why we use 2α in the definition of the functions denoted by $r_{2\alpha}$ and $r_{2\alpha,\gamma}$ is that in this way the expression for the Korobov norm in terms of derivatives in Propositions 2.2, 2.4, and 2.17 becomes simpler. In principle (weighted) Korobov spaces can also be introduced for $\alpha \in [0, 1/2]$ as the space of all functions with finite norm, with the same norm and inner product as introduced in Section 2.4 (see, for example, [143] and the "Notes and Remarks" Section at the end of Chapter 11). Then, for $\alpha = 0$ the Korobov space is the same as the L_2-space of square integrable functions, i.e., $\mathcal{H}_{\mathrm{kor},d,\alpha} = L_2([0,1]^d)$. For $\alpha > 0$, the Korobov space is a proper subspace of $L_2([0,1]^d)$. However, for $\alpha \in [0, 1/2]$ the (weighted) Korobov spaces are no reproducing kernel Hilbert spaces anymore. In the literature, Korobov spaces are also referred to as *periodic Sobolev spaces of dominating mixed smoothness* or *functions with bounded mixed derivatives*. A detailed introduction to the concept of Korobov spaces can be found in [210, Appendix A.1].

The proof of Bakhvalov's lower bound in Theorem 2.10 follows the exposition of Temlyakov in [250, Lemma 3.1].

In Section 2.3 we have presented a proof that QMC integration in the unweighted Korobov space suffers from the curse of dimensionality, and this negative result holds true for any algorithm. The general proof uses a suitably constructed Sobolev space for which numerical integration is not harder than in the unweighted Korobov space; this construction is due to Hickernell and Woźniakowski [106]. Then, one can show with the method of decomposable kernels that integration in this Sobolev space suffers from the curse of dimensionality and hence the same holds true for the unweighted Korobov space. For details we refer to [106] and to [211, Section 16.8].

Also in the weighted case we have restricted ourselves to QMC rules when considering necessary conditions for tractability of integration in weighted Korobov spaces. However, as shown by Hickernell and Woźniakowski in [106], the same results again hold true for arbitrary algorithms. Proofs of these results for (strong) polynomial tractability are given in [106] and in [211, Theorem 16.16]. A proof of the necessary condition for weak tractability can be found in [72].

Chapter 3
Constructions of Lattice Rules

As outlined in Chapter 2, we have results based on averaging arguments which guarantee the existence of lattice rules with excellent behavior regarding a given quality criterion.

For dimension $d = 1$, a good lattice point set consisting of N points is given by the set

$$\left\{0, \frac{1}{N}, \frac{2}{N}, \ldots, \frac{N-1}{N}\right\}.$$

For $d = 2$ one can use a relation to Diophantine approximation to find explicit constructions, see Sections 1.8 and 5.1. Particularly prominent examples of two-dimensional lattice rules are Fibonacci lattice rules as outlined in Example 1.54.

Though Fibonacci rules are elegant examples of good lattice rules in dimension $d = 2$, there are no explicit constructions known of good lattice rules with respect to any of the common figures of merit for $d > 2$. Consequently, it is necessary to resort to computer search algorithms, which shall be discussed in the present chapter.

3.1 Exhaustive Search for Generating Vectors

Recall that (rank-1) lattice point sets $\mathcal{P}(\boldsymbol{g}, N)$ are, for fixed N and d, fully characterized by their generating vector $\boldsymbol{g} \in \mathbb{Z}^d$. Since the elements of $\mathcal{P}(\boldsymbol{g}, N)$ are of the form $\boldsymbol{x}_k = \{(k/N)\boldsymbol{g}\}$, it is clear that two generating vectors $\boldsymbol{g}, \boldsymbol{g}' \in \mathbb{Z}^d$ with $\boldsymbol{g} \equiv \boldsymbol{g}' \pmod{N}$, where the congruence is assumed to hold component-wise, yield the same lattice point set. Therefore it suffices to search for lattice points $\boldsymbol{g}$ whose components belong to a complete residue system modulo N, and thus in most cases we will restrict the search to the set

$$G_d(N) := \{0, 1, \ldots, N-1\}^d,$$

which consists of N^d elements.

J. Dick et al., *Lattice Rules*, Springer Series in Computational Mathematics 58,
https://doi.org/10.1007/978-3-031-09951-9_3

A further restriction that is usually made is that every projection of a lattice point set has N distinct values. This is equivalent to requiring that the components of the generating vector $\boldsymbol{g}$ are all coprime with N, i.e., $\boldsymbol{g}$ is chosen from the set

$$G_d^{\varphi}(N) := \{g \in \{1, 2, \ldots, N\} \ : \ \gcd(g, N) = 1\}^d . \tag{3.1}$$

We obviously have $|G_d^{\varphi}(N)| = (\varphi(N))^d$, where φ denotes Euler's totient function. It is known that

$$\liminf_{N \to \infty} \frac{\varphi(N) \log \log N}{N} = \mathrm{e}^{-\gamma_{\mathrm{EM}}},$$

where $\gamma_{\mathrm{EM}} = 0.57721\ldots$ is the Euler–Mascheroni constant, so $\varphi(N)$ can be said to practically grow like N (see, for example, [93, Theorem 328]).

For the sake of simplicity, we will often restrict ourselves to prime numbers N in the following sections. In this case we have $G_d^{\varphi}(N) = (G_1(N) \setminus \{0\})^d$ and $\varphi(N) = N - 1$. Hence, for fixed d and prime N, an—at the first glance—obvious approach to identifying a generating vector that yields a lattice rule with good properties would be to perform an exhaustive search, going through all possible candidates in $G_d(N)$ or $G_d^{\varphi}(N)$ with N^d or $(\varphi(N))^d$ elements, respectively. It is obvious, however, that a full search is infeasible even for moderate values of N and d, which is particularly relevant for modern applications where N and d may be very large.

There have been attempts to reduce the size of the search space for good lattice points for higher rank lattice rules, as it was, e.g., outlined in [236, 237] (see also [230]) for rank-2 lattice rules with small values of the quality criterion P_α (see Definition 1.14). Restricting one's attention to relatively coprime invariants m and n allows for writing a rank-2 lattice rule in a very symmetric three-sum form. This representation makes it easier to eliminate geometrically equivalent rules from the search. Since P_α is invariant for geometrically equivalent rules it suffices to calculate P_α only for one element of each geometrically equivalent family. This may shrink the search space since there may be a huge number of rules which are distinct but geometrically equivalent. However, in general this method is still not satisfying for practically relevant numbers N and d, which is why an exhaustive search is not feasible in most cases.

The nowadays most common and most effective search algorithms for generating vectors of good lattice rules are designed for rank-1 lattice rules, on which we will focus in the following.

3.2 Korobov Type Generating Vectors

One way to reduce the size of the search space for good generating vectors is a method that goes back to Korobov [139]. Let us again assume that N is prime. Korobov suggested considering generating vectors of a special form, depending on only one parameter $g \in G_1(N) = \{0, 1, \ldots, N-1\}$.

Definition 3.1 (Korobov type lattice point) For $g \in G_1(N)$ the lattice point

$$\boldsymbol{g}_d(g) = (1, g, g^2, \ldots, g^{d-1})$$

in $\mathbb{Z}^d$ is called a *Korobov type lattice point* or *Korobov type generating vector*. The corresponding rank-1 lattice rule is called a *Korobov type lattice rule*.

In searching for good Korobov type lattice points, this restriction reduces the size of the search space from N^d to N, since for given N one only has to search for g in the set $G_1(N)$. The limitation to Korobov type lattice points is in many cases justified by averaging results that are basically of the same quality as the averaging results over all lattice points in $G_d(N)$.

Let us, as an example, consider the use of lattice rules generated by Korobov type lattice points $\boldsymbol{g}_d(g)$ for integration in the weighted Korobov space. As in Section 2.5, let us denote the worst-case error of a lattice rule using $\mathcal{P}(\boldsymbol{g}, N)$ in the Korobov space $\mathcal{H}_{\mathrm{kor},d,\alpha,\boldsymbol{\gamma}}$ by $\mathrm{err}_{N,d,\alpha,\boldsymbol{\gamma}}(\boldsymbol{g})$. We can formulate the following algorithm to identify a candidate for a good Korobov type generating vector.

Algorithm 3.2 (Finding a Korobov type generating vector) Let N be a prime number and let $d \in \mathbb{N}$. The optimal Korobov type generating vector is found by minimizing $[\mathrm{err}_{N,d,\alpha,\boldsymbol{\gamma}}(\boldsymbol{g}_d(g))]^2$ with respect to $g \in G_1(N)$.

Recall from Remark 2.9 that for $\alpha \in \mathbb{N}$ the worst-case error in the Korobov space $\mathcal{H}_{\mathrm{kor},d,\alpha,\boldsymbol{\gamma}}$ can be computed in $O(d\,N)$ elementary operations. In Algorithm 3.2 the worst-case error has to be computed N times. Hence, the number of operations needed to find the optimal Korobov type lattice rule for $\alpha \in \mathbb{N}$ and a dimension d is $O(d\,N^2)$. This is a remarkable improvement compared to the order $O(N^d)$ for a full search. Surprisingly, the lattice rule obtained by using Algorithm 3.2 is still of good quality. The following result for product weights was first shown in [259].

Theorem 3.3 *Let N be a prime number, let $d \in \mathbb{N}$, and let $\boldsymbol{\gamma} = (\gamma_j)_{j\ge 1}$ be product weights. Assume that $\boldsymbol{g}_d(g_*)$ has been found by Algorithm 3.2. Then we have, for arbitrary $\tau \in [1/2, \alpha)$,*

$$\mathrm{err}_{N,d,\alpha,\boldsymbol{\gamma}}(\boldsymbol{g}_d(g_*)) \le \frac{d^\tau}{N^\tau} \prod_{j=1}^{d} \left(1 + 2\gamma_j^{1/(2\tau)} \zeta\left(\frac{\alpha}{\tau}\right)\right)^\tau. \tag{3.2}$$

Proof According to Corollary 2.21 we have

$$\begin{aligned}\frac{1}{N}\sum_{g=0}^{N-1} [\mathrm{err}_{N,d,\alpha,\boldsymbol{\gamma}}(\boldsymbol{g}_d(g))]^2 &= \frac{1}{N}\sum_{g=0}^{N-1} \sum_{\substack{\boldsymbol{h}\in\mathbb{Z}^d\setminus\{\boldsymbol{0}\}\\ \boldsymbol{h}\cdot\boldsymbol{g}_d(g)\equiv 0 \pmod{N}}} \frac{1}{r_{2\alpha,\boldsymbol{\gamma}}(\boldsymbol{h})}\\ &= \frac{1}{N}\sum_{\boldsymbol{h}\in\mathbb{Z}^d\setminus\{\boldsymbol{0}\}} \frac{1}{r_{2\alpha,\boldsymbol{\gamma}}(\boldsymbol{h})} \sum_{\substack{g=0\\ \boldsymbol{h}\cdot\boldsymbol{g}_d(g)\equiv 0 \pmod{N}}}^{N-1} 1\end{aligned}$$

$$= \frac{1}{N} \sum_{\substack{\boldsymbol{h} \in \mathbb{Z}^d \setminus \{\boldsymbol{0}\} \\ \boldsymbol{h} \equiv \boldsymbol{0} \pmod{N}}} \frac{1}{r_{2\alpha,\boldsymbol{\gamma}}(\boldsymbol{h})} \sum_{\substack{g=0 \\ \boldsymbol{h} \cdot \boldsymbol{g}_d(g) \equiv 0 \pmod{N}}}^{N-1} 1$$

$$+ \frac{1}{N} \sum_{\substack{\boldsymbol{h} \in \mathbb{Z}^d \setminus \{\boldsymbol{0}\} \\ \boldsymbol{h} \not\equiv \boldsymbol{0} \pmod{N}}} \frac{1}{r_{2\alpha,\boldsymbol{\gamma}}(\boldsymbol{h})} \sum_{\substack{g=0 \\ \boldsymbol{h} \cdot \boldsymbol{g}_d(g) \equiv 0 \pmod{N}}}^{N-1} 1,$$

where $\boldsymbol{h} \equiv \boldsymbol{0} \pmod{N}$ indicates summation over all $\boldsymbol{h} \neq \boldsymbol{0}$ that are multiples of N and $\boldsymbol{h} \not\equiv \boldsymbol{0} \pmod{N}$ indicates summation over all $\boldsymbol{h} \neq \boldsymbol{0}$ that have at least one component that is not a multiple of N.

For given $\boldsymbol{h} = (h_1, \ldots, h_d) \in \mathbb{Z}^d$ the innermost sums in the two summands of the latter expression count the number of solutions $g \in G_1(N)$ of the polynomial congruence

$$h_1 + h_2 g + \cdots + h_d g^{d-1} \equiv 0 \pmod{N}. \tag{3.3}$$

If $\boldsymbol{h} \equiv \boldsymbol{0} \pmod{N}$, then $\boldsymbol{h} \cdot \boldsymbol{g}_d(g) \equiv 0 \pmod{N}$ is satisfied for all $g \in G_1(N)$, and hence we have

$$\begin{aligned} \frac{1}{N} \sum_{\substack{\boldsymbol{h} \in \mathbb{Z}^d \setminus \{\boldsymbol{0}\} \\ \boldsymbol{h} \equiv \boldsymbol{0} \pmod{N}}} \frac{1}{r_{2\alpha,\boldsymbol{\gamma}}(\boldsymbol{h})} \sum_{\substack{g=0 \\ \boldsymbol{h} \cdot \boldsymbol{g}_d(g) \equiv 0 \pmod{N}}}^{N-1} 1 &= \sum_{\substack{\boldsymbol{h} \in \mathbb{Z}^d \setminus \{\boldsymbol{0}\} \\ \boldsymbol{h} \equiv \boldsymbol{0} \pmod{N}}} \frac{1}{r_{2\alpha,\boldsymbol{\gamma}}(\boldsymbol{h})} \\ &= \sum_{\boldsymbol{h} \in \mathbb{Z}^d \setminus \{\boldsymbol{0}\}} \frac{1}{r_{2\alpha,\boldsymbol{\gamma}}(N\boldsymbol{h})} \\ &\le \frac{1}{N} \sum_{\boldsymbol{h} \in \mathbb{Z}^d \setminus \{\boldsymbol{0}\}} \frac{1}{r_{2\alpha,\boldsymbol{\gamma}}(\boldsymbol{h})}, \end{aligned}$$

where in the last estimate we used that at least one component of $\boldsymbol{h}$ is different from zero, and so the term N appears at least with multiplicity one in $r_{2\alpha,\boldsymbol{\gamma}}(N\boldsymbol{h})$.

If $\boldsymbol{h} \not\equiv \boldsymbol{0} \pmod{N}$, the number of solutions $g \in G_1(N)$ of the polynomial congruence (3.3) is at most $d-1$, since any polynomial of degree $d-1$ can have at most $d-1$ distinct roots. Thus,

$$\frac{1}{N} \sum_{\substack{\boldsymbol{h} \in \mathbb{Z}^d \setminus \{\boldsymbol{0}\} \\ \boldsymbol{h} \not\equiv \boldsymbol{0} \pmod{N}}} \frac{1}{r_{2\alpha,\boldsymbol{\gamma}}(\boldsymbol{h})} \sum_{\substack{g=0 \\ \boldsymbol{h} \cdot \boldsymbol{g}_d(g) \equiv 0 \pmod{N}}}^{N-1} 1 \le \frac{d-1}{N} \sum_{\boldsymbol{h} \in \mathbb{Z}^d \setminus \{\boldsymbol{0}\}} \frac{1}{r_{2\alpha,\boldsymbol{\gamma}}(\boldsymbol{h})}.$$

Altogether we obtain

$$\frac{1}{N} \sum_{g=0}^{N-1} [\mathrm{err}_{N,d,\alpha,\boldsymbol{\gamma}}(\boldsymbol{g}_d(g))]^2 \le \frac{d}{N} \sum_{\boldsymbol{h} \in \mathbb{Z}^d \setminus \{\boldsymbol{0}\}} \frac{1}{r_{2\alpha,\boldsymbol{\gamma}}(\boldsymbol{h})}$$

$$= \frac{d}{N}\left(-1+\prod_{j=1}^{d}\left(\sum_{h\in\mathbb{Z}}\frac{1}{r_{2\alpha,\gamma_j}(h)}\right)\right)$$

$$= \frac{d}{N}\left(-1+\prod_{j=1}^{d}\left(1+2\gamma_j\sum_{h=1}^{\infty}\frac{1}{h^{2\alpha}}\right)\right)$$

$$\le \frac{d}{N}\prod_{j=1}^{d}\left(1+2\gamma_j\zeta(2\alpha)\right).$$

Since there must be a g that is at least as good as average (see the standard averaging argument in Remark 2.14), we therefore obtain the existence of a $\widetilde{g} \in G_1(N)$ such that

$$[\mathrm{err}_{N,d,\alpha,\boldsymbol{\gamma}}(\boldsymbol{g}_d(\widetilde{g}))]^2 \le \frac{d}{N}\prod_{j=1}^{d}\left(1+2\gamma_j\zeta(2\alpha)\right). \tag{3.4}$$

Next, we apply this existence result to the worst-case error in $\mathcal{H}_{\mathrm{kor},d,\alpha\lambda,\boldsymbol{\gamma}^\lambda}$ for $\lambda \in (1/(2\alpha), 1]$. This yields the existence of a $\widetilde{g} \in G_1(N)$ satisfying

$$[\mathrm{err}_{N,d,\alpha\lambda,\boldsymbol{\gamma}^\lambda}(\boldsymbol{g}_d(\widetilde{g}))]^2 \le \frac{d}{N}\prod_{j=1}^{d}\left(1+2\gamma_j^\lambda\zeta(2\alpha\lambda)\right).$$

Then, using (2.26) we see that there exists a $\widetilde{g} \in G_1(N)$ such that

$$[\mathrm{err}_{N,d,\alpha,\boldsymbol{\gamma}}(\boldsymbol{g}_d(\widetilde{g}))]^{2\lambda} \le \frac{d}{N}\prod_{j=1}^{d}\left(1+2\gamma_j^\lambda\zeta(2\alpha\lambda)\right).$$

Note that $\widetilde{g}$ may depend on λ here. Putting $\tau = 1/(2\lambda)$, such that $\tau \in [1/2, \alpha)$, we therefore obtain the existence of $\widetilde{g}_\tau \in G_1(N)$ such that

$$\mathrm{err}_{N,d,\alpha,\boldsymbol{\gamma}}(\boldsymbol{g}_d(\widetilde{g}_\tau)) \le \left(\frac{d}{N}\right)^\tau \prod_{j=1}^{d}\left(1+2\gamma_j^{1/(2\tau)}\zeta\left(\frac{\alpha}{\tau}\right)\right)^\tau. \tag{3.5}$$

For the optimal Korobov type generating vector $\boldsymbol{g}(g_*)$ we have

$$\mathrm{err}_{N,d,\alpha,\boldsymbol{\gamma}}(\boldsymbol{g}_d(g_*)) \le \mathrm{err}_{N,d,\alpha,\boldsymbol{\gamma}}(\boldsymbol{g}_d(\widetilde{g}_\tau))$$

for all $\tau \in [1/2, \alpha)$ and so the result follows. □

Remark 3.4 As pointed out in [259], the cost of Algorithm 3.2 can be reduced if one modifies the search to be applicable to composite N. In particular, one can obtain a construction cost of order arbitrarily close to $O(d\,N)$ if one chooses N

as the product of sufficiently many distinct primes. However, there is a trade-off in that the error convergence and the dependence on the dimension d will change in a disadvantageous way if one improves on the cost of the algorithm like this.

Even though restricting oneself to considering only Korobov type generating vectors drastically reduces the size of the search space for good generating vectors, there is one drawback that should be highlighted. Considering the bound in Equation (3.2), we see that there is polynomial dependence on d that cannot be avoided even if the weights γ_j decay fast. This currently renders it impossible to obtain strong polynomial tractability results by using Korobov type generating vectors, which may be a considerable disadvantage for very high-dimensional problems. It may be suspected that this phenomenon is inherent to the structure of Korobov type generating vectors.

3.3 Component-By-Component Constructions

The nowadays most widely used variants of computer search algorithms for lattice rule generating vectors are *component-by-component*, or for short *CBC*, constructions. CBC algorithms are greedy algorithms, where the components g_j of a generating vector $\boldsymbol{g}$ are chosen one at a time. Already as early as 1963, Korobov [140] suggested such algorithms to search for good generating vectors of rank-1 lattice rules. However, this method fell into oblivion and it was only in 2002 that the component-by-component construction was reinvented by Sloan and Reztsov [235]. At this time, this result was considered a big surprise. Quoting Sloan and Reztsov [235, p. 263], *"The results may be thought surprising, since it is generally accepted that knowledge of a good lattice rule in s dimensions does not help in finding a good rule in s + 1 dimensions."*

The general principle

We now state the component-by-component construction principle for a general error measure $E_{N,d}(\boldsymbol{g})$ for a lattice point set $\mathcal{P}(\boldsymbol{g}, N)$. Later, we will adapt this principle to various concrete figures of merit, such as the worst-case error in a Korobov space or the criterion R.

Algorithm 3.5 (CBC construction principle) Let N and d be given. Construct a generating vector $\boldsymbol{g} = (g_1, \ldots, g_d) \in G_d^{\varphi}(N)$, where $G_d^{\varphi}(N)$ is defined in (3.1), as follows.

(1) Choose a sufficiently good one-dimensional generator $g_1 \in G_1^{\varphi}(N)$.
(2) For s from 1 to $d-1$:
Assume that $g_1, \ldots, g_s \in G_1^{\varphi}(N)$ have already been found. Consider them as fixed and choose $g_{s+1} \in G_1^{\varphi}(N)$ to minimize $E_{N,s+1}(g_1, \ldots, g_s, g)$ as a function in g, i.e., as

$$g_{s+1} := \underset{g \in G_1^{\varphi}(N)}{\operatorname{argmin}} \; E_{N,s+1}((g_1, \ldots, g_s, g)).$$

If the error criterion has the same minimal value for several distinct elements in a step of the algorithm, it is—if not explicitly stated otherwise—allowed to choose any of them.

End for.

(3) Set $\boldsymbol{g} = (g_1, \ldots, g_d)$.

Note that the CBC construction as outlined in Algorithm 3.5 is extensible in the dimension, i.e., if one would like to add further components to the obtained vector later, this can be done by simply running the loop in the algorithm as often as required. Furthermore, note that the size of the search space for $\boldsymbol{g}$ is reduced to $d\,\varphi(N)$ by the CBC algorithm, since for each component $j \in [d]$ there are only $|G_1^{\varphi}(N)| = \varphi(N)$ possible candidates g_j. This is a huge advantage over a full search where the search space is of size $(\varphi(N))^d$. However, note that one also needs to take into account the cost of optimizing $E_{N,s+1}$ in each step for determining the overall cost of the CBC construction.

The CBC construction with respect to the worst-case error in the Korobov space

Let us now discuss the CBC construction principle in greater detail for the worst-case error in the weighted Korobov space $\mathcal{H}_{\mathrm{kor},d,\alpha,\boldsymbol{\gamma}}$. In this case, we try to minimize the worst-case error of a rank-1 lattice rule in each step of the algorithm, and the construction has the following form.

Algorithm 3.6 (CBC construction for the weighted Korobov space) Let N and d be given. Construct a generating vector $\boldsymbol{g} = (g_1, \ldots, g_d) \in G_d^{\varphi}(N)$ as follows.

(1) Choose $g_1 = 1$.

(2) For s from 1 to $d-1$:

Assume that $g_1, \ldots, g_s \in G_1^{\varphi}(N)$ have already been found. Choose $g_{s+1} \in G_1^{\varphi}(N)$ as

$$g_{s+1} := \underset{g \in G_1^{\varphi}(N)}{\operatorname{argmin}} \; \mathrm{err}_{N,s+1,\alpha,\boldsymbol{\gamma}}((g_1, \ldots, g_s, g)).$$

End for.

(3) Set $\boldsymbol{g} = (g_1, \ldots, g_d)$.

Algorithm 3.6 can be practically implemented, in particular for product weights $(\gamma_j)_{j \geq 1}$, as we have the closed error formula for the (squared) worst-case error of rank-1 lattice rules in weighted Korobov spaces, similarly to Remark 2.9,

$$[\mathrm{err}_{N,d,\alpha,\boldsymbol{\gamma}}(\boldsymbol{g})]^2 = -1 + \frac{1}{N} \sum_{k=0}^{N-1} \prod_{j=1}^{d} \left(1 + \gamma_j \sum_{h \in \mathbb{Z} \setminus \{0\}} \frac{\mathrm{e}^{2\pi \mathrm{i} h k g_j / N}}{|h|^{2\alpha}} \right). \tag{3.6}$$

Let us write, for real x,

$$\varphi_\alpha(x) := \sum_{h \in \mathbb{Z}\setminus\{0\}} \frac{\mathrm{e}^{2\pi \mathrm{i} h x}}{|h|^{2\alpha}}. \tag{3.7}$$

Note that for evaluating (3.6), it is sufficient to know the function values of φ_α for all arguments of the form k/N with $k \in \{0, 1, \ldots, N-1\}$. These N values of φ_α can be reasonably well approximated; in particular, they can be precomputed before running the actual CBC algorithm and stored in a look-up table of size N.

In the special case where α is a positive integer, we can express φ_α as the Bernoulli polynomial $B_{2\alpha}$ of degree 2α,

$$\varphi_\alpha\left(\frac{kg_j}{N}\right) = \varphi_\alpha\left(\left\{\frac{kg_j}{N}\right\}\right) = \frac{(-1)^{\alpha+1}(2\pi)^{2\alpha}}{(2\alpha)!} B_{2\alpha}\left(\left\{\frac{kg_j}{N}\right\}\right),$$

see again Remark 2.9.

Using these facts, we can analyze the construction cost of a straightforward implementation of Algorithm 3.6 as follows. Let us assume that the values of φ_α have been precomputed. Computing the value of $e_{N,s,\alpha,\gamma}((g_1, \ldots, g_s, g))$ has a cost of at most dN operations for a $g \in G_1^\varphi(N)$. This has to be done for each element of $G_1^\varphi(N)$ in each of the d steps of the algorithm, which gives a total cost of order $O(d^2 N\varphi(N))$.

This order can, with respect to the dependence on d, be easily reduced to a total cost of $O(d\, N\varphi(N))$ by storing the value of the product in each step. Indeed, we can write in step s,

$$[\mathrm{err}_{N,s+1,\alpha,\gamma}(\boldsymbol{g})]^2 = -1 + \frac{1}{N}\sum_{k=0}^{N-1}\left(1 + \gamma_{s+1}\varphi_\alpha\left(\frac{kg_{s+1}}{N}\right)\right)\prod_{j=1}^{s}\left(1 + \gamma_j\varphi_\alpha\left(\frac{kg_j}{N}\right)\right),$$

and put

$$\eta_s(k) := \prod_{j=1}^{s}\left(1 + \gamma_j\varphi_\alpha\left(\frac{kg_j}{N}\right)\right)$$

for $k \in \{0, 1, \ldots, N-1\}$. Note that the N values of $\eta_s(k)$ do not depend on g_{s+1}, so they can be stored at the expense of $O(N)$, which yields the desired reduction in the required operations. (Observe that the values $\eta_s(k)$, $k \in \{0, 1, \ldots, N-1\}$, have already been computed when searching for g_s and therefore computing them does not incur additional computation in the search for g_{s+1}.)

We shall discuss further significant reductions of this computation cost in Sections 3.4 and 4.1 below. Before doing so, let us first check that the lattice rules found using Algorithm 3.6 actually yield a low integration error. Indeed, we will show next that the error convergence rate of such lattice rules can be arbitrarily close to the optimal rate in the Korobov space.

Error analysis for prime N and product weights

In order to present the basic idea of the error analysis, we consider, as a first step, the easiest case where the weights are of product form and where N is a prime number. This result was first shown by Kuo [156, Theorem 4].

Theorem 3.7 *Let N be a prime number, let $d \in \mathbb{N}$, and let $\boldsymbol{\gamma} = (\gamma_j)_{j\geq 1}$ be product weights. Assume that $\boldsymbol{g} = (g_1, \ldots, g_d)$ has been found by Algorithm 3.6. Then for arbitrary $\tau \in [1/2, \alpha)$ and for any $s \in [d]$ we have*

$$\mathrm{err}_{N,s,\alpha,\boldsymbol{\gamma}}((g_1, \ldots, g_s)) \leq \frac{2^\tau}{N^\tau} \prod_{j=1}^{s} \left(1 + 2\gamma_j^{1/(2\tau)} \zeta\left(\frac{\alpha}{\tau}\right)\right)^\tau . \tag{3.8}$$

Remark 3.8 Theorem 3.7 is formulated for product weights. A corresponding result for general weights follows as a special case of Theorem 3.9. Moreover, Theorem 3.7 deals with the case of prime N, which is technically easier to handle than general values of N, and for this reason we illustrate this result and its proof first. Note that in the case of prime N the set $G_1^\varphi(N)$ equals $\{1, 2, \ldots, N-1\}$. For a similar result for composite N and general weights, see Theorem 3.9 below.

Proof of Theorem 3.7 The result is shown by induction on s.

For $s = 1$, we have, due to (3.6),

$$\begin{aligned}
[\mathrm{err}_{N,1,\alpha,\gamma_1}(1)]^2 &= -1 + \frac{1}{N}\sum_{k=0}^{N-1}\left(1 + \gamma_1 \sum_{h\in\mathbb{Z}\setminus\{0\}} \frac{\mathrm{e}^{2\pi \mathrm{i} hk/N}}{|h|^{2\alpha}}\right) \\
&= \frac{\gamma_1}{N}\sum_{k=0}^{N-1}\sum_{h\in\mathbb{Z}\setminus\{0\}} \frac{\mathrm{e}^{2\pi \mathrm{i} hk/N}}{|h|^{2\alpha}} \\
&= \frac{\gamma_1}{N}\sum_{h\in\mathbb{Z}\setminus\{0\}} \frac{1}{|h|^{2\alpha}} \sum_{k=0}^{N-1} \mathrm{e}^{2\pi \mathrm{i} hk/N}.
\end{aligned}$$

If $h \equiv 0 \pmod{N}$,

$$\sum_{k=0}^{N-1} \mathrm{e}^{2\pi \mathrm{i} hk/N} = \sum_{k=0}^{N-1} 1 = N.$$

If, however, $h \not\equiv 0 \pmod{N}$, then we obtain from the geometric sum formula that

$$\sum_{k=0}^{N-1} \mathrm{e}^{2\pi \mathrm{i} hk/N} = \sum_{k=0}^{N-1}\left(\mathrm{e}^{2\pi \mathrm{i} h/N}\right)^k = \frac{\mathrm{e}^{2\pi \mathrm{i} h} - 1}{\mathrm{e}^{2\pi \mathrm{i} h/N} - 1} = 0.$$

This yields

$$\begin{aligned}
[\mathrm{err}_{N,1,\alpha,\gamma_1}(1)]^2 &= \gamma_1 \sum_{\substack{h\in\mathbb{Z}\setminus\{0\}\\ h\equiv 0 \pmod N}} \frac{1}{|h|^{2\alpha}} \\
&= \gamma_1 \sum_{h\in\mathbb{Z}\setminus\{0\}} \frac{1}{|Nh|^{2\alpha}} \\
&= \frac{\gamma_1}{N^{2\alpha}} \sum_{h\in\mathbb{Z}\setminus\{0\}} \frac{1}{|h|^{2\alpha}} \\
&= \frac{\gamma_1}{N^{2\alpha}} 2\zeta(2\alpha). \qquad (3.9)
\end{aligned}$$

Note that $\alpha > 1/2$ and hence $\zeta(2\alpha) < \infty$. Using (2.26) we obtain from (3.9) for $\lambda \in (1/(2\alpha), 1]$ that

$$\begin{aligned}
[\mathrm{err}_{N,1,\alpha,\gamma_1}(1)]^{2\lambda} &\le [\mathrm{err}_{N,1,\alpha\lambda,\gamma_1^\lambda}(1)]^2 \\
&\le \frac{\gamma_1^\lambda}{N^{2\alpha\lambda}} 2\zeta(2\alpha\lambda) \le \frac{1}{N^{2\alpha\lambda}} \left(1 + 2\gamma_1^\lambda \zeta(2\alpha\lambda)\right).
\end{aligned}$$

Setting $\tau = 1/(2\lambda)$, which implies $\tau \in [1/2, \alpha)$, we therefore obtain

$$\mathrm{err}_{N,1,\alpha,\gamma_1}(1) \le \frac{1}{N^\alpha} \left(1 + 2\gamma_1^{1/(2\tau)} \zeta\left(\frac{\alpha}{\tau}\right)\right)^\tau,$$

and this yields the result for $s = 1$.

For the induction step, let $\boldsymbol{g}^{(s)} = (g_1, \ldots, g_s) \in G_s^\varphi(N)$ denote the s-dimensional generating vector that has been found in the first s steps of the algorithm, and suppose that the claimed error bound holds for the lattice rule generated by $\boldsymbol{g}^{(s)}$. In the following we will write, with some abuse of notation, $(\boldsymbol{g}^{(s)}, g)$ for the vector $(g_1, \ldots, g_s, g) \in G_{s+1}^\varphi(N)$.

We will now deduce that the error bound holds analogously by choosing the $(s+1)$-st component according to Algorithm 3.6. To this end, we start again from Equation (3.6),

$$\begin{aligned}
&[\mathrm{err}_{N,s+1,\alpha,\boldsymbol{\gamma}}((\boldsymbol{g}^{(s)}, g))]^2 \\
&= -1 + \frac{1}{N} \sum_{k=0}^{N-1} \left(\prod_{j=1}^{s} \left(1 + \gamma_j \sum_{h\in\mathbb{Z}\setminus\{0\}} \frac{\mathrm{e}^{2\pi\mathrm{i}hkg_j/N}}{|h|^{2\alpha}}\right)\right) \left(1 + \gamma_{s+1} \sum_{h\in\mathbb{Z}\setminus\{0\}} \frac{\mathrm{e}^{2\pi\mathrm{i}hkg/N}}{|h|^{2\alpha}}\right) \\
&= -1 + \frac{1}{N} \sum_{k=0}^{N-1} \prod_{j=1}^{s} \left(1 + \gamma_j \sum_{h\in\mathbb{Z}\setminus\{0\}} \frac{\mathrm{e}^{2\pi\mathrm{i}hkg_j/N}}{|h|^{2\alpha}}\right) \\
&\quad + \frac{\gamma_{s+1}}{N} \sum_{k=0}^{N-1} \left(\prod_{j=1}^{s} \left(1 + \gamma_j \sum_{h\in\mathbb{Z}\setminus\{0\}} \frac{\mathrm{e}^{2\pi\mathrm{i}hkg_j/N}}{|h|^{2\alpha}}\right)\right) \left(\sum_{h\in\mathbb{Z}\setminus\{0\}} \frac{\mathrm{e}^{2\pi\mathrm{i}hkg/N}}{|h|^{2\alpha}}\right)
\end{aligned}$$

$$= [\mathrm{err}_{N,s,\alpha,\boldsymbol{\gamma}}(\boldsymbol{g}^{(s)})]^2 + \theta_{N,s,\alpha,\boldsymbol{\gamma}}(\boldsymbol{g}^{(s)}, g),$$

where we write

$$\theta_{N,s,\alpha,\boldsymbol{\gamma}}(\boldsymbol{g}^{(s)}, g) := \frac{\gamma_{s+1}}{N} \sum_{k=0}^{N-1} \left(\prod_{j=1}^{s} \left(1 + \gamma_j \sum_{h \in \mathbb{Z} \setminus \{0\}} \frac{\mathrm{e}^{2\pi \mathrm{i} h k g_j / N}}{|h|^{2\alpha}} \right) \right) \left(\sum_{h \in \mathbb{Z} \setminus \{0\}} \frac{\mathrm{e}^{2\pi \mathrm{i} h k g / N}}{|h|^{2\alpha}} \right).$$

Note that $\theta_{N,s,\alpha,\boldsymbol{\gamma}}(\boldsymbol{g}^{(s)}, g)$ is the only term in $[\mathrm{err}_{N,s+1,\alpha,\boldsymbol{\gamma}}((\boldsymbol{g}^{(s)}, g))]^2$ that depends on g.

We now consider the average of $\theta_{N,s,\alpha,\boldsymbol{\gamma}}(\boldsymbol{g}^{(s)}, g)$ over all possible values of $g \in \{1, 2, \ldots, N-1\}$,

$$\begin{aligned}
\Theta_{N,s,\alpha,\boldsymbol{\gamma}}(\boldsymbol{g}^{(s)}) &:= \frac{1}{N-1} \sum_{g=1}^{N-1} \theta_{N,s,\alpha,\boldsymbol{\gamma}}(\boldsymbol{g}^{(s)}, g) \\
&= \frac{\gamma_{s+1}}{N} \sum_{k=0}^{N-1} \left(\prod_{j=1}^{s} \left(1 + \gamma_j \sum_{h \in \mathbb{Z} \setminus \{0\}} \frac{\mathrm{e}^{2\pi \mathrm{i} h k g_j / N}}{|h|^{2\alpha}} \right) \right) \\
&\qquad \times \frac{1}{N-1} \sum_{g=1}^{N-1} \sum_{h \in \mathbb{Z} \setminus \{0\}} \frac{\mathrm{e}^{2\pi \mathrm{i} h k g / N}}{|h|^{2\alpha}} \\
&\le \frac{\gamma_{s+1}}{N} \sum_{k=0}^{N-1} \left(\prod_{j=1}^{s} \left(1 + \gamma_j \sum_{h \in \mathbb{Z} \setminus \{0\}} \frac{1}{|h|^{2\alpha}} \right) \right) \\
&\qquad \times \left| \frac{1}{N-1} \sum_{g=1}^{N-1} \sum_{h \in \mathbb{Z} \setminus \{0\}} \frac{\mathrm{e}^{2\pi \mathrm{i} h k g / N}}{|h|^{2\alpha}} \right| \\
&= \frac{\gamma_{s+1}}{N} \left(\prod_{j=1}^{s} (1 + 2\gamma_j \zeta(2\alpha)) \right) \sum_{k=0}^{N-1} \left| \frac{1}{N-1} \sum_{g=1}^{N-1} \sum_{h \in \mathbb{Z} \setminus \{0\}} \frac{\mathrm{e}^{2\pi \mathrm{i} h k g / N}}{|h|^{2\alpha}} \right|.
\end{aligned}$$

For short, we use the notation

$$T_{N,\alpha}(k) := \sum_{g=1}^{N-1} \sum_{h \in \mathbb{Z} \setminus \{0\}} \frac{\mathrm{e}^{2\pi \mathrm{i} h k g / N}}{|h|^{2\alpha}},$$

and obtain

$$\Theta_{N,s,\alpha,\boldsymbol{\gamma}}(\boldsymbol{g}^{(s)}) \le \frac{\gamma_{s+1}}{N} \left(\prod_{j=1}^{s} (1 + 2\gamma_j \zeta(2\alpha)) \right) \frac{1}{N-1} \sum_{k=0}^{N-1} \left| T_{N,\alpha}(k) \right|.$$

If $k = 0$, we have

$$T_{N,\alpha}(0) = \sum_{g=1}^{N-1} \sum_{h \in \mathbb{Z}\setminus\{0\}} \frac{1}{|h|^{2\alpha}} = 2\zeta(2\alpha)(N-1).$$

On the other hand, if $k \in \{1, 2, \ldots, N-1\}$, we have

$$\begin{aligned} T_{N,\alpha}(k) &= \sum_{g=1}^{N-1} \sum_{h \in \mathbb{Z}\setminus\{0\}} \frac{\mathrm{e}^{2\pi \mathrm{i} hkg/N}}{|h|^{2\alpha}} \\ &= \sum_{h \in \mathbb{Z}\setminus\{0\}} \frac{1}{|h|^{2\alpha}} \sum_{g=1}^{N-1} \mathrm{e}^{2\pi \mathrm{i} hkg/N} \\ &= \sum_{\substack{h \in \mathbb{Z}\setminus\{0\} \\ h \equiv 0 \pmod{N}}} \frac{1}{|h|^{2\alpha}} \sum_{g=1}^{N-1} \mathrm{e}^{2\pi \mathrm{i} hkg/N} \\ &\quad + \sum_{\substack{h \in \mathbb{Z}\setminus\{0\} \\ h \not\equiv 0 \pmod{N}}} \frac{1}{|h|^{2\alpha}} \sum_{g=1}^{N-1} \mathrm{e}^{2\pi \mathrm{i} hkg/N}. \end{aligned}$$

We treat the two sums in the latter expression separately. First,

$$\begin{aligned} T_{N,\alpha}^{(1)}(k) &:= \sum_{\substack{h \in \mathbb{Z}\setminus\{0\} \\ h \equiv 0 \pmod{N}}} \frac{1}{|h|^{2\alpha}} \sum_{g=1}^{N-1} \mathrm{e}^{2\pi \mathrm{i} hkg/N} \\ &= \sum_{h \in \mathbb{Z}\setminus\{0\}} \frac{1}{|Nh|^{2\alpha}} \sum_{g=1}^{N-1} 1 \\ &= \frac{N-1}{N^{2\alpha}} 2\zeta(2\alpha). \end{aligned}$$

Moreover, using the fact that

$$\sum_{g=1}^{N-1} \mathrm{e}^{2\pi \mathrm{i} hkg/N} = -1 + \sum_{g=0}^{N-1} \mathrm{e}^{2\pi \mathrm{i} hkg/N} = -1$$

for $h \not\equiv 0 \pmod{N}$, where in the last step we used the geometric sum formula, we obtain

$$T_{N,\alpha}^{(2)}(k) := \sum_{\substack{h \in \mathbb{Z}\setminus\{0\} \\ h \not\equiv 0 \pmod{N}}} \frac{1}{|h|^{2\alpha}} \sum_{g=1}^{N-1} \mathrm{e}^{2\pi \mathrm{i} hkg/N}$$

$$
\begin{aligned}
&= - \sum_{\substack{h \in \mathbb{Z}\setminus\{0\} \\ h \not\equiv 0 \pmod N}} \frac{1}{|h|^{2\alpha}} \\
&= \sum_{\substack{h \in \mathbb{Z}\setminus\{0\} \\ h \equiv 0 \pmod N}} \frac{1}{|h|^{2\alpha}} - \sum_{h \in \mathbb{Z}\setminus\{0\}} \frac{1}{|h|^{2\alpha}} \\
&= \frac{2\zeta(2\alpha)}{N^{2\alpha}} - 2\zeta(2\alpha).
\end{aligned}
$$

Adding $T_{N,\alpha}^{(1)}(k)$ and $T_{N,\alpha}^{(2)}(k)$ thus yields

$$
T_{N,\alpha}(k) = \begin{cases} 2\zeta(2\alpha)(N-1) & \text{if } k = 0, \\ 2\zeta(2\alpha)\left(N^{1-2\alpha} - 1\right) & \text{otherwise.} \end{cases}
$$

Note that the right-hand side of the latter equation is the same for all $k \in \{1, 2, \ldots, N-1\}$. Consequently,

$$
\frac{1}{N-1} \sum_{k=0}^{N-1} \left|T_{N,\alpha}(k)\right| = 2\zeta(2\alpha) + 2\zeta(2\alpha)\left(1 - \frac{1}{N^{2\alpha-1}}\right) \le 4\zeta(2\alpha).
$$

This yields

$$
\Theta_{N,s,\alpha,\boldsymbol{\gamma}}(\boldsymbol{g}^{(s)}) \le \frac{\gamma_{s+1} 4\zeta(2\alpha)}{N} \prod_{j=1}^{s} (1 + 2\gamma_j \zeta(2\alpha)),
$$

and hence, by the standard averaging argument (see Remark 2.14), there must exist a $\widetilde{g}_{s+1} \in \{1, 2, \ldots, N-1\}$, which may depend on $\alpha, \boldsymbol{\gamma}$, and $\boldsymbol{g}^{(s)}$, such that

$$
\theta_{N,s,\alpha,\boldsymbol{\gamma}}(\boldsymbol{g}^{(s)}, \widetilde{g}_{s+1}) \le \frac{\gamma_{s+1} 4\zeta(2\alpha)}{N} \prod_{j=1}^{s} (1 + 2\gamma_j \zeta(2\alpha)). \tag{3.10}
$$

For the next step in the proof, we again use the function $r_{2\alpha,\gamma_j}$, as defined in (2.22), to rewrite $\theta_{N,s,\alpha,\boldsymbol{\gamma}}(\boldsymbol{g}^{(s)}, g)$, for $g \in \{1, 2, \ldots, N-1\}$, in the form

$$
\begin{aligned}
\theta_{N,s,\alpha,\boldsymbol{\gamma}}(\boldsymbol{g}^{(s)}, g) &= \frac{\gamma_{s+1}}{N} \sum_{k=0}^{N-1} \left(\prod_{j=1}^{s} \left(\sum_{h \in \mathbb{Z}} \frac{\mathrm{e}^{2\pi \mathrm{i} h k g_j / N}}{r_{2\alpha,\gamma_j}(h)} \right) \right) \left(\sum_{h \in \mathbb{Z}\setminus\{0\}} \frac{\mathrm{e}^{2\pi \mathrm{i} h k g / N}}{|h|^{2\alpha}} \right) \\
&= \frac{\gamma_{s+1}}{N} \sum_{k=0}^{N-1} \sum_{\substack{\boldsymbol{h} \in \mathbb{Z}^{s+1} \\ h_{s+1} \neq 0}} \frac{\mathrm{e}^{2\pi \mathrm{i} k \boldsymbol{h} \cdot (\boldsymbol{g}^{(s)}, g)/N}}{|h_{s+1}|^{2\alpha} \prod_{j=1}^{s} r_{2\alpha,\gamma_j}(h_j)}.
\end{aligned}
$$

Note that the latter expression is zero if $\boldsymbol{h} \cdot (\boldsymbol{g}^{(s)}, g) \not\equiv 0 \pmod N$, so

$$\theta_{N,s,\alpha,\gamma}(\boldsymbol{g}^{(s)},g) = \frac{\gamma_{s+1}}{N}\sum_{k=0}^{N-1}\sum_{\substack{\boldsymbol{h}\in\mathbb{Z}^{s+1}\\ h_{s+1}\neq 0\\ \boldsymbol{h}\cdot(\boldsymbol{g}^{(s)},g)\equiv 0 \pmod N}} \frac{1}{|h_{s+1}|^{2\alpha}\prod_{j=1}^{s} r_{2\alpha,\gamma_j}(h_j)}$$

$$= \sum_{\substack{\boldsymbol{h}\in\mathbb{Z}^{s+1}\\ h_{s+1}\neq 0\\ \boldsymbol{h}\cdot(\boldsymbol{g}^{(s)},g)\equiv 0 \pmod N}} \frac{\gamma_{s+1}}{|h_{s+1}|^{2\alpha}}\prod_{j=1}^{s}\frac{1}{r_{2\alpha,\gamma_j}(h_j)}. \tag{3.11}$$

Recall that for $\lambda\in(1/(2\alpha),1]$ we have $(r_{2\alpha,\gamma_j}(h_j))^{\lambda} = r_{2\alpha\lambda,\gamma_j^{\lambda}}(h_j)$. Thus, applying Jensen's inequality (see Lemma 2.25) to (3.11) yields

$$\theta_{N,s,\alpha,\gamma}(\boldsymbol{g}^{(s)},g) \le \left(\sum_{\substack{\boldsymbol{h}\in\mathbb{Z}^{s+1}\\ h_{s+1}\neq 0\\ \boldsymbol{h}\cdot(\boldsymbol{g}^{(s)},g)\equiv 0 \pmod N}} \frac{\gamma_{s+1}^{\lambda}}{|h_{s+1}|^{2\alpha\lambda}}\prod_{j=1}^{s}\frac{1}{r_{2\alpha\lambda,\gamma_j^{\lambda}}(h_j)}\right)^{1/\lambda}$$

$$= \left(\theta_{N,s,\alpha\lambda,\gamma^{\lambda}}(\boldsymbol{g}^{(s)},g)\right)^{1/\lambda}. \tag{3.12}$$

We can replace α by $\alpha\lambda$ and $\boldsymbol{\gamma}$ by $\boldsymbol{\gamma}^{\lambda}$ in (3.10), and hence deduce the existence of a $\widetilde{g}_{s+1}\in\{1,2,\ldots,N-1\}$, which may depend on $\alpha\lambda$, $\boldsymbol{\gamma}^{\lambda}$, and $\boldsymbol{g}^{(s)}$, such that

$$\theta_{N,s,\alpha\lambda,\gamma^{\lambda}}(\boldsymbol{g}^{(s)},\widetilde{g}_{s+1}) \le \frac{\gamma_{s+1}^{\lambda}4\zeta(2\alpha\lambda)}{N}\prod_{j=1}^{s}(1+2\gamma_j^{\lambda}\zeta(2\alpha\lambda)).$$

Then it follows from (3.12) that this $\widetilde{g}_{s+1}$ satisfies

$$\theta_{N,s,\alpha,\gamma}(\boldsymbol{g}^{(s)},\widetilde{g}_{s+1}) \le \left(\frac{\gamma_{s+1}^{\lambda}4\zeta(2\alpha\lambda)}{N}\right)^{1/\lambda}\prod_{j=1}^{s}(1+2\gamma_j^{\lambda}\zeta(2\alpha\lambda))^{1/\lambda}.$$

Setting again $\tau = 1/(2\lambda)$ this bound can be rewritten in the form

$$\theta_{N,s,\alpha,\gamma}(\boldsymbol{g}^{(s)},\widetilde{g}_{s+1}) \le \left(\frac{\gamma_{s+1}^{1/(2\tau)}4\zeta(\alpha/\tau)}{N}\right)^{2\tau}\prod_{j=1}^{s}\left(1+2\gamma_j^{1/(2\tau)}\zeta\left(\frac{\alpha}{\tau}\right)\right)^{2\tau},$$

and the same bound also applies to $\theta_{N,s,\alpha,\gamma}(\boldsymbol{g}^{(s)},g_{s+1})$, where g_{s+1} denotes a minimizer of $\theta_{N,s,\alpha,\gamma}(\boldsymbol{g}^{(s)},g)$ over $g\in G_1^{\varphi}(N)$ obtained in Algorithm 3.6. Note that g_{s+1} may depend on α, $\boldsymbol{\gamma}$ and $\boldsymbol{g}^{(s)}$, but is independent of τ.

Hence, for a minimizer g_{s+1} we obtain, using the induction assumption,

$$\begin{aligned}
&[\mathrm{err}_{N,s+1,\alpha,\boldsymbol{\gamma}}((\boldsymbol{g}^{(s)}, g_{s+1}))]^2 \\
&\leq [\mathrm{err}_{N,s,\alpha,\boldsymbol{\gamma}}(\boldsymbol{g}^{(s)})]^2 + \left(\frac{\gamma_{s+1}^{1/(2\tau)} 4\zeta(\alpha/\tau)}{N}\right)^{2\tau} \prod_{j=1}^{s} \left(1 + 2\gamma_j^{1/(2\tau)} \zeta\left(\frac{\alpha}{\tau}\right)\right)^{2\tau} \\
&\leq \frac{2^{2\tau}}{N^{2\tau}} \prod_{j=1}^{s} \left(1 + 2\gamma_j^{1/(2\tau)} \zeta\left(\frac{\alpha}{\tau}\right)\right)^{2\tau} \\
&\quad + \left(\frac{\gamma_{s+1}^{1/(2\tau)} 4\zeta(\alpha/\tau)}{N}\right)^{2\tau} \prod_{j=1}^{s} \left(1 + 2\gamma_j^{1/(2\tau)} \zeta\left(\frac{\alpha}{\tau}\right)\right)^{2\tau} \\
&= \frac{2^{2\tau}}{N^{2\tau}} \left(\prod_{j=1}^{s} \left(1 + 2\gamma_j^{1/(2\tau)} \zeta\left(\frac{\alpha}{\tau}\right)\right)^{2\tau}\right) \left(1 + 2^{2\tau} \gamma_{s+1} \left(\zeta\left(\frac{\alpha}{\tau}\right)\right)^{2\tau}\right) \\
&\leq \frac{2^{2\tau}}{N^{2\tau}} \prod_{j=1}^{s+1} \left(1 + 2\gamma_j^{1/(2\tau)} \zeta\left(\frac{\alpha}{\tau}\right)\right)^{2\tau}.
\end{aligned}$$

Taking the square root yields the desired result. □

Error analysis for the general case

In Theorem 3.7 and its proof we have shown the error bound for rank-1 lattice rules obtained via the CBC construction for the case of prime N and product weights. In the subsequent theorem, we show a corresponding bound for the most general case, namely arbitrary $N \geq 2$ and general weights.

Theorem 3.9 *Let $N \geq 2$ be an arbitrary integer, let $d \in \mathbb{N}$, and let $\boldsymbol{\gamma} = \{\gamma_{\mathfrak{u}}\}_{\mathfrak{u} \subseteq [d]}$ be general weights. Assume that $\boldsymbol{g} = (g_1, \ldots, g_d)$ has been found by Algorithm 3.6. Then for arbitrary $\tau \in [1/2, \alpha)$ and for any $s \in [d]$ we have*

$$\mathrm{err}_{N,s,\alpha,\boldsymbol{\gamma}}((g_1, \ldots, g_s)) \leq \left(\frac{1}{\varphi(N)} \sum_{\emptyset \neq \mathfrak{u} \subseteq [s]} \gamma_{\mathfrak{u}}^{1/(2\tau)} \left(2\zeta\left(\frac{\alpha}{\tau}\right)\right)^{|\mathfrak{u}|}\right)^{\tau}. \tag{3.13}$$

Proof The result is again shown by induction on s. For $s = 1$ and arbitrary $\lambda \in (1/(2\alpha), 1]$ we have by (2.26), Theorem 2.19, and since $g_1 = 1$,

$$\begin{aligned}
[\mathrm{err}_{N,1,\alpha,\gamma_{\{1\}}}(1)]^{2\lambda} &\leq [\mathrm{err}_{N,1,\alpha\lambda,\gamma_{\{1\}}^{\lambda}}(1)]^2 \\
&= \gamma_{\{1\}}^{\lambda} \sum_{\substack{h \in \mathbb{Z} \setminus \{0\} \\ h \equiv 0 \pmod{N}}} \frac{1}{|h|^{2\alpha\lambda}}
\end{aligned}$$

$$
\begin{aligned}
&= \gamma_{\{1\}}^{\lambda} \sum_{h \in \mathbb{Z} \setminus \{0\}} \frac{1}{|Nh|^{2\alpha\lambda}} \\
&= \gamma_{\{1\}}^{\lambda} \frac{2\zeta(2\alpha\lambda)}{N^{2\alpha\lambda}} \\
&\le \frac{\gamma_{\{1\}}^{\lambda} 2\zeta(2\alpha\lambda)}{(\varphi(N))^{2\alpha\lambda}}.
\end{aligned}
$$

Setting $\tau = 1/(2\lambda)$ and taking the root gives the result for $s = 1$.

In the induction step, let again $\boldsymbol{g}^{(s)} = (g_1, \ldots, g_s)$ be the s-dimensional generating vector that has been chosen in the first s steps of Algorithm 3.6, and suppose that (3.13) holds for the rank-1 lattice rule generated by $\boldsymbol{g}^{(s)}$. We will show that (3.13) holds analogously for the $(s+1)$-dimensional case by choosing the $(s+1)$-st component according to Algorithm 3.6. To this end, we consider

$$
\begin{aligned}
[\mathrm{err}_{N,s+1,\alpha,\boldsymbol{\gamma}}((\boldsymbol{g}^{(s)}, g))]^2 &= \sum_{\substack{\boldsymbol{h} \in \mathbb{Z}^{s+1} \setminus \{\boldsymbol{0}\} \\ \boldsymbol{h} \cdot (\boldsymbol{g}^{(s)}, g) \equiv 0 \pmod{N}}} \frac{1}{r_{2\alpha,\boldsymbol{\gamma}}(\boldsymbol{h})} \\
&= \sum_{\substack{\boldsymbol{h} \in \mathbb{Z}^{s} \setminus \{\boldsymbol{0}\} \\ \boldsymbol{h} \cdot \boldsymbol{g}^{(s)} \equiv 0 \pmod{N}}} \frac{1}{r_{2\alpha,\boldsymbol{\gamma}}(\boldsymbol{h})} + \sum_{\substack{\boldsymbol{h} \in \mathbb{Z}^{s+1} \setminus \{\boldsymbol{0}\} \\ h_{s+1} \neq 0 \\ \boldsymbol{h} \cdot (\boldsymbol{g}^{(s)}, g) \equiv 0 \pmod{N}}} \frac{1}{r_{2\alpha,\boldsymbol{\gamma}}(\boldsymbol{h})} \\
&= [\mathrm{err}_{N,s,\alpha,\boldsymbol{\gamma}}(\boldsymbol{g}^{(s)})]^2 + \theta_{N,s,\alpha,\boldsymbol{\gamma}}(\boldsymbol{g}^{(s)}, g),
\end{aligned}
$$

where we write

$$
\theta_{N,s,\alpha,\boldsymbol{\gamma}}(\boldsymbol{g}^{(s)}, g) = \sum_{\substack{\boldsymbol{h} \in \mathbb{Z}^{s+1} \setminus \{\boldsymbol{0}\} \\ h_{s+1} \neq 0 \\ \boldsymbol{h} \cdot (\boldsymbol{g}^{(s)}, g) \equiv 0 \pmod{N}}} \frac{1}{r_{2\alpha,\boldsymbol{\gamma}}(\boldsymbol{h})}. \tag{3.14}
$$

Note that $\theta_{N,s,\alpha,\boldsymbol{\gamma}}(\boldsymbol{g}^{(s)}, g)$ is the only term in $[\mathrm{err}_{N,s+1,\alpha,\boldsymbol{\gamma}}((\boldsymbol{g}^{(s)}, g))]^2$ that depends on g. We can rewrite $\theta_{N,s,\alpha,\boldsymbol{\gamma}}(\boldsymbol{g}^{(s)}, g)$ as

$$
\begin{aligned}
\theta_{N,s,\alpha,\boldsymbol{\gamma}}(\boldsymbol{g}^{(s)}, g) &= \sum_{\substack{\mathfrak{u} \subseteq [s+1] \\ s+1 \in \mathfrak{u}}} \sum_{\substack{\boldsymbol{h}_{\mathfrak{u}} \in (\mathbb{Z} \setminus \{0\})^{|\mathfrak{u}|} \\ \boldsymbol{h}_{\mathfrak{u}} \cdot \boldsymbol{g}_{\mathfrak{u}} \equiv 0 \pmod{N}}} \frac{1}{\prod_{j \in \mathfrak{u}} |h_j|^{2\alpha}} \\
&= \gamma_{s+1} \sum_{\substack{h_{s+1} \in \mathbb{Z} \setminus \{0\} \\ h_{s+1} g \equiv 0 \pmod{N}}} \frac{1}{|h_{s+1}|^{2\alpha}} \\
&\quad + \sum_{\substack{\mathfrak{u} \subseteq [s+1] \\ |\mathfrak{u}| \ge 2 \\ s+1 \in \mathfrak{u}}} \gamma_{\mathfrak{u}} \sum_{h_{s+1} \in \mathbb{Z} \setminus \{0\}} \frac{1}{|h_{s+1}|^{2\alpha}} \sum_{\substack{\boldsymbol{h}_{\mathfrak{u} \setminus \{s+1\}} \in (\mathbb{Z} \setminus \{0\})^{|\mathfrak{u}|-1} \\ \boldsymbol{h}_{\mathfrak{u} \setminus \{s+1\}} \cdot \boldsymbol{g}_{\mathfrak{u} \setminus \{s+1\}} \equiv -h_{s+1} g \pmod{N}}} \frac{1}{\prod_{j \in \mathfrak{u} \setminus \{s+1\}} |h_j|^{2\alpha}}.
\end{aligned}
$$

Note that, as $g \in G_1^{\varphi}(N)$, we have

$$\sum_{\substack{h_{s+1} \in \mathbb{Z}\setminus\{0\} \\ h_{s+1} g \equiv 0 \pmod N}} \frac{1}{|h_{s+1}|^{2\alpha}} = \sum_{\substack{h_{s+1} \in \mathbb{Z}\setminus\{0\} \\ h_{s+1} \equiv 0 \pmod N}} \frac{1}{|h_{s+1}|^{2\alpha}} = \frac{2\zeta(2\alpha)}{N^{2\alpha}},$$

and so

$$\begin{aligned}
&\theta_{N,s,\alpha,\boldsymbol{\gamma}}(\boldsymbol{g}^{(s)}, g) \\
&= \gamma_{s+1} \frac{2\zeta(2\alpha)}{N^{2\alpha}} \\
&\quad + \sum_{\substack{\mathfrak{u} \subseteq [s+1] \\ |\mathfrak{u}| \geq 2 \\ s+1 \in \mathfrak{u}}} \gamma_{\mathfrak{u}} \sum_{h_{s+1} \in \mathbb{Z}\setminus\{0\}} \frac{1}{|h_{s+1}|^{2\alpha}} \sum_{\substack{\boldsymbol{h}_{\mathfrak{u}\setminus\{s+1\}} \in (\mathbb{Z}\setminus\{0\})^{|\mathfrak{u}|-1} \\ \boldsymbol{h}_{\mathfrak{u}\setminus\{s+1\}} \cdot \boldsymbol{g}_{\mathfrak{u}\setminus\{s+1\}} \equiv -h_{s+1} g \pmod N}} \frac{1}{\prod_{j \in \mathfrak{u}\setminus\{s+1\}} |h_j|^{2\alpha}}.
\end{aligned}$$

Let now $\lambda \in (1/(2\alpha), 1]$. We consider the average of $(\theta_{N,s,\alpha,\boldsymbol{\gamma}}(\boldsymbol{g}^{(s)}, g))^{\lambda}$ over all possible values of $g \in G_1^{\varphi}(N)$,

$$\begin{aligned}
\Theta_{N,s,\alpha,\boldsymbol{\gamma},\lambda}(\boldsymbol{g}^{(s)}) &:= \frac{1}{\varphi(N)} \sum_{g \in G_1^{\varphi}(N)} (\theta_{N,s,\alpha,\boldsymbol{\gamma}}(\boldsymbol{g}^{(s)}, g))^{\lambda} \\
&\leq \gamma_{s+1}^{\lambda} \frac{2\zeta(2\alpha\lambda)}{N^{2\alpha\lambda}} \\
&\quad + \frac{1}{\varphi(N)} \sum_{g \in G_1^{\varphi}(N)} \sum_{\substack{\mathfrak{u} \subseteq [s+1] \\ |\mathfrak{u}| \geq 2 \\ s+1 \in \mathfrak{u}}} \gamma_{\mathfrak{u}}^{\lambda} \sum_{h_{s+1} \in \mathbb{Z}\setminus\{0\}} \frac{1}{|h_{s+1}|^{2\alpha\lambda}} \\
&\quad \times \sum_{\substack{\boldsymbol{h}_{\mathfrak{u}\setminus\{s+1\}} \in (\mathbb{Z}\setminus\{0\})^{|\mathfrak{u}|-1} \\ \boldsymbol{h}_{\mathfrak{u}\setminus\{s+1\}} \cdot \boldsymbol{g}_{\mathfrak{u}\setminus\{s+1\}} \equiv -h_{s+1} g \pmod N}} \frac{1}{\prod_{j \in \mathfrak{u}\setminus\{s+1\}} |h_j|^{2\alpha\lambda}},
\end{aligned}$$

where we again used Jensen's inequality (Lemma 2.25). Then,

$$\begin{aligned}
&\Theta_{N,s,\alpha,\boldsymbol{\gamma},\lambda}(\boldsymbol{g}^{(s)}) \\
&\leq \gamma_{s+1}^{\lambda} \frac{2\zeta(2\alpha\lambda)}{N^{2\alpha\lambda}} \\
&\quad + \frac{1}{\varphi(N)} \sum_{g \in G_1^{\varphi}(N)} \sum_{\substack{\mathfrak{u} \subseteq [s+1] \\ |\mathfrak{u}| \geq 2 \\ s+1 \in \mathfrak{u}}} \gamma_{\mathfrak{u}}^{\lambda} \sum_{\substack{h_{s+1} \in \mathbb{Z}\setminus\{0\} \\ h_{s+1} \equiv 0 \pmod N}} \frac{1}{|h_{s+1}|^{2\alpha\lambda}} \\
&\quad \times \sum_{\substack{\boldsymbol{h}_{\mathfrak{u}\setminus\{s+1\}} \in (\mathbb{Z}\setminus\{0\})^{|\mathfrak{u}|-1} \\ \boldsymbol{h}_{\mathfrak{u}\setminus\{s+1\}} \cdot \boldsymbol{g}_{\mathfrak{u}\setminus\{s+1\}} \equiv 0 \pmod N}} \frac{1}{\prod_{j \in \mathfrak{u}\setminus\{s+1\}} |h_j|^{2\alpha\lambda}}
\end{aligned}$$

$$
\begin{aligned}
&+\frac{1}{\varphi(N)}\sum_{g\in G_1^{\varphi}(N)}\sum_{\substack{\mathfrak{u}\subseteq[s+1]\\|\mathfrak{u}|\geq 2\\ s+1\in\mathfrak{u}}}\gamma_{\mathfrak{u}}^{\lambda}\sum_{\substack{h_{s+1}\in\mathbb{Z}\setminus\{0\}\\ h_{s+1}\not\equiv 0 \pmod N}}\frac{1}{|h_{s+1}|^{2\alpha\lambda}}\\
&\times\sum_{\substack{\boldsymbol{h}_{\mathfrak{u}\setminus\{s+1\}}\in(\mathbb{Z}\setminus\{0\})^{|\mathfrak{u}|-1}\\ \boldsymbol{h}_{\mathfrak{u}\setminus\{s+1\}}\cdot\boldsymbol{g}_{\mathfrak{u}\setminus\{s+1\}}\equiv -h_{s+1}g \pmod N}}\frac{1}{\prod_{j\in\mathfrak{u}\setminus\{s+1\}}|h_j|^{2\alpha\lambda}}\\
&=\gamma_{s+1}^{\lambda}\frac{2\zeta(2\alpha\lambda)}{N^{2\alpha\lambda}}\\
&+\sum_{\substack{\mathfrak{u}\subseteq[s+1]\\|\mathfrak{u}|\geq 2\\ s+1\in\mathfrak{u}}}\gamma_{\mathfrak{u}}^{\lambda}\sum_{h\in\mathbb{Z}\setminus\{0\}}\frac{1}{|Nh|^{2\alpha\lambda}}\sum_{\substack{\boldsymbol{h}_{\mathfrak{u}\setminus\{s+1\}}\in(\mathbb{Z}\setminus\{0\})^{|\mathfrak{u}|-1}\\ \boldsymbol{h}_{\mathfrak{u}\setminus\{s+1\}}\cdot\boldsymbol{g}_{\mathfrak{u}\setminus\{s+1\}}\equiv 0 \pmod N}}\frac{1}{\prod_{j\in\mathfrak{u}\setminus\{s+1\}}|h_j|^{2\alpha\lambda}}\\
&+\frac{1}{\varphi(N)}\sum_{g\in G_1^{\varphi}(N)}\sum_{\substack{\mathfrak{u}\subseteq[s+1]\\|\mathfrak{u}|\geq 2\\ s+1\in\mathfrak{u}}}\sum_{c=1}^{N-1}\gamma_{\mathfrak{u}}^{\lambda}\sum_{\substack{h_{s+1}\in\mathbb{Z}\setminus\{0\}\\ h_{s+1}\equiv -cg^{-1} \pmod N}}\frac{1}{|h_{s+1}|^{2\alpha\lambda}}\\
&\times\sum_{\substack{\boldsymbol{h}_{\mathfrak{u}\setminus\{s+1\}}\in(\mathbb{Z}\setminus\{0\})^{|\mathfrak{u}|-1}\\ \boldsymbol{h}_{\mathfrak{u}\setminus\{s+1\}}\cdot\boldsymbol{g}_{\mathfrak{u}\setminus\{s+1\}}\equiv c \pmod N}}\frac{1}{\prod_{j\in\mathfrak{u}\setminus\{s+1\}}|h_j|^{2\alpha\lambda}}\\
&=\gamma_{s+1}^{\lambda}\frac{2\zeta(2\alpha\lambda)}{N^{2\alpha\lambda}}\\
&+\sum_{\substack{\mathfrak{u}\subseteq[s+1]\\|\mathfrak{u}|\geq 2\\ s+1\in\mathfrak{u}}}\gamma_{\mathfrak{u}}^{\lambda}\frac{2\zeta(2\alpha\lambda)}{N^{2\alpha\lambda}}\sum_{\substack{\boldsymbol{h}_{\mathfrak{u}\setminus\{s+1\}}\in(\mathbb{Z}\setminus\{0\})^{|\mathfrak{u}|-1}\\ \boldsymbol{h}_{\mathfrak{u}\setminus\{s+1\}}\cdot\boldsymbol{g}_{\mathfrak{u}\setminus\{s+1\}}\equiv 0 \pmod N}}\frac{1}{\prod_{j\in\mathfrak{u}\setminus\{s+1\}}|h_j|^{2\alpha\lambda}}\\
&+\frac{1}{\varphi(N)}\sum_{g\in G_1^{\varphi}(N)}\sum_{\substack{\mathfrak{u}\subseteq[s+1]\\|\mathfrak{u}|\geq 2\\ s+1\in\mathfrak{u}}}\sum_{c=1}^{N-1}\gamma_{\mathfrak{u}}^{\lambda}\sum_{\substack{h_{s+1}\in\mathbb{Z}\setminus\{0\}\\ h_{s+1}\equiv -cg^{-1} \pmod N}}\frac{1}{|h_{s+1}|^{2\alpha\lambda}}\\
&\times\sum_{\substack{\boldsymbol{h}_{\mathfrak{u}\setminus\{s+1\}}\in(\mathbb{Z}\setminus\{0\})^{|\mathfrak{u}|-1}\\ \boldsymbol{h}_{\mathfrak{u}\setminus\{s+1\}}\cdot\boldsymbol{g}_{\mathfrak{u}\setminus\{s+1\}}\equiv c \pmod N}}\frac{1}{\prod_{j\in\mathfrak{u}\setminus\{s+1\}}|h_j|^{2\alpha\lambda}},
\end{aligned}
$$

where by g^{-1} we denote the multiplicative inverse of g in $G_1^{\varphi}(N)$.

For $c \in \{1, 2, \ldots, N-1\}$, let $z := \gcd(c, N)$. Then $\gcd(c/z, N/z) = 1$. Furthermore, note that

$$\{c\, g^{-1} \pmod N \;:\; g \in G_1^{\varphi}(N)\} = \{c\, g \pmod N \;:\; g \in G_1^{\varphi}(N)\}.$$

Hence,

$$
\begin{aligned}
\sum_{g\in G_1^{\varphi}(N)} & \sum_{\substack{h_{s+1}\in\mathbb{Z}\setminus\{0\}\\ h_{s+1}\equiv -cg^{-1} \pmod N}} \frac{1}{|h_{s+1}|^{2\alpha\lambda}} \\
&= \sum_{g\in G_1^{\varphi}(N)} \sum_{\substack{h_{s+1}\in\mathbb{Z}\setminus\{0\}\\ h_{s+1}\equiv -cg \pmod N}} \frac{1}{|h_{s+1}|^{2\alpha\lambda}} \\
&= \sum_{g\in G_1^{\varphi}(N)} \sum_{m\in\mathbb{Z}} \frac{1}{|mN-cg|^{2\alpha\lambda}} \\
&= \frac{1}{z^{2\alpha\lambda}} \sum_{g\in G_1^{\varphi}(N)} \sum_{m\in\mathbb{Z}} \frac{1}{|m(N/z)-(c/z)g|^{2\alpha\lambda}} \\
&= \frac{1}{z^{2\alpha\lambda}} \sum_{g\in G_1^{\varphi}(N)} \sum_{\substack{h\in\mathbb{Z}\setminus\{0\}\\ h\equiv -(c/z)g \pmod{N/z}}} \frac{1}{|h|^{2\alpha\lambda}} \\
&\le \frac{z}{z^{2\alpha\lambda}} \sum_{a=1}^{N/z-1} \sum_{\substack{h\in\mathbb{Z}\setminus\{0\}\\ h\equiv a \pmod{N/z}}} \frac{1}{|h|^{2\alpha\lambda}}.
\end{aligned}
$$

Note that

$$
\begin{aligned}
\sum_{a=1}^{N/z-1} \sum_{\substack{h\in\mathbb{Z}\setminus\{0\}\\ h\equiv a \pmod{N/z}}} \frac{1}{|h|^{2\alpha\lambda}} &= \sum_{a=0}^{N/z-1} \sum_{\substack{h\in\mathbb{Z}\setminus\{0\}\\ h\equiv a \pmod{N/z}}} \frac{1}{|h|^{2\alpha\lambda}} - \sum_{\substack{h\in\mathbb{Z}\setminus\{0\}\\ h\equiv 0 \pmod{N/z}}} \frac{1}{|h|^{2\alpha\lambda}} \\
&= \sum_{h\in\mathbb{Z}\setminus\{0\}} \frac{1}{|h|^{2\alpha\lambda}} - \sum_{h\in\mathbb{Z}\setminus\{0\}} \frac{1}{|hN/z|^{2\alpha\lambda}} \\
&= 2\zeta(2\alpha\lambda) - \left(\frac{z}{N}\right)^{2\alpha\lambda} 2\zeta(2\alpha\lambda).
\end{aligned}
$$

Consequently,

$$
\begin{aligned}
\sum_{g\in G_1^{\varphi}(N)} \sum_{\substack{h_{s+1}\in\mathbb{Z}\setminus\{0\}\\ h_{s+1}\equiv -cg^{-1} \pmod N}} \frac{1}{|h_{s+1}|^{2\alpha\lambda}} &\le \frac{z}{z^{2\alpha\lambda}} 2\zeta(2\alpha\lambda) \left(1 - \left(\frac{z}{N}\right)^{2\alpha\lambda}\right) \\
&\le 2\zeta(2\alpha\lambda),
\end{aligned}
$$

since $z \ge 1$ and $\lambda > 1/(2\alpha)$. So we have shown

$$
\begin{aligned}
&\Theta_{N,s,\alpha,\boldsymbol{\gamma},\lambda}(\boldsymbol{g}^{(s)}) \\
&\le \gamma_{s+1}^{\lambda}\frac{2\zeta(2\alpha\lambda)}{N^{2\alpha\lambda}} \\
&\quad + \sum_{\substack{\mathfrak{u}\subseteq[s+1]\\ |\mathfrak{u}|\ge 2\\ s+1\in\mathfrak{u}}} \gamma_{\mathfrak{u}}^{\lambda}\frac{2\zeta(2\alpha\lambda)}{N^{2\alpha\lambda}} \sum_{\substack{\boldsymbol{h}_{\mathfrak{u}\setminus\{s+1\}}\in(\mathbb{Z}\setminus\{0\})^{|\mathfrak{u}|-1}\\ \boldsymbol{h}_{\mathfrak{u}\setminus\{s+1\}}\cdot\boldsymbol{g}_{\mathfrak{u}\setminus\{s+1\}}\equiv 0 \pmod N}} \frac{1}{\prod_{j\in\mathfrak{u}\setminus\{s+1\}}|h_j|^{2\alpha\lambda}} \\
&\quad + \frac{1}{\varphi(N)}\sum_{\substack{\mathfrak{u}\subseteq[s+1]\\ |\mathfrak{u}|\ge 2\\ s+1\in\mathfrak{u}}} \gamma_{\mathfrak{u}}^{\lambda}\, 2\zeta(2\alpha\lambda) \sum_{\substack{\boldsymbol{h}_{\mathfrak{u}\setminus\{s+1\}}\in(\mathbb{Z}\setminus\{0\})^{|\mathfrak{u}|-1}\\ \boldsymbol{h}_{\mathfrak{u}\setminus\{s+1\}}\cdot\boldsymbol{g}_{\mathfrak{u}\setminus\{s+1\}}\not\equiv 0 \pmod N}} \frac{1}{\prod_{j\in\mathfrak{u}\setminus\{s+1\}}|h_j|^{2\alpha\lambda}} \\
&\le \gamma_{s+1}^{\lambda}\frac{2\zeta(2\alpha\lambda)}{\varphi(N)} \\
&\quad + \frac{1}{\varphi(N)}\sum_{\substack{\mathfrak{u}\subseteq[s+1]\\ |\mathfrak{u}|\ge 2\\ s+1\in\mathfrak{u}}} \gamma_{\mathfrak{u}}^{\lambda}\, 2\zeta(2\alpha\lambda) \sum_{\boldsymbol{h}_{\mathfrak{u}\setminus\{s+1\}}\in(\mathbb{Z}\setminus\{0\})^{|\mathfrak{u}|-1}} \frac{1}{\prod_{j\in\mathfrak{u}\setminus\{s+1\}}|h_j|^{2\alpha\lambda}} \\
&= \frac{1}{\varphi(N)}\sum_{\substack{\mathfrak{u}\subseteq[s+1]\\ s+1\in\mathfrak{u}}} \gamma_{\mathfrak{u}}^{\lambda}\,(2\zeta(2\alpha\lambda))^{|\mathfrak{u}|}.
\end{aligned} \tag{3.15}
$$

By construction, the g_{s+1} chosen by Algorithm 3.6 must satisfy

$$
(\theta_{N,s,\alpha,\boldsymbol{\gamma}}(\boldsymbol{g}^{(s)},g_{s+1}))^{\lambda} \le \Theta_{N,s,\alpha,\boldsymbol{\gamma},\lambda}(\boldsymbol{g}^{(s)}) \le \frac{1}{\varphi(N)}\sum_{\substack{\mathfrak{u}\subseteq[s+1]\\ s+1\in\mathfrak{u}}} \gamma_{\mathfrak{u}}^{\lambda}\,(2\zeta(2\alpha\lambda))^{|\mathfrak{u}|}. \tag{3.16}
$$

We then obtain

$$
\begin{aligned}
&[\mathrm{err}_{N,s+1,\alpha,\boldsymbol{\gamma}}((\boldsymbol{g}^{(s)},g_{s+1}))]^{2\lambda} \\
&\le \left([\mathrm{err}_{N,s,\alpha,\boldsymbol{\gamma}}(\boldsymbol{g}^{(s)})]^2 + \theta_{N,s,\alpha,\boldsymbol{\gamma}}(\boldsymbol{g}^{(s)},g_{s+1})\right)^{\lambda} \\
&\le [\mathrm{err}_{N,s,\alpha,\boldsymbol{\gamma}}(\boldsymbol{g}^{(s)})]^{2\lambda} + (\theta_{N,s,\alpha,\boldsymbol{\gamma}}(\boldsymbol{g}^{(s)},g_{s+1}))^{\lambda} \\
&\le [\mathrm{err}_{N,s,\alpha,\boldsymbol{\gamma}}(\boldsymbol{g}^{(s)})]^{2\lambda} + \frac{1}{\varphi(N)}\sum_{\substack{\mathfrak{u}\subseteq[s+1]\\ s+1\in\mathfrak{u}}} \gamma_{\mathfrak{u}}^{\lambda}\,(2\zeta(2\alpha\lambda))^{|\mathfrak{u}|},
\end{aligned} \tag{3.17}
$$

where we used Lemma (2.25) once again, and also (3.16). Setting $\tau = 1/(2\lambda)$ we get

$$
\begin{aligned}
&[\mathrm{err}_{N,s+1,\alpha,\boldsymbol{\gamma}}((\boldsymbol{g}^{(s)},g_{s+1}))]^{1/\tau} \\
&\le [\mathrm{err}_{N,s,\alpha,\boldsymbol{\gamma}}(\boldsymbol{g}^{(s)})]^{1/\tau} + \frac{1}{\varphi(N)}\sum_{\substack{\mathfrak{u}\subseteq[s+1]\\ s+1\in\mathfrak{u}}} \gamma_{\mathfrak{u}}^{1/(2\tau)}\left(2\zeta\left(\frac{\alpha}{\tau}\right)\right)^{|\mathfrak{u}|}.
\end{aligned}
$$

Using the induction assumption gives

$$\begin{aligned}
&[\mathrm{err}_{N,s+1,\alpha,\boldsymbol{\gamma}}((\boldsymbol{g}^{(s)},g_{s+1}))]^{1/\tau} \\
&\le \frac{1}{\varphi(N)}\sum_{\emptyset\neq\mathfrak{u}\subseteq[s]}\gamma_{\mathfrak{u}}^{1/(2\tau)}\left(2\zeta\left(\frac{\alpha}{\tau}\right)\right)^{|\mathfrak{u}|}+\frac{1}{\varphi(N)}\sum_{\substack{\mathfrak{u}\subseteq[s+1]\\ s+1\in\mathfrak{u}}}\gamma_{\mathfrak{u}}^{1/(2\tau)}\left(2\zeta\left(\frac{\alpha}{\tau}\right)\right)^{|\mathfrak{u}|} \\
&=\frac{1}{\varphi(N)}\sum_{\emptyset\neq\mathfrak{u}\subseteq[s+1]}\gamma_{\mathfrak{u}}^{1/(2\tau)}\left(2\zeta\left(\frac{\alpha}{\tau}\right)\right)^{|\mathfrak{u}|},
\end{aligned}$$

which yields the final result. □

3.4 The Fast CBC Construction for Product Weights

As pointed out in Section 3.3, a more or less straightforward implementation of Algorithm 3.6 needs $O(d^2N^2)$ or, when using a storage cost of order $O(N)$, $O(d\,N^2)$ operations. This order of magnitude is prohibitive when one is in need of constructing a large number of integration nodes, which is often necessary to obtain sufficiently high accuracy.

The problem of the rather high computational complexity of an implementation of Algorithm 3.6 was significantly mitigated by a benchmark result of Nuyens and Cools (see [213, 214, 215]) who showed that by a clever implementation the cost of Algorithm 3.6 can be reduced to $O(d\,N\log N)$ operations under the assumption of product weights, using the fast Fourier transform (FFT). We will now outline how this reduction of the computation time can be obtained following the ideas of Nuyens and Cools.

Let us assume that we have product weights $\boldsymbol{\gamma}=(\gamma_j)_{j\ge 1}$. Recall from Equation (3.6) that we can write

$$[\mathrm{err}_{N,d,\alpha,\boldsymbol{\gamma}}(\boldsymbol{g})]^2=-1+\frac{1}{N}\sum_{k=0}^{N-1}\prod_{j=1}^{d}\left(1+\gamma_j\varphi_\alpha\left(\left\{\frac{kg_j}{N}\right\}\right)\right),$$

where φ_α is defined in (3.7). If $\alpha\in\mathbb{N}$, we can express φ_α in terms of the Bernoulli polynomial $B_{2\alpha}$. If $\alpha>1/2$ is not an integer, we do in general not have an explicit formula for φ_α, but the function values may nevertheless be approximated numerically using the (inverse) fast Fourier transform.

Approximating φ_α

Computing approximate values of φ_α is possible making use of $O(N\log N)$ operations, at the expense of $O(N)$ storage. Indeed, we can approximate φ_α by

$$\widetilde{\varphi}_\alpha(x) := \sum_{\substack{h \in \mathbb{Z} \\ 0<|h|<N}} \frac{e^{2\pi i h x}}{|h|^{2\alpha}} \quad \text{for} \quad x \in [0,1].$$

For any $x \in [0, 1]$, the absolute error of this approximation satisfies

$$|\varphi_\alpha(x) - \widetilde{\varphi}_\alpha(x)| \le 2 \sum_{h=N}^{\infty} \frac{1}{h^{2\alpha}} \le 2 \int_{N-1}^{\infty} \frac{1}{h^{2\alpha}} \le \frac{2}{2\alpha - 1} \frac{1}{(N-1)^{2\alpha-1}}.$$

Now the values $\widetilde{\varphi}_\alpha(k/N)$ for $k \in \{0, 1, \ldots, N-1\}$ can be computed using only $O(N \log N)$ operations. To this end consider the column vector

$$\boldsymbol{\phi}_N := \left(\widetilde{\varphi}_\alpha\left(\frac{0}{N}\right), \widetilde{\varphi}_\alpha\left(\frac{1}{N}\right), \ldots, \widetilde{\varphi}_\alpha\left(\frac{N-1}{N}\right)\right)^\top,$$

where

$$\begin{aligned} \widetilde{\varphi}_\alpha\left(\frac{k}{N}\right) &= \sum_{h=1}^{N-1} \frac{e^{2\pi i hk/N}}{h^{2\alpha}} + \sum_{h=-(N-1)}^{-1} \frac{e^{2\pi i hk/N}}{|h|^{2\alpha}} \\ &= \sum_{h=1}^{N-1} \left(\frac{1}{h^{2\alpha}} + \frac{1}{(N-h)^{2\alpha}} \right) e^{2\pi i hk/N} \end{aligned}$$

for $k \in \{0, 1, \ldots, N-1\}$. Let $t(h) := h^{-2\alpha} + (N-h)^{-2\alpha}$ for $h \in \{1, 2, \ldots, N-1\}$ and put $t(0) := 0$.

For $m \in \mathbb{N}$, let $\omega_m := e^{2\pi i/m}$ and let

$$F_m := \frac{1}{\sqrt{m}} \left(\omega_m^{k\ell}\right)_{k,\ell=0,1,\ldots,m-1} \tag{3.18}$$

be the Fourier matrix of order m. Note that F_m is symmetric and that $F_m\overline{F}_m = U_m$, where U_m denotes the $m \times m$ identity matrix, and where $\overline{F}_m$ denotes the complex conjugate of the matrix F_m.

With this notation we can write

$$\boldsymbol{\phi}_N = \sqrt{N} F_N \boldsymbol{x},$$

where

$$\boldsymbol{x} := (t(0), t(1), t(2), \ldots, t(N-1))^\top.$$

Using FFT, the matrix-vector product $F_N\boldsymbol{x}$ in the computation of $\boldsymbol{\phi}_N$ requires only $O(N \log N)$ operations. The method of FFT will be explained later in this section (see Theorem 3.11 below). Note that for the storage of the vector $\boldsymbol{\phi}_N$ we require memory space of size $O(N)$.

The CBC algorithm and efficient computation of matrix-vector products

Let us return to the actual CBC algorithm. For the sake of simplicity, we assume prime N in the following.

Suppose that we have precomputed the values

$$\varphi_\alpha(0), \varphi_\alpha\left(\frac{1}{N}\right), \ldots, \varphi_\alpha\left(\frac{N-1}{N}\right).$$

Then, we set $g_1 := 1$ and we write

$$\eta_1(k) := 1 + \gamma_1 \varphi_\alpha\left(\frac{k}{N}\right).$$

For $s \in \{1, 2, \ldots, d-1\}$, we describe how to perform one step in the CBC algorithm in an efficient way. Suppose that we have already chosen $g_1, \ldots, g_s$. According to Algorithm 3.6, we now have to find $g \in G_1^\varphi(N)$ that minimizes $\mathrm{err}_{N,s+1,\alpha,\boldsymbol{\gamma}}((g_1, \ldots, g_s, g))$, which is equivalent to minimizing

$$\sum_{k=0}^{N-1} \left(1 + \gamma_{s+1} \varphi_\alpha\left(\left\{\frac{kg}{N}\right\}\right)\right) \eta_s(k) = \sum_{k=0}^{N-1} \eta_s(k) + \gamma_{s+1} \sum_{k=0}^{N-1} \varphi_\alpha\left(\left\{\frac{kg}{N}\right\}\right) \eta_s(k),$$

where, as above,

$$\eta_s(k) := \prod_{j=1}^{s} \left(1 + \gamma_j \varphi_\alpha\left(\left\{\frac{kg_j}{N}\right\}\right)\right)$$

for $s \in \mathbb{N}$. Thus, minimizing $\mathrm{err}_{N,s+1,\alpha,\boldsymbol{\gamma}}((g_1, \ldots, g_s, g))$ with respect to g is equivalent to minimizing

$$T_{s+1}(g) := \sum_{k=0}^{N-1} \varphi_\alpha\left(\left\{\frac{kg}{N}\right\}\right) \eta_s(k). \tag{3.19}$$

A crucial observation is that, using (3.19), we can write the vector

$$\boldsymbol{T}_{s+1} := (T_{s+1}(1), \ldots, T_{s+1}(N-1))^\top$$

as the product of the $(N-1) \times N$ matrix

$$\boldsymbol{\Omega}_N := \left(\varphi_\alpha\left(\left\{\frac{kg}{N}\right\}\right)\right)_{g \in G_1^\varphi(N), k \in G_1(N)}$$

with the column vector

$$\boldsymbol{\eta}_s := (\eta_s(0), \ldots, \eta_s(N-1))^\top,$$

i.e.,

$$\boldsymbol{T}_{s+1} = \boldsymbol{\Omega}_N \boldsymbol{\eta}_s.$$

Note that Algorithm 3.6 will choose

$$g_{s+1} = \underset{g \in G_1^{\varphi}(N)}{\operatorname{argmin}} \; T_{s+1}(g).$$

Let us analyze the matrix-vector product $\boldsymbol{\Omega}_N \boldsymbol{\eta}_s$. In general the cost of a matrix-vector multiplication of this size is quadratic in N, but due to the special structure of $\boldsymbol{\Omega}_N$, this cost can be reduced to $O(N \log N)$ operations. Indeed, the entries of $\boldsymbol{\Omega}_N$ are of the form $\varphi_\alpha(\{(k/N)g\})$. As N is a prime number, there exists a primitive root modulo N, i.e., there exists $q \in \{1, 2, \ldots, N-1\}$ such that

$$\{q^k \pmod{N} \; : \; 0 \le k \le N-2\} = \{1, 2, \ldots, N-1\}.$$

Thus, any product of nonzero integers k and g can be written as a power of q modulo N. The key trick is now to reorder the indices $g \in G_1^{\varphi}(N)$ and $k \in G_1(N)$ (or, the rows and columns of $\boldsymbol{\Omega}_N$, respectively) in a way that allows fast matrix-vector multiplication. This is done such that $\boldsymbol{\Omega}_N$ has a circulant submatrix C_N of size $(N-1) \times (N-1)$, and it is achieved by using the so-called Rader transform. For the primitive root q define an $(N-1) \times (N-1)$ permutation matrix $\Pi(q) = (\pi_{i,j}(q))_{i,j \in \{1,2,\ldots,N-1\}}$ by

$$\pi_{i,j}(q) = \begin{cases} 1 & \text{if } i \equiv q^j \pmod{N}, \\ 0 & \text{otherwise.} \end{cases}$$

The matrix $\Pi(q)$ is indeed a permutation matrix, since q is a primitive root modulo N. Note that the first column of $\boldsymbol{\Omega}_N$ has $N-1$ identical entries, which are all $\varphi_\alpha(0)$. Let $\boldsymbol{\Omega}_N^{(N-1)}$ be the $(N-1) \times (N-1)$ matrix obtained by deleting the first column of $\boldsymbol{\Omega}_N$. Then we define $C_{N-1} = (c_{i,j}^{(N-1)})_{i,j \in \{1,2,\ldots,N-1\}}$ by

$$C_{N-1} := \Pi(q) \boldsymbol{\Omega}_N^{(N-1)} \Pi(q^{-1}), \tag{3.20}$$

where q^{-1} denotes the multiplicative inverse of q. We then have

$$c_{i,j}^{(N-1)} = \sum_{u,v=1}^{N-1} \pi_{u,i}(q)\, \varphi_\alpha\left(\left\{\frac{uv}{N}\right\}\right) \pi_{v,j}(q^{-1}) = \varphi_\alpha\left(\left\{\frac{q^i q^{-j}}{N}\right\}\right).$$

We write for short $c_\ell := \varphi_\alpha(\{q^\ell/N\})$, and since q is a primitive root modulo N we obviously have $c_\ell = c_{\ell'}$ for integers ℓ, ℓ' with $\ell \equiv \ell' \pmod{N-1}$. This implies $c_{i,j}^{(N-1)} = c_{i-j \pmod{N-1}}$, and hence

$$C_{N-1} = \begin{pmatrix} c_0 & c_{N-2} & \cdots & \cdots & c_2 & c_1 \\ c_1 & c_0 & c_{N-2} & \cdots & \cdots & c_2 \\ \vdots & \ddots & \ddots & \ddots & & \vdots \\ \vdots & & \ddots & \ddots & \ddots & \vdots \\ c_{N-3} & \cdots & \cdots & c_1 & c_0 & c_{N-2} \\ c_{N-2} & c_{N-3} & \cdots & \cdots & c_1 & c_0 \end{pmatrix}. \tag{3.21}$$

A matrix with a structure like C_{N-1} is called a *circulant matrix*. Note that a circulant matrix is completely determined by its first row or column.

Next, we need a technical lemma which shows how a circulant matrix can be expressed in terms of a Fourier matrix and a diagonal matrix. We remind the reader of the definition of a Fourier matrix F_m in (3.18). Furthermore, we denote, for $m \in \mathbb{N}$, by $\mathrm{diag}(a_1, a_2, \ldots, a_m)$ the diagonal matrix

$$(A_{i,j})_{i,j \in \{1,2,\ldots,m\}} \quad \text{with} \quad A_{i,j} = \begin{cases} a_i & \text{if } i = j, \\ 0 & \text{otherwise.} \end{cases}$$

What is more, for $m \in \mathbb{N}$ and a real or complex vector $\boldsymbol{v} = (v_0, v_1, \ldots, v_{m-1})$ of length m, we write $p_{\boldsymbol{v}}(z) := v_0 + v_1 z + \cdots + v_{m-1} z^{m-1}$ for $z \in \mathbb{C}$.

Using this notation, we can formulate the following lemma for general circulant matrices.

Lemma 3.10 *Let C_m be a circulant $m \times m$ matrix,*

$$C_m = \begin{pmatrix} c_0 & c_{m-1} & \cdots & \cdots & c_2 & c_1 \\ c_1 & c_0 & c_{m-1} & \cdots & \cdots & c_2 \\ \vdots & \ddots & \ddots & \ddots & & \vdots \\ \vdots & & \ddots & \ddots & \ddots & \vdots \\ c_{m-2} & \cdots & \cdots & c_1 & c_0 & c_{m-1} \\ c_{m-1} & c_{m-2} & \cdots & \cdots & c_1 & c_0 \end{pmatrix}.$$

Then C_m can be expressed in terms of a similarity transformation, $C_m = \overline{F}_m D F_m$, where F_m is the Fourier matrix of order m, and where

$$D = \mathrm{diag}\left(p_{\boldsymbol{c}}(1), p_{\boldsymbol{c}}(\omega_m), p_{\boldsymbol{c}}(\omega_m^2), \ldots, p_{\boldsymbol{c}}(\omega_m^{m-1})\right),$$

where $p_{\boldsymbol{c}}(z)$ is defined as outlined above, and where $\boldsymbol{c} = (c_0, c_1, c_2, \ldots, c_{m-1})$.

Proof Let $D = (d_{k,\ell})_{k,\ell \in \{0,1,\ldots,m-1\}}$ be given by

$$D = F_m C_m \overline{F}_m.$$

Then we have

$$\begin{aligned} d_{k,\ell} &= \frac{1}{m} \sum_{u,v=0}^{m-1} \omega_m^{ku} \, c_{u-v \pmod m} \, \omega_m^{-\ell v} \\ &= \frac{1}{m} \sum_{u=0}^{m-1} \omega_m^{u(k-\ell)} \sum_{v=0}^{m-1} c_{u-v \pmod m} \, \omega_m^{\ell(u-v)}. \end{aligned}$$

It is easily checked that $\sum_{v=0}^{m-1} c_{u-v \pmod m} \, \omega_m^{\ell(u-v)} = p_{\boldsymbol{c}}(\omega_m^{\ell})$, independently of u, and therefore

$$d_{k,\ell} = p_{\boldsymbol{c}}(\omega_m^{\ell}) \, \frac{1}{m} \sum_{u=0}^{m-1} \omega_m^{u(\ell-k)}.$$

The result now follows by noting that $(1/m) \sum_{u=0}^{m-1} \omega_m^{u(\ell-k)} = 1$ if $\ell = k$, and 0 otherwise. □

Now, combining (3.20) and Lemma 3.10, we obtain

$$\boldsymbol{\Omega}_N^{(N-1)} = (\Pi(q))^{\top} \overline{F}_{N-1} D F_{N-1} (\Pi(q^{-1}))^{\top},$$

for a certain diagonal matrix D.

Let $\boldsymbol{x} \in \mathbb{C}^{N-1}$ be an arbitrary column vector. Then it is obvious that the matrix-vector products $(\Pi(q))^{\top}\boldsymbol{x}$, $D\boldsymbol{x}$, and $(\Pi(q^{-1}))^{\top}\boldsymbol{x}$, respectively, can all be calculated using $O(N)$ operations.

We are going to outline next how the matrix-vector products $F_{N-1}\boldsymbol{x}$ and $\overline{F}_{N-1}\boldsymbol{x}$ can each be computed using $O(N \log N)$ operations. As $\overline{F}_{N-1}\boldsymbol{x} = \overline{F_{N-1}\overline{\boldsymbol{x}}}$, we can restrict ourselves to discussing only $F_{N-1}\boldsymbol{x}$. The desired property of this matrix-vector product can be obtained using FFT, which we will outline now.

The method of FFT

The method of FFT is particularly efficient for Fourier matrices whose size is a power of 2, while the Fourier matrix F_{N-1} is of size $N-1$, which in general is not a power of 2. However, this technical issue can be removed easily, by relating the matrix-vector product $F_{N-1}\boldsymbol{x}$ to one in which the size of the Fourier matrix involved is indeed a power of 2. To this end, let $h \in \mathbb{N}$ be such that $2^{h-1} < N-1 \leq 2^h$ and let $v \in \{0, 1, \ldots, 2^{h-1}-1\}$, such that $N-1+v = 2^h$. We extend the matrix C_{N-1} in (3.21) by v rows and v columns, to obtain a matrix T, given by

$$T = \begin{pmatrix}
c_0 & c_{N-2} & \cdots & \cdots & c_2 & c_1 & 0 & 0 & \cdots & 0 \\
c_1 & c_0 & c_{N-2} & \cdots & \cdots & c_2 & c_1 & 0 & \cdots & 0 \\
\vdots & \ddots & \ddots & \ddots & & \vdots & c_2 & c_1 & \cdots & 0 \\
\vdots & & \ddots & \ddots & \ddots & \vdots & \vdots & \vdots & \ddots & \ddots \\
c_{N-3} & \cdots & \cdots & c_1 & c_0 & c_{N-2} & \vdots & \vdots & \ddots & \ddots \\
c_{N-2} & c_{N-3} & \cdots & \cdots & c_1 & c_0 & c_{N-2} & c_{N-3} & \ddots & \ddots \\
0 & c_{N-2} & c_{N-3} & \cdots & \cdots & c_1 & c_0 & c_{N-2} & \ddots & \ddots \\
0 & 0 & c_{N-2} & c_{N-3} & \cdots & \cdots & c_1 & c_0 & \ddots & \ddots \\
\vdots & \ddots & \ddots & & & & \ddots & \ddots & \ddots & \ddots \\
0 & \ddots & \ddots & \ddots & & & \ddots & \ddots & c_1 & c_0
\end{pmatrix}.$$

Now, T is a $2^h \times 2^h$ matrix, which is no longer a circulant matrix, but a so-called *Toeplitz matrix* . In order to obtain a circulant matrix, we define another $2^h \times 2^h$ matrix

$$R := \begin{pmatrix}
0 & \cdots & 0 & c_{N-2} & \cdots & \cdots & c_2 & c_1 \\
\vdots & \ddots & \ddots & \ddots & \ddots & \ddots & \ddots & \vdots \\
0 & \cdots & 0 & \cdots & 0 & c_{N-2} & \ddots & \vdots \\
c_1 & \ddots & \ddots & \ddots & \ddots & \ddots & \ddots & \vdots \\
\vdots & \ddots & \ddots & \ddots & \ddots & \ddots & \ddots & c_{N-2} \\
\vdots & \ddots & c_1 & \ddots & \ddots & \ddots & \ddots & 0 \\
c_{N-3} & \ddots & \ddots & \ddots & \ddots & \ddots & \ddots & \vdots \\
c_{N-2} & c_{N-3} & \cdots & \cdots & c_1 & 0 & \cdots & 0
\end{pmatrix},$$

and then put

$$C'_{N-1} := \begin{pmatrix} T & R \\ R & T \end{pmatrix},$$

which is now a circulant $2^{h+1} \times 2^{h+1}$ matrix.

Suppose now that we want to compute $C_{N-1}\boldsymbol{x}$ for a complex vector $\boldsymbol{x} = (x_1, \ldots, x_{N-1})^\top$. This can also be achieved by computing $C'_{N-1}\boldsymbol{x}'$, where $\boldsymbol{x}'$ is a column vector of length 2^{h+1} given by

$$\boldsymbol{x}' = (x_1, \ldots, x_{N-1}, \underbrace{0, \ldots, 0}_{v+2^h \text{ components}})^\top.$$

Indeed, $C_{N-1}\boldsymbol{x}$ is the projection of $C'_{N-1}\boldsymbol{x}'$ onto the first $N-1$ components. Note that Lemma 3.10 can also be applied to C'_{N-1}, as C'_{N-1} is circulant, and hence C'_{N-1} can be represented by the product of Fourier matrices and a diagonal matrix, where the matrices involved all are of size 2^{h+1}. Now, applying the following classical result by Cooley and Tukey (see [26]) yields the fast matrix-vector multiplication that we desire.

Theorem 3.11 *Let $F_{2^{h+1}}$ be the Fourier matrix* (3.18) *of order 2^{h+1}. Let $\boldsymbol{u}$ be a complex column vector of length 2^{h+1}. Then the matrix-vector product $F_{2^{h+1}}\boldsymbol{u}$ can be computed using $O((h+1)2^{h+1})$ operations.*

Proof Let $\boldsymbol{z} = (z_0, \ldots, z_{2^{h+1}-1})^\top = F_{2^{h+1}}\boldsymbol{u}$, where $\boldsymbol{u} = (u_0, \ldots, u_{2^{h+1}-1})^\top$. Then

$$z_k = 2^{-(h+1)/2} \sum_{l=0}^{2^{h+1}-1} \omega_{2^{h+1}}^{kl} u_l \quad \text{for all } k \in \{0, 1, \ldots, 2^{h+1}-1\}.$$

We show how to compute the above sum recursively. Let k and l have binary expansions

$$k = \kappa_0 + \kappa_1 2 + \cdots + \kappa_h 2^h \quad \text{and} \quad l = \lambda_0 + \lambda_1 2 + \cdots + \lambda_h 2^h,$$

with digits $\kappa_0, \ldots, \kappa_h, \lambda_0, \ldots, \lambda_h \in \{0, 1\}$, respectively. Put

$$G_0(\lambda_0, \ldots, \lambda_{h-1}, \kappa_0) := \sum_{\lambda_h=0}^{1} \omega_2^{\kappa_0\lambda_h} u_{\lambda_0+\lambda_1 2+\cdots+\lambda_h 2^h}, \tag{3.22}$$

and, for $r \in [h]$, put

$$\begin{aligned} &G_r(\lambda_0, \ldots, \lambda_{h-r-1}, \kappa_0, \ldots, \kappa_r) \\ &:= \sum_{\lambda_{h-r}=0}^{1} \omega_{2^{r+1}}^{(\kappa_0+\kappa_1 2+\cdots+\kappa_r 2^r)\lambda_{h-r}} G_{r-1}(\lambda_0, \ldots, \lambda_{h-r}, \kappa_0, \ldots, \kappa_{r-1}). \end{aligned} \tag{3.23}$$

Now we show by induction on r that, for all $r \in [h]$, we have

$$\begin{aligned} &G_r(\lambda_0, \ldots, \lambda_{h-r-1}, \kappa_0, \ldots, \kappa_r) \\ &= \sum_{\lambda_{h-r}=0}^{1} \cdots \sum_{\lambda_h=0}^{1} \omega_{2^{r+1}}^{(\kappa_0+\kappa_1 2+\cdots+\kappa_r 2^r)(\lambda_{h-r}+\cdots+\lambda_h 2^r)} u_{\lambda_0+\lambda_1 2+\cdots+\lambda_h 2^h}. \end{aligned} \tag{3.24}$$

For $r = 1$, we obtain

$$G_1(\lambda_0, \ldots, \lambda_{h-2}, \kappa_0, \kappa_1) = \sum_{\lambda_{h-1}=0}^{1} \omega_4^{(\kappa_0+\kappa_1 2)\lambda_{h-1}} \sum_{\lambda_h=0}^{1} \omega_2^{\kappa_0\lambda_h} u_{\lambda_0+\lambda_1 2+\cdots+\lambda_h 2^h}.$$

Note that $\omega_2^{\kappa_0\lambda_h} = \omega_4^{2\kappa_0\lambda_h} = \omega_4^{2\kappa_0\lambda_h+4\kappa_1\lambda_h} = \omega_4^{(\kappa_0+2\kappa_1)2\lambda_h}$ and hence

$$G_1(\lambda_0, \ldots, \lambda_{h-2}, \kappa_0, \kappa_1) = \sum_{\lambda_{h-1}=0}^{1} \sum_{\lambda_h=0}^{1} \omega_4^{(\kappa_0+\kappa_1 2)(\lambda_{h-1}+\lambda_h 2)} u_{\lambda_0+\lambda_1 2+\cdots+\lambda_h 2^h}.$$

Thus (3.24) holds for $r = 1$. Assume that (3.24) holds for the index r. Then

$$\begin{aligned} G_{r+1}&(\lambda_0, \ldots, \lambda_{h-r-2}, \kappa_0, \ldots, \kappa_{r+1}) \\ &= \sum_{\lambda_{h-r-1}=0}^{1} \omega_{2^{r+2}}^{(\kappa_0+\cdots+\kappa_{r+1}2^{r+1})\lambda_{h-r-1}} G_r(\lambda_0, \ldots, \lambda_{h-r-1}, \kappa_0, \ldots, \kappa_r) \\ &= \sum_{\lambda_{h-r-1}=0}^{1} \omega_{2^{r+2}}^{(\kappa_0+\kappa_1 2+\cdots+\kappa_{r+1}2^{r+1})\lambda_{h-r-1}} \\ &\quad \times \sum_{\lambda_{h-r}=0}^{1} \cdots \sum_{\lambda_h=0}^{1} \omega_{2^{r+1}}^{(\kappa_0+\kappa_1 2+\cdots+\kappa_r 2^r)(\lambda_{h-r}+\cdots+\lambda_h 2^r)} u_{\lambda_0+\lambda_1 2+\cdots+\lambda_h 2^h}. \end{aligned}$$

Since

$$\begin{aligned} \omega_{2^{r+1}}^{(\kappa_0+\kappa_1 2+\cdots+\kappa_r 2^r)(\lambda_{h-r}+\cdots+\lambda_h 2^r)} &= \omega_{2^{r+1}}^{(\kappa_0+\kappa_1 2+\cdots+\kappa_r 2^r+\kappa_{r+1}2^{r+1})(\lambda_{h-r}+\cdots+\lambda_h 2^r)} \\ &= \omega_{2^{r+2}}^{(\kappa_0+\kappa_1 2+\cdots+\kappa_{r+1}2^{r+1})(\lambda_{h-r}2+\cdots+\lambda_h 2^{r+1})} \end{aligned}$$

we obtain

$$\begin{aligned} G_{r+1}&(\lambda_0, \ldots, \lambda_{h-r-2}, \kappa_0, \ldots, \kappa_{r+1}) \\ &= \sum_{\lambda_{h-r-1}=0}^{1} \cdots \sum_{\lambda_h=0}^{1} \omega_{2^{r+2}}^{(\kappa_0+\kappa_1 2+\cdots+\kappa_{r+1}2^{r+1})(\lambda_{h-r-1}+\cdots+\lambda_h 2^{r+1})} u_{\lambda_0+\lambda_1 2+\cdots+\lambda_h 2^h}, \end{aligned}$$

and hence (3.24) is shown.

In particular, with the choice $r = h$, (3.24) leads to the desired formula

$$z_k = 2^{-(h+1)/2} G_h(\kappa_0, \ldots, \kappa_h),$$

for all $k = \kappa_0 + \kappa_1 2 + \cdots + \kappa_h 2^h \in \{0, 1, \ldots, 2^{h+1} - 1\}$.

For the calculation of the z_k we compute G_r recursively. For $r \in \{0, 1, \ldots, h\}$ compute

$$G_r(\lambda_0, \ldots, \lambda_{m-r-1}, \kappa_0, \ldots, \kappa_r)$$

for all $\lambda_0, \ldots, \lambda_{h-r-1}, \kappa_0, \ldots, \kappa_r \in \{0, 1\}$ using (3.22) and the recursion formula (3.23). For each r this requires $O(2^{h+1})$ operations.

Overall we require $O((h+1)2^{h+1})$ operations to compute G_h and hence we can compute z in

$$O((h+1)2^{h+1})$$

operations. □

Theorem 3.11 can now be directly applied to computing $C'_{N-1}\boldsymbol{x}'$. Since we chose h such that $2^{h-1} < N-1 \leq 2^h$, we have $h < 1+\log_2 N$, and so we see that $C_{N-1}\boldsymbol{x}$ (and therefore also $\boldsymbol{T}_{s+1} = \boldsymbol{\Omega}_N \boldsymbol{\eta}_s$) can be computed using $O(N \log N)$ operations. In total, this results in a computational cost of $O(d N \log N)$ operations for Algorithm 3.6, if it is implemented accordingly. These observations are the keystones of the results by Nuyens and Cools, and these made the CBC algorithm feasible for the generation of high-dimensional lattice rules with many points, which is needed in practice.

We further remark that the order $O(d\,N \log N)$ in the construction cost for product weights can—if we assume sufficiently fast decreasing weights—be reduced further using the so-called reduced CBC construction, see Section 4.2.

3.5 The Fast CBC Construction for POD Weights

Other than the aforementioned product weights there is another type of weights that is important for practical applications of lattice rules (especially in the context of PDEs with random coefficients, see Appendix A). These weights are so-called product and order dependent (or, for short, POD) weights, which take the form

$$\gamma_{\mathfrak{u}} = \Gamma_{|\mathfrak{u}|} \prod_{j \in \mathfrak{u}} \gamma_j, \tag{3.25}$$

for $\mathfrak{u} \subseteq [d]$, where the γ_j have a role similar to product weights, and where $\Gamma_{|\mathfrak{u}|}$ may depend on $|\mathfrak{u}|$ but not on the elements of $\mathfrak{u}$. For $\mathfrak{u} = \emptyset$, we set $\Gamma_{|\mathfrak{u}|} = 1$ and hence (since we define the empty product to be one), we have $\gamma_\emptyset = 1$.

A fast implementation of the CBC construction for arbitrary general weights currently seems out of reach, as one would have to take into account 2^d different weights $\gamma_{\mathfrak{u}}$. However, for the special case of POD weights there is a relatively fast implementation of the CBC construction available.

We again start with the general error formula for the squared worst-case error in Corollary 2.8,

$$\begin{aligned}
[\mathrm{err}_{N,d,\alpha,\boldsymbol{\gamma}}(\boldsymbol{g})]^2 &= -1 + \sum_{\boldsymbol{h} \in \mathbb{Z}^d} \frac{1}{r_{2\alpha,\boldsymbol{\gamma}}(\boldsymbol{h})} \frac{1}{N} \sum_{k=0}^{N-1} \mathrm{e}^{2\pi \mathrm{i} k \boldsymbol{g}\cdot \boldsymbol{h}/N} \\
&= -1 + \frac{1}{N} \sum_{k=0}^{N-1} \sum_{\mathfrak{u} \subseteq [d]} \gamma_{\mathfrak{u}} \prod_{j \in \mathfrak{u}} \sum_{h \in \mathbb{Z}\setminus\{0\}} \frac{\mathrm{e}^{2\pi \mathrm{i} h k g_j/N}}{|h|^{2\alpha}} \\
&= -1 + \frac{1}{N} \sum_{k=0}^{N-1} \sum_{\mathfrak{u} \subseteq [d]} \Gamma_{|\mathfrak{u}|} \prod_{j \in \mathfrak{u}} \gamma_j \varphi_\alpha\left(\left\{\frac{k g_j}{N}\right\}\right),
\end{aligned}$$

where φ_α is given by (3.7). Since, in the previous expression, the summand for $\mathfrak{u} = \emptyset$ is one, we can rewrite this expression further as

$$[\mathrm{err}_{N,d,\alpha,\boldsymbol{\gamma}}(\boldsymbol{g})]^2 = \frac{1}{N} \sum_{k=0}^{N-1} \sum_{\emptyset \neq \mathfrak{u} \subseteq [d]} \Gamma_{|\mathfrak{u}|} \prod_{j \in \mathfrak{u}} \gamma_j \varphi_\alpha\left(\left\{\frac{k g_j}{N}\right\}\right)$$

$$= \frac{1}{N} \sum_{k=0}^{N-1} \sum_{\ell=1}^{d} \sum_{\substack{\mathfrak{u} \subseteq [d] \\ |\mathfrak{u}|=\ell}} \Gamma_\ell \prod_{j \in \mathfrak{u}} \gamma_j \varphi_\alpha \left(\left\{ \frac{k g_j}{N} \right\} \right). \tag{3.26}$$

Let again $s \in [d-1]$, and suppose that $g_1, \ldots, g_s$ have already been chosen by Algorithm 3.6. Let us write $\boldsymbol{g}^{(s+1)} := (\boldsymbol{g}^{(s)}, g_{s+1})$, where $\boldsymbol{g}^{(s)}$ denotes the s-dimensional vector found in the first s steps of the CBC construction, and g_{s+1} is the component selected in step $s+1$. For selecting g_{s+1}, we can split the innermost sum in (3.26) into a sum over all subsets $\mathfrak{u}$ with $(s+1) \notin \mathfrak{u}$ and a sum over $\mathfrak{u}$ with $(s+1) \in \mathfrak{u}$. This yields, after short calculation,

$$\begin{aligned}
&[\mathrm{err}_{N,s+1,\alpha,\boldsymbol{\gamma}}(\boldsymbol{g}^{(s+1)})]^2 \\
&= \frac{1}{N} \sum_{k=0}^{N-1} \sum_{\ell=1}^{s+1} \left(\sum_{\substack{\mathfrak{u} \subseteq [s] \\ |\mathfrak{u}|=\ell}} \Gamma_\ell \prod_{j \in \mathfrak{u}} \gamma_j \varphi_\alpha \left(\left\{ \frac{k g_j}{N} \right\} \right) \right. \\
&\quad \left. + \frac{\Gamma_\ell}{\Gamma_{\ell-1}} \gamma_{s+1} \varphi_\alpha \left(\left\{ \frac{k g_{s+1}}{N} \right\} \right) \sum_{\substack{\mathfrak{u} \subseteq [s] \\ |\mathfrak{u}|=\ell-1}} \Gamma_{\ell-1} \prod_{j \in \mathfrak{u}} \gamma_j \varphi_\alpha \left(\left\{ \frac{k g_j}{N} \right\} \right) \right).
\end{aligned}$$

We write, for short,

$$\Omega_N(g,k) := \varphi_\alpha \left(\left\{ \frac{kg}{N} \right\} \right) \quad \text{for } g \in G_1^{\varphi}(N) \text{ and } k \in G_1(N),$$

and, for $k \in G_1(N)$,

$$p_{s,\ell}(k) := \sum_{\substack{\mathfrak{u} \subseteq [s] \\ |\mathfrak{u}|=\ell}} \Gamma_\ell \prod_{j \in \mathfrak{u}} \gamma_j \varphi_\alpha \left(\left\{ \frac{k g_j}{N} \right\} \right) \quad \text{for } \ell \in \{0, 1, \ldots, s\}. \tag{3.27}$$

Using this notation, we can write

$$\begin{aligned}
&[\mathrm{err}_{N,s+1,\alpha,\boldsymbol{\gamma}}((\boldsymbol{g}^{(s)}, g_{s+1}))]^2 \\
&= [\mathrm{err}_{N,s,\alpha,\boldsymbol{\gamma}}(\boldsymbol{g}^{(s)})]^2 + \frac{\gamma_{s+1}}{N} \sum_{k=0}^{N-1} \Omega_N(g_{s+1}, k) \sum_{\ell=1}^{s+1} \frac{\Gamma_\ell}{\Gamma_{\ell-1}} p_{s,\ell-1}(k) \\
&= [\mathrm{err}_{N,s,\alpha,\boldsymbol{\gamma}}(\boldsymbol{g}^{(s)})]^2 + \frac{\gamma_{s+1}}{N} \boldsymbol{\Omega}_N(g_{s+1}) \left(\sum_{\ell=1}^{s+1} \frac{\Gamma_\ell}{\Gamma_{\ell-1}} \boldsymbol{p}_{s,\ell-1} \right)^{\top},
\end{aligned}$$

where

$$\boldsymbol{\Omega}_N(g_{s+1}) := (\Omega_N(g_{s+1}, 0), \ldots, \Omega_N(g_{s+1}, N-1)) \quad \text{and}$$

$$\boldsymbol{p}_{s,\ell-1} := (p_{s,\ell-1}(0), \dots, p_{s,\ell-1}(N-1)),$$

and where $\boldsymbol{\Omega}_N(g_{s+1})\,(\sum_{\ell=1}^{s+1}(\Gamma_\ell/\Gamma_{\ell-1})\boldsymbol{p}_{s,\ell-1})^\top$ is considered as the product of the row vector $\boldsymbol{\Omega}_N(g_{s+1})$ with the column vector $(\sum_{\ell=1}^{s+1}(\Gamma_\ell/\Gamma_{\ell-1})\boldsymbol{p}_{s,\ell-1})^\top$.

Let us furthermore write $\boldsymbol{E}_{N,s+1,\alpha,\boldsymbol{\gamma}}(\boldsymbol{g}^{(s)})$ to denote the column vector with entries $[\mathrm{err}_{N,s+1,\alpha,\boldsymbol{\gamma}}((\boldsymbol{g}^{(s)}, g))]^2$ for $g \in G_1^\varphi(N)$, and let $\boldsymbol{\Omega}_N$ be the $\varphi(N) \times N$ matrix with row vectors $\boldsymbol{\Omega}_N(g)$ for $g \in G_1^\varphi(N)$. Then we can write

$$\boldsymbol{E}_{N,s+1,\alpha,\boldsymbol{\gamma}}(\boldsymbol{g}^{(s)}) = [\mathrm{err}_{N,s,\alpha,\boldsymbol{\gamma}}(\boldsymbol{g}^{(s)})]^2\,\mathbf{1}_{\varphi(N)} + \frac{\gamma_{s+1}}{N}\,\boldsymbol{\Omega}_N \left(\sum_{\ell=1}^{s+1} \frac{\Gamma_\ell}{\Gamma_{\ell-1}} \boldsymbol{p}_{s,\ell-1}\right)^\top, \tag{3.28}$$

where $\mathbf{1}_{\varphi(N)}$ denotes the all-1 vector of length $\varphi(N)$, and where one uses the update formula

$$\boldsymbol{p}_{s+1,\ell} = \boldsymbol{p}_{s,\ell} + \frac{\Gamma_\ell}{\Gamma_{\ell-1}} \gamma_{s+1}\,\boldsymbol{\Omega}_N(g_{s+1}) * \boldsymbol{p}_{s,\ell-1}$$

for $\ell \in [s+1]$, with "$*$" denoting component-wise multiplication of vectors in $\mathbb{R}^N$, with the initial values

$$\boldsymbol{p}_{s+1,0} = \mathbf{1}_N \quad \text{and} \quad \boldsymbol{p}_{s+1,\ell} = \mathbf{0}_N \quad \text{for } s \geq 0 \text{ and } \ell > s+1,$$

where $\mathbf{0}_N$ denotes the zero vector of length N. Regarding the cost of this implementation, the matrix-vector multiplication in (3.28) can be carried out in a similar way as for product weights, resulting in a computational cost of order $O(N \log N)$ in each step. Updating $\boldsymbol{p}_{s+1,\ell}$ needs N operations for each ℓ, and thus $O(Ns)$ operations in total in each step. This gives a total order of

$$O(d\,N \log N + d^2 N)$$

for the number of elementary operations needed in the fast CBC algorithm. Since each of the vectors $\boldsymbol{p}_{s+1,\ell}$ has length N, and there are $s+1$ such vectors, the storage cost is $O(d\,N)$, and the vectors can be overwritten in the subsequent step.

We remark that the order $O(d\,N \log N + d^2 N)$ in the construction cost for POD weights can—for sufficiently fast decreasing weights—be reduced further using the so-called reduced CBC construction, see Section 4.2.

3.6 A CBC Algorithm Based on the Quality Criterion *R*

In Section 3.3 we outlined the classical CBC algorithm (Algorithm 3.6) for the Korobov space with smoothness α. This construction is based on the worst-case error. For its application we have to know the smoothness-parameter α and the specific weights $\boldsymbol{\gamma}$ of the function space for which the so-obtained lattice rules satisfy the desired error bounds. This, however, is often a problem in applications, since these parameters are often not known for a given integrand f. Most desirable would be a construction of lattice rules which work well simultaneously for all choices of the parameters α and $\boldsymbol{\gamma}$. We have already seen in Theorem 2.15 that such lattice rules exist, at least in the unweighted case, where we would like to find lattice rules which work well for several choices of α simultaneously. These existence results are based on the quality criterion R, but the efficient construction remained open in Section 2.3. In the present section, we introduce a CBC algorithm that is also based on the figure of merit R.

The question regarding algorithms that work simultaneously for different kinds of weights will be discussed in Chapter 12.

We consider the weighted setting and define a generalization of the classical quality criterion R in Definition 1.48.

Definition 3.12 For $\tau \geq 1$, given weights $\boldsymbol{\gamma}$, $N \in \mathbb{N}$, $N \geq 2$, and $\boldsymbol{g} \in \mathbb{Z}^d$, define

$$R_{N,d,\tau,\boldsymbol{\gamma}}(\boldsymbol{g}) := \sum_{\substack{\boldsymbol{h} \in C_d^*(N) \\ \boldsymbol{h}\cdot\boldsymbol{g} \equiv 0 \pmod{N}}} \frac{1}{r_{\tau,\boldsymbol{\gamma}}(\boldsymbol{h})},$$

where $C_d^*(N)$ is as introduced in Section 1.8 and $r_{\tau,\boldsymbol{\gamma}}$ is as in (2.17).

For $\tau = 1$ and the unweighted case, i.e., $\gamma_{\mathfrak{u}} = 1$ for all $\mathfrak{u} \subseteq [d]$, Definition 3.12 recovers the classical criterion R introduced in Definition 1.48.

Using Jensen's inequality (Lemma 2.25) we obtain

$$R_{N,d,\tau,\boldsymbol{\gamma}}(\boldsymbol{g}) \leq \left(\sum_{\substack{\boldsymbol{h} \in C_d^*(N) \\ \boldsymbol{h}\cdot\boldsymbol{g} \equiv 0 \pmod{N}}} \frac{1}{r_{1,\boldsymbol{\gamma}^{1/\tau}}(\boldsymbol{h})} \right)^{\tau} = (R_{N,d,1,\boldsymbol{\gamma}^{1/\tau}}(\boldsymbol{g}))^{\tau}, \tag{3.29}$$

where $\boldsymbol{\gamma}^{1/\tau}$ denotes the collection of the weights $\boldsymbol{\gamma}_{\mathfrak{u}}$ raised to the power $1/\tau$, i.e.,

$$\boldsymbol{\gamma}^{1/\tau} = \{\gamma_{\mathfrak{u}}^{1/\tau} \ : \ \mathfrak{u} \subseteq [d]\}.$$

Re-ordering the summation in the definition of $R_{N,d,1,\boldsymbol{\gamma}}$ according to the groups of nonzero components of the summation index $\boldsymbol{h}$ we can write

$$R_{N,d,1,\boldsymbol{\gamma}}(\boldsymbol{g}) = \sum_{\emptyset \neq \mathfrak{u} \subseteq [d]} \gamma_{\mathfrak{u}} \widetilde{R}_N(\boldsymbol{g}_{\mathfrak{u}}),$$

where

$$\widetilde{R}_N(\boldsymbol{g}_{\mathfrak{u}}) := \sum_{\substack{\boldsymbol{h}_{\mathfrak{u}} \in (C_1^*(N))^{|\mathfrak{u}|} \\ \boldsymbol{h}_{\mathfrak{u}}\cdot\boldsymbol{g}_{\mathfrak{u}} \equiv 0 \pmod{N}}} \frac{1}{r_1(\boldsymbol{h}_{\mathfrak{u}})}. \tag{3.30}$$

Observe that $\widetilde{R}_N(\boldsymbol{g}_{\mathfrak{u}}) = 0$ if $|\mathfrak{u}| = 1$ and $\boldsymbol{g}_{\mathfrak{u}} \in G_{|\mathfrak{u}|}^{\varphi}(N)$ since then the range of summation is empty. Note, furthermore, that we may write

$$\widetilde{R}_N(\boldsymbol{g}_{\mathfrak{u}}) = \frac{1}{N}\sum_{k=0}^{N-1}\prod_{j\in\mathfrak{u}}\left(\sum_{h\in C_1^*(N)}\frac{\mathrm{e}^{2\pi \mathrm{i} hkg_j/N}}{|h|}\right) = \frac{1}{N}\sum_{k=0}^{N-1}\prod_{j\in\mathfrak{u}} S_{k,N}(g_j),$$

where, for $z \in \mathbb{Z}$,

$$S_{k,N}(z) := \sum_{h\in C_1^*(N)}\frac{\mathrm{e}^{2\pi \mathrm{i} hkz/N}}{|h|} \quad \text{for } k \in \{0, 1, \ldots, N-1\}. \tag{3.31}$$

Note that for $k = 0$ we have

$$S_{0,N} = S_N = \sum_{h\in C_1^*(N)}\frac{1}{|h|},$$

which we already encountered in (2.15).

A relation to the worst-case error in the weighted Korobov space

We now show a proposition implying that the squared worst-case error of integration using a rank-1 lattice rule in the Korobov space with smoothness $\alpha > 1/2$ is essentially characterized by $R_{N,d,2\alpha,\boldsymbol{\gamma}}$, which in turn can be bounded in terms of $R_{N,d,1,\boldsymbol{\gamma}^{1/(2\alpha)}}$. This result is a weighted version of Theorem 2.11.

Proposition 3.13 *Let $\alpha > 1/2$, let $\boldsymbol{\gamma} = \{\gamma_{\mathfrak{u}}\}_{\mathfrak{u}\subseteq[d]}$ be general weights, and let $\boldsymbol{g} \in G_d^{\varphi}(N)$. Then,*

$$[\mathrm{err}_{N,d,\alpha,\boldsymbol{\gamma}}(\boldsymbol{g})]^2 \le R_{N,d,2\alpha,\boldsymbol{\gamma}}(\boldsymbol{g}) + \frac{2^{2\alpha}}{N^{2\alpha}}\sum_{\emptyset\neq\mathfrak{u}\subseteq[d]}\gamma_{\mathfrak{u}}(4\zeta(2\alpha))^{|\mathfrak{u}|}.$$

Moreover,

$$[\mathrm{err}_{N,d,\alpha,\boldsymbol{\gamma}}(\boldsymbol{g})]^2 \le (R_{N,d,1,\boldsymbol{\gamma}^{1/(2\alpha)}}(\boldsymbol{g}))^{2\alpha} + \frac{2^{2\alpha}}{N^{2\alpha}}\sum_{\emptyset\neq\mathfrak{u}\subseteq[d]}\gamma_{\mathfrak{u}}(4\zeta(2\alpha))^{|\mathfrak{u}|}.$$

Proof We study the difference

$$[\mathrm{err}_{N,d,\alpha,\boldsymbol{\gamma}}(\boldsymbol{g})]^2 - R_{N,d,2\alpha,\boldsymbol{\gamma}}(\boldsymbol{g}),$$

which can be written as

$$\sum_{\emptyset\neq\mathfrak{u}\subseteq[d]}\left(\sum_{\substack{\boldsymbol{h}_{\mathfrak{u}}\in(\mathbb{Z}\setminus\{0\})^{|\mathfrak{u}|}\\ \boldsymbol{h}_{\mathfrak{u}}\cdot\boldsymbol{g}_{\mathfrak{u}}\equiv 0 \pmod N}}\frac{1}{r_{2\alpha,\gamma_{\mathfrak{u}}}(\boldsymbol{h}_{\mathfrak{u}})} - \sum_{\substack{\boldsymbol{h}_{\mathfrak{u}}\in(C_1^*(N))^{|\mathfrak{u}|}\\ \boldsymbol{h}_{\mathfrak{u}}\cdot\boldsymbol{g}_{\mathfrak{u}}\equiv 0 \pmod N}}\frac{1}{r_{2\alpha,\gamma_{\mathfrak{u}}}(\boldsymbol{h}_{\mathfrak{u}})}\right),$$

motivating us to define, for $\emptyset \neq \mathfrak{u} \subseteq [d]$,

$$T_{\mathfrak{u}} := \sum_{\substack{\boldsymbol{h}_{\mathfrak{u}} \in (\mathbb{Z} \setminus \{0\})^{|\mathfrak{u}|} \\ \boldsymbol{h}_{\mathfrak{u}} \cdot \boldsymbol{g}_{\mathfrak{u}} \equiv 0 \pmod{N}}} \frac{1}{r_{2\alpha,\gamma_{\mathfrak{u}}}(\boldsymbol{h}_{\mathfrak{u}})} - \sum_{\substack{\boldsymbol{h}_{\mathfrak{u}} \in (C_1^*(N))^{|\mathfrak{u}|} \\ \boldsymbol{h}_{\mathfrak{u}} \cdot \boldsymbol{g}_{\mathfrak{u}} \equiv 0 \pmod{N}}} \frac{1}{r_{2\alpha,\gamma_{\mathfrak{u}}}(\boldsymbol{h}_{\mathfrak{u}})}.$$

We now distinguish two cases.

Case 1: Suppose that $|\mathfrak{u}| = 1$ such that $\mathfrak{u} = \{j\}$ for some $j \in [d]$. Then,

$$\begin{aligned} T_{\{j\}} &= \sum_{\substack{h_j \in \mathbb{Z} \setminus \{0\} \\ h_j g_j \equiv 0 \pmod{N}}} \frac{1}{r_{2\alpha,\gamma_{\{j\}}}(h_j)} - \sum_{\substack{h_j \in C_1^*(N) \\ h_j g_j \equiv 0 \pmod{N}}} \frac{1}{r_{2\alpha,\gamma_{\{j\}}}(h_j)} \\ &= \sum_{\substack{h_j \in \mathbb{Z} \setminus \{0\} \\ h_j g_j \equiv 0 \pmod{N}}} \frac{\gamma_{\{j\}}}{|h_j|^{2\alpha}} \\ &= 2\gamma_{\{j\}} \sum_{t=1}^{\infty} \frac{1}{(tN)^{2\alpha}} \\ &= \frac{2\gamma_{\{j\}}\zeta(2\alpha)}{N^{2\alpha}}, \end{aligned}$$

which follows since $\gcd(g_j, N) = 1$ and thus $h_j g_j \equiv 0 \pmod{N}$ if and only if g_j is a multiple of N, but $C_1^*(N)$ does not contain any multiple of N.

Case 2: Suppose that $|\mathfrak{u}| \geq 2$. For such a $\mathfrak{u}$ and $i \in \mathfrak{u}$, and $\boldsymbol{g}, \boldsymbol{h} \in \mathbb{Z}^{|\mathfrak{u}|}$, we write for short $\boldsymbol{g}_{\mathfrak{u}\setminus\{i\}}$ and $\boldsymbol{h}_{\mathfrak{u}\setminus\{i\}} \in \mathbb{Z}^{|\mathfrak{u}|-1}$ to denote the projections onto those components with indices in $\mathfrak{u} \setminus \{i\}$.

Note that

$$T_{\mathfrak{u}} = \sum_{\substack{\boldsymbol{h}_{\mathfrak{u}} \in (\mathbb{Z} \setminus \{0\})^{|\mathfrak{u}|} \setminus (C_1^*(N))^{|\mathfrak{u}|} \\ \boldsymbol{h}_{\mathfrak{u}} \cdot \boldsymbol{g}_{\mathfrak{u}} \equiv 0 \pmod{N}}} \frac{1}{r_{2\alpha,\gamma_{\mathfrak{u}}}(\boldsymbol{h}_{\mathfrak{u}})}.$$

Hence we can estimate

$$T_{\mathfrak{u}} \leq \sum_{i \in \mathfrak{u}} \sum_{\boldsymbol{h}_{\mathfrak{u}\setminus\{i\}} \in (\mathbb{Z} \setminus \{0\})^{|\mathfrak{u}|-1}} \sum_{\substack{h_i \in \mathbb{Z} \setminus C_1(N) \\ \boldsymbol{h}_{\mathfrak{u}\setminus\{i\}} \cdot \boldsymbol{g}_{\mathfrak{u}\setminus\{i\}} + h_i g_i \equiv 0 \pmod{N}}} \frac{1}{r_{2\alpha,\gamma_{\mathfrak{u}}}(\boldsymbol{h}_{\mathfrak{u}})}.$$

For $\boldsymbol{h}_{\mathfrak{u}\setminus\{i\}} \in (\mathbb{Z} \setminus \{0\})^{|\mathfrak{u}|-1}$, we write $b := \boldsymbol{h}_{\mathfrak{u}\setminus\{i\}} \cdot \boldsymbol{g}_{\mathfrak{u}\setminus\{i\}}$ and consider the expression

$$\begin{aligned} &\sum_{\substack{h_i \in \mathbb{Z} \setminus C_1(N) \\ b + h_i g_i \equiv 0 \pmod{N}}} \frac{1}{r_{2\alpha,\gamma_{\mathfrak{u}}}(\boldsymbol{h}_{\mathfrak{u}})} \\ &= \gamma_{\mathfrak{u}} \sum_{\substack{h_i \in \mathbb{Z} \setminus C_1(N) \\ b + h_i g_i \equiv 0 \pmod{N}}} \prod_{j \in \mathfrak{u}} \frac{1}{|h_j|^{2\alpha}} \end{aligned}$$

$$
\begin{aligned}
&= \gamma_{\mathfrak{u}}\left(\prod_{j\in\mathfrak{u}\setminus\{i\}}\frac{1}{|h_j|^{2\alpha}}\right)\sum_{\substack{h_i\in\mathbb{Z}\setminus C_1(N)\\ b+h_ig_i\equiv 0 \pmod N}}\frac{1}{|h_i|^{2\alpha}}\\
&= \gamma_{\mathfrak{u}}\left(\prod_{j\in\mathfrak{u}\setminus\{i\}}\frac{1}{|h_j|^{2\alpha}}\right)\left(\sum_{\substack{h_i=\lfloor N/2\rfloor+1\\ b+h_ig_i\equiv 0 \pmod N}}^{\infty}\frac{1}{|h_i|^{2\alpha}}+\sum_{\substack{h_i=\lceil N/2\rceil\\ -b+h_ig_i\equiv 0 \pmod N}}^{\infty}\frac{1}{|h_i|^{2\alpha}}\right)\\
&= \gamma_{\mathfrak{u}}\left(\prod_{j\in\mathfrak{u}\setminus\{i\}}\frac{1}{|h_j|^{2\alpha}}\right)\sum_{t=1}^{\infty}\left(\sum_{\substack{h_i=t(\lfloor N/2\rfloor+1)\\ b+h_ig_i\equiv 0 \pmod N}}^{(t+1)(\lfloor N/2\rfloor+1)-1}\frac{1}{|h_i|^{2\alpha}}+\sum_{\substack{h_i=t\lceil N/2\rceil\\ -b+h_ig_i\equiv 0 \pmod N}}^{(t+1)\lceil N/2\rceil-1}\frac{1}{|h_i|^{2\alpha}}\right)\\
&\le \gamma_{\mathfrak{u}}\left(\prod_{j\in\mathfrak{u}\setminus\{i\}}\frac{1}{|h_j|^{2\alpha}}\right)\sum_{t=1}^{\infty}\frac{2^{2\alpha}}{(tN)^{2\alpha}}\left(\sum_{\substack{h_i=t(\lfloor N/2\rfloor+1)\\ b+h_ig_i\equiv 0 \pmod N}}^{(t+1)(\lfloor N/2\rfloor+1)-1}1+\sum_{\substack{h_i=t\lceil N/2\rceil\\ -b+h_ig_i\equiv 0 \pmod N}}^{(t+1)\lceil N/2\rceil-1}1\right)\\
&\le \gamma_{\mathfrak{u}}\left(\prod_{j\in\mathfrak{u}\setminus\{i\}}\frac{1}{|h_j|^{2\alpha}}\right)\frac{2\cdot 2^{2\alpha}\zeta(2\alpha)}{N^{2\alpha}},
\end{aligned}
$$

which holds since for any $b, g \in \mathbb{Z}$ with $\gcd(g, N) = 1$ the congruence $b + hg \equiv 0 \pmod N$ has at most one solution h in $\{t(\lfloor N/2\rfloor + 1), \ldots, (t + 1)(\lfloor N/2\rfloor + 1) - 1\}$ and at most one solution h in $\{t\lceil N/2\rceil, \ldots, (t + 1)\lceil N/2\rceil - 1\}$. Consequently, we can bound $T_{\mathfrak{u}}$ for $|\mathfrak{u}| \ge 2$ by

$$
\begin{aligned}
T_{\mathfrak{u}} &\le \gamma_{\mathfrak{u}}\frac{2\cdot 2^{2\alpha}\zeta(2\alpha)}{N^{2\alpha}}\sum_{i\in\mathfrak{u}}\sum_{\boldsymbol{h}_{\mathfrak{u}\setminus\{i\}}\in(\mathbb{Z}\setminus\{0\})^{|\mathfrak{u}|-1}}\prod_{j\in\mathfrak{u}\setminus\{i\}}\frac{1}{|h_j|^{2\alpha}}\\
&= \gamma_{\mathfrak{u}}\frac{2\cdot 2^{2\alpha}\zeta(2\alpha)}{N^{2\alpha}}\sum_{i\in\mathfrak{u}}\left(2\sum_{m=1}^{\infty}\frac{1}{m^{2\alpha}}\right)^{|\mathfrak{u}|-1}\\
&= \gamma_{\mathfrak{u}}\frac{2^{2\alpha}}{N^{2\alpha}}(2\zeta(2\alpha))^{|\mathfrak{u}|}|\mathfrak{u}|\\
&\le \gamma_{\mathfrak{u}}\frac{2^{2\alpha}}{N^{2\alpha}}(4\zeta(2\alpha))^{|\mathfrak{u}|}.
\end{aligned}
$$

In summary, we obtain, using the results for both cases from above,

$$
\sum_{\emptyset\neq\mathfrak{u}\subseteq[d]}T_{\mathfrak{u}} \le \frac{2^{2\alpha}}{N^{2\alpha}}\sum_{\emptyset\neq\mathfrak{u}\subseteq[d]}\gamma_{\mathfrak{u}}(4\zeta(2\alpha))^{|\mathfrak{u}|},
$$

as claimed.

The second estimate in terms of $R_{N,d,1,\boldsymbol{\gamma}^{1/(2\alpha)}}$ follows from the first one after an application of Inequality (3.29). □

A CBC construction based on $R_{N,d,1,\gamma}$

Next, we introduce a CBC algorithm that is based on the figure of merit $R_{N,d,1,\gamma}$, which is independent of α. In the following approach due to Sinescu and Joe [228] we restrict ourselves to prime N.

Algorithm 3.14 (CBC construction based on $R_{N,d,1,\gamma}$) Let N be a prime number and let d be a positive integer. Construct a generating vector $\boldsymbol{g} = (g_1, \dots, g_d) \in G_d(N)$ as follows.

(1) Choose $g_1 = 1$.
(2) For s from 1 to $d-1$:
Assume that $g_1, \dots, g_s \in G_1(N)$ have already been found. Choose $g_{s+1} \in G_1(N)$ as

$$g_{s+1} := \underset{g \in G_1(N)}{\operatorname{argmin}} \ R_{N,s+1,1,\gamma}((g_1, \dots, g_s, g)).$$

End for.
(3) Set $\boldsymbol{g} = (g_1, \dots, g_d)$.

The following theorem shows that the generating vectors returned by Algorithm 3.14 indeed yield low values of $R_{N,d,1,\gamma}$.

Theorem 3.15 *Let N be a prime number, let $d \in \mathbb{N}$, and let $\boldsymbol{\gamma} = \{\gamma_{\mathfrak{u}}\}_{\mathfrak{u} \subseteq [d]}$ be general weights. Assume that $\boldsymbol{g} = (g_1, \dots, g_d)$ has been constructed according to Algorithm 3.14. Then, for any $s \in [d]$, we have*

$$R_{N,s,1,\gamma}((g_1, \dots, g_s)) \le \frac{1}{N} \sum_{\emptyset \ne \mathfrak{u} \subseteq [s]} \gamma_{\mathfrak{u}} S_N^{|\mathfrak{u}|},$$

where $S_N = \sum_{\boldsymbol{h} \in C_1^*(N)} |h|^{-1}$ as in (2.15). *In particular, for product weights $\gamma_{\mathfrak{u}} = \prod_{j \in \mathfrak{u}} \gamma_j$ for $\mathfrak{u} \subseteq [d]$, we have*

$$R_{N,s,1,\gamma}((g_1, \dots, g_s)) \le \frac{1}{N} \prod_{j=1}^{s} (1 + \gamma_j S_N),$$

and in the unweighted case, i.e., when all weights are equal to one, it is true that

$$R((g_1, \dots, g_s), N) = R_{N,s,1,\gamma}((g_1, \dots, g_s)) \le \frac{(1 + S_N)^s}{N}.$$

Remark 3.16 Recall from (2.16) that $S_N \le 2(\log N + 1 - \log 2)$. Hence for the vector $(g_1, \dots, g_s)$ in Theorem 3.15 we obtain

$$R_{N,s,1,\gamma}((g_1, \dots, g_s)) \le \frac{1}{N} \sum_{\emptyset \ne \mathfrak{u} \subseteq [s]} \gamma_{\mathfrak{u}} (2(\log N + 1 - \log 2))^{|\mathfrak{u}|}.$$

Note, furthermore, that in the unweighted case the constructive result in Theorem 3.15 matches the average type result in (2.14).

Proof of Theorem 3.15 We prove the result by induction on s. For $s = 1$ we have $R_{N,1,1,\gamma}(g) = 0$ for any $g \in G_1(N)$, and so the desired bound holds trivially.

Assume that we have already constructed a vector $\boldsymbol{g}^{(s)} = (g_1, \ldots, g_s) \in G_s(N)$ satisfying

$$R_{N,s,1,\gamma}(\boldsymbol{g}^{(s)}) \le \frac{1}{N} \sum_{\emptyset \neq \mathfrak{u} \subseteq [s]} \gamma_{\mathfrak{u}} S_N^{|\mathfrak{u}|}. \tag{3.32}$$

For arbitrary $g \in G_1(N)$ we write $(\boldsymbol{g}^{(s)}, g) := (g_1, \ldots, g_s, g)$. Recall the definition of $\widetilde{R}_N(\boldsymbol{g}_{\mathfrak{u}})$ from (3.30). Then we have

$$\begin{aligned} R_{N,s+1,1,\gamma}((\boldsymbol{g}^{(s)}, g)) &= \sum_{\emptyset \neq \mathfrak{u} \subseteq [s+1]} \gamma_{\mathfrak{u}} \widetilde{R}_N((\boldsymbol{g}^{(s)}, g)_{\mathfrak{u}}) \\ &= \sum_{\emptyset \neq \mathfrak{u} \subseteq [s]} \gamma_{\mathfrak{u}} \widetilde{R}_N(\boldsymbol{g}_{\mathfrak{u}}^{(s)}) + \sum_{\substack{\mathfrak{u} \subseteq [s+1] \\ s+1 \in \mathfrak{u}}} \gamma_{\mathfrak{u}} \widetilde{R}_N((\boldsymbol{g}^{(s)}, g)_{\mathfrak{u}}) \\ &= R_{N,s,1,\gamma}(\boldsymbol{g}^{(s)}) + \sum_{\substack{\mathfrak{u} \subseteq [s+1] \\ s+1 \in \mathfrak{u}}} \gamma_{\mathfrak{u}} \widetilde{R}_N((\boldsymbol{g}^{(s)}, g)_{\mathfrak{u}}). \end{aligned}$$

The term $\widetilde{R}_N((\boldsymbol{g}^{(s)}, g)_{\mathfrak{u}})$ that appears in the latter sum, where $\mathfrak{u} \subseteq [s+1]$ with $s+1 \in \mathfrak{u}$, is zero if $|\mathfrak{u}| = 1$, and if $|\mathfrak{u}| \ge 2$ it can be rewritten in the form

$$\begin{aligned} \widetilde{R}_N((\boldsymbol{g}^{(s)}, g)_{\mathfrak{u}}) &= \frac{1}{N} \sum_{k=0}^{N-1} \left(\prod_{j \in \mathfrak{u} \setminus \{s+1\}} S_{k,N}(g_j) \right) S_{k,N}(g) \\ &= \frac{S_N^{|\mathfrak{u}|}}{N} + \frac{1}{N} \sum_{k=1}^{N-1} \left(\prod_{j \in \mathfrak{u} \setminus \{s+1\}} S_{k,N}(g_j) \right) S_{k,N}(g), \end{aligned}$$

where $S_{k,N}$ is defined as in (3.31). Therefore we obtain

$$\begin{aligned} R_{N,s+1,1,\gamma}((\boldsymbol{g}^{(s)}, g)) = R_{N,s,1,\gamma}(\boldsymbol{g}^{(s)}) &+ \frac{1}{N} \sum_{\substack{\mathfrak{u} \subseteq [s+1] \\ |\mathfrak{u}| \ge 2 \\ s+1 \in \mathfrak{u}}} \gamma_{\mathfrak{u}} S_N^{|\mathfrak{u}|} \\ &+ \frac{1}{N} \sum_{\substack{\mathfrak{u} \subseteq [s+1] \\ |\mathfrak{u}| \ge 2 \\ s+1 \in \mathfrak{u}}} \gamma_{\mathfrak{u}} \sum_{k=1}^{N-1} \left(\prod_{j \in \mathfrak{u} \setminus \{s+1\}} S_{k,N}(g_j) \right) S_{k,N}(g). \end{aligned} \tag{3.33}$$

Now assume that $g_{s+1} \in G_1(N)$ has been chosen according to Algorithm 3.14. Let $(\boldsymbol{g}^{(s)}, g_{s+1}) = (g_1, \ldots, g_s, g_{s+1})$. Again we use the familiar standard averaging argument (see Remark 2.14). Since the minimum is not larger than the average, we have

$$R_{N,s+1,1,\gamma}((\boldsymbol{g}^{(s)}, g_{s+1})) \le \frac{1}{N} \sum_{g=0}^{N-1} R_{N,s+1,1,\gamma}((\boldsymbol{g}^{(s)}, g)).$$

Using (3.33) we get

$$\begin{aligned} & R_{N,s+1,1,\gamma}((\boldsymbol{g}^{(s)}, g_{s+1})) \\ & \le R_{N,s,1,\gamma}(\boldsymbol{g}^{(s)}) + \frac{1}{N} \sum_{\substack{\mathfrak{u} \subseteq [s+1] \\ |\mathfrak{u}| \ge 2 \\ s+1 \in \mathfrak{u}}} \gamma_{\mathfrak{u}} \, S_N^{|\mathfrak{u}|} \\ & + \frac{1}{N} \sum_{\substack{\mathfrak{u} \subseteq [s+1] \\ |\mathfrak{u}| \ge 2 \\ s+1 \in \mathfrak{u}}} \gamma_{\mathfrak{u}} \sum_{k=1}^{N-1} \left(\prod_{j \in \mathfrak{u} \setminus \{s+1\}} S_{k,N}(g_j) \right) \frac{1}{N} \sum_{g=0}^{N-1} S_{k,N}(g). \end{aligned}$$

Now we use that N is a prime number. For $k \in \{1, 2, \dots, N-1\}$ we have

$$\sum_{g=0}^{N-1} S_{k,N}(g) = \sum_{h \in C_1^*(N)} \frac{1}{|h|} \sum_{g=0}^{N-1} \left(\mathrm{e}^{2\pi \mathrm{i} hk/N} \right)^g = \sum_{\substack{h \in C_1^*(N) \\ hk \equiv 0 \pmod{N}}} \frac{N}{|h|} = 0,$$

since no $h \in C_1^*(N)$ satisfies $hk \equiv 0 \pmod{N}$. Thus,

$$R_{N,s+1,1,\gamma}((\boldsymbol{g}^{(s)}, g_{s+1})) \le R_{N,s,1,\gamma}(\boldsymbol{g}^{(s)}) + \frac{1}{N} \sum_{\substack{\mathfrak{u} \subseteq [s+1] \\ |\mathfrak{u}| \ge 2 \\ s+1 \in \mathfrak{u}}} \gamma_{\mathfrak{u}} \, S_N^{|\mathfrak{u}|}.$$

Using the induction hypothesis (3.32) we obtain

$$\begin{aligned} R_{N,s+1,1,\gamma}((\boldsymbol{g}^{(s)}, g_{s+1})) & \le \frac{1}{N} \sum_{\emptyset \ne \mathfrak{u} \subseteq [s]} \gamma_{\mathfrak{u}} \, S_N^{|\mathfrak{u}|} + \frac{1}{N} \sum_{\substack{\mathfrak{u} \subseteq [s+1] \\ |\mathfrak{u}| \ge 2 \\ s+1 \in \mathfrak{u}}} \gamma_{\mathfrak{u}} \, S_N^{|\mathfrak{u}|} \\ & \le \frac{1}{N} \sum_{\emptyset \ne \mathfrak{u} \subseteq [s+1]} \gamma_{\mathfrak{u}} \, S_N^{|\mathfrak{u}|}, \end{aligned}$$

and this finishes the proof of the first claim in the theorem. The other claims follow by straightforward reasoning. □

Combining Theorem 3.15 with Proposition 3.13 we can formulate the following theorem.

Theorem 3.17 *Let $\alpha > 1/2$, let $d \in \mathbb{N}$, and let $\boldsymbol{\gamma} = \{\gamma_{\mathfrak{u}}\}_{\mathfrak{u} \subseteq [d]}$ be general weights. Then the following two statements hold.*

1. *Let N be a prime number and assume that $\boldsymbol{g}$ has been constructed according to Algorithm 3.14 using the weights $\{\boldsymbol{\gamma}_{\mathfrak{u}}^{1/(2\alpha)}\}_{\mathfrak{u} \subseteq [d]}$. Then we have*

$$\mathrm{err}_{N,d,\alpha,\boldsymbol{\gamma}}(\boldsymbol{g}) \le \frac{1}{N^\alpha} \left(\left(\sum_{\emptyset \ne \mathfrak{u} \subseteq [d]} \gamma_{\mathfrak{u}}^{1/(2\alpha)} (2(\log N + 1 - \log 2))^{|\mathfrak{u}|} \right)^{2\alpha} \right.$$
$$\left. + 2^{2\alpha} \sum_{\emptyset \ne \mathfrak{u} \subseteq [d]} \gamma_{\mathfrak{u}} (4\zeta(2\alpha))^{|\mathfrak{u}|} \right)^{1/2}.$$

2. *Let N be a prime number and assume that $\boldsymbol{g}$ has been constructed according to Algorithm 3.14 using the weights $\{\boldsymbol{\gamma}_{\mathfrak{u}}\}_{\mathfrak{u} \subseteq [d]}$. Then we have*

$$\mathrm{err}_{N,d,\alpha,\boldsymbol{\gamma}^{2\alpha}}(\boldsymbol{g}) \le \frac{1}{N^\alpha} \left(\left(\sum_{\emptyset \ne \mathfrak{u} \subseteq [d]} \gamma_{\mathfrak{u}} (2(\log N + 1 - \log 2))^{|\mathfrak{u}|} \right)^{2\alpha} \right.$$
$$\left. + 2^{2\alpha} \sum_{\emptyset \ne \mathfrak{u} \subseteq [d]} \gamma_{\mathfrak{u}}^{2\alpha} (4\zeta(2\alpha))^{|\mathfrak{u}|} \right)^{1/2}.$$

Remark 3.18 Note that there is a kind of trade-off to be observed in Theorem 3.17. If we would like to have a good error bound on $\mathrm{err}_{N,d,\alpha,\boldsymbol{\gamma}}(\boldsymbol{g})$, then we need to run Algorithm 3.14 using the weights $\{\boldsymbol{\gamma}_{\mathfrak{u}}^{1/(2\alpha)}\}_{\mathfrak{u} \subseteq [d]}$, which implies that the algorithm is not independent of the smoothness parameter α.

However, if we are content with an error bound for the smaller weights $\boldsymbol{\gamma}^{2\alpha}$, i.e., a bound for $\mathrm{err}_{N,d,\alpha,\boldsymbol{\gamma}^{2\alpha}}(\boldsymbol{g})$, then we can run Algorithm 3.14 using just the weights $\{\boldsymbol{\gamma}_{\mathfrak{u}}\}_{\mathfrak{u} \subseteq [d]}$, which implies that the algorithm is indeed independent of the smoothness parameter α. Theorem 3.17 can also be combined with Theorem 12.1 to obtain a result for more general weights.

Remark 3.19 For the unweighted case, i.e., $\gamma_{\mathfrak{u}} = 1$ for all $\mathfrak{u} \subseteq [d]$, we obviously have $\gamma_{\mathfrak{u}}^{1/(2\alpha)} = \gamma_{\mathfrak{u}} = 1$. Hence, running Algorithm 3.14, which is independent of the parameter α, we get the following error bound for the output generating vector $\boldsymbol{g}$,

$$\mathrm{err}_{N,d,\alpha,\boldsymbol{\gamma}}(\boldsymbol{g}) \le \frac{1}{N^\alpha} \left((2(1 + \log N))^{2\alpha d} + 2^{2\alpha} (1 + 4\zeta(2\alpha))^d \right)^{1/2},$$

simultaneously for every $\alpha > 1/2$. This is essentially the bound from the existence result in Theorem 2.15.

In summary, we have, at least to a certain extent, achieved the goal we set ourselves at the beginning of this section.

Error bounds independent of the dimension

We aim at finding conditions on the weights which guarantee that the bounds in Remark 3.16 and Theorem 3.17 can be made independent of the dimension. In the following we use the notation $\{\gamma_{\mathfrak{u}}\}_{\mathfrak{u}\subseteq\mathbb{N},|\mathfrak{u}|<\infty}$, which indicates that we are not considering d as fixed, but assume that—in the sense of Information-Based Complexity—the dimension d tends to infinity. Hence we have to assume that there is a weight $\gamma_{\mathfrak{u}}$ for every finite $\mathfrak{u} \subseteq \mathbb{N}$.

To start, we state a lemma which was first shown in [69] and which will be crucial in several instances in this book. For this lemma we need to assume positive weights.

Lemma 3.20 *Let* $\{\gamma_{\mathfrak{u}}\}_{\mathfrak{u}\subseteq\mathbb{N},|\mathfrak{u}|<\infty}$ *be arbitrary but positive weights with* $\gamma_\emptyset = 1$ *such that*

$$\sum_{j=1}^{\infty} \max_{\mathfrak{v}\subseteq[j-1]} \frac{\gamma_{\mathfrak{v}\cup\{j\}}}{\gamma_{\mathfrak{v}}} < \infty. \tag{3.34}$$

Then, for any $\delta > 0$*, there exists a* $C(\boldsymbol{\gamma},\delta) > 0$*, which is independent of* d *and* N*, such that for all* $N \in \mathbb{N}$ *we have*

$$\sum_{\substack{\mathfrak{u}\subseteq\mathbb{N}\\|\mathfrak{u}|<\infty}} \gamma_{\mathfrak{u}} (\log N)^{|\mathfrak{u}|} \le C(\boldsymbol{\gamma},\delta)\, N^{\delta}.$$

In the case of product weights $\gamma_{\mathfrak{u}} = \prod_{j\in\mathfrak{u}} \gamma_j$ *for* $\mathfrak{u} \subseteq \mathbb{N}, |\mathfrak{u}| < \infty$*, Condition* (3.34) *is equivalent to* $\sum_{j=1}^{\infty} \gamma_j < \infty$ *and the assertion reads*

$$\prod_{j=1}^{\infty}(1+\gamma_j \log N) \le C(\boldsymbol{\gamma},\delta) N^{\delta}.$$

Proof We first recall the proof for product weights from [105, Lemma 3]. In this case Condition (3.34) boils down to $\sum_{j=1}^{\infty} \gamma_j < \infty$. We need to show then that for any $\delta > 0$ there exists a $C(\boldsymbol{\gamma},\delta) > 0$ such that for all $N \in \mathbb{N}$ we have

$$\sum_{\substack{\mathfrak{u}\subseteq\mathbb{N}\\|\mathfrak{u}|<\infty}} \gamma_{\mathfrak{u}} (\log N)^{|\mathfrak{u}|} = \prod_{j=1}^{\infty}(1+\gamma_j \log N) \le C(\boldsymbol{\gamma},\delta) N^{\delta}. \tag{3.35}$$

Let

$$S(\boldsymbol{\gamma},N) := \prod_{j=1}^{\infty}(1+\gamma_j \log N),$$

and define $\sigma_\ell := \sum_{j=\ell+1}^{\infty} \gamma_j$ for $\ell \in \mathbb{N}_0$. If $\sum_{j=1}^{\infty} \gamma_j < \infty$, it is obviously true that $\lim_{\ell\to\infty} \sigma_\ell = 0$. Then,

$$\begin{aligned}
\log S(\boldsymbol{\gamma}, N) &= \sum_{j=1}^{\infty} \log(1 + \gamma_j \log N) \\
&\leq \sum_{j=1}^{\ell} \log(1 + \sigma_\ell^{-1} + \gamma_j \log N) + \sum_{j=\ell+1}^{\infty} \log(1 + \gamma_j \log N) \\
&\leq \ell \log(1 + \sigma_\ell^{-1}) + \sum_{j=1}^{\ell} \log(1 + \gamma_j \sigma_\ell \log N) \\
&\quad + \sum_{j=\ell+1}^{\infty} \log(1 + \gamma_j \log N) \\
&\leq \ell \log(1 + \sigma_\ell^{-1}) + \sigma_\ell (\log N) \sum_{j=1}^{\ell} \gamma_j + \sigma_\ell \log N \\
&\leq \ell \log(1 + \sigma_\ell^{-1}) + \sigma_\ell (\sigma_0 + 1) \log N.
\end{aligned}$$

Hence we obtain

$$S(\boldsymbol{\gamma}, N) \leq (1 + \sigma_\ell^{-1})^\ell N^{(\sigma_0+1)\sigma_\ell}.$$

For $\delta > 0$ choose ℓ sufficiently large to make $\sigma_\ell \leq \delta/(\sigma_0 + 1)$. Then we get (3.35) for a suitably chosen positive real $C(\boldsymbol{\gamma}, \delta)$.

Now we deal with the general case. We start by considering a finite sum over all subsets of $[d]$ for, without loss of generality, $d \geq 4$,

$$\begin{aligned}
\sum_{\mathfrak{u} \subseteq [d]} \gamma_{\mathfrak{u}} (\log N)^{|\mathfrak{u}|} &= \sum_{\mathfrak{u} \subseteq [d-1]} \gamma_{\mathfrak{u}} (\log N)^{|\mathfrak{u}|} + \sum_{\mathfrak{u} \subseteq [d-1]} \gamma_{\mathfrak{u} \cup \{d\}} (\log N)^{|\mathfrak{u}|+1} \\
&= \sum_{\mathfrak{u} \subseteq [d-1]} \gamma_{\mathfrak{u}} (\log N)^{|\mathfrak{u}|} \left(1 + \frac{\gamma_{\mathfrak{u} \cup \{d\}}}{\gamma_{\mathfrak{u}}} \log N\right) \\
&= \sum_{d-1 \notin \mathfrak{u} \subseteq [d-1]} \gamma_{\mathfrak{u}} (\log N)^{|\mathfrak{u}|} \left(1 + \frac{\gamma_{\mathfrak{u} \cup \{d\}}}{\gamma_{\mathfrak{u}}} \log N\right) \\
&\quad + \sum_{d-1 \in \mathfrak{u} \subseteq [d-1]} \gamma_{\mathfrak{u}} (\log N)^{|\mathfrak{u}|} \left(1 + \frac{\gamma_{\mathfrak{u} \cup \{d\}}}{\gamma_{\mathfrak{u}}} \log N\right) \\
&= \sum_{\mathfrak{u} \subseteq [d-2]} \gamma_{\mathfrak{u}} (\log N)^{|\mathfrak{u}|} \left(1 + \frac{\gamma_{\mathfrak{u} \cup \{d\}}}{\gamma_{\mathfrak{u}}} \log N\right) \\
&\quad + \sum_{\mathfrak{u} \subseteq [d-2]} \gamma_{\mathfrak{u} \cup \{d-1\}} (\log N)^{|\mathfrak{u}|+1} \left(1 + \frac{\gamma_{\mathfrak{u} \cup \{d-1\} \cup \{d\}}}{\gamma_{\mathfrak{u} \cup \{d-1\}}} \log N\right) \\
&\leq \sum_{\mathfrak{u} \subseteq [d-2]} \gamma_{\mathfrak{u}} (\log N)^{|\mathfrak{u}|} \left(1 + \frac{\gamma_{\mathfrak{u} \cup \{d-1\}}}{\gamma_{\mathfrak{u}}} \log N\right) \left(1 + \max_{\mathfrak{v} \subseteq \{d-1\}} \frac{\gamma_{\mathfrak{u} \cup \mathfrak{v} \cup \{d\}}}{\gamma_{\mathfrak{u} \cup \mathfrak{v}}} \log N\right)
\end{aligned}$$

$$\leq \sum_{\mathfrak{u}\subseteq[d-3]} \gamma_{\mathfrak{u}} (\log N)^{|\mathfrak{u}|} \left(1 + \frac{\gamma_{\mathfrak{u}\cup\{d-2\}}}{\gamma_{\mathfrak{u}}} \log N\right)$$
$$\times \left(1 + \max_{\mathfrak{v}\subseteq\{d-2\}} \frac{\gamma_{\mathfrak{u}\cup\mathfrak{v}\cup\{d-1\}}}{\gamma_{\mathfrak{u}\cup\mathfrak{v}}} \log N\right) \left(1 + \max_{\mathfrak{v}\subseteq\{d-2,d-1\}} \frac{\gamma_{\mathfrak{u}\cup\mathfrak{v}\cup\{d\}}}{\gamma_{\mathfrak{u}\cup\mathfrak{v}}} \log N\right).$$

Repeatedly applying this argument, we obtain

$$\sum_{\mathfrak{u}\subseteq[d]} \gamma_{\mathfrak{u}} (\log N)^{|\mathfrak{u}|} \leq \prod_{j=1}^{d} \left(1 + \widetilde{\gamma}_j \log N\right), \qquad \text{where } \widetilde{\gamma}_j := \max_{\mathfrak{v}\subseteq[j-1]} \frac{\gamma_{\mathfrak{v}\cup\{j\}}}{\gamma_{\mathfrak{v}}}.$$

Now the general result follows similarly as (3.35). □

We can combine the previous results to the following theorem.

Theorem 3.21 *Let $N > 2$ be a prime number. Then the following statements hold true.*

1. *Let $\boldsymbol{\gamma} = \{\gamma_{\mathfrak{u}}\}_{\mathfrak{u}\subseteq[d]}$ be positive weights satisfying*

$$\sum_{j=1}^{\infty} \max_{\mathfrak{v}\subseteq[j-1]} \frac{\gamma_{\mathfrak{v}\cup\{j\}}}{\gamma_{\mathfrak{v}}} < \infty.$$

Then for any $\delta > 0$ there exists a positive real $C(\boldsymbol{\gamma}, \delta)$, that depends only on the weights $\boldsymbol{\gamma}$ and on δ, with the following property. For any $d \in \mathbb{N}$, if $\boldsymbol{g} = (g_1, \ldots, g_d)$ has been constructed according to Algorithm 3.14 using the weights $\{\boldsymbol{\gamma}_{\mathfrak{u}}\}_{\mathfrak{u}\subseteq[d]}$, we have

$$R_{N,d,1,\boldsymbol{\gamma}}(\boldsymbol{g}) \leq \frac{C(\boldsymbol{\gamma}, \delta)}{N^{1-\delta}}.$$

2. *Let $\boldsymbol{\gamma} = \{\gamma_{\mathfrak{u}}\}_{\mathfrak{u}\subseteq[d]}$ be positive weights satisfying*

$$\sum_{j=1}^{\infty} \max_{\mathfrak{v}\subseteq[j-1]} \frac{\gamma_{\mathfrak{v}\cup\{j\}}}{\gamma_{\mathfrak{v}}} < \infty.$$

Then for any $\alpha > 1/2$ and any $\delta > 0$ there exists a positive real $\widetilde{C}(\boldsymbol{\gamma}, \delta)$, that depends only on the weights $\boldsymbol{\gamma}$ and on δ, with the following property. For any $d \in \mathbb{N}$, if $\boldsymbol{g} = (g_1, \ldots, g_d)$ has been constructed according to Algorithm 3.14 using the weights $\{\boldsymbol{\gamma}_{\mathfrak{u}}\}_{\mathfrak{u}\subseteq[d]}$, we have

$$\mathrm{err}_{N,d,\alpha,\boldsymbol{\gamma}^{2\alpha}}(\boldsymbol{g}) \leq \frac{\widetilde{C}(\boldsymbol{\gamma}, \delta)}{N^{\alpha-\delta}}.$$

3. *Let $\boldsymbol{\gamma} = \{\gamma_{\mathfrak{u}}\}_{\mathfrak{u}\subseteq[d]}$ be positive weights satisfying*

$$\sum_{j=1}^{\infty} \max_{\mathfrak{v}\subseteq[j-1]} \frac{\gamma_{\mathfrak{v}\cup\{j\}}^{1/(2\alpha)}}{\gamma_{\mathfrak{v}}^{1/(2\alpha)}} < \infty.$$

Then for any $\alpha > 1/2$ *and any* $\delta > 0$ *there exists a positive real* $\widetilde{C}(\boldsymbol{\gamma}^{1/(2\alpha)}, \delta)$, *that depends only on the weights* $\boldsymbol{\gamma}^{1/(2\alpha)}$ *and on* δ, *with the following property. For any* $d \in \mathbb{N}$, *if* $\boldsymbol{g} = (g_1, \ldots, g_d)$ *has been constructed according to Algorithm 3.14 using the weights* $\{\gamma_{\mathfrak{u}}^{1/(2\alpha)}\}_{\mathfrak{u} \subseteq [d]}$, *we have*

$$\mathrm{err}_{N,d,\alpha,\boldsymbol{\gamma}}(\boldsymbol{g}) \leq \frac{\widetilde{C}(\boldsymbol{\gamma}^{1/(2\alpha)}, \delta)}{N^{\alpha-\delta}}.$$

Proof Regarding Item 1, we have, according to Remark 3.16,

$$\begin{aligned} R_{N,d,1,\boldsymbol{\gamma}}(\boldsymbol{g}) &\leq \frac{1}{N} \sum_{\emptyset \neq \mathfrak{u} \subseteq [d]} \gamma_{\mathfrak{u}} (2 \log N + 2 - \log 4)^{|\mathfrak{u}|} \\ &\leq \frac{1}{N} \sum_{\emptyset \neq \mathfrak{u} \subseteq [d]} \gamma_{\mathfrak{u}} (3 \log N)^{|\mathfrak{u}|}. \end{aligned}$$

A simple adaption of Lemma 3.20 yields the result.

Regarding Item 2, this result can be shown by combining Item 2 of Theorem 3.17 with Lemma 3.20, and noting that the assumptions of that lemma imply that the second sum in the bound in Item 2 of Theorem 3.17 can be bounded by a constant that is independent of d.

Regarding Item 3, the reasoning is analogous to Item 2. □

Notes and Remarks

For further information on the relation of two-dimensional lattice point sets to Diophantine approximation we refer to [6, 9, 199, 230] or [204, Example 4.3.15].

Exhaustive searches for generating vectors of lattice rules were undertaken in several papers, see, for example, [89, 90, 186, 187, 223].

Averaging results for Korobov type lattice points can be found, for example, in [139, 199, 204].

The original version of Theorem 3.7 according to Kuo [156, Theorem 4] is slightly more general. She considered a Korobov space with a reproducing kernel depending on a sequence of additional parameters $(\beta_j)_{j \geq 1}$. For the sake of simplicity, we choose all β_j equal to 1 in the present book, as it has been common during the last years. For further details on the case of general β_j, we refer to [156].

The generalization of [156, Theorem 4] to arbitrary $N \in \mathbb{N}$ was done in [35]. The proof of the general error bound in Theorem 3.9 follows the presentation in [48, Proof of Theorem 5.8].

As already mentioned, the fast CBC approach was developed by Nuyens and Cools (see [213, 214, 215]). Related survey articles are [31, 32]. In the outline of the fast matrix-vector product by means of FFT we followed the book [180, Section 4.2]. For the presentation of the fast CBC construction we restricted ourselves to the case of prime N for the sake of simplicity. The general case can be handled analogously; see [215]. Note also that for the case of prime-power N the fast implementation of the CBC algorithm is a special case of the fast implementation of the so-called reduced CBC algorithm, which we shall outline in Section 4.2. Indeed, by setting all reduction indices w_j equal to zero there, we obtain a fast implementation of Algorithm 3.6 for prime-power N. In the present description of the fast CBC construction for POD weights we followed [48].

For implementations of the fast CBC construction, we refer to the websites of various active researchers in this field, and to several ongoing projects that provide software for generating QMC point sets. We also refer to Appendix B of the present book for numerical experiments corresponding to the construction algorithms introduced in this chapter.

Lemma 3.20 is a generalization of [105, Lemma 3] and of [160, Lemma 4.4] that hold for product weights.

It should also be noted that a CBC algorithm similar to Algorithm 3.14 with similar error bounds on the obtained lattice rules has recently been analyzed in [69], where it was also shown that it can be implemented using the same order of magnitude of operations as the fast implementation of Algorithm 3.6, but without the need to use the fast Fourier transform. Furthermore, numerical results indicate that both algorithms yield lattice rules of similar quality.

Chapter 4
Modified Construction Schemes

In the previous chapter we have introduced the basic concept of the (fast) CBC construction for lattice rules together with an analysis of the quality of the resulting generating vectors and an analysis of the runtime of the algorithms. In this chapter we will continue to study this important topic and will present options to fine-tune the CBC construction for various situations and applications. These include the reduced CBC construction, the successive coordinate search construction, the projection-corrected CBC construction, and the component-by-component digit-by-digit construction, which all shall be presented here.

This chapter contains advanced material that can be skipped by beginners.

Let us begin with the reduced CBC construction.

4.1 The Reduced CBC Construction

We have outlined in Section 3.4 how the CBC construction can be implemented using only $O(d\,N \log N)$ operations. In modern applications, however, the values of N and d might be extremely large simultaneously, which will make even this order of magnitude infeasible for practical use of the CBC construction. Hence, it would be beneficial to further reduce the runtime of the CBC construction, at least for certain, suitably chosen function classes.

A second motivation stems from tractability results for integration in weighted function spaces as discussed, for example, in Section 2.6. It is known that, in order to obtain strong polynomial tractability of the integration problem in the weighted Korobov space $\mathcal{H}_{\mathrm{kor},d,\alpha,\boldsymbol{\gamma}}$ with product weights $\boldsymbol{\gamma} = (\gamma_j)_{j\ge 1}$, it is necessary and sufficient (see Corollary 2.28) that $\boldsymbol{\gamma}$ satisfies

$$\sum_{j=1}^{\infty} \gamma_j < \infty.$$

J. Dick et al., *Lattice Rules*, Springer Series in Computational Mathematics 58,
https://doi.org/10.1007/978-3-031-09951-9_4

If even

$$\sum_{j=1}^{\infty} \gamma_j^{1/(2\tau)} < \infty \tag{4.1}$$

for some $\tau \in [1/2, \alpha)$, then one can choose the ε-exponent to be $1/\tau$ (see again Corollary 2.28 or Theorem 3.7). Assume, on the other hand, that (4.1) holds for some $\tau > \alpha$. Then no further advantage is obtained over the results from Chapter 3, since one still gets strong polynomial tractability with the optimal ε-exponent and the construction cost of the lattice rule is independent of the choice of weights. However, since (4.1) implies that the importance of coordinates with bigger index is much smaller than that of earlier ones, it seems unreasonable to spend the same amount of work to search for the corresponding component of the generating vector. This is exactly the key idea that is discussed in this section. We will see below that in a situation where (4.1) holds for $\tau > \alpha$, we can reduce the construction cost of the generating vector of a lattice rule by making the search space for later components smaller than for earlier ones, while still achieving strong polynomial tractability with the optimal ε-exponent.

The reduced fast CBC construction

Our approach is the following. In this section, we do not assume that N is a prime number, but a prime power of the form $N = b^m$, where b is prime and $m \in \mathbb{N}$. Furthermore, let $w_1, w_2, \ldots, w_d \in \mathbb{N}_0$ with $w_1 \leq w_2 \leq \cdots \leq w_d$. We will sometimes refer to the w_j as *reduction indices* in the following. The most interesting case is when $w_1 = 0$, since otherwise the construction below results in each point just being counted b^{w_1} times.

The reduced fast CBC construction is based on the idea to shrink the search spaces for later components g_j of the generating vector $\boldsymbol{g} = (g_1, \ldots, g_d)$ of a rank-1 lattice rule. While in the common CBC construction (cf. Algorithms 3.5 and 3.6) g_j is chosen from $G_1^{\varphi}(N)$ for each $j \in [d]$, we introduce search spaces $Z_{w_j}(N)$ here, which depend on the choice of the w_j and are in general different for each j. To be more precise, we define

$$Z_{w_j}(N) := \begin{cases} \{g \in \{1, 2, \ldots, b^{m-w_j} - 1\} : \gcd(g, b) = 1\} & \text{if } w_j < m, \\ \{1\} & \text{if } w_j \geq m. \end{cases} \tag{4.2}$$

Note that then

$$|Z_{w_j}(N)| = \begin{cases} b^{m-w_j-1}(b-1) & \text{if } w_j < m, \\ 1 & \text{if } w_j \geq m. \end{cases}$$

In the following, we write

$$Y_j := b^{w_j} \quad \text{for } j \in [d]. \tag{4.3}$$

We can now formulate the reduced fast CBC construction algorithm for generating vectors $\boldsymbol{g}$ with respect to general weights $\boldsymbol{\gamma} = \{\gamma_{\mathfrak{u}}\}_{\mathfrak{u}\subseteq[d]}$.

Algorithm 4.1 (Reduced CBC construction for the weighted Korobov space) Let b be a prime number, let $m \in \mathbb{N}_0$ and $N = b^m$. Let $\boldsymbol{\gamma} = \{\gamma_{\mathfrak{u}}\}_{\mathfrak{u}\subseteq[d]}$ be general weights, and let $w_1, w_2, \ldots, w_d \in \mathbb{N}_0$ with $w_1 \leq w_2 \leq \cdots \leq w_d$. Furthermore, let $Y_1, Y_2, \ldots, Y_d$ be as in (4.3). Construct $\boldsymbol{g} = (Y_1 g_1, \ldots, Y_d g_d)$ as follows.

(1) Choose $g_1 = 1$.
(2) For s from 1 to $d-1$:
Assume that $g_j \in Z_{w_j}(N)$, $j \in [s]$, have already been found. Choose $g_{s+1} \in Z_{w_{s+1}}(N)$ as

$$g_{s+1} := \underset{g \in Z_{w_{s+1}}(N)}{\operatorname{argmin}} \ \operatorname{err}_{N,s+1,\alpha,\boldsymbol{\gamma}}((Y_1 g_1, \ldots, Y_s g_s, Y_{s+1} g)).$$

End for.
(3) Set $\boldsymbol{g} = (Y_1 g_1, \ldots, Y_d g_d)$.

Just as the classical CBC construction (see Algorithms 3.5 and 3.6), also the reduced CBC construction is extensible in the dimension. If one would like to add further components to the obtained vector later, this can be done by choosing further reduction indices $w_{d+1}, w_{d+2}, \ldots$ as required, and then running the loop in the algorithm as often as necessary.

Remark 4.2 Note that when setting $w_1 = w_2 = \cdots = w_d = 0$, Algorithm 4.1 coincides with Algorithm 3.6, hence the reduced CBC construction is a generalization of the usual CBC construction.

Error analysis

We show the following theorem from [45], which states that Algorithm 4.1 yields generating vectors $\boldsymbol{g}$ with a small integration error.

Theorem 4.3 *Let b be a prime number, let $m \in \mathbb{N}_0$ and $N = b^m$. Let $\boldsymbol{\gamma} = \{\gamma_{\mathfrak{u}}\}_{\mathfrak{u}\subseteq[d]}$ be general weights, and let $w_1, w_2, \ldots, w_d \in \mathbb{N}_0$ with $w_1 \leq w_2 \leq \cdots \leq w_d$. Furthermore, let $Y_1, Y_2, \ldots, Y_d$ be as in* (4.3). *Assume that $\boldsymbol{g} = (Y_1 g_1, \ldots, Y_d g_d) \in \mathbb{Z}^d$ has been constructed according to Algorithm 4.1. Then for arbitrary $\tau \in [1/2, \alpha)$ and any $s \in [d]$ we have*

$$\operatorname{err}_{N,s,\alpha,\boldsymbol{\gamma}}((Y_1 g_1, \ldots, Y_s g_s)) \leq \left(\sum_{\emptyset \neq \mathfrak{u} \subseteq [s]} \gamma_{\mathfrak{u}}^{1/(2\tau)} \frac{2(2\zeta(\alpha/\tau))^{|\mathfrak{u}|}}{b^{\max(0, m - \max_{j\in\mathfrak{u}} w_j)}} \right)^{\tau}. \tag{4.4}$$

Proof We recall that we denote the dual lattice of an integration lattice $\mathcal{L}$ by $\mathcal{L}^{\perp}$. Furthermore, for $\emptyset \neq \mathfrak{u} \subseteq [d]$ and a lattice $\mathcal{L}$ with generating vector $\boldsymbol{g}$, we write

$$\mathcal{L}_{\mathfrak{u}}^{\perp} := \{\boldsymbol{h}_{\mathfrak{u}} \in (\mathbb{Z} \setminus \{0\})^{|\mathfrak{u}|} : \boldsymbol{h}_{\mathfrak{u}} \cdot \boldsymbol{g}_{\mathfrak{u}} \equiv 0 \pmod{N}\}. \tag{4.5}$$

The result is shown by induction on s.

For $s = 1$, we have $g_1 = 1$. Thus we have, using (2.23) in Theorem 2.19,

$$[\mathrm{err}_{N,1,\alpha,\gamma_{\{1\}}}(Y_1)]^2 = \gamma_{\{1\}} \sum_{\substack{h\in\mathbb{Z}\setminus\{0\}\\ hY_1\equiv 0 \pmod{N}}} \frac{1}{r_{2\alpha}(h)}.$$

Let now $\lambda \in (1/(2\alpha), 1]$. Applying Jensen's inequality (Lemma 2.25) to the squared error $[\mathrm{err}_{N,1,\alpha,\gamma_{\{1\}}}(Y_1)]^2$, and noting that $(r_{2\alpha}(h))^{\lambda} = r_{2\alpha\lambda}(h)$ for $h \in \mathbb{Z}$, we obtain

$$[\mathrm{err}_{N,1,\alpha,\gamma_{\{1\}}}(Y_1)]^{2\lambda} \le \gamma_{\{1\}}^{\lambda} \sum_{\substack{h\in\mathbb{Z}\setminus\{0\}\\ hY_1\equiv 0 \pmod{N}}} \frac{1}{r_{2\alpha\lambda}(h)}.$$

If $w_1 \ge m$, we have $N|Y_1$, and the condition $hY_1 \equiv 0 \pmod{N}$ is satisfied for any integer h. Consequently,

$$[\mathrm{err}_{N,1,\alpha,\gamma_{\{1\}}}(Y_1)]^{2\lambda} \le \gamma_{\{1\}}^{\lambda} \sum_{h\in\mathbb{Z}\setminus\{0\}} \frac{1}{r_{2\alpha\lambda}(h)} = \gamma_{\{1\}}^{\lambda} 2\zeta(2\alpha\lambda).$$

If $w_1 < m$, then $hY_1 \equiv 0 \pmod{N}$ is equivalent to $b^{m-w_1}|h$, and

$$[\mathrm{err}_{N,1,\alpha,\gamma_{\{1\}}}(Y_1)]^{2\lambda} \le \gamma_{\{1\}}^{\lambda} \sum_{\substack{h\in\mathbb{Z}\setminus\{0\}\\ b^{m-w_1}|h}} \frac{1}{r_{2\alpha\lambda}(h)} = \gamma_{\{1\}}^{\lambda} \sum_{h\in\mathbb{Z}\setminus\{0\}} \frac{1}{r_{2\alpha\lambda}(hb^{m-w_1})}.$$

For $h \in \mathbb{Z} \setminus \{0\}$ we have $r_{2\alpha\lambda}(hb^{m-w_1}) = b^{-2\alpha\lambda(m-w_1)} r_{2\alpha\lambda}(h)$, and we also have $2\alpha\lambda > 1$. Consequently, we obtain

$$[\mathrm{err}_{N,1,\alpha,\gamma_{\{1\}}}(Y_1)]^{2\lambda} \le \frac{\gamma_{\{1\}}^{\lambda}}{b^{2\alpha\lambda(m-w_1)}} \sum_{h\in\mathbb{Z}\setminus\{0\}} \frac{1}{r_{2\alpha\lambda}(h)} = \gamma_{\{1\}}^{\lambda} \frac{2\zeta(2\alpha\lambda)}{b^{2\alpha\lambda(m-w_1)}}.$$

In any case,

$$[\mathrm{err}_{N,1,\alpha,\gamma_{\{1\}}}(Y_1)]^{2\lambda} \le \gamma_{\{1\}}^{\lambda} \frac{2\zeta(2\alpha\lambda)}{b^{2\alpha\lambda \max(0,m-w_1)}} \le \gamma_{\{1\}}^{\lambda} \frac{4\zeta(2\alpha\lambda)}{b^{2\alpha\lambda \max(0,m-w_1)}}.$$

Taking the left-hand side and the right-hand side of the previous inequality to the power $1/(2\lambda)$ and setting $\tau = 1/(2\lambda)$, which implies $\tau \in [1/2, \alpha)$, we obtain

$$\mathrm{err}_{N,1,\alpha,\gamma_{\{1\}}}(Y_1) \le \gamma_{\{1\}}^{1/2} \frac{(4\zeta(\alpha/\tau))^{\tau}}{b^{\alpha \max(0,m-w_1)}} \le \left(\gamma_{\{1\}}^{1/(2\tau)} \frac{4\zeta(\alpha/\tau)}{b^{\max(0,m-w_1)}}\right)^{\tau},$$

and this is the desired result (4.4) for the case $s = 1$.

Assume now that we have shown the result for some fixed s, i.e., the generating vector $\boldsymbol{g}^{(s)} = (Y_1 g_1, \ldots, Y_s g_s)$ satisfies

$$\mathrm{err}_{N,s,\alpha,\boldsymbol{\gamma}}(\boldsymbol{g}^{(s)}) \leq \left(\sum_{\emptyset \neq \mathfrak{u} \subseteq [s]} \gamma_{\mathfrak{u}}^{1/(2\tau)} \frac{2(2\zeta(\alpha/\tau))^{|\mathfrak{u}|}}{b^{\max(0, m - \max_{j \in \mathfrak{u}} w_j)}} \right)^{\tau}. \tag{4.6}$$

Furthermore, assume that $g_{s+1} \in Z_{w_{s+1}}(N)$ has been chosen according to Algorithm 4.1. Then, using (2.23) in Theorem 2.19,

$$\begin{aligned}
&[\mathrm{err}_{N,s+1,\alpha,\boldsymbol{\gamma}}((\boldsymbol{g}^{(s)}, Y_{s+1} g_{s+1}))]^2 \\
&= \sum_{\emptyset \neq \mathfrak{u} \subseteq [s+1]} \gamma_{\mathfrak{u}} \sum_{\boldsymbol{h}_{\mathfrak{u}} \in \mathcal{L}_{\mathfrak{u}}^{\perp}} \frac{1}{r_{2\alpha}(\boldsymbol{h}_{\mathfrak{u}})} \\
&= \sum_{\emptyset \neq \mathfrak{u} \subseteq [s]} \gamma_{\mathfrak{u}} \sum_{\boldsymbol{h}_{\mathfrak{u}} \in \mathcal{L}_{\mathfrak{u}}^{\perp}} \frac{1}{r_{2\alpha}(\boldsymbol{h}_{\mathfrak{u}})} + \sum_{\substack{\mathfrak{u} \subseteq [s+1] \\ s+1 \in \mathfrak{u}}} \gamma_{\mathfrak{u}} \sum_{\boldsymbol{h}_{\mathfrak{u}} \in \mathcal{L}_{\mathfrak{u}}^{\perp}} \frac{1}{r_{2\alpha}(\boldsymbol{h}_{\mathfrak{u}})} \\
&= [\mathrm{err}_{N,s,\alpha,\boldsymbol{\gamma}}(\boldsymbol{g}^{(s)})]^2 + \theta_{N,s,\alpha,\boldsymbol{\gamma}}(\boldsymbol{g}^{(s)}, g_{s+1}),
\end{aligned} \tag{4.7}$$

where we write

$$\theta_{N,s,\alpha,\boldsymbol{\gamma}}(\boldsymbol{g}^{(s)}, g_{s+1}) = \sum_{\substack{\mathfrak{u} \subseteq [s+1] \\ s+1 \in \mathfrak{u}}} \gamma_{\mathfrak{u}} \sum_{\boldsymbol{h}_{\mathfrak{u}} \in \mathcal{L}_{\mathfrak{u}}^{\perp}} \frac{1}{r_{2\alpha}(\boldsymbol{h}_{\mathfrak{u}})}, \tag{4.8}$$

which is the only term in $[\mathrm{err}_{N,s+1,\alpha,\boldsymbol{\gamma}}((\boldsymbol{g}^{(s)}, Y_{s+1} g_{s+1}))]^2$ that depends on g_{s+1} (note that the dependence on g_{s+1} is in $\mathcal{L}_{\mathfrak{u}}^{\perp}$).

Let $\lambda \in (1/(2\alpha), 1]$. We now analyze the average of $(\theta_{N,s,\alpha,\boldsymbol{\gamma}}(\boldsymbol{g}^{(s)}, g))^{\lambda}$ over all possible values $g \in Z_{w_{s+1}}(N)$,

$$\Theta_{N,s,\alpha,\boldsymbol{\gamma},\lambda}(\boldsymbol{g}^{(s)}) := \frac{1}{|Z_{w_{s+1}}(N)|} \sum_{g \in Z_{w_{s+1}}(N)} (\theta_{N,s,\alpha,\boldsymbol{\gamma}}(\boldsymbol{g}^{(s)}, g))^{\lambda},$$

where $\theta_{N,s,\alpha,\boldsymbol{\gamma}}(\boldsymbol{g}^{(s)}, g)$ is the analogue of (4.8) for $g \in Z_{w_{s+1}}(N)$. We now have, using Jensen's inequality (Lemma 2.25) twice,

$$\begin{aligned}
\Theta_{N,s,\alpha,\boldsymbol{\gamma},\lambda}(\boldsymbol{g}^{(s)}) &\leq \frac{1}{|Z_{w_{s+1}}(N)|} \sum_{g \in Z_{w_{s+1}}(N)} \sum_{\substack{\mathfrak{u} \subseteq [s+1] \\ s+1 \in \mathfrak{u}}} \gamma_{\mathfrak{u}}^{\lambda} \sum_{\boldsymbol{h}_{\mathfrak{u}} \in \mathcal{L}_{\mathfrak{u}}^{\perp}} \frac{1}{r_{2\alpha\lambda}(\boldsymbol{h}_{\mathfrak{u}})} \\
&= \frac{1}{|Z_{w_{s+1}}(N)|} \sum_{g \in Z_{w_{s+1}}(N)} \gamma_{\{s+1\}}^{\lambda} \sum_{\substack{h_{s+1} \in \mathbb{Z} \setminus \{0\} \\ h_{s+1} Y_{s+1} g \equiv 0 \pmod{N}}} \frac{1}{r_{2\alpha\lambda}(h_{s+1})} \\
&\quad + \frac{1}{|Z_{w_{s+1}}(N)|} \sum_{g \in Z_{w_{s+1}}(N)} \sum_{\emptyset \neq \mathfrak{v} \subseteq [s]} \gamma_{\mathfrak{v} \cup \{s+1\}}^{\lambda} \sum_{h_{s+1} \in \mathbb{Z} \setminus \{0\}} \frac{1}{r_{2\alpha\lambda}(h_{s+1})}
\end{aligned}$$

$$\times \sum_{\substack{\boldsymbol{h}_{\mathfrak{v}} \in (\mathbb{Z}\setminus\{0\})^{|\mathfrak{v}|} \\ \sum_{j\in\mathfrak{v}} h_j Y_j g_j \equiv -h_{s+1}Y_{s+1}g \pmod{N}}} \frac{1}{r_{2\alpha\lambda}(\boldsymbol{h}_{\mathfrak{v}})}$$

$$= T_1 + T_2,$$

where

$$T_1 := \frac{1}{|Z_{w_{s+1}}(N)|} \sum_{g\in Z_{w_{s+1}}(N)} \gamma^{\lambda}_{\{s+1\}} \sum_{\substack{h_{s+1}\in\mathbb{Z}\setminus\{0\} \\ h_{s+1}Y_{s+1}g\equiv 0 \pmod{N}}} \frac{1}{r_{2\alpha\lambda}(h_{s+1})},$$

and

$$T_2 := \frac{1}{|Z_{w_{s+1}}(N)|} \sum_{g\in Z_{w_{s+1}}(N)} \sum_{\emptyset\neq\mathfrak{v}\subseteq[s]} \gamma^{\lambda}_{\mathfrak{v}\cup\{s+1\}} \sum_{h_{s+1}\in\mathbb{Z}\setminus\{0\}} \frac{1}{r_{2\alpha\lambda}(h_{s+1})}$$
$$\times \sum_{\substack{\boldsymbol{h}_{\mathfrak{v}} \in (\mathbb{Z}\setminus\{0\})^{|\mathfrak{v}|} \\ \sum_{j\in\mathfrak{v}} h_j Y_j g_j \equiv -h_{s+1}Y_{s+1}g \pmod{N}}} \frac{1}{r_{2\alpha\lambda}(\boldsymbol{h}_{\mathfrak{v}})}.$$

We first consider T_1. Since $\gcd(g, N) = 1$, the congruence $h_{s+1}Y_{s+1}g \equiv 0 \pmod{N}$ is equivalent to $h_{s+1}Y_{s+1} \equiv 0 \pmod{N}$. Thus by the same arguments as for the case $s = 1$ we obtain

$$T_1 \leq \frac{4\gamma^{\lambda}_{\{s+1\}}\zeta(2\alpha\lambda)}{b^{\max(0,m-w_{s+1})}}. \tag{4.9}$$

Regarding T_2, we distinguish the two cases $w_{s+1} \geq m$ and $w_{s+1} < m$.

If $w_{s+1} \geq m$, we have $Z_{w_{s+1}}(N) = \{1\}$, $g = 1$, and $h_{s+1}Y_{s+1}g \equiv 0 \pmod{N}$, since $N|Y_{s+1}$ and therefore $N|h_{s+1}Y_{s+1}g$. Consequently, T_2 simplifies to

$$\begin{aligned}
T_2 &= \sum_{\emptyset\neq\mathfrak{v}\subseteq[s]} \gamma^{\lambda}_{\mathfrak{v}\cup\{s+1\}} \sum_{h_{s+1}\in\mathbb{Z}\setminus\{0\}} \frac{1}{r_{2\alpha\lambda}(h_{s+1})} \sum_{\substack{\boldsymbol{h}_{\mathfrak{v}} \in (\mathbb{Z}\setminus\{0\})^{|\mathfrak{v}|} \\ \sum_{j\in\mathfrak{v}} h_j Y_j g_j \equiv 0 \pmod{N}}} \frac{1}{r_{2\alpha\lambda}(\boldsymbol{h}_{\mathfrak{v}})} \\
&= 2\zeta(2\alpha\lambda) \sum_{\emptyset\neq\mathfrak{v}\subseteq[s]} \gamma^{\lambda}_{\mathfrak{v}\cup\{s+1\}} \sum_{\substack{\boldsymbol{h}_{\mathfrak{v}} \in (\mathbb{Z}\setminus\{0\})^{|\mathfrak{v}|} \\ \sum_{j\in\mathfrak{v}} h_j Y_j g_j \equiv 0 \pmod{N}}} \frac{1}{r_{2\alpha\lambda}(\boldsymbol{h}_{\mathfrak{v}})} \\
&\leq 2\zeta(2\alpha\lambda) \sum_{\emptyset\neq\mathfrak{v}\subseteq[s]} \gamma^{\lambda}_{\mathfrak{v}\cup\{s+1\}} \sum_{\boldsymbol{h}_{\mathfrak{v}} \in (\mathbb{Z}\setminus\{0\})^{|\mathfrak{v}|}} \frac{1}{r_{2\alpha\lambda}(\boldsymbol{h}_{\mathfrak{v}})} \\
&= 2\zeta(2\alpha\lambda) \sum_{\emptyset\neq\mathfrak{v}\subseteq[s]} \gamma^{\lambda}_{\mathfrak{v}\cup\{s+1\}} (2\zeta(2\alpha\lambda))^{|\mathfrak{v}|} \\
&= \sum_{\emptyset\neq\mathfrak{v}\subseteq[s]} \gamma^{\lambda}_{\mathfrak{v}\cup\{s+1\}} (2\zeta(2\alpha\lambda))^{|\mathfrak{v}|+1}.
\end{aligned}$$

If $w_{s+1} < m$, we split T_2 into two parts,

$$T_2 = T_{2,1} + T_{2,2},$$

where

$$T_{2,1} := \frac{1}{|Z_{w_{s+1}}(N)|} \sum_{g \in Z_{w_{s+1}}(N)} \sum_{\emptyset \neq \mathfrak{v} \subseteq [s]} \gamma^{\lambda}_{\mathfrak{v} \cup \{s+1\}} \sum_{\substack{h_{s+1} \in \mathbb{Z} \setminus \{0\} \\ h_{s+1} \equiv 0 \ (b^{m-w_{s+1}})}} \frac{1}{r_{2\alpha\lambda}(h_{s+1})} \times \sum_{\substack{\boldsymbol{h}_{\mathfrak{v}} \in (\mathbb{Z}\setminus\{0\})^{|\mathfrak{v}|} \\ \sum_{j \in \mathfrak{v}} h_j Y_j g_j \equiv 0 \ (\mathrm{mod}\ N)}} \frac{1}{r_{2\alpha\lambda}(\boldsymbol{h}_{\mathfrak{v}})}$$

and

$$T_{2,2} := \frac{1}{|Z_{w_{s+1}}(N)|} \sum_{g \in Z_{w_{s+1}}(N)} \sum_{\emptyset \neq \mathfrak{v} \subseteq [s]} \gamma^{\lambda}_{\mathfrak{v} \cup \{s+1\}} \sum_{\substack{h_{s+1} \in \mathbb{Z} \setminus \{0\} \\ h_{s+1} \not\equiv 0 \ (\mathrm{mod}\ b^{m-w_{s+1}})}} \frac{1}{r_{2\alpha\lambda}(h_{s+1})} \times \sum_{\substack{\boldsymbol{h}_{\mathfrak{v}} \in (\mathbb{Z}\setminus\{0\})^{|\mathfrak{v}|} \\ \sum_{j \in \mathfrak{v}} h_j Y_j g_j \equiv -h_{s+1} Y_{s+1} g \ (\mathrm{mod}\ N)}} \frac{1}{r_{2\alpha\lambda}(\boldsymbol{h}_{\mathfrak{v}})}.$$

For the term $T_{2,1}$, note that if $h_{s+1} \equiv 0 \pmod{b^{m-w_{s+1}}}$, then $h_{s+1} Y_{s+1} g \equiv 0 \pmod N$ for any $g \in Z_{w_{s+1}}(N)$, so we obtain

$$\begin{aligned}
T_{2,1} &= \sum_{\emptyset \neq \mathfrak{v} \subseteq [s]} \gamma^{\lambda}_{\mathfrak{v} \cup \{s+1\}} \sum_{\substack{\boldsymbol{h}_{\mathfrak{v}} \in (\mathbb{Z}\setminus\{0\})^{|\mathfrak{v}|} \\ \sum_{j \in \mathfrak{v}} h_j Y_j g_j \equiv 0 \ (\mathrm{mod}\ N)}} \frac{1}{r_{2\alpha\lambda}(\boldsymbol{h}_{\mathfrak{v}})} \sum_{\substack{h_{s+1} \in \mathbb{Z} \setminus \{0\} \\ h_{s+1} \equiv 0 \ (b^{m-w_{s+1}})}} \frac{1}{r_{2\alpha\lambda}(h_{s+1})} \\
&= \frac{2\zeta(2\alpha\lambda)}{b^{(m-w_{s+1})2\alpha\lambda}} \sum_{\emptyset \neq \mathfrak{v} \subseteq [s]} \gamma^{\lambda}_{\mathfrak{v} \cup \{s+1\}} \sum_{\substack{\boldsymbol{h}_{\mathfrak{v}} \in (\mathbb{Z}\setminus\{0\})^{|\mathfrak{v}|} \\ \sum_{j \in \mathfrak{v}} h_j Y_j g_j \equiv 0 \ (\mathrm{mod}\ N)}} \frac{1}{r_{2\alpha\lambda}(\boldsymbol{h}_{\mathfrak{v}})} \\
&\le \frac{2\zeta(2\alpha\lambda)}{b^{m-w_{s+1}}} \sum_{\emptyset \neq \mathfrak{v} \subseteq [s]} \gamma^{\lambda}_{\mathfrak{v} \cup \{s+1\}} \sum_{\substack{\boldsymbol{h}_{\mathfrak{v}} \in (\mathbb{Z}\setminus\{0\})^{|\mathfrak{v}|} \\ \sum_{j \in \mathfrak{v}} h_j Y_j g_j \equiv 0 \ (\mathrm{mod}\ N)}} \frac{1}{r_{2\alpha\lambda}(\boldsymbol{h}_{\mathfrak{v}})},
\end{aligned}$$

where we used that $2\alpha\lambda > 1$ in the last line.

For $T_{2,2}$ we obtain

$$T_{2,2} = \frac{1}{|Z_{w_{s+1}}(N)|} \sum_{g \in Z_{w_{s+1}}(N)} \sum_{\emptyset \neq \mathfrak{v} \subseteq [s]} \gamma^{\lambda}_{\mathfrak{v} \cup \{s+1\}} \times \sum_{c=1}^{b^{m-w_{s+1}}-1} \sum_{\substack{h_{s+1} \in \mathbb{Z} \setminus \{0\} \\ h_{s+1} \equiv -cg^{-1} \ (\mathrm{mod}\ b^{m-w_{s+1}})}} \frac{1}{r_{2\alpha\lambda}(h_{s+1})}$$

$$\times \sum_{\substack{\boldsymbol{h}_{\mathfrak{v}} \in (\mathbb{Z}\setminus\{0\})^{|\mathfrak{v}|} \\ \sum_{j\in\mathfrak{v}} h_j Y_j g_j \equiv c Y_{s+1} \pmod{N}}} \frac{1}{r_{2\alpha\lambda}(\boldsymbol{h}_{\mathfrak{v}})},$$

where g^{-1} denotes the multiplicative inverse of g in $Z_{w_{s+1}}(N)$. For fixed $c \in \{1, 2, \ldots, b^{m-w_{s+1}} - 1\}$ let $z := \gcd(c, b^{m-w_{s+1}})$. Then $\gcd(c/z, b^{m-w_{s+1}}/z) = 1$. Furthermore, note that

$$\{c\, g^{-1} \pmod{b^{m-w_{s+1}}} \; : \; g \in Z_{w_{s+1}}(N)\} = \{c\, g \pmod{b^{m-w_{s+1}}} \; : \; g \in Z_{w_{s+1}}(N)\}.$$

Hence

$$\begin{aligned}
&\sum_{g\in Z_{w_{s+1}}(N)} \sum_{\substack{h_{s+1}\in\mathbb{Z}\setminus\{0\} \\ h_{s+1}\equiv -cg^{-1} \pmod{b^{m-w_{s+1}}}}} \frac{1}{r_{2\alpha\lambda}(h_{s+1})} \\
&= \sum_{g\in Z_{w_{s+1}}(N)} \sum_{\substack{h_{s+1}\in\mathbb{Z}\setminus\{0\} \\ h_{s+1}\equiv -cg \pmod{b^{m-w_{s+1}}}}} \frac{1}{|h_{s+1}|^{2\alpha\lambda}} \\
&= \sum_{g\in Z_{w_{s+1}}(N)} \sum_{k\in\mathbb{Z}} \frac{1}{|k b^{m-w_{s+1}} - cg|^{2\alpha\lambda}} \\
&= \frac{1}{z^{2\alpha\lambda}} \sum_{g\in Z_{w_{s+1}}(N)} \sum_{k\in\mathbb{Z}} \frac{1}{|k(b^{m-w_{s+1}}/z) - (c/z)g|^{2\alpha\lambda}} \\
&= \frac{1}{z^{2\alpha\lambda}} \sum_{g\in Z_{w_{s+1}}(N)} \sum_{\substack{h\in\mathbb{Z}\setminus\{0\} \\ h\equiv -(c/z)g \pmod{b^{m-w_{s+1}}/z}}} \frac{1}{|h|^{2\alpha\lambda}} \\
&\le \frac{z}{z^{2\alpha\lambda}} \sum_{t=1}^{b^{m-w_{s+1}}/z-1} \sum_{\substack{h\in\mathbb{Z}\setminus\{0\} \\ h\equiv t \pmod{b^{m-w_{s+1}}/z}}} \frac{1}{|h|^{2\alpha\lambda}} \\
&\le \sum_{h\in\mathbb{Z}\setminus\{0\}} \frac{1}{|h|^{2\alpha\lambda}} \\
&= 2\zeta(2\alpha\lambda),
\end{aligned} \tag{4.10}$$

since $z \ge 1$ and $\lambda > 1/(2\alpha)$.

It follows that

$$\begin{aligned}
T_{2,2} &\le \frac{2\zeta(2\alpha\lambda)}{|Z_{w_{s+1}}(N)|} \sum_{\emptyset\neq\mathfrak{v}\subseteq[s]} \gamma_{\mathfrak{v}\cup\{s+1\}}^{\lambda} \sum_{c=1}^{b^{m-w_{s+1}}-1} \sum_{\substack{\boldsymbol{h}_{\mathfrak{v}} \in (\mathbb{Z}\setminus\{0\})^{|\mathfrak{v}|} \\ \sum_{j\in\mathfrak{v}} h_j Y_j g_j \equiv c Y_{s+1} \pmod{N}}} \frac{1}{r_{2\alpha\lambda}(\boldsymbol{h}_{\mathfrak{v}})} \\
&\le \frac{2\zeta(2\alpha\lambda)}{|Z_{w_{s+1}}(N)|} \sum_{\emptyset\neq\mathfrak{v}\subseteq[s]} \gamma_{\mathfrak{v}\cup\{s+1\}}^{\lambda} \sum_{\substack{\boldsymbol{h}_{\mathfrak{v}} \in (\mathbb{Z}\setminus\{0\})^{|\mathfrak{v}|} \\ \sum_{j\in\mathfrak{v}} h_j Y_j g_j \not\equiv 0 \pmod{N}}} \frac{1}{r_{2\alpha\lambda}(\boldsymbol{h}_{\mathfrak{v}})}.
\end{aligned}$$

Recall that $|Z_{w_{s+1}}(N)| = b^{m-w_{s+1}-1}(b-1) \geq b^{m-w_{s+1}}/2$. Therefore we get

$$\begin{aligned}
T_{2,2} &\leq \frac{4\zeta(2\alpha\lambda)}{b^{m-w_{s+1}}} \sum_{\emptyset \neq \mathfrak{v} \subseteq [s]} \gamma_{\mathfrak{v}\cup\{s+1\}}^{\lambda} \sum_{\substack{\boldsymbol{h}_{\mathfrak{v}} \in (\mathbb{Z}\setminus\{0\})^{|\mathfrak{v}|} \\ \sum_{j\in\mathfrak{v}} h_j Y_j g_j \not\equiv 0 \pmod{N}}} \frac{1}{r_{2\alpha\lambda}(\boldsymbol{h}_{\mathfrak{v}})} \\
&= \frac{4\zeta(2\alpha\lambda)}{b^{m-w_{s+1}}} \sum_{\emptyset \neq \mathfrak{v} \subseteq [s]} \gamma_{\mathfrak{v}\cup\{s+1\}}^{\lambda} \\
&\quad \times \left(\sum_{\boldsymbol{h}_{\mathfrak{v}} \in (\mathbb{Z}\setminus\{0\})^{|\mathfrak{v}|}} \frac{1}{r_{2\alpha\lambda}(\boldsymbol{h}_{\mathfrak{v}})} - \sum_{\substack{\boldsymbol{h}_{\mathfrak{v}} \in (\mathbb{Z}\setminus\{0\})^{|\mathfrak{v}|} \\ \sum_{j\in\mathfrak{v}} h_j Y_j g_j \equiv 0 \pmod{N}}} \frac{1}{r_{2\alpha\lambda}(\boldsymbol{h}_{\mathfrak{v}})} \right).
\end{aligned}$$

Thus, we obtain for $T_2 = T_{2,1} + T_{2,2}$,

$$\begin{aligned}
T_2 &\leq \frac{4\zeta(2\alpha\lambda)}{b^{m-w_{s+1}}} \sum_{\emptyset \neq \mathfrak{v} \subseteq [s]} \gamma_{\mathfrak{v}\cup\{s+1\}}^{\lambda} \sum_{\boldsymbol{h}_{\mathfrak{v}} \in (\mathbb{Z}\setminus\{0\})^{|\mathfrak{v}|}} \frac{1}{r_{2\alpha\lambda}(\boldsymbol{h}_{\mathfrak{v}})} \\
&= \sum_{\emptyset \neq \mathfrak{v} \subseteq [s]} \gamma_{\mathfrak{v}\cup\{s+1\}}^{\lambda} \frac{2(2\zeta(2\alpha\lambda))^{|\mathfrak{v}|+1}}{b^{m-w_{s+1}}}.
\end{aligned}$$

In any case, when $w_{s+1} \geq m$ and when $w_{s+1} < m$, we have

$$T_2 \leq \sum_{\emptyset \neq \mathfrak{v} \subseteq [s]} \gamma_{\mathfrak{v}\cup\{s+1\}}^{\lambda} \frac{2(2\zeta(2\alpha\lambda))^{|\mathfrak{v}|+1}}{b^{\max(0,m-w_{s+1})}}. \tag{4.11}$$

Combining (4.9) and (4.11) yields

$$\Theta_{N,s,\alpha,\boldsymbol{\gamma},\lambda}(\boldsymbol{g}^{(s)}) \leq \sum_{\substack{\mathfrak{u} \subseteq [s+1] \\ s+1 \in \mathfrak{u}}} \gamma_{\mathfrak{u}}^{\lambda} \frac{2(2\zeta(2\alpha\lambda))^{|\mathfrak{u}|}}{b^{\max(0,m-w_{s+1})}}.$$

By construction, the g_{s+1} chosen by Algorithm 4.1 must satisfy

$$(\theta_{N,s,\alpha,\boldsymbol{\gamma}}(\boldsymbol{g}^{(s)}, g_{s+1}))^{\lambda} \leq \Theta_{N,s,\alpha,\boldsymbol{\gamma},\lambda}(\boldsymbol{g}^{(s)}) \leq \sum_{\substack{\mathfrak{u} \subseteq [s+1] \\ s+1 \in \mathfrak{u}}} \gamma_{\mathfrak{u}}^{\lambda} \frac{2(2\zeta(2\alpha\lambda))^{|\mathfrak{u}|}}{b^{\max(0,m-w_{s+1})}}, \tag{4.12}$$

where we again used the standard averaging argument in Remark 2.14.

We then obtain from (4.7) that

$$\begin{aligned}
&[\mathrm{err}_{N,s+1,\alpha,\boldsymbol{\gamma}}((\boldsymbol{g}^{(s)}, Y_{s+1}g_{s+1}))]^{2\lambda} \\
&= \left([\mathrm{err}_{N,s,\alpha,\boldsymbol{\gamma}}(\boldsymbol{g}^{(s)})]^2 + \theta_{N,s,\alpha,\boldsymbol{\gamma}}(\boldsymbol{g}^{(s)}, g_{s+1}) \right)^{\lambda} \\
&\leq [\mathrm{err}_{N,s,\alpha,\boldsymbol{\gamma}}(\boldsymbol{g}^{(s)})]^{2\lambda} + (\theta_{N,s,\alpha,\boldsymbol{\gamma}}(\boldsymbol{g}^{(s)}, g_{s+1}))^{\lambda}
\end{aligned}$$

$$\le [\mathrm{err}_{N,s,\alpha,\boldsymbol{\gamma}}(\boldsymbol{g}^{(s)})]^{2\lambda} + \sum_{\substack{\mathfrak{u}\subseteq[s+1]\\ s+1\in\mathfrak{u}}} \gamma_{\mathfrak{u}}^{\lambda} \frac{2(2\zeta(2\alpha\lambda))^{|\mathfrak{u}|}}{b^{\max(0,m-w_{s+1})}},$$

where we once again used Lemma 2.25, and also (4.12). Setting again $\tau = 1/(2\lambda)$ we arrive at

$$\begin{aligned}
&[\mathrm{err}_{N,s+1,\alpha,\boldsymbol{\gamma}}((\boldsymbol{g}^{(s)}, Y_{s+1}g_{s+1}))]^{1/\tau}\\
&\le [\mathrm{err}_{N,s,\alpha,\boldsymbol{\gamma}}(\boldsymbol{g}^{(s)})]^{1/\tau} + \sum_{\substack{\mathfrak{u}\subseteq[s+1]\\ s+1\in\mathfrak{u}}} \gamma_{\mathfrak{u}}^{1/(2\tau)} \frac{2(2\zeta(\alpha/\tau))^{|\mathfrak{u}|}}{b^{\max(0,m-w_{s+1})}}.
\end{aligned}$$

Using the induction hypothesis (4.6) we get

$$\begin{aligned}
&[\mathrm{err}_{N,s+1,\alpha,\boldsymbol{\gamma}}((\boldsymbol{g}^{(s)}, Y_{s+1}g_{s+1}))]^{1/\tau}\\
&\le \sum_{\emptyset\neq\mathfrak{u}\subseteq[s]} \gamma_{\mathfrak{u}}^{1/(2\tau)} \frac{2(2\zeta(\alpha/\tau))^{|\mathfrak{u}|}}{b^{\max(0,m-\max_{j\in\mathfrak{u}} w_j)}} + \sum_{\substack{\mathfrak{u}\subseteq[s+1]\\ s+1\in\mathfrak{u}}} \gamma_{\mathfrak{u}}^{1/(2\tau)} \frac{2(2\zeta(\alpha/\tau))^{|\mathfrak{u}|}}{b^{\max(0,m-w_{s+1})}}\\
&= \sum_{\emptyset\neq\mathfrak{u}\subseteq[s+1]} \gamma_{\mathfrak{u}}^{1/(2\tau)} \frac{2(2\zeta(\alpha/\tau))^{|\mathfrak{u}|}}{b^{\max(0,m-\max_{j\in\mathfrak{u}} w_j)}},
\end{aligned}$$

which yields the final result. □

Tractability

Recall that we motivated the study of the reduced CBC construction by sufficient conditions for (strong polynomial) tractability. We formulate two corollaries to Theorem 4.3 as follows, and use the terminology introduced in Section 1.7.

Corollary 4.4 *Let b be a prime number, let $m \in \mathbb{N}_0$ and $N = b^m$. Let $\boldsymbol{\gamma} = \{\gamma_{\mathfrak{u}}\}_{\mathfrak{u}\subseteq[d]}$ be general weights, and let $w_1, w_2, \ldots, w_d \in \mathbb{N}_0$ with $w_1 \le w_2 \le \cdots \le w_d$. Moreover, let $Y_1, Y_2, \ldots, Y_d$ be as in (4.3). Assume that $\boldsymbol{g} = (Y_1 g_1, \ldots, Y_d g_d) \in \mathbb{Z}^d$ has been constructed according to Algorithm 4.1. Then,*

$$\mathrm{err}_{N,d,\alpha,\boldsymbol{\gamma}}(\boldsymbol{g}) \le \frac{c_{d,\alpha,\boldsymbol{\gamma},\delta,\boldsymbol{w}}}{N^{\alpha-\delta}} \quad \text{for all } \delta \in (0, \alpha - 1/2]\,, \tag{4.13}$$

where

$$c_{d,\alpha,\boldsymbol{\gamma},\delta,\boldsymbol{w}} := \left(2 \sum_{\emptyset\neq\mathfrak{u}\subseteq[d]} \gamma_{\mathfrak{u}}^{1/(2\alpha-2\delta)} \left(2\zeta\left(\frac{\alpha}{\alpha-\delta}\right)\right)^{|\mathfrak{u}|} b^{\max_{j\in\mathfrak{u}} w_j}\right)^{\alpha-\delta}.$$

Furthermore, for $\delta \in (0, \alpha - 1/2]$ and $q \ge 0$ define

$$C_{\delta,q} := \sup_{d\in\mathbb{N}}\left(\frac{1}{d^q}\sum_{\emptyset\neq\mathfrak{u}\subseteq[d]}\gamma_{\mathfrak{u}}^{1/(2\alpha-2\delta)}\left(2\zeta\left(\frac{\alpha}{2\alpha-2\delta}\right)\right)^{|\mathfrak{u}|} b^{\max_{j\in\mathfrak{u}} w_j}\right).$$

With this notation, the following holds.

1. *If*
$$C_{\delta,q}<\infty \text{ for some } \delta\in(0,\alpha-1/2] \text{ and a nonnegative } q,$$
then
$$\mathrm{err}_{N,d,\alpha,\boldsymbol{\gamma}}(\mathbf{g}) \le \frac{(2d^q C_{\delta,q})^{\alpha-\delta}}{N^{\alpha-\delta}},$$
and hence the information complexity depends only polynomially on d and ε^{-1}. In particular, this implies polynomial tractability.
2. *If*
$$C_{\delta,0}<\infty \text{ for some } \delta\in(0,\alpha-1/2],$$
then
$$\mathrm{err}_{N,d,\alpha,\boldsymbol{\gamma}}(\mathbf{g}) \le \frac{(2C_{\delta,0})^{\alpha-\delta}}{N^{\alpha-\delta}},$$
and hence the information complexity depends only polynomially on ε^{-1} and is independent of d. In particular, this implies strong polynomial tractability with an ε-exponent of at most $1/(\alpha-\delta)$.
3. *If*
$$\lim_{d\to\infty}\frac{\log\left(\sum_{\emptyset\neq\mathfrak{u}\subseteq[d]}\gamma_{\mathfrak{u}}(2\zeta(2\alpha))^{|\mathfrak{u}|}b^{\max_{j\in\mathfrak{u}} w_j}\right)}{d}=0,$$
then we obtain weak tractability.

Proof The proof of (4.13) follows by setting $\tau := \alpha-\delta$ in Theorem 4.3. The proof of Items 1–3 follows from an adaption of the proof of the corresponding assertions in Corollary 2.27. □

Corollary 4.5 *Let b be a prime number, let $m\in\mathbb{N}_0$ and $N=b^m$. Let $\boldsymbol{\gamma}=(\gamma_j)_{j\ge1}$ be positive product weights, and let $w_1,w_2,\ldots,w_d\in\mathbb{N}_0$ with $w_1\le w_2\le\cdots\le w_d$. Furthermore, let $Y_1,Y_2,\ldots,Y_d$ be as in (4.3). Assume that $\mathbf{g}=(Y_1g_1,\ldots,Y_dg_d)\in\mathbb{Z}^d$ has been constructed according to Algorithm 4.1. Then the following holds true.*

1. *The term $c_{d,\alpha,\boldsymbol{\gamma},\delta,\mathbf{w}}$ in Corollary 4.4 satisfies*
$$c_{d,\alpha,\boldsymbol{\gamma},\delta,\mathbf{w}} \le \left(2\prod_{j=1}^{d}\left(1+\gamma_j^{1/(2\alpha-2\delta)}2\zeta\left(\frac{\alpha}{\alpha-\delta}\right)b^{w_j}\right)\right)^{\alpha-\delta}.$$
2. *If*
$$A:=\limsup_{d\to\infty}\frac{\sum_{j=1}^{d}\gamma_j^{1/(2\alpha-2\delta)}b^{w_j}}{\log d}<\infty,$$
then for every $\eta>0$ there exists a $c_\eta>0$ such that

$$c_{d,\alpha,\gamma,\delta,\boldsymbol{w}} \le c_\eta 2^{\alpha-\delta} d^{\zeta(\alpha/(\alpha-\delta))(A+\eta)(2\alpha-2\delta)},$$

and thus we obtain polynomial tractability.

3. *If*

$$B := \sum_{j=1}^{\infty} \gamma_j^{1/(2\alpha-2\delta)} b^{w_j} < \infty,$$

then

$$c_{d,\alpha,\gamma,\delta,\boldsymbol{w}} \le 2^{\alpha-\delta} \mathrm{e}^{(2\alpha-2\delta)\zeta(\alpha/(\alpha-\delta))B}$$

and hence the information complexity depends only polynomially on ε^{-1} and is independent of d. This implies strong polynomial tractability with an ε-exponent of at most $1/(\alpha-\delta)$.

4. *If*

$$\lim_{d\to\infty} \frac{1}{d} \sum_{j=1}^{d} \gamma_j b^{w_j} = 0,$$

then we have weak tractability.

Proof The results follow from Corollary 4.4 by using similar arguments as in the proof of the corresponding results in Corollary 2.28. □

Let us briefly return to the main idea of the reduced CBC construction, which we illustrate using Item 3 of Corollary 4.5. Assume that we want to obtain strong polynomial tractability with an ε-exponent arbitrarily close to $1/\alpha$. Further assume that $\sum_{j=1}^{\infty} \gamma_j^{1/(2\tau)} < \infty$ for some $\tau > \alpha$. It would be sufficient if the weights satisfy $\sum_{j=1}^{\infty} \gamma_j^{1/(2\alpha)} < \infty$ to achieve this (see Corollary 2.28). If we use the standard (fast) CBC approach, in some sense we would "waste" the additional faster decay rate of the weights. However, the reduced CBC construction allows us to take advantage of the faster decay rate, since we can introduce reduction indices $(w_j)_{j\ge 1}$ with $0 \le w_1 \le w_2 \le \cdots$ such that $\sum_{j=1}^{\infty} \gamma_j^{1/(2\alpha)} b^{w_j} < \infty$. As we have seen above, Algorithm 4.1 allows us to use those $w_1, w_2, \ldots$ to reduce the construction cost of the fast CBC algorithm. How much reduction in the construction cost can be achieved in this way will be discussed in the next section.

4.2 The Reduced Fast CBC Construction for Product and POD Weights

Let us now discuss how the reduced CBC construction in Algorithm 4.1 can be implemented efficiently for product weights and for POD weights. We start the discussion with the case of product weights.

Product weights

Let again $s \in [d-1]$, and suppose that $g_1, \dots, g_s$ have already been chosen. We describe how to perform the step of selecting g_{s+1} in a fast way. If $w_{s+1} \geq m$, then we set $g_{s+1} = 1$ and no computation is necessary. Thus we assume now that $w_{s+1} < m$. Then, according to Algorithm 4.1, we need to find a $g \in Z_{w_{s+1}}(N)$ which minimizes the squared error $[\mathrm{err}_{N,s+1,\alpha,\boldsymbol{\gamma}}((Y_1 g_1, \dots, Y_s g_s, Y_{s+1} g))]^2$ with respect to g, which is equivalent to minimizing

$$\sum_{k=0}^{N-1} \left(1 + \gamma_j \varphi_\alpha\left(\frac{kY_{s+1}g}{N}\right)\right) \eta_s(k) = \sum_{k=0}^{N-1} \eta_s(k) + \gamma_j \sum_{k=0}^{N-1} \varphi_\alpha\left(\frac{kY_{s+1}g}{N}\right) \eta_s(k),$$

where $\eta_0(k) := 1$ for all $k \in \{0, 1, \dots, N-1\}$, and, for $s \in \mathbb{N}$,

$$\eta_s(k) := \prod_{j=1}^{s} \left(1 + \gamma_j \varphi_\alpha\left(\frac{kY_j g_j}{N}\right)\right).$$

Thus, like in (3.19), minimizing $[\mathrm{err}_{N,s+1,\alpha,\boldsymbol{\gamma}}((Y_1 g_1, \dots, Y_s g_s, Y_{s+1} g))]^2$ with respect to g is done by minimizing

$$T_{s+1}(g) := \sum_{k=0}^{N-1} \varphi_\alpha\left(\frac{kY_{s+1}g}{N}\right) \eta_s(k).$$

Now the key observations are that the vector $\boldsymbol{T}_{s+1} := (T_{s+1}(g))^\top_{g \in Z_{w_{s+1}}(N)}$ is the product of a special $(b^{m-w_{s+1}} - 1) \times N$ matrix

$$\boldsymbol{A}_N := \left(\varphi_\alpha\left(\frac{kY_{s+1}g}{N}\right)\right)_{g \in Z_{w_{s+1}}(N), k \in G_1(N)}$$

with the vector $\boldsymbol{\eta}_s := (\eta_s(0), \eta_s(1), \dots, \eta_s(N-1))^\top$, and that this matrix-vector product can be computed very efficiently, as we will show in the following.

Note that the rows of $\boldsymbol{A}_N$ are periodic with period $b^{m-w_{s+1}}$, since $Y_{s+1} = b^{w_{s+1}}$, and $N = b^m$, and therefore

$$(k + b^{m-w_{s+1}}) Y_{s+1} g \equiv k Y_{s+1} g \pmod{b^m}.$$

More specifically, $\boldsymbol{A}_N$ is a block matrix

$$\boldsymbol{A}_N = \underbrace{\left(\boldsymbol{\Omega}_{b^{m-w_{s+1}}}, \dots, \boldsymbol{\Omega}_{b^{m-w_{s+1}}}\right)}_{b^{w_{s+1}} \text{ times}},$$

where

$$\boldsymbol{\Omega}_{b^\ell} := \left(\varphi_\alpha\left(\frac{kg}{b^\ell}\right)\right)_{g \in Z_0(b^\ell), k \in G_1(b^\ell)}.$$

If $\boldsymbol{\xi} = (\xi_1, \ldots, \xi_N)$ is a vector of length $N = b^m$, we compute

$$A_N \boldsymbol{\xi}^\top = \boldsymbol{\Omega}_{b^{m-w_{s+1}}} \boldsymbol{\xi}_1^\top + \cdots + \boldsymbol{\Omega}_{b^{m-w_{s+1}}} \boldsymbol{\xi}_{b^{w_{s+1}}}^\top = \boldsymbol{\Omega}_{b^{m-w_{s+1}}} (\boldsymbol{\xi}_1 + \cdots + \boldsymbol{\xi}_{b^{w_{s+1}}})^\top,$$

where $\boldsymbol{\xi}_1$ consists of the first $b^{m-w_{s+1}}$ coordinates of $\boldsymbol{\xi}$, where $\boldsymbol{\xi}_2$ consists of the next $b^{m-w_{s+1}}$ coordinates of $\boldsymbol{\xi}$, and so forth.

We have already seen in the previous chapter (see Section 3.4) how to multiply a vector of length b^ℓ with $\boldsymbol{\Omega}_{b^\ell}$ using at most $O(kb^\ell)$ elementary operations employing the method of Nuyens and Cools. Addition of the vectors $\boldsymbol{\xi}_1, \ldots, \boldsymbol{\xi}_{b^{w_{s+1}}}$ uses b^m single additions. Thus multiplication of a vector of length b^m with A_N uses $O(b^m + (m - w_{s+1})b^{m-w_{s+1}})$ operations.

It is crucial to note that Algorithm 4.1 only needs actual computations as long as the reduction indices are strictly less than m. As soon as we have reached the stage where the reduction indices are at least m, the corresponding components of the generating vector are set equal to one, and no further computations are necessary.

Furthermore, we also need to take into account that the precomputation of the values of φ_α requires $O(N \log N) = O(mb^m)$ operations, as mentioned in Section 3.4. Therefore, the computational cost of Algorithm 4.1 is of order

$$O\left(mb^m + b^m \min(d, d_*) + \sum_{s=1}^{\min(d,d_*)} (m - w_s) b^{m-w_s}\right),$$

where $d_* := \max\{s \in \mathbb{N}_0 \ : \ w_s < m\}$. Hence, if the reduction indices are chosen such that they increase sufficiently fast, the computational cost of Algorithm 4.1 can be independent of d. However, one should keep in mind that large reduction indices have a negative influence on the error bound in Theorem 4.3. So, in general, the choice of the w_j needs to be balanced with the weights γ_j in order to achieve best possible results.

POD weights

We outlined in Algorithm 4.1 how we can use reduction indices to modify the CBC construction for general weights, and we outlined the fast implementation of this reduced CBC construction for the case of product weights above. Also in the case of POD weights (see (3.25)), which have gained much attention due to their use in applications in QMC for PDEs with random coefficients (see Appendix A), we can use the reduced construction to improve on the runtime of the CBC algorithm. As seen in Section 3.5, the runtime of the CBC construction for POD weights with no reduction indices is of order $O(d N \log N + d^2 N)$. We now show how to decrease this runtime by using reduction indices.

Recall that according to (3.26) the worst-case error of integration using a rank-1 lattice rule for POD weights is given by

$$[\mathrm{err}_{N,d,\alpha,\boldsymbol{\gamma}}(\boldsymbol{g})]^2 = \frac{1}{N}\sum_{k=0}^{N-1}\sum_{\ell=1}^{d}\sum_{\substack{\mathfrak{u}\subseteq[d]\\|\mathfrak{u}|=\ell}}\Gamma_\ell\prod_{j\in\mathfrak{u}}\gamma_j\varphi_\alpha\left(\frac{kg_j}{N}\right).$$

Let $s \in [d-1]$, and suppose that $g_1, \ldots, g_s$ have already been chosen by the reduced CBC algorithm. As before we write $\boldsymbol{g}^{(s)} := (Y_1 g_1, \ldots, Y_s g_s)$. We describe how to select g_{s+1} in a fast way. If $w_{s+1} > m$, then we set $g_{s+1} = 1$ and no computation is necessary. Thus we assume now that $w_{s+1} \leq m$. Then, according to Algorithm 4.1, we need to find a $g \in Z_{w_{s+1}}(N)$ which minimizes the error $[\mathrm{err}_{N,s+1,\alpha,\boldsymbol{\gamma}}((\boldsymbol{g}^{(s)}, Y_{s+1}g))]^2$ with respect to g.

Then, again similar to Section 3.5, we can write

$$\begin{aligned}
&[\mathrm{err}_{N,s+1,\alpha,\boldsymbol{\gamma}}((\boldsymbol{g}^{(s)}, Y_{s+1}g_{s+1}))]^2\\
&= [\mathrm{err}_{N,s,\alpha,\boldsymbol{\gamma}}(\boldsymbol{g}^{(s)})]^2 + \frac{1}{b^m}\sum_{k=0}^{b^m-1}\sum_{\ell=1}^{s+1}\frac{\Gamma_\ell}{\Gamma_{\ell-1}}\sum_{\substack{\mathfrak{u}\subseteq[s]\\|\mathfrak{u}|=\ell-1}}\Gamma_{\ell-1}\\
&\times\left(\prod_{j\in\mathfrak{u}}\gamma_j\varphi_\alpha\left(\frac{kg_j \pmod{b^{m-w_j}}}{b^{m-w_j}}\right)\right)\gamma_{s+1}\varphi_\alpha\left(\frac{kg_{s+1} \pmod{b^{m-w_{s+1}}}}{b^{m-w_{s+1}}}\right).
\end{aligned}$$

We now represent every $k \in \{0, 1, \ldots, b^m - 1\}$ in the form $k = \kappa + tb^{m-w_{s+1}}$ with $\kappa \in \{0, 1, \ldots, b^{m-w_{s+1}} - 1\}$ and $t \in \{0, 1, \ldots, b^{w_{s+1}} - 1\}$ to obtain

$$\begin{aligned}
&[\mathrm{err}_{N,s+1,\alpha,\boldsymbol{\gamma}}((\boldsymbol{g}^{(s)}, Y_{s+1}g_{s+1}))]^2\\
&= [\mathrm{err}_{N,s,\alpha,\boldsymbol{\gamma}}(\boldsymbol{g}^{(s)})]^2 + \frac{1}{b^m}\sum_{\kappa=0}^{b^{m-w_{s+1}}-1}\sum_{\ell=1}^{s+1}\frac{\Gamma_\ell}{\Gamma_{\ell-1}}\sum_{t=0}^{b^{w_{s+1}}-1}\sum_{\substack{\mathfrak{u}\subseteq[s]\\|\mathfrak{u}|=\ell-1}}\Gamma_{\ell-1}\\
&\times\left(\prod_{j\in\mathfrak{u}}\gamma_j\varphi_\alpha\left(\frac{(\kappa+tb^{m-w_{s+1}})g_j \pmod{b^{m-w_j}}}{b^{m-w_j}}\right)\right)\\
&\times\gamma_{s+1}\varphi_\alpha\left(\frac{(\kappa+tb^{m-w_{s+1}})g_{s+1} \pmod{b^{m-w_{s+1}}}}{b^{m-w_{s+1}}}\right)\\
&= [\mathrm{err}_{N,s,\alpha,\boldsymbol{\gamma}}(\boldsymbol{g}^{(s)})]^2 + \frac{1}{b^m}\sum_{\kappa=0}^{b^{m-w_{s+1}}-1}\sum_{\ell=1}^{s+1}\frac{\Gamma_\ell}{\Gamma_{\ell-1}}\sum_{t=0}^{b^{w_{s+1}}-1}\sum_{\substack{\mathfrak{u}\subseteq[s]\\|\mathfrak{u}|=\ell-1}}\Gamma_{\ell-1}\\
&\times\left(\prod_{j\in\mathfrak{u}}\gamma_j\varphi_\alpha\left(\frac{(\kappa+tb^{m-w_{s+1}})g_j \pmod{b^{m-w_j}}}{b^{m-w_j}}\right)\right)\\
&\times\gamma_{s+1}\varphi_\alpha\left(\frac{\kappa g_{s+1} \pmod{b^{m-w_{s+1}}}}{b^{m-w_{s+1}}}\right).
\end{aligned}$$

We write, for short, for $w \in \{0, 1, \ldots, m\}$ and $\kappa \in \{0, 1, \ldots, b^{m-w} - 1\}$,

$$p_{s,\ell,w}(\kappa) := \sum_{t=0}^{b^w-1} \sum_{\substack{\mathfrak{u}\subseteq[s] \\ |\mathfrak{u}|=\ell}} \Gamma_\ell \prod_{j\in\mathfrak{u}} \gamma_j \varphi_\alpha\left(\frac{(\kappa+tb^{m-w})g_j \pmod{b^{m-w_j}}}{b^{m-w_j}}\right).$$

Additionally, we put $p_{s,0,w} := 1$. Note that if $w = 0$, the quantity $p_{s,\ell,w}$ corresponds to the quantity $p_{s,\ell}$ in (3.27) for $N = b^{m-w_j}$. Hence we can write

$$\begin{aligned}
&[\mathrm{err}_{N,s+1,\alpha,\boldsymbol{\gamma}}((\boldsymbol{g}^{(s)}, Y_{s+1}g_{s+1}))]^2 \\
&= [\mathrm{err}_{N,s,\alpha,\boldsymbol{\gamma}}(\boldsymbol{g}^{(s)})]^2 + \frac{1}{b^m} \sum_{\kappa=0}^{b^{m-w_{s+1}}-1} \sum_{\ell=1}^{s+1} \frac{\Gamma_\ell}{\Gamma_{\ell-1}} p_{s,\ell-1,w_{s+1}}(\kappa) \\
&\times \gamma_{s+1}\varphi_\alpha\left(\frac{\kappa g_{s+1} \pmod{b^{m-w_{s+1}}}}{b^{m-w_{s+1}}}\right).
\end{aligned}$$

Now consider $w_s \in \{0, 1, \ldots, w_{s+1}\}$. Every $t \in \{0, 1, \ldots, b^{w_{s+1}} - 1\}$ can be represented in the form $t = t' + t'' b^{w_{s+1}-w_s}$ with $t' \in \{0, 1, \ldots, b^{w_{s+1}-w_s} - 1\}$ and $t'' \in \{0, 1, \ldots, b^{w_s} - 1\}$. Consequently,

$$\begin{aligned}
p_{s,\ell-1,w_{s+1}}(\kappa) &= \sum_{t=0}^{b^{w_{s+1}}-1} \sum_{\substack{\mathfrak{u}\subseteq[s] \\ |\mathfrak{u}|=\ell-1}} \Gamma_{\ell-1} \prod_{j\in\mathfrak{u}} \gamma_j \varphi_\alpha\left(\frac{(\kappa+tb^{m-w_{s+1}})g_j \pmod{b^{m-w_j}}}{b^{m-w_j}}\right) \\
&= \sum_{t'=0}^{b^{w_{s+1}-w_s}-1} \sum_{t''=0}^{b^{w_s}-1} \sum_{\substack{\mathfrak{u}\subseteq[s] \\ |\mathfrak{u}|=\ell-1}} \Gamma_{\ell-1} \\
&\quad \times \prod_{j\in\mathfrak{u}} \gamma_j \varphi_\alpha\left(\frac{(\kappa+(t'+t''b^{w_{s+1}-w_s})b^{m-w_{s+1}})g_j \pmod{b^{m-w_j}}}{b^{m-w_j}}\right) \\
&= \sum_{t'=0}^{b^{w_{s+1}-w_s}-1} p_{s,\ell-1,w_s}(\kappa + t' b^{m-w_{s+1}}). \qquad (4.14)
\end{aligned}$$

Next, we define a "fold-and-sum" operator which will be very useful for the further considerations. For $m \in \mathbb{N}$ and $w', w'' \in \{0, 1, \ldots, m\}$ with $w' \leq w''$, and given base b, let

$$P^m_{w'',w'} : \mathbb{R}^{b^{m-w'}} \to \mathbb{R}^{b^{m-w''}}, \quad P^m_{w'',w'}\boldsymbol{v} = \underbrace{(U_{b^{m-w''}} | \ldots | U_{b^{m-w''}})}_{b^{w''-w'} \text{ times}} \boldsymbol{v}^\top, \qquad (4.15)$$

where $U_{b^{m-w''}}$ is the $b^{m-w''} \times b^{m-w''}$ identity matrix. This operator divides a real vector of length $b^{m-w'}$ into blocks of equal length $b^{m-w''}$ and sums them up. For example, for $b = 2$, $m = 4$, $w' = 1$, and $w'' = 3$, the operator $P^4_{3,1} : \mathbb{R}^8 \to \mathbb{R}^2$ applied to $\boldsymbol{v} = (v_1, \ldots, v_8) \in \mathbb{R}^8$ yields

$$P_{3,1}^4 \boldsymbol{v} = \begin{pmatrix} 1 & 0 & 1 & 0 & 1 & 0 & 1 & 0 \\ 0 & 1 & 0 & 1 & 0 & 1 & 0 & 1 \end{pmatrix} \boldsymbol{v}^\top = \begin{pmatrix} v_1 + v_3 + v_5 + v_7 \\ v_2 + v_4 + v_6 + v_8 \end{pmatrix}.$$

Note that the computational cost of applying the operator $P^m_{w'',w'}$ is the length of the input vector, i.e., it is of order $O(b^{m-w'})$. Furthermore, the operator has a kind of a transitivity property, i.e., for $0 \le w' \le w'' \le w''' \le m$ it holds that

$$P^m_{w''',w''} P^m_{w'',w'} \boldsymbol{v} = P^m_{w''',w'} \boldsymbol{v}, \quad \text{for } \boldsymbol{v} \in \mathbb{R}^{b^{m-w'}}.$$

Using the vector notation

$$\boldsymbol{p}_{s,\ell-1,w} = (p_{s,\ell-1,w}(0), p_{s,\ell-1,w}(1), \ldots, p_{s,\ell-1,w}(b^{m-w}-1))^\top$$

for integers $w \in \{0, 1, \ldots, m\}$, we can rewrite (4.14) in vector form as

$$\boldsymbol{p}_{s,\ell-1,w_{s+1}} = P^m_{w_{s+1},w_s} \boldsymbol{p}_{s,\ell-1,w_s},$$

where $P^m_{w_{s+1},w_s}$ is the "fold-and-sum" operator in (4.15) for w_s and w_{s+1}.

Similarly to Section 3.5 we use $\boldsymbol{E}_{N,s+1,\alpha,\gamma}(\boldsymbol{g}^{(s)})$ to denote the column vector with entries $[\mathrm{err}_{N,s+1,\alpha,\gamma}((\boldsymbol{g}^{(s)}, Y_{s+1}g))]^2$ for $g \in Z_{w_{s+1}}(N)$. Then we can write

$$\begin{aligned} \boldsymbol{E}_{N,s+1,\alpha,\gamma}(\boldsymbol{g}^{(s)}) &= [\mathrm{err}_{N,s,\alpha,\gamma}(\boldsymbol{g}^{(s)})]^2 \, \mathbf{1}_{|Z_{w_{s+1}}(N)|} \\ &\quad + \frac{\gamma_{s+1}}{N} \boldsymbol{\Omega}_{b^{m-w_{s+1}}} \left(\sum_{\ell=1}^{s+1} \frac{\Gamma_\ell}{\Gamma_{\ell-1}} P^m_{w_{s+1},w_s} \boldsymbol{p}_{s-1,\ell-1,w_s} \right)^\top, \end{aligned} \tag{4.16}$$

where again $\boldsymbol{\Omega}_{b^\ell} := (\varphi_\alpha(\{(k/b^\ell)g\}))_{g \in Z_0(b^\ell), k \in G_1(b^\ell)}$. Denote by $\boldsymbol{\Omega}_{b^\ell}(g)$, $g \in Z_0(b^\ell)$, the rows of the matrix $\boldsymbol{\Omega}_{b^\ell}$.

Once g_{s+1} has been selected, we can use the update formula

$$\boldsymbol{p}_{s+1,\ell,w_{s+1}} = P^m_{w_{s+1},w_s} \boldsymbol{p}_{s,\ell,w_s} + \frac{\Gamma_\ell}{\Gamma_{\ell-1}} \gamma_{s+1} \boldsymbol{\Omega}_{b^m}(g_{s+1}) * P^m_{w_{s+1},w_s} \boldsymbol{p}_{s,\ell-1,w_s}$$

for $\ell \in [s+1]$, with "$*$" denoting component-wise multiplication of vectors in $\mathbb{R}^N$, with the initial values

$$\boldsymbol{p}_{s+1,0,w_{s+1}} = \mathbf{1}_{b^{m-w_{s+1}}} \quad \text{and} \quad \boldsymbol{p}_{s+1,\ell,w_{s+1}} = \mathbf{0}_{b^{m-w_{s+1}}} \quad \text{for } \ell > s+1.$$

Let us discuss the computational cost of this procedure. For the sake of simplicity, we assume $w_1 = 0$, as the case $w_1 > 0$ can easily be reduced to the case $w_1 = 0$. For computing $\boldsymbol{E}_{N,s+1,\alpha,\gamma}(\boldsymbol{g}^{(s)})$ in (4.16), the following steps have to be taken. In order to compute the quantity

$$\sum_{\ell=1}^{s+1} \frac{\Gamma_\ell}{\Gamma_{\ell-1}} P^m_{w_{s+1},w_s} \boldsymbol{p}_{s-1,\ell-1,w_s}$$

we need $O((s+1)b^{m-w_s})$ operations, according to our previous observation regarding application of the "fold-and-sum" operator. Performing the matrix-vector multiplication in (4.16) can be done by exploiting the block-circulant structure of the matrix, using FFT, at a cost of $O((m-w_{s+1})b^{m-w_{s+1}})$ operations. Updating $\boldsymbol{p}_{s,\ell,w_{s+1}}$ needs b^{m-w_s} operations for each ℓ, and thus $O(sb^{m-w_s})$ operations in total in each step. As pointed out in [68], this cost can be reduced to $O((s+1)b^{m-w_{s+1}})$ by a slight rearrangement of the steps of the algorithm, by applying the "fold-and-sum" operator one iteration earlier. This yields that the construction cost of the reduced fast CBC construction for POD weights is of order

$$O\left(\sum_{s=1}^{\min(d,d_*)} (m - w_s + s)b^{m-w_s}\right),$$

where again $d_* := \max\{s \in \mathbb{N}_0 : w_s < m\}$. Furthermore, the memory cost of the implementation is of order $O(\sum_{s=1}^{\min(d,d_*)} b^{m-w_s})$.

4.3 The Successive Coordinate Search Construction

It is inherent to the CBC constructions outlined in Sections 3.3, 3.6, and 4.1 that these are greedy algorithms, which construct one component of the generating vector $\boldsymbol{g}$ at a time. Even though it is known that the error bounds are asymptotically optimal, it cannot be expected that CBC algorithms yield the generating vectors actually minimizing the respective error criteria under consideration.

In order to address this problem, a so-called successive coordinate search (for short, SCS) construction was introduced in [71]. The basic idea of the SCS construction is that the algorithm is initialized by an existing generating vector $\boldsymbol{g}$ of a rank-1 lattice point set, which is then updated in a component-wise fashion. In this way, it is possible that the worst-case error of a lattice rule with underlying generating vector $\boldsymbol{g}$ is improved by updating $\boldsymbol{g}$. In particular, it is possible to use the output vector of the CBC algorithm as the initial vector for the SCS algorithm, and then to possibly improve on the quality with respect to the error criterion. Moreover, the SCS construction can be applied as an iterative method.

In this section, we present the precise formulation of the SCS algorithm and derive an error estimate for the lattice rules obtained for the weighted Korobov space.

We start by formalizing the SCS construction algorithm. We illustrate this by the algorithm using the worst-case error in the weighted Korobov space as the error criterion, but the basic principle could also be used for other error criteria.

Algorithm 4.6 (SCS construction for the weighted Korobov space) Let positive integers N and d, and an initial vector $\boldsymbol{g}^{(0)} = (g_1^{(0)}, \ldots, g_d^{(0)}) \in G_d^{\varphi}(N)$ be given. Update $\boldsymbol{g}^{(0)}$ to a generating vector $\boldsymbol{g} = (g_1, \ldots, g_d) \in G_d^{\varphi}(N)$ as follows.

(1) For s from 1 to $d-1$:
Assume that $g_1, \ldots, g_s \in G_1^{\varphi}(N)$ have already been found. Choose $g_{s+1} \in G_1^{\varphi}(N)$ as

$$g_{s+1} := \underset{g \in G_1^{\varphi}(N)}{\operatorname{argmin}} \operatorname{err}_{N,d,\alpha,\boldsymbol{\gamma}}((g_1, \ldots, g_s, g, g_{s+2}^{(0)}, \ldots, g_d^{(0)})),$$

with the obvious adaptions if $s = d-1$.
End for.

(2) Set $\boldsymbol{g} = (g_1, \ldots, g_d)$.

Note that Algorithm 4.6 is, as opposed to CBC construction algorithms, not extensible in the dimension.

The following proposition is obvious from the formulation of the algorithm.

Proposition 4.7 *Let $\boldsymbol{g}^{(0)} \in G_d^{\varphi}(N)$ be an arbitrary initial vector for Algorithm 4.6, and let $\boldsymbol{g}$ be the corresponding output vector. Then it is true that*

$$\operatorname{err}_{N,d,\alpha,\boldsymbol{\gamma}}(\boldsymbol{g}) \le \operatorname{err}_{N,d,\alpha,\boldsymbol{\gamma}}(\boldsymbol{g}^{(0)}).$$

In the special case of product weights, the following theorem relates the SCS construction, i.e., Algorithm 4.6, to the usual CBC construction for the Korobov space, i.e., Algorithm 3.6. In order to guarantee a meaningful statement, we assume that both algorithms select the same instance of g if there are multiple minimizers in the minimization step. To this end, we extend the domain of the starting vector $\boldsymbol{g}^{(0)}$ of the SCS construction to allow zero components, i.e., $\boldsymbol{g}^{(0)} \in (G_1^{\varphi}(N) \cup \{0\})^d$. This, however, is just a technical extension causing no problems.

The following result is due to [71].

Theorem 4.8 *In the special case of positive product weights $(\gamma_j)_{j \ge 1}$, Algorithm 3.6 and Algorithm 4.6 both yield the same generating vector if Algorithm 4.6 is initialized with the all-zero vector $\boldsymbol{g}^{(0)} = (0, \ldots, 0)$.*

Proof Let $s \in [d-1]$, assume that $\boldsymbol{g} = (g_1, \ldots, g_{s+1}) \in (G_1^{\varphi}(N) \cup \{0\})^{s+1}$, and define $\widetilde{\boldsymbol{g}} = (\widetilde{g}_1, \ldots, \widetilde{g}_d) = (\boldsymbol{g}, \boldsymbol{0}_{d-s-1}) := (g_1, \ldots, g_{s+1}, 0, 0, \ldots, 0)$. Then we have, using the worst-case error formula (3.6),

$$\begin{aligned}
[\operatorname{err}_{N,d,\alpha,\boldsymbol{\gamma}}(\widetilde{\boldsymbol{g}})]^2 &= -1 + \frac{1}{N} \sum_{k=0}^{N-1} \prod_{j=1}^{d} \left(1 + \gamma_j \varphi_\alpha\left(\frac{k\widetilde{g}_j}{N}\right)\right) \\
&= -1 + \frac{1}{N} \sum_{k=0}^{N-1} \left(\prod_{j=1}^{s+1} \left(1 + \gamma_j \varphi_\alpha\left(\frac{k\widetilde{g}_j}{N}\right)\right)\right)\left(\prod_{j=s+2}^{d} (1 + \gamma_j \varphi_\alpha(0))\right).
\end{aligned}$$

By definition we have $\varphi_\alpha(0) = 2\zeta(2\alpha)$. Using the notation

$$C_{s+2} := \prod_{j=s+2}^{d} (1 + \gamma_j 2\zeta(2\alpha))$$

we can rewrite the previous formula as

$$
\begin{aligned}
[\mathrm{err}_{N,d,\alpha,\boldsymbol{\gamma}}(\widetilde{\boldsymbol{g}})]^2 &= -1 + \frac{C_{s+2}}{N} \sum_{k=0}^{N-1} \prod_{j=1}^{s+1} \left(1 + \gamma_j \varphi_\alpha\left(\frac{k\widetilde{g}_j}{N}\right)\right) \\
&= -1 + C_{s+2}([\mathrm{err}_{N,s+1,\alpha,\boldsymbol{\gamma}}(\boldsymbol{g})]^2 + 1).
\end{aligned}
$$

Now suppose that, for $s \in [d-1]$, the SCS Algorithm 4.6 has already selected $g_1, \ldots, g_s$ as the first s components of the generating vector. Then, according to the argument above, the algorithm selects g_{s+1} as the minimizer of $[\mathrm{err}_{N,s+1,\alpha,\boldsymbol{\gamma}}((g_1, \ldots, g_s, g))]^2$. By an inductive argument, this means that the SCS algorithm with initial vector $\boldsymbol{g}^{(0)} = (0, \ldots, 0)$ and the CBC algorithm coincide. □

Remark 4.9 Note that an analogue of Theorem 4.8 does, in general, not hold for arbitrary weights $\boldsymbol{\gamma} = \{\gamma_{\mathfrak{u}}\}_{\mathfrak{u} \subseteq [d]}$.

We now have the following theorem which gives an error bound for the lattice rules constructed by the SCS algorithm. The result is taken from [67, Theorem 1].

Theorem 4.10 *Let N be a prime power and let $\boldsymbol{g} \in G_d^{\varphi}(N)$ be constructed by Algorithm 4.6 with some initial vector $\boldsymbol{g}^{(0)}$. Then, for $\tau \in [1/2, \alpha)$, the worst-case error of the lattice rule generated by $\boldsymbol{g}$ satisfies*

$$
\mathrm{err}_{N,d,\alpha,\boldsymbol{\gamma}}(\boldsymbol{g}) \le \frac{1}{N^\tau} \left(\sum_{s=1}^{d} \sum_{\substack{\mathfrak{u} \subseteq [d] \\ s \in \mathfrak{u}}} \gamma_{\mathfrak{u}}^{1/(2\tau)} 2 \left(2\zeta\left(\frac{\alpha}{\tau}\right)\right)^{|\mathfrak{u}|} \right)^\tau .
$$

Proof The result follows from setting $w_1 = w_2 = \cdots = w_d = 0$ in Theorem 4.12 below. □

In [71], the authors discuss a fast implementation of the SCS algorithm for product weights and prime N that takes $O(d N \log N)$ operations. We will outline how this implementation works for the reduced fast SCS construction in Section 4.4 below for more general choices than prime N, and by allowing additional reduction indices. Setting all reduction indices equal to zero then yields a fast implementation for the nonreduced case outlined in the present section.

4.4 The Reduced Fast SCS Construction

Similar to the CBC construction, there also exists a reduced version of the SCS construction, which, for sufficiently fast decaying weights, can help in lowering the construction cost of the algorithm. This approach was studied in [67], and again the underlying idea is to shrink the search spaces for the components g_j of the lattice generating vector $\boldsymbol{g}$ for which the corresponding weights are small. To this end, again assume in this section that $N = b^m$ is a prime power, and let $w_1, w_2, \ldots, w_d \in \mathbb{N}_0$ with $w_1 \leq w_2 \leq \cdots \leq w_d$ be reduction indices. Similarly to the reduced CBC construction, we are mostly interested in the case where $w_1 = 0$, as for $w_1 > 0$ the construction results in each point being counted b^{w_1} times. We again use the sets $Z_{w_j}(N)$ defined in (4.2) and the notation $Y_j = b^{w_j}$ for $j \in [d]$ as in (4.3).

Algorithm 4.11 (Reduced SCS construction for the weighted Korobov space) Let $d \in \mathbb{N}$ and let b be a prime number, let $m \in \mathbb{N}_0$ and $N = b^m$. Let $\boldsymbol{\gamma} = \{\gamma_{\mathfrak{u}}\}_{\mathfrak{u} \subseteq [d]}$ be general weights, and let $w_1, w_2, \ldots, w_d \in \mathbb{N}_0$ with $w_1 \leq w_2 \leq \cdots \leq w_d$. Furthermore, let an initial vector $\boldsymbol{g}^{(0)} = (g_1^{(0)}, \ldots, g_d^{(0)}) \in G_d^{\varphi}(N)$ be given. Update $\boldsymbol{g}^{(0)}$ to a generating vector $\boldsymbol{g} = (Y_1 g_1, \ldots, Y_d g_d)$ with $g_j \in Z_{w_j}(N)$ for $j \in [d]$ as follows.

(1) For s from 1 to $d-1$:
Assume that $g_j \in Z_{w_j}(N)$ for $j \in [s]$ have already been found. Choose $g_{s+1} \in Z_{w_{s+1}}(N)$ as

$$g_{s+1} := \underset{g \in Z_{w_{s+1}}(N)}{\operatorname{argmin}} \ \operatorname{err}_{N,d,\alpha,\boldsymbol{\gamma}}((Y_1 g_1, \ldots, Y_s g_s, Y_{s+1} g, g_{s+2}^{(0)}, \ldots, g_d^{(0)})),$$

with the obvious adaptions if $s = d-1$.
End for.

(2) Set $\boldsymbol{g} = (Y_1 g_1, \ldots, Y_d g_d)$.

Error analysis

The following theorem provides an error bound for the lattice rules based on the generating vectors found by Algorithm 4.11.

Theorem 4.12 *Let $d \in \mathbb{N}$ and let b be a prime number, let $m \in \mathbb{N}_0$ and $N = b^m$. Let $\boldsymbol{\gamma} = \{\gamma_{\mathfrak{u}}\}_{\mathfrak{u} \subseteq [d]}$ be general weights, and let $w_1, w_2, \ldots, w_d \in \mathbb{N}_0$ with $w_1 \leq w_2 \leq \cdots \leq w_d$. Assume that $\boldsymbol{g}$ has been constructed by Algorithm 4.11 with some initial vector $\boldsymbol{g}^{(0)}$. Then, for $\tau \in [1/2, \alpha)$, the worst-case error of the lattice rule generated by $\boldsymbol{g}$ satisfies*

$$\operatorname{err}_{N,d,\alpha,\boldsymbol{\gamma}}(\boldsymbol{g}) \leq \left(\sum_{s=1}^{d} \sum_{\substack{\mathfrak{u} \subseteq [d] \\ s \in \mathfrak{u}}} \gamma_{\mathfrak{u}}^{1/(2\tau)} \frac{2(2\zeta(\alpha/\tau))^{|\mathfrak{u}|}}{b^{\max(0, m - w_s)}} \right)^{\tau}.$$

Proof Let $\boldsymbol{g} = (Y_1 g_1, \ldots, Y_d g_d)$ be the vector found by Algorithm 4.11 with initial vector $\boldsymbol{g}^{(0)}$. In the following, we will again make use of the dual lattice for projections of a lattice. In order to display the dependence of the dual of a lattice on the generating vector, we write, for $\emptyset \neq \mathfrak{u} \subseteq [d]$, $\mathcal{L}_{\mathfrak{u}}^{\perp}(\boldsymbol{g}_{\mathfrak{u}}) = \mathcal{L}_{\mathfrak{u}}^{\perp}$, where $\mathcal{L}_{\mathfrak{u}}^{\perp}$ is defined in (4.5).

By (2.23) in Theorem 2.19, we have for our $\boldsymbol{g}$,

$$[\mathrm{err}_{N,d,\alpha,\boldsymbol{\gamma}}(\boldsymbol{g})]^2 = \sum_{\emptyset \neq \mathfrak{u} \subseteq [d]} \gamma_{\mathfrak{u}} \sum_{\boldsymbol{h}_{\mathfrak{u}} \in \mathcal{L}_{\mathfrak{u}}^{\perp}(\boldsymbol{g}_{\mathfrak{u}})} \frac{1}{r_{2\alpha}(\boldsymbol{h}_{\mathfrak{u}})}$$
$$= \sum_{s=1}^{d} \sum_{\substack{\mathfrak{u} \subseteq [s] \\ s \in \mathfrak{u}}} \gamma_{\mathfrak{u}} \sum_{\boldsymbol{h}_{\mathfrak{u}} \in \mathcal{L}_{\mathfrak{u}}^{\perp}(\boldsymbol{g}_{\mathfrak{u}})} \frac{1}{r_{2\alpha}(\boldsymbol{h}_{\mathfrak{u}})}.$$

For $s \in [d]$ we write

$$\boldsymbol{g}^{(s)} := (Y_1 g_1, \ldots, Y_{s-1} g_{s-1}, Y_s g_s, g_{s+1}^{(0)}, \ldots, g_d^{(0)}),$$

and

$$\theta_s(\boldsymbol{g}^{(s)}) := \sum_{\substack{\mathfrak{u} \subseteq [d] \\ s \in \mathfrak{u}}} \gamma_{\mathfrak{u}} \sum_{\boldsymbol{h}_{\mathfrak{u}} \in \mathcal{L}_{\mathfrak{u}}^{\perp}(\boldsymbol{g}_{\mathfrak{u}}^{(s)})} \frac{1}{r_{2\alpha}(\boldsymbol{h}_{\mathfrak{u}})}.$$

Then we have

$$[\mathrm{err}_{N,d,\alpha,\boldsymbol{\gamma}}(\boldsymbol{g})]^2 \leq \sum_{s=1}^{d} \theta_s(\boldsymbol{g}^{(s)}).$$

Using Jensen's inequality (Lemma 2.25) we obtain, for $\lambda \in (1/(2\alpha), 1]$,

$$[\mathrm{err}_{N,d,\alpha,\boldsymbol{\gamma}}(\boldsymbol{g})]^{2\lambda} \leq \left(\sum_{s=1}^{d} \theta_s(\boldsymbol{g}^{(s)})\right)^{\lambda} \leq \sum_{s=1}^{d} (\theta_s(\boldsymbol{g}^{(s)}))^{\lambda}.$$

By the standard averaging argument (see Remark 2.14) we obtain that

$$(\theta_s(\boldsymbol{g}^{(s)}))^{\lambda} = (\theta_s((Y_1 g_1, \ldots, Y_s g_s, g_{s+1}^{(0)}, \ldots, g_d^{(0)})))^{\lambda}$$
$$\leq \frac{1}{|Z_{w_s}(N)|} \sum_{g \in Z_{w_s}(N)} (\theta_s((Y_1 g_1, \ldots, Y_{s-1} g_{s-1}, Y_s g, g_{s+1}^{(0)}, \ldots, g_d^{(0)}))^{\lambda}.$$

We now use the notation

$$\widehat{\boldsymbol{g}}^{(s)} = \widehat{\boldsymbol{g}}^{(s)}(g) = (\widehat{g}_1, \ldots, \widehat{g}_d) := (Y_1 g_1, \ldots, Y_{s-1} g_{s-1}, Y_s g, g_{s+1}^{(0)}, \ldots, g_d^{(0)}).$$

Next, we establish an upper bound on the quantity $(\theta_s(\widehat{\boldsymbol{g}}^{(s)}))^{\lambda}$ for each $s \in [d]$. We have, by using Jensen's inequality twice,

$$(\theta_s(\widehat{\boldsymbol{g}}^{(s)}))^{\lambda} \leq \sum_{\substack{\mathfrak{u} \subseteq [d] \\ s \in \mathfrak{u}}} \gamma_{\mathfrak{u}}^{\lambda} \sum_{\boldsymbol{h}_{\mathfrak{u}} \in \mathcal{L}_{\mathfrak{u}}^{\perp}(\widehat{\boldsymbol{g}}_{\mathfrak{u}}^{(s)})} \frac{1}{r_{2\alpha\lambda}(\boldsymbol{h}_{\mathfrak{u}})}$$

$$= \gamma_{\{s\}}^{\lambda} \sum_{h_s \in \mathcal{L}_{\{s\}}^{\perp}(Y_s g)} \frac{1}{r_{2\alpha\lambda}(h_s)}$$

$$+ \sum_{\substack{\emptyset \neq \mathfrak{v} \subseteq [d] \\ s \notin \mathfrak{v}}} \gamma_{\mathfrak{v} \cup \{s\}}^{\lambda} \sum_{h_s \in \mathbb{Z} \setminus \{0\}} \frac{1}{r_{2\alpha\lambda}(h_s)} \sum_{\substack{\boldsymbol{h}_{\mathfrak{v}} \in (\mathbb{Z} \setminus \{0\})^{|\mathfrak{v}|} \\ \sum_{j \in \mathfrak{v}} h_j \widehat{g}_j \equiv -h_s Y_s g \pmod{N}}} \frac{1}{r_{2\alpha\lambda}(\boldsymbol{h}_{\mathfrak{v}})}.$$

This implies in turn that

$$(\theta_s(\boldsymbol{g}^{(s)}))^{\lambda} \leq \frac{1}{|Z_{w_s}(N)|} \sum_{g \in Z_{w_s}(N)} (\theta_s(\widehat{\boldsymbol{g}}^{(s)}))^{\lambda} \leq T_1 + T_2,$$

where

$$T_1 := \frac{1}{|Z_{w_s}(N)|} \sum_{g \in Z_{w_s}(N)} \gamma_{\{s\}}^{\lambda} \sum_{h_s \in \mathcal{L}_{\{s\}}^{\perp}(Y_s g)} \frac{1}{r_{2\alpha\lambda}(h_s)}$$

and

$$T_2 := \frac{1}{|Z_{w_s}(N)|} \sum_{g \in Z_{w_s}(N)} \sum_{\substack{\emptyset \neq \mathfrak{v} \subseteq [d] \\ s \notin \mathfrak{v}}} \gamma_{\mathfrak{v} \cup \{s\}}^{\lambda} \sum_{h_s \in \mathbb{Z} \setminus \{0\}} \frac{1}{r_{2\alpha\lambda}(h_s)}$$
$$\times \sum_{\substack{\boldsymbol{h}_{\mathfrak{v}} \in (\mathbb{Z} \setminus \{0\})^{|\mathfrak{v}|} \\ \sum_{j \in \mathfrak{v}} h_j \widehat{g}_j \equiv -h_s Y_s g \pmod{N}}} \frac{1}{r_{2\alpha\lambda}(\boldsymbol{h}_{\mathfrak{v}})}.$$

Regarding T_1, we distinguish the cases $w_s \geq m$ and $w_s < m$.

If $w_s \geq m$, then $g \in Z_{w_s}(N) = \{1\}$ and $N = b^m$ is a divisor of b^{w_s} such that

$$T_1 = \gamma_{\{s\}}^{\lambda} \sum_{h_s \in \mathbb{Z} \setminus \{0\}} \frac{1}{r_{2\alpha\lambda}(h_s)} = \gamma_{\{s\}}^{\lambda}\, 2\zeta(2\alpha\lambda).$$

If $w_s < m$, then $h_s b^{w_s} g \equiv 0 \pmod{N}$ is equivalent to $h_s g \equiv 0 \pmod{b^{m-w_s}}$. Since $g \in Z_{w_s}(N)$, and in particular $\gcd(g, b) = 1$,

$$h_s b^{w_s} g \equiv 0 \pmod{N} \text{ if and only if } h_s \equiv 0 \pmod{b^{m-w_s}}.$$

Hence we obtain

$$T_1 = \frac{1}{|Z_{w_s}(N)|} \sum_{g \in Z_{w_s}(N)} \gamma_{\{s\}}^{\lambda} \sum_{\substack{h_s \in \mathbb{Z} \setminus \{0\} \\ h_s \equiv 0 \pmod{b^{m-w_s}}}} \frac{1}{r_{2\alpha\lambda}(h_s)}$$
$$= \gamma_{\{s\}}^{\lambda} \sum_{h \in \mathbb{Z} \setminus \{0\}} \frac{1}{r_{2\alpha\lambda}(b^{m-w_s} h)}$$

$$= \gamma_{\{s\}}^{\lambda} \frac{2\zeta(2\alpha\lambda)}{b^{2\alpha\lambda(m-w_s)}}$$
$$\leq \gamma_{\{s\}}^{\lambda} \frac{2\zeta(2\alpha\lambda)}{b^{m-w_s}},$$

since $2\alpha\lambda > 1$.

Therefore, in both possible cases, it holds that

$$T_1 \leq \gamma_{\{s\}}^{\lambda} \frac{2\zeta(2\alpha\lambda)}{b^{\max(0,m-w_s)}}.$$

Similarly, we study the term T_2 for the cases $w_s \geq m$ and $w_s < m$.

If $w_s \geq m$, then $g \in Z_{w_s}(N) = \{1\}$ and so

$$\begin{aligned}
T_2 &= \sum_{\substack{\emptyset \neq \mathfrak{v} \subseteq [d] \\ s \notin \mathfrak{v}}} \gamma_{\mathfrak{v}\cup\{s\}}^{\lambda} \sum_{h_s \in \mathbb{Z}\setminus\{0\}} \frac{1}{r_{2\alpha\lambda}(h_s)} \sum_{\substack{\boldsymbol{h}_{\mathfrak{v}} \in (\mathbb{Z}\setminus\{0\})^{|\mathfrak{v}|} \\ \sum_{j\in\mathfrak{v}} h_j \widehat{g}_j \equiv -h_s Y_s \pmod N}} \frac{1}{r_{\alpha\lambda}(\boldsymbol{h}_{\mathfrak{v}})} \\
&\leq \sum_{h_s \in \mathbb{Z}\setminus\{0\}} \frac{1}{r_{2\alpha\lambda}(h_s)} \sum_{\substack{\emptyset \neq \mathfrak{v} \subseteq [d] \\ s \notin \mathfrak{v}}} \gamma_{\mathfrak{v}\cup\{s\}}^{\lambda} \sum_{\boldsymbol{h}_{\mathfrak{v}} \in (\mathbb{Z}\setminus\{0\})^{|\mathfrak{v}|}} \frac{1}{r_{2\alpha\lambda}(\boldsymbol{h}_{\mathfrak{v}})} \\
&= 2\zeta(2\alpha\lambda) \sum_{\substack{\emptyset \neq \mathfrak{v} \subseteq [d] \\ s \notin \mathfrak{v}}} \gamma_{\mathfrak{v}\cup\{s\}}^{\lambda} (2\zeta(2\alpha\lambda))^{|\mathfrak{v}|} \\
&= \sum_{\substack{\emptyset \neq \mathfrak{v} \subseteq [d] \\ s \notin \mathfrak{v}}} \gamma_{\mathfrak{v}\cup\{s\}}^{\lambda} (2\zeta(2\alpha\lambda))^{|\mathfrak{v}|+1} \\
&= \sum_{\substack{\{s\} \neq \mathfrak{u} \subseteq [d] \\ s \in \mathfrak{u}}} \gamma_{\mathfrak{u}}^{\lambda} (2\zeta(2\alpha\lambda))^{|\mathfrak{u}|}.
\end{aligned}$$

If $w_s < m$, we write

$$T_2 = T_{2,1} + T_{2,2},$$

where

$$\begin{aligned}
T_{2,1} := \frac{1}{|Z_{w_s}(N)|} \sum_{g \in Z_{w_s}(N)} \sum_{\substack{\emptyset \neq \mathfrak{v} \subseteq [d] \\ s \notin \mathfrak{v}}} \gamma_{\mathfrak{v}\cup\{s\}}^{\lambda} \sum_{\substack{h_s \in \mathbb{Z}\setminus\{0\} \\ h_s \equiv 0 \pmod{b^{m-w_s}}}} \frac{1}{r_{2\alpha\lambda}(h_s)} \\
\times \sum_{\substack{\boldsymbol{h}_{\mathfrak{v}} \in (\mathbb{Z}\setminus\{0\})^{|\mathfrak{v}|} \\ \sum_{j\in\mathfrak{v}} h_j \widehat{g}_j \equiv -h_s Y_s g \pmod N}} \frac{1}{r_{2\alpha\lambda}(\boldsymbol{h}_{\mathfrak{v}})}
\end{aligned}$$

and

$$T_{2,2} := \frac{1}{|Z_{w_s}(N)|} \sum_{g \in Z_{w_s}(N)} \sum_{\substack{\emptyset \neq \mathfrak{v} \subseteq [d] \\ s \notin \mathfrak{v}}} \gamma_{\mathfrak{v} \cup \{s\}}^{\lambda} \sum_{\substack{h_s \in \mathbb{Z} \setminus \{0\} \\ h_s \not\equiv 0 \pmod{b^{m-w_s}}}} \frac{1}{r_{2\alpha\lambda}(h_s)} \times \sum_{\substack{\boldsymbol{h}_{\mathfrak{v}} \in (\mathbb{Z} \setminus \{0\})^{|\mathfrak{v}|} \\ \sum_{j \in \mathfrak{v}} h_j \widehat{g}_j \equiv -h_s Y_s g \pmod{N}}} \frac{1}{r_{2\alpha\lambda}(\boldsymbol{h}_{\mathfrak{v}})}.$$

For $T_{2,1}$ we see that if $h_s \equiv 0 \ (b^{m-w_s})$ then $h_s Y_s g \equiv 0 \pmod{N}$. Thus

$$\begin{aligned}
T_{2,1} &= \frac{1}{|Z_{w_s}(N)|} \sum_{g \in Z_{w_s}(N)} \sum_{\substack{\emptyset \neq \mathfrak{v} \subseteq [d] \\ s \notin \mathfrak{v}}} \gamma_{\mathfrak{v} \cup \{s\}}^{\lambda} \sum_{\substack{h_s \in \mathbb{Z} \setminus \{0\} \\ h_s \equiv 0 \pmod{b^{m-w_s}}}} \frac{1}{r_{2\alpha\lambda}(h_s)} \\
&\qquad \times \sum_{\substack{\boldsymbol{h}_{\mathfrak{v}} \in (\mathbb{Z} \setminus \{0\})^{|\mathfrak{v}|} \\ \sum_{j \in \mathfrak{v}} h_j \widehat{g}_j \equiv 0 \pmod{N}}} \frac{1}{r_{2\alpha\lambda}(\boldsymbol{h}_{\mathfrak{v}})} \\
&= \sum_{\substack{\emptyset \neq \mathfrak{v} \subseteq [d] \\ s \notin \mathfrak{v}}} \gamma_{\mathfrak{v} \cup \{s\}}^{\lambda} \sum_{h \in \mathbb{Z} \setminus \{0\}} \frac{1}{r_{2\alpha\lambda}(b^{m-w_s} h)} \sum_{\substack{\boldsymbol{h}_{\mathfrak{v}} \in (\mathbb{Z} \setminus \{0\})^{|\mathfrak{v}|} \\ \sum_{j \in \mathfrak{v}} h_j \widehat{g}_j \equiv 0 \pmod{N}}} \frac{1}{r_{2\alpha\lambda}(\boldsymbol{h}_{\mathfrak{v}})} \\
&= \frac{2\zeta(2\alpha\lambda)}{b^{(m-w_s)2\alpha\lambda}} \sum_{\substack{\emptyset \neq \mathfrak{v} \subseteq [d] \\ s \notin \mathfrak{v}}} \gamma_{\mathfrak{v} \cup \{s\}}^{\lambda} \sum_{\substack{\boldsymbol{h}_{\mathfrak{v}} \in (\mathbb{Z} \setminus \{0\})^{|\mathfrak{v}|} \\ \sum_{j \in \mathfrak{v}} h_j \widehat{g}_j \equiv 0 \pmod{N}}} \frac{1}{r_{2\alpha\lambda}(\boldsymbol{h}_{\mathfrak{v}})} \\
&\leq \frac{2\zeta(2\alpha\lambda)}{b^{m-w_s}} \sum_{\substack{\emptyset \neq \mathfrak{v} \subseteq [d] \\ s \notin \mathfrak{v}}} \gamma_{\mathfrak{v} \cup \{s\}}^{\lambda} \sum_{\substack{\boldsymbol{h}_{\mathfrak{v}} \in (\mathbb{Z} \setminus \{0\})^{|\mathfrak{v}|} \\ \sum_{j \in \mathfrak{v}} h_j \widehat{g}_j \equiv 0 \pmod{N}}} \frac{1}{r_{2\alpha\lambda}(\boldsymbol{h}_{\mathfrak{v}})}.
\end{aligned}$$

For $T_{2,2}$ we obtain

$$T_{2,2} = \frac{1}{|Z_{w_s}(N)|} \sum_{g \in Z_{w_s}(N)} \sum_{\substack{\emptyset \neq \mathfrak{v} \subseteq [d] \\ s \notin \mathfrak{v}}} \gamma_{\mathfrak{v} \cup \{s\}}^{\lambda} \sum_{c=1}^{b^{m-w_s}-1} \sum_{\substack{h_s \in \mathbb{Z} \setminus \{0\} \\ h_s \equiv -c g^{-1} \pmod{b^{m-w_s}}}} \frac{1}{r_{2\alpha\lambda}(h_s)} \times \sum_{\substack{\boldsymbol{h}_{\mathfrak{v}} \in (\mathbb{Z} \setminus \{0\})^{|\mathfrak{v}|} \\ \sum_{j \in \mathfrak{v}} h_j \widehat{g}_j \equiv c Y_s \pmod{N}}} \frac{1}{r_{2\alpha\lambda}(\boldsymbol{h}_{\mathfrak{v}})},$$

where g^{-1} denotes the multiplicative inverse of g in $Z_{w_s}(N)$. As in (4.10) we have

$$\sum_{g \in Z_{w_s}(N)} \sum_{\substack{h_s \in \mathbb{Z} \setminus \{0\} \\ h_s \equiv -c g^{-1} \pmod{b^{m-w_s}}}} \frac{1}{r_{2\alpha\lambda}(h_s)} \leq 2\zeta(2\alpha\lambda),$$

and so

$$T_{2,2} \le \frac{2\zeta(2\alpha\lambda)}{|Z_{w_s}(N)|} \sum_{\substack{\emptyset \neq \mathfrak{v} \subseteq [d] \\ s \notin \mathfrak{v}}} \gamma_{\mathfrak{v}\cup\{s\}}^{\lambda} \sum_{c=1}^{b^{m-w_s}-1} \sum_{\substack{\boldsymbol{h}_{\mathfrak{v}} \in (\mathbb{Z}\setminus\{0\})^{|\mathfrak{v}|} \\ \sum_{j\in\mathfrak{v}} h_j \widehat{g}_j \equiv cY_s \pmod N}} \frac{1}{r_{2\alpha\lambda}(\boldsymbol{h}_{\mathfrak{v}})}$$

$$\le \frac{2\zeta(2\alpha\lambda)}{|Z_{w_s}(N)|} \sum_{\substack{\emptyset \neq \mathfrak{v} \subseteq [d] \\ s \notin \mathfrak{v}}} \gamma_{\mathfrak{v}\cup\{s\}}^{\lambda} \sum_{\substack{\boldsymbol{h}_{\mathfrak{v}} \in (\mathbb{Z}\setminus\{0\})^{|\mathfrak{v}|} \\ \sum_{j\in\mathfrak{v}} h_j \widehat{g}_j \not\equiv 0 \pmod N}} \frac{1}{r_{2\alpha\lambda}(\boldsymbol{h}_{\mathfrak{v}})}.$$

Recall that $|Z_{w_s}(N)| = b^{m-w_s-1}(b-1) \ge b^{m-w_s}/2$. Therefore we obtain

$$T_{2,2} \le \frac{4\zeta(2\alpha\lambda)}{b^{m-w_s}} \sum_{\substack{\emptyset \neq \mathfrak{v} \subseteq [d] \\ s \notin \mathfrak{v}}} \gamma_{\mathfrak{v}\cup\{s\}}^{\lambda} \sum_{\substack{\boldsymbol{h}_{\mathfrak{v}} \in (\mathbb{Z}\setminus\{0\})^{|\mathfrak{v}|} \\ \sum_{j\in\mathfrak{v}} h_j \widehat{g}_j \not\equiv 0 \pmod N}} \frac{1}{r_{2\alpha\lambda}(\boldsymbol{h}_{\mathfrak{v}})}$$

$$= \frac{4\zeta(2\alpha\lambda)}{b^{m-w_s}} \sum_{\substack{\emptyset \neq \mathfrak{v} \subseteq [d] \\ s \notin \mathfrak{v}}} \gamma_{\mathfrak{v}\cup\{s\}}^{\lambda}$$

$$\times \left(\sum_{\boldsymbol{h}_{\mathfrak{v}} \in (\mathbb{Z}\setminus\{0\})^{|\mathfrak{v}|}} \frac{1}{r_{2\alpha\lambda}(\boldsymbol{h}_{\mathfrak{v}})} - \sum_{\substack{\boldsymbol{h}_{\mathfrak{v}} \in (\mathbb{Z}\setminus\{0\})^{|\mathfrak{v}|} \\ \sum_{j\in\mathfrak{v}} h_j \widehat{g}_j \equiv 0 \pmod N}} \frac{1}{r_{2\alpha\lambda}(\boldsymbol{h}_{\mathfrak{v}})} \right).$$

Thus, we get for $T_2 = T_{2,1} + T_{2,2}$,

$$T_2 \le \frac{4\zeta(2\alpha\lambda)}{b^{m-w_s}} \sum_{\substack{\emptyset \neq \mathfrak{v} \subseteq [d] \\ s \notin \mathfrak{v}}} \gamma_{\mathfrak{v}\cup\{s\}}^{\lambda} \sum_{\boldsymbol{h}_{\mathfrak{v}} \in (\mathbb{Z}\setminus\{0\})^{|\mathfrak{v}|}} \frac{1}{r_{2\alpha\lambda}(\boldsymbol{h}_{\mathfrak{v}})}$$

$$= \sum_{\substack{\emptyset \neq \mathfrak{v} \subseteq [d] \\ s \notin \mathfrak{v}}} \gamma_{\mathfrak{v}\cup\{s\}}^{\lambda} \frac{2(2\zeta(2\alpha\lambda))^{|\mathfrak{v}|+1}}{b^{m-w_s}}$$

$$= \sum_{\substack{\{s\} \neq \mathfrak{u} \subseteq [d] \\ s \in \mathfrak{u}}} \gamma_{\mathfrak{u}}^{\lambda} \frac{2(2\zeta(2\alpha\lambda))^{|\mathfrak{u}|}}{b^{m-w_s}}.$$

Combining both cases, T_2 is always bounded by

$$T_2 \le \sum_{\substack{\{s\} \neq \mathfrak{u} \subseteq [d] \\ s \in \mathfrak{u}}} \gamma_{\mathfrak{u}}^{\lambda} \frac{2(2\zeta(2\alpha\lambda))^{|\mathfrak{u}|}}{b^{\max(0,m-w_s)}}.$$

Hence we see that

$$(\theta_s(\boldsymbol{g}^{(s)}))^{\lambda} \le T_1 + T_2$$

$$\le \gamma_{\{s\}}^{\lambda} \frac{2\zeta(2\alpha\lambda)}{b^{\max(0,m-w_s)}} + \sum_{\substack{\{s\} \neq \mathfrak{u} \subseteq [d] \\ s \in \mathfrak{u}}} \gamma_{\mathfrak{u}}^{\lambda} \frac{2(2\zeta(2\alpha\lambda))^{|\mathfrak{u}|}}{b^{\max(0,m-w_s)}}$$

$$\leq \sum_{\substack{\mathfrak{u}\subseteq[d]\\ s\in\mathfrak{u}}} \gamma_{\mathfrak{u}}^{\lambda} \frac{2(2\zeta(2\alpha\lambda))^{|\mathfrak{u}|}}{b^{\max(0,m-w_s)}},$$

and so

$$[\mathrm{err}_{N,d,\alpha,\boldsymbol{\gamma}}(\boldsymbol{g})]^{2\lambda} \leq \sum_{s=1}^{d}(\theta_s(\boldsymbol{g}^{(s)}))^{\lambda} \leq \sum_{s=1}^{d} \sum_{\substack{\mathfrak{u}\subseteq[d]\\ s\in\mathfrak{u}}} \gamma_{\mathfrak{u}}^{\lambda} \frac{2(2\zeta(2\alpha\lambda))^{|\mathfrak{u}|}}{b^{\max(0,m-w_s)}}.$$

Taking the left-hand side and the right-hand side of the latter inequality to the power $1/(2\lambda)$ and setting $\tau = 1/(2\lambda)$, which implies $\tau \in [1/2, \alpha)$, yields

$$\mathrm{err}_{N,d,\alpha,\boldsymbol{\gamma}}(\boldsymbol{g}) \leq \left(\sum_{s=1}^{d} \sum_{\substack{\mathfrak{u}\subseteq[d]\\ s\in\mathfrak{u}}} \gamma_{\mathfrak{u}}^{1/(2\tau)} \frac{2(2\zeta(\alpha/\tau))^{|\mathfrak{u}|}}{b^{\max(0,m-w_s)}} \right)^{\tau}$$

as claimed. □

Fast implementation of the reduced SCS algorithm

Let us now discuss the fast implementation of the reduced SCS algorithm for product weights and certain types of initial vectors $\boldsymbol{g}^{(0)}$. We require for a fast implementation that the initial vector $\boldsymbol{g}^{(0)}$ is of the form

$$\boldsymbol{g}^{(0)} = (g_1^{(0)}, \ldots, g_d^{(0)}) = (Y_1\overline{g}_1, \ldots, Y_d\overline{g}_d) \tag{4.17}$$

with $\overline{g}_j \in Z_{w_j}(N)$ for $j \in [d]$. Note that the components $Y_j\overline{g}_j$ satisfy $Y_j\overline{g}_j \equiv 0 \pmod N$ for $j > d_*$, where again $d_* := \max\{s \in \mathbb{N}_0 \,:\, w_s < m\}$.

Our starting point is the worst-case error formula (3.6). Let $N = b^m$, let $s \in \{0, 1, \ldots, d-1\}$ be arbitrarily chosen, and assume that, for $s \neq 0$, $g_1, \ldots, g_s$ have already been selected. According to Algorithm 4.11, we now need to find a minimizer in $Z_{w_{s+1}}(N)$ such that the squared error

$$[\mathrm{err}_{N,s+1,\alpha,\boldsymbol{\gamma}}((Y_1 g_1, \ldots, Y_s g_s, Y_{s+1} g, g_{s+2}^{(0)}, \ldots, g_d^{(0)}))]^2 \tag{4.18}$$

is minimized as a function of g, which means that we need to minimize

$$\sum_{k=0}^{N-1}\left(1+\gamma_j\varphi_\alpha\left(\frac{kY_{s+1}g}{N}\right)\right)\widetilde{\eta}_s(k) = \sum_{k=0}^{N-1}\widetilde{\eta}_s(k) + \gamma_j \sum_{k=0}^{N-1}\varphi_\alpha\left(\frac{kY_{s+1}g}{N}\right)\widetilde{\eta}_s(k),$$

where

$$\widetilde{\eta}_s(k) := \left(\prod_{j=1}^{s}\left(1+\gamma_j\varphi_\alpha\left(\frac{kY_j g_j}{N}\right)\right)\right)\prod_{j=s+2}^{d}\left(1+\gamma_j\varphi_\alpha\left(\frac{k g_j^{(0)}}{N}\right)\right).$$

The empty products occurring for $s = 0$ or $s = d - 1$, respectively, are considered as equal to one. Accordingly, minimizing the squared error (4.18) with respect to g is equivalent to minimizing

$$\widetilde{T}_{s+1}(g) := \sum_{k=0}^{N-1} \varphi_\alpha\left(\frac{kY_{s+1}g}{N}\right)\widetilde{\eta}_s(k).$$

Using this notation, we can now outline the idea of the fast implementation of Algorithm 4.11 and derive its cost.

As outlined in Section 3.4, we can assume that the function values of φ_α for arguments k/b^m with $k \in \{0, 1, \ldots, b^m - 1\}$, can be precomputed using $O(N \log N)$ operations, at the expense of $O(N)$ storage.

In the next step we initialize a vector $\widetilde{\boldsymbol{\eta}} = (\widetilde{\eta}(0), \ldots, \widetilde{\eta}(N-1))^\top$ by

$$\widetilde{\eta}(k) := \prod_{j=1}^{d}\left(1 + \gamma_j\varphi_\alpha\left(\frac{kg_j^{(0)}}{N}\right)\right) \quad \text{for} \quad k \in \{0, 1, \ldots, N-1\}.$$

This initialization of $\widetilde{\boldsymbol{\eta}}$ can be done using $O(N \min(d, d_*))$ operations, as (4.17) actually implies

$$\widetilde{\eta}(k) = \left(\prod_{j=1}^{d_*}\left(1 + \gamma_j\varphi_\alpha\left(\frac{kg_j^{(0)}}{N}\right)\right)\right)\prod_{j=d_*+1}^{d}(1 + \gamma_j\varphi_\alpha(0)).$$

After the initialization, one can use a loop over increasing $s \in \{2, 3, \ldots, \min(d, d_*) - 1\}$ as follows. Assume that $g_1, \ldots, g_s$ have already been chosen in the previous steps (for $s = 0$, no previous components have been chosen). Then update the vector $\widetilde{\boldsymbol{\eta}}_s = (\widetilde{\eta}_s(0), \ldots, \widetilde{\eta}_s(N-1))^\top$ using $\widetilde{\boldsymbol{\eta}}$, by dividing out the term corresponding to $g_{s+1}^{(0)}$, i.e., put

$$\widetilde{\eta}_s(k) := \left(\prod_{j=1}^{s}\left(1 + \gamma_j\varphi_\alpha\left(\frac{kY_jg_j}{N}\right)\right)\right)\prod_{j=s+2}^{d}\left(1 + \gamma_j\varphi_\alpha\left(\frac{kg_j^{(0)}}{N}\right)\right)$$

for $k \in \{0, 1, \ldots, N-1\}$. Like the initialization, this step can be done in $O(N \min(d, d_*))$ operations in total for all steps in the loop over s.

Next, divide $\widetilde{\boldsymbol{\eta}}_s$ into $b^{w_{s+1}}$ vectors $\widetilde{\boldsymbol{\eta}}_s^{(1)}, \ldots, \widetilde{\boldsymbol{\eta}}_s^{(b^{w_{s+1}})}$ of length $b^{m-w_{s+1}}$ each, where

$$\widetilde{\boldsymbol{\eta}}_s^{(\ell)} = (\widetilde{\eta}_s(1 + (\ell - 1)b^{m-w_{s+1}}), \ldots, \widetilde{\eta}_s(\ell\, b^{m-w_{s+1}}))^\top \quad \text{for } \ell \in [b^{w_{s+1}}],$$

and put $\widetilde{\boldsymbol{\eta}}_s' := \widetilde{\boldsymbol{\eta}}_s^{(1)} + \cdots + \widetilde{\boldsymbol{\eta}}_s^{(b^{w_{s+1}})}$. These additions again require $O(N \min(d, d_*))$ operations in total for all steps in the loop.

As in Section 4.2, let

$$\boldsymbol{A}_N := \left(\varphi_\alpha\left(\frac{kY_{s+1}g}{N}\right)\right)_{g\in Z_{w_{s+1}}(N),k\in G_1(N)}.$$

Then, we see again that $\boldsymbol{A}_N$ is a block matrix,

$$\boldsymbol{A}_N = \left(\underbrace{\boldsymbol{\Omega}_{b^{m-w_{s+1}}},\ldots,\boldsymbol{\Omega}_{b^{m-w_{s+1}}}}_{b^{w_{s+1}}\text{ times}}\right),$$

and that multiplication of $\boldsymbol{A}_N$ with $\widetilde{\boldsymbol{\eta}}_s$ boils down to multiplication of $\boldsymbol{\Omega}_{b^{m-w_{s+1}}}$ (defined as in Section 4.2) with $\widetilde{\boldsymbol{\eta}}'_s$. Using FFT, this can again be done in $O((m-w_{s+1})b^{m-w_{s+1}})$ operations.

Finally, update $\widetilde{\boldsymbol{\eta}}$ using $\widetilde{\boldsymbol{\eta}}_s$, by

$$\widetilde{\eta}(k) := \left(\prod_{j=1}^{s+1}\left(1+\gamma_j\varphi_\alpha\left(\frac{kY_jg_j}{N}\right)\right)\right)\prod_{j=s+2}^{d}\left(1+\gamma_j\varphi_\alpha\left(\frac{kg_j^{(0)}}{N}\right)\right)$$

for $k\in\{0,1,\ldots,N-1\}$, and this again requires $O(N\min(d,d_*))$ operations in total for all steps in the loop.

Then, increase s by one and repeat the loop until s has reached $\min(d,d_*)$. For $s\geq d_*$, all components g_s are chosen as 1.

This shows that the total number of operations required for a fast implementation of Algorithm 4.11 is, as for the reduced fast CBC construction outlined in Section 4.1, of order

$$O\left(mb^m + b^m\min(d,d_*) + \sum_{s=1}^{\min(d,d_*)}(m-w_s)b^{m-w_s}\right),$$

where $d_* := \max\{s\in\mathbb{N}_0 : w_s < m\}$. Furthermore, if the reduction indices are chosen such that they increase sufficiently fast, the computational cost of Algorithm 4.11 can be independent of d, and again one needs to carefully balance the choice of the reduction indices with the coordinate weights.

4.5 Projection-Corrected Constructions

Let us now address another issue that may be of relevance in the context of CBC constructions. Several authors (see, e.g., [77]) report that in running the CBC construction or variants thereof it may happen that components of the generating vector $\boldsymbol{g}=(g_1,\ldots,g_d)$ repeat themselves, i.e., there are $i,j\in[d], i\neq j$, such that $g_i=g_j$.

To alleviate this problem, Gantner and Schwab [77] introduced the method of "pruning". Here, we outline the method for the worst-case error in weighted Korobov spaces, but the general idea also works for other quality measures for lattice rules.

Assume that one would like to construct a generating vector $\boldsymbol{g}$ of a d-dimensional rank-1 lattice rule with $N \in \mathbb{N}$ points by the usual CBC construction, as given in Algorithm 3.6, but that one would like to avoid repeated components. Suppose that for $s \in [d-1]$ the components $g_1, \ldots, g_s \in G_1^{\varphi}(N)$ of the generating vector $\boldsymbol{g}$ have already been chosen. Then, choose $g_{s+1} \in G_1^{\varphi}(N) \setminus \{g_1, \ldots, g_s\}$. Obeying this rule naturally avoids repeated components in $\boldsymbol{g}$. The only additional assumption to be made in this procedure is that the cardinality of $G_1^{\varphi}(N)$ needs to be (substantially) larger than d, which, however, is no big restriction, as it is common that the number of points significantly exceeds the dimension.

We now formulate the so-called projection-corrected CBC algorithm.

Algorithm 4.13 (Projection-corrected CBC construction) Let N and d be given. Construct a generating vector $\boldsymbol{g} = (g_1, \ldots, g_d) \in G_d^{\varphi}(N)$ as follows.

(1) Choose $g_1 = 1$ and $\mathcal{E}_1 = \emptyset$.
(2) For s from 1 to $d-1$:
Assume that $g_1, \ldots, g_s \in G_1^{\varphi}(N)$ have already been found, and choose $\mathcal{E}_{s+1} \subsetneq G_1^{\varphi}(N)$. Choose $g_{s+1} \in G_1^{\varphi}(N) \setminus \mathcal{E}_{s+1}$ as

$$g_{s+1} := \underset{g \in G_1^{\varphi}(N) \setminus \mathcal{E}_{s+1}}{\operatorname{argmin}} \; \mathrm{err}_{N,s+1,\alpha,\boldsymbol{\gamma}}((g_1, \ldots, g_s)).$$

End for.
(3) Set $\boldsymbol{g} = (g_1, \ldots, g_d)$.

Remark 4.14 Note that Algorithm 4.13 generalizes the idea of pruning in the sense that it allows arbitrary exclusion sets $\mathcal{E}_s$ in step s. If no elements of $G_1^{\varphi}(N)$ are to be excluded in step s, choose $\mathcal{E}_s = \emptyset$. In particular, by choosing $\mathcal{E}_s = \emptyset$ for all $s \in \{2, 3, \ldots, d\}$, one obtains Algorithm 3.6. Note, furthermore, that for $s = 1$ we do not use any exclusions since in our setting all choices of g_1 would yield the same one-dimensional point set.

The following result, which is a generalization of Theorem 3.9, was shown in [43], but only for product weights. Here we state it for general weights.

Theorem 4.15 *Let $N \geq 2$ be an arbitrary integer and let $\boldsymbol{\gamma} = \{\gamma_{\mathfrak{u}}\}_{\mathfrak{u} \subseteq [d]}$ be general weights. Moreover, assume that $\boldsymbol{g} = (g_1, \ldots, g_d)$ and exclusion sets $\mathcal{E}_1, \ldots, \mathcal{E}_d \subsetneq G_1^{\varphi}(N)$ have been chosen in Algorithm 4.13. Then for arbitrary $\tau \in [1/2, \alpha)$ and for any $s \in [d]$ we have*

$$\mathrm{err}_{N,s,\alpha,\boldsymbol{\gamma}}((g_1, \ldots, g_s)) \leq \left(\frac{1}{\varphi(N)} \sum_{\emptyset \neq \mathfrak{u} \subseteq [s]} \gamma_{\mathfrak{u}}^{1/(2\tau)} \left(2\zeta\left(\frac{\alpha}{\tau}\right)\right)^{|\mathfrak{u}|} \prod_{j \in \mathfrak{u}} \frac{\varphi(N)}{\varphi(N) - |\mathcal{E}_j|} \right)^{\tau}. \tag{4.19}$$

Proof The proof is a modification of the proof of Theorem 3.9 in Section 3.3, and we also use the same notation as there. Again, the proof is based on induction on s.

For $s = 1$ the argumentation is exactly the same as in the proof of Theorem 3.9.

We now explain the changes in the induction step. Assume that the bound (4.19) holds for s and assume further that g_{s+1} and $\mathcal{E}_{s+1}$ are chosen according to Algorithm 4.13. Let $\boldsymbol{g}^{(s)} := (g_1, \ldots, g_s)$ be the vector found in the first s steps of Algorithm 4.13, and, for $g \in G_1^{\varphi}(N)$, let $(\boldsymbol{g}^{(s)}, g) := (g_1, \ldots, g_s, g)$. For $\lambda \in (1/(2\alpha), 1]$ and $c \geq 1$ define

$$G_c := \{g \in G_1^{\varphi}(N) \; : \; (\theta_{N,s,\alpha,\boldsymbol{\gamma}}(\boldsymbol{g}^{(s)}, g))^{\lambda} \leq c\, \Theta_{N,s,\alpha,\boldsymbol{\gamma},\lambda}(\boldsymbol{g}^{(s)})\}.$$

We employ Markov's inequality, which states that for a nonnegative random variable X and any real number $c \geq 1$ we have that $\mathbb{P}[X < c\mathbb{E}[X]] > 1 - c^{-1}$. We use the normalized counting measure μ on $G_1^{\varphi}(N)$ as the probability measure. Then

$$\frac{|G_c|}{\varphi(N)} = \mu(G_c) > 1 - \frac{1}{c}.$$

In particular, for any $c \geq 1$ there is a subset $G_c \subseteq G_1^{\varphi}(N)$ of size bigger than $\varphi(N)(1 - c^{-1})$ such that for all g in this subset we have

$$(\theta_{N,s,\alpha,\boldsymbol{\gamma}}(\boldsymbol{g}^{(s)}, g))^{\lambda} \leq c\, \Theta_{N,s,\alpha,\boldsymbol{\gamma},\lambda}(\boldsymbol{g}^{(s)}).$$

By choosing $c \geq 1$ such that

$$\varphi(N)\left(1 - \frac{1}{c}\right) = |\mathcal{E}_{s+1}|,$$

it follows that the set $G_c \setminus \mathcal{E}_{s+1}$ is nonempty. This condition is satisfied for

$$c = \frac{\varphi(N)}{\varphi(N) - |\mathcal{E}_{s+1}|}.$$

In particular, if $\mathcal{E}_{s+1} = \emptyset$, then $c = 1$. Since g_{s+1} is a minimizer of the error we obtain for $g \in G_c \setminus \mathcal{E}_{s+1}$ that

$$\begin{aligned}(\theta_{N,s,\alpha,\boldsymbol{\gamma}}(\boldsymbol{g}^{(s)}, g_{s+1}))^{\lambda} &\leq (\theta_{N,s,\alpha,\boldsymbol{\gamma}}(\boldsymbol{g}^{(s)}, g))^{\lambda} \\ &\leq c\, \Theta_{N,s,\alpha,\boldsymbol{\gamma},\lambda}(\boldsymbol{g}^{(s)}) \\ &\leq \frac{\varphi(N)}{\varphi(N) - |\mathcal{E}_{s+1}|} \frac{1}{\varphi(N)} \sum_{\substack{\mathfrak{u} \subseteq [s+1] \\ s+1 \in \mathfrak{u}}} \gamma_{\mathfrak{u}}^{\lambda} \, (2\zeta(2\alpha\lambda))^{|\mathfrak{u}|},\end{aligned}$$

where we used (3.15) in the last estimate. Since $\varphi(N)/(\varphi(N) - |\mathcal{E}_j|) \geq 1$ for every $j \in [d]$, we get

$$(\theta_{N,s,\alpha,\boldsymbol{\gamma}}(\boldsymbol{g}^{(s)}, g_{s+1}))^{\lambda} \leq \frac{1}{\varphi(N)} \sum_{\substack{\mathfrak{u} \subseteq [s+1] \\ s+1 \in \mathfrak{u}}} \gamma_{\mathfrak{u}}^{\lambda} \, (2\zeta(2\alpha\lambda))^{|\mathfrak{u}|} \prod_{j \in \mathfrak{u}} \frac{\varphi(N)}{\varphi(N) - |\mathcal{E}_j|}. \tag{4.20}$$

Now we proceed as in the final part of the proof of Theorem 3.9. From (3.17) and (4.20) we obtain

$$
\begin{aligned}
&[\mathrm{err}_{N,s+1,\alpha,\boldsymbol{\gamma}}((\boldsymbol{g}^{(s)}, g_{s+1}))]^{2\lambda} \\
&\le [\mathrm{err}_{N,s,\alpha,\boldsymbol{\gamma}}(\boldsymbol{g}^{(s)})]^{2\lambda} + (\theta_{N,s,\alpha,\boldsymbol{\gamma}}(\boldsymbol{g}^{(s)}, g_{s+1}))^{\lambda} \\
&\le [\mathrm{err}_{N,s,\alpha,\boldsymbol{\gamma}}(\boldsymbol{g}^{(s)})]^{2\lambda} + \frac{1}{\varphi(N)} \sum_{\substack{\mathfrak{u}\subseteq[s+1]\\ s+1\in\mathfrak{u}}} \gamma_{\mathfrak{u}}^{\lambda} (2\zeta(2\alpha\lambda))^{|\mathfrak{u}|} \prod_{j\in\mathfrak{u}} \frac{\varphi(N)}{\varphi(N)-|\mathcal{E}_j|}.
\end{aligned}
$$

Setting again $\tau = 1/(2\lambda)$ we obtain

$$
\begin{aligned}
&[\mathrm{err}_{N,s+1,\alpha,\boldsymbol{\gamma}}((\boldsymbol{g}^{(s)}, g_{s+1}))]^{1/\tau} \\
&\le [\mathrm{err}_{N,s,\alpha,\boldsymbol{\gamma}}(\boldsymbol{g}^{(s)})]^{1/\tau} + \frac{1}{\varphi(N)} \sum_{\substack{\mathfrak{u}\subseteq[s+1]\\ s+1\in\mathfrak{u}}} \gamma_{\mathfrak{u}}^{1/(2\tau)} \left(2\zeta\left(\frac{\alpha}{\tau}\right)\right)^{|\mathfrak{u}|} \prod_{j\in\mathfrak{u}} \frac{\varphi(N)}{\varphi(N)-|\mathcal{E}_j|}.
\end{aligned}
$$

Using the induction assumption yields

$$
\begin{aligned}
&[\mathrm{err}_{N,s+1,\alpha,\boldsymbol{\gamma}}((\boldsymbol{g}^{(s)}, g_{s+1}))]^{1/\tau} \\
&\le \frac{1}{\varphi(N)} \sum_{\emptyset\neq\mathfrak{u}\subseteq[s]} \gamma_{\mathfrak{u}}^{1/(2\tau)} \left(2\zeta\left(\frac{\alpha}{\tau}\right)\right)^{|\mathfrak{u}|} \prod_{j\in\mathfrak{u}} \frac{\varphi(N)}{\varphi(N)-|\mathcal{E}_j|} \\
&\quad + \frac{1}{\varphi(N)} \sum_{\substack{\mathfrak{u}\subseteq[s+1]\\ s+1\in\mathfrak{u}}} \gamma_{\mathfrak{u}}^{1/(2\tau)} \left(2\zeta\left(\frac{\alpha}{\tau}\right)\right)^{|\mathfrak{u}|} \prod_{j\in\mathfrak{u}} \frac{\varphi(N)}{\varphi(N)-|\mathcal{E}_j|} \\
&= \frac{1}{\varphi(N)} \sum_{\emptyset\neq\mathfrak{u}\subseteq[s+1]} \gamma_{\mathfrak{u}}^{1/(2\tau)} \left(2\zeta\left(\frac{\alpha}{\tau}\right)\right)^{|\mathfrak{u}|} \prod_{j\in\mathfrak{u}} \frac{\varphi(N)}{\varphi(N)-|\mathcal{E}_j|},
\end{aligned}
$$

which gives the final result. □

The following corollary, which is a straightforward consequence of Theorem 4.15, deals with the case when the relative sizes of the exclusion sets are uniformly bounded.

Corollary 4.16 *Let sequences of positive integers $(N_k)_{k\in\mathbb{N}}$ and $(d_k)_{k\in\mathbb{N}}$ be given. Assume that $\boldsymbol{g}_k = (g_{1,k}, \ldots, g_{d_k,k})$ have been constructed by Algorithm 4.13, using N_k, d_k and the exclusion sets $\mathcal{E}_{1,k}, \ldots, \mathcal{E}_{d_k,k}$ for $k \in \mathbb{N}$. Assume that there is a $\delta \in (0, 1)$ such that*

$$
\sup_{k\in\mathbb{N}} \frac{\max_{j\in[d_k]} |\mathcal{E}_{j,k}|}{\varphi(N_k)} \le \delta. \tag{4.21}
$$

Then for any $\tau \in [1/2, \alpha)$ and any $k \in \mathbb{N}$ we have

$$
\mathrm{err}_{N_k,d_k,\alpha,\boldsymbol{\gamma}}(\boldsymbol{g}_k) \le \left(\frac{1}{\varphi(N_k)} \sum_{\emptyset\neq\mathfrak{u}\subseteq[d_k]} \gamma_{\mathfrak{u}}^{1/(2\tau)} \left(\frac{2\zeta(\alpha/\tau)}{1-\delta}\right)^{|\mathfrak{u}|}\right)^{\tau}.
$$

Corollary 4.16 implies that as long as the relative sizes of the exclusion sets are uniformly bounded as in (4.21), tractability results are not affected. In other words, if, e.g., we obtain strong polynomial tractability for the integration problem in the weighted Korobov space using the standard CBC algorithm (Algorithm 3.6), then the same is possible with the use of the modified CBC algorithm using uniformly bounded exclusion sets.

Although Algorithm 4.13, Theorem 4.15, and Corollary 4.16 apply to arbitrary exclusion sets, some choices are of particular interest. Let us discuss some of them in the following.

Repeated components

If, as in the motivation for this section, the aim is simply to avoid repeated components as observed in some numerical experiments, and as used in [77] for pruning, one can choose the exclusion sets as

$$\mathcal{E}_{s+1} = \{g_1, \ldots, g_s\} \qquad \text{for } s \in [d-1],$$

where the $g_1, \ldots, g_s$ are the components chosen in the first steps of Algorithm 4.13. This guarantees that there are no two-dimensional projections of the lattice point set whose points all lie on the main diagonal $\{(x, x) : 0 \le x \le 1\}$.

Avoiding diagonals

To also exclude having two-dimensional projections where all the points of the lattice point set lie on an antidiagonal $\{(x, 1-x) : 0 \le x \le 1\}$, one can additionally exclude the components $N - g_1, N - g_2, \ldots, N - g_s$ in step $s + 1$ of Algorithm 4.13. This suggests using the exclusion sets

$$\mathcal{E}_{s+1} = \{g_1, N - g_1, g_2, N - g_2, \ldots, g_s, N - g_s\} \qquad \text{for } s \in [d-1]. \tag{4.22}$$

Note that this is only possible as long as $2(d-1) < \varphi(N)$. Even so, if, say, $|\mathcal{E}_j| = \varphi(N) - \ell$ for some small $\ell \in \mathbb{N}$, then the factor $\varphi(N)/(\varphi(N) - |\mathcal{E}_j|) = \varphi(N)/\ell$ becomes large, in which case the bound in Theorem 4.15 becomes meaningless. So one still wants to impose a restriction, e.g., of the form $\max_{j \in [d]} |\mathcal{E}_j| \le \delta\varphi(N)$ for some "reasonable choice" (depending on the application) of $\delta < 1$. This and (4.22) imply $2(d-1) \le \delta\varphi(N)$.

Avoiding diagonals in smaller dimension

Under certain circumstances a condition of the form $2(d-1) \le \delta\varphi(N)$ cannot be satisfied. For example, when considering tractability questions one wants to study the dependence on the dimension as d tends to infinity. Another case in which problems can arise is when N and d need to be increased simultaneously, as for example in applications related to PDEs with random coefficients (see, e.g., [160], and Appendix A). In this case one can, for instance, choose $\mathcal{E}_{s+1}$ as in (4.22) for $s \in [d^* - 1]$ for some fixed d^* (independent of N and d) and set $\mathcal{E}_{s+1} = \emptyset$ for $s \in \{d^*, \dots, d-1\}$. As long as $2(d^* - 1) \leq \delta\varphi(N)$, Corollary 4.16 applies since the relative sizes of the exclusion sets are uniformly bounded, and therefore strong polynomial tractability results can still be obtained. The particular choice of d^* will depend on the problem under consideration.

Reduced fast CBC construction

Another instance where a particular type of exclusion sets has been considered is the reduced (fast) CBC construction outlined in Section 4.1. There, however, the original aim is slightly different, namely, to reduce the search space in each coordinate such that one obtains a speedup of the fast CBC algorithm. In this case, instead of having to do additional computational work, the computational work actually decreases. We refer to the preceding sections for further details.

4.6 The Component-By-Component Digit-By-Digit Construction

In the final section of this chapter, we present a construction algorithm for generating vectors of rank-1 lattice rules that is motivated by earlier work of Korobov from 1963 and 1982, more specifically [140] and [141]. To be more precise, we will consider an algorithm that constructs the generating vector $\boldsymbol{g}$ in a CBC fashion in which each component g_j is assembled digit-by-digit (DBD), that is, for a total number $N = 2^m$, $m \in \mathbb{N}$, of points we greedily construct the components g_j bit-by-bit starting from the least significant bit. We call this the component-by-component digit-by-digit (CBC-DBD) algorithm. This construction method was presented for the weighted Korobov space setting in [69], and we follow this reference in what we outline below.

Throughout this section we restrict our considerations to product weights $\boldsymbol{\gamma} = (\gamma_j)_{j \geq 1}$. Our starting point is Proposition 3.13, which implies that

$$[\mathrm{err}_{N,d,\alpha,\boldsymbol{\gamma}}(\boldsymbol{g})]^2 \leq R_{N,d,2\alpha,\boldsymbol{\gamma}}(\boldsymbol{g}) + \frac{2^{2\alpha}}{N^{2\alpha}} \left(-1 + \prod_{j=1}^{d} (1 + \gamma_j 4\zeta(2\alpha)) \right),$$

and also

$$[\mathrm{err}_{N,d,\alpha,\boldsymbol{\gamma}}(\boldsymbol{g})]^2 \leq (R_{N,d,1,\boldsymbol{\gamma}^{1/(2\alpha)}}(\boldsymbol{g}))^{2\alpha} + \frac{2^{2\alpha}}{N^{2\alpha}} \left(-1 + \prod_{j=1}^{d} (1 + \gamma_j 4\zeta(2\alpha)) \right)$$

for any $\boldsymbol{g} \in G_d^{\varphi}(N)$, where the function $R_{N,d,\tau,\boldsymbol{\gamma}}$ for $\tau \geq 1$ is as in Definition 3.12. I.e.,

$$R_{N,d,\tau,\boldsymbol{\gamma}}(\boldsymbol{g}) = \sum_{\substack{\boldsymbol{h} \in C_d^*(N) \\ \boldsymbol{h} \cdot \boldsymbol{g} \equiv 0 \pmod{N}}} \frac{1}{r_{\tau,\boldsymbol{\gamma}}(\boldsymbol{h})},$$

where $C_d^*(N)$ is defined as in Section 1.8 and $r_{\tau,\boldsymbol{\gamma}}$ is given in (2.17). Note that, since we assume that $N = 2^m$ with $m \geq 1$, we have $N/2 \in \mathbb{Z}$, so we obtain

$$C_1 = \left(-\frac{N}{2}, \frac{N}{2}\right] \cap \mathbb{Z} = \{-2^{m-1} + 1, -2^{m-1} + 2, \ldots, 2^{m-1}\}.$$

Similarly to Section 3.6 we use the fact that, for finding a rank-1 lattice rule with a small worst-case error, it is essentially sufficient to make sure that the quantity $R_{N,d,2\alpha,\boldsymbol{\gamma}}(\boldsymbol{g})$ is small for the generating vector $\boldsymbol{g}$. In Section 3.6, this was achieved by using a CBC construction (see Algorithm 3.14), where the target function is $R_{N,s,1,\boldsymbol{\gamma}}$ for $s \in [d]$. In this section, we will modify this approach by constructing the components of $\boldsymbol{g}$ digit-wise. This alternative construction is not only interesting from a mathematical point of view, but it has the advantage that there is a fast implementation without the need to employ the fast Fourier transform, as we will see below.

A new bound on $R_{N,d,1,\boldsymbol{\gamma}}$

The approach presented below is based on a suitable general estimate of $R_{N,d,1,\boldsymbol{\gamma}}$ that relates it to another quantity, which is then used as a new figure of merit better suited for the present purpose.

We first show the following bound on the quantity $R_{N,d,1,\boldsymbol{\gamma}}$, which already indicates the target function to be minimized in the CBC-DBD algorithm below.

Proposition 4.17 *Let $N = 2^m$, $m \in \mathbb{N}$, and let $\boldsymbol{\gamma} = (\gamma_j)_{j \geq 1}$ be positive product weights. Furthermore, let $\boldsymbol{g} = (g_1, \ldots, g_d) \in G_d^{\varphi}(N)$. Then,*

$$R_{N,d,1,\boldsymbol{\gamma}}(\boldsymbol{g}) \leq \frac{2 + 6 \log N}{N} \left(-1 + \prod_{j=1}^{d} (1 + \gamma_j 2(6 + 4 \log N))\right)$$
$$+ \frac{1}{N} H_{d,m,\boldsymbol{\gamma}}(\boldsymbol{g}) - \left(-1 + \prod_{j=1}^{d} (1 + \gamma_j \log 4)\right),$$

where

$$H_{d,m,\boldsymbol{\gamma}} = H_{d,m,\boldsymbol{\gamma}}(\boldsymbol{g}) := \sum_{k=1}^{2^m - 1} \left(-1 + \prod_{j=1}^{d} \left(1 + \gamma_j \log\left(\frac{1}{\sin^2(\pi k g_j / 2^m)}\right)\right)\right).$$

For the proof we need two auxiliary results.

Lemma 4.18 *Let $N \in \mathbb{N}$. Then for any $x \in (0,1)$ there exists a $\tau(x) \in \mathbb{R}$ with $|\tau(x)| \le 3$ such that*

$$\log\left(\frac{1}{\sin^2(\pi x)}\right) = \log 4 + \sum_{h \in C_1^*(N)} \frac{\mathrm{e}^{2\pi \mathrm{i} h x}}{|h|} + \frac{\tau(x)}{N\,\|x\|},$$

where $\|x\|$ denotes the distance to the nearest integer of x.

Proof Let $x \in (0,1)$. For $\sigma \in \{-1,1\}$, Euler's formula yields the identity

$$\begin{aligned}
\log(\sin(\pi x)) &= \log\left(\frac{\mathrm{e}^{\pi \mathrm{i} x} - \mathrm{e}^{-\pi \mathrm{i} x}}{2\mathrm{i}}\right)\\
&= \log\left(\mathrm{e}^{\sigma\pi \mathrm{i} x}\sigma\left(\frac{1-\mathrm{e}^{-2\sigma\pi \mathrm{i} x}}{2\mathrm{i}}\right)\right)\\
&= \sigma\pi \mathrm{i} x - \log(2\sigma\mathrm{i}) + \log\left(1-\mathrm{e}^{-2\sigma\pi \mathrm{i} x}\right)\\
&= \sigma\pi \mathrm{i} x - \log 2 - \sigma\frac{\pi\mathrm{i}}{2} + \log\left(1-\mathrm{e}^{-2\sigma\pi \mathrm{i} x}\right).
\end{aligned}$$

We recall that the Maclaurin series of $\log(1-y)$ equals $-\sum_{h=1}^{\infty} y^h/h$, which converges to $\log(1-y)$ for $y = \mathrm{e}^{2\pi \mathrm{i} x}$ provided that $x \notin \mathbb{Z}$. Then, averaging over both choices of $\sigma \in \{-1,1\}$ yields

$$\log(\sin(\pi x)) = -\log 2 - \frac{1}{2}\sum_{h=1}^{\infty}\left(\frac{\mathrm{e}^{-2\pi \mathrm{i} h x} + \mathrm{e}^{2\pi \mathrm{i} h x}}{h}\right) = -\log 2 - \frac{1}{2}\sum_{h \in \mathbb{Z}\setminus\{0\}} \frac{\mathrm{e}^{2\pi \mathrm{i} h x}}{|h|},$$

and we note that the series is convergent for $x \in (0,1)$. Now we have

$$\begin{aligned}
\log\left(\frac{1}{\sin^2(\pi x)}\right) &= -2\log(\sin(\pi x))\\
&= \log 4 + \sum_{h \in \mathbb{Z}\setminus\{0\}} \frac{\mathrm{e}^{2\pi \mathrm{i} h x}}{|h|}\\
&= \log 4 + \sum_{h \in C_1^*(N)} \frac{\mathrm{e}^{2\pi \mathrm{i} h x}}{|h|} + r(x),
\end{aligned} \tag{4.23}$$

where the remainder $r(x)$ satisfies

$$\begin{aligned}
|r(x)| &= \left|\sum_{h=N/2+1}^{\infty} \frac{\mathrm{e}^{2\pi \mathrm{i} h x}}{h} + \sum_{h=N/2}^{\infty} \frac{\mathrm{e}^{-2\pi \mathrm{i} h x}}{h}\right|\\
&\le \frac{2}{N} + 2\left|\sum_{h=N/2+1}^{\infty} \frac{\mathrm{e}^{2\pi \mathrm{i} h x}}{h}\right|.
\end{aligned}$$

If we make use of the identity

$$\frac{\mathrm{e}^{2\pi \mathrm{i} h x}}{h} = \frac{1}{\mathrm{e}^{2\pi \mathrm{i} x} - 1} \left(\frac{\mathrm{e}^{2\pi \mathrm{i}(h+1)x}}{h+1} - \frac{\mathrm{e}^{2\pi \mathrm{i} h x}}{h} + \frac{\mathrm{e}^{2\pi \mathrm{i}(h+1)x}}{h(h+1)} \right),$$

summing over the $h \geq N/2 + 1$ yields a telescoping sum such that

$$\begin{aligned}
\left| \sum_{h=N/2+1}^{\infty} \frac{\mathrm{e}^{2\pi \mathrm{i} h x}}{h} \right| &= \frac{1}{|\mathrm{e}^{2\pi \mathrm{i} x} - 1|} \left| \sum_{h=N/2+1}^{\infty} \left(\frac{\mathrm{e}^{2\pi \mathrm{i}(h+1)x}}{h+1} - \frac{\mathrm{e}^{2\pi \mathrm{i} h x}}{h} + \frac{\mathrm{e}^{2\pi \mathrm{i}(h+1)x}}{h(h+1)} \right) \right| \\
&= \frac{1}{|\mathrm{e}^{2\pi \mathrm{i} x} - 1|} \left| - \frac{\mathrm{e}^{2\pi \mathrm{i}(N/2+1)x}}{N/2+1} + \sum_{h=N/2+1}^{\infty} \frac{\mathrm{e}^{2\pi \mathrm{i}(h+1)x}}{h(h+1)} \right| \\
&\leq \frac{1}{2\sin(\pi x)} \left(\frac{1}{N/2+1} + \sum_{h=N/2+1}^{\infty} \frac{1}{h(h+1)} \right) \\
&= \frac{1}{(N/2+1)\sin(\pi x)},
\end{aligned}$$

where we used that

$$\sum_{h=N/2+1}^{\infty} \frac{1}{h(h+1)} = \sum_{h=N/2+1}^{\infty} \left(\frac{1}{h} - \frac{1}{h+1} \right) = \frac{1}{N/2+1}$$

and that

$$|\mathrm{e}^{2\pi \mathrm{i} x} - 1| = |\mathrm{e}^{\pi \mathrm{i} x}| \cdot |\mathrm{e}^{\pi \mathrm{i} x} - \mathrm{e}^{-\pi \mathrm{i} x}| = |2\mathrm{i} \sin(\pi x)| = 2|\sin(\pi x)|,$$

where the latter absolute value can be ignored since $x \in (0, 1)$.

Since for $x \in [0, 1/2]$ we have that $\sin(\pi x) \geq 2x = 2\|x\|$, the symmetry of $\sin(\pi x)$ and $\|x\|$ about $1/2$ implies that $\sin(\pi x) \geq 2\|x\|$ for all $x \in [0, 1]$. This then yields that

$$|r(x)| \leq \frac{2}{N} + \frac{2}{(N/2+1)\sin(\pi x)} \leq \frac{3}{N\,\|x\|} \quad \text{and thus} \quad r(x) = \frac{\tau(x)}{N\,\|x\|}$$

for some $\tau(x) \in \mathbb{R}$ with $|\tau(x)| \leq 3$. This together with the expression in (4.23) yields the claim. □

The following lemma provides a result regarding the difference of two products. Variants of such a result can be found in the literature (see also, e.g., Lemma 5.2), but, as the lemma is crucial in showing the main result of this section, we provide a proof for the sake of completeness.

Lemma 4.19 *For $j \in [d]$, let $u_j, \bar{u}_j, v_j$, and t_j be real numbers satisfying*

$$(a) \quad u_j = v_j + t_j, \quad (b) \quad |u_j| \le \bar{u}_j, \quad (c) \quad \bar{u}_j \ge 1.$$

Then, for any set $\mathfrak{u}$, $\emptyset \neq \mathfrak{u} \subseteq [d]$, there exists a $\theta_{\mathfrak{u}}$ with $|\theta_{\mathfrak{u}}| \le 1$ such that

$$\prod_{j\in\mathfrak{u}} u_j = \prod_{j\in\mathfrak{u}} v_j + \theta_{\mathfrak{u}} \left(\prod_{j\in\mathfrak{u}} (\bar{u}_j + |t_j|)\right) \sum_{j\in\mathfrak{u}} |t_j|. \tag{4.24}$$

Proof We prove the statement by induction on $|\mathfrak{u}|$. For $|\mathfrak{u}| = 1$ we have that $\mathfrak{u} = \{j\}$ for some $j \in [d]$ and obtain, since $\bar{u}_j \ge 1$, that

$$u_j = v_j + t_j \le v_j + |t_j|(\bar{u}_j + |t_j|),$$

and thus $u_j = v_j + \theta_{\mathfrak{u}}|t_j|(\bar{u}_j + |t_j|)$ for some $\theta_{\mathfrak{u}}$ with $|\theta_{\mathfrak{u}}| \le 1$. Indeed, it is sufficient to choose $\theta_{\mathfrak{u}} = t_j/(|t_j|(\bar{u}_j + |t_j|))$.

Consider then $|\mathfrak{u}| \ge 2$ and assume that (4.24) holds for sets of cardinality $|\mathfrak{u}| - 1$. Writing $\mathfrak{u} = \{j_1, \ldots, j_{|\mathfrak{u}|}\} \subseteq [d]$ and using Properties (a) and (b), this yields

$$\begin{aligned}
\left|\prod_{j\in\mathfrak{u}} u_j - \prod_{j\in\mathfrak{u}} v_j\right| &= \left|\left(\prod_{i=1}^{|\mathfrak{u}|-1} u_{j_i} - \prod_{i=1}^{|\mathfrak{u}|-1} v_{j_i}\right) v_{j_{|\mathfrak{u}|}} + t_{j_{|\mathfrak{u}|}} \prod_{i=1}^{|\mathfrak{u}|-1} u_{j_i}\right| \\
&\le \left|\prod_{i=1}^{|\mathfrak{u}|-1} u_{j_i} - \prod_{i=1}^{|\mathfrak{u}|-1} v_{j_i}\right| |u_{j_{|\mathfrak{u}|}} - t_{j_{|\mathfrak{u}|}}| + |t_{j_{|\mathfrak{u}|}}| \prod_{i=1}^{|\mathfrak{u}|-1} \bar{u}_{j_i} \\
&\le \left|\theta_{\mathfrak{u}\setminus\{j_{|\mathfrak{u}|}\}}\right| (\bar{u}_{j_{|\mathfrak{u}|}} + |t_{j_{|\mathfrak{u}|}}|) \left(\prod_{i=1}^{|\mathfrak{u}|-1} (\bar{u}_{j_i} + |t_{j_i}|)\right) \sum_{i=1}^{|\mathfrak{u}|-1} |t_{j_i}| \\
&\quad + |t_{j_{|\mathfrak{u}|}}| \prod_{i=1}^{|\mathfrak{u}|-1} \bar{u}_{j_i} \\
&= |\theta_{\mathfrak{u}\setminus\{j_{|\mathfrak{u}|}\}}| \left(\prod_{i=1}^{|\mathfrak{u}|} (\bar{u}_{j_i} + |t_{j_i}|)\right) \sum_{i=1}^{|\mathfrak{u}|-1} |t_{j_i}| + |t_{j_{|\mathfrak{u}|}}| \prod_{i=1}^{|\mathfrak{u}|-1} \bar{u}_{j_i} \\
&\le \widetilde{\theta}_{\mathfrak{u}} \left(\prod_{j\in\mathfrak{u}} (\bar{u}_j + |t_j|)\right) \sum_{j\in\mathfrak{u}} |t_j|,
\end{aligned}$$

where $< \theta_{\mathfrak{u}\setminus\{j_{|\mathfrak{u}|}\}}|, |\widetilde{\theta}_{\mathfrak{u}}| \le 1$. Indeed, the last inequality follows by noting that we can write

$$\prod_{i=1}^{|\mathfrak{u}|-1} \bar{u}_{j_i} = \widetilde{\theta} \prod_{i=1}^{|\mathfrak{u}|} (\bar{u}_{j_i} + |t_{j_i}|)$$

for some $\widetilde{\theta}$ with $|\widetilde{\theta}| \le 1$ because of Property (c) and then setting $\widetilde{\theta}_{\mathfrak{u}} = \max(|\theta_{\mathfrak{u}\setminus\{j_{|\mathfrak{u}|}\}}|, \widetilde{\theta})$. This then implies the existence of a $\theta_{\mathfrak{u}}$ with $|\theta_{\mathfrak{u}}| \le 1$ such that (4.24) holds. □

We are now ready to state the proof of Proposition 4.17.

Proof of Proposition 4.17 For $\boldsymbol{g} \in G_d^{\varphi}(N)$ we have

$$\begin{aligned} R_{N,d,1,\boldsymbol{\gamma}}(\boldsymbol{g}) &= \sum_{\substack{\boldsymbol{h} \in C_d^*(N) \\ \boldsymbol{h}\cdot\boldsymbol{g} \equiv 0 \pmod{N}}} \frac{1}{r_{1,\boldsymbol{\gamma}}(\boldsymbol{h})} \\ &= \sum_{\emptyset \neq \mathfrak{u} \subseteq [d]} \gamma_{\mathfrak{u}} \sum_{\substack{\boldsymbol{h}_{\mathfrak{u}} \in (C_1^*(N))^{|\mathfrak{u}|} \\ \boldsymbol{h}_{\mathfrak{u}}\cdot\boldsymbol{g}_{\mathfrak{u}} \equiv 0 \pmod{N}}} \prod_{j \in \mathfrak{u}} \frac{1}{|h_j|} \\ &\le \sum_{\emptyset \neq \mathfrak{u} \subseteq [d]} \gamma_{\mathfrak{u}} \sum_{\substack{\boldsymbol{h}_{\mathfrak{u}} \in C_{|\mathfrak{u}|}^*(N) \\ \boldsymbol{h}_{\mathfrak{u}}\cdot\boldsymbol{g}_{\mathfrak{u}} \equiv 0 \pmod{N}}} \prod_{j \in \mathfrak{u}} \frac{1}{b(h_j)}, \end{aligned}$$

where we use the abbreviation

$$b(h) := \begin{cases} |h| & \text{if } h \neq 0, \\ 1/(\log 4) & \text{if } h = 0, \end{cases}$$

for $h \in \mathbb{Z}$. Now we write, employing Remark 1.10,

$$\sum_{\substack{\boldsymbol{h}_{\mathfrak{u}} \in C_{|\mathfrak{u}|}^*(N) \\ \boldsymbol{h}_{\mathfrak{u}}\cdot\boldsymbol{g}_{\mathfrak{u}} \equiv 0 \pmod{N}}} \prod_{j \in \mathfrak{u}} \frac{1}{b(h_j)} = \frac{1}{N} \sum_{k=0}^{N-1} \left(\sum_{\boldsymbol{h}_{\mathfrak{u}} \in C_{|\mathfrak{u}|}(N)} \frac{\mathrm{e}^{2\pi \mathrm{i} k \boldsymbol{h}_{\mathfrak{u}}\cdot\boldsymbol{g}_{\mathfrak{u}}/N}}{\prod_{j \in \mathfrak{u}} b(h_j)} - (\log 4)^{|\mathfrak{u}|} \right).$$

Using this identity we continue estimating $R_{N,d,1,\boldsymbol{\gamma}}(\boldsymbol{g})$. We have

$$\begin{aligned} &R_{N,d,1,\boldsymbol{\gamma}}(\boldsymbol{g}) \\ &\le \sum_{\emptyset \neq \mathfrak{u} \subseteq [d]} \frac{\gamma_{\mathfrak{u}}}{N} \sum_{k=0}^{N-1} \left(\sum_{\boldsymbol{h}_{\mathfrak{u}} \in C_{|\mathfrak{u}|}(N)} \frac{\mathrm{e}^{2\pi \mathrm{i} k \boldsymbol{h}_{\mathfrak{u}}\cdot\boldsymbol{g}_{\mathfrak{u}}/N}}{\prod_{j \in \mathfrak{u}} b(h_j)} - (\log 4)^{|\mathfrak{u}|} \right) \\ &= \sum_{\emptyset \neq \mathfrak{u} \subseteq [d]} \frac{\gamma_{\mathfrak{u}}}{N} \left(\sum_{\boldsymbol{h}_{\mathfrak{u}} \in C_{|\mathfrak{u}|}(N)} \frac{1}{\prod_{j \in \mathfrak{u}} b(h_j)} + \sum_{k=1}^{N-1} \prod_{j \in \mathfrak{u}} \left(\log 4 + \sum_{h \in C_1^*(N)} \frac{\mathrm{e}^{2\pi \mathrm{i} k h g_j/N}}{|h|} \right) \right) \\ &\quad - \sum_{\emptyset \neq \mathfrak{u} \subseteq [d]} \gamma_{\mathfrak{u}} (\log 4)^{|\mathfrak{u}|} \\ &= \frac{1}{N} \left(-1 + \prod_{j=1}^{d} \left(1 + \gamma_j \sum_{h \in C_1(N)} \frac{1}{b(h)} \right) \right) - \left(-1 + \prod_{j=1}^{d} (1 + \gamma_j \log 4) \right) \\ &\quad + \sum_{\emptyset \neq \mathfrak{u} \subseteq [d]} \frac{\gamma_{\mathfrak{u}}}{N} \sum_{k=1}^{N-1} \prod_{j \in \mathfrak{u}} \left(\log 4 + \sum_{h \in C_1^*(N)} \frac{\mathrm{e}^{2\pi \mathrm{i} k h g_j/N}}{|h|} \right). \end{aligned} \tag{4.25}$$

Note that $\sum_{h \in C_1(N)} (b(h))^{-1} = \log 4 + S_N$, where, as before, $S_N := \sum_{h \in C_1^*(N)} |h|^{-1}$.

Hence it remains to estimate (4.25) to complete the proof. To this end we employ Lemma 4.19 with

$$u_j = u_j(k) := \log(\sin^{-2}(\pi k g_j/N)), \qquad \bar{u}_j = \bar{u}_j(k) := 2\log N,$$
$$v_j = v_j(k) := \log 4 + \sum_{h\in C_1^*(N)} \frac{e^{2\pi i k h g_j/N}}{|h|}, \qquad t_j = t_j(k) := \frac{\tau_j(k)}{N\,\|k g_j/N\|},$$

where the numbers $\tau_j(k)$ are a shorthand for $\tau(\{kg_j/N\})$ with τ defined as in Lemma 4.18. We then obtain

$$\begin{aligned}
&\sum_{\emptyset\neq \mathfrak{u}\subseteq[d]} \frac{\gamma_{\mathfrak{u}}}{N}\sum_{k=1}^{N-1}\prod_{j\in\mathfrak{u}}\left(\log 4 + \sum_{h\in C_1^*(N)}\frac{e^{2\pi i k h g_j/N}}{|h|}\right)\\
&= \sum_{\emptyset\neq \mathfrak{u}\subseteq[d]} \frac{\gamma_{\mathfrak{u}}}{N}\sum_{k=1}^{N-1}\left(\left(\prod_{j\in\mathfrak{u}} v_j(k) - \prod_{j\in\mathfrak{u}} u_j(k)\right) + \prod_{j\in\mathfrak{u}} u_j(k)\right)\\
&= \sum_{\emptyset\neq \mathfrak{u}\subseteq[d]} \frac{\gamma_{\mathfrak{u}}}{N}\sum_{k=1}^{N-1}\left(\theta_{\mathfrak{u}}(k)\left(\prod_{j\in\mathfrak{u}}(\bar{u}_j(k)+|t_j(k)|)\right)\sum_{j\in\mathfrak{u}}|t_j(k)| + \prod_{j\in\mathfrak{u}} u_j(k)\right),
\end{aligned} \tag{4.26}$$

where all $|\theta_{\mathfrak{u}}(k)|$ are at most 1. Recall that we are considering product weights. Therefore we have

$$\begin{aligned}
\sum_{\emptyset\neq \mathfrak{u}\subseteq[d]} \frac{\gamma_{\mathfrak{u}}}{N}\sum_{k=1}^{N-1}\prod_{j\in\mathfrak{u}} u_j(k) &= \frac{1}{N}\sum_{k=1}^{N-1}\left(-1+\prod_{j=1}^{d}\left(1+\gamma_j\log\left(\frac{1}{\sin^2(\pi k g_j/N)}\right)\right)\right)\\
&= \frac{1}{N} H_{d,m,\boldsymbol{\gamma}}(\boldsymbol{g}).
\end{aligned} \tag{4.27}$$

Due to Lemma 4.18, Condition (a) in Lemma 4.19 is fulfilled. Furthermore, we have for $k \in [N-1]$ and $g \in G_1^{\varphi}(N)$ that

$$\sin^2\left(\frac{\pi k g}{N}\right) \geq \sin^2\left(\frac{\pi}{N}\right) \geq \left(\frac{2}{N}\right)^2 \geq \frac{1}{N^2},$$

where we used that for $x \in [0,\pi/2]$ the estimate $\sin x \geq (2/\pi)x$ holds. This implies that Conditions (b) and (c) in Lemma 4.19 are fulfilled since

$$|u_j| = \log\left(\frac{1}{\sin^2(\pi k g_j/N)}\right) \leq \log(N^2) = 2\log N = \bar{u}_j,$$

and the latter expression is not smaller than one as long as $N \geq 2$.

We now show how to bound the sum

$$\sum_{\emptyset\neq\mathfrak{u}\subseteq[d]} \frac{\gamma_{\mathfrak{u}}}{N} \sum_{k=1}^{N-1} \left(\theta_{\mathfrak{u}}(k) \left(\prod_{j\in\mathfrak{u}} \left(\bar{u}_j(k) + |t_j(k)|\right) \right) \sum_{j\in\mathfrak{u}} |t_j(k)| \right)$$

independently of the choice of $\boldsymbol{g}$. Inserting the definitions of $\bar{u}_j(k)$ and $t_j(k)$, and using the estimates $|\theta_{\mathfrak{u}}(k)| \leq 1$, $|\tau_j(k)| \leq 3$, and $N\|kg_j/N\| \geq N\|1/N\| = 1$, we obtain

$$\begin{aligned}
&\sum_{\emptyset\neq\mathfrak{u}\subseteq[d]} \frac{\gamma_{\mathfrak{u}}}{N} \sum_{k=1}^{N-1} \theta_{\mathfrak{u}}(k) \left(\prod_{j\in\mathfrak{u}} \left(\bar{u}_j(k) + |t_j(k)|\right) \right) \sum_{j\in\mathfrak{u}} |t_j(k)| \\
&\leq \sum_{\emptyset\neq\mathfrak{u}\subseteq[d]} \frac{\gamma_{\mathfrak{u}}}{N} \sum_{k=1}^{N-1} \left(\prod_{j\in\mathfrak{u}} \left(2\log N + \frac{|\tau_j(k)|}{N\,\|kg_j/N\|} \right) \right) \sum_{j\in\mathfrak{u}} \frac{|\tau_j(k)|}{N\,\|kg_j/N\|} \\
&\leq \sum_{\emptyset\neq\mathfrak{u}\subseteq[d]} \frac{\gamma_{\mathfrak{u}}}{N} \left(\prod_{j\in\mathfrak{u}} (3 + 2\log N) \right) \sum_{j\in\mathfrak{u}} \sum_{k=1}^{N-1} \frac{3}{N\,\|kg_j/N\|}.
\end{aligned}$$

Next we use a well-known estimate for the innermost sum in the latter expression, which is a Diophantine sum. As $\gcd(g_j, N) = 1$, we have $\{kg_j \pmod N : k \in \{1, 2, \ldots, N-1\}\} = \{1, 2, \ldots, N-1\}$, and consequently

$$\begin{aligned}
\sum_{k=1}^{N-1} \frac{1}{N\,\|kg_j/N\|} &= \frac{1}{N} \sum_{k=1}^{N-1} \frac{1}{\|k/N\|} \\
&= \frac{1}{N} \left(\sum_{k=1}^{N/2} \frac{1}{k/N} + \sum_{k=N/2+1}^{N-1} \frac{1}{1-k/N} \right) \\
&= \sum_{k=1}^{N/2} \frac{1}{k} + \sum_{k=1}^{N-N/2-1} \frac{1}{k} \\
&= S_N,
\end{aligned}$$

where we recall that $N = 2^m$. So we obtain

$$\begin{aligned}
&\sum_{\emptyset\neq\mathfrak{u}\subseteq[d]} \frac{\gamma_{\mathfrak{u}}}{N} \sum_{k=1}^{N-1} \theta_{\mathfrak{u}}(k) \left(\prod_{j\in\mathfrak{u}} \left(\bar{u}_j(k) + |t_j(k)|\right) \right) \sum_{j\in\mathfrak{u}} |t_j(k)| \\
&\leq \sum_{\emptyset\neq\mathfrak{u}\subseteq[d]} \frac{\gamma_{\mathfrak{u}}}{N} \left(\prod_{j\in\mathfrak{u}} (3 + 2\log N) \right) 3|\mathfrak{u}| S_N \\
&\leq \frac{3S_N}{N} \left(-1 + \prod_{j=1}^{d} \left(1 + \gamma_j 2(3 + 2\log N)\right) \right), \qquad (4.28)
\end{aligned}$$

where we used that $|\mathfrak{u}| \leq 2^{|\mathfrak{u}|}$ for $\emptyset \neq \mathfrak{u} \subseteq [d]$. Employing (4.25)–(4.28) yields

$$\begin{aligned} R_{N,d,1,\gamma}(\boldsymbol{g}) \leq & \frac{1}{N}\left(-1+\prod_{j=1}^{d}(1+\gamma_j(\log 4+S_N))\right) \\ & +\frac{3S_N}{N}\left(-1+\prod_{j=1}^{d}(1+\gamma_j 2(3+2\log N))\right) \\ & +\frac{1}{N}H_{d,m,\gamma}(\boldsymbol{g})-\left(-1+\prod_{j=1}^{d}(1+\gamma_j\log 4)\right). \end{aligned}$$

From (2.16) we know that $S_N \leq 2(\log N + 1 - \log 2)$ and hence $\log 4 + S_N \leq 2(\log N + 1)$. This yields

$$\begin{aligned} R_{N,d,1,\gamma}(\boldsymbol{g}) \leq & \frac{2+6\log N}{N}\left(-1+\prod_{j=1}^{d}(1+\gamma_j(6+4\log N))\right) \\ & +\frac{1}{N}H_{d,m,\gamma}(\boldsymbol{g})-\left(-1+\prod_{j=1}^{d}(1+\gamma_j\log 4)\right), \end{aligned}$$

as desired. □

The CBC-DBD construction

Proposition 4.17 implies that if we can find a generating vector $\boldsymbol{g}$ of a lattice rule with $N = 2^m$ points for which a good bound on

$$\frac{1}{N}H_{d,m,\gamma}(\boldsymbol{g})-\left(-1+\prod_{j=1}^{d}(1+\gamma_j\log 4)\right) \tag{4.29}$$

holds, then for this generating vector we also have a good bound on $R_{N,d,1,\gamma}(\boldsymbol{g})$. Indeed, we will show next that we can find such a generating vector in a component-by-component digit-by-digit construction. Note, however, that the CBC-DBD algorithm introduced below does not directly optimize the function $H_{d,m,\gamma}(\boldsymbol{g})$ in terms of $\boldsymbol{g}$, but, digit-by-digit, a function that is closely related to $H_{d,m,\gamma}$, in the sense that it corresponds to the increment of a function like $H_{d,m,\gamma}$ when the number m of digits of a single component of $\boldsymbol{g}$ is increased. This relation is made precise in Item 3 of Lemma 4.22 below.

Next, we introduce a digit-wise quality function and state the construction algorithm, which is based on a double loop, namely an outer loop for the components and an inner loop for the digits of the single components of $\boldsymbol{g}$.

In the following, let $x \in \mathbb{N}$ be an odd integer, let $m, d \in \mathbb{N}$, and let $\boldsymbol{\gamma} = (\gamma_j)_{j\geq 1}$ be product weights. For $v \in [m]$, $s \in [d]$, and positive integers $a_{1,m}, \ldots, a_{s-1,m}$ we define the quality function $h_{s,v,\boldsymbol{\gamma}} : \mathbb{N} \to \mathbb{R}$ as

$$h_{s,v,\boldsymbol{\gamma}}(x) := \sum_{t=v}^{m} \frac{1}{2^{t-v}} \sum_{\substack{k=1 \\ k \equiv 1 \pmod 2}}^{2^t} \left(-1 + \left(1 + \gamma_s \log\left(\frac{1}{\sin^2(\pi k x/2^v)}\right)\right) \right.$$
$$\left. \times \prod_{j=1}^{s-1} \left(1 + \gamma_j \log\left(\frac{1}{\sin^2(\pi k a_{j,m}/2^t)}\right)\right)\right). \tag{4.30}$$

We are now ready to outline the CBC-DBD construction algorithm.

Algorithm 4.20 (CBC-DBD construction for the weighted Korobov space) Let $N = 2^m$, product weights $\boldsymbol{\gamma} = (\gamma_j)_{j\geq 1}$, and $d \in \mathbb{N}$ be given. Construct a generating vector $\boldsymbol{g} = (g_1, \ldots, g_d) \in G_d^{\varphi}(N)$ as follows.

(1) Set $a_{1,m} = 1$ and $a_{2,1} = \cdots = a_{d,1} = 1$.
(2) For s from 1 to $d-1$:
Assume that $a_{1,m}, \ldots, a_{s,m}$ have already been found. Choose $a_{s+1,m}$ as follows.

(2a) For v from 1 to $m-1$:
Assume that $a_{s+1,1}, \ldots, a_{s+1,v}$ have already been found. Choose

$$g^* = \underset{g \in \{0,1\}}{\operatorname{argmin}}\, h_{s+1,v+1,\boldsymbol{\gamma}}(a_{s+1,v} + 2^v g).$$

(2b) Set

$$a_{s+1,v+1} = a_{s+1,v} + 2^v g^*.$$

End for.

End for.
(3) Set $g_s := a_{s,m}$ for $s \in [d]$ and $\boldsymbol{g} := (g_1, \ldots, g_d)$.

Note that the generating vector $\boldsymbol{g} = (g_1, \ldots, g_d)$ constructed by Algorithm 4.20 satisfies that $g_j \in G_1^{\varphi}(N) = G_1^{\varphi}(2^m)$ for all $j \in [d]$.

The quality of generating vectors constructed by the CBC-DBD algorithm

In the following, we assess the quality of generating vectors found by the CBC-DBD algorithm. We show that Algorithm 4.20 yields low values of $R_{N,d,1,\boldsymbol{\gamma}}(\boldsymbol{g})$ (and, hence, also of the corresponding worst-case integration error in the Korobov space). The next theorem gives an estimate on the quantity $H_{d,m,\boldsymbol{\gamma}} = H_{d,m,\boldsymbol{\gamma}}(\boldsymbol{g})$, which shows that the term (4.29) is sufficiently small. This will be a key observation in showing that a generating vector $\boldsymbol{g}$ constructed by Algorithm 4.20 yields low values of $R_{N,d,1,\boldsymbol{\gamma}}(\boldsymbol{g})$.

Theorem 4.21 *Let $m, d \in \mathbb{N}$, let $N = 2^m$, and let $\boldsymbol{\gamma} = (\gamma_j)_{j \geq 1}$ be product weights. Furthermore, assume that the integers $a_{s,v}$ with $s \in [d]$ and $v \in [m]$ and the corresponding generating vector $\mathbf{g}$ have been constructed by Algorithm 4.20. Then,*

$$H_{d,m,\boldsymbol{\gamma}} = H_{d,m,\boldsymbol{\gamma}}(\mathbf{g}) \leq N\left(-1 + \prod_{j=1}^{d}(1 + \gamma_j \log 4)\right).$$

For the proof of this result we need to define some additional quantities and auxiliary results for them.

Assume that the integers $a_{s,v}$ with $s \in [d]$ and $v \in [m]$ have been constructed by Algorithm 4.20. For $s \in [d]$ and $v \in [m]$ we define the additional quantities

$$H_{s,v,\boldsymbol{\gamma}} := \sum_{k=1}^{2^v-1}\left(-1 + \prod_{j=1}^{s}\left(1 + \gamma_j \log\left(\frac{1}{\sin^2(\pi k a_{j,v}/2^v)}\right)\right)\right), \tag{4.31}$$

$$h^*_{s,v,\boldsymbol{\gamma}} := h_{s,v,\boldsymbol{\gamma}}(a_{s,v}).$$

Note that, for $s = d$ and $v = m$, the definition of $H_{s,v,\boldsymbol{\gamma}}$ corresponds to the analogous quantity $H_{d,m,\boldsymbol{\gamma}}$ defined in Proposition 4.17.

For the quantities $h^*_{s,v,\boldsymbol{\gamma}}$ and $H_{s,v,\boldsymbol{\gamma}}$, the following identities hold.

Lemma 4.22 *Let $m, d \in \mathbb{N}$, $m \geq 2$, let $\boldsymbol{\gamma} = (\gamma_j)_{j\geq 1}$ be given product weights, and assume that the integers $a_{s,v}$ with $s \in [d]$ and $v \in [m]$ have been constructed by Algorithm 4.20. Then,*

1. $h^*_{1,m,\boldsymbol{\gamma}} = \gamma_1(2^{m-1} - 1)\log 4$,
2. $H_{1,v,\boldsymbol{\gamma}} = \gamma_1(2^v - v - 1)\log 4$, *and*
3. $H_{s,m,\boldsymbol{\gamma}} = h^*_{s,m,\boldsymbol{\gamma}} + H_{s,m-1,\boldsymbol{\gamma}}$.

Proof We prove the claimed identities one by one.

Let us start with Item 1. For $s = 1$ and $v = m$ we obtain that

$$\begin{aligned}
h^*_{1,m,\boldsymbol{\gamma}} &= h_{1,m,\boldsymbol{\gamma}}(a_{1,m}) \\
&= \sum_{\substack{k=1 \\ k\equiv 1 \pmod 2}}^{2^m} \gamma_1 \log\left(\frac{1}{\sin^2(\pi k/2^m)}\right) \\
&= \sum_{k=1}^{2^m-1} \gamma_1 \log\left(\frac{1}{\sin^2(\pi k/2^m)}\right) - \sum_{\substack{k=1 \\ k\equiv 0 \pmod 2}}^{2^m-1} \gamma_1 \log\left(\frac{1}{\sin^2(\pi k/2^m)}\right) \\
&= -2\gamma_1\left(\sum_{k=1}^{2^m-1}\log\left(\sin\left(\frac{\pi k}{2^m}\right)\right) - \sum_{k=1}^{2^{m-1}-1}\log\left(\sin\left(\frac{\pi k}{2^{m-1}}\right)\right)\right) \\
&= -2\gamma_1\left(\log\left(\prod_{k=1}^{2^m-1}\sin\left(\frac{\pi k}{2^m}\right)\right) - \log\left(\prod_{k=1}^{2^{m-1}-1}\sin\left(\frac{\pi k}{2^{m-1}}\right)\right)\right).
\end{aligned}$$

We now use the well-known identity

$$\prod_{k=1}^{n-1} 2\sin\left(\frac{\pi k}{n}\right) = n \quad \text{or, equivalently,} \quad \prod_{k=1}^{n-1} \sin\left(\frac{\pi k}{n}\right) = \frac{n}{2^{n-1}} \tag{4.32}$$

for any $n \in \mathbb{N}$ (see, e.g., [3, Lemma 3], [140, p. 99], or [190]).

Applying (4.32) to the previous expression gives

$$\begin{aligned}
h^*_{1,m,\gamma} &= -2\gamma_1 \left(\log\left(\frac{2^m}{2^{2^m-1}}\right) - \log\left(\frac{2^{m-1}}{2^{2^{m-1}-1}}\right)\right) \\
&= -2\gamma_1(m\log 2 - (2^m-1)\log 2 - (m-1)\log 2 + (2^{m-1}-1)\log 2) \\
&= -2\gamma_1(1 - 2^m + 2^{m-1})\log 2 \\
&= \gamma_1(2^{m-1}-1)\log 4.
\end{aligned}$$

We proceed with showing Item 2. Again using (4.32), we obtain that

$$\begin{aligned}
H_{1,v,\gamma} &= \sum_{k=1}^{2^v-1} \gamma_1 \log\left(\frac{1}{\sin^2(\pi k/2^v)}\right) \\
&= -2\gamma_1 \sum_{k=1}^{2^v-1} \log\left(\sin\left(\frac{\pi k}{2^v}\right)\right) \\
&= -2\gamma_1 \log\left(\prod_{k=1}^{2^v-1} \sin\left(\frac{\pi k}{2^v}\right)\right) \\
&= -2\gamma_1 \log\left(\frac{2^v}{2^{2^v-1}}\right) \\
&= -2\gamma_1(v\log 2 - (2^v-1)\log 2) \\
&= \gamma_1(2^v - v - 1)\log 4,
\end{aligned}$$

so Item 2 is shown.

Regarding Item 3, from the definition of $h^*_{s,m,\gamma}$ and $H_{s,m,\gamma}$ we see that $H_{s,m,\gamma}$ can be rewritten as

$$\begin{aligned}
H_{s,m,\gamma} = &\sum_{\substack{k=1 \\ k\equiv 1 \pmod 2}}^{2^m} \left(-1 + \prod_{j=1}^{s}\left(1 + \gamma_j \log\left(\frac{1}{\sin^2(\pi k a_{j,m}/2^m)}\right)\right)\right) \\
&+ \sum_{\substack{k=1 \\ k\equiv 0 \pmod 2}}^{2^m-2} \left(-1 + \prod_{j=1}^{s}\left(1 + \gamma_j \log\left(\frac{1}{\sin^2(\pi k a_{j,m}/2^m)}\right)\right)\right) \\
= &\, h_{s,m,\gamma}(a_{s,m}) + H_{s,m-1,\gamma} \\
= &\, h^*_{s,m,\gamma} + H_{s,m-1,\gamma}.
\end{aligned}$$

□

The next lemma will allow us to quantify and rewrite the average over the choices $g \in \{0, 1\}$ for the base 2 digits in Step (2a) of Algorithm 4.20.

Lemma 4.23 *Let a and k be odd integers and let $v \geq 2$ be an integer. Then it holds true that*

$$\sum_{g=0}^{1} \log\left(\frac{1}{\sin^2(\pi k(a + 2^{v-1}g)/2^v)}\right) = \log 4 + \log\left(\frac{1}{\sin^2(\pi ka/2^{v-1})}\right).$$

Proof Since k is odd, we have that

$$\sin^2\left(\frac{\pi k(a + 2^{v-1})}{2^v}\right) = \sin^2\left(\frac{\pi ka}{2^v} + \frac{\pi(k-1)}{2} + \frac{\pi}{2}\right) = \cos^2\left(\frac{\pi ka}{2^v}\right),$$

and therefore, using that $\sin(x)\cos(x) = \sin(2x)/2$, we obtain

$$\begin{aligned}\sum_{g=0}^{1} \log\left(\frac{1}{\sin^2(\pi k(a + 2^{v-1}g)/2^v)}\right) &= \log\left(\frac{1}{\sin^2(\pi ka/2^v)}\right) + \log\left(\frac{1}{\cos^2(\pi ka/2^v)}\right) \\ &= \log\left(\frac{1}{\sin^2(\pi ka/2^v)\cos^2(\pi ka/2^v)}\right) \\ &= \log\left(\frac{4}{\sin^2(\pi ka/2^{v-1})}\right) \\ &= \log 4 + \log\left(\frac{1}{\sin^2(\pi ka/2^{v-1})}\right).\end{aligned}$$

□

Now we can state the proof of Theorem 4.21.

Proof of Theorem 4.21 By the standard averaging argument (see Remark 2.14), by which the best choice for $g \in \{0, 1\}$ is at least as good as the average over all possible choices, and by using Lemma 4.23, we obtain

$$\begin{aligned}&h^*_{s,v,\gamma} \\ &\leq \frac{1}{2}\sum_{g=0}^{1} h_{s,v,\gamma}(a_{s,v-1} + 2^{v-1}g) \\ &= \frac{1}{2}\sum_{g=0}^{1}\sum_{t=v}^{m}\frac{1}{2^{t-v}}\sum_{\substack{k=1 \\ k\equiv 1 \pmod 2}}^{2^t}\left(-1 + \left(1 + \gamma_s \log\left(\frac{1}{\sin^2(\pi k(a_{s,v-1} + 2^{v-1}g)/2^v)}\right)\right)\right. \\ &\qquad \left.\times \prod_{j=1}^{s-1}\left(1 + \gamma_j \log\left(\frac{1}{\sin^2(\pi ka_{j,m}/2^t)}\right)\right)\right)\end{aligned}$$

$$= \sum_{t=v}^{m} \frac{1}{2^{t-v+1}} \sum_{\substack{k=1 \\ k\equiv 1 \pmod 2}}^{2^t} \left(-2 + \left(2 + \gamma_s \sum_{g=0}^{1} \log \left(\frac{1}{\sin^2(\pi k (a_{s,v-1} + 2^{v-1} g)/2^v)} \right) \right) \right.$$
$$\left. \times \prod_{j=1}^{s-1} \left(1 + \gamma_j \log \left(\frac{1}{\sin^2(\pi k a_{j,m}/2^t)} \right) \right) \right)$$
$$= \sum_{t=v}^{m} \frac{1}{2^{t-v+1}} \sum_{\substack{k=1 \\ k\equiv 1 \pmod 2}}^{2^t} \left(-2 + \left(2 + \gamma_s \left(\log 4 + \log \left(\frac{1}{\sin^2(\pi k a_{s,v-1}/2^{v-1}} \right) \right) \right) \right.$$
$$\left. \times \prod_{j=1}^{s-1} \left(1 + \gamma_j \log \left(\frac{1}{\sin^2(\pi k a_{j,m}/2^t)} \right) \right) \right).$$

Comparing the last expression to $h^*_{s,v-1,\gamma}$ we see that

$$\begin{aligned} h^*_{s,v,\gamma} \le h^*_{s,v-1,\gamma} & \\ - \sum_{\substack{k=1 \\ k\equiv 1 \pmod 2}}^{2^{v-1}} & \left(-1 + \left(1 + \gamma_s \log \frac{1}{\sin^2(\pi k a_{s,v-1}/2^{v-1})} \right) \right. \\ & \left. \times \prod_{j=1}^{s-1} \left(1 + \gamma_j \log \left(\frac{1}{\sin^2(\pi k a_{j,m}/2^{v-1})} \right) \right) \right) \\ + \sum_{t=v}^{m} \frac{1}{2^{t-(v-1)}} & \sum_{\substack{k=1 \\ k\equiv 1 \pmod 2}}^{2^t} \left(-1 + (1 + \gamma_s \log 4) \right. \\ & \left. \times \prod_{j=1}^{s-1} \left(1 + \gamma_j \log \left(\frac{1}{\sin^2(\pi k a_{j,m}/2^t)} \right) \right) \right). \end{aligned}$$

Repeatedly applying this inequality to $h^*_{s,\ell,\gamma}$ for $\ell \in [v-1]$, we further obtain

$$\begin{aligned} h^*_{s,v,\gamma} \le h^*_{s,1,\gamma} & \\ - \sum_{\ell=1}^{v-1} \sum_{\substack{k=1 \\ k\equiv 1 \pmod 2}}^{2^\ell} & \left(-1 + \left(1 + \gamma_s \log \left(\frac{1}{\sin^2(\pi k a_{s,\ell}/2^\ell)} \right) \right) \right. \\ & \left. \times \prod_{j=1}^{s-1} \left(1 + \gamma_j \log \left(\frac{1}{\sin^2(\pi k a_{j,m}/2^\ell)} \right) \right) \right) \end{aligned}$$

$$
\begin{aligned}
&+\sum_{\ell=1}^{v-1}\sum_{t=\ell+1}^{m}\frac{1}{2^{t-\ell}}\sum_{\substack{k=1\\k\equiv 1 \pmod 2}}^{2^t}\left(-1+(1+\gamma_s\log 4)\right.\\
&\qquad\left.\times\prod_{j=1}^{s-1}\left(1+\gamma_j\log\left(\frac{1}{\sin^2(\pi k a_{j,m}/2^t)}\right)\right)\right)\\
&\le h^*_{s,1,\gamma}-\sum_{k=1}^{2^{v-1}-1}\left(-1+\prod_{j=1}^{s}\left(1+\gamma_j\log\left(\frac{1}{\sin^2(\pi k a_{j,m}/2^{v-1})}\right)\right)\right)\\
&+\sum_{\ell=1}^{v-1}\sum_{t=\ell+1}^{m}\frac{1}{2^{t-\ell}}\sum_{\substack{k=1\\k\equiv 1 \pmod 2}}^{2^t}\left(-1+(1+\gamma_s\log 4)\right.\\
&\qquad\left.\times\prod_{j=1}^{s-1}\left(1+\gamma_j\log\left(\frac{1}{\sin^2(\pi k a_{j,m}/2^t)}\right)\right)\right),
\end{aligned}
\tag{4.33}
$$

where the last inequality follows since $a_{s,\ell} \equiv a_{s,m} \pmod{2^\ell}$ for $\ell \in [m]$ and $s \in [d]$, and thus

$$
\begin{aligned}
\sin^2\left(\pi\frac{ka_{s,\ell}}{2^\ell}\right) &= \sin^2\left(\pi\frac{ka_{s,\ell} \pmod{2^\ell}}{2^\ell}\right)\\
&= \sin^2\left(\pi\frac{ka_{s,m} \pmod{2^\ell}}{2^\ell}\right) = \sin^2\left(\pi\frac{ka_{s,m}}{2^\ell}\right),
\end{aligned}
$$

and the fact that in general, for $p \in \mathbb{N}$,

$$
\sum_{\ell=1}^{p}\sum_{\substack{k=1\\k\equiv 1 \pmod 2}}^{2^\ell} f\left(\frac{k}{2^\ell}\right) = \sum_{k=1}^{2^p-1} f\left(\frac{k}{2^p}\right). \tag{4.34}
$$

Note that we have

$$
\begin{aligned}
H_{s,v-1,\gamma} &= \sum_{k=1}^{2^{v-1}-1}\left(-1+\prod_{j=1}^{s}\left(1+\gamma_j\log\left(\frac{1}{\sin^2(\pi k a_{j,v-1}/2^{v-1})}\right)\right)\right)\\
&= \sum_{k=1}^{2^{v-1}-1}\left(-1+\prod_{j=1}^{s}\left(1+\gamma_j\log\left(\frac{1}{\sin^2(\pi k a_{j,m}/2^{v-1})}\right)\right)\right).
\end{aligned}
$$

Since by definition $h^*_{s,1,\gamma}$ contains terms of the form $\log(\sin^{-2}(\pi k a_{s,1}/2))$ with $a_{s,1} = 1$ and k an odd integer, which equal zero, we obtain that

$$h^*_{s,1,\boldsymbol{\gamma}} = \sum_{t=1}^{m} \frac{1}{2^{t-1}} \sum_{\substack{k=1 \\ k \equiv 1 \pmod 2}}^{2^t} \left(-1 + \prod_{j=1}^{s-1} \left(1 + \gamma_j \log \left(\frac{1}{\sin^2(\pi k a_{j,m}/2^t)} \right) \right) \right).$$

In order to derive an upper bound on $h^*_{s,m,\boldsymbol{\gamma}}$, we restrict ourselves to the case where $v = m$, and obtain, based on (4.33), that

$$\begin{aligned}
h^*_{s,m,\boldsymbol{\gamma}} \le{}& \sum_{t=1}^{m} \frac{1}{2^{t-1}} \sum_{\substack{k=1 \\ k \equiv 1 \pmod 2}}^{2^t} \left(-1 + \prod_{j=1}^{s-1} \left(1 + \gamma_j \log \left(\frac{1}{\sin^2(\pi k a_{j,m}/2^t)} \right) \right) \right) \\
&+ \sum_{\ell=1}^{m-1} \sum_{t=\ell+1}^{m} \frac{1}{2^{t-\ell}} \sum_{\substack{k=1 \\ k \equiv 1 \pmod 2}}^{2^t} \Bigg(-1 + (1 + \gamma_s \log 4) \\
&\qquad \times \prod_{j=1}^{s-1} \left(1 + \gamma_j \log \left(\frac{1}{\sin^2(\pi k a_{j,m}/2^t)} \right) \right) \Bigg) \\
&- H_{s,m-1,\boldsymbol{\gamma}} \\
\le{}& \sum_{t=1}^{m} \frac{1}{2^{t-1}} \sum_{\substack{k=1 \\ k \equiv 1 \pmod 2}}^{2^t} \left(-1 + \prod_{j=1}^{s-1} \left(1 + \gamma_j \log \left(\frac{1}{\sin^2(\pi k a_{j,m}/2^t)} \right) \right) \right) \\
&+ \sum_{t=1}^{m} \left(1 - \frac{1}{2^{t-1}} \right) \sum_{\substack{k=1 \\ k \equiv 1 \pmod 2}}^{2^t} \Bigg(-1 + (1 + \gamma_s \log 4) \\
&\qquad \times \prod_{j=1}^{s-1} \left(1 + \gamma_j \log \left(\frac{1}{\sin^2(\pi k a_{j,m}/2^t)} \right) \right) \Bigg) \\
&- H_{s,m-1,\boldsymbol{\gamma}},
\end{aligned}$$

where we used the fact that for a mapping $f : \mathbb{N} \to \mathbb{R}_+$ we have the general relation

$$\sum_{\ell=1}^{m-1} \sum_{t=\ell+1}^{m} \frac{1}{2^{t-\ell}} f(t) = \sum_{t=2}^{m} \left(1 - \frac{1}{2^{t-1}} \right) f(t) = \sum_{t=1}^{m} \left(1 - \frac{1}{2^{t-1}} \right) f(t).$$

Based on this, we then estimate further, again using (4.34), that

$$\begin{aligned}
h^*_{s,m,\boldsymbol{\gamma}} \le{}& \sum_{t=1}^{m} \sum_{\substack{k=1 \\ k \equiv 1 \pmod 2}}^{2^t} \left(-1 + (1 + \gamma_s \log 4) \prod_{j=1}^{s-1} \left(1 + \gamma_j \log \left(\frac{1}{\sin^2(\pi k a_{j,m}/2^t)} \right) \right) \right) \\
&- H_{s,m-1,\boldsymbol{\gamma}}
\end{aligned}$$

$$
\begin{aligned}
&= \sum_{k=1}^{2^m-1}\left(-1+(1+\gamma_s\log 4)\prod_{j=1}^{s-1}\left(1+\gamma_j\log\left(\frac{1}{\sin^2(\pi k a_{j,m}/2^m)}\right)\right)\right)\\
&\quad -H_{s,m-1,\gamma}\\
&= H_{s-1,m,\gamma}+\gamma_s(2^m-1+H_{s-1,m,\gamma})\log 4 - H_{s,m-1,\gamma}. \qquad (4.35)
\end{aligned}
$$

We then use Item 3 of Lemma 4.22 and (4.35) to see that $H_{s,m,\gamma}$ satisfies the upper bound

$$
\begin{aligned}
H_{s,m,\gamma} &= H_{s,m-1,\gamma}+h^*_{s,m,\gamma}\\
&\le H_{s,m-1,\gamma}+H_{s-1,m,\gamma}+\gamma_s(2^m-1+H_{s-1,m,\gamma})\log 4 - H_{s,m-1,\gamma}\\
&= H_{s-1,m,\gamma}+\gamma_s(2^m-1+H_{s-1,m,\gamma})\log 4\\
&\le (1+\gamma_s\log 4)H_{s-1,m,\gamma}+\gamma_s 2^m\log 4. \qquad (4.36)
\end{aligned}
$$

The inequality (4.36) holds for all $s\in\{2,3,\ldots,d\}$, and applying it recursively we obtain

$$
\begin{aligned}
H_{d,m,\gamma} &\le (1+\gamma_d\log 4)H_{d-1,m,\gamma}+\gamma_d 2^m\log 4\\
&\le (1+\gamma_d\log 4)\left((1+\gamma_{d-1}\log 4)H_{d-2,m,\gamma}+\gamma_{d-1}2^m\log 4\right)+\gamma_d 2^m\log 4\\
&= H_{d-2,m,\gamma}\prod_{j=d-1}^{d}(1+\gamma_j\log 4)+2^m\left(-1+\prod_{j=d-1}^{d}(1+\gamma_j\log 4)\right)\\
&\le H_{1,m,\gamma}\prod_{j=2}^{d}(1+\gamma_j\log 4)+2^m\left(-1+\prod_{j=2}^{d}(1+\gamma_j\log 4)\right). \qquad (4.37)
\end{aligned}
$$

According to Algorithm 4.20 we have chosen $a_{1,m}=1$ and hence, using also (4.32), $H_{1,m,\gamma}$ equals

$$
\begin{aligned}
H_{1,m,\gamma} &= \sum_{k=1}^{2^m-1}\gamma_1\log\left(\frac{1}{\sin^2(\pi k a_{1,m}/2^m)}\right)\\
&= -2\gamma_1\log\left(\prod_{k=1}^{2^m-1}\sin\left(\frac{\pi k}{2^m}\right)\right)\\
&= -2\gamma_1\log\left(\frac{2^m}{2^{2^m-1}}\right)\\
&= \gamma_1(2^m-m-1)\log 4.
\end{aligned}
$$

Combining the obtained expression with (4.37) finally gives

$$
H_{d,m,\gamma}(\boldsymbol{g}) \le 2^m\gamma_1(\log 4)\prod_{j=2}^{d}(1+\gamma_j\log 4)+2^m\left(-1+\prod_{j=2}^{d}(1+\gamma_j\log 4)\right)
$$

$$= 2^m \left(-1 + \prod_{j=1}^{d} (1 + \gamma_j \log 4) \right),$$

as claimed. □

We are now able to show the main result regarding the CBC-DBD construction.

Theorem 4.24 *Let $N = 2^m$, with $m \in \mathbb{N}$, and let $\boldsymbol{\gamma} = (\gamma_j)_{j \geq 1}$ be positive product weights. Furthermore, let $\boldsymbol{g} = (g_1, \ldots, g_d)$ be a generating vector constructed by Algorithm 4.20. Then we have*

$$R_{N,d,1,\boldsymbol{\gamma}}(\boldsymbol{g}) \leq \frac{2 + 6 \log N}{N} \left(-1 + \prod_{j=1}^{d} (1 + \gamma_j (6 + 4 \log N)) \right). \tag{4.38}$$

Moreover, if the weights satisfy

$$\sum_{j=1}^{\infty} \gamma_j < \infty,$$

then $R_{N,d,1,\boldsymbol{\gamma}}(\boldsymbol{g})$ can be bounded independently of the dimension. To be more precise, for any $\delta > 0$, there exists a positive quantity $C(\boldsymbol{\gamma}, \delta)$ independent of d such that

$$R_{N,d,1,\boldsymbol{\gamma}}(\boldsymbol{g}) \leq \frac{C(\boldsymbol{\gamma}, \delta)}{N^{1-\delta}}.$$

Proof Combining the bound on $R_{N,d,1,\boldsymbol{\gamma}}(\boldsymbol{g})$ in Proposition 4.17 with $\boldsymbol{g}$ being a generating vector obtained from Algorithm 4.20, and then using the bound on $H_{d,m,\boldsymbol{\gamma}}$ in Theorem 4.21 directly yields (4.38). The second part of the theorem follows in exactly the same way as in the proof of Theorem 3.21, using Lemma 3.20. □

Finally, we can deduce the following result from Theorem 4.24, which implies a similar result as Theorem 3.17.

Theorem 4.25 *Let $N = 2^m$, with $m \in \mathbb{N}$, and let $\boldsymbol{\gamma} = (\gamma_j)_{j \geq 1}$ be positive product weights. If $\boldsymbol{g} = (g_1, \ldots, g_d)$ is a generating vector constructed by Algorithm 4.20 for the weights $\boldsymbol{\gamma} = (\gamma_j)_{j \geq 1}$, then*

$$\mathrm{err}_{N,d,\alpha,\boldsymbol{\gamma}^{2\alpha}}(\boldsymbol{g}) \leq \frac{1}{N^\alpha}$$
$$\times \left(\left((2 + 6 \log N) \prod_{j=1}^{d} (1 + \gamma_j (6 + 4 \log N)) \right)^{2\alpha} + 2^{2\alpha} \prod_{j=1}^{d} (1 + \gamma_j 4 \zeta(2\alpha)) \right)^{1/2}.$$

Moreover, if the weights satisfy

$$\sum_{j=1}^{\infty} \gamma_j < \infty,$$

then $\mathrm{err}_{N,d,\alpha,\gamma^{2\alpha}}(\mathbf{g})$ *can be bounded independently of the dimension, with an error convergence rate arbitrarily close to* $O(N^{-\alpha})$.

Furthermore, let $\mathbf{g} = (g_1, \ldots, g_d)$ *be a generating vector constructed by Algorithm 4.20 for the weights* $(\gamma_j^{1/(2\alpha)})_{j\geq 1}$. *Then it is true that*

$$\mathrm{err}_{N,d,\alpha,\gamma}(\mathbf{g}) \leq \frac{1}{N^\alpha}$$

$$\times\left(\left((2+6\log N)\prod_{j=1}^{d}(1+\gamma_j^{1/(2\alpha)}(6+4\log N))\right)^{2\alpha} + 2^{2\alpha}\prod_{j=1}^{d}(1+\gamma_j 4\zeta(2\alpha))\right)^{1/2}.$$

Moreover, if the weights satisfy

$$\sum_{j=1}^{\infty}\gamma_j^{1/(2\alpha)} < \infty,$$

then $\mathrm{err}_{N,d,\alpha,\gamma}(\mathbf{g})$ *can be bounded independently of the dimension, with an error convergence rate arbitrarily close to* $O(N^{-\alpha})$.

Proof The result follows by combining Theorem 4.24 with Proposition 3.13, and by adjusting the weights accordingly. □

Notes and Remarks

The idea of the reduced CBC construction was first presented in the paper [45]. The outline of the reduced CBC construction for POD weights in Section 4.2 follows the paper [68]. It can be shown that this approach, in the special case of product weights (note that product weights can be viewed as a special case of POD weights), yields a construction cost of order

$$O\left(\sum_{s=1}^{\min(d,d_*)}(m-w_s)b^{m-w_s}\right),$$

with a required memory of order $O(b^{m-w_1})$, see [68]. Note that this result for product weights is a slight improvement over the results outlined in the first part of Section 4.2.

The presentation of the SCS construction in Section 4.3 follows [71]. Theorem 4.10 is a generalization of an upper bound on the error occurring in [71, Proof of Theorem 2]. The theoretical results presented show that the rules constructed by means of the SCS construction are at least not worse than rules constructed component-by-component. This is also confirmed by the numerical results presented in Appendix B. Whether the SCS construction really can significantly outperform the CBC construction is not yet clear and may be questioned.

The idea of "pruning" was introduced in [77] and then analyzed from a theoretical viewpoint in [43], where it was shown how the worst-case error of the rule constructed in this fashion, in a slightly more general setting, can be bounded from above. The projection-corrected CBC algorithm presented as Algorithm 4.13 is according to [43].

It is possible to combine the ideas of Algorithms 4.1 and 4.13, which has been outlined in [169], and even further combinations of algorithms presented in this chapter could be considered.

In Section 4.6 we had to restrict the considerations to product weights. However, the only passage in that section where this restriction is necessary are the very last few steps in the proof of Theorem 4.21 after (4.36). It is open how to generalize the result to a broader class of weights. See [69] for further information.

Chapter 5
Discrepancy of Lattice Point Sets

The discrepancy of the quadrature points is an important quality criterion for QMC integration rules, as we have seen in Section 1.6, where we have studied discrepancy and the Koksma–Hlawka inequality. This of course also applies to lattice rules and the underlying lattice point sets. In this chapter we study the (extreme) discrepancy of lattice point sets and show its relation to the figure of merit R. Based on the latter we obtain efficient constructions of lattice point sets with the almost optimal order of magnitude of discrepancy.

However, classical discrepancy bounds have a rather poor dependence on the dimension d. To circumvent the this problem and to obtain better behavior in high dimensions, we introduce a weighted discrepancy. Depending on the decay rate of the weights in this weighted discrepancy, several notions of tractability can be shown to hold by means of lattice point sets.

In the last part of this chapter we study the isotropic discrepancy of lattice point sets. Low isotropic discrepancy is of advantage when one aims at transferring lattice points onto the sphere.

5.1 Extreme Discrepancy

For a lattice point set $\mathcal{P}(\mathcal{L}) = \{\boldsymbol{x}_0, \boldsymbol{x}_1, \ldots, \boldsymbol{x}_{N-1}\}$ it follows from the fact that the group $\mathcal{L}/\mathbb{Z}^d$ has order N that $N\boldsymbol{x}_k \in \mathbb{Z}^d$ for all $k \in \{0, 1, \ldots, N-1\}$. Hence all components of any point of $\mathcal{P}(\mathcal{L})$ are rational numbers with a denominator of at most N.

In [193] Niederreiter proved a discrepancy estimate for point sets with such a property, which can be seen as a rational version of the inequality of Erdős, Turán, and Koksma (for the original version of the inequality, we refer, e.g., to [66, 155]). This estimate is the basis for most other discrepancy estimates for lattice point sets presented in this book. In order to state Niederreiter's result we first need to introduce some notation.

J. Dick et al., *Lattice Rules*, Springer Series in Computational Mathematics 58,
https://doi.org/10.1007/978-3-031-09951-9_5

For $M \in \mathbb{N}$, $M \geq 2$, we use the sets $C(M)$, $C_d(M)$, and $C_d^*(M)$ as defined in Section 1.8. For $h \in C(M)$ put

$$r(h, M) = \begin{cases} M \sin(\pi|h|/M) & \text{if } h \neq 0, \\ 1 & \text{if } h = 0, \end{cases}$$

and for $\boldsymbol{h} = (h_1, \ldots, h_d) \in C_d(M)$ put

$$r(\boldsymbol{h}, M) = \prod_{j=1}^{d} r(h_j, M).$$

We now state Niederreiter's discrepancy bound in terms of the extreme discrepancy as given in Definition 1.31. Recall from Chapter 1 that the extreme discrepancy always dominates the star-discrepancy.

Theorem 5.1 *For $M \in \mathbb{N}$, $M \geq 2$, and $\boldsymbol{y}_0, \boldsymbol{y}_1, \ldots, \boldsymbol{y}_{N-1} \in \mathbb{Z}^d$, let $\mathcal{P} = \{\boldsymbol{x}_0, \boldsymbol{x}_1, \ldots, \boldsymbol{x}_{N-1}\}$ be the point set consisting of the fractional parts $\boldsymbol{x}_k = \{\boldsymbol{y}_k/M\}$ for $k \in \{0, 1, \ldots, N-1\}$. Then we have*

$$D_N(\mathcal{P}) \leq 1 - \left(1 - \frac{1}{M}\right)^d + \sum_{\boldsymbol{h} \in C_d^*(M)} \frac{1}{r(\boldsymbol{h}, M)} \left| \frac{1}{N} \sum_{k=0}^{N-1} \mathrm{e}^{2\pi \mathrm{i} \boldsymbol{h} \cdot \boldsymbol{y}_k / M} \right|.$$

For the proof of this result we need the following lemma, which actually is a special, but more refined, case of Lemma 4.19.

Lemma 5.2 *For $j \in [d]$ let $u_j, v_j \in [0, 1]$ with $|u_j - v_j| \leq \delta$. Then we have*

$$\left| \prod_{j=1}^{d} u_j - \prod_{j=1}^{d} v_j \right| \leq 1 - (1 - \delta)^d \leq d\,\delta.$$

Proof We prove the result using induction on d. Trivially, the assertion holds true for $d = 1$. Assume that the result holds true for a fixed $d \in \mathbb{N}$. Without loss of generality we may assume that $u_{d+1} \geq v_{d+1}$. Then we have, using the induction assumption,

$$\begin{aligned} \left| \prod_{j=1}^{d+1} u_j - \prod_{j=1}^{d+1} v_j \right| &= \left| (u_{d+1} - v_{d+1}) \prod_{j=1}^{d} u_j + v_{d+1} \left(\prod_{j=1}^{d} u_j - \prod_{j=1}^{d} v_j \right) \right| \\ &\leq |u_{d+1} - v_{d+1}| + v_{d+1}(1 - (1 - \delta)^d) \\ &= u_{d+1} - v_{d+1}(1 - \delta)^d \\ &= u_{d+1}(1 - (1 - \delta)^d) + (u_{d+1} - v_{d+1})(1 - \delta)^d \\ &\leq 1 - (1 - \delta)^d + \delta(1 - \delta)^d \\ &= 1 - (1 - \delta)^{d+1}. \end{aligned}$$

Hence the first inequality in the lemma is shown.

According to the mean value theorem we have for all $y, z \in \mathbb{R}$ with $z \leq y$ that $y^d - z^d = d\,\xi^{d-1}(y-z)$ for some $\xi \in (z, y)$. Choosing $y = 1$ and $z = 1-\delta$ yields the second inequality in the lemma. □

We now give the proof of Theorem 5.1.

Proof of Theorem 5.1 For $\boldsymbol{\ell} = (\ell_1, \ldots, \ell_d) \in \mathbb{Z}^d$ let

$$A(\boldsymbol{\ell}) := |\{k \in \{0, 1, \ldots, N-1\} \ : \ \boldsymbol{y}_k \equiv \boldsymbol{\ell} \pmod{M}\}|,$$

where a congruence of vectors is meant to be component-wise. Then we have

$$\begin{aligned}
\sum_{k=0}^{N-1} \frac{1}{M^d} \sum_{\boldsymbol{h} \in C_d(M)} \mathrm{e}^{2\pi \mathrm{i} \boldsymbol{h} \cdot (\boldsymbol{y}_k - \boldsymbol{\ell})/M}
&= \sum_{k=0}^{N-1} \prod_{j=1}^{d} \left(\frac{1}{M} \sum_{h_j \in C(M)} \mathrm{e}^{2\pi \mathrm{i} h_j (y_{k,j} - \ell_j)/M} \right) \\
&= A(\boldsymbol{\ell}),
\end{aligned}$$

where we used that

$$\frac{1}{M} \sum_{h_j \in C(M)} \mathrm{e}^{2\pi \mathrm{i} h_j (y_{k,j} - \ell_j)/M} = \begin{cases} 1 & \text{if } y_{k,j} \equiv \ell_j \pmod{M}, \\ 0 & \text{otherwise.} \end{cases}$$

Consequently,

$$A(\boldsymbol{\ell}) - \frac{N}{M^d} = \frac{1}{M^d} \sum_{\boldsymbol{h} \in C_d^*(M)} \mathrm{e}^{-2\pi \mathrm{i} \boldsymbol{h} \cdot \boldsymbol{\ell}/M} \sum_{k=0}^{N-1} \mathrm{e}^{2\pi \mathrm{i} \boldsymbol{h} \cdot \boldsymbol{y}_k / M}.$$

Let now $J = \prod_{j=1}^{d} [u_j, v_j)$ be an arbitrary subinterval of $[0,1)^d$. For $j \in [d]$ let $a_j \in \mathbb{Z}$ be minimal such that $u_j \leq a_j/M$ and let $b_j \in \mathbb{Z}$ be maximal such that $b_j/M < v_j$. In particular, we have $[a_j/M, b_j/M] \subseteq [u_j, v_j)$. In the following, it is convenient to write

$$A(J, \mathcal{P}, N) := |\{k \in \{0, 1, \ldots, N-1\} \ : \ \boldsymbol{x}_k \in J\}|.$$

If $[a_j/M, b_j/M] = \emptyset$, i.e., $a_j > b_j$, for some $j \in [d]$, then J does not contain any point of $\mathcal{P}$ and $v_j - u_j < 1/M$. Therefore we have

$$\left| \frac{A(J, \mathcal{P}, N)}{N} - \lambda_d(J) \right| = \lambda_d(J) < \frac{1}{M} \leq 1 - \left(1 - \frac{1}{M}\right)^d.$$

Now assume that $[a_j/M, b_j/M] \neq \emptyset$ for all $j \in [d]$. Then,

$$A(J,\mathcal{P},N) = \sum_{\boldsymbol{a} \le \boldsymbol{\ell} \le \boldsymbol{b}} A(\boldsymbol{\ell}) \quad \text{and} \quad \sum_{\boldsymbol{a} \le \boldsymbol{\ell} \le \boldsymbol{b}} \frac{1}{M^d} = \frac{1}{M^d} \prod_{j=1}^{d} (b_j - a_j + 1),$$

where the summation range $\boldsymbol{a} \le \boldsymbol{\ell} \le \boldsymbol{b}$ means summation over all $\boldsymbol{\ell} \in \mathbb{Z}^d$ for which $a_j \le \ell_j \le b_j$ for all $j \in [d]$, and hence

$$\begin{aligned}
&\frac{A(J,\mathcal{P},N)}{N} - \lambda_d(J) \\
&= \sum_{\boldsymbol{a} \le \boldsymbol{\ell} \le \boldsymbol{b}} \left(\frac{A(\boldsymbol{\ell})}{N} - \frac{1}{M^d} \right) + \frac{1}{M^d} \prod_{j=1}^{d} (b_j - a_j + 1) - \lambda_d(J) \\
&= \frac{1}{M^d} \sum_{\boldsymbol{h} \in C_d^*(M)} \left(\sum_{\boldsymbol{a} \le \boldsymbol{\ell} \le \boldsymbol{b}} \mathrm{e}^{-2\pi \mathrm{i} \boldsymbol{h} \cdot \boldsymbol{\ell}/M} \right) \frac{1}{N} \sum_{k=0}^{N-1} \mathrm{e}^{2\pi \mathrm{i} \boldsymbol{h} \cdot \boldsymbol{y}_k/M} \\
&\quad + \prod_{j=1}^{d} \frac{b_j - a_j + 1}{M} - \prod_{j=1}^{d} (v_j - u_j).
\end{aligned}$$

For all $j \in [d]$ we have

$$\left| \frac{b_j - a_j + 1}{M} - (v_j - u_j) \right| < \frac{1}{M},$$

and so it follows from Lemma 5.2 that

$$\left| \prod_{j=1}^{d} \frac{b_j - a_j + 1}{M} - \prod_{j=1}^{d} (v_j - u_j) \right| \le 1 - \left(1 - \frac{1}{M}\right)^d.$$

Therefore we obtain

$$\begin{aligned}
&\left| \frac{A(J,\mathcal{P},N)}{N} - \lambda_d(J) \right| \\
&\le 1 - \left(1 - \frac{1}{M}\right)^d + \frac{1}{M^d} \sum_{\boldsymbol{h} \in C_d^*(M)} \left| \sum_{\boldsymbol{a} \le \boldsymbol{\ell} \le \boldsymbol{b}} \mathrm{e}^{-2\pi \mathrm{i} \boldsymbol{h} \cdot \boldsymbol{\ell}/M} \right| \cdot \left| \frac{1}{N} \sum_{k=0}^{N-1} \mathrm{e}^{2\pi \mathrm{i} \boldsymbol{h} \cdot \boldsymbol{y}_k/M} \right|.
\end{aligned}$$

For short, we write

$$r^*(\boldsymbol{h}, M) := \left| \sum_{\boldsymbol{a} \le \boldsymbol{\ell} \le \boldsymbol{b}} \mathrm{e}^{-2\pi \mathrm{i} \boldsymbol{h} \cdot \boldsymbol{\ell}/M} \right|.$$

Since for any complex number z we have $|z| = |\overline{z}|$, and since for any real number t it holds that $|\mathrm{e}^{2\pi \mathrm{i} t}| = 1$, we obtain

$$r^*(\boldsymbol{h}, M) = \prod_{j=1}^{d} \left| \sum_{\ell_j = a_j}^{b_j} \mathrm{e}^{2\pi \mathrm{i} h_j \ell_j/M} \right|$$

$$= \prod_{j=1}^{d} \left| \sum_{\ell_j=0}^{b_j-a_j} \mathrm{e}^{2\pi \mathrm{i} h_j \ell_j/M} \mathrm{e}^{2\pi \mathrm{i} h_j a_j/M} \right|$$

$$= \prod_{j=1}^{d} \left| \sum_{\ell_j=0}^{b_j-a_j} \mathrm{e}^{2\pi \mathrm{i} h_j \ell_j/M} \right|.$$

If $h_j = 0$, we have

$$\left| \sum_{\ell_j=0}^{b_j-a_j} \mathrm{e}^{2\pi \mathrm{i} h_j \ell_j/M} \right| = b_j - a_j + 1 \leq M = \frac{M}{r(h_j, M)}.$$

If $h_j \in C^*(M)$, we have

$$\begin{aligned} \left| \sum_{\ell_j=0}^{b_j-a_j} \mathrm{e}^{2\pi \mathrm{i} h_j \ell_j/M} \right| &= \left| \frac{\mathrm{e}^{2\pi \mathrm{i} h_j (b_j-a_j+1)/M} - 1}{\mathrm{e}^{2\pi \mathrm{i} h_j/M} - 1} \right| \\ &= \left| \frac{\sin(\pi h_j (b_j - a_j + 1)/M)}{\sin(\pi h_j/M)} \right| \\ &\leq \frac{1}{\sin(\pi |h_j|/M)} \\ &= \frac{M}{r(h_j, M)}. \end{aligned}$$

In either case we obtain

$$r^*(\boldsymbol{h}, M) \leq \prod_{j=1}^{d} \frac{M}{r(h_j, M)} = \frac{M^d}{r(\boldsymbol{h}, M)},$$

and therefore

$$\left| \frac{A(J, \mathcal{P}, N)}{N} - \lambda_d(J) \right| \leq 1 - \left(1 - \frac{1}{M}\right)^d + \sum_{\boldsymbol{h} \in C_d^*(M)} \frac{1}{r(\boldsymbol{h}, M)} \left| \frac{1}{N} \sum_{k=0}^{N-1} \mathrm{e}^{2\pi \mathrm{i} \boldsymbol{h} \cdot \boldsymbol{y}_k/M} \right|.$$

The right-hand side of this inequality is independent of the specific choice of the interval J and thus the result follows. □

A general discrepancy bound for lattice point sets

Let $\mathcal{P}(\mathcal{L}) = \{\boldsymbol{x}_0, \boldsymbol{x}_1, \ldots, \boldsymbol{x}_{N-1}\}$ be an N-element lattice point set in $[0, 1)^d$. As already discussed, $N\boldsymbol{x}_k \in \mathbb{Z}^d$ for all $k \in \{0, 1, \ldots, N-1\}$, and hence we can apply Theorem 5.1 in order to obtain

$$D(\mathcal{P}(\mathcal{L})) \leq 1 - \left(1 - \frac{1}{N}\right)^d + \sum_{\boldsymbol{h} \in C_d^*(N)} \frac{1}{r(\boldsymbol{h}, N)} \left| \frac{1}{N} \sum_{k=0}^{N-1} \mathrm{e}^{2\pi \mathrm{i} \boldsymbol{h} \cdot \boldsymbol{x}_k} \right|.$$

Using the character property of lattice point sets stated in Lemma 1.9 we obtain

$$D_N(\mathcal{P}(\mathcal{L})) \leq 1 - \left(1 - \frac{1}{N}\right)^d + \sum_{\boldsymbol{h} \in C_d^*(N) \cap \mathcal{L}^\perp} \frac{1}{r(\boldsymbol{h}, N)}. \tag{5.1}$$

Recalling the definition of r_1 in (1.12) in Section 1.4, we see that we have $r(\boldsymbol{h}, N) \geq 2r_1(\boldsymbol{h})$ for any $\boldsymbol{h} \in C_d^*(N)$. This inequality follows from the fact that $\sin(\pi t) \geq 2t$ for $t \in [0, 1/2]$. Hence

$$\sum_{\boldsymbol{h} \in C_d^*(N) \cap \mathcal{L}^\perp} \frac{1}{r(\boldsymbol{h}, N)} \leq \frac{R(\mathcal{L})}{2},$$

where R is as in Definition 1.48. Summing up, we obtain the following estimate for the discrepancy of a lattice point set, which yields the promised relation to R.

Proposition 5.3 *For the extreme discrepancy of an N-element lattice point set $\mathcal{P}(\mathcal{L})$ in $[0, 1)^d$ with $N \geq 2$ we have*

$$D_N(\mathcal{P}(\mathcal{L})) \leq 1 - \left(1 - \frac{1}{N}\right)^d + \frac{R(\mathcal{L})}{2} \leq \frac{d}{N} + \frac{R(\mathcal{L})}{2}.$$

Proposition 5.3 gives a bound on the extreme discrepancy of lattice point sets which is easier to handle than the estimate (5.1). Let us now restrict ourselves to rank-1 lattice point sets. For given N, we aim at finding a generating vector $\boldsymbol{g}$ with a low value of $R(\boldsymbol{g}, N)$ as given in Definition 1.48, and hence, according to Proposition 5.3, with a low value of the extreme discrepancy of the corresponding lattice point set.

If N is a prime number, then we obtain from Lemma 2.13 that the average of $R(\boldsymbol{g}, N)$ over all generating vectors in $G_d(N) = \{0, 1, \ldots, N-1\}^d$ is not larger than

$$\frac{2^d}{N}(\log N + 1)^d. \tag{5.2}$$

As a consequence, there always exists a lattice point $\boldsymbol{g} \in G_d(N)$ such that

$$D_N(\mathcal{P}(\boldsymbol{g}, N)) \leq \frac{d}{N} + \frac{2^{d-1}}{N}(\log N + 1)^d = O\left(\frac{(\log N)^d}{N}\right).$$

Remark 5.4 The previous averaging argument can be extended to arbitrary integers $N \geq 2$. Niederreiter proved in [194, Theorem 1] that

$$\frac{1}{|C_d^{\varphi}(N)|} \sum_{\boldsymbol{g} \in C_d^{\varphi}(N)} R(\boldsymbol{g}, N) \leq \frac{1}{N}\left(2 \log N + \frac{7}{5}\right)^d \quad \text{for all } d \geq 2 \text{ and all } N \geq 2, \tag{5.3}$$

where

$$C_d^{\varphi}(N) := \{\boldsymbol{g} = (g_1, \ldots, g_d) \in C_d(N) \ : \ \gcd(g_j, N) = 1 \ \text{ for all } \ j \in [d]\}.$$

An improved, but less explicit version of this result was provided by Niederreiter in [199, Theorem 5.10]. For arbitrary integers $d \geq 2$ and $N \geq 2$ we have

$$\frac{1}{|C_d^{\varphi}(N)|} \sum_{\boldsymbol{g} \in C_d^{\varphi}(N)} R(\boldsymbol{g}, N) \leq \frac{(2 \log N + c)^d - 2d \log N}{N} + O\left(\frac{(\log \log N)^2}{N}\right),$$

with $c = 2\gamma_{\mathrm{E-M}} - \log 4 + 1 = 0.768\ldots$, where $\gamma_{\mathrm{E-M}}$ is the Euler–Mascheroni constant, and where the implied constant in the O-notation depends only on d. The proof of this result is technically much more involved than that of the corresponding result for prime N.

Again, it is clear that there must exist at least one generating vector which is at least as good as the average (see Remark 2.14). Hence, for any $d, N \geq 2$, there exists a generating vector $\boldsymbol{g} \in C_d^{\varphi}(N)$ such that

$$\begin{aligned} D_N(\mathcal{P}(\boldsymbol{g}, N)) &\leq \frac{(2 \log N + c)^d + 2d - 2d \log N}{2N} + O\left(\frac{(\log \log N)^2}{N}\right) \\ &= O\left(\frac{(\log N)^d}{N}\right). \end{aligned}$$

These existence results for generating vectors $\boldsymbol{g}$ for which $R(\boldsymbol{g}, N)$ is of order $O((\log N)^d/N)$ are best possible with respect to the order of magnitude in N, which was shown by Larcher [171] in the following result.

Theorem 5.5 *For every $d \geq 2$ there exists a $c_d > 0$, depending only on d, such that for all $\boldsymbol{g} = (g_1, \ldots, g_d) \in \mathbb{Z}^d$ and every $N \in \mathbb{N}$ we have*

$$R(\boldsymbol{g}, N) > c_d \frac{\ell(\log(N/\ell))^d}{N},$$

with $\ell = \gcd(g_1, \ldots, g_d, N)$.

The currently best existence result for lattice point sets with low extreme discrepancy is a result by Bykovskii [22] who directly studied the discrepancy rather than the upper bound in terms of $R(\cdot, N)$. The corresponding result for the special case $d = 2$ was already obtained earlier by Larcher [170]. Bykovskii's result reads as follows.

Theorem 5.6 *For every choice of integers $d \geq 2$ and $N \geq 3$, there exists a generating vector $\boldsymbol{g} \in \mathbb{Z}^d$ for which*

$$D_N(\mathcal{P}(\boldsymbol{g}, N)) = O\left(\frac{(\log N)^{d-1} \log\log N}{N}\right), \tag{5.4}$$

where the implied multiplicative factor may depend on d but is independent of N.

This result is close to the best known upper bounds on the discrepancy for point sets with N elements in $[0, 1)^d$, however, there is a gap to the general lower bound for the discrepancy of arbitrary N-element point sets in the d-dimensional unit cube in (1.20).

Remark 5.7 A special instance of Theorem 5.6 is the two-dimensional case, because here we can employ the theory of continued fractions. Let $N \geq 2$ and $g \in G_1^{\varphi}(N)$, and let $a_1, a_2, \ldots, a_\ell$ be the partial quotients in the continued fraction expansion of g/N as in (1.38). Then it can be shown (see [199, Eq. (5.39)]) that

$$D_N(\mathcal{P}((1, g), N)) \leq \frac{1}{N}\left(\sum_{j=1}^{\ell} a_j + 1\right).$$

It is known that $\ell \leq c \log N$ with an absolute constant $c > 0$. Hence we have

$$\sum_{j=1}^{\ell} a_j \leq \ell K\left(\frac{g}{N}\right) \leq cK\left(\frac{g}{N}\right) \log N,$$

where $K(g/N)$ is defined in (1.39). Therefore,

$$D_N(\mathcal{P}((1, g), N)) \leq \frac{cK(g/N) \log N + 1}{N}.$$

E.g., for the Fibonacci lattice $\mathcal{P}((1, F_{n-1}), F_n)$ for some integer $n \geq 3$ (see Example 1.54) we have $K(F_{n-1}/F_n) = 1$ and so

$$D_N(\mathcal{P}((1, F_{n-1}), F_n)) \leq \frac{c \log N + 1}{N}, \quad \text{where } N = F_n.$$

This order of magnitude is best possible for the discrepancy of N-element point sets in the unit square (see (1.21)).

5.2 CBC Construction of Low Discrepancy Lattice Point Sets

We now aim at constructing generating vectors $\boldsymbol{g}$ of rank-1 lattice point sets which yield a low extreme discrepancy. The basic idea is to use the relation of the discrepancy to the figure of merit R in Proposition 5.3, and to construct generating vectors $\boldsymbol{g}$ with small values of $R(\boldsymbol{g}, N)$. Here we can again employ the CBC construction in Algorithm 3.14 (see Section 3.6) for the unweighted case.

According to the unweighted case in Theorem 3.15, the resulting generating vector $\boldsymbol{g} = (g_1, \dots, g_d) \in G_d(N)$ satisfies, for prime N,

$$R(\boldsymbol{g}^{(s)}, N) \leq \frac{1}{N}(1 + S_N)^s$$

for all $s \in [d]$, where, as before, $S_N = \sum_{h \in C_1^*(N)} |h|^{-1}$, and where $\boldsymbol{g}^{(s)} = (g_1, \dots, g_s)$. This result was first shown by Joe [120] (see also [229] for the case of composite N).

Using (2.16), which states that $S_N \leq 2(\log N + 1 - \log 2)$, we obtain

$$R(\boldsymbol{g}^{(s)}, N) \leq \frac{1}{N}(1 + 2(\log N + 1 - \log 2))^s \leq \frac{2^s}{N}(\log N + 1)^s,$$

which exactly matches the bound (5.2) according to the average type result. Thus we obtain the following corollary.

Corollary 5.8 *Let N be a prime number. If the lattice point $\boldsymbol{g} = (g_1, \dots, g_d)$ has been constructed according to Algorithm 3.14 in Section 3.6, then, for all $s \in [d]$, we have*

$$D_N(\mathcal{P}(\boldsymbol{g}^{(s)}, N)) \leq \frac{s}{N} + \frac{2^{s-1}}{N} (\log N + 1)^s,$$

where $\boldsymbol{g}^{(s)} = (g_1, \dots, g_s)$.

5.3 Weighted Star-Discrepancy

Similar to weighted function spaces (see Section 2.4), there is also the concept of weighted discrepancy. Also in this case one tries to adapt to situations where different dependencies on variables or groups of variables have to be taken into account. Weighted discrepancy is a generalization of the classical notion of discrepancy, and it should be seen in the context of weighted numerical integration and a weighted version of the Koksma–Hlawka inequality (see Remark 5.11). The concept was first introduced by Sloan and Woźniakowski [239] in connection with weighted function spaces, and since then it has become a very useful figure of merit for sets of QMC integration nodes and, in particular, for lattice point sets.

We outline the idea of weighted discrepancy for the case of the star-discrepancy, as this is most common in the literature. However, one could similarly define other types of weighted discrepancy.

Definition 5.9 (Weighted star-discrepancy) Let $\boldsymbol{\gamma} = \{\gamma_{\mathfrak{u}}\}_{\mathfrak{u}\subseteq[d]}$ be given weights. For an N-point set $\mathcal{P}$ in $[0,1)^d$ the $\boldsymbol{\gamma}$*-weighted star-discrepancy* is defined as

$$D^*_{N,\boldsymbol{\gamma}}(\mathcal{P}) := \sup_{\boldsymbol{t}\in[0,1]^d} \max_{\emptyset\neq\mathfrak{u}\subseteq[d]} \gamma_{\mathfrak{u}} |\Delta_{\mathcal{P}}(\boldsymbol{t}_{\mathfrak{u}},\boldsymbol{1})|,$$

where for $\boldsymbol{t} = (t_1,\ldots,t_d) \in [0,1]^d$ and for $\mathfrak{u} \subseteq [d]$ we put $(\boldsymbol{t}_{\mathfrak{u}},\boldsymbol{1}) := (y_1,\ldots,y_d)$ with $y_j = t_j$ if $j \in \mathfrak{u}$ and $y_j = 1$ if $j \notin \mathfrak{u}$.

Remark 5.10 The $\boldsymbol{\gamma}$-weighted star-discrepancy is sometimes alternatively defined as

$$D^*_{N,\boldsymbol{\gamma}}(\mathcal{P}) = \max_{\emptyset\neq\mathfrak{u}\subseteq[d]} \gamma_{\mathfrak{u}}\, D^*_N(\mathcal{P}_{\mathfrak{u}}), \tag{5.5}$$

where the sets $\mathcal{P}_{\mathfrak{u}}$ consist of the points of $\mathcal{P}$ projected onto the components with indices in $\mathfrak{u}$. As shown in [218] this definition is equivalent to Definition 5.9. Indeed, we have on the one hand $|\Delta_{\mathcal{P}}(\boldsymbol{t}_{\mathfrak{u}},\boldsymbol{1})| \leq D^*_N(\mathcal{P}_{\mathfrak{u}})$ for any $\mathfrak{u} \subseteq [d]$ and any $\boldsymbol{t} \in [0,1]^d$, and hence

$$D^*_{N,\boldsymbol{\gamma}}(\mathcal{P}_d) \leq \max_{\emptyset\neq\mathfrak{u}\subseteq[d]} \gamma_{\mathfrak{u}}\, D^*_N(\mathcal{P}_{\mathfrak{u}}).$$

On the other hand, for every $\emptyset \neq \mathfrak{u} \subseteq [d]$ we have

$$D^*_{N,\boldsymbol{\gamma}}(\mathcal{P}) \geq \sup_{\boldsymbol{t}\in[0,1]^d} \gamma_{\mathfrak{u}}\, |\Delta_{\mathcal{P}}(\boldsymbol{t}_{\mathfrak{u}},\boldsymbol{1})| = \gamma_{\mathfrak{u}}\, D^*_N(\mathcal{P}_{\mathfrak{u}}).$$

Remark 5.11 The $\boldsymbol{\gamma}$-weighted star-discrepancy of an N-element point set $\mathcal{P} = \{\boldsymbol{x}_0,\boldsymbol{x}_1,\ldots,\boldsymbol{x}_{N-1}\}$ in $[0,1)^d$ is intimately linked to the worst-case integration error of QMC rules

$$\frac{1}{N}\sum_{k=0}^{N-1} f(\boldsymbol{x}_k)$$

for functions f in the weighted function class $\mathcal{F}_{d,1,\boldsymbol{\gamma}}$, which is a weighted version of $\mathcal{F}_{d,q}$ from (1.27) in Section 1.6 for $q = 1$. It is defined as follows. Let $\mathcal{W}_1^{(1,1,\ldots,1)}([0,1]^d)$ be the Sobolev space of functions defined on $[0,1]^d$ that are once differentiable in each variable, and whose derivatives have finite L_1-norm. Then

$$\mathcal{F}_{d,1,\boldsymbol{\gamma}} = \{f \in \mathcal{W}_1^{(1,1,\ldots,1)}([0,1]^d) \ : \ \|f\|_{d,1,\boldsymbol{\gamma}} < \infty\},$$

where

$$\|f\|_{d,1,\boldsymbol{\gamma}} = |f(\boldsymbol{1})| + \sum_{\emptyset\neq\mathfrak{u}\subseteq[d]} \frac{1}{\gamma_{\mathfrak{u}}} \int_{[0,1]^{|\mathfrak{u}|}} \left|\frac{\partial^{|\mathfrak{u}|}}{\partial \boldsymbol{x}_{\mathfrak{u}}} f(\boldsymbol{x}_{\mathfrak{u}},\boldsymbol{1})\right| \mathrm{d}\boldsymbol{x}_{\mathfrak{u}}.$$

A fundamental error estimate is a weighted version of the Koksma–Hlawka inequality in Corollary 1.36. In fact, the worst-case error of a QMC rule in $\mathcal{F}_{d,1,\boldsymbol{\gamma}}$ is exactly the $\boldsymbol{\gamma}$-weighted star-discrepancy of the point set used in the QMC rule, i.e.,

$$\mathrm{err}_{N,d}(\mathcal{F}_{d,1,\gamma},\mathcal{P}) = D^*_{N,\gamma}(\mathcal{P}_d).$$

In other words, for functions $f \in \mathcal{F}_{d,1,\gamma}$ we have

$$|\mathrm{err}_{N,d}(f,\mathcal{P})| \le \|f\|_{d,1,\gamma}\, D^*_{N,\gamma}(\mathcal{P}_d).$$

This error estimate is often called the *weighted version of the Koksma–Hlawka inequality*. We remark that one can also define a weighted L_p-discrepancy with a corresponding relation to the worst-case error of QMC rules in weighted spaces $\mathcal{F}_{d,q,\gamma}$ based on the L_q-norm, where $1/p + 1/q = 1$. For further information see [211, Chapter 9] or [239].

The weighted star-discrepancy of lattice point sets

Next, we consider the weighted star-discrepancy of rank-1 lattice point sets $\mathcal{P}(\boldsymbol{g}, N)$. Note that the projection of the elements of $\mathcal{P}(\boldsymbol{g}, N)$ onto the components with indices in $\mathfrak{u} \subseteq [d]$, $\mathfrak{u} \neq \emptyset$, yields the rank-1 lattice point set $\mathcal{P}(\boldsymbol{g}_{\mathfrak{u}}, N)$ with the $|\mathfrak{u}|$-dimensional generating vector $\boldsymbol{g}_{\mathfrak{u}}$ consisting of the components of $\boldsymbol{g}$ whose indices belong to $\mathfrak{u}$. Combining (5.5) with Proposition 5.3 gives

$$D^*_{N,\gamma}(\mathcal{P}(\boldsymbol{g}, N)) \le \max_{\emptyset \neq \mathfrak{u} \subseteq [d]} \gamma_{\mathfrak{u}} \left(1 - \left(1 - \frac{1}{N}\right)^{|\mathfrak{u}|} + \frac{R(\boldsymbol{g}_{\mathfrak{u}}, N)}{2}\right). \tag{5.6}$$

Regarding the first term in the parentheses in (5.6), the following auxiliary result holds.

Lemma 5.12 *Let $d \in \mathbb{N}$. For general weights $\{\gamma_{\mathfrak{u}}\}_{\mathfrak{u} \subseteq [d]}$ we have, for any $N \in \mathbb{N}$,*

$$\max_{\emptyset \neq \mathfrak{u} \subseteq [d]} \gamma_{\mathfrak{u}} \left(1 - \left(1 - \frac{1}{N}\right)^{|\mathfrak{u}|}\right) \le \frac{1}{N} \max_{\emptyset \neq \mathfrak{u} \subseteq [d]} |\mathfrak{u}|\, \gamma_{\mathfrak{u}}.$$

For product weights $(\gamma_j)_{j \ge 1}$ we have, for any $N \in \mathbb{N}$,

$$\max_{\emptyset \neq \mathfrak{u} \subseteq [d]} \gamma_{\mathfrak{u}} \left(1 - \left(1 - \frac{1}{N}\right)^{|\mathfrak{u}|}\right) \le \frac{1}{N} \prod_{j=1}^{d} (1 + 2\gamma_j).$$

Proof For general weights the result follows from an application of Bernoulli's inequality which implies

$$\left(1 - \frac{1}{N}\right)^{|\mathfrak{u}|} \ge 1 - \frac{|\mathfrak{u}|}{N}.$$

For product weights we use $|\mathfrak{u}| \le 2^{|\mathfrak{u}|}$. Then we obtain from the estimate for general weights,

$$\max_{\emptyset \neq \mathfrak{u} \subseteq [d]} \gamma_{\mathfrak{u}} \left(1 - \left(1 - \frac{1}{N}\right)^{|\mathfrak{u}|}\right) \leq \frac{1}{N} \max_{\emptyset \neq \mathfrak{u} \subseteq [d]} \prod_{j \in \mathfrak{u}} (2\gamma_j)$$
$$\leq \frac{1}{N} \sum_{\mathfrak{u} \subseteq [d]} \prod_{j \in \mathfrak{u}} (2\gamma_j)$$
$$= \frac{1}{N} \prod_{j=1}^{d} (1 + 2\gamma_j),$$

as desired. □

Let us now consider the term $R(\boldsymbol{g}_{\mathfrak{u}}, N)$ occurring in (5.6). From the form of R in Definition 1.48 we obtain

$$R(\boldsymbol{g}_{\mathfrak{u}}, N) = \sum_{\substack{\boldsymbol{h} \in C^*_{|\mathfrak{u}|}(N) \\ \boldsymbol{h} \cdot \boldsymbol{g}_{\mathfrak{u}} \equiv 0 \pmod N}} \frac{1}{r_1(\boldsymbol{h})}$$
$$= \sum_{\boldsymbol{h} \in C^*_{|\mathfrak{u}|}(N)} \frac{1}{r_1(\boldsymbol{h})} \frac{1}{N} \sum_{k=0}^{N-1} \mathrm{e}^{2\pi \mathrm{i} k \boldsymbol{h} \cdot \boldsymbol{g}_{\mathfrak{u}}/N}$$
$$= -1 + \frac{1}{N} \sum_{k=0}^{N-1} \prod_{j \in \mathfrak{u}} \left(1 + \sum_{h \in C^*_1(N)} \frac{\mathrm{e}^{2\pi \mathrm{i} k h g_j/N}}{|h|}\right)$$
$$= \frac{1}{N} \sum_{k=0}^{N-1} \left(-1 + \prod_{j \in \mathfrak{u}} \left(1 + S_{k,N}(g_j)\right)\right),$$

where $S_{k,N}$ is defined as in (3.31). Expanding the above product we obtain

$$R(\boldsymbol{g}_{\mathfrak{u}}, N) = \frac{1}{N} \sum_{k=0}^{N-1} \sum_{\emptyset \neq \mathfrak{v} \subseteq \mathfrak{u}} \prod_{j \in \mathfrak{v}} S_{k,N}(g_j)$$
$$= \sum_{\emptyset \neq \mathfrak{v} \subseteq \mathfrak{u}} \frac{1}{N} \sum_{k=0}^{N-1} \prod_{j \in \mathfrak{v}} S_{k,N}(g_j)$$
$$= \sum_{\emptyset \neq \mathfrak{v} \subseteq \mathfrak{u}} \widetilde{R}_N(\boldsymbol{g}_{\mathfrak{v}}),$$

where $\widetilde{R}_N$ is defined as in (3.30).

In the following we shall sometimes make the practically reasonable assumption

$$\gamma_{\mathfrak{u}} \leq \gamma_{\mathfrak{v}} \quad \text{for all } \mathfrak{v} \subseteq \mathfrak{u} \text{ and all } \mathfrak{u} \subseteq [d]. \tag{5.7}$$

Note that this assumption is automatically satisfied for product weights $\gamma_{\mathfrak{u}} = \prod_{j \in \mathfrak{u}} \gamma_j$ for $\mathfrak{u} \subseteq [d]$, with $\gamma_j \in (0, 1]$ for all $j \in \mathbb{N}$. Under this assumption we have

$$\begin{aligned}\max_{\emptyset\neq\mathfrak{u}\subseteq[d]} \gamma_{\mathfrak{u}} R(\boldsymbol{g}_{\mathfrak{u}}, N) &= \max_{\emptyset\neq\mathfrak{u}\subseteq[d]} \gamma_{\mathfrak{u}} \sum_{\emptyset\neq\mathfrak{v}\subseteq\mathfrak{u}} \widetilde{R}_N(\boldsymbol{g}_{\mathfrak{v}}) \\ &\le \sum_{\emptyset\neq\mathfrak{u}\subseteq[d]} \gamma_{\mathfrak{u}} \widetilde{R}_N(\boldsymbol{g}_{\mathfrak{u}}) \\ &= R_{N,d,1,\boldsymbol{\gamma}}(\boldsymbol{g}),\end{aligned}$$

where $R_{N,d,1,\boldsymbol{\gamma}}(\boldsymbol{g})$ is given in Definition 3.12.

We can now formulate a theorem which follows from (5.6) and our observations above.

Theorem 5.13 *For general weights* $\boldsymbol{\gamma} = \{\gamma_{\mathfrak{u}}\}_{\mathfrak{u}\subseteq[d]}$ *satisfying* (5.7) *we have*

$$D^*_{N,\boldsymbol{\gamma}}(\mathcal{P}(\boldsymbol{g}, N)) \le \frac{1}{N} \max_{\emptyset\neq\mathfrak{u}\subseteq[d]} |\mathfrak{u}|\,\gamma_{\mathfrak{u}} + \frac{R_{N,d,1,\boldsymbol{\gamma}}(\boldsymbol{g})}{2}.$$

We have discussed in Section 3.6 how to construct generating vectors with a small value of $R_{N,d,1,\boldsymbol{\gamma}}$. The lattice point sets constructed in this way have a low weighted star-discrepancy due to Theorem 5.13. We summarize this in the following corollary.

Corollary 5.14 *Let N be a prime number, let $d \in \mathbb{N}$, and let $\boldsymbol{\gamma} = \{\gamma_{\mathfrak{u}}\}_{\mathfrak{u}\subseteq[d]}$ be general weights satisfying Condition* (5.7). *Assume that $\boldsymbol{g} = (g_1, \ldots, g_d)$ has been constructed according to Algorithm 3.14. Then for any $s \in [d]$ we have*

$$D^*_{N,\boldsymbol{\gamma}}(\mathcal{P}((g_1, \ldots, g_s), N)) \le \frac{1}{N}\left(\max_{\emptyset\neq\mathfrak{u}\subseteq[s]} |\mathfrak{u}|\,\gamma_{\mathfrak{u}} + \frac{1}{2}\sum_{\emptyset\neq\mathfrak{u}\subseteq[s]} \gamma_{\mathfrak{u}}\, S_N^{|\mathfrak{u}|}\right),$$

where S_N is defined as in (2.15). *In particular, for product weights $\gamma_{\mathfrak{u}} = \prod_{j\in\mathfrak{u}} \gamma_j$, for $\mathfrak{u} \subseteq [d]$, we have*

$$D^*_{N,\boldsymbol{\gamma}}(\mathcal{P}((g_1, \ldots, g_s), N)) \le \frac{1}{N}\left(\prod_{j=1}^{s}(1 + 2\gamma_j) + \frac{1}{2}\prod_{j=1}^{s}(1 + \gamma_j S_N)\right).$$

Recall that $S_N = O(\log N)$, which implies that the discrepancy bounds in Corollary 5.14 are both of order

$$O\left(\frac{(\log N)^s}{N}\right),$$

where the implied multiplicative factor depends only on s and $\boldsymbol{\gamma}$. Under suitable assumptions on the weights this implied factor is independent of the dimension. This will be discussed in the subsequent section.

5.4 Tractability of the Weighted Star-Discrepancy

In accordance with Definition 1.40 we define the N-th minimal weighted star-discrepancy.

Definition 5.15 For $d, N \in \mathbb{N}$ and weights $\boldsymbol{\gamma} = \{\gamma_{\mathfrak{u}}\}_{\mathfrak{u} \subseteq [d]}$ the *N-th minimal weighted star-discrepancy* is defined as

$$\mathrm{disc}^*_{\boldsymbol{\gamma}}(N, d) = \inf_{\mathcal{P}} D^*_{N,\boldsymbol{\gamma}}(\mathcal{P}),$$

where the infimum is extended over all N-element point sets $\mathcal{P}$ in $[0, 1)^d$.

Definition 5.16 For $d \in \mathbb{N}$, $\varepsilon \in (0, 1)$, and $\boldsymbol{\gamma} = \{\gamma_{\mathfrak{u}}\}_{\mathfrak{u} \subseteq [d]}$ the *inverse of the weighted star-discrepancy* is defined as

$$N^*_{\boldsymbol{\gamma}}(\varepsilon, d) = \min\{N \in \mathbb{N} \,:\, \mathrm{disc}^*_{\boldsymbol{\gamma}}(N, d) \le \varepsilon\}.$$

Note that the inverse of the weighted star-discrepancy is a special instance of the information complexity as given in Definition 1.42.

A further analysis of the discrepancy bounds in Corollary 5.14, similar to what was shown in Section 5.2, leads to the following estimates.

Corollary 5.17 *Let $\boldsymbol{\gamma} = \{\gamma_{\mathfrak{u}}\}_{\mathfrak{u} \subseteq [d]}$ be positive weights satisfying Condition* (5.7) *and the condition*

$$\sum_{j=1}^{\infty} \max_{\mathfrak{v} \subseteq [j-1]} \frac{\gamma_{\mathfrak{v} \cup \{j\}}}{\gamma_{\mathfrak{v}}} < \infty. \tag{5.8}$$

Then for any $\delta > 0$ there exists a positive real $C(\boldsymbol{\gamma}, \delta)$, that depends only on δ and on the weights $\boldsymbol{\gamma}$, with the following property. For any $d, N \in \mathbb{N}$, and for a $\boldsymbol{g} \in G_d(N)$ that has been constructed according to Algorithm 3.14, we have

$$D^*_{N,\boldsymbol{\gamma}}(\mathcal{P}(\boldsymbol{g}, N)) \le \frac{C(\boldsymbol{\gamma}, \delta)}{N^{1-\delta}}.$$

Assume that the weights $\boldsymbol{\gamma} = \{\gamma_{\mathfrak{u}}\}_{\mathfrak{u} \subseteq [d]}$ satisfy Conditions (5.7) and (5.8). Fix $\delta > 0$, and let N be the smallest prime number greater than or equal to

$$M := \left\lceil (C(\boldsymbol{\gamma}, \delta)\varepsilon^{-1})^{1/(1-\delta)} \right\rceil,$$

where $C(\boldsymbol{\gamma}, \delta)$ is as in Corollary 5.17. Then for the generating vector $\boldsymbol{g}$ in Corollary 5.17 we have

$$\mathrm{disc}^*_{\boldsymbol{\gamma}}(N, d) \le D^*_{N,\boldsymbol{\gamma}}(\mathcal{P}(\boldsymbol{g}, N)) \le \varepsilon.$$

Hence it follows that

$$N^*_{\boldsymbol{\gamma}}(\varepsilon, d) \le N \le 2M = 2\left\lceil (C(\boldsymbol{\gamma}, \delta)\varepsilon^{-1})^{1/(1-\delta)} \right\rceil \le \widetilde{C}(\boldsymbol{\gamma}, \delta)\varepsilon^{-1/(1-\delta)},$$

with a suitably chosen positive real $\widetilde{C}(\boldsymbol{\gamma}, \delta)$ that is independent of ε, d, and N. This means that the weighted star-discrepancy is strongly polynomially tractable. Since $\delta > 0$ can be chosen arbitrarily close to zero, the ε-exponent of strong polynomial tractability equals 1. We summarize this result in the following corollary.

Corollary 5.18 *Assume that the weights $\boldsymbol{\gamma} = \{\gamma_{\mathfrak{u}}\}_{\mathfrak{u} \subseteq [d]}$ satisfy Conditions* (5.7) *and* (5.8). *Then the weighted star-discrepancy is strongly polynomially tractable with an ε-exponent equal to 1, which can be obtained using lattice point sets.*

For product weights Condition (5.8) is equivalent to $\sum_{j=1}^{\infty} \gamma_j < \infty$. In this case it is an easy task, using methods already introduced in this book, to also find conditions on the weights for polynomial or weak tractability of the weighted star-discrepancy.

5.5 Korobov Type Lattice Point Sets With Low Weighted Star-Discrepancy

For $g \in \mathbb{Z}$ let $\boldsymbol{g}_d(g) = (1, g, g^2, \ldots, g^{d-1})$ be a Korobov type generating vector in $\mathbb{Z}^d$ (see also Section 3.2). The existence of good Korobov type generating vectors with respect to the weighted star-discrepancy can be shown by averaging $R_{N,d,1,\boldsymbol{\gamma}}(\boldsymbol{g}_d(g))$ over all $g \in G_1(N)$.

Theorem 5.19 *Let N be a prime number. Then*

$$\frac{1}{N} \sum_{g=0}^{N-1} R_{N,d,1,\boldsymbol{\gamma}}(\boldsymbol{g}_d(g)) \leq \frac{d-1}{N} \sum_{\emptyset \neq \mathfrak{u} \subseteq [d]} \gamma_{\mathfrak{u}} \, S_N^{|\mathfrak{u}|},$$

where S_N is defined as in (2.15). *Furthermore, for every $\theta \in [0, 1)$ there are more than θN elements g in $G_1(N)$ such that*

$$R_{N,d,1,\boldsymbol{\gamma}}(\boldsymbol{g}_d(g)) \leq \frac{1}{1-\theta} \frac{d-1}{N} \sum_{\emptyset \neq \mathfrak{u} \subseteq [d]} \gamma_{\mathfrak{u}} \, S_N^{|\mathfrak{u}|}.$$

Proof We have

$$\begin{aligned} \frac{1}{N} \sum_{g=0}^{N-1} R_{N,d,1,\boldsymbol{\gamma}}(\boldsymbol{g}_d(g)) &= \frac{1}{N} \sum_{g=0}^{N-1} \sum_{\substack{\boldsymbol{h} \in C_d^*(N) \\ \boldsymbol{h} \cdot \boldsymbol{g}_d(g) \equiv 0 \pmod{N}}} \frac{1}{r_{1,\boldsymbol{\gamma}}(\boldsymbol{h})} \\ &= \frac{1}{N} \sum_{\boldsymbol{h} \in C_d^*(N)} \frac{1}{r_{1,\boldsymbol{\gamma}}(\boldsymbol{h})} \sum_{\substack{g=0 \\ \boldsymbol{h} \cdot \boldsymbol{g}_d(g) \equiv 0 \pmod{N}}}^{N-1} 1. \end{aligned}$$

For given $\boldsymbol{h} = (h_1, \ldots, h_d) \neq \boldsymbol{0}$ the innermost sum of the latter term counts the number of solutions $g \in \{0, 1, \ldots, N-1\}$ of the polynomial congruence

$$h_1 + h_2 g + \cdots + h_d g^{d-1} \equiv 0 \pmod{N}.$$

This number is, however, at most $d-1$, since any polynomial of degree $d-1$ can have at most $d-1$ distinct roots. Hence we obtain

$$\frac{1}{N}\sum_{g=0}^{N-1} R_{N,d,1,\boldsymbol{\gamma}}(\boldsymbol{g}_d(g)) \le \frac{d-1}{N} \sum_{\boldsymbol{h}\in C_d^*(N)} \frac{1}{r_{1,\boldsymbol{\gamma}}(\boldsymbol{h})}.$$

Now the first inequality in the theorem follows since

$$\begin{aligned}\sum_{\boldsymbol{h}\in C_d^*(N)} \frac{1}{r_{1,\boldsymbol{\gamma}}(\boldsymbol{h})} &= \sum_{\boldsymbol{h}\in C_d^*(N)} \gamma_{\mathfrak{u}(\boldsymbol{h})} \prod_{j\in\mathfrak{u}(\boldsymbol{h})} \frac{1}{|h_j|}\\ &= \sum_{\emptyset\neq\mathfrak{u}\subseteq[d]} \gamma_{\mathfrak{u}} \sum_{\boldsymbol{h}\in(C_1^*(N))^{|\mathfrak{u}|}} \prod_{j\in\mathfrak{u}} \frac{1}{|h_j|}\\ &= \sum_{\emptyset\neq\mathfrak{u}\subseteq[d]} \gamma_{\mathfrak{u}}\, S_N^{|\mathfrak{u}|}.\end{aligned}$$

In order to prove the second inequality we use the abbreviation

$$T(d,N,\boldsymbol{\gamma}) := \frac{d-1}{N} \sum_{\emptyset\neq\mathfrak{u}\subseteq[d]} \gamma_{\mathfrak{u}}\, S_N^{|\mathfrak{u}|}.$$

We then have

$$\begin{aligned} N\,T(d,N,\boldsymbol{\gamma}) &\ge \sum_{g=0}^{N-1} R_{N,d,1,\boldsymbol{\gamma}}(\boldsymbol{g}_d(g))\\ &> \frac{T(d,N,\boldsymbol{\gamma})}{1-\theta}\left|\left\{g\in G_1(N)\ :\ R_{N,d,1,\boldsymbol{\gamma}}(\boldsymbol{g}_d(g)) > \frac{T(d,N,\boldsymbol{\gamma})}{1-\theta}\right\}\right|.\end{aligned}$$

Thus, the number of $g\in G_1(N)$ that satisfy $R_{N,d,1,\boldsymbol{\gamma}}(\boldsymbol{g}_d(g)) > (1-\theta)^{-1}T(d,N,\boldsymbol{\gamma})$ is less than $(1-\theta)\,N$. This implies that the number of $g\in G_1(N)$ that satisfy

$$R_{N,d,1,\boldsymbol{\gamma}}(\boldsymbol{g}_d(g)) \le \frac{T(d,N,\boldsymbol{\gamma})}{1-\theta} = \frac{1}{1-\theta}\frac{d-1}{N} \sum_{\emptyset\neq\mathfrak{u}\subseteq[d]} \gamma_{\mathfrak{u}}\, S_N^{|\mathfrak{u}|}$$

is greater than θN. □

Formally, we state the following search algorithm.

Algorithm 5.20 (Korobov type lattice rule with low $R_{N,d,1,\boldsymbol{\gamma}}$) Let N be a prime number and let $d\in\mathbb{N}$. The optimal Korobov type generating vector is found by minimizing $R_{N,d,1,\boldsymbol{\gamma}}(\boldsymbol{g}_d(g))$ with respect to $g\in G_1(N)$.

From Theorems 5.13 and 5.19 we deduce the subsequent corollary.

Corollary 5.21 *Let N be a prime number and let $\boldsymbol{\gamma} = \{\gamma_{\mathfrak{u}}\}_{\mathfrak{u}\subseteq[d]}$ be general weights satisfying* (5.7). *For $g \in G_1(N)$ found by Algorithm 5.20 we have*

$$D^*_{N,\gamma}(\mathcal{P}(\boldsymbol{g}_d(g), N)) \leq \frac{1}{N}\left(\max_{\emptyset\neq\mathfrak{u}\subseteq[d]} |\mathfrak{u}|\,\gamma_{\mathfrak{u}} + \frac{d-1}{2}\sum_{\emptyset\neq\mathfrak{u}\subseteq[d]} \gamma_{\mathfrak{u}}\, S_N^{|\mathfrak{u}|}\right).$$

If in addition the weights satisfy Condition (5.8), *then for every $\delta > 0$ there exists a positive $C(\boldsymbol{\gamma},\delta)$, which is independent of d and N, such that*

$$D^*_{N,\gamma}(\mathcal{P}(\boldsymbol{g}_d(g), N)) \leq C(\boldsymbol{\gamma},\delta)\,\frac{d}{N^{1-\delta}}.$$

Observe that in the above estimate we have linear dependence on the dimension d. This means that we cannot guarantee strong polynomial tractability. However, we can obtain polynomial tractability by means of optimal Korobov type generating vectors. In analogy to what we remarked at the end of Section 3.2, it is not clear whether the linear dependence on d can be avoided at all if we restrict ourselves to Korobov type generating vectors.

5.6 Isotropic Discrepancy and Lattice Point Sets on the Sphere

The usual (star-) discrepancy uses axes-parallel boxes or anchored axes-parallel boxes as test sets for the uniformity of point sets in the unit cube. This type of discrepancy is important because of its occurrence in the original Koksma–Hlawka inequality for the absolute integration error of QMC rules. However, there are several extensions of discrepancies to various other classes of test sets. One prominent example is the so-called isotropic discrepancy, which uses all convex subsets of the unit cube as test sets.

Definition 5.22 For a finite point set $\mathcal{P} = \{\boldsymbol{x}_0, \boldsymbol{x}_1, \ldots, \boldsymbol{x}_{N-1}\}$ in the d-dimensional unit cube $[0,1)^d$ the *isotropic discrepancy* is defined as

$$J_N(\mathcal{P}) := \sup_C \left|\frac{|\{n \in \{0,1,\ldots,N-1\} \,:\, \boldsymbol{x}_n \in C\}|}{N} - \lambda_d(C)\right|,$$

where the supremum is extended over all convex subsets C of $[0,1]^d$.

The isotropic discrepancy is an important quantitative measure for the irregularity of distribution of $\mathcal{P}$, see, e.g., [66, 155]. According to [2], point sets in $[0,1)^2$ with a low isotropic discrepancy can be used to generate point sets on the sphere $\mathbb{S}^2$ in $\mathbb{R}^3$ with a small spherical cap discrepancy (see Definition 5.23 below) by means of the *Lambert cylindrical equal-area projection*,

$$\boldsymbol{\Phi}(\alpha,\tau) := \left(2\sqrt{\tau-\tau^2}\cos(2\pi\alpha), 2\sqrt{\tau-\tau^2}\sin(2\pi\alpha), 1-2\tau\right) \quad \text{for } \alpha,\tau \in [0,1].$$

Figure 5.1 illustrates the image of a Fibonacci lattice point set $\mathcal{P}((1, F_{m-1}), F_m)$ as defined in Example 1.54 with $m = 15$, using $F_{14} = 377$ and $F_{15} = 610$, under Lambert's cylindrical equal-area projection on the sphere $\mathbb{S}^2$.

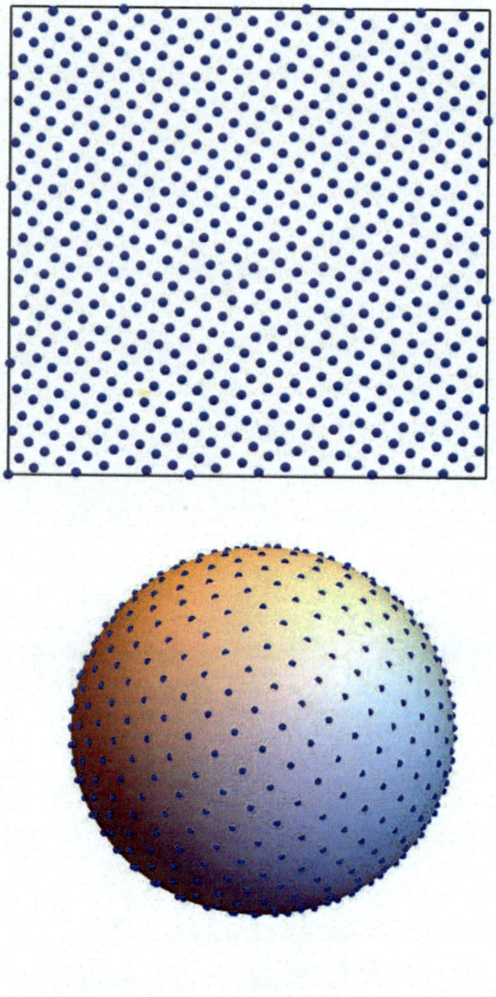

Fig. 5.1: The Fibonacci lattice point set $\mathcal{P}((1, 377), 610)$ and its image under Lambert's cylindrical equal-area projection on the sphere $\mathbb{S}^2$.

Definition 5.23 The *spherical cap discrepancy* of a point set $Q = \{z_0, z_1, \ldots, z_{N-1}\}$ on $\mathbb{S}^2$ uses spherical caps as test sets and is defined as

$$D_N^{\mathbb{S}^2}(Q) := \sup_{w,t} \left| \frac{|\{k \in \{0,1,\ldots,N-1\} \ : \ z_k \in C(w,t)\}|}{N} - \varsigma(C(w,t)) \right|,$$

where the supremum is extended over all $w \in \mathbb{S}^2$ and all $t \in [-1,1]$, where the $C(w,t)$ are spherical caps of the form

$$C(w,t) := \{x \in \mathbb{S}^2 \ : \ w \cdot x > t\},$$

and where ς is the Lebesgue surface area measure on $\mathbb{S}^2$ normalized to a probability measure.

If $\mathcal{P} = \{x_0, x_1, \ldots, x_{N-1}\}$ is a point set in $[0,1]^2$, then

$$Q = \Phi(\mathcal{P}) = \{\Phi(x_0), \Phi(x_1), \ldots, \Phi(x_{N-1})\}$$

is a point set on $\mathbb{S}^2$ whose spherical cap discrepancy satisfies

$$D_N^{\mathbb{S}^2}(Q) \leq 11\, J_N(\mathcal{P}). \tag{5.9}$$

This estimate, shown in [2, Theorem 6], provides an avenue to construct point sets on the sphere with low spherical cap discrepancy and is one motivation for finding point sets with low isotropic discrepancy.

We summarize some well-known bounds on the isotropic discrepancy. A famous result of Schmidt [225, Theorem 1] gives a general lower bound for arbitrary point sets, which we state in the following theorem.

Theorem 5.24 *For every $d \in \mathbb{N}$ and every $N \in \mathbb{N}$ there exists a positive quantity c_d such that for all N-element point sets $\mathcal{P}$ in $[0,1)^d$ we have*

$$J_N(\mathcal{P}) \geq \frac{c_d}{N^{2/(d+1)}}.$$

This result is essentially (up to log factors) best possible as shown by Beck [11] for $d = 2$, and Stute [246] for $d \geq 3$ using probabilistic methods. However, this optimal order cannot be attained by the means of lattice point sets as we will show in the following.

Isotropic discrepancy of lattice point sets

The subsequent theorem shows that the isotropic discrepancy of a lattice point set is, up to a factor depending only on d, the same as the spectral test of the corresponding integration lattice that was introduced in Definition 1.64. Recall that for an integration lattice $\mathcal{L}$ the spectral test is defined as

$$\sigma(\mathcal{L}) := \frac{1}{\min\{\|h\|_2 \ : \ h \in \mathcal{L}^\perp \setminus \{\mathbf{0}\}\}}.$$

Theorem 5.25 *Let $\mathcal{P}(\mathcal{L})$ be an N-element lattice point set in $[0,1)^d$. Then we have*

$$\frac{\sigma(\mathcal{L})}{\sqrt{d}+\sigma(\mathcal{L})} \leq J_N(\mathcal{P}(\mathcal{L})) \leq d\, 2^{2(d+1)}\sigma(\mathcal{L}).$$

If $\sigma(\mathcal{L}) \leq 1/2$, then the lower bound can be replaced by $c\,\sigma(\mathcal{L})$, where $c > 0$ is an absolute constant.

Proof For the first part of the proof we follow [242]. Consider a family $\mathcal{H}^*$ of parallel hyperplanes covering the lattice $\mathcal{L}$, where the distance between adjacent hyperplanes is maximal and equal to the spectral test. It follows from the geometric interpretation of the spectral test that such a hyperplane covering of $\mathcal{L}$ exists (see p. 51). Since the unit cube $[0,1)^d$ has diameter $\sqrt{d}$, it cannot be intersected by more than $\sqrt{d}/\sigma(\mathcal{L})+1$ hyperplanes from this covering $\mathcal{H}^*$. Assume that any of these intersections contains strictly less than $\sigma(\mathcal{L})N/(\sqrt{d}+\sigma(\mathcal{L}))$ lattice points. Since every point of $\mathcal{P}(\mathcal{L})$ lies on exactly one hyperplane of $\mathcal{H}^*$ we then would have

$$N = |\mathcal{P}(\mathcal{L})| = \sum_{H \in \mathcal{H}^*} |\mathcal{P}(\mathcal{L}) \cap H| < \frac{\sigma(\mathcal{L})N}{\sqrt{d}+\sigma(\mathcal{L})}\left(\frac{\sqrt{d}}{\sigma(\mathcal{L})}+1\right) = N,$$

where the summation is extended over all parallel hyperplanes H of the considered covering $\mathcal{H}^*$ of $\mathcal{P}(\mathcal{L})$. This obviously is a contradiction. So, there exists at least one hyperplane that contains at least $\sigma(\mathcal{L})N/(\sqrt{d}+\sigma(\mathcal{L}))$ lattice points. Since the intersection of a hyperplane with the unit cube is convex and has measure zero, this gives the desired lower bound.

Suppose now that additionally $\sigma(\mathcal{L}) \leq 1/2$. Then we show the lower bound on the isotropic discrepancy by finding an empty convex set $C \subseteq [0,1)^d$ of volume at least $c\sigma(\mathcal{L})$ for a suitably chosen absolute constant $c > 0$. This convex set will be constructed by intersecting the slab between two suitable hyperplanes with the unit cube. To this end, consider again the covering $\mathcal{H}^*$ of $\mathcal{P}(\mathcal{L})$ from above and consider the one-dimensional space orthogonal to all hyperplanes in $\mathcal{H}^*$ which is spanned by some $\boldsymbol{h} \in \mathcal{L}^\perp$. The rays emanating from the center of the cube into the directions $\pm\boldsymbol{h}$ hit a hyperplane of $\mathcal{H}^*$ at a distance of at most $\sigma(\mathcal{L})$ from the center of the cube. In this way, we get a pair of adjacent hyperplanes $H_1, H_2 \in \mathcal{H}^*$ sandwiching the center of the cube, with possibly one of them containing the center, and having distance $\sigma(\mathcal{L})$. Denote the collection of all hyperplanes parallel to the family $\mathcal{H}^*$ which lie between H_1 and H_2 by $\widetilde{\mathcal{H}}$ and define the open convex set

$$C := \mathrm{int}(\mathrm{conv}(H_1 \cup H_2) \cap [0,1)^d)$$

(here, by "int" we mean the interior, and by "conv" the convex hull), which does not contain any point of $\mathcal{P}(\mathcal{L})$. The volume of C can be bounded from below by

$$\lambda_d(C) \geq \sigma(\mathcal{L}) \inf_{H \in \widetilde{\mathcal{H}}} \lambda_{d-1}(H \cap [0,1)^d).$$

In order to bound the infimum on the right-hand side of this inequality from below, we make use of [136, Theorem 1.1] due to König and Rudelson, which establishes the existence of a $c > 0$ such that for any $d \in \mathbb{N}$ the $(d-1)$-dimensional volume of the intersection of the cube $[0,1)^d$ with any hyperplane having distance at most $1/2$ from its center is bounded from below by c. However, since in our case all hyperplanes in $\widetilde{\mathcal{H}}$ have a distance of at most $\sigma(\mathcal{L}) \le 1/2$ from the center, this lower bound of König and Rudelson yields that the infimum is bounded from below by some $c > 0$. Thus, we have found an empty convex set C with $\lambda_d(C) \ge c\,\sigma(\mathcal{L})$ and hence also

$$J_N(\mathcal{P}(\mathcal{L})) \ge \lambda_d(C) \ge c\,\sigma(\mathcal{L}).$$

Consequently, in any case the lower bound is proven.

We now prove the upper bound. By means of the LLL-algorithm, see, e.g., [76, Chapter 17], we find a reduced basis $\{\boldsymbol{b}_1, \boldsymbol{b}_2, \ldots, \boldsymbol{b}_d\}$ of the lattice $\mathcal{L}$ containing short near-orthogonal vectors. The definition of a reduced basis requires the Gram-Schmidt orthogonalization $\{\boldsymbol{b}_1^*, \boldsymbol{b}_2^*, \ldots, \boldsymbol{b}_d^*\}$ which is obtained from the lattice basis by setting $\boldsymbol{b}_1^* = \boldsymbol{b}_1$ and

$$\boldsymbol{b}_i^* = \boldsymbol{b}_i - \sum_{j=1}^{i-1} \mu_{i,j} \boldsymbol{b}_j^* \quad \text{for } i \in \{2, 3, \ldots, d\},$$

where

$$\mu_{i,j} = \frac{\boldsymbol{b}_i \cdot \boldsymbol{b}_j^*}{\|\boldsymbol{b}_j^*\|_2^2}.$$

Then, from the properties of a reduced basis it can easily be deduced, see, e.g., [76, Lemma 17.2.8], that

(a) $\|\boldsymbol{b}_j^*\|_2^2 \le 2^{i-j} \|\boldsymbol{b}_i^*\|_2^2$ for $i, j \in [d]$, where $j \le i$, and
(b) $\|\boldsymbol{b}_j\|_2^2 \le 2^{d-1} \|\boldsymbol{b}_j^*\|_2^2$ for $j \in [d]$.

Together these properties imply

$$\|\boldsymbol{b}_d^*\|_2 \ge \frac{1}{2^{(d-1)/2}} \max_{j \in [d]} \|\boldsymbol{b}_j^*\|_2 \ge \frac{1}{2^{d-1}} \max_{j \in [d]} \|\boldsymbol{b}_j\|_2.$$

Consider now the unit cell associated with the lattice basis $\{\boldsymbol{b}_1, \boldsymbol{b}_2, \ldots, \boldsymbol{b}_d\}$, i.e.,

$$\mathcal{U} := \left\{ t_1 \boldsymbol{b}_1 + t_2 \boldsymbol{b}_2 + \cdots + t_d \boldsymbol{b}_d \ : \ t_j \in [0,1) \text{ for all } j \in [d] \right\}, \tag{5.10}$$

which induces a partition of $\mathbb{R}^d$ into disjoint cells $\boldsymbol{x} + \mathcal{U}$, where $\boldsymbol{x} \in \mathcal{L}$. Each of these translated unit cells contains only the lattice point $\boldsymbol{x}$ and has a diameter satisfying

$$\operatorname{diam}(\mathcal{U}) \le \sum_{j=1}^{d} \|\boldsymbol{b}_j\|_2 \le d \max_{j \in [d]} \|\boldsymbol{b}_j\|_2 \le d\, 2^{d-1} \|\boldsymbol{b}_d^*\|_2.$$

By construction we have

$$\|\boldsymbol{b}_d^*\|_2 = \left\| \boldsymbol{b}_d - \sum_{j=1}^{d-1} \mu_{d,j} \boldsymbol{b}_j^* \right\|_2 .$$

That is, the length of the last vector in the Gram-Schmidt orthogonalization is the length of the projection of the vector $\boldsymbol{b}_d$ onto the orthogonal complement of the subspace $\mathrm{span}\{\boldsymbol{b}_1^*, \ldots, \boldsymbol{b}_{d-1}^*\} = \mathrm{span}\{\boldsymbol{b}_1, \ldots, \boldsymbol{b}_{d-1}\}$ spanned by the other basis vectors. However, this is exactly the distance between two adjacent hyperplanes of the family of parallel hyperplanes

$$k\boldsymbol{b}_d + \mathrm{span}\{\boldsymbol{b}_1, \ldots, \boldsymbol{b}_{d-1}\}, \text{ where } k \in \mathbb{Z},$$

which covers the entire lattice $\mathcal{L}$. Therefore, since this distance cannot be larger than the spectral test, we have

$$d\, 2^{d-1}\, \sigma(\mathcal{L}) \geq d\, 2^{d-1}\, \|\boldsymbol{b}_d^*\|_2 \geq \mathrm{diam}(\mathcal{U}). \tag{5.11}$$

Next, we bound the local discrepancy

$$\Delta_{\mathcal{P}(\mathcal{L})}(C) := \frac{|\{n \in \{0, 1, \ldots, N-1\} \,:\, \boldsymbol{x}_n \in C\}|}{N} - \lambda_d(C)$$

with respect to an arbitrary convex set $C \subseteq [0,1]^d$ in terms of the diameter of the unit cell $\mathcal{U}$ given in (5.10). To this end, consider the collection of translated unit cells $\boldsymbol{x} + \mathcal{U}$ with $\boldsymbol{x} \in \mathcal{L}$ that are fully contained in C, and denote their union by W°. Likewise, denote the union of translated unit cells having nonempty intersection with C by $\overline{W}$. Clearly,

$$W^\circ \subseteq C \subseteq \overline{W}.$$

Since $\mathcal{L}$ is an integration lattice, we have, according to what is outlined in Section 1.2 and in particular in Theorem 1.4, that

$$\lambda_d(\mathcal{U}) = \det(\mathcal{L}) = \frac{1}{\det(\mathcal{L}^\perp)} = \frac{1}{N}.$$

Furthermore, every translated unit cell $\boldsymbol{x} + \mathcal{U}$ with $\boldsymbol{x} \in \mathcal{L}$ contains only the lattice point $\boldsymbol{x}$. Hence we have

$$\Delta_{\mathcal{P}(\mathcal{L})}(W^\circ) = \Delta_{\mathcal{P}(\mathcal{L})}(\overline{W}) = 0.$$

This implies

$$\Delta_{\mathcal{P}(\mathcal{L})}(C) \leq \Delta_{\mathcal{P}(\mathcal{L})}(\overline{W}) + \lambda_d(\overline{W} \setminus C) = \lambda_d(\overline{W} \setminus C)$$

and

$$\Delta_{\mathcal{P}(\mathcal{L})}(C) \geq \Delta_{\mathcal{P}(\mathcal{L})}(W^{\circ}) - \lambda_d(C \setminus W^{\circ}) = -\lambda_d(C \setminus W^{\circ}).$$

Therefore we get the fundamental insight that the local discrepancy for C is only influenced by cells intersecting the boundary of C and satisfies

$$|\Delta_{\mathcal{P}(\mathcal{L})}(C)| \leq \max(\lambda_d(\overline{W} \setminus C), \lambda_d(C \setminus W^{\circ})). \tag{5.12}$$

It remains to estimate the maximum of $\lambda_d(\overline{W} \setminus C)$ and $\lambda_d(C \setminus W^{\circ})$. Let, for $\rho > 0$,

$$C_{\rho}^{+} := \{\boldsymbol{x} \in C^c \ : \ \mathrm{dist}(\boldsymbol{x}, C) \leq \rho\}$$

and

$$C_{\rho}^{-} := \{\boldsymbol{x} \in C \ : \ \mathrm{dist}(\boldsymbol{x}, C^c) \leq \rho\}.$$

Here, $\mathrm{dist}(\boldsymbol{x}, A) := \inf_{\boldsymbol{y} \in A} \|\boldsymbol{x} - \boldsymbol{y}\|_2$ for $\boldsymbol{x} \in \mathbb{R}^d$ and $A \subseteq \mathbb{R}^d$, and $\|\cdot\|_2$ is the Euclidean norm on $\mathbb{R}^d$. A helpful side note for better understanding is that

$$C_{\rho}^{+} \cup C_{\rho}^{-} = \{\boldsymbol{x} \in \mathbb{R}^d \ : \ \mathrm{disc}(\boldsymbol{x}, \partial C) \leq \rho\}.$$

We now use a result which states that for $C \subseteq [0, 1]^d$ and $\rho \in [0, 1]$ we have

$$\max(\lambda_d(C_{\rho}^{+}), \lambda_d(C_{\rho}^{-})) \leq 2^{d+3}\rho. \tag{5.13}$$

A proof of this result uses machinery from convex geometry (for example, Steiner's formula for the volume of the Minkowski sum of convex sets) and is beyond the scope of this book. For details we refer to [243].

With the choice $\rho = \rho_{\mathcal{L}} := \mathrm{diam}(\mathcal{U})$ we obviously have

$$\overline{W} \setminus C \subseteq C_{\rho_{\mathcal{L}}}^{+} \quad \text{and} \quad C \setminus W^{\circ} \subseteq C_{\rho_{\mathcal{L}}}^{-}.$$

Hence we obtain from (5.12), (5.13), and (5.11) that

$$|\Delta_{\mathcal{P}(\mathcal{L})}(C)| \leq 2^{d+3}\mathrm{diam}(\mathcal{U}) \leq d\, 2^{2(d+1)}\sigma(\mathcal{L}).$$

Since the convex set C has been chosen arbitrarily, the same inequality also holds for the supremum over all convex sets C. □

Theorem 5.25 shows that in order to bound the isotropic discrepancy of a lattice point set $\mathcal{P}(\mathcal{L})$ it suffices to bound the spectral test of the lattice $\mathcal{L}$. In order to deduce a general lower bound we use Theorem 1.65.

Theorem 5.26 *Let $\mathcal{P}(\mathcal{L})$ be an N-element lattice point set in $[0, 1)^d$. Then we have*

$$J_N(\mathcal{P}(\mathcal{L})) \geq \min\left(\frac{1}{2\sqrt{d}+1}, \frac{c_d}{N^{1/d}}\right),$$

where

$$c_d := \frac{\sqrt{\pi}}{2\sqrt{d}+1}\left(\Gamma\left(\frac{d}{2}+1\right)\right)^{-1/d},$$

and where Γ *denotes the Gamma function.*

If additionally $\sigma(\mathcal{L}) \leq 1/2$, *then* c_d *may be replaced by* $\widetilde{c}_d = (c\sqrt{\pi}/2)\,(\Gamma(d/2+1))^{-1/d}$, *where* $c > 0$ *is the absolute constant in Theorem 5.25.*

Proof If $\sigma(\mathcal{L}) > 1/2$ we obtain from Theorem 5.25 together with the fact that $x \mapsto x/(\sqrt{d}+x)$ is increasing for $x \geq 1/2$, that

$$J_N(\mathcal{P}(\mathcal{L})) \geq \frac{\sigma(\mathcal{L})}{\sqrt{d}+\sigma(\mathcal{L})} \geq \frac{1}{2\sqrt{d}+1}.$$

If $\sigma(\mathcal{L}) \leq 1/2$, then it suffices to combine Theorems 5.25 and 1.65. □

We remark that it is easily seen by means of Stirling's formula for the Gamma function that

$$c_d \sim \sqrt{\frac{\pi\,\mathrm{e}}{2}\frac{1}{d}} \quad \text{as } d \to \infty.$$

In view of Proposition 1.66 it is clear that the order of magnitude $N^{-1/d}$ is best possible for the isotropic discrepancy of integration lattices in dimension d, and this order can even be attained by rank-1 lattice point sets.

Corollary 5.27 *For every dimension d there exists a positive number* C_d, *depending only on d, with the following property. For every prime number N there exists a lattice point* $\boldsymbol{g} \in G_d(N)$ *such that*

$$J(\mathcal{P}(\boldsymbol{g}, N)) \leq \frac{C_d}{N^{1/d}}.$$

Corollary 5.27 is a pure existence result, and explicit constructions are still missing. Only in dimension $d = 2$ an explicit example of a lattice point set with isotropic discrepancy of order $O(N^{-1/2})$ is known, which is once again the Fibonacci lattice point set.

Theorem 5.28 *For* $m \in \mathbb{N}$ *let* F_m *denote the m-th Fibonacci number. Then we have*

$$J_{F_m}(\mathcal{P}((1, F_{m-1}), F_m)) \leq \begin{cases} 4\sqrt{2/F_m} & \text{if } m \text{ is odd,} \\ 4\sqrt{8/F_m} & \text{if } m \text{ is even.} \end{cases}$$

For a proof of this result we refer to [2].

Notes and Remarks

The results on the discrepancy of lattice point sets presented in Section 5.1 are classical. We mainly followed [199, Chapter 5] in our presentation. This material is also discussed in the book [204] by Niederreiter and Winterhof. The relation of two-dimensional lattice point sets to the theory of Diophantine approximation is well known and studied in many papers, see again [199, pp. 121–124] for further information. An extension of Theorem 5.5 to rank-2 lattice rules can be found in [217].

The CBC construction of lattice point sets with a low figure of merit R and therefore with low star-discrepancy was first considered by Joe [120].

Weighted L_p-discrepancy was introduced by Sloan and Woźniakowski in [239], and was first studied in greater detail in [179]; for general information, in particular regarding the relation to QMC integration, we refer to [211, Chapter 9] and [52, Section 3.6]. The study of CBC constructions of lattice point sets with low weighted star-discrepancy was initiated by Joe [121] and continued by Sinescu and Joe [228, 229] and Sinescu [227].

Tractability of the weighted star-discrepancy is studied in greater detail in [1, 40, 55, 111]. See also [211, Chapter 9] and Section 10.2 of the present book.

The isotropic discrepancy is intimately related to numerical integration of functions over convex subsets of the d-dimensional unit cube. The corresponding Koksma–Hlawka type inequality is based on the isotropic discrepancy of the point set used in an integration rule. A result of this flavor was shown by Zaremba in [267, Proposition 2.2]. A general estimate of the isotropic discrepancy in terms of the extreme discrepancy states that for any N-element point set $\mathcal{P}$ in $[0, 1)^d$ we have $D_N(\mathcal{P}) \le J_N(\mathcal{P}) \le 4\, d\, D_N(\mathcal{P})^{1/d}$. The second, nontrivial inequality follows from a result of Niederreiter and Wills [202, 203]; see also [195, pp. 981–982] for a further discussion. The results on isotropic discrepancy of lattice point sets in Section 5.6 are taken from the papers [220, 243]. In these papers appears a slight inaccuracy in the presented lower bounds. The corrected lower bounds in Theorems 5.25 and 5.26 are due to Sonnleitner [242]. Theorem 5.28 is taken from [2]. We further remark that for the special case of rank-1 lattice point sets (in particular for the Fibonacci lattice) the lower bound for isotropic discrepancy of order of magnitude $N^{-1/d}$ (see Theorem 5.26) can already be deduced from a result by Larcher [172] on initial segments of Kronecker sequences, which, however, is proven with different methods and shows that there are large empty rotated boxes.

The relation (5.9) between the spherical cap discrepancy and the isotropic discrepancy was shown in [2]. We already have pointed out that this offers a way to construct point sets on the sphere with low spherical cap discrepancy via point sets in the unit square with low isotropic discrepancy. Note, though, that it is not possible to obtain point sets with optimal spherical cap discrepancy on $\mathbb{S}^2$ via this inequality in the sense that the optimal convergence rate of the spherical cap discrepancy on $\mathbb{S}^2$ is, up to $\log N$ factors, $N^{-3/4}$ (see [10]), whereas the optimal rate of convergence for the isotropic discrepancy in dimension 2 is, as already mentioned, $N^{-2/3}$, also up to $\log N$ factors. Explicit constructions of point sets on the sphere with optimal spherical discrepancy are not known. However, numerical evidence suggests that Fibonacci lattice point sets projected onto the sphere via the Lambert transform yield point sets with optimal spherical cap discrepancy (possibly up to $\log N$ factors).

Chapter 6
Extensible Lattice Point Sets

Let us go back to the definition of a rank-1 lattice point set $\mathcal{P}(\boldsymbol{g}, N)$ with $N \in \mathbb{N}$ and $\boldsymbol{g} \in \mathbb{Z}^d$, which is

$$\mathcal{P}(\boldsymbol{g}, N) = \left\{ \left\{ \frac{k}{N} \boldsymbol{g} \right\} \; : \; k \in \{0, 1, \ldots, N-1\} \right\}. \tag{6.1}$$

By using CBC constructions, as outlined in Chapters 3 and 4, we know that we can extend the dimension of such a point set by concatenating the generating vector $\boldsymbol{g}$ with suitable further components, and it is possible to still obtain point sets with good quality. However, the situation is different if we ask for an extension with respect to the number of points, N. Extending the number of points may be of practical relevance if one wants to improve the accuracy of approximation of an integral by increasing the number of integration nodes without having to discard previously computed function values. In the first place, it is visible from (6.1) that the construction of $\mathcal{P}(\boldsymbol{g}, N)$ crucially depends on the choice of N, so it is not obvious how extensibility with respect to the number of points could be reached in a useful fashion. Secondly, even if this issue is resolved, the question is whether one could extend N in a way such that one obtains larger point sets with good quality, e.g., with respect to their performance in integration rules.

We shall address this issue in this chapter and outline how lattice point sets can be extended in N. Such rules are then usually referred to as extensible lattice point sets or lattice sequences.

6.1 The Definition of Extensible Lattice Point Sets

The question of how to adapt the construction scheme (6.1) such that it can be used for varying values of N was addressed for the first time by Hickernell and Hong in [102], where the following modification was suggested.

J. Dick et al., *Lattice Rules*, Springer Series in Computational Mathematics 58,
https://doi.org/10.1007/978-3-031-09951-9_6

Let $b \geq 2$ be an integer. For any nonnegative integer $k \in \mathbb{N}_0$, we use the base b representation of k, i.e.

$$k = \kappa_0 + \kappa_1 b + \kappa_2 b^2 + \cdots ,$$

where $\kappa_i \in \{0, 1, \ldots, b-1\}$, $i \in \mathbb{N}_0$, are the base b digits of k. We then define the well-known *radical inverse function* in base b, ϕ_b, as $\phi_b : \mathbb{N}_0 \to [0, 1)$,

$$\phi_b(k) := \sum_{i=0}^{\infty} \frac{\kappa_i}{b^{i+1}}.$$

The radical inverse function is prominent in the theory of uniform distribution and numerical integration, as it forms the basis of two classes of infinite point sets, namely *van der Corput sequences* and *Halton sequences* (we refer to, e.g., [52] or [74] for further information). The radical inverse function has the convenient property that, for any $m \in \mathbb{N}_0$,

$$\{\phi_b(k) \ : \ k \in \{0, 1, \ldots, b^m - 1\}\} = \left\{ \frac{k}{b^m} \ : \ k \in \{0, 1, \ldots, b^m - 1\} \right\}. \tag{6.2}$$

In this way, we can construct an extensible lattice point set.

Definition 6.1 (Extensible lattice point set) For a lattice point $\boldsymbol{g} \in \mathbb{Z}^d$ the corresponding *extensible lattice point set* (also called (infinite) lattice sequence) is given by

$$\mathcal{P}(\boldsymbol{g}) := \{\{\phi_b(k)\boldsymbol{g}\} \ : \ k \in \mathbb{N}_0\}. \tag{6.3}$$

If we consider $N = b^m$ for $m \in \mathbb{N}_0$ and fix $\boldsymbol{g} \in \mathbb{Z}^d$, by (6.2), the first N points of $\mathcal{P}(\boldsymbol{g})$ in (6.3) and the point set $\mathcal{P}(\boldsymbol{g}, N)$ in (6.1) coincide. Furthermore, note that, if we consider the base b digits of the components of $\boldsymbol{g}$, only the m least significant digits influence the first b^m points of $\mathcal{P}(\boldsymbol{g})$ since all other digits only contribute integers to $\phi_b(k)\boldsymbol{g}$. Consequently, the first b^m terms of $\mathcal{P}(\boldsymbol{g})$ indeed form a rank-1 lattice point set, and succeeding runs of b^m elements of $\mathcal{P}(\boldsymbol{g})$ form shifted copies of the b^m initial elements (i.e., they are shifted by a constant modulo one). We refer to Section 7.1 for further information on the concept of shifted lattice rules. We summarize these easy observations in the following proposition (see also [103]).

Proposition 6.2 *Let $\ell \in \mathbb{N}_0$ and $m \in \mathbb{N}$, and let $\boldsymbol{g} \in \mathbb{Z}^d$. Moreover, let*

$$\mathcal{P}_{\ell,m} := \{\{\phi_b(\ell b^m + k)\boldsymbol{g}\} \ : \ k \in \{0, 1, \ldots, b^m - 1\}\}$$

be the $(\ell + 1)$-st run of b^m points of the extensible lattice point set $\mathcal{P}(\boldsymbol{g})$ in (6.3). *Then it is true that*

$$\mathcal{P}_{\ell,m} = \left\{\{\phi_b(k)\boldsymbol{g} + \phi_b(\ell) b^{-m-1}\boldsymbol{g}\} \ : \ k \in \{0, 1, \ldots, b^m - 1\}\right\},$$

i.e., $\mathcal{P}_{\ell,m}$ is $\mathcal{P}_{0,m}$ shifted modulo one by $\phi_b(\ell) b^{-m-1}\boldsymbol{g}$.

Proof The result follows from the fact that, as can be checked easily, for the radical inverse function ϕ_b and $\ell \in \mathbb{N}_0$, $m \in \mathbb{N}$, and $k \in \{0, 1, \dots, b^m - 1\}$ it holds that

$$\phi_b(\ell b^m + k) = \phi_b(k) + \frac{\phi_b(\ell)}{b^{m+1}}.$$

□

Remark 6.3 Theoretically, it is possible to modify the definition of rank-r lattice point sets (see Section 1.3) in a way that they can be made extensible. For practical purposes, however, this is not extremely useful, as we then would need to consider r indices $k_1, \dots, k_r$, each of which could or could not tend to infinity, as pointed out in [103]. Therefore, we will restrict ourselves to rank-1 lattice point sets in this chapter.

Having defined extensible lattice point sets in Definition 6.1, the crucial question remains how (and even if) one can find a generating vector $\boldsymbol{g}$ such that at least certain parts of this infinite sequence can be used as integration node sets in QMC rules with reasonably low integration errors. A second question is how to estimate the errors of integration rules using certain parts of consecutive points of an extensible lattice point set. As it turns out, this question is far from trivial. In [102, 103], it is shown how errors can be estimated in certain instances, and practical examples of searching for generating vectors guaranteeing suitable performance of at least some finite lattice point sets contained in extensible lattices are shown. A theoretical breakthrough was made in [105], where it was shown that generating vectors $\boldsymbol{g}$ of extensible rank-1 lattices exist such that common figures of merit like R, P_α, or the discrepancy are reasonably small for the first N points with $N = b, b^2, b^3, \dots$ (for a base $b \geq 2$) and dimensions $d = 1, 2, \dots$. Due to the construction method in (6.3) it is natural to suspect that a usual integer vector $\boldsymbol{g}$, for which each component has a finite base b expansion, is not going to suffice for providing a good lattice sequence if we consider portions of N points with N tending to infinity. This is why the authors of [105] considered the problem of showing the existence of a good generating vector $\boldsymbol{g}$ which lies in a superset of the integers, namely so-called b-adic numbers.

Definition 6.4 For an integer base $b \geq 2$, the set of b-adic numbers (sometimes also referred to as b-adic integers) is defined as the set of formal sums

$$\mathbb{Z}_b := \left\{ z = \sum_{r=0}^{\infty} z_r b^r \ : \ z_r \in \{0, 1, \dots, b-1\} \text{ for } r \in \mathbb{N}_0 \right\}.$$

We call the $z_r \in \{0, 1, \dots, b-1\}$ the digits of $z \in \mathbb{Z}_b$.

Note that $\mathbb{N}_0 \subsetneq \mathbb{Z}_b$, and elements of $\mathbb{N}_0$ are characterized as numbers with at most finitely many digits different from 0. For two nonnegative integers, $z, y \in \mathbb{Z}_b$, we can define $z + y$ in $\mathbb{Z}_b$ like for normal integers, and the operation can be extended to all b-adic numbers, with addition carried out in the usual way. E.g., the inverse of $1 \in \mathbb{Z}_b$ is given by the formal sum

$$\sum_{r=0}^{\infty}(b-1)b^r$$

as

$$\begin{aligned}1 + ((b-1) + (b-1)b + (b-1)b^2 + \cdots) &= 0 + (1 + (b-1))b + (b-1)b^2 + \cdots \\ &= 0 + 0 + (1 + (b-1))b^2 + \cdots \\ &= 0 + 0 + 0 + \cdots .\end{aligned}$$

Using this addition in $\mathbb{Z}_b$, the set forms an abelian group, with 0 as the neutral element, and we also have $\mathbb{Z} \subsetneq \mathbb{Z}_b$.

Remark 6.5 (Measurable structure of $\mathbb{Z}_b$) For $z \in \mathbb{Z}_b \setminus \{0\}$ the b-adic order $\mathrm{ord}_b(z)$ is the smallest index r such that $z_r \neq 0$. The order of 0 is defined as ∞. Then, for $z \in \mathbb{Z}_b \setminus \{0\}$, the b-adic absolute value is $|z|_b := b^{-\mathrm{ord}_b(z)}$ and $|0|_b := 0$. This b-adic absolute value induces a metric, which in turn induces a topology on $\mathbb{Z}_b$. In this way, $\mathbb{Z}_b$ becomes a compact group. Then, let $\widetilde{\mu}_b$ denote the uniquely determined normalized Haar measure on $\mathbb{Z}_b$. We remark that $\widetilde{\mu}_b$ has the following rather simple form. For fixed $m \in \mathbb{N}_0$ and $c_0, \ldots, c_{m-1} \in \{0, 1, \ldots, b-1\}$ the so-called cylinder set $C(c_0, \ldots, c_{m-1}) := \{z \in \mathbb{Z}_b \ : \ z_r = c_r \ \text{ for } \ r \in \{0, 1, \ldots, m-1\}\}$ has measure $\widetilde{\mu}_b(C(c_0, \ldots, c_{m-1})) = b^{-m}$. See also [87, Section 2.4] and the references therein for further information.

We now define

$$\mathcal{Z}_b := \{z \in \mathbb{Z}_b \ : \ \gcd(z, b) = 1\}$$

and will consider the ∞-fold cartesian product $\mathcal{Z}_b^\infty := \mathcal{Z}_b \times \mathcal{Z}_b \times \ldots$ as the candidate set for an ∞-dimensional generating vector $\boldsymbol{g}$ of an ∞-dimensional extensible lattice point set.

We then define another probability measure μ_b on $\mathbb{Z}_b$ by

$$\mu_b(A) := \widetilde{\mu}_b(A | \mathcal{Z}_b),$$

and the corresponding product measure μ_b^∞ on $\mathbb{Z}_b^\infty$. We will use the measure μ_b^∞ in order to show the existence of a generating vector $\boldsymbol{g} \in \mathcal{Z}_b^\infty$ of an ∞-dimensional extensible lattice point set with good properties.

6.2 Existence of Extensible Lattice Point Sets With Good Properties

We will now show that there exist extensible lattice point sets with desirable properties, where we first study the weighted quality measure R given in Definition 3.12. Throughout this section we restrict our considerations to product weights.

Remark 6.6 Before we go into the technical details, we describe a simplified version of the underlying idea. In its simplest form the idea is the following. Let X be a finite set and let $A, B \subseteq X$ be two subsets with $|A|, |B| > |X|/2$. Then it is clear that $A \cap B \neq \emptyset$. Analogously, if $|A^c|, |B^c| < |X|/2$, where $A^c = X \setminus A$, $B^c = X \setminus B$ are the complements of A and B, respectively, then $A \cap B \neq \emptyset$. This can be extended to more than two sets, even a sequence of sets. For instance, for $A, B, C \subseteq X$ with $|A|, |B|, |C| > 2|X|/3$ (or analogously $|A^c|, |B^c|, |C^c| < |X|/3$), then $A \cap B \cap C \neq \emptyset$.

In our context, if X is a set of generating vectors, A is a subset of generating vectors yielding a good error bound for a given number of points b^{m_1} and B is the subset of generating vectors satisfying a good error bound using b^{m_2} points, and each subset contains more than half of all generating vectors in X, then there is at least one generating vector which works for both b^{m_1} and b^{m_2} points.

This idea can be extended to infinite sets X where the number of elements in a set is replaced by a probability measure, say μ, on X. For instance, for two subsets $A, B \subseteq X$ with $\mu(A), \mu(B) > 1/2$ (or analogously $\mu(A^c), \mu(B^c) < 1/2$), we again have $A \cap B \neq \emptyset$. More generally, if $A_1, A_2, \ldots \subseteq X$ are a sequence of subsets such that $\mu(A_1^c) + \mu(A_2^c) + \cdots < 1$, then there is at least one element which lies in all sets A_r, $r \in \mathbb{N}$.

We will use these ideas in the following to show the existence of extensible lattice point sets with good properties.

For an ∞-dimensional vector $\boldsymbol{g} = (g_1, g_2, \ldots) \in \mathcal{Z}_b^\infty$ and a finite, nonempty subset $\mathfrak{u}$ of $\mathbb{N}$ we denote by $\boldsymbol{g}_\mathfrak{u}$ the $|\mathfrak{u}|$-dimensional projection onto those components of $\boldsymbol{g}$ whose indices belong to $\mathfrak{u}$, i.e., $\boldsymbol{g}_\mathfrak{u} = (g_j)_{j \in \mathfrak{u}}$.

The following basic result is due to Hickernell and Niederreiter [105, Lemma 1].

Lemma 6.7 *Let $b \in \mathbb{N}$, $b \geq 2$, be fixed, let $N = b^m$ with $m \in \mathbb{N}$, and let $\boldsymbol{\gamma} = \mathbf{1}$ be product weights which are all equal to 1, i.e., $\gamma_j = 1$ for all $j \in \mathbb{N}$. For every finite, nonempty subset $\mathfrak{u}$ of $\mathbb{N}$ we have*

$$\int_{\mathcal{Z}_b^\infty} R_{N,|\mathfrak{u}|,1,\mathbf{1}}(\boldsymbol{g}_\mathfrak{u}) \,\mathrm{d}\mu_b^\infty(\boldsymbol{g}) \leq \frac{1}{N}(\beta_1 + \beta_2 \log N)^{|\mathfrak{u}|},$$

for absolute constants $\beta_1, \beta_2 > 0$, which are independent of N, $\mathfrak{u}$, and $|\mathfrak{u}|$.

Proof Let $N = b^m$. We have

$$\begin{aligned}\int_{\mathcal{Z}_b^\infty} R_{N,|\mathfrak{u}|,1,\mathbf{1}}(\boldsymbol{g}_\mathfrak{u}) \,\mathrm{d}\mu_b^\infty(\boldsymbol{g}) &= \int_{\mathcal{Z}_b^{|\mathfrak{u}|}} \sum_{\substack{\boldsymbol{h}_\mathfrak{u} \in C_{|\mathfrak{u}|}^*(N) \\ \boldsymbol{h}_\mathfrak{u} \cdot \boldsymbol{g}_\mathfrak{u} \equiv 0 \pmod{N}}} \frac{1}{r_1(\boldsymbol{h}_\mathfrak{u})} \,\mathrm{d}\mu_b^{|\mathfrak{u}|}(\boldsymbol{g}_\mathfrak{u}) \\ &= \frac{1}{|G_{|\mathfrak{u}|}^\varphi(b^m)|} \sum_{\boldsymbol{g}_\mathfrak{u} \in G_{|\mathfrak{u}|}^\varphi(b^m)} R(\boldsymbol{g}_\mathfrak{u}, b^m),\end{aligned}$$

where in the last step we used the simple form of the measure for cylinder sets as discussed in Remark 6.5. From here the claim follows from a classical result according to Niederreiter which was already presented in Remark 5.4 (Equation (5.3)). Here, however, we average over $G^{\varphi}_{|\mathfrak{u}|}(b^m)$ and not over $C^{\varphi}_{|\mathfrak{u}|}(b^m)$ as in Remark 5.4, which only changes the values of the implied constants denoted by β_1 and β_2 here. □

Extensible lattice point sets with low $R_{N,d,\tau,\gamma}$

Lemma 6.7 implies that there must exist some generating vector(s) $\boldsymbol{g} \in \mathcal{Z}_b^{\infty}$ for which the values of $R_{N,|\mathfrak{u}|,1,\mathbf{1}}(\boldsymbol{g}_{\mathfrak{u}})$ are in a certain sense low. The following result, which is one of the main results in [105], implies an analogous existence result for larger smoothness parameters $\tau \geq 1$ and more general product weights, i.e., for $R_{N,d,\tau,\gamma}$.

Theorem 6.8 *Let $b \in \mathbb{N}$, $b \geq 2$, be fixed, and let $\boldsymbol{\gamma} = (\gamma_j)_{j\geq 1}$ be a sequence of product weights. Let $\tau \geq 1$ and $\varepsilon > 0$. Then there exists a μ_b^{∞}-measurable set $B_b \subseteq \mathcal{Z}_b^{\infty}$ of measure strictly less than 1 such that for all $\boldsymbol{g} \in \mathcal{Z}_b^{\infty} \setminus B_b$ we have*

$$R_{N,d,\tau,\gamma}(\boldsymbol{g}_{[d]}) \leq C_{\tau,\gamma,\varepsilon,d,b} \frac{(\log N)^{\tau(d+1)}(\log\log(N+1))^{\tau(1+\varepsilon)}}{N^{\tau}} \tag{6.4}$$

for all $d \in \mathbb{N}$ and all $N = b^m$ with $m \in \mathbb{N}$, where $C_{\tau,\gamma,\varepsilon,d,b}$ is a positive real independent of m. The measure of B_b, $\mu_b^{\infty}(B_b)$, can be made arbitrarily small by choosing $C_{\tau,\gamma,\varepsilon,d,b}$ sufficiently large.

Proof We first define the auxiliary quantity

$$\widehat{R}_{N,d,\tau,\gamma}(\boldsymbol{g}_{[d]}) := \sum_{\emptyset\neq\mathfrak{u}\subseteq[d]} \gamma_{\mathfrak{u}} R_{N,|\mathfrak{u}|,\tau,\mathbf{1}}(\boldsymbol{g}_{\mathfrak{u}}),$$

and show that this quantity dominates $R_{N,d,\tau,\gamma}(\boldsymbol{g}_{[d]})$. Indeed, we have

$$\begin{aligned} R_{N,d,\tau,\gamma}(\boldsymbol{g}_{[d]}) &= \sum_{\substack{\boldsymbol{h}\in C_d^*(N)\\ \boldsymbol{h}\cdot\boldsymbol{g}_{[d]}\equiv 0 \pmod N}} \frac{1}{r_{\tau,\gamma}(\boldsymbol{h})} \\ &= \sum_{\substack{\boldsymbol{h}\in C_d^*(N)\\ \boldsymbol{h}\cdot\boldsymbol{g}_{[d]}\equiv 0 \pmod N}} \gamma_{\mathfrak{u}(\boldsymbol{h})} \prod_{j\in\mathfrak{u}(\boldsymbol{h})} \frac{1}{|h_j|^{\tau}}, \end{aligned}$$

where we remind the reader of the notation $\mathfrak{u}(\boldsymbol{h}) := \{j \in [d] : h_j \neq 0\}$, for $\boldsymbol{h} \in \mathbb{Z}^d$. Consequently,

$$\begin{aligned}
R_{N,d,\tau,\boldsymbol{\gamma}}(\boldsymbol{g}_{[d]}) &= \sum_{\emptyset\neq\mathfrak{u}\subseteq[d]} \gamma_{\mathfrak{u}} \sum_{\substack{\boldsymbol{h}\in C_d^*(N)\\ \boldsymbol{h}\cdot\boldsymbol{g}_{[d]}\equiv 0 \pmod N\\ \mathfrak{u}(\boldsymbol{h})=\mathfrak{u}}} \prod_{j\in\mathfrak{u}} \frac{1}{|h_j|^\tau}\\
&= \sum_{\emptyset\neq\mathfrak{u}\subseteq[d]} \gamma_{\mathfrak{u}} \sum_{\substack{\boldsymbol{h}_{\mathfrak{u}}\in(C_1^*(N))^{|\mathfrak{u}|}\\ \boldsymbol{h}_{\mathfrak{u}}\cdot\boldsymbol{g}_{\mathfrak{u}}\equiv 0 \pmod N}} \prod_{j\in\mathfrak{u}} \frac{1}{|h_j|^\tau}\\
&\le \sum_{\emptyset\neq\mathfrak{u}\subseteq[d]} \gamma_{\mathfrak{u}} R_{N,|\mathfrak{u}|,\tau,\mathbf{1}}(\boldsymbol{g}_{\mathfrak{u}})\\
&= \widehat{R}_{N,d,\tau,\boldsymbol{\gamma}}(\boldsymbol{g}_{[d]}).
\end{aligned}$$

Hence it suffices to show the upper bound in (6.4) for $\widehat{R}_{N,d,\tau,\boldsymbol{\gamma}}(\boldsymbol{g})$.

Note that, for any finite, nonempty $\mathfrak{u}\subseteq\mathbb{N}$, and $\tau\ge 1$, we have

$$\begin{aligned}
R_{N,|\mathfrak{u}|,\tau,\mathbf{1}}(\boldsymbol{g}_{\mathfrak{u}}) &= \sum_{\substack{\boldsymbol{h}_{\mathfrak{u}}\in C_{|\mathfrak{u}|}^*(N)\\ \boldsymbol{h}_{\mathfrak{u}}\cdot\boldsymbol{g}_{\mathfrak{u}}\equiv 0 \pmod N}} \left(\frac{1}{r_{1,\mathbf{1}}(\boldsymbol{h}_{\mathfrak{u}})}\right)^\tau\\
&\le \left(\sum_{\substack{\boldsymbol{h}_{\mathfrak{u}}\in C_{|\mathfrak{u}|}^*(N)\\ \boldsymbol{h}_{\mathfrak{u}}\cdot\boldsymbol{g}_{\mathfrak{u}}\equiv 0 \pmod N}} \frac{1}{r_{1,\mathbf{1}}(\boldsymbol{h}_{\mathfrak{u}})}\right)^\tau\\
&= \left(R_{N,|\mathfrak{u}|,1,\mathbf{1}}(\boldsymbol{g}_{\mathfrak{u}})\right)^\tau,
\end{aligned}$$

where we used the simple fact that $\sum a_j^\tau \le \left(\sum a_j\right)^\tau$ for positive a_j and $\tau\ge 1$. Applying this principle once more leads to

$$\begin{aligned}
\widehat{R}_{N,d,\tau,\boldsymbol{\gamma}}(\boldsymbol{g}_{[d]}) &= \sum_{\emptyset\neq\mathfrak{u}\subseteq[d]} \gamma_{\mathfrak{u}} R_{N,|\mathfrak{u}|,\tau,\mathbf{1}}(\boldsymbol{g}_{\mathfrak{u}})\\
&\le \sum_{\emptyset\neq\mathfrak{u}\subseteq[d]} (\gamma_{\mathfrak{u}}^{1/\tau})^\tau \left(R_{N,|\mathfrak{u}|,1,\mathbf{1}}(\boldsymbol{g}_{\mathfrak{u}})\right)^\tau\\
&\le \left(\sum_{\emptyset\neq\mathfrak{u}\subseteq[d]} \gamma_{\mathfrak{u}}^{1/\tau} R_{N,|\mathfrak{u}|,1,\mathbf{1}}(\boldsymbol{g}_{\mathfrak{u}})\right)^\tau\\
&= \left(\widehat{R}_{N,d,1,\boldsymbol{\gamma}^{1/\tau}}(\boldsymbol{g}_{[d]})\right)^\tau. \qquad (6.5)
\end{aligned}$$

In the next step of the proof, we introduce a further auxiliary quantity, namely

$$\widetilde{R}_{N,d,\tau,\boldsymbol{\gamma}}(\boldsymbol{g}_{[d]}) := \sum_{\substack{\mathfrak{u}\subseteq[d]\\ d\in\mathfrak{u}}} \gamma_{\mathfrak{u}} R_{N,|\mathfrak{u}|,\tau,\mathbf{1}}(\boldsymbol{g}_{\mathfrak{u}}).$$

Using Lemma 6.7 we obtain

$$
\begin{aligned}
&\int_{\mathcal{Z}_b^\infty} \widetilde{R}_{N,d,1,\gamma^{1/\tau}}(\boldsymbol{g}_{[d]})\,\mathrm{d}\mu_b^\infty(\boldsymbol{g})\\
&= \sum_{\substack{\mathfrak{u}\subseteq[d]\\ d\in\mathfrak{u}}} \gamma_{\mathfrak{u}}^{1/\tau} \int_{\mathcal{Z}_b^\infty} R_{N,|\mathfrak{u}|,1,\mathbf{1}}(\boldsymbol{g}_{\mathfrak{u}})\,\mathrm{d}\mu_b^\infty(\boldsymbol{g})\\
&\le \sum_{\substack{\mathfrak{u}\subseteq[d]\\ d\in\mathfrak{u}}} \gamma_{\mathfrak{u}}^{1/\tau} \frac{1}{N}(\beta_1+\beta_2\log N)^{|\mathfrak{u}|}\\
&= \frac{\gamma_d^{1/\tau}}{N}(\beta_1+\beta_2\log N) \sum_{\mathfrak{u}\subseteq[d-1]} \gamma_{\mathfrak{u}}^{1/\tau}(\beta_1+\beta_2\log N)^{|\mathfrak{u}|}\\
&= \frac{\gamma_d^{1/\tau}}{N}(\beta_1+\beta_2\log N)\prod_{j=1}^{d-1}(1+\gamma_j^{1/\tau}(\beta_1+\beta_2\log N)).
\end{aligned}
$$

For short, we write

$$
M_{N,\gamma^{1/\tau},d} := \frac{\gamma_d^{1/\tau}}{N}(\beta_1+\beta_2\log N)\prod_{j=1}^{d-1}(1+\gamma_j^{1/\tau}(\beta_1+\beta_2\log N)).
$$

For a given $\varepsilon > 0$, let $c_0(\varepsilon)$ be chosen sufficiently large such that

$$
\sum_{j=1}^{\infty}\frac{1}{j(\log(j+1))^{1+\varepsilon}} < c_0(\varepsilon),
$$

and define, for $j \in \mathbb{N}$, $c_j = c_j(\varepsilon) := c_0(\varepsilon)j(\log(j+1))^{1+\varepsilon}$. Note that we always have $c_j \ge 1$.

Now, for given N of the form $N = b^m$, $m \in \mathbb{N}$, and given $d \in \mathbb{N}$, we define a set of "bad" generating vectors as

$$
\widetilde{B}_{b,m,d,\tau,\gamma} := \left\{\boldsymbol{g}\in\mathcal{Z}_b^\infty \;:\; \widetilde{R}_{N,d,1,\gamma^{1/\tau}}(\boldsymbol{g}_{[d]}) > c_m c_d M_{N,\gamma^{1/\tau},d}\right\}.
$$

We can bound the measure of $\widetilde{B}_{b,m,d,\tau,\gamma}$ by observing that

$$
\begin{aligned}
\mu_b^\infty(\widetilde{B}_{b,m,d,\tau,\gamma^{1/\tau}})c_m c_d M_{N,\gamma,d} &\le \int_{\widetilde{B}_{b,m,d,\tau,\gamma^{1/\tau}}} \widetilde{R}_{N,d,1,\gamma^{1/\tau},d}(\boldsymbol{g}_{[d]})\,\mathrm{d}\mu_b^\infty(\boldsymbol{g})\\
&\le \int_{\mathcal{Z}_b^\infty} \widetilde{R}_{N,d,1,\gamma^{1/\tau}}(\boldsymbol{g}_{[d]})\,\mathrm{d}\mu_b^\infty(\boldsymbol{g})\\
&\le M_{N,\gamma^{1/\tau},d},
\end{aligned}
$$

which implies $\mu_b^\infty(\widetilde{B}_{b,m,d,\tau,\boldsymbol{\gamma}}) \le 1/(c_m c_d)$. We now define B_b as the union of all "bad" sets, namely

$$B_b = \bigcup_{m\in\mathbb{N}} \bigcup_{d\in\mathbb{N}} \widetilde{B}_{b,m,d,\tau,\boldsymbol{\gamma}}.$$

For the measure of B_b we obtain

$$\begin{aligned}
\mu_b^\infty(B_b) &\le \sum_{m\in\mathbb{N}} \sum_{d\in\mathbb{N}} \mu_b^\infty(\widetilde{B}_{b,m,d,\tau,\boldsymbol{\gamma}}) \\
&\le \sum_{m\in\mathbb{N}} \sum_{d\in\mathbb{N}} \frac{1}{c_m c_d} \\
&= \left(\frac{1}{c_0(\varepsilon)} \sum_{j=1}^{\infty} \frac{1}{j(\log(j+1))^{1+\varepsilon}} \right)^2 \\
&< 1.
\end{aligned}$$

This implies that $\mu_b^\infty(\mathcal{Z}_b^\infty \setminus B_b) > 0$, and in particular the set $\mathcal{Z}_b^\infty \setminus B_b$ is nonempty. Note that $\mathcal{Z}_b^\infty \setminus B_b$ contains those $\boldsymbol{g} \in \mathcal{Z}_b^\infty$ with $\boldsymbol{g} \notin \widetilde{B}_{b,m,d,\tau,\boldsymbol{\gamma}}$ for all $m, d \in \mathbb{N}$. By increasing the value of $c_0(\varepsilon)$ we can even shrink the measure of the "bad" set B_b to an arbitrarily small positive value.

Now, for $\boldsymbol{g} \in \mathcal{Z}_b^\infty \setminus B_b$ and $m, d \in \mathbb{N}$, where $N = b^m$, we obtain

$$\begin{aligned}
\widehat{R}_{N,d,1,\boldsymbol{\gamma}^{1/\tau}}(\boldsymbol{g}_{[d]}) &= \sum_{s=1}^{d} \widetilde{R}_{N,s,1,\boldsymbol{\gamma}^{1/\tau}}(\boldsymbol{g}_{[s]}) \\
&\le c_m \sum_{s=1}^{d} c_s M_{b^m,\boldsymbol{\gamma}^{1/\tau},s} \\
&= c_0(\varepsilon) m (\log(m+1))^{1+\varepsilon} \sum_{s=1}^{d} c_s M_{b^m,\boldsymbol{\gamma}^{1/\tau},s} \\
&\le c_0(\varepsilon) m (\log(m+1))^{1+\varepsilon} \\
&\quad \times \frac{1}{N} \sum_{s=1}^{d} \Bigg(c_0(\varepsilon) s (\log(s+1))^{1+\varepsilon} \gamma_s^{1/\tau} (\beta_1 + \beta_2 \log N) \\
&\quad \times \prod_{j=1}^{s-1} (1 + \gamma_j^{1/\tau} (\beta_1 + \beta_2 \log N)) \Bigg)
\end{aligned}$$

$$
\begin{aligned}
&\leq c_0(\varepsilon)m(\log(m+1))^{1+\varepsilon} \\
&\quad\times\frac{1}{N}\sum_{s=1}^{d}\left(c_0(\varepsilon)s(\log(s+1))^{1+\varepsilon}\gamma_s^{1/\tau}(\beta_1+\beta_2\log N)\right. \\
&\quad\left.\times\prod_{j=1}^{s-1}(1+\gamma_j^{1/\tau}c_0(\varepsilon)j(\log(j+1))^{1+\varepsilon}(\beta_1+\beta_2\log N))\right) \\
&\leq c_0(\varepsilon)m(\log(m+1))^{1+\varepsilon} \\
&\quad\times\frac{1}{N}\prod_{j=1}^{d}(1+\gamma_j^{1/\tau}c_0(\varepsilon)j(\log(j+1))^{1+\varepsilon}(\beta_1+\beta_2\log N)),
\end{aligned}
\tag{6.6}
$$

where we made use of the definitions of the sets $\widetilde{B}_{b,m,j}$, the c_j, and the quantities $M_{b^m,\gamma^{1/\tau},j}$. Recalling that $m = \log_b N$, this implies the existence of a suitable positive real $\widetilde{C}_{\gamma^{1/\tau},\varepsilon,d,b}$, which is independent of N, such that

$$
\widehat{R}_{N,d,1,\gamma^{1/\tau}}(\boldsymbol{g}_{[d]}) \leq \widetilde{C}_{\gamma^{1/\tau},\varepsilon,d,b}\,\frac{(\log N)^{d+1}(\log\log(N+1))^{1+\varepsilon}}{N}
$$

for $N = b^m$ and $m, d \in \mathbb{N}$. Using Inequality (6.5) yields

$$
\begin{aligned}
R_{N,d,\tau,\gamma}(\boldsymbol{g}_{[d]}) &\leq \left(R_{N,d,1,\gamma^{1/\tau}}(\boldsymbol{g}_{[d]})\right)^{\tau} \\
&\leq \left(\widetilde{C}_{\gamma^{1/\tau},\varepsilon,d,b}\,\frac{(\log N)^{d+1}(\log\log(N+1))^{1+\varepsilon}}{N}\right)^{\tau},
\end{aligned}
$$

and this proves the desired result with $C_{\tau,\gamma,\varepsilon,d,b} := \widetilde{C}^{\tau}_{\gamma^{1/\tau},\varepsilon,d,b}$. □

In Theorem 6.8 we still have an unfavorable dependence of the bound on the dimension d. Our next goal is to show that under suitable conditions on the weight sequence $\boldsymbol{\gamma}$, there exist generating vectors of extensible lattice rules for which $\widehat{R}_{N,d,\tau,\gamma,d}$ can be bounded independently of the dimension d, and for which the upper bound converges to zero reasonably fast.

The following theorem is also due to [105].

Theorem 6.9 *Let $b \in \mathbb{N}$, $b \geq 2$, be fixed, and let $\boldsymbol{\gamma} = (\gamma_j)_{j\geq 1}$ be a sequence of product weights. Let $\tau \geq 1$ and $\varepsilon > 0$. If*

$$
\sum_{j=1}^{\infty}\gamma_j^{\lambda/\tau}j(\log(j+1))^{1+\varepsilon} < \infty \quad \textit{for some } \lambda \in [1,\tau], \tag{6.7}
$$

then for any fixed $\delta > 0$ there exists a positive real $C_{\delta,\tau,\lambda,\gamma,\varepsilon}$, which is independent of d and N, and a set $B_{b,\lambda} \subseteq \mathcal{Z}_b^{\infty}$ such that for all $\boldsymbol{g} \in \mathcal{Z}_b^{\infty} \setminus B_{b,\lambda}$ we have

$$R_{N,d,\tau,\gamma}(\boldsymbol{g}_{[d]}) \le \frac{C_{\delta,\tau,\lambda,\gamma,\varepsilon}}{N^{\tau/\lambda-\delta}} \tag{6.8}$$

for all $d \in \mathbb{N}$ and all $N = b^m$ with $m \in \mathbb{N}$. The measure of $B_{b,\lambda}$, $\mu_b^\infty(B_{b,\lambda})$, can be made arbitrarily small by choosing $C_{\delta,\tau,\lambda,\gamma,\varepsilon}$ sufficiently large.

Proof First of all it is easily checked that for $\lambda \ge 1$ we have

$$\widehat{R}_{N,d,\lambda,\gamma^{\lambda/\tau}}(\boldsymbol{g}_{[d]}) \le \widehat{R}_{N,d,1,\gamma^{\lambda/\tau}}(\boldsymbol{g}_{[d]}). \tag{6.9}$$

Similar to the derivation of (6.6), we can identify a "bad set" $B_{b,\lambda} \subseteq \mathcal{Z}_b^\infty$ such that for every $\boldsymbol{g} \in \mathcal{Z}_b^\infty \setminus B_{b,\lambda}$ and $\varepsilon > 0$ we have

$$\begin{aligned}\widehat{R}_{N,d,1,\gamma^{\lambda/\tau}}(\boldsymbol{g}_{[d]}) &\le c_0(\varepsilon)m(\log(m+1))^{1+\varepsilon} \\ &\quad\times \frac{1}{N}\prod_{j=1}^{d}(1+\gamma_j^{\lambda/\tau}c_0(\varepsilon)j(\log(j+1))^{1+\varepsilon}(\beta_1+\beta_2\log N))\end{aligned} \tag{6.10}$$

for $m, d \in \mathbb{N}$ and all N of the form $N = b^m$.

Next we use a slight generalization of (6.5). For $\lambda \in [1, \tau]$ it is again easily checked that

$$\widehat{R}_{N,d,\tau,\gamma}(\boldsymbol{g}_{[d]}) \le \left(\widehat{R}_{N,d,\lambda,\gamma^{\lambda/\tau}}(\boldsymbol{g}_{[d]})\right)^{\tau/\lambda}. \tag{6.11}$$

Now, using (6.10) we obtain for all $\boldsymbol{g} \in \mathcal{Z}_b^\infty \setminus B_{b,\lambda}$ that

$$\begin{aligned}\widehat{R}_{N,d,\tau,\gamma}(\boldsymbol{g}_{[d]}) \le \Bigg(&\frac{c_0(\varepsilon)m(\log(m+1))^{1+\varepsilon}}{N} \\ &\times\prod_{j=1}^{d}(1+\gamma_j^{\lambda/\tau}c_0(\varepsilon)j(\log(j+1))^{1+\varepsilon}(\beta_1+\beta_2\log N))\Bigg)^{\tau/\lambda}.\end{aligned}$$

Let $\delta > 0$ be chosen arbitrarily small but fixed. Then there exists a positive real $\widetilde{c}(\varepsilon, \delta)$ such that

$$c_0(\varepsilon)m(\log(m+1))^{1+\varepsilon} \le \widetilde{c}(\varepsilon,\delta)N^{\delta\lambda/(2\tau)} \quad \text{for all } m \in \mathbb{N}.$$

Furthermore, since (6.7) is assumed to hold, we obtain from Lemma 3.20 that

$$\prod_{j=1}^{d}\left(1+\gamma_j^{\lambda/\tau}c_0(\varepsilon)j(\log(j+1))^{1+\varepsilon}(\beta_1+\beta_2\log N)\right) \le \widehat{c}(\lambda,\tau,\boldsymbol{\gamma},\varepsilon,\delta)N^{\delta\lambda/(2\tau)}$$

for all $d \in \mathbb{N}$ and all $N = b^m$ with $m \in \mathbb{N}$. Here the positive real $\widehat{c}(\lambda,\tau,\boldsymbol{\gamma},\varepsilon,\delta)$ is again independent of m (and also of d). Combining the latter two estimates now yields

$$\widehat{R}_{N,d,\tau,\boldsymbol{\gamma}}(\boldsymbol{g}_{[d]}) \leq \left(\frac{\widetilde{c}(\varepsilon,\delta) N^{\delta\lambda/(2\tau)}\, \widehat{c}(\lambda,\tau,\boldsymbol{\gamma},\varepsilon,\delta) N^{\delta\lambda/(2\tau)}}{N} \right)^{\tau/\lambda}.$$

Since $\widehat{R}_{N,d,\tau,\boldsymbol{\gamma}}$ dominates $R_{N,d,\tau,\boldsymbol{\gamma}}$, as shown in the proof of Theorem 6.8, this yields the desired result with $C_{\delta,\tau,\lambda,\boldsymbol{\gamma},\varepsilon} := (\widetilde{c}(\varepsilon,\delta)\, \widehat{c}(\lambda,\tau,\boldsymbol{\gamma},\varepsilon,\delta))^{\tau/\lambda}$. □

The summability condition (6.7) contains the term $j(\log(j+1))^{1+\varepsilon}$, which somewhat weakens the result. The following theorem, again taken from [105], shows how we can get rid of this extra term in the summability condition on the weights.

Theorem 6.10 *Let $b \in \mathbb{N}$, $b \geq 2$, be fixed, and let $\boldsymbol{\gamma} = (\gamma_j)_{j\geq 1}$ be a sequence of product weights. Let $\tau \geq 1$. Suppose that*

$$\sum_{j=1}^{\infty} \gamma_j^{\lambda/\tau} < \infty \quad \textit{for some } \lambda \in [1,\tau]. \tag{6.12}$$

Then for any fixed $\delta > 0$ there exists a μ_b^∞-measurable set $B_{b,\lambda,\delta} \subseteq \mathcal{Z}_b^\infty$ of measure strictly less than 1, such that for all $\boldsymbol{g} \in \mathcal{Z}_b^\infty \setminus B_{b,\lambda,\delta}$ we have

$$R_{N,d,\tau,\boldsymbol{\gamma}}(\boldsymbol{g}_{[d]}) \leq \frac{C_{\delta,\tau,\lambda,\boldsymbol{\gamma}}}{N^{\tau/\lambda-\delta}} \tag{6.13}$$

for all $d \in \mathbb{N}$ and all $N = b^m$ with $m \in \mathbb{N}$.

Again, the measure of $B_{b,\lambda,\delta}$, $\mu_b^\infty(B_{b,\lambda,\delta})$, can be made arbitrarily small.

Remark 6.11 As already indicated, in Theorem 6.10 we have a less demanding condition on the weights as compared to Theorem 6.9 in order to get an error bound that holds uniformly in d. The price that we have to pay for this improvement is that now the "bad" set $B_{b,\lambda,\delta}$, and therefore also the vectors $\boldsymbol{g} \in \mathcal{Z}_b^\infty \setminus B_{b,\lambda,\delta}$, depend on the chosen value of δ. This is in contrast to Theorem 6.9, where the corresponding set does not depend on δ and where the respective bounds hold simultaneously for all positive δ.

Proof of Theorem 6.10 Let $N = b^m$ for some $m \in \mathbb{N}$. Then, by Lemma 6.7, we have for $d \in \mathbb{N}$, and any finite, nonempty $\mathfrak{u} \subseteq \mathbb{N}$ that

$$\int_{\mathcal{Z}_b^\infty} R_{N,|\mathfrak{u}|,1,\mathbf{1}}(\boldsymbol{g}_{\mathfrak{u}})\, \mathrm{d}\mu_b^\infty(\boldsymbol{g}) \leq \frac{1}{N}(\beta_1 + \beta_2 \log N)^{|\mathfrak{u}|}.$$

Consequently,

$$\begin{aligned}
\int_{\mathcal{Z}_b^\infty} \widehat{R}_{N,d,1,\boldsymbol{\gamma}^{\lambda/\tau}}(\boldsymbol{g}_{[d]})\, \mathrm{d}\mu_b^\infty(\boldsymbol{g}) &= \sum_{\emptyset\neq\mathfrak{u}\subseteq[d]} \gamma_{\mathfrak{u}}^{\lambda/\tau} \int_{\mathcal{Z}_b^\infty} R_{N,|\mathfrak{u}|,1,\mathbf{1}}(\boldsymbol{g}_{\mathfrak{u}})\, \mathrm{d}\mu_b^\infty(\boldsymbol{g}) \\
&\leq \sum_{\emptyset\neq\mathfrak{u}\subseteq[d]} \gamma_{\mathfrak{u}}^{\lambda/\tau} \frac{1}{N}(\beta_1 + \beta_2 \log N)^{|\mathfrak{u}|}
\end{aligned}$$

$$\leq \frac{1}{N} \prod_{j=1}^{d} (1 + \gamma_j^{\lambda/\tau} (\beta_1 + \beta_2 \log N)).$$

Let $\delta > 0$. Since (6.12) holds, and according to Lemma 3.20, there exists a positive real $\widehat{c}_{\delta,\boldsymbol{\gamma}^{\lambda/\tau}}$, which is independent of N and d, such that

$$\prod_{j=1}^{d} \left((1 + \gamma_j^{\lambda/\tau} (\beta_1 + \beta_2 \log N)) \right) < \widehat{c}_{\delta,\boldsymbol{\gamma}^{\lambda/\tau}} N^{\delta\lambda/(2\tau)}$$

for all $d \in \mathbb{N}$. For given and fixed $\varepsilon > 0$ let now $c_j = c_j(\varepsilon)$, $j \in \mathbb{N}$, and $c_0 = c_0(\varepsilon)$ be defined as in the proof of Theorem 6.8. We choose ε equal to 1 from here on. We again define a set of "bad" generating vectors by

$$\widehat{B}_{b,m,d,\lambda,\delta} := \left\{ \boldsymbol{g} \in \mathcal{Z}_b^\infty \ : \ \widehat{R}_{N,d,1,\boldsymbol{\gamma}^{\lambda/\tau}}(\boldsymbol{g}_{[d]}) > c_m M_{N,\boldsymbol{\gamma},\lambda,\delta} \right\}$$

for $m, d \in \mathbb{N}$, where we write $M_{N,\boldsymbol{\gamma},\lambda,\delta} := \widehat{c}_{\delta,\boldsymbol{\gamma}^{\lambda/\tau}} N^{\delta\lambda/(2\tau)-1}$. Analogously to the proof of Theorem 6.8 we have

$$\mu_b^\infty(\widehat{B}_{b,m,d,\lambda,\delta}) c_m M_{N,\boldsymbol{\gamma}} \leq \int_{\widehat{B}_{b,m,d,\lambda,\delta}} \widehat{R}_{N,d,1,\boldsymbol{\gamma}^{\lambda/\tau}}(\boldsymbol{g}_{[d]}) \, \mathrm{d}\mu_b^\infty(\boldsymbol{g}) \leq M_{N,\boldsymbol{\gamma},\lambda,\delta},$$

which then implies $\mu_b^\infty(\widehat{B}_{b,m,d,\lambda,\delta}) \leq 1/c_m$ for all $m, d \in \mathbb{N}$. Note that, by definition, $\widehat{R}_{N,d,1,\boldsymbol{\gamma}^{\lambda/\tau}}$ is nondecreasing in d. So it follows that $\widehat{B}_{b,m,1,\lambda,\delta} \subseteq \widehat{B}_{b,m,2,\lambda,\delta} \subseteq \ldots$. Define

$$\widehat{B}_{b,m,\lambda,\delta} := \bigcup_{d=1}^{\infty} \widehat{B}_{b,m,d,\lambda,\delta}.$$

Due to the monotonicity of the $\widehat{B}_{b,m,d,\lambda,\delta}$ and the fact that $\mu_b^\infty(\widehat{B}_{b,m,d,\lambda,\delta}) \leq 1/c_m$ for all $d \in \mathbb{N}$, it follows that also $\mu_b^\infty(\widehat{B}_{b,m,\lambda,\delta}) \leq 1/c_m$. Now, let

$$B_{b,\lambda,\delta} := \bigcup_{m=1}^{\infty} \widehat{B}_{b,m,\lambda,\delta},$$

then

$$\mu_b^\infty(B_{b,\lambda,\delta}) \leq \sum_{m=1}^{\infty} \mu_b^\infty(\widehat{B}_{b,m,\lambda,\delta}) = \sum_{m=1}^{\infty} \frac{1}{c_m} = \sum_{m=1}^{\infty} \frac{1}{c_0 \, m (\log(1+m))^2} < 1,$$

by the definition of the c_j for $j \in \mathbb{N}$ and c_0, respectively. This implies that $\mu_b^\infty(\mathcal{Z}_b^\infty \setminus B_{b,\lambda,\delta}) > 0$. In particular, $\mathcal{Z}_b^\infty \setminus B_{b,\lambda,\delta}$ is nonempty.

For $\boldsymbol{g} \in \mathcal{Z}_b^\infty \setminus B_{b,\lambda,\delta}$ we therefore have

$$\widehat{R}_{N,d,1,\boldsymbol{\gamma}^{\lambda/\tau}}(\boldsymbol{g}_{[d]}) \leq \frac{c_0 m (\log(m+1))^2 \, \widehat{c}_{\delta,\boldsymbol{\gamma}^{\lambda/\tau}}}{N^{1-\delta\lambda/(2\tau)}}$$

for all $d, m \in \mathbb{N}$ and $N = b^m$. Again, there also exists a positive real $\widetilde{c}_\delta$ such that

$$c_0 m(\log(m+1))^2 \leq \widetilde{c}_\delta N^{\delta\lambda/(2\tau)} \quad \text{for all } m \in \mathbb{N}.$$

Hence, for $\boldsymbol{g} \in \mathcal{Z}_b^\infty \setminus B_{b,\lambda,\delta}$ we have

$$\widehat{R}_{N,d,1,\gamma^{\lambda/\tau}}(\boldsymbol{g}_{[d]}) \leq \frac{\widetilde{c}_\delta\, \widehat{c}_{\delta,\gamma^{\lambda/\tau}}}{N^{1-\delta\lambda/\tau}}$$

for all $d, m \in \mathbb{N}$ and $N = b^m$.

Now we finish the proof like the proof of Theorem 6.9. Using (6.9) and (6.11) we obtain

$$\widehat{R}_{N,d,\tau,\gamma}(\boldsymbol{g}_{[d]}) \leq \left(\widehat{R}_{N,d,1,\gamma^{\lambda/\tau}}(\boldsymbol{g}_{[d]})\right)^{\tau/\lambda} \leq \frac{(\widetilde{c}_\delta\, \widehat{c}_{\delta,\gamma^{\lambda/\tau}})^{\tau/\lambda}}{N^{\tau/\lambda-\delta}}.$$

This proves the desired result with $C_{\delta,\tau,\lambda,\gamma} = (\widetilde{c}_\delta\, \widehat{c}_{\delta,\gamma^{\lambda/\tau}})^{\tau/\lambda}$. □

Extensible lattice point sets with low worst-case error in Korobov spaces

Theorems 6.8–6.10 show the existence of extensible lattice rules for which the quantity $R_{N,d,\tau,\gamma}$ is small for all $d \in \mathbb{N}$ and all N of the form $N = b^m$ with $m \in \mathbb{N}$. Under certain conditions on the weights these bounds hold uniformly in d. A similar result can be obtained in terms of the worst-case error in Korobov spaces of smoothness $\alpha > 1/2$. To this end, we use Proposition 3.13 in Chapter 3, which implies, for product weights, that

$$[\mathrm{err}_{N,d,\alpha,\gamma}(\boldsymbol{g}_{[d]})]^2 \leq R_{N,d,2\alpha,\gamma}(\boldsymbol{g}_{[d]}) + \frac{2^{2\alpha}}{N^{2\alpha}} \prod_{j=1}^{d} (1 + \gamma_j 4\zeta(2\alpha))$$

for all $d \in \mathbb{N}$. This immediately yields the following result.

Corollary 6.12 *By setting $\tau = 2\alpha$, the results in Theorems 6.8–6.10 also hold true, except for an adaption of the respective constants, if $R_{N,d,2\alpha,\gamma,[d]}$ is replaced by $[\mathrm{err}_{N,d,\alpha,\gamma}(\boldsymbol{g}_{[d]})]^2$.*

Extensible lattice point sets with low discrepancy

We can also relate $R_{N,d,1,\mathbf{1}}$ to the (unweighted) star-discrepancy of extensible lattice points. Indeed, using Proposition 5.3 we have

$$D_N^*(\mathcal{P}(\boldsymbol{g}_{\mathfrak{u}}, N)) \leq 1 - \left(1 - \frac{1}{N}\right)^{|\mathfrak{u}|} + \frac{R_{N,|\mathfrak{u}|,1,\mathbf{1}}(\boldsymbol{g}_{\mathfrak{u}})}{2}.$$

Using this estimate, we obtain the following result for the weighted star-discrepancy (this should also be compared with the results outlined in Section 5.3).

Corollary 6.13 *Let $b \in \mathbb{N}$, $b \geq 2$, be fixed, let $\boldsymbol{\gamma} = (\gamma_j)_{j\geq 1}$ be a sequence of product weights, and let $\varepsilon > 0$. Then the following assertions hold.*

1. *There exists a μ_b^∞-measurable set $B_b \subseteq \mathcal{Z}_b^\infty$ of measure strictly less than 1 such that for all $\boldsymbol{g} \in \mathcal{Z}_b^\infty \setminus B_b$ it is true that*

$$D^*_{N,\boldsymbol{\gamma}}(\mathcal{P}(\boldsymbol{g}_{[d]}, N)) \leq C_{\boldsymbol{\gamma},\varepsilon,d} \frac{(\log N)^{d+1}(\log\log(N+1))^{1+\varepsilon}}{N} \tag{6.14}$$

for all $d \in \mathbb{N}$ and all $N = b^m$ with $m \in \mathbb{N}$, where $C_{\boldsymbol{\gamma},\varepsilon,d}$ is a positive real independent of N. The measure of B_b, $\mu_b^\infty(B_b)$, can be made arbitrarily small by choosing $C_{\boldsymbol{\gamma},\varepsilon,d}$ sufficiently large.

2. *If $\sum_{j=1}^\infty \gamma_j(\log(j+1))^{1+\varepsilon} < \infty$, then for any $\delta > 0$ there exists a positive real $C_{\delta,\boldsymbol{\gamma},\varepsilon}$, independent of N and d, such that for all $\boldsymbol{g} \in \mathcal{Z}_b^\infty \setminus B_b$ we have*

$$D^*_{N,\boldsymbol{\gamma}}(\mathcal{P}(\boldsymbol{g}_{[d]}, N)) \leq \frac{C_{\delta,\boldsymbol{\gamma},\varepsilon}}{N^{1-\delta}} \tag{6.15}$$

for all $d \in \mathbb{N}$ and all $N = b^m$ with $m \in \mathbb{N}$. The measure of B_b, $\mu_b^\infty(B_b)$, can be made arbitrarily small by choosing $C_{\delta,\boldsymbol{\gamma},\varepsilon}$ sufficiently large.

3. *If $\sum_{j=1}^\infty \gamma_j < \infty$, then for any fixed $\delta > 0$ there exists a μ_b^∞-measurable set $B_{b,\delta} \subseteq \mathcal{Z}_b^\infty$ such that for all $\boldsymbol{g} \in \mathcal{Z}_b^\infty \setminus B_{b,\delta}$ the bound (6.15) holds. Again, the measure of $B_{b,\delta}$, $\mu_b^\infty(B_{b,\delta})$, can be made arbitrarily small by choosing $C_{\delta,\boldsymbol{\gamma},\varepsilon}$ sufficiently large.*

Why do we need N of the form b^m?

As a final remark in this section, we briefly discuss the question why results like Theorem 6.8–6.10 "only" hold for N of the form $N = b^m$, i.e., for a geometric progression of N, and not, e.g., for an arithmetic progression or even all sufficiently large N? The answer to this question is surprisingly simple. As already pointed out by Sobol' in [241] (see also [216] for a more detailed discussion), a sequence of equal-weight quadrature rules that are embedded into each other can—at least in nontrivial settings—not have a better error convergence rate than N^{-1} for an arithmetic sequence of N.

To be more precise, following [104], suppose that we have a Hilbert space $\mathcal{H}$ of d-variate functions defined on $[0,1]^d$ with the following property. For any fixed number M, an M-point QMC rule using any point set $\mathcal{P}$ has a positive worst-case error, and this error is bounded below away from zero no matter how well one chooses the set $\mathcal{P}$, i.e.,

$$\begin{aligned}\inf_{\substack{\mathcal{P}\subseteq[0,1]^d\\|\mathcal{P}|=M}} \mathrm{err}_{M,d}(\mathcal{H},\mathcal{P}) &= \inf_{\substack{\mathcal{P}\subseteq[0,1]^d\\|\mathcal{P}|=M}} \sup_{\substack{f\in\mathcal{H}\\\|f\|_{\mathcal{H}}\le 1}} \left|\int_{[0,1]^d} f(\boldsymbol{x})\,\mathrm{d}\boldsymbol{x} - \frac{1}{M}\sum_{\boldsymbol{y}\in\mathcal{P}} f(\boldsymbol{y})\right| \\ &\ge c_{\mathcal{H},M} \\ &> 0, \end{aligned} \tag{6.16}$$

where $c_{\mathcal{H},M}$ may naturally depend on $\mathcal{H}$ and M. This assumption is not very restrictive, since if it is violated one can get an arbitrarily accurate approximation of an integral by evaluating the integrand at a finite number of points.

Now the following proposition holds.

Proposition 6.14 *Let $\mathcal{S} = (\boldsymbol{x}_k)_{k\ge 0}$ be any sequence of points in $[0,1]^d$, and let $\mathcal{S}_N$ denote the point set consisting of the first N terms of $\mathcal{S}$.*

- *If assumption* (6.16) *holds for $M = 1$, then the sequence $(\mathrm{err}_{N,d}(\mathcal{H},\mathcal{S}_N))_{N\ge 1}$ cannot converge to zero faster than $O(N^{-1})$.*
- *If assumption* (6.16) *holds for some $M > 1$, then the subsequence of errors $(\mathrm{err}_{N_0+nM,d}(\mathcal{H},\mathcal{S}_{N_0+nM}))_{n\ge 0}$ cannot converge to zero faster than $O(N^{-1})$, with N of the form $N = N_0 + nM$, for any fixed $N_0 \in \mathbb{N}$.*

Proof The proof for $M = 1$ and arbitrary $M \ge 1$ are the same. For any f in the unit ball of $\mathcal{H}$ and any $N \in \mathbb{N}$, the error of the QMC rule using the point set $\mathcal{S}_N = \{\boldsymbol{x}_0, \boldsymbol{x}_1, \ldots, \boldsymbol{x}_{N-1}\}$ is

$$\mathrm{err}_{N,d}(f,\mathcal{S}_N) = \int_{[0,1]^d} f(\boldsymbol{x})\,\mathrm{d}\boldsymbol{x} - \frac{1}{N}\sum_{k=0}^{N-1} f(\boldsymbol{x}_k).$$

Thus, for any $N_0, n \in \mathbb{N}$, and $N = N_0 + nM$, we have

$$\begin{aligned} &M\left|\int_{[0,1]^d} f(\boldsymbol{x})\,\mathrm{d}\boldsymbol{x} - \frac{1}{M}\sum_{k=N}^{N+M-1} f(\boldsymbol{x}_k)\right| \\ &= |(N+M)\mathrm{err}_{N+M,d}(f,\mathcal{S}_{N+M}) - N\mathrm{err}_{N,d}(f,\mathcal{S}_N)| \\ &\le (N+M)|\mathrm{err}_{N+M,d}(f,\mathcal{S}_{N+M})| + N|\mathrm{err}_{N,d}(f,\mathcal{S}_N)|, \end{aligned}$$

which leads to

$$\begin{aligned} &M \sup_{\substack{f\in\mathcal{H}\\\|f\|_{\mathcal{H}}\le 1}} \left|\int_{[0,1]^d} f(\boldsymbol{x})\,\mathrm{d}\boldsymbol{x} - \frac{1}{M}\sum_{k=N}^{N+M-1} f(\boldsymbol{x}_k)\right| \\ &\le (N+M)\,\mathrm{err}_{N+M,d}(\mathcal{H},\mathcal{S}_{N+M}) + N\,\mathrm{err}_{N,d}(\mathcal{H},\mathcal{S}_N). \end{aligned}$$

Suppose there exist constants $C > 0$ and $\beta > 0$ such that

$$\mathrm{err}_{N,d}(\mathcal{H},\mathcal{S}_N) \le \frac{C}{N^\beta} \qquad \text{for all} \quad N = N_0 + nM \quad \text{with} \quad n \in \mathbb{N}_0.$$

Then assumption (6.16) implies that

$$
\begin{aligned}
M\, c_{\mathcal{H},M} &\le M \sup_{\substack{f\in\mathcal{H} \\ \|f\|_{\mathcal{H}}\le 1}} \left| \int_{[0,1]^d} f(\boldsymbol{x})\,\mathrm{d}\boldsymbol{x} - \frac{1}{M}\sum_{k=N}^{N+M-1} f(\boldsymbol{x}_k) \right| \\
&\le \frac{C}{(N+M)^{\beta-1}} + \frac{C}{N^{\beta-1}}.
\end{aligned}
$$

Since the left-hand side of the above inequality cannot converge to 0 as N increases, it is necessary that $\beta \le 1$. □

Due to Proposition 6.14, it is, for common spaces of interest, useless to hope to find a sequence of embedded quadrature rules that yield a better convergence order than $O(N^{-1})$ for an arithmetic progression of N. One remedy to solve this issue is to modify the integration rules by changing the integration weights, and to move away from equal-weight rules. This procedure is outlined in detail in [104], but is beyond the scope of this chapter.

6.3 Constructions of Extensible Lattice Rules—Embedded Lattice Rules

The results in Theorems 6.8–6.10 and Corollaries 6.12–6.13 are of great significance from a theoretical point of view. However, they do not imply any clue on how good generating vectors for extensible lattice rules can practically be found. In particular, it is a challenging problem that, as pointed out above, a generating vector with finite base b expansions of its components cannot be expected to yield extensible lattice rules of good quality.

Therefore, various authors have resorted to constructing not extensible (rank-1) lattice rules, but a suitable substitute which is referred to as embedded lattice rules. The idea of an embedded rule is that one tries to find a finite sequence of integration lattices with sizes b^m where m is in a certain range $\{\underline{m}, \underline{m}+1, \ldots, \overline{m}\}$. Furthermore, for practical constructions, one considers only a finite range of components, i.e., d is limited to the set $[d_{\max}]$ for some $d_{\max} \in \mathbb{N}$. To be more precise, an embedded lattice is not only a finite sequence of integration lattices, but—as the name suggests—a sequence of integration lattices of different sizes, where smaller lattices are embedded in larger ones. To this end, for $m \in \mathbb{N}_0$ and $\boldsymbol{g} \in \mathbb{Z}^d$, and for a base $b \in \mathbb{N}$, $b \ge 2$, we define

$$
\mathcal{P}_m := \mathcal{P}(\boldsymbol{g}, b^m) = \left\{ \left\{ \frac{k}{b^m}\boldsymbol{g} \right\} \,:\, k \in \{0, 1, \ldots, b^m - 1\} \right\}. \tag{6.17}
$$

Then, by definition, each $\mathcal{P}_m$ is a rank-1 lattice point set of size b^m, and we have the relation $\mathcal{P}_m \subseteq \mathcal{P}_{m+1}$. However, there is even more structure, namely $\mathcal{P}_m$ contains exactly those points of $\mathcal{P}_{m+1}$ whose indices $k \in \{0, 1, \ldots, b^{m+1} - 1\}$ in (6.17) are multiples of b. Furthermore, if we restrict our consideration to one $\mathcal{P}_m$ for a fixed

$m \in \mathbb{N}_0$, it is sufficient to consider instead of $\boldsymbol{g}$ the vector $\boldsymbol{g}^{(m)}$, where all components of $\boldsymbol{g}$ are reduced modulo b^m. In general, one can of course not assume that, if $\mathcal{P}_m$ is of high quality with respect to a certain figure of merit for a certain $m \in \mathbb{N}_0$, the $\mathcal{P}_k$ contained in $\mathcal{P}_m$ with $k < m$ are of high quality as well. This is exactly one of the crucial ideas of finding good embedded lattice rules, namely to find a whole sequence of lattices $(\mathcal{P}_{\underline{m}}, \mathcal{P}_{\underline{m}+1}, \ldots, \mathcal{P}_{\overline{m}})$ of the form (6.17) such that there is a certain high quality for all m between $\underline{m}$ and $\overline{m}$.

A first attempt to find good embedded lattice rules was presented by Hickernell, Hong, L'Ecuyer, and Lemieux [103], who restricted themselves to Korobov type lattice points of the form $\boldsymbol{g}_d(g) = (1, g, g^2, \ldots, g^{d-1})$ for $d \in [d_{\max}]$ and a suitable integer g. They designed a quality measure (based on a kind of L_2-discrepancy) whose minimizer over all integers g in a finite range is guaranteed to be of excellent quality for all $m \in \{\underline{m}, \underline{m}+1, \ldots, \overline{m}\}$ and $d \in [d_{\max}]$. This exactly corresponds to the idea of embedded lattice rules, as outlined above.

Later, Cools, Kuo, and Nuyens [27] extended this approach. They used the same idea as in [103] in order to construct a quality measure—now based on the worst-case error—but they used the CBC-construction for general rank-1 lattice point sets instead of Korobov type lattice points only. For the sake of simplicity and to make the results comparable to those in earlier chapters, we restrict ourselves to outlining the theory only for the weighted Korobov space with smoothness parameter α and weights $\boldsymbol{\gamma}$, i.e., we consider embedded lattice rules that can particularly be used for numerically integrating periodic functions. Similar results to the ones we will outline below exist for nonperiodic functions in certain Sobolev spaces, where one adds a random shift to the lattice rules under consideration (see Section 7.1).

CBC construction of embedded rules

For fixed $m \in \{\underline{m}, \underline{m}+1, \ldots, \overline{m}\}$ let $\boldsymbol{g}_0^{(m)}$ be the generating vector found by the usual CBC algorithm, Algorithm 3.6. The worst-case errors $\mathrm{err}_{b^m,d,\alpha,\boldsymbol{\gamma}}(\boldsymbol{g}_0^{(m)})$ for $d \in [d_{\max}]$ can be stored in a look-up table. Based on the idea in [103], we can then define, for a generating vector $\boldsymbol{g} \in \mathbb{Z}^d$, the quantity

$$\widetilde{D}_{\underline{m},\overline{m},d}(\boldsymbol{g}) := \max_{\underline{m} \le m \le \overline{m}} \frac{\mathrm{err}_{b^m,d,\alpha,\boldsymbol{\gamma}}(\boldsymbol{g})}{\mathrm{err}_{b^m,d,\alpha,\boldsymbol{\gamma}}(\boldsymbol{g}_0^{(m)})}. \tag{6.18}$$

Based on this quantity, we can now formulate a CBC algorithm for the construction of embedded lattice rules in the weighted Korobov space.

Algorithm 6.15 (CBC construction of embedded rules) Let b and $d_{\max}$ be given, and let the values $\mathrm{err}_{N,d,\alpha,\boldsymbol{\gamma}}(\boldsymbol{g}_0^{(m)})$ for $m \in \{\underline{m}, \underline{m}+1, \ldots, \overline{m}\}$ and $d \in [d_{\max}]$ be stored in a look-up table. Construct a generating vector $\boldsymbol{g} = (g_1, \ldots, g_{d_{\max}}) \in G_{d_{\max}}^{\varphi}(b^{\overline{m}})$ as follows.

(1) Choose $g_1 = 1$.
(2) For s from 1 to $d_{\max} - 1$:
Assume that $g_1, \ldots, g_s \in G_1^{\varphi}(b^{\overline{m}})$ have already been found. Choose $g_{s+1} \in G_1^{\varphi}(b^{\overline{m}})$ as

$$g_{s+1} := \underset{g \in G_1^{\varphi}(b^{\overline{m}})}{\operatorname{argmin}} \; \widetilde{D}_{\underline{m},\overline{m},s+1}((g_1, \ldots, g_s, g)).$$

End for.
(3) Set $\boldsymbol{g} = (g_1, \ldots, g_{d_{\max}})$.

Remark 6.16 We remark that, as long as the precomputed data of $\mathrm{err}_{N,d,\alpha,\boldsymbol{\gamma}}(\boldsymbol{g}_0^{(m)})$ are available, Algorithm 6.15 can be extended in the dimension $d_{\max}$, i.e., the limitation to $d \in [d_{\max}]$ is of practical, but not of theoretical relevance.

Computational cost

We analyze the computational cost of Algorithm 6.15 for the most common form of weights, namely product weights. As outlined in Section 3.4, the cost of running Algorithm 3.6 once is of order $O(d\, N \log N)$ for given N. For the precomputations necessary to run Algorithm 6.15, we therefore need $O(d\, b^m m)$ operations for each instance of $m \in \{\underline{m}, \underline{m}+1, \ldots, \overline{m}\}$, so in total we require $O(d\, b^{\overline{m}} \overline{m}^2)$ operations for the precomputations for Algorithm 6.15. Next, we analyze the cost of actually running Algorithm 6.15.

To this end, it is necessary to consider the fast CBC construction of generating vectors for a composite number N of points in greater detail. We will outline some crucial ideas for the special case where N is the power of a prime b here, as this is the case relevant for the construction of embedded lattice rules. We closely study the matrix

$$\boldsymbol{\Omega}_N := \left(\varphi_\alpha\left(\left\{\frac{kg}{N}\right\}\right)\right)_{g \in G_1^{\varphi}(N), k \in G_1(N)},$$

where φ_α is defined as in (3.7). While we assumed N to be a prime in Section 3.4, we are now in the situation that $N = b^m$ is a prime power. We start our discussion by giving an example, inspired by [27]. To this end, we use the same notation as in Section 3.4.

Example 6.17 Let $N = 2^4 = 16$, i.e., $b = 2$, and $m = 4$. Hence, in our example the cardinality of $G_1^{\varphi}(N) = G_1^{\varphi}(16)$ is 8, so $\boldsymbol{\Omega}_N$ is an 8×16 matrix (in the case where N is a prime, $\boldsymbol{\Omega}_N$ is always an $(N-1) \times N$ matrix). Like in Section 3.4, we will use a special circulant structure of $\boldsymbol{\Omega}_N$ which allows for fast matrix-vector multiplication. Again, we will use the Rader transform, so we first require a primitive root of the multiplicative group of integers modulo N. In the case of $b = 2$, it is known that we can express this group in terms of the generator 5, to be more precise, the group can be written by separating it into two subgroups,

$$
\begin{aligned}
G_1^{\varphi}(16) &= \{1,3,5,7,9,11,13,15\} \\
&= \{5^i \pmod{16} \ : \ i \in \{0,1,2,3\}\} \cup \{-5^i \pmod{16} \ : \ i \in \{0,1,2,3\}\} \\
&= \{1,5,9,13\} \cup \{15,11,7,3\}.
\end{aligned}
$$

We can take the ordering of the elements of $G_1^{\varphi}(16)$ as in the last row of the latter expression for reordering the columns and rows of $\mathbf{\Omega}_{16}$. Obviously, the first column of $\mathbf{\Omega}_{16}$ consists only of the entries $\varphi_\alpha(0)$ and can therefore (trivially) be neglected. As the second column, we obtain

$$
\left(\varphi_\alpha\left(\frac{1}{16}\right), \varphi_\alpha\left(\frac{5}{16}\right), \varphi_\alpha\left(\frac{9}{16}\right), \varphi_\alpha\left(\frac{13}{16}\right), \right.
\left. \varphi_\alpha\left(\frac{15}{16}\right), \varphi_\alpha\left(\frac{11}{16}\right), \varphi_\alpha\left(\frac{7}{16}\right), \varphi_\alpha\left(\frac{3}{16}\right)\right)^\top .
$$

Let us now consider the submatrix Ξ_{16} of $\mathbf{\Omega}_{16}$ that corresponds to the first 8 nontrivial columns (Columns 2-9) of $\mathbf{\Omega}_{16}$. This matrix can be arranged such that it has, though no circulant structure, a "block-circulant-with-circulant-block" structure, namely Ξ_{16} can be written in the form

$$
\Xi_{16} = \left(\begin{array}{c|c} X_{16}^{(1)} & X_{16}^{(2)} \\ \hline X_{16}^{(2)} & X_{16}^{(1)} \end{array}\right),
$$

where

$$
X_{16}^{(1)} = \begin{pmatrix}
\varphi_\alpha(1/16) & \varphi_\alpha(13/16) & \varphi_\alpha(9/16) & \varphi_\alpha(5/16) \\
\varphi_\alpha(5/16) & \varphi_\alpha(1/16) & \varphi_\alpha(13/16) & \varphi_\alpha(9/16) \\
\varphi_\alpha(9/16) & \varphi_\alpha(5/16) & \varphi_\alpha(1/16) & \varphi_\alpha(13/16) \\
\varphi_\alpha(13/16) & \varphi_\alpha(9/16) & \varphi_\alpha(5/16) & \varphi_\alpha(1/16)
\end{pmatrix}
$$

and

$$
X_{16}^{(2)} = \begin{pmatrix}
\varphi_\alpha(15/16) & \varphi_\alpha(3/16) & \varphi_\alpha(7/16) & \varphi_\alpha(11/16) \\
\varphi_\alpha(11/16) & \varphi_\alpha(15/16) & \varphi_\alpha(3/16) & \varphi_\alpha(7/16) \\
\varphi_\alpha(7/16) & \varphi_\alpha(11/16) & \varphi_\alpha(15/16) & \varphi_\alpha(3/16) \\
\varphi_\alpha(3/16) & \varphi_\alpha(7/16) & \varphi_\alpha(11/16) & \varphi_\alpha(15/16)
\end{pmatrix}.
$$

As we can see, both $X_{16}^{(1)}$ and $X_{16}^{(2)}$ are circulant matrices. Thus, Ξ_{16} has much structure that can be made use of in calculating the matrix product of $\mathbf{\Omega}_{16}$ and the vector $\boldsymbol{\eta}_s$ in the s-th step of the CBC algorithm (see Section 3.4). However, there is even more structure that can be exploited. A convenient property of the function φ_α defined in (3.7) is that it is symmetric in the sense that

$$
\varphi_\alpha(x) = \varphi_\alpha(1-x) \quad \text{for any } x \in [0,1].
$$

This implies that $\varphi_\alpha(k/N) = \varphi_\alpha((N-k)/N)$ for $k \in \{0,1,\ldots,N\}$, which in turn implies $\varphi_\alpha(k/N) = \varphi_\alpha(\min(k, N-k)/N)$. Hence, instead of studying Ξ_{16}, we can restrict ourselves to considering

$$\widehat{\Xi}_{16} = \begin{pmatrix} \widehat{X}_{16} \mid \widehat{X}_{16} \\ \hline \widehat{X}_{16} \mid \widehat{X}_{16} \end{pmatrix},$$

where

$$\widehat{X}_{16} = \begin{pmatrix} \varphi_\alpha(1/16) & \varphi_\alpha(3/16) & \varphi_\alpha(7/16) & \varphi_\alpha(5/16) \\ \varphi_\alpha(5/16) & \varphi_\alpha(1/16) & \varphi_\alpha(3/16) & \varphi_\alpha(7/16) \\ \varphi_\alpha(7/16) & \varphi_\alpha(5/16) & \varphi_\alpha(1/16) & \varphi_\alpha(3/16) \\ \varphi_\alpha(3/16) & \varphi_\alpha(7/16) & \varphi_\alpha(5/16) & \varphi_\alpha(1/16) \end{pmatrix}. \tag{6.19}$$

Note that $\widehat{\Xi}_{16}$ is a circulant matrix.

We have now described how the first 8 nontrivial columns (Columns 2-9) of $\mathbf{\Omega}_{16}$ can be structured. The analogous question regarding the remaining columns will be answered in Theorem 6.18 in a general result.

Structure analysis of $\mathbf{\Omega}_N$ if N is a power of 2

Returning to the general case, we first consider the case where N is a power of 2. Note that it is well known from number theory that the group of primitive residue classes modulo a power of 2, say 2^m for $m \geq 3$, is not cyclic anymore. However, the powers $5, 5^2, 5^3, \ldots, 5^{2^{m-2}}$ constitute half of a reduced residue system modulo 2^m, namely all integers which are congruent to $-1 \pmod 4$. The missing residue classes are represented by $-5, -5^2, -5^3, \ldots, -5^{2^{m-2}}$ (see, e.g., [181, Theorem 4.6]).

The following theorem is due to Nuyens and Cools (see [215] and also [27]).

Theorem 6.18 *Let $N = 2^m$ with $m \in \mathbb{N}$, and let*

$$\mathbf{\Omega}_{2^m} = \left(\varphi_\alpha\left(\left\{\frac{kg}{N}\right\}\right)\right)_{g \in G_1^\varphi(N), k \in G_1(N)}.$$

Then the rows and columns of $\mathbf{\Omega}_{2^m}$ can be reordered such that we obtain a matrix

$$\widehat{\mathbf{\Omega}}_{2^m} = \left(\mathbf{1}_{2^0} \otimes \widehat{\Xi}_{2^m} | \mathbf{1}_{2^1} \otimes \widehat{\Xi}_{2^{m-1}} | \cdots | \mathbf{1}_{2^{m-1}} \otimes \widehat{\Xi}_2 | \mathbf{1}_{2^{m-1}} \otimes \widehat{\Xi}_1\right).$$

Here, the notation $\mathbf{1}_\ell \otimes A$ means the stacking of ℓ instances of a matrix A. Furthermore, $\widehat{\Xi}_1 = (\varphi_\alpha(0))$, $\widehat{\Xi}_2 = (\varphi_\alpha(1/2))$, and

$$\widehat{\Xi}_{2^\ell} := \begin{pmatrix} \widehat{X}_{2^\ell} \mid \widehat{X}_{2^\ell} \\ \hline \widehat{X}_{2^\ell} \mid \widehat{X}_{2^\ell} \end{pmatrix}$$

for $\ell \geq 2$, with

$$\widehat{X}_{2^\ell} = \left(\varphi_\alpha\left(\left\{\frac{kg}{N}\right\}\right)\right)_{g \in \langle 5 \rangle_{2^\ell}, k \in \langle -5 \rangle_{2^\ell}},$$

where

$$\langle 5 \rangle_{2^\ell} = \left\{5^i \pmod{2^\ell} \;:\; 0 \leq i \leq 2^{\ell-2} - 1\right\}$$

and

$$\langle -5\rangle_{2^\ell} = \left\{-5^i \pmod{2^\ell} \; : \; 0 \le i \le 2^{\ell-2}-1\right\}.$$

In particular, each of the matrices $\widehat{\boldsymbol{\Xi}}_{2^\ell}$ *is circulant.*

Remark 6.19 Note that what we considered as the first (trivial) column of $\boldsymbol{\Omega}_{2^m}$ (the column containing only $\varphi_\alpha(0)$) is written as the last column of $\widehat{\boldsymbol{\Omega}}_{2^m}$ in Theorem 6.18 above. This is merely for technical reasons. Note furthermore that the number of matrices $\widehat{\boldsymbol{\Xi}}_{2^\ell}$ corresponds to the number of divisors of 2^m, which is $m+1$.

Using Theorem 6.18, we can return to our previous example of $\boldsymbol{\Omega}_{16}$.

Example 6.20 (Example 6.17 revisited) The matrix $\boldsymbol{\Omega}_{16}$ can be restructured to

$$\widehat{\boldsymbol{\Omega}}_{16} = \left(\begin{array}{c|c|c|c|c} & & & \widehat{\boldsymbol{\Xi}}_2 & \widehat{\boldsymbol{\Xi}}_1 \\ & & & \widehat{\boldsymbol{\Xi}}_2 & \widehat{\boldsymbol{\Xi}}_1 \\ & & \widehat{\boldsymbol{\Xi}}_4 & \widehat{\boldsymbol{\Xi}}_2 & \widehat{\boldsymbol{\Xi}}_1 \\ & \widehat{\boldsymbol{\Xi}}_8 & \widehat{\boldsymbol{\Xi}}_4 & \widehat{\boldsymbol{\Xi}}_2 & \widehat{\boldsymbol{\Xi}}_1 \\ \widehat{\boldsymbol{\Xi}}_{16} & \widehat{\boldsymbol{\Xi}}_8 & \widehat{\boldsymbol{\Xi}}_4 & \widehat{\boldsymbol{\Xi}}_2 & \widehat{\boldsymbol{\Xi}}_1 \\ & & \widehat{\boldsymbol{\Xi}}_4 & \widehat{\boldsymbol{\Xi}}_2 & \widehat{\boldsymbol{\Xi}}_1 \\ & & & \widehat{\boldsymbol{\Xi}}_2 & \widehat{\boldsymbol{\Xi}}_1 \\ & & & \widehat{\boldsymbol{\Xi}}_2 & \widehat{\boldsymbol{\Xi}}_1 \end{array}\right).$$

Here,

$$\widehat{\boldsymbol{\Xi}}_{16} = \begin{pmatrix} \widehat{X}_{16} & | & \widehat{X}_{16} \\ \widehat{X}_{16} & | & \widehat{X}_{16} \end{pmatrix}, \quad \text{with} \quad \widehat{X}_{16} \text{ as in (6.19)},$$

$$\widehat{\boldsymbol{\Xi}}_8 = \begin{pmatrix} \widehat{X}_8 & | & \widehat{X}_8 \\ \widehat{X}_8 & | & \widehat{X}_8 \end{pmatrix}, \quad \text{with} \quad \widehat{X}_8 = \begin{pmatrix} \varphi_\alpha(1/8) & \varphi_\alpha(5/8) \\ \varphi_\alpha(5/8) & \varphi_\alpha(1/8) \end{pmatrix} = \begin{pmatrix} \varphi_\alpha(2/16) & \varphi_\alpha(6/16) \\ \varphi_\alpha(6/16) & \varphi_\alpha(2/16) \end{pmatrix},$$

where we remind the reader that $\varphi_\alpha(10/16) = \varphi_\alpha(6/16)$ due to the symmetry of φ_α,

$$\widehat{\boldsymbol{\Xi}}_4 = \begin{pmatrix} \widehat{X}_4 & | & \widehat{X}_4 \\ \widehat{X}_4 & | & \widehat{X}_4 \end{pmatrix}, \quad \text{with} \quad \widehat{X}_4 = \left(\varphi_\alpha(1/4)\right) = \left(\varphi_\alpha(4/16)\right),$$

$$\widehat{\boldsymbol{\Xi}}_2 = \left(\varphi_\alpha(1/2)\right) = \left(\varphi_\alpha(8/16)\right),$$

and

$$\widehat{\boldsymbol{\Xi}}_1 = \left(\varphi_\alpha(0)\right).$$

Theorem 6.18 shows how $\boldsymbol{\Omega}_{2^m}$ can be reordered such that it has the aforementioned "block-circulant-with-circulant-block" structure. For $N = b^m$ being a prime power with $b \neq 2$ an analogous theorem holds (see Theorem 6.21 below). The case $b = 2$ has a special role, since in this case the group $G_1^\varphi(2^m)$ can be expressed in terms of the union of two subgroups which are obtained by considering suitable powers of the primitive root 5, as pointed out above.

Structure analysis of $\boldsymbol{\Omega}_N$ if N is an odd prime power

For $b \neq 2$, the following result holds, which is slightly different to Theorem 6.18. For a proof of this result we refer to [215].

Theorem 6.21 *Let b be an odd prime number and let $N = b^m$ with $m \in \mathbb{N}$. Let furthermore*

$$\boldsymbol{\Omega}_{b^m} = \left(\varphi_\alpha\left(\left\{\frac{kg}{N}\right\}\right)\right)_{g \in G_1^\varphi(N), k \in G_1(N)}.$$

Then the rows and columns of $\boldsymbol{\Omega}_{b^m}$ can be reordered such that we obtain a matrix

$$\widehat{\boldsymbol{\Omega}}_{b^m} = \left(\mathbf{1}_{b^0} \otimes \widehat{\Xi}_{b^m} | \mathbf{1}_{b^1} \otimes \widehat{\Xi}_{b^{m-1}} | \cdots | \mathbf{1}_{b^{m-1}} \otimes \widehat{\Xi}_b | \mathbf{1}_{b^{m-1}(b-1)} \otimes \widehat{\Xi}_1\right).$$

Again, the notation $\mathbf{1}_\ell \otimes A$ means the stacking of ℓ instances of a matrix A. Furthermore, $\widehat{\Xi}_1 = (\varphi_\alpha(0))$, and

$$\widehat{\Xi}_{b^\ell} := \begin{pmatrix} \widehat{X}_{b^\ell} & | & \widehat{X}_{b^\ell} \\ \hline \widehat{X}_{b^\ell} & | & \widehat{X}_{b^\ell} \end{pmatrix}$$

for $\ell \in \mathbb{N}$, with

$$\widehat{X}_{b^\ell} = \left(\varphi_\alpha\left(\left\{\frac{kg}{N}\right\}\right)\right)_{g \in \langle t \rangle_{b^\ell}, k \in \langle t^{-1} \rangle_{b^\ell}},$$

where t is a generator of the group $G_1^\varphi(b^m)$, and where

$$\langle t \rangle_{b^\ell} = \left\{t^i \pmod{b^\ell} \ : \ 0 \le i \le \varphi(b^\ell)/2 - 1\right\}$$

and

$$\langle t^{-1} \rangle_{2^\ell} = \left\{t^{-i} \pmod{b^\ell} \ : \ 0 \le i \le \varphi(b^\ell)/2 - 1\right\}.$$

In particular, each of the matrices $\widehat{\Xi}_{b^\ell}$ is circulant.

Remark 6.22 Similar to Theorem 6.18, the last column of $\widehat{\boldsymbol{\Omega}}_{b^m}$ in Theorem 6.21 is the column that contains only $\varphi_\alpha(0)$ as entries. Also, note again that the number of matrices $\widehat{\Xi}_{b^\ell}$ corresponds to the number of divisors of b^m, which is $m + 1$.

Regarding the question of how to find generators of the groups $G_1^\varphi(b^m)$ for general prime b, we refer the interested reader to [27, Section 4.3]. Indeed, it is essential that a generator g of $G_1^\varphi(b^m)$ is also a generator of $G_1^\varphi(b^\ell)$ for all $\ell \in [m]$. Furthermore, as discussed in [27], for each prime $b \in \{3, 5, 7, \ldots, 40487\}$, the smallest generator g of $G_1^\varphi(b)$ is also a generator of $G_1^\varphi(b^\ell)$ for any $\ell \in \mathbb{N}$. This is sufficient for practical purposes.

Theorems 6.18 and 6.21, respectively, imply that we can reorder $\boldsymbol{\Omega}_{b^m}$ for prime b in a way that the newly obtained matrix consists of circulant blocks, and itself has a block-circulant structure. Moreover, the number of distinct blocks corresponds to the number of divisors of b^m and is ordered in terms of the size of the divisors. To be more precise, this ordering even facilitates the extension of the reordered matrices

$\widehat{\mathbf{\Omega}}_{b^m}$ to the "next step", namely a matrix $\widehat{\mathbf{\Omega}}_{b^{m+1}}$, which is then just the analogously reordered matrix corresponding to $\mathbf{\Omega}_{b^{m+1}}$. This feature is particularly important for embedded lattice rules, and the larger matrices can be obtained by means of the smaller ones simply by noting that

$$\widehat{\mathbf{\Omega}}_{b^{m+1}} = (\widehat{\mathbf{\Xi}}_{b^{m+1}} | \mathbf{1}_b \otimes \widehat{\mathbf{\Omega}}_{b^m}).$$

Example 6.23 (Example 6.17 revisited) Let us return to our example of $\widehat{\mathbf{\Omega}}_{16}$. In this case, $\widehat{\mathbf{\Omega}}_{32}$ is given by

$$\widehat{\mathbf{\Omega}}_{32} = \left(\widehat{\mathbf{\Xi}}_{32} \middle| \begin{matrix} \widehat{\mathbf{\Omega}}_{16} \\ \widehat{\mathbf{\Omega}}_{16} \end{matrix} \right),$$

where $\widehat{\mathbf{\Xi}}_{32}$ has a form as outlined in Theorem 6.18.

Theorems 6.18 and 6.21 imply that also for N of the form $N = b^m$ (with prime b) the matrix-vector product

$$\widehat{\mathbf{\Omega}}_{b^m} \boldsymbol{\eta}$$

for a vector $\boldsymbol{\eta}$ of length N can be performed in $O(N \log N)$ operations. Indeed, also the reordering of $\mathbf{\Omega}_{b^m}$ to $\widehat{\mathbf{\Omega}}_{b^m}$ can be expressed in terms of permutation matrices. Furthermore, multiplying $\widehat{\mathbf{\Omega}}_{b^m}$ with $\boldsymbol{\eta}$ can be done block-wise; this is done by starting with the smallest block. The result of this computation is then replicated $b-1$ times and added to the result corresponding to the next block (with one exception for the case $b = 2$, where we have two smallest blocks, which can be handled by making the obvious adaptions). Then, the result of this addition is replicated $b-1$ times and added to the result corresponding to the next block, etc. Since all blocks of $\widehat{\mathbf{\Omega}}_{b^m}$ are themselves circulant, and there are $m+1$ different block sizes, the cost of computing the matrix-vector products for all blocks is

$$\sum_{\ell=0}^{m} O\left(\varphi(b^{m-\ell}) \log(\varphi(b^{m-\ell}))\right) = O(N \log N).$$

Moreover, the cost for the summation of the results of the different blocks is of order $O(N)$, and can be neglected, and the summation needs only additional memory of order $O(m)$.

In total we therefore see that the matrix-vector multiplication that is needed for efficiently implementing the CBC algorithm for prime power N can be done in $O(N \log N)$ operations, and, consequently, also for the case of prime power N the CBC algorithm can be implemented in a fast way.

Finally, we return to the computational cost of Algorithm 6.15 to find generating vectors for embedded lattice rules. The answer to this question is surprisingly simple, due to the embedded structure not only of the lattice rules to be constructed but also of the matrices $\mathbf{\Omega}_{b^m}$ for $m \in \{\underline{m}, \underline{m}+1, \ldots, \overline{m}\}$. Since all of these matrices are essentially embedded into each other, we can calculate and keep track of $\mathrm{err}_{b^m,d,\alpha,\gamma}$ for any $m \in \{\underline{m}, \underline{m}+1, \ldots, \overline{m}\}$ "along the way" when calculating $\mathrm{err}_{b^{\overline{m}},d,\alpha,\gamma}$, and this does not incur any extra cost.

Hence, the total cost of Algorithm 6.15 is $O(d\, b^{\overline{m}}\, \overline{m}^2)$, namely, as already discussed, $O(d\, b^{\overline{m}}\, \overline{m}^2)$ operations for the precomputations and $O(d\, b^{\overline{m}}\, \overline{m})$ operations for actually running the algorithm.

So far, we have not presented theoretical error bounds for the embedded lattice rules whose construction we have considered before. The reason is that it is actually difficult to give good error bounds for these point sets in the sense that they would hold for a whole range of m. There are some remarks on how to give an estimate for the error in [27, 103], which are based on observations like in Proposition 6.2, namely that an embedded lattice rule can be seen as the union of instances of a lattice point set that has been shifted a (large) number of times. However, these observations, though intuitively clear, do not yield a theoretical error bound, e.g., for the rules constructed by Algorithm 6.15. We will present another way of constructing embedded lattice rules, which also allows for error bounds, in the next section.

6.4 A Sieve Principle for Constructing Embedded Lattice Rules

As outlined in Remark 6.6, the idea behind the proof of the existence of good extensible lattice rules in [105] is based on finding large subsets of generating vectors such that the underlying lattice rules satisfy good error bounds for given parameters, and then choosing a generating vector from the intersection of those subsets to obtain a generating vector which satisfies all these error bounds simultaneously. In the following we call this a sieve principle. In [58] this idea was turned into a fast algorithm to find good embedded lattice rules.

In the following let $b \geq 2$ be a prime. The lattice rules under consideration will again have $N = b^m$ points for some positive exponent m. Additionally to the set $G_d^{\varphi}(N)$ that has been used before, we introduce the set

$$G_d^{\varphi}(b, \infty) := \{g \in \mathbb{N} \,:\, \gcd(g, b) = 1\}^d . \tag{6.20}$$

A few comments will be helpful for the following. We recall that $|G_d^{\varphi}(b^m)| = (\varphi(b^m))^d$, but $G_d^{\varphi}(b, \infty)$ has infinite cardinality. The difference between $G_1^{\varphi}(b, \infty)$ and $\mathcal{Z}_b$ in Section 6.1 is that the former set is a subset of $\mathbb{N}$ whereas the latter contains elements of $\mathbb{Z}_b$. Note, furthermore, that for each $\boldsymbol{g} \in G_d^{\varphi}(b, \infty)$ and $m \in \mathbb{N}$ there exists a $\overline{\boldsymbol{g}} \in G_d^{\varphi}(b^m)$ such that $\boldsymbol{g} \equiv \overline{\boldsymbol{g}} \pmod{b^m}$, where, as usual, the equivalence of vectors is meant component-wise. It is important to note that

$$\mathrm{err}_{b^m,d,\alpha,\gamma}(\boldsymbol{g}) = \mathrm{err}_{b^m,d,\alpha,\gamma}(\overline{\boldsymbol{g}}) \quad \text{if } \boldsymbol{g} \equiv \overline{\boldsymbol{g}} \pmod{b^m}.$$

We now define a probability measure $\widehat{\mu}_{b,d}$ on $G_d^{\varphi}(b, \infty)$ which will play a role analogous to the probability measure μ_b^{∞} in Section 6.1. For $m \in \mathbb{N}$ we denote the uniform measure on $G_d^{\varphi}(b^m)$ by $\widehat{\mu}_{b,d,m}$, i.e.,

$$\widehat{\mu}_{b,d,m}(A) = \frac{|A|}{(\varphi(b^m))^d} \qquad \text{for } A \subseteq G_d^{\varphi}(b^m).$$

Next, we aim at extending this measure to a measure on $G_d^{\varphi}(b,\infty)$. We say that a set $A \subseteq G_d^{\varphi}(b,\infty)$ is of finite type if there exists an integer $m = m(A) \in \mathbb{N}$, and a set $A' \subseteq G_d^{\varphi}(b^m)$ such that

$$A = \left\{ \boldsymbol{g} \in G_d^{\varphi}(b,\infty) \ : \ \boldsymbol{g} \ (\mathrm{mod}\ b^m) \in A' \right\}.$$

It should be noted that not every set $A \subseteq G_d^{\varphi}(b,\infty)$ is of finite type. Indeed, suppose that A_0 contains a representative of each primitive residue class modulo b^m except zero for all $m \in \mathbb{N}$, but that $A_0 \neq G_d^{\varphi}(b,\infty)$ (for instance, for $d = 1$, $b = 2$ consider $A_0 = \{k \in \mathbb{N} : k \text{ odd}\} \setminus \{5\}$). Then the corresponding A_0' can only be constructed such that

$$\left\{ \boldsymbol{g} \in G_d^{\varphi}(b,\infty) \ : \ \boldsymbol{g} \ (\mathrm{mod}\ b^m) \in A_0' \right\} = G_d^{\varphi}(b,\infty) \neq A_0,$$

hence A_0 is not of finite type.

If, however, A is of finite type, then we define the measure of A as

$$\widehat{\mu}_{b,d}(A) := \widehat{\mu}_{b,d,m}(A') = \frac{|A'|}{(\varphi(b^m))^d}, \quad \text{where } m = m(A).$$

Naturally, the number $m = m(A)$ is not uniquely determined by A, since if m works, then also any number larger than m works in the definition of a finite type subset. It is easy to see, though, that the definition of the measure $\widehat{\mu}_{b,d}$ does not depend on the specific choice of m.

Sets of good generating vectors

We will now define sets of "good generating vectors". As a benchmark for a fair understanding of "good" we use our previous error bounds for lattice rules with a fixed number of points from Chapters 2–4. For short, we write

$$E_{b^m,d,\alpha,\boldsymbol{\gamma}}(\tau) := \frac{1}{b^{\tau m}} \left(\prod_{j=1}^{d} \left(1 + 4\gamma_j^{1/(2\tau)} \zeta\left(\frac{\alpha}{\tau}\right)\right) - 1 \right)^{\tau} \quad \text{for } \tau \in [1/2, \alpha).$$

Now, for given $N = b^m$ and a given constant $c \geq 1$, let

$$C_{b^m,d,\alpha,\boldsymbol{\gamma}}(c) := \left\{ \boldsymbol{g} \in G_d^{\varphi}(b,\infty) \ : \ \mathrm{err}_{b^m,d,\alpha,\boldsymbol{\gamma}}(\boldsymbol{g}) \leq c^{1/2} E_{b^m,d,\alpha,\boldsymbol{\gamma}}(1/2) \right\}, \tag{6.21}$$

and, more generally,

$$\begin{aligned}&\widetilde{C}_{b^m,d,\alpha,\boldsymbol{\gamma}}(c) \\ &:= \left\{ \boldsymbol{g} \in G_d^{\varphi}(b,\infty) \ : \ \mathrm{err}_{b^m,d,\alpha,\boldsymbol{\gamma}}(\boldsymbol{g}) \leq c^{\tau} E_{b^m,d,\alpha,\boldsymbol{\gamma}}(\tau) \quad \forall \tau \in [1/2,\alpha) \right\}.\end{aligned}$$

With respect to the definition in (6.21), note that, if we define

$$C'_{b^m,d,\alpha,\gamma}(c) := \left\{ \boldsymbol{g} \in G_d^{\varphi}(b^m) \ : \ \mathrm{err}_{b^m,d,\alpha,\gamma}(\boldsymbol{g}) \le c^{1/2} E_{b^m,d,\alpha,\gamma}(1/2) \right\},$$

then

$$C_{b^m,d,\alpha,\gamma}(c) = \left\{ \boldsymbol{g} \in G_d^{\varphi}(b,\infty) \ : \ \boldsymbol{g} \ (\mathrm{mod}\ b^m) \in C'_{b^m,d,\alpha,\gamma}(c) \right\},$$

and hence

$$\widehat{\mu}_{b,d}(C_{b^m,d,\alpha,\gamma}(c)) = \widehat{\mu}_{b,d,m}(C'_{b^m,d,\alpha,\gamma}(c)). \tag{6.22}$$

We can formulate the following result.

Lemma 6.24 *Let b be a prime number and let $m \in \mathbb{N}$. Then, for every $c \ge 1$, we have*

$$\widehat{\mu}_{b,d}(C_{b^m,d,\alpha,\gamma}(c)) > 1 - \frac{1}{c} \tag{6.23}$$

and

$$\widehat{\mu}_{b,d}(\widetilde{C}_{b^m,d,\alpha,\gamma}(c)) > 1 - \frac{1}{c}. \tag{6.24}$$

Proof First we recall from Theorem 2.24 and its proof that, for prime N,

$$\frac{1}{N^d} \sum_{\boldsymbol{g} \in G_d^{\varphi}(N)} [\mathrm{err}_{N,d,\alpha,\gamma}(\boldsymbol{g})]^2 \le \frac{1}{N} \left(\prod_{j=1}^{d} (1 + 4\gamma_j \zeta(2\alpha)) - 1 \right).$$

This result can be extended to N of the form $N = b^m$, e.g., by methods outlined in the proof of Theorem 4.3 or using [158], where it is shown that

$$\frac{1}{(\varphi(b^m))^d} \sum_{\boldsymbol{g} \in G_d^{\varphi}(b^m)} [\mathrm{err}_{b^m,d,\alpha,\gamma}(\boldsymbol{g})]^2 \le \frac{1}{N} \left(\prod_{j=1}^{d} (1 + 4\gamma_j \zeta(2\alpha)) - 1 \right). \tag{6.25}$$

Next, we show (6.23). We know from (6.22) that

$$\begin{aligned} &\widehat{\mu}_{b,d}(C_{b^m,d,\alpha,\gamma}(c)) \\ &= \widehat{\mu}_{b,d,m}\left(\{ \boldsymbol{g} \in G_d^{\varphi}(b^m) \ : \ \mathrm{err}_{b^m,d,\alpha,\gamma}(\boldsymbol{g}) \le c^{1/2} E_{b^m,d,\alpha,\gamma}(1/2) \} \right). \end{aligned}$$

However, by (6.25), $E_{b^m,d,\alpha,\gamma}(1/2)$ is just an upper bound on the root mean square value of $\mathrm{err}_{b^m,d,\alpha,\gamma}(\boldsymbol{g})$ for $\boldsymbol{g} \in G_d^{\varphi}(b^m)$. Hence, an application of Markov's inequality yields the result in (6.23).

Regarding (6.24), we first note that

$$\lim_{\tau \to \alpha} c^{\tau} E_{b^m,d,\alpha,\gamma}(\tau) = \infty.$$

Thus, there exists $\tau^* \in [1/2, \alpha)$ such that

$$c^{\tau^*} E_{b^m,d,\alpha,\gamma}(\tau^*) \le c^{\tau} E_{b^m,d,\alpha,\gamma}(\tau) \quad \forall \tau \in [1/2, \alpha).$$

We fix such a τ^* and note that due to (6.23) we have

$$\widehat{\mu}_{b,d}(C_{b^m,d,\alpha/(2\tau^*),\gamma^{1/(2\tau^*)}}(c)) > 1 - \frac{1}{c}. \tag{6.26}$$

For $\boldsymbol{g} \in C_{b^m,d,\alpha/(2\tau^*),\gamma^{1/(2\tau^*)}}(c)$ we obviously have

$$\mathrm{err}_{b^m,d,\alpha/(2\tau^*),\gamma^{1/(2\tau^*)}}(\boldsymbol{g}) \le c^{1/2} E_{b^m,d,\alpha/(2\tau^*),\gamma^{1/(2\tau^*)}}(1/2).$$

An application of Jensen's inequality (cf. Lemma 2.25) then yields

$$\begin{aligned}\left([\mathrm{err}_{b^m,d,\alpha,\gamma}(\boldsymbol{g})]^2\right)^{1/(2\tau^*)} &\le [\mathrm{err}_{b^m,d,\alpha/(2\tau^*),\gamma^{1/(2\tau^*)}}(\boldsymbol{g})]^2 \\ &\le c[E_{b^m,d,\alpha/(2\tau^*),\gamma^{1/(2\tau^*)}}(1/2)]^2.\end{aligned}$$

Consequently,

$$\mathrm{err}_{b^m,d,\alpha,\gamma}(\boldsymbol{g}) \le c^{\tau^*} [E_{b^m,d,\alpha/(2\tau^*),\gamma^{1/(2\tau^*)}}(1/2)]^{2\tau^*} = c^{\tau^*} E_{b^m,d,\alpha,\gamma}(\tau^*),$$

and this yields $\boldsymbol{g} \in \widetilde{C}_{b^m,d,\alpha,\gamma}(c)$. Consequently,

$$C_{b^m,d,\alpha/(2\tau^*),\gamma^{1/(2\tau^*)}}(c) \subseteq \widetilde{C}_{b^m,d,\alpha,\gamma}(c),$$

which implies, due to (6.26),

$$\widehat{\mu}_{b,d}(\widetilde{C}_{b^m,d,\alpha,\gamma}(c)) \ge \widehat{\mu}_{b,d}(C_{b^m,d,\alpha/(2\tau^*),\gamma^{1/(2\tau^*)}}(c)) > 1 - \frac{1}{c},$$

yielding (6.24). □

The existence of good embedded lattice rules

We now go on to consider embedded lattice rules, again with the goal to find lattice rules of high quality, in the sense that we have good embedded lattice rules with b^m points for m in a finite range $\{\underline{m}, \underline{m}+1, \ldots, \overline{m}\}$. We first show the following existence result, which is similar to Theorems 6.8–6.10, and Corollaries 6.12 and 6.13.

Theorem 6.25 *Let b be a prime number and let d, $\underline{m}$, and $\overline{m} \in \mathbb{N}$ be such that $\underline{m} \le \overline{m}$. Choose positive c_m for $m \in \{\underline{m}, \underline{m}+1, \ldots, \overline{m}\}$ such that $\sum_{m=\underline{m}}^{\overline{m}} c_m^{-1} \le 1$. Then there exists a $\boldsymbol{g} \in G_d^{\varphi}(b, \infty)$ such that*

$$\mathrm{err}_{b^m,d,\alpha,\gamma}(\boldsymbol{g}) \le \frac{c_m^{\tau}}{b^{\tau m}} \left(\prod_{j=1}^{d} \left(1 + 4\gamma_j^{1/(2\tau)} \zeta\left(\frac{\alpha}{\tau}\right)\right) - 1 \right)^{\tau}$$

for all $\tau \in [1/2, \alpha)$ *and all* $m \in \{\underline{m}, \underline{m}+1, \ldots, \overline{m}\}$.

Proof The proof is done by showing that

$$\widehat{\mu}_{b,d}\left(\bigcap_{m=\underline{m}}^{\overline{m}} \widetilde{C}_{b^m,d,\alpha,\gamma}(c_m)\right) > 0.$$

Indeed, by denoting the complement of $\widetilde{C}_{b^m,d,\alpha,\gamma}(c_m)$ by $\widetilde{C}^c_{b^m,d,\alpha,\gamma}(c_m)$, we obtain

$$\begin{aligned}\widehat{\mu}_{b,d}\left(\bigcap_{m=\underline{m}}^{\overline{m}} \widetilde{C}_{b^m,d,\alpha,\gamma}(c_m)\right) &= 1 - \widehat{\mu}_{b,d}\left(\bigcup_{m=\underline{m}}^{\overline{m}} \widetilde{C}^c_{b^m,d,\alpha,\gamma}(c_m)\right)\\ &\geq 1 - \sum_{m=\underline{m}}^{\overline{m}} \widehat{\mu}_{b,d}\left(\widetilde{C}^c_{b^m,d,\alpha,\gamma}(c_m)\right)\\ &> 1 - \sum_{m=\underline{m}}^{\overline{m}} \frac{1}{c_m}\\ &\geq 0,\end{aligned}$$

as claimed. □

Remark 6.26 Note that in Theorem 6.25 we can replace the set $\{\underline{m}, \underline{m}+1, \ldots, \overline{m}\}$ by some other subset of $\mathbb{N}$, or indeed the entire set $\mathbb{N}$. For the case of the entire set of natural numbers, we can for instance choose $c_m = b^{m\delta}/(1-b^{-\delta})$ for $m \in \mathbb{N}$ and for some $\delta > 0$, in which case we obtain $\sum_{m=1}^{\infty} c_m^{-1} = 1$. The error bound is then

$$\mathrm{err}_{b^m,d,\alpha,\gamma}(\boldsymbol{g}) \leq \frac{1}{b^{\tau m(1-\delta)}(1-b^{-\delta})}\left(\prod_{j=1}^{d}\left(1 + 4\gamma_j^{1/(2\tau)}\zeta\left(\frac{\alpha}{\tau}\right)\right) - 1\right)^{\tau}$$

for all $\tau \in [1/2, \alpha)$ and all $m \in \mathbb{N}$.

By choosing for instance $c_m = \kappa b^{m\delta}/(1-b^{-\delta})$ for some $\kappa \in (0,1)$ we obtain that $1 - \sum_{m=1}^{\infty} c_m^{-1} = 1 - \kappa > 0$. In this way we can also increase the measure of the set of extensible lattice rules satisfying certain error bounds.

A general search principle

Since we know by Theorem 6.25 that a good $\boldsymbol{g}$ for all $m \in \{\underline{m}, \underline{m}+1, \ldots, \overline{m}\}$ must exist, we now go on to describe how such a $\boldsymbol{g}$ can be found. The general search principle, which will then be slightly modified and optimized, is a sieve algorithm that works by starting from $m = \underline{m}$ and proceeding, step by step, to $m = \overline{m}$.

In the first step, for $m = \underline{m}$, we define

$$T_{\underline{m}} := \left\{\boldsymbol{g} \in G_d^{\varphi}(b^{\underline{m}}) \;:\; \mathrm{err}_{b^{\underline{m}},d,\alpha,\gamma}(\boldsymbol{g}) \leq c_{\underline{m}}^{\tau} E_{b^{\underline{m}},d,\alpha,\gamma}(\tau) \quad \forall \tau \in [1/2, \alpha)\right\}.$$

By Lemma 6.24 we know that

$$|T_{\underline{m}}| \geq \left\lfloor \left(1 - \frac{1}{c_{\underline{m}}}\right) (\varphi(b^{\underline{m}}))^d \right\rfloor + 1.$$

In the next step, we consider $m = \underline{m} + 1$, and define

$$S_{\underline{m}+1} := \left\{ \boldsymbol{g} \in G_d^{\varphi}(b^{\underline{m}+1}) \; : \; \exists \overline{\boldsymbol{g}} \in T_{\underline{m}} \quad \text{such that} \quad \boldsymbol{g} \equiv \overline{\boldsymbol{g}} \pmod{b^{\underline{m}}} \right\}.$$

Furthermore, we set

$$T_{\underline{m}+1} := \left\{ \boldsymbol{g} \in S_{\underline{m}+1} \; : \; \mathrm{err}_{b^{\underline{m}+1},d,\alpha,\gamma}(\boldsymbol{g}) \leq c_{\underline{m}+1}^{\tau} E_{b^{\underline{m}+1},d,\alpha,\gamma}(\tau) \quad \forall \tau \in [1/2, \alpha) \right\}.$$

We will now show that

$$|T_{\underline{m}+1}| \geq \left\lfloor \left(1 - \frac{1}{c_{\underline{m}}} - \frac{1}{c_{\underline{m}+1}}\right) (\varphi(b^{\underline{m}+1}))^d \right\rfloor + 1.$$

Indeed, note first that $|S_{\underline{m}+1}| = |T_{\underline{m}}| b^d$, and define

$$\widetilde{T}_{\underline{m}+1} := \left\{ \boldsymbol{g} \in G_d^{\varphi}(b^{\underline{m}+1}) \; : \; \mathrm{err}_{b^{\underline{m}+1},d,\alpha,\gamma}(\boldsymbol{g}) \leq c_{\underline{m}+1}^{\tau} E_{b^{\underline{m}+1},d,\alpha,\gamma}(\tau) \quad \forall \tau \in [1/2, \alpha) \right\}.$$

Then, again by Lemma 6.24,

$$|\widetilde{T}_{\underline{m}+1}| > \left(1 - \frac{1}{c_{\underline{m}+1}}\right) (\varphi(b^{\underline{m}+1}))^d.$$

As $T_{\underline{m}+1} = S_{\underline{m}+1} \cap \widetilde{T}_{\underline{m}+1}$, we have, by again denoting the complement of a set A by A^c,

$$\begin{aligned}
|T_{\underline{m}+1}| &= |S_{\underline{m}+1} \cap \widetilde{T}_{\underline{m}+1}| \\
&= (\varphi(b^{\underline{m}+1}))^d - |(S_{\underline{m}+1} \cap \widetilde{T}_{\underline{m}+1})^c| \\
&= (\varphi(b^{\underline{m}+1}))^d - |S_{\underline{m}+1}^c \cup \widetilde{T}_{\underline{m}+1}^c| \\
&\geq (\varphi(b^{\underline{m}+1}))^d - |S_{\underline{m}+1}^c| - |\widetilde{T}_{\underline{m}+1}^c| \\
&= (\varphi(b^{\underline{m}+1}))^d - \left((\varphi(b^{\underline{m}+1}))^d - |S_{\underline{m}+1}|\right) - \left((\varphi(b^{\underline{m}+1}))^d - |\widetilde{T}_{\underline{m}+1}|\right) \\
&= |S_{\underline{m}+1}| + |\widetilde{T}_{\underline{m}+1}| - (\varphi(b^{\underline{m}+1}))^d \\
&> \left(1 - \frac{1}{c_{\underline{m}}}\right) (\varphi(b^{\underline{m}}))^d b^d + \left(1 - \frac{1}{c_{\underline{m}+1}}\right) (\varphi(b^{\underline{m}+1}))^d - (\varphi(b^{\underline{m}+1}))^d \\
&= \left(1 - \frac{1}{c_{\underline{m}}} - \frac{1}{c_{\underline{m}+1}}\right) (\varphi(b^{\underline{m}+1}))^d,
\end{aligned}$$

where we used that

$$\varphi(b^{\underline{m}})\, b = b\, b^{\underline{m}}\left(1 - \frac{1}{b}\right) = b^{\underline{m}+1}\left(1 - \frac{1}{b}\right) = \varphi(b^{\underline{m}+1}).$$

This implies that indeed

$$|T_{\underline{m}+1}| \geq \left\lfloor \left(1 - \frac{1}{c_{\underline{m}}} - \frac{1}{c_{\underline{m}+1}}\right) (\varphi(b^{\underline{m}+1}))^d \right\rfloor + 1.$$

We can continue this procedure, and construct then, step by step, $S_{\underline{m}+2}, T_{\underline{m}+2}, \ldots$ until we arrive at $S_{\overline{m}}$ and $T_{\overline{m}}$. Eventually, also $T_{\overline{m}}$ contains at least one element. Hence, any vector in this set has the desired property stated in Theorem 6.25.

In analogy to what we stated above, we again remark that, theoretically speaking, the value of $\overline{m}$ in this procedure is not restricted to being finite. In principle, as long as $\sum_{m=\underline{m}}^{\overline{m}} c_m^{-1} \leq 1$, we can also allow $\overline{m} = \infty$. Furthermore, it is not necessary to consider only successive values of $m \in \{\underline{m}, \underline{m}+1, \ldots, \overline{m}\}$, but we could allow "gaps" in the set of indices m considered. Even though the procedure just outlined yields generating vectors of embedded (or, if we choose $\overline{m} = \infty$, even infinitely extensible) lattice point sets, the algorithm of how to find such a good generating vector $\boldsymbol{g}$ will in general be rather slow. We shall address this issue in the next section.

6.5 The CBC Sieve Algorithm

In order to make the construction outlined in the previous section practically feasible, it was shown in [58] how to find a good generating vector of lattice point sets with b^m points, with $m \in \{\underline{m}, \underline{m}+1, \ldots, \overline{m}\}$ by using a combination of the sieve principle and a component-by-component algorithm. However, for the CBC sieve algorithm, it will no longer be possible to choose $\overline{m} = \infty$ and thereby make the constructed lattice rules infinitely extensible with respect to their numbers of points. In that sense, the generating vectors found in the present section are "only" generating vectors of embedded lattice rules for fixed numbers of b^m points, with $m \in \{\underline{m}, \underline{m}+1, \ldots, \overline{m}\}$, where $\underline{m}$ and $\overline{m}$ with $\underline{m} \leq \overline{m}$ are both finite and fixed beforehand. The idea of the CBC sieve algorithm is based on similar principles as the ordinary CBC algorithm.

We start by considering a quantity that has already been studied in Section 3.3 (though, there, for prime N). Suppose that we are given $N = b^m$, with $m \in \mathbb{N}$ and prime b, $\alpha > 1/2$, product weights $\boldsymbol{\gamma} = (\gamma_j)_{j\geq 1}$, and an index $s \in [d-1]$. Furthermore, assume that we are given an s-dimensional vector $\boldsymbol{g}^{(s)} \in G_s^{\varphi}(b^m)$ and a $g_{s+1} \in G_1^{\varphi}(b^m)$. Then we define

$$\theta_{b^m,s,\alpha,\boldsymbol{\gamma}}(\boldsymbol{g}^{(s)}, g_{s+1}) := [\mathrm{err}_{b^m,s+1,\alpha,\boldsymbol{\gamma}}((\boldsymbol{g}^{(s)}, g_{s+1}))]^2 - [\mathrm{err}_{b^m,s,\alpha,\boldsymbol{\gamma}}(\boldsymbol{g}^{(s)})]^2.$$

As a special instance of (3.15) in Chapter 3 we have, for fixed $\boldsymbol{g}^{(s)} \in G_s^{\varphi}(b^m)$, that

$$\frac{1}{\varphi(b^m)} \sum_{g_{s+1} \in G_1^{\varphi}(b^m)} \theta_{b^m,s,\alpha,\boldsymbol{\gamma}}(\boldsymbol{g}^{(s)}, g_{s+1}) \leq \frac{4}{b^m} \gamma_{s+1} \zeta(2\alpha) \prod_{j=1}^{s} (1 + 2\gamma_j \zeta(2\alpha)). \tag{6.27}$$

To simplify notation, we write

$$\theta_{b^m,s,\alpha,\boldsymbol{\gamma}}^{\text{avg-bound}} := \frac{4}{b^m} \gamma_{s+1} \zeta(2\alpha) \prod_{j=1}^{s} (1 + 2\gamma_j \zeta(2\alpha))$$

in the following, and we define a set analogous to $C_{b^m,d,\alpha,\boldsymbol{\gamma}}(c)$. For $c \geq 1$ and given $\boldsymbol{g}^{(s)} \in G_s^{\varphi}(b, \infty)$, let

$$\mathcal{B}_{b^m,s,\alpha,\boldsymbol{\gamma}}(c, \boldsymbol{g}^{(s)}) := \left\{ g_{s+1} \in G_1^{\varphi}(b, \infty) \,:\, \theta_{b^m,s,\alpha,\boldsymbol{\gamma}}(\boldsymbol{g}^{(s)}, g_{s+1}) \leq c\, \theta_{b^m,s,\alpha,\boldsymbol{\gamma}}^{\text{avg-bound}} \right\},$$

and, more specifically,

$$\begin{aligned} &\widetilde{\mathcal{B}}_{b^m,s,\alpha,\boldsymbol{\gamma}}(c, \boldsymbol{g}^{(s)}) \\ &:= \Bigg\{ g_{s+1} \in G_1^{\varphi}(b, \infty) \,:\, \theta_{b^m,s,\alpha,\boldsymbol{\gamma}}(\boldsymbol{g}^{(s)}, g_{s+1}) \leq c^{2\tau} \left(\theta_{b^m,s,\alpha/(2\tau),\boldsymbol{\gamma}^{1/(2\tau)}}^{\text{avg-bound}} \right)^{2\tau} \\ &\qquad \forall \tau \in [1/2, \alpha) \Bigg\}. \end{aligned}$$

The following lemma is similar to Lemma 6.24.

Lemma 6.27 *Let b be a prime number and let $s, m \in \mathbb{N}$. Then, for every $c \geq 1$ and every $\boldsymbol{g}^{(s)} \in G_s^{\varphi}(b, \infty)$, we have*

$$\widehat{\mu}_{b,1}\left(\mathcal{B}_{b^m,s,\alpha,\boldsymbol{\gamma}}(c, \boldsymbol{g}^{(s)}) \right) > 1 - \frac{1}{c},$$

and

$$\widehat{\mu}_{b,1}\left(\widetilde{\mathcal{B}}_{b^m,s,\alpha,\boldsymbol{\gamma}}(c, \boldsymbol{g}^{(s)}) \right) > 1 - \frac{1}{c}.$$

Proof The proof of the theorem is analogous to that of Lemma 6.24. We remark that Markov's inequality can be applied since the value of $\theta_{b^m,s,\alpha,\boldsymbol{\gamma}}(\boldsymbol{g}^{(s)}, g_{s+1})$ is always nonnegative (see (3.14)). For a full proof, see [58]. □

The following theorem can be shown using Lemma 6.27.

Theorem 6.28 *Let b be a prime number and let s, $\underline{m}$, and $\overline{m} \in \mathbb{N}$ be such that $\underline{m} \leq \overline{m}$. Choose positive c_m for $m \in \{\underline{m}, \underline{m}+1, \ldots, \overline{m}\}$ such that $\sum_{m=\underline{m}}^{\overline{m}} c_m^{-1} \leq 1$. Then, for any $\boldsymbol{g}^{(s)} \in G_s^{\varphi}(b, \infty)$, there exists $g_{s+1} \in G_1^{\varphi}(b, \infty)$ such that*

$$\theta_{b^m,s,\alpha,\boldsymbol{\gamma}}(\boldsymbol{g}^{(s)}, g_{s+1}) \leq c^{2\tau} \left(\theta_{b^m,s,\alpha/(2\tau),\boldsymbol{\gamma}^{1/(2\tau)}}^{\text{avg-bound}} \right)^{2\tau}$$

for all $\tau \in [1/2, \alpha)$ *and all* $m \in \{\underline{m}, \ldots, \overline{m}\}$.

Proof The proof of the theorem works in the same way as that of Theorem 6.25. For a full proof, we refer to [58]. □

We need one more technical estimate before coming to one of the main results in this section.

Proposition 6.29 *Let b be a prime, let* $s, m \in \mathbb{N}$, *and let* $c_m \geq 1$. *Let* $\boldsymbol{g}^{(s)} \in G_s^{\varphi}(b, \infty)$ *be such that*

$$\mathrm{err}_{b^m,s,\alpha,\gamma}(\boldsymbol{g}^{(s)}) \leq c_m^{\tau} E_{b^m,s,\alpha,\gamma}(\tau) \quad \textit{for all } \tau \in [1/2, \alpha),$$

and let $g_{s+1} \in G_1^{\varphi}(b, \infty)$ *be such that*

$$\theta_{b^m,s,\alpha,\gamma}(\boldsymbol{g}^{(s)}, g_{s+1}) \leq c^{2\tau} \left(\theta^{\text{avg-bound}}_{b^m,s,\alpha/(2\tau),\gamma^{1/(2\tau)}}\right)^{2\tau} \quad \textit{for all } \tau \in [1/2, \alpha).$$

Then it is true that

$$\mathrm{err}_{b^m,s+1,\alpha,\gamma}(\boldsymbol{g}^{(s)}, g_{s+1}) \leq c_m^{\tau} E_{b^m,s+1,\alpha,\gamma}(\tau) \quad \textit{for all } \tau \in [1/2, \alpha).$$

Proof The proof is done by standard methods, analogously to what we showed in Chapter 3. For a full proof, see again [58]. □

Having collected all necessary estimates, we are now ready to formulate the CBC sieve algorithm, which constructs, in a component-wise fashion, a good generating vector of an embedded lattice rule which has a low worst-case error for all instances of $N = b^m$ with $m \in \{\underline{m}, \underline{m} + 1, \ldots, \overline{m}\}$.

Algorithm 6.30 (CBC sieve construction) Let b be a prime number and let d, $\underline{m}$, and $\overline{m} \in \mathbb{N}$ be such that $\underline{m} \leq \overline{m}$. For $m \in \{\underline{m}, \underline{m} + 1, \ldots, \overline{m}\}$ let c_m be positive reals such that $\sum_{m=\underline{m}}^{\overline{m}} c_m^{-1} \leq 1$. Construct a generating vector $\boldsymbol{g} = (g_1, \ldots, g_d) \in G_d^{\varphi}(b, \infty)$ as follows.

(1) Set $g_1 = 1$.
(2) For s from 1 to $d - 1$:
Assume that $\boldsymbol{g}^{(s)} := (g_1, \ldots, g_s) \in G_s^{\varphi}(b, \infty)$ has already been found. Consider this vector as fixed and choose $g_{s+1} \in G_1^{\varphi}(b, \infty)$ as follows.

(2a) Find $\lfloor (1 - c_{\underline{m}})\varphi(b^{\underline{m}}) \rfloor + 1$ elements $g_{s+1} \in G_d^{\varphi}(b^{\underline{m}})$ to populate the set

$$T_{\underline{m},s+1} \subseteq \Big\{ g \in G_1^{\varphi}(b^{\underline{m}}) : \\ \mathrm{err}_{b^{\underline{m}},s+1,\alpha,\gamma}(\boldsymbol{g}^{(s)}, g) \leq c_{\underline{m}}^{\tau} E_{b^{\underline{m}},s+1,\alpha,\gamma}(\tau) \; \forall \tau \in [1/2, \alpha) \Big\}.$$

(2b) For m from $\underline{m} + 1$ to $\overline{m}$:
Define

$$S_{m,s+1} := \left\{ g \in G_1^{\varphi}(b^m) \; : \; \exists \overline{g} \in T_{m-1,s+1} \text{ such that } g \equiv \overline{g} \pmod{b^{m-1}} \right\}.$$

Find $\lfloor (1 - \sum_{i=\underline{m}}^{m} c_i^{-1}) \varphi(b^m) \rfloor + 1$ elements to populate the set

$$T_{m,s+1} \subseteq \big\{ g \in S_{m,s+1} : \\ \mathrm{err}_{b^m,s+1,\alpha,\boldsymbol{\gamma}}(\boldsymbol{g}^{(s)}, g) \le c_m^\tau E_{b^m,s+1,\alpha,\boldsymbol{\gamma}}(\tau) \quad \forall \tau \in [1/2, \alpha) \big\}.$$

End for.

(2c) Choose $g_{s+1} \in T_{\overline{m},s+1}$.

End for.

(3) Set $\boldsymbol{g} = (g_1, \ldots, g_d)$.

The following theorem summarizes the properties of the generating vectors $\boldsymbol{g}$ constructed by Algorithm 6.30.

Theorem 6.31 *Let b be a prime number and let $d, \underline{m}$, and $\overline{m} \in \mathbb{N}$ be such that $\underline{m} \le \overline{m}$. For $m \in \{\underline{m}, \underline{m}+1, \ldots, \overline{m}\}$ let c_m be positive reals such that $\sum_{m=\underline{m}}^{\overline{m}} c_m^{-1} \le 1$. Assume that $\boldsymbol{g} = (g_1, \ldots, g_d) \in G_d^\varphi(b, \infty)$ has been constructed according to Algorithm 6.30. Then it holds that*

$$\mathrm{err}_{b^m,s,\alpha,\boldsymbol{\gamma}}((g_1, \ldots, g_s)) \le \frac{c_m^\tau}{b^{\tau m}} \left(\prod_{j=1}^{s} \left(1 + 4\gamma_j^{1/(2\tau)} \zeta\left(\frac{\alpha}{\tau}\right)\right) - 1 \right)^\tau$$

for all $s \in [d]$, for all $\tau \in [1/2, \alpha)$, and for all $m \in \{\underline{m}, \underline{m}+1, \ldots, \overline{m}\}$.

Proof Theorem 6.28 and Proposition 6.29 imply that the sets to be populated in Algorithm 6.30 are nonempty and have the proper cardinalities, respectively. This immediately yields the result. □

While Algorithm 6.30 is a theoretical algorithm yielding a generating vector that is, by Theorem 6.31, guaranteed to satisfy an excellent error estimate, it cannot be expected to be particularly fast if implemented on a computer. Obviously, one would hope to find a CBC sieve construction that has a runtime comparable to that of the fast CBC algorithm in Section 3.4 or that of Algorithm 6.15 outlined above. Indeed, this is possible, but—to this end—one has to modify the search criterion.

6.6 The Fast CBC Sieve Algorithm

In order to make the CBC sieve algorithm accessible to a fast implementation in the sense of the fast CBC construction (see Section 3.4), it needs to be slightly modified, namely one has to compute the error for all elements of $G_d^\varphi(b^{\underline{m}})$ for $m \in \{\underline{m}, \underline{m}+1, \ldots, \overline{m}\}$ and not only those which satisfy a certain criterion. This can be achieved by modifying the search and simultaneously keeping track of the "good" elements of $G_d^\varphi(b^{\underline{m}})$ by storing and finally selecting one of them. We formulate the algorithm as follows.

Algorithm 6.32 (CBC sieve construction, fast version) Let b be a prime number and let $d, \underline{m}$, and $\overline{m} \in \mathbb{N}$ be such that $\underline{m} \le \overline{m}$. For $m \in \{\underline{m}, \underline{m}+1, \ldots, \overline{m}\}$ let c_m be positive reals such that $\sum_{m=\underline{m}}^{\overline{m}} c_m^{-1} \le 1$. Construct a generating vector $\boldsymbol{g} = (g_1, \ldots, g_d) \in G_d^{\varphi}(b, \infty)$ as follows.

(1) Set $g_1 = 1$.
(2) For s from 1 to $d-1$:
Assume that $\boldsymbol{g}^{(s)} := (g_1, \ldots, g_s) \in G_s^{\varphi}(b, \infty)$ has already been found. Consider this vector as fixed and choose $g_{s+1} \in G_1^{\varphi}(b, \infty)$ as follows.

(2a) For m from $\underline{m}$ to $\overline{m}$:
Compute $\tau^* \in [1/2, \alpha)$ such that

$$\tau^* = \underset{\tau \in [1/2, \alpha)}{\operatorname{argmin}}\ c^{\tau} E_{b^m, d, \alpha, \boldsymbol{\gamma}}(\tau).$$

The existence of such a τ^* follows as outlined in the proof of Lemma 6.24.
(2b) Compute, for each $g \in G_1^{\varphi}(b^m)$,

$$\frac{\mathrm{err}_{b^m, s+1, \alpha, \boldsymbol{\gamma}}((\boldsymbol{g}^{(s)}, g))}{c^{\tau} E_{b^m, d, \alpha, \boldsymbol{\gamma}}(\tau)}.$$

End for.

(3) Define

$$T_{s+1} := \left\{ g \in G_1^{\varphi}(b^{\overline{m}}) \ :\ \max_{\underline{m} \le m \le \overline{m}} \frac{\mathrm{err}_{b^m, s+1, \alpha, \boldsymbol{\gamma}}((\boldsymbol{g}^{(s)}, g))}{c^{\tau} E_{b^m, d, \alpha, \boldsymbol{\gamma}}(\tau)} \le 1 \right\}.$$

Select g_{s+1} as

$$g_{s+1} = \underset{g \in T_{s+1}}{\operatorname{argmin}} \sum_{m=\underline{m}}^{\overline{m}} \frac{[\mathrm{err}_{b^m, s+1, \alpha, \boldsymbol{\gamma}}((\boldsymbol{g}^{(s)}, g))]^2}{c^{2\tau} [E_{b^m, d, \alpha, \boldsymbol{\gamma}}(\tau)]^2}. \tag{6.28}$$

End for.
(4) Set $\boldsymbol{g} = (g_1, \ldots, g_d)$.

The following theorem is the analogue of Theorem 6.31 and states that the generating vectors constructed by Algorithm 6.32 indeed have a low worst-case error for all instances of $m \in \{\underline{m}, \underline{m}+1, \ldots, \overline{m}\}$.

Theorem 6.33 *Let b be a prime number and let $d, \underline{m}$, and $\overline{m} \in \mathbb{N}$ be such that $\underline{m} \le \overline{m}$. For $m \in \{\underline{m}, \underline{m}+1, \ldots, \overline{m}\}$ let c_m be positive reals such that $\sum_{m=\underline{m}}^{\overline{m}} c_m^{-1} \le 1$. Assume that $\boldsymbol{g} = (g_1, \ldots, g_d) \in G_d^{\varphi}(b, \infty)$ has been constructed according to Algorithm 6.32. Then it holds that*

$$\mathrm{err}_{b^m,s,\alpha,\boldsymbol{\gamma}}((g_1,\ldots,g_s)) \le \frac{c_m^\tau}{b^{\tau m}}\left(\prod_{j=1}^{s}\left(1+4\gamma_j^{1/(2\tau)}\zeta\left(\frac{\alpha}{\tau}\right)\right)-1\right)^\tau$$

for all $s \in [d]$, *for all* $\tau \in [1/2, \alpha)$, *and for all* $m \in \{\underline{m}, \underline{m}+1, \ldots, \overline{m}\}$.

Proof Theorem 6.28 and Proposition 6.29 imply that the sets to be populated in Algorithm 6.30 are nonempty and have the proper cardinalities, respectively. This immediately yields the result. □

Remark 6.34 It should be noted that selecting g_{s+1} as in (6.28) in Step (3) of Algorithm 6.32 is not absolutely necessary. Indeed, Theorem 6.33 would also hold if we would choose any $g \in T_{s+1}$ as g_{s+1}. However, from the computational point of view, choosing g_{s+1} as in (6.28) does not cause significant additional cost, and yields better results.

The construction cost of Algorithm 6.32

The core of the algorithm is computing, in Step (2b), the quantity

$$\frac{\mathrm{err}_{b^m,s+1,\alpha,\boldsymbol{\gamma}}((\boldsymbol{g}^{(s)},g))}{c^\tau E_{b^m,d,\alpha,\boldsymbol{\gamma}}(\tau)},$$

but this can be done for each m using $O(b^m \log b^m)$ operations, employing the fast construction method outlined in Chapter 3. Hence, in total we need $O(b^{\overline{m}}(\overline{m} - \underline{m}) \log b^{\overline{m}})$ operations. Computing the maximum and the minimum in Step (3) does not cause computational cost of higher order than $O(b^{\overline{m}}(\overline{m} - \underline{m}) \log b^{\overline{m}})$. Finding the optimal τ^* in Step (2a) can be done using standard software packages, and therefore can be neglected. In total, we see that Algorithm 6.32 can be run using $O(b^{\overline{m}}(\overline{m} - \underline{m}) \log b^{\overline{m}}) = O(\overline{m}^2 b^{\overline{m}})$ operations.

6.7 A Digit-By-Digit Construction

Another, maybe even more natural idea for constructing extensible lattice rules is based on a digit-wise approach. In a nutshell, this construction works as follows. Each component of a generating vector $\boldsymbol{g}$ is considered in its b-adic expansion, where b is a prime. If we have already constructed the first m digits of each component such that the generating vector yields good results for $N = b, b^2, \ldots, b^m$, we search, in each component, for the $(m+1)$-st digit such that the generating vector yields good results also for $N = b^{m+1}$. In this way we find, digit-by-digit, a good generating vector for all $N = b, b^2, b^3, \ldots$. In contrast to other algorithms, we do not have to stop at some fixed $b^{\overline{m}}$.

As the relevant quality measure we again choose the worst-case integration error in the weighted Korobov space with smoothness α and product weights. We present an algorithm which constructs, digit-by-digit, a generating vector which is good with respect to the worst-case error $\mathrm{err}_{N,d,\alpha,\boldsymbol{\gamma}}$ for all $N = b, b^2, b^3, \ldots$.

Algorithm 6.35 (Digit-by-digit construction of extensible lattice point sets) Let b be a prime number and let $G_d(b) := \{0, 1, \ldots, b-1\}^d$.

(1) Find $\boldsymbol{g}_1$ by minimizing $[\mathrm{err}_{b,d,\alpha,\boldsymbol{\gamma}}(\boldsymbol{g})]^2$ over all $\boldsymbol{g} \in G_d(b)$, i.e.,

$$\boldsymbol{g}_1 := \operatorname*{argmin}_{\boldsymbol{g} \in G_d(b)} [\mathrm{err}_{b,d,\alpha,\boldsymbol{\gamma}}(\boldsymbol{g})]^2.$$

For $b = 2$ one may simply choose $\boldsymbol{g}_1 = \mathbf{1} \in \mathbb{Z}^d$.

(2) For $m = 2, 3, \ldots$:
Find $\boldsymbol{g}_m$ by minimizing $[\mathrm{err}_{b^m,d,\alpha,\boldsymbol{\gamma}}(\boldsymbol{g}_{m-1} + b^{m-1}\boldsymbol{g})]^2$ over all $\boldsymbol{g} \in G_d(b)$, i.e.,

$$\boldsymbol{g}_m := \operatorname*{argmin}_{\boldsymbol{g} \in G_d(b)} [\mathrm{err}_{b^m,d,\alpha,\boldsymbol{\gamma}}(\boldsymbol{g}_{m-1} + b^{m-1}\boldsymbol{g})]^2.$$

Error analysis

The lattice points $\boldsymbol{g}_m$ constructed in Algorithm 6.35 satisfy the following error bound.

Theorem 6.36 *Let b be a prime number and let $m, d \in \mathbb{N}$. Furthermore, let $\alpha > 1/2$ and let $\boldsymbol{\gamma} = (\gamma_j)_{j \geq 1}$ be product weights. Assume that $\boldsymbol{g}_m \in \mathbb{Z}^d$ has been constructed according to Algorithm 6.35. Then we have*

$$\mathrm{err}_{b^m,d,\alpha,\boldsymbol{\gamma}}(\boldsymbol{g}_m) \leq \frac{1}{b^{m/2}} \left(2 \left(\prod_{j=1}^{d} (1 + 2\gamma_j \zeta(2\alpha)) - 1 \right) \min \left(m, \frac{b^{2\alpha-1}}{b^{2\alpha-1} - 1} \right) \right)^{1/2}.$$

In the case where $b = 2$ and $\gamma_j \in (0, 1]$ for all $j \in \mathbb{N}$ the factor $\sqrt{2}$ in the upper bound can be omitted.

Remark 6.37 The upper bound in Theorem 6.36 essentially matches the upper bound in Theorem 6.25 obtained for the sieve algorithm, but only for $\tau = 1/2$. This is a theoretical disadvantage. With the proof method for the digit-by-digit construction we cannot theoretically guarantee better convergence rates such as $O(N^{-\alpha})$, although these can be observed in numerical experiments.

Remark 6.38 The search for $\boldsymbol{g}_1$ in the first step of Algorithm 6.35 takes—at least for integer α—as many as $O(d\, b^{d+1})$ operations, which can be reduced to $O(d^2 b^2)$ operations by using a CBC construction. In this case one gets a slightly weaker error bound in Theorem 6.36 (the term "-1" after the product must be dropped).

Alternatively, one can choose $\boldsymbol{g}_1 = \mathbf{1}$ in all cases, but then the upper bound in Theorem 6.36 has to be replaced by

$$\mathrm{err}_{b^m,d,\alpha,\boldsymbol{\gamma}}(\boldsymbol{g}_m) \le \frac{1}{b^{m/2}}\left(b\left(\prod_{j=1}^{d}(1+2\gamma_j\zeta(2\alpha))-1\right)\min\left(m, \frac{b^{2\alpha-1}}{b^{2\alpha-1}-1}\right)\right)^{1/2},$$

except for the case $b = 2$ and $\gamma_j \in (0, 1]$ for all $j \in \mathbb{N}$, where the bound in Theorem 6.36 still holds and the factor $\sqrt{2}$ can again be omitted. This follows immediately from the subsequent proof of Theorem 6.36.

Proof of Theorem 6.36 We first show the result for $m = 1$. Employing again the idea that the minimum is at least as good as the average (see the standard averaging argument in Remark 2.14), we have

$$\begin{aligned}
[\mathrm{err}_{b,d,\alpha,\boldsymbol{\gamma}}(\boldsymbol{g}_1)]^2 &\le \frac{1}{b^d}\sum_{\boldsymbol{g}\in G_d(b)}[\mathrm{err}_{b,d,\alpha,\boldsymbol{\gamma}}(\boldsymbol{g})]^2 \\
&= \frac{1}{b^d}\sum_{\boldsymbol{h}\in\mathbb{Z}^d\setminus\{\boldsymbol{0}\}}\frac{1}{r_{2\alpha,\boldsymbol{\gamma}}(\boldsymbol{h})}\sum_{\substack{\boldsymbol{g}\in G_d(b)\\ \boldsymbol{h}\cdot\boldsymbol{g}\equiv 0 \pmod{b}}} 1,
\end{aligned}$$

where in the second step we used the worst-case error formula in Corollary 2.21 and changed the order of summation. Recall that b is a prime number, so

$$\sum_{\substack{\boldsymbol{g}\in G_d(b)\\ \boldsymbol{h}\cdot\boldsymbol{g}\equiv 0 \pmod{b}}} 1 = |\{\boldsymbol{g}\in G_d(b)\ :\ \boldsymbol{h}\cdot\boldsymbol{g}\equiv 0 \pmod{b}\}| = \begin{cases} b^d & \text{if } \boldsymbol{h}\equiv\boldsymbol{0} \pmod{b},\\ b^{d-1} & \text{if } \boldsymbol{h}\not\equiv\boldsymbol{0} \pmod{b}.\end{cases}$$

Hence we obtain

$$\begin{aligned}
[\mathrm{err}_{b,d,\alpha,\boldsymbol{\gamma}}(\boldsymbol{g}_1)]^2 &\le \sum_{\substack{\boldsymbol{h}\in\mathbb{Z}^d\setminus\{\boldsymbol{0}\}\\ \boldsymbol{h}\equiv\boldsymbol{0} \pmod{b}}}\frac{1}{r_{2\alpha,\boldsymbol{\gamma}}(\boldsymbol{h})} + \frac{1}{b}\sum_{\substack{\boldsymbol{h}\in\mathbb{Z}^d\setminus\{\boldsymbol{0}\}\\ \boldsymbol{h}\not\equiv\boldsymbol{0} \pmod{b}}}\frac{1}{r_{2\alpha,\boldsymbol{\gamma}}(\boldsymbol{h})} \\
&= \left(1-\frac{1}{b}\right)\sum_{\boldsymbol{h}\in\mathbb{Z}^d\setminus\{\boldsymbol{0}\}}\frac{1}{r_{2\alpha,\boldsymbol{\gamma}}(b\boldsymbol{h})} + \frac{1}{b}\sum_{\boldsymbol{h}\in\mathbb{Z}^d\setminus\{\boldsymbol{0}\}}\frac{1}{r_{2\alpha,\boldsymbol{\gamma}}(\boldsymbol{h})} \\
&\le \frac{2}{b}\sum_{\boldsymbol{h}\in\mathbb{Z}^d\setminus\{\boldsymbol{0}\}}\frac{1}{r_{2\alpha,\boldsymbol{\gamma}}(\boldsymbol{h})} \\
&= \frac{2}{b}\left(\prod_{j=1}^{d}(1+2\gamma_j\zeta(2\alpha))-1\right).
\end{aligned}$$

For $b = 2$ and $\gamma_j \in (0, 1]$ for all $j \in \mathbb{N}$, and with the choice $\boldsymbol{g}_1 = \mathbf{1}$, we can bound $[\mathrm{err}_{2,d,\alpha,\boldsymbol{\gamma}}(\mathbf{1})]^2$ directly. In this case we obtain with Corollary 2.21,

$$
\begin{aligned}
&[\mathrm{err}_{2,d,\alpha,\boldsymbol{\gamma}}(\mathbf{1})]^2 \\
&= -1 + \sum_{h_1,\ldots,h_{d-1}=-\infty}^{\infty} \prod_{j=1}^{d-1} \frac{1}{r_{2\alpha,\gamma_j}(h_j)} \sum_{\substack{h=-\infty \\ h\equiv h_1+\cdots+h_{d-1} \pmod 2}}^{\infty} \frac{1}{r_{2\alpha,\gamma_d}(h)}.
\end{aligned}
$$

We denote the innermost sum in the previous expression by Σ_1. If $h_1 + \cdots + h_{d-1} \equiv 0 \pmod 2$, then we have

$$
\Sigma_1 = \sum_{h=-\infty}^{\infty} \frac{1}{r_{2\alpha,\gamma_d}(2h)} = 1 + \frac{2}{2^{2\alpha}} \gamma_d \zeta(2\alpha).
$$

If $h_1 + \cdots + h_{s-1} \equiv 1 \pmod 2$, we get

$$
\begin{aligned}
\Sigma_1 &= \sum_{h=-\infty}^{\infty} \frac{1}{r_{2\alpha,\gamma_d}(2h+1)} \\
&= 2\sum_{h=0}^{\infty} \frac{1}{r_{2\alpha,\gamma_d}(2h+1)} \\
&= 2\gamma_d \left(\sum_{h=1}^{\infty} \frac{1}{h^{2\alpha}} - \sum_{h=1}^{\infty} \frac{1}{(2h)^{2\alpha}} \right) \\
&= 2\gamma_d \zeta(2\alpha) - \frac{2}{2^{2\alpha}} \gamma_d \zeta(2\alpha).
\end{aligned}
$$

Altogether we obtain

$$
\Sigma_1 = \frac{1}{2} + \gamma_d \zeta(2\alpha) + (-1)^{h_1+\cdots+h_{d-1}} \left(\frac{1}{2} - \gamma_d \zeta(2\alpha) + \frac{2}{2^{2\alpha}} \gamma_d \zeta(2\alpha) \right).
$$

Therefore we get

$$
\begin{aligned}
[\mathrm{err}_{2,d,\alpha,\boldsymbol{\gamma}}(\mathbf{1})]^2 &= -1 + \left(\frac{1}{2} + \gamma_d \zeta(2\alpha) \right) \prod_{j=1}^{d-1} \left(\sum_{h=-\infty}^{\infty} \frac{1}{r_{2\alpha,\gamma_j}(h)} \right) \\
&\quad + \left(\frac{1}{2} - \gamma_d \zeta(2\alpha) + \frac{2}{2^{2\alpha}} \gamma_d \zeta(2\alpha) \right) \prod_{j=1}^{d-1} \left(\sum_{h=-\infty}^{\infty} \frac{(-1)^h}{r_{2\alpha,\gamma_j}(h)} \right) \\
&= -1 + \frac{1}{2} \prod_{j=1}^{d} (1 + 2\gamma_j \zeta(2\alpha)) \\
&\quad + \frac{1}{2} \left(1 + \frac{4}{2^{2\alpha}} \gamma_d \zeta(2\alpha) - 2\gamma_d \zeta(2\alpha) \right) \prod_{j=1}^{d-1} \left(\sum_{h=-\infty}^{\infty} \frac{(-1)^h}{r_{2\alpha,\gamma_j}(h)} \right).
\end{aligned}
$$

Since

$$\sum_{h=-\infty}^{\infty} \frac{(-1)^h}{r_{2\alpha,\gamma_j}(h)} = 1 + \frac{4}{2^{2\alpha}}\gamma_j\zeta(2\alpha) - 2\gamma_j\zeta(2\alpha),$$

we obtain

$$[\mathrm{err}_{2,d,\alpha,\boldsymbol{\gamma}}(\mathbf{1})]^2 = -1 + \frac{1}{2}\prod_{j=1}^{d}(1 + 2\gamma_j\zeta(2\alpha)) + \frac{1}{2}\prod_{j=1}^{d}\left(1 - \left(2 - \frac{4}{2^{2\alpha}}\right)\gamma_j\zeta(2\alpha)\right). \tag{6.29}$$

Next we claim that for any $\alpha > 1/2$ the assertion

$$1 < \left(2 - \frac{4}{2^{2\alpha}}\right)\zeta(2\alpha) < 2 \tag{6.30}$$

holds, which is equivalent to

$$\frac{1}{2} \cdot \frac{1}{1 - 2/2^{2\alpha}} < \zeta(2\alpha) < \frac{1}{1 - 2/2^{2\alpha}},$$

which in turn is equivalent to

$$\frac{1}{2}\sum_{i=0}^{\infty}\left(\frac{2}{2^{2\alpha}}\right)^i < \sum_{i=1}^{\infty}\frac{1}{i^{2\alpha}} < \sum_{i=0}^{\infty}\left(\frac{2}{2^{2\alpha}}\right)^i.$$

The latter chain of inequalities is shown by comparing the three series above by suitably grouping the summands. For instance, to show the upper bound on $\zeta(2\alpha)$, we compare the first term of the series for $\zeta(2\alpha)$ with the first term of the last series, the sum of the second and third terms of the series for $\zeta(2\alpha)$ with the second term of the last series, the sum of the fourth, fifth, sixth, and seventh terms of the series for $\zeta(2\alpha)$ with the third term of the last series, and so on. In this way we find that (6.30) is indeed correct.

As $\gamma_j \in (0, 1]$ for all $j \in \mathbb{N}$, (6.29) and (6.30) yield

$$[\mathrm{err}_{2,d,\alpha,\boldsymbol{\gamma}}(\mathbf{1})]^2 = \frac{1}{2}\prod_{j=1}^{d}(1+2\gamma_j\zeta(2\alpha)) - 1 + \frac{1}{2}\prod_{j=1}^{d}\theta_j(\alpha) \quad \text{with} \quad -1 < \theta_j(\alpha) < 1,$$

from which we obtain

$$[\mathrm{err}_{2,d,\alpha,\boldsymbol{\gamma}}(\mathbf{1})]^2 \le \frac{1}{2}\left(\prod_{j=1}^{d}(1 + 2\gamma_j\zeta(2\alpha)) - 1\right).$$

Hence the result follows for $m = 1$ and any prime number b.

Let now $m \geq 2$. Like for $m = 1$ we get

$$[\mathrm{err}_{b^m,d,\alpha,\gamma}(\boldsymbol{g}_m)]^2 \leq \frac{1}{b^d} \sum_{\boldsymbol{g} \in G_d(b)} [\mathrm{err}_{b^m,d,\alpha,\gamma}(\boldsymbol{g}_{m-1} + b^{m-1}\boldsymbol{g})]^2$$
$$= \frac{1}{b^d} \sum_{\boldsymbol{h} \in \mathbb{Z}^d \setminus \{\boldsymbol{0}\}} \frac{1}{r_{2\alpha,\gamma}(\boldsymbol{h})} \sum_{\substack{\boldsymbol{g} \in G_d(b) \\ \boldsymbol{h} \cdot (\boldsymbol{g}_{m-1} + b^{m-1}\boldsymbol{g}) \equiv 0 \pmod{b^m}}} 1.$$

The inner sum of the previous term is equal to the number of $\boldsymbol{g} \in G_d(b)$ satisfying $b^{m-1}\boldsymbol{h} \cdot \boldsymbol{g} \equiv -\boldsymbol{h} \cdot \boldsymbol{g}_{m-1} \pmod{b^m}$, which implies $\boldsymbol{h} \cdot \boldsymbol{g}_{m-1} \equiv 0 \pmod{b^{m-1}}$, and then $\boldsymbol{h} \cdot \boldsymbol{g} \equiv -b^{1-m}\boldsymbol{h} \cdot \boldsymbol{g}_{m-1} \pmod{b}$. Thus,

$$[\mathrm{err}_{b^m,d,\alpha,\gamma}(\boldsymbol{g}_m)]^2 \leq \frac{1}{b^d} \sum_{\substack{\boldsymbol{h} \in \mathbb{Z}^d \setminus \{\boldsymbol{0}\} \\ \boldsymbol{h} \cdot \boldsymbol{g}_{m-1} \equiv 0 \pmod{b^{m-1}}}} \frac{1}{r_{2\alpha,\gamma}(\boldsymbol{h})} \sum_{\substack{\boldsymbol{g} \in G_d(b) \\ \boldsymbol{h} \cdot \boldsymbol{g} \equiv -b^{1-m}\boldsymbol{h} \cdot \boldsymbol{g}_{m-1} \pmod{b}}} 1.$$

Consider the inner sum of the latter expression. If $\boldsymbol{h} \not\equiv \boldsymbol{0} \pmod{b}$, then the sum is equal to b^{d-1}. If $\boldsymbol{h} \equiv \boldsymbol{0} \pmod{b}$, the sum equals 0 if $\boldsymbol{h} \cdot \boldsymbol{g}_{m-1} \not\equiv 0 \pmod{b^m}$ and b^d if $\boldsymbol{h} \cdot \boldsymbol{g}_{m-1} \equiv 0 \pmod{b^m}$. Consequently,

$$[\mathrm{err}_{b^m,d,\alpha,\gamma}(\boldsymbol{g}_m)]^2$$
$$\leq \frac{1}{b} \sum_{\substack{\boldsymbol{h} \in \mathbb{Z}^d \\ \boldsymbol{h} \not\equiv \boldsymbol{0} \pmod{b} \\ \boldsymbol{h} \cdot \boldsymbol{g}_{m-1} \equiv 0 \pmod{b^{m-1}}}} \frac{1}{r_{2\alpha,\gamma}(\boldsymbol{h})} + \sum_{\substack{\boldsymbol{h} \in \mathbb{Z}^d \setminus \{\boldsymbol{0}\} \\ \boldsymbol{h} \equiv \boldsymbol{0} \pmod{b} \\ \boldsymbol{h} \cdot \boldsymbol{g}_{m-1} \equiv 0 \pmod{b^m}}} \frac{1}{r_{2\alpha,\gamma}(\boldsymbol{h})}$$
$$= \frac{1}{b}[\mathrm{err}_{b^{m-1},d,\alpha,\gamma}(\boldsymbol{g}_{m-1})]^2 - \frac{1}{b} \sum_{\substack{\boldsymbol{h} \in \mathbb{Z}^d \setminus \{\boldsymbol{0}\} \\ \boldsymbol{h} \equiv \boldsymbol{0} \pmod{b} \\ \boldsymbol{h} \cdot \boldsymbol{g}_{m-1} \equiv 0 \pmod{b^{m-1}}}} \frac{1}{r_{2\alpha,\gamma}(\boldsymbol{h})}$$
$$+ \sum_{\substack{\boldsymbol{h} \in \mathbb{Z}^d \setminus \{\boldsymbol{0}\} \\ \boldsymbol{h} \equiv \boldsymbol{0} \pmod{b} \\ \boldsymbol{h} \cdot \boldsymbol{g}_{m-1} \equiv 0 \pmod{b^m}}} \frac{1}{r_{2\alpha,\gamma}(\boldsymbol{h})}.$$

We proceed inductively and insert the latter bound analogously for $[\mathrm{err}_{b^{m-1},d,\alpha,\gamma}(\boldsymbol{g}_{m-1})]^2$. Then we get

$$[\mathrm{err}_{b^m,d,\alpha,\gamma}(\boldsymbol{g}_m)]^2 \leq \frac{1}{b^2}[\mathrm{err}_{b^{m-2},d,\alpha,\gamma}(\boldsymbol{g}_{m-2})]^2$$
$$-\frac{1}{b^2} \sum_{\substack{\boldsymbol{h} \in \mathbb{Z}^d \setminus \{\boldsymbol{0}\} \\ \boldsymbol{h} \equiv \boldsymbol{0} \pmod{b} \\ \boldsymbol{h} \cdot \boldsymbol{g}_{m-2} \equiv 0 \pmod{b^{m-2}}}} \frac{1}{r_{2\alpha,\gamma}(\boldsymbol{h})} + \frac{1}{b} \sum_{\substack{\boldsymbol{h} \in \mathbb{Z}^d \setminus \{\boldsymbol{0}\} \\ \boldsymbol{h} \equiv \boldsymbol{0} \pmod{b} \\ \boldsymbol{h} \cdot \boldsymbol{g}_{m-2} \equiv 0 \pmod{b^{m-1}}}} \frac{1}{r_{2\alpha,\gamma}(\boldsymbol{h})}$$

$$-\frac{1}{b}\sum_{\substack{\boldsymbol{h}\in\mathbb{Z}^d\setminus\{\boldsymbol{0}\}\\ \boldsymbol{h}\equiv\boldsymbol{0}\ (\mathrm{mod}\ b)\\ \boldsymbol{h}\cdot\boldsymbol{g}_{m-1}\equiv 0\ (\mathrm{mod}\ b^{m-1})}}\frac{1}{r_{2\alpha,\boldsymbol{\gamma}}(\boldsymbol{h})}+\sum_{\substack{\boldsymbol{h}\in\mathbb{Z}^d\setminus\{\boldsymbol{0}\}\\ \boldsymbol{h}\equiv\boldsymbol{0}\ (\mathrm{mod}\ b)\\ \boldsymbol{h}\cdot\boldsymbol{g}_{m-1}\equiv 0\ (\mathrm{mod}\ b^{m})}}\frac{1}{r_{2\alpha,\boldsymbol{\gamma}}(\boldsymbol{h})}.$$

Assume that $\boldsymbol{h}\in\mathbb{Z}^d$ with $\boldsymbol{h}\equiv\boldsymbol{0}\ (\mathrm{mod}\ b)$, i.e., $\boldsymbol{h}=b\widetilde{\boldsymbol{h}}$ for some $\widetilde{\boldsymbol{h}}\in\mathbb{Z}^d$, and $\boldsymbol{h}\cdot\boldsymbol{g}_{m-2}\equiv 0\ (\mathrm{mod}\ b^{m-1})$. Then we have

$$\boldsymbol{h}\cdot\boldsymbol{g}_{m-1}=\boldsymbol{h}\cdot(\boldsymbol{g}_{m-2}+b^{m-2}\boldsymbol{g})=\boldsymbol{h}\cdot\boldsymbol{g}_{m-2}+b^{m-1}\widetilde{\boldsymbol{h}}\cdot\boldsymbol{g}\equiv 0\ (\mathrm{mod}\ b^{m-1})$$

for some $\boldsymbol{g}\in G_d(b)$. Therefore we obtain

$$[\mathrm{err}_{b^m,d,\alpha,\boldsymbol{\gamma}}(\boldsymbol{g}_m)]^2\le\frac{1}{b^2}[\mathrm{err}_{b^{m-2},d,\alpha,\boldsymbol{\gamma}}(\boldsymbol{g}_{m-2})]^2$$
$$-\frac{1}{b^2}\sum_{\substack{\boldsymbol{h}\in\mathbb{Z}^d\setminus\{\boldsymbol{0}\}\\ \boldsymbol{h}\equiv\boldsymbol{0}\ (\mathrm{mod}\ b)\\ \boldsymbol{h}\cdot\boldsymbol{g}_{m-2}\equiv 0\ (\mathrm{mod}\ b^{m-2})}}\frac{1}{r_{2\alpha,\boldsymbol{\gamma}}(\boldsymbol{h})}+\sum_{\substack{\boldsymbol{h}\in\mathbb{Z}^d\setminus\{\boldsymbol{0}\}\\ \boldsymbol{h}\equiv\boldsymbol{0}\ (\mathrm{mod}\ b)\\ \boldsymbol{h}\cdot\boldsymbol{g}_{m-1}\equiv 0\ (\mathrm{mod}\ b^{m})}}\frac{1}{r_{2\alpha,\boldsymbol{\gamma}}(\boldsymbol{h})}.$$

Repeating this argument, we get

$$[\mathrm{err}_{b^m,d,\alpha,\boldsymbol{\gamma}}(\boldsymbol{g}_m)]^2\le\frac{1}{b^{m-1}}[\mathrm{err}_{b,d,\alpha,\boldsymbol{\gamma}}(\boldsymbol{g}_1)]^2$$
$$-\frac{1}{b^{m-1}}\sum_{\substack{\boldsymbol{h}\in\mathbb{Z}^d\setminus\{\boldsymbol{0}\}\\ \boldsymbol{h}\equiv\boldsymbol{0}\ (\mathrm{mod}\ b)\\ \boldsymbol{h}\cdot\boldsymbol{g}_{1}\equiv 0\ (\mathrm{mod}\ b)}}\frac{1}{r_{2\alpha,\boldsymbol{\gamma}}(\boldsymbol{h})}+\sum_{\substack{\boldsymbol{h}\in\mathbb{Z}^d\setminus\{\boldsymbol{0}\}\\ \boldsymbol{h}\equiv\boldsymbol{0}\ (\mathrm{mod}\ b)\\ \boldsymbol{h}\cdot\boldsymbol{g}_{m-1}\equiv 0\ (\mathrm{mod}\ b^{m})}}\frac{1}{r_{2\alpha,\boldsymbol{\gamma}}(\boldsymbol{h})}.$$

For the second sum in the previous expression we have

$$\begin{aligned}\sum_{\substack{\boldsymbol{h}\in\mathbb{Z}^d\setminus\{\boldsymbol{0}\}\\ \boldsymbol{h}\equiv\boldsymbol{0}\ (\mathrm{mod}\ b)\\ \boldsymbol{h}\cdot\boldsymbol{g}_{m-1}\equiv 0\ (\mathrm{mod}\ b^{m})}}\frac{1}{r_{2\alpha,\boldsymbol{\gamma}}(\boldsymbol{h})}&=\sum_{\substack{\boldsymbol{h}\in\mathbb{Z}^d\setminus\{\boldsymbol{0}\}\\ b\boldsymbol{h}\cdot\boldsymbol{g}_{m-1}\equiv 0\ (\mathrm{mod}\ b^{m})}}\frac{1}{r_{2\alpha,\boldsymbol{\gamma}}(b\boldsymbol{h})}\\&=\sum_{\substack{\boldsymbol{h}\in\mathbb{Z}^d\setminus\{\boldsymbol{0}\}\\ \boldsymbol{h}\cdot\boldsymbol{g}_{m-1}\equiv 0\ (\mathrm{mod}\ b^{m-1})}}\frac{1}{r_{2\alpha,\boldsymbol{\gamma}}(b\boldsymbol{h})}\\&\le\frac{1}{b^{2\alpha}}\sum_{\substack{\boldsymbol{h}\in\mathbb{Z}^d\setminus\{\boldsymbol{0}\}\\ \boldsymbol{h}\cdot\boldsymbol{g}_{m-1}\equiv 0\ (\mathrm{mod}\ b^{m-1})}}\frac{1}{r_{2\alpha,\boldsymbol{\gamma}}(\boldsymbol{h})}\\&=\frac{1}{b^{2\alpha}}[\mathrm{err}_{b^{m-1},d,\alpha,\boldsymbol{\gamma}}(\boldsymbol{g}_{m-1})]^2.\end{aligned}$$

With this upper bound, we obtain

$$[\mathrm{err}_{b^m,d,\alpha,\boldsymbol{\gamma}}(\boldsymbol{g}_m)]^2\le\frac{1}{b^{m-1}}[\mathrm{err}_{b,d,\alpha,\boldsymbol{\gamma}}(\boldsymbol{g}_1)]^2+\frac{1}{b^{2\alpha}}[\mathrm{err}_{b^{m-1},d,\alpha,\boldsymbol{\gamma}}(\boldsymbol{g}_{m-1})]^2.$$

With backward induction on m and invoking the upper bound on $[\mathrm{err}_{b,d,\alpha,\boldsymbol{\gamma}}(\boldsymbol{g}_1)]^2$ we get

$$\begin{aligned}
&[\mathrm{err}_{b^m,d,\alpha,\boldsymbol{\gamma}}(\boldsymbol{g}_m)]^2 \\
&\le [\mathrm{err}_{b,d,\alpha,\boldsymbol{\gamma}}(\boldsymbol{g}_1)]^2 \left(\frac{1}{b^{m-1}} + \frac{1}{b^{m-2+2\alpha}}\right) + \frac{1}{b^{4\alpha}} [\mathrm{err}_{b^{m-2},d,\alpha,\boldsymbol{\gamma}}(\boldsymbol{g}_{m-2})]^2 \\
&\ \vdots \\
&\le [\mathrm{err}_{b,d,\alpha,\boldsymbol{\gamma}}(\boldsymbol{g}_1)]^2 \sum_{k=0}^{m-2} \frac{1}{b^{m-1-k+k2\alpha}} + \frac{1}{b^{(m-1)2\alpha}} [\mathrm{err}_{b,d,\alpha,\boldsymbol{\gamma}}(\boldsymbol{g}_1)]^2 \\
&= [\mathrm{err}_{b,d,\alpha,\boldsymbol{\gamma}}(\boldsymbol{g}_1)]^2 \sum_{k=0}^{m-1} \frac{1}{b^{m-1-k+k2\alpha}} \\
&\le \left(\prod_{j=1}^{d} (1 + 2\gamma_j \zeta(2\alpha)) - 1\right) \min\left(m, \frac{b^{2\alpha-1}}{b^{2\alpha-1} - 1}\right) \frac{2}{b^m},
\end{aligned}$$

and, for $b = 2$ and $\boldsymbol{g}_1 = \mathbf{1}$,

$$[\mathrm{err}_{2^m,d,\alpha,\boldsymbol{\gamma}}(\boldsymbol{g}_m)]^2 \le \left(\prod_{j=1}^{d} (1 + 2\gamma_j \zeta(2\alpha)) - 1\right) \min\left(m, \frac{2^{2\alpha-1}}{2^{2\alpha-1} - 1}\right) \frac{1}{2^m},$$

as long as $\gamma_j \in (0, 1]$ for all $j \in \mathbb{N}$. Taking the square root yields the desired result. □

Notes and Remarks

The existence results for extensible lattice rules in Section 6.2 are taken from [105]. Regarding the result in Proposition 6.14, we remark that Algorithm 6.15 was used in [104] as a basis to construct sequences of embedded nonequal-weight quadrature rules that may yield convergence rates better than N^{-1} for arithmetic sequences of N.

The presentation in Section 6.3 concerning embedded lattice rules mainly follows [27]. Theorems 6.18 and 6.21 are due to Nuyens and Cools; we have omitted their proofs in the present outline and refer to the original paper [215].

The results presented in Section 6.4 are taken from [58]. For the presentation we have restricted ourselves to the case of prime numbers b but it should be remarked that the results in [58] even hold for arbitrary choices of $b \ge 2$. We note that there are some obvious similarities between Algorithm 6.32 presented in Section 6.6 and Algorithm 6.15 presented in Section 6.3. Indeed, with some modifications Algorithm 6.32 can be reformulated such that it yields Algorithm 6.15, and one can then also show efficient error bounds for the generating vectors constructed by this algorithm, which was still missing in [27]. For further details we refer to [58, Theorem 13].

The digit-wise construction in Section 6.7 is based on an idea of Korobov [141] (see also Section 4.6) and was worked out for the present problem in [201]. Observe that Algorithm 6.35 is designed for a fixed dimension d and therefore is not extensible in the dimension. On the other hand the number of points N is allowed to be extended in Algorithm 6.35, and the construction yields—to some extent—good results for consecutive prime powers. Note that this is exactly the "opposite" as for Algorithm 4.20 in Section 4.6, which is extensible in d but not in the number of points. Whether or not one can find a digit-by-digit algorithm that is extensible in both the dimension and the number of points remains an open question.

Finally it should be also remarked that the idea of finding good extensible lattice rules by means of a sieve method or a digit-by-digit approach can also be applied to the setting where the figure of merit is given in terms of the quality measure R, which is related to the discrepancy of the lattice point sets. For details, see [58, Section 6] (for the sieve method) and [201, Section 2.2] (for the digit-wise approach).

Chapter 7
Lattice Rules for Nonperiodic Integrands

We have already seen that lattice rules perform well for the numerical integration of functions belonging to Korobov spaces, which are spaces of periodic functions. If we are given a not necessarily periodic, but still smooth integrand, then the theory we have discussed so far does not apply. We have several options to resolve this situation. A classical approach is to use periodization, which means that we transform a sufficiently smooth nonperiodic integrand f into a periodic integrand $\overline{f}$ without changing the value of the integral. A problem with this method is that the periodic extensions $\overline{f}$ often do not behave nicely at the boundaries of $[0, 1]^d$. In particular, the norm of the transformed integrand can depend exponentially on the dimension d, which implies that periodization is only feasible for small dimensions d. We will not consider periodization methods here since this is classical knowledge that is available in the existing literature (see the "Notes and Remarks" Section at the end of this chapter).

In this chapter we present three methods allowing us to obtain a higher convergence rate for smooth, nonperiodic functions using lattice rules (i.e., under suitable smoothness assumptions we obtain a bound on the worst-case error tending to zero faster than $N^{-1/2}$), namely

1. shifted lattice rules,
2. folded or tent-transformed lattice rules,
3. symmetrized lattice rules.

As the underlying function spaces we consider Sobolev spaces of a given smoothness, and cosine spaces. As we will see, all of these methods relate the nonperiodic integrands back to periodic integrands in some way. However, in the function spaces we consider, they avoid the problems associated with some classical periodization techniques.

J. Dick et al., *Lattice Rules*, Springer Series in Computational Mathematics 58,
https://doi.org/10.1007/978-3-031-09951-9_7

7.1 Shifted Lattice Rules and Integration in Weighted Sobolev Spaces

One way to treat nonperiodic integrands is to use shifted lattice rules, which are based on a geometric shift modulo one of the underlying integration nodes.

Definition 7.1 For an N-element point set $\mathcal{P} = \{\boldsymbol{x}_0, \boldsymbol{x}_1, \ldots, \boldsymbol{x}_{N-1}\}$ in $[0,1)^d$ and a point $\boldsymbol{\Delta} \in [0,1]^d$ the *shifted point set* $\mathcal{P}_{\boldsymbol{\Delta}}$ is defined as

$$\mathcal{P}_{\boldsymbol{\Delta}} := \{\{\boldsymbol{x}_k + \boldsymbol{\Delta}\} \ : \ k \in \{0, 1, \ldots, N-1\}\},$$

where $\{\boldsymbol{x}_k + \boldsymbol{\Delta}\}$ means component-wise addition modulo one (we remind the reader that by $\{x\}$ we mean the fractional part of a real number $x \in \mathbb{R}$, i.e., $\{x\} = x - \lfloor x \rfloor$. For vectors, the fractional part is defined component-wise). In this context, the point $\boldsymbol{\Delta}$ is called the *shift*. For a QMC rule $Q_{N,d}$ based on a point set $\mathcal{P}$ and for a shift $\boldsymbol{\Delta}$ the QMC rule

$$Q_{N,d}(f, \boldsymbol{\Delta}) := \frac{1}{N} \sum_{k=0}^{N-1} f\left(\{\boldsymbol{x}_k + \boldsymbol{\Delta}\}\right)$$

is called a *shifted QMC rule*.

We are particularly interested in the case where $\mathcal{P} = \mathcal{P}(\boldsymbol{g}, N)$ is the node set underlying a rank-1 lattice rule. In this case we write $\mathcal{P}_{\boldsymbol{\Delta}}(\boldsymbol{g}, N)$ for the shifted point set

$$\left\{\left\{\frac{k}{N}\boldsymbol{g} + \boldsymbol{\Delta}\right\} \ : \ k \in \{0, 1, \ldots, N-1\}\right\}.$$

Figure 7.1 shows an example of a shifted lattice point set.

Furthermore, we call a shifted QMC rule that is based on $\mathcal{P}_{\boldsymbol{\Delta}}(\boldsymbol{g}, N)$, i.e.,

$$Q_{N,d}(f, \boldsymbol{\Delta}) = \frac{1}{N} \sum_{k=0}^{N-1} f\left(\left\{\frac{k}{N}\boldsymbol{g} + \boldsymbol{\Delta}\right\}\right),$$

a *shifted lattice rule*.

In the study of nonperiodic functions it is useful to choose the shift randomly, where the random shift is a vector whose components are i.i.d. (independent and identically distributed) and uniformly distributed in the unit cube. Since we then have

$$\begin{aligned}
\mathbb{E}\left[\frac{1}{N} \sum_{k=0}^{N-1} f(\{\boldsymbol{x}_k + \boldsymbol{\Delta}\})\right] &= \frac{1}{N} \sum_{k=0}^{N-1} \mathbb{E}[f(\{\boldsymbol{x}_k + \boldsymbol{\Delta}\})] \\
&= \frac{1}{N} \sum_{k=0}^{N-1} \int_{[0,1]^d} f(\{\boldsymbol{x}_k + \boldsymbol{\Delta}\}) \, \mathrm{d}\boldsymbol{\Delta} \\
&= \int_{[0,1]^d} f(\boldsymbol{\Delta}) \, \mathrm{d}\boldsymbol{\Delta},
\end{aligned}$$

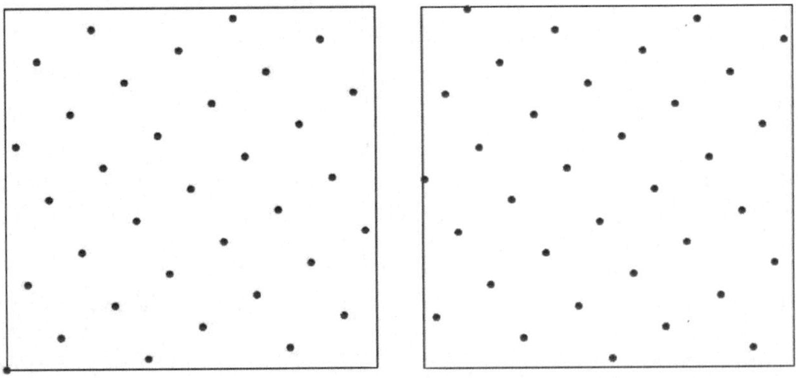

Fig. 7.1: The Fibonacci lattice point set $\mathcal{P}((1, 21), 34)$ (left) and the shifted version thereof with shift $\boldsymbol{\Delta} = (0.3, 0.7)$ (right).

it follows that the estimator $Q_{N,d}(f, \boldsymbol{\Delta})$ is an *unbiased* estimator, which simply means that the expected value of the random variable $Q_{N,d}(f, \boldsymbol{\Delta})$ is the exact value of the integral of f. Since $Q_{N,d}(f, \boldsymbol{\Delta})$ is unbiased, we study the variance (or equivalently, the standard deviation) of the worst-case integration error. (Later, in Chapter 11, we will consider a different criterion.)

Recall that the integration error using a QMC rule based on the point set $\mathcal{P}$ in (reproducing kernel) Hilbert spaces $(\mathcal{H}, \|\cdot\|)$ of functions on $[0, 1]^d$ is, according to Definition 1.26, given by

$$\mathrm{err}_{N,d}(\mathcal{H}, \mathcal{P}) = \sup_{\substack{f \in \mathcal{H} \\ \|f\| \le 1}} |\mathrm{err}_{N,d}(f, \mathcal{P})|.$$

Definition 7.2 Let $\mathrm{err}_{N,d}(\mathcal{H}, \mathcal{P})$ be the worst-case error of a QMC rule based on the point set $\mathcal{P}$ in $[0, 1)^d$ applied to functions from a Hilbert space $(\mathcal{H}, \|\cdot\|)$. Assume that the components of the shift $\boldsymbol{\Delta}$ are i.i.d. uniformly distributed on $[0, 1]$. Then the root mean square worst-case error of the shifted rule,

$$\mathrm{err}^{\mathrm{sh}}_{N,d}(\mathcal{H}, \mathcal{P}) := \sqrt{\mathbb{E}\left[[\mathrm{err}_{N,d}(\mathcal{H}, \mathcal{P}_{\boldsymbol{\Delta}})]^2\right]}$$

is referred to as the *shift-averaged worst-case error*. Here, the expectation is taken with respect to the shift $\boldsymbol{\Delta}$.

Remark 7.3

1. Since the components of $\boldsymbol{\Delta}$ are i.i.d. uniformly distributed on $[0, 1]^d$, we have

$$\mathbb{E}\left[[\mathrm{err}_{N,d}(\mathcal{H}, \mathcal{P}_{\boldsymbol{\Delta}})]^2\right] = \int_{[0,1]^d} [\mathrm{err}_{N,d}(\mathcal{H}, \mathcal{P}_{\boldsymbol{\Delta}})]^2 \, \mathrm{d}\boldsymbol{\Delta}.$$

2. According to the averaging argument (see Remark 2.14), there exists at least one shift $\boldsymbol{\Delta} \in [0,1]^d$ for every N-element point set $\mathcal{P}$ such that

$$\mathrm{err}_{N,d}(\mathcal{H}, \mathcal{P}_{\boldsymbol{\Delta}}) \leq \mathrm{err}^{\mathrm{sh}}_{N,d}(\mathcal{H}, \mathcal{P}).$$

The shift-averaged worst-case error will be used as a quality criterion for shifted lattice rules. In the context of numerical integration in reproducing kernel Hilbert spaces $\mathcal{H}(K)$, Hickernell [100] introduced the notion of a shift-invariant kernel, which is an important tool to represent the shift-averaged worst-case error.

Definition 7.4 For a reproducing kernel $K : [0,1]^d \times [0,1]^d \to \mathbb{C}$, the *shift-invariant kernel* is defined as

$$K^{\mathrm{sh}}(\boldsymbol{x}, \boldsymbol{y}) := \int_{[0,1]^d} K(\{\boldsymbol{x} + \boldsymbol{\Delta}\}, \{\boldsymbol{y} + \boldsymbol{\Delta}\}) \, \mathrm{d}\boldsymbol{\Delta} \quad \text{for } \boldsymbol{x}, \boldsymbol{y} \in [0,1]^d.$$

We discuss some essential properties of the shift-invariant kernel. It can be checked that the function K^{sh} in Definition 7.4 is indeed a reproducing kernel. Further, using the transformation of variables $\boldsymbol{z} = \boldsymbol{y} + \boldsymbol{\Delta}$, we obtain

$$\int_{[0,1]^d} K(\{\boldsymbol{x} + \boldsymbol{\Delta}\}, \{\boldsymbol{y} + \boldsymbol{\Delta}\}) \, \mathrm{d}\boldsymbol{\Delta} = \int_{[0,1]^d} K(\{\boldsymbol{x} - \boldsymbol{y} + \boldsymbol{z}\}, \boldsymbol{z}) \, \mathrm{d}\boldsymbol{z}.$$

This implies that

$$K^{\mathrm{sh}}(\boldsymbol{x}, \boldsymbol{y}) = K^{\mathrm{sh}}(\{\boldsymbol{x} - \boldsymbol{y}\}, \boldsymbol{0}).$$

Thus K^{sh} only depends on the component-wise fractional part of the difference $\boldsymbol{x} - \boldsymbol{y}$ in each coordinate. Define the function $\Theta : \mathbb{R}^d \to \mathbb{R}$ by

$$\Theta(\boldsymbol{z}) = K^{\mathrm{sh}}(\{\boldsymbol{z}\}, \boldsymbol{0}).$$

Since we only use the fractional part of each component of $\boldsymbol{z}$, it follows that Θ is periodic in each variable with period 1.

We restrict ourselves to reproducing kernels K for which the associated shift-invariant kernel can be represented pointwise by a Fourier series, that is, the function Θ can be written as

$$\Theta(\boldsymbol{z}) = \sum_{\boldsymbol{k} \in \mathbb{Z}^d} \widehat{\Theta}(\boldsymbol{k}) \, \mathrm{e}^{2\pi \mathrm{i} \boldsymbol{k} \cdot \boldsymbol{z}}.$$

Hence we can write

$$K^{\mathrm{sh}}(\boldsymbol{x}, \boldsymbol{y}) = \sum_{\boldsymbol{k} \in \mathbb{Z}^d} \widehat{\Theta}(\boldsymbol{k}) \, \mathrm{e}^{2\pi \mathrm{i} \boldsymbol{k} \cdot (\boldsymbol{x} - \boldsymbol{y})}.$$

Consequently, the reproducing kernel Hilbert space associated with the shift-invariant kernel is a space of 1-periodic Fourier series, similar to a Korobov space, with the only difference being the precise definition of the Fourier coefficients. As we will see below, this relation will allow us to carry over results for Korobov spaces to some reproducing kernel Hilbert spaces containing also nonperiodic functions.

The Fourier coefficients $\widehat{\Theta}(\boldsymbol{k})$, $\boldsymbol{k} \in \mathbb{Z}^d$, of K^{sh} are given by

$$\begin{aligned}
\widehat{\Theta}(\boldsymbol{k}) &= \int_{[0,1]^d} K^{\mathrm{sh}}(\boldsymbol{z}, \boldsymbol{0})\, \mathrm{e}^{-2\pi \mathrm{i} \boldsymbol{k}\cdot\boldsymbol{z}}\, \mathrm{d}\boldsymbol{z} \\
&= \int_{[0,1]^d} \int_{[0,1]^d} K^{\mathrm{sh}}(\boldsymbol{x}, \boldsymbol{y})\, \mathrm{e}^{-2\pi \mathrm{i} \boldsymbol{k}\cdot(\boldsymbol{x}-\boldsymbol{y})}\, \mathrm{d}\boldsymbol{x}\, \mathrm{d}\boldsymbol{y} \\
&= \int_{[0,1]^d} \int_{[0,1]^d} \int_{[0,1]^d} K(\{\boldsymbol{x}+\boldsymbol{\Delta}\}, \{\boldsymbol{y}+\boldsymbol{\Delta}\})\, \mathrm{e}^{-2\pi \mathrm{i} \boldsymbol{k}\cdot(\boldsymbol{x}-\boldsymbol{y})}\, \mathrm{d}\boldsymbol{\Delta}\, \mathrm{d}\boldsymbol{x}\, \mathrm{d}\boldsymbol{y} \\
&= \int_{[0,1]^d} \int_{[0,1]^d} \int_{[0,1]^d} K(\{\boldsymbol{x}+\boldsymbol{\Delta}\}, \{\boldsymbol{y}+\boldsymbol{\Delta}\})\, \mathrm{e}^{-2\pi \mathrm{i} \boldsymbol{k}\cdot(\{\boldsymbol{x}+\boldsymbol{\Delta}\}-\{\boldsymbol{y}+\boldsymbol{\Delta}\})}\, \mathrm{d}\boldsymbol{x}\, \mathrm{d}\boldsymbol{y}\, \mathrm{d}\boldsymbol{\Delta} \\
&= \int_{[0,1]^d} \int_{[0,1]^d} \int_{[0,1]^d} K(\boldsymbol{x}, \boldsymbol{y})\, \mathrm{e}^{-2\pi \mathrm{i} \boldsymbol{k}\cdot(\boldsymbol{x}-\boldsymbol{y})}\, \mathrm{d}\boldsymbol{x}\, \mathrm{d}\boldsymbol{y}\, \mathrm{d}\boldsymbol{\Delta} \\
&= \int_{[0,1]^d} \int_{[0,1]^d} K(\boldsymbol{x}, \boldsymbol{y})\, \mathrm{e}^{-2\pi \mathrm{i} \boldsymbol{k}\cdot(\boldsymbol{x}-\boldsymbol{y})}\, \mathrm{d}\boldsymbol{x}\, \mathrm{d}\boldsymbol{y}.
\end{aligned}$$

This illustrates another property of the shift-invariant kernel. Assume that we can represent the reproducing kernel K pointwise by

$$K(\boldsymbol{x}, \boldsymbol{y}) = \sum_{\boldsymbol{k}, \boldsymbol{\ell} \in \mathbb{Z}^d} \widehat{K}(\boldsymbol{k}, \boldsymbol{\ell})\, \mathrm{e}^{2\pi \mathrm{i} \boldsymbol{k}\cdot\boldsymbol{x}} \mathrm{e}^{-2\pi \mathrm{i} \boldsymbol{\ell}\cdot\boldsymbol{y}},$$

where

$$\widehat{K}(\boldsymbol{k}, \boldsymbol{\ell}) = \int_{[0,1]^d} \int_{[0,1]^d} K(\boldsymbol{x}, \boldsymbol{y})\, \mathrm{e}^{-2\pi \mathrm{i} \boldsymbol{k}\cdot\boldsymbol{x}} \mathrm{e}^{2\pi \mathrm{i} \boldsymbol{\ell}\cdot\boldsymbol{y}}\, \mathrm{d}\boldsymbol{x}\, \mathrm{d}\boldsymbol{y}.$$

Then the associated shift-invariant kernel just contains the diagonal elements of this Fourier series representation, i.e.,

$$K^{\mathrm{sh}}(\boldsymbol{x}, \boldsymbol{y}) = \sum_{\boldsymbol{k} \in \mathbb{Z}^d} \widehat{K}(\boldsymbol{k}, \boldsymbol{k})\, \mathrm{e}^{2\pi \mathrm{i} \boldsymbol{k}\cdot(\boldsymbol{x}-\boldsymbol{y})}. \tag{7.1}$$

The following theorem shows that the worst-case error of a QMC rule in the reproducing kernel Hilbert space $\mathcal{H}(K^{\mathrm{sh}})$ equals the shift-averaged worst-case error of the corresponding shifted QMC rule in $\mathcal{H}(K)$.

Theorem 7.5 *The shift-averaged worst-case error of a QMC rule based on an N-element point set $\mathcal{P} = \{\boldsymbol{x}_0, \boldsymbol{x}_1, \ldots, \boldsymbol{x}_{N-1}\}$ in $[0,1)^d$ in a reproducing kernel Hilbert space $\mathcal{H}(K)$ is given as*

$$[\mathrm{err}^{\mathrm{sh}}_{N,d}(\mathcal{H}(K),\mathcal{P})]^2 = -\int_{[0,1]^d}\int_{[0,1]^d} K(\boldsymbol{x},\boldsymbol{y})\,\mathrm{d}\boldsymbol{x}\,\mathrm{d}\boldsymbol{y} + \frac{1}{N^2}\sum_{k,\ell=0}^{N-1} K^{\mathrm{sh}}(\boldsymbol{x}_k,\boldsymbol{x}_\ell)$$

$$= \sum_{\boldsymbol{k}\in\mathbb{Z}^d\setminus\{\boldsymbol{0}\}} \widehat{K}(\boldsymbol{k},\boldsymbol{k}) \left|\frac{1}{N}\sum_{k=0}^{N-1} \mathrm{e}^{2\pi\mathrm{i}\boldsymbol{k}\cdot\boldsymbol{x}_k}\right|^2,$$

where

$$\widehat{K}(\boldsymbol{k},\boldsymbol{k}) = \int_{[0,1]^d}\int_{[0,1]^d} K(\boldsymbol{x},\boldsymbol{y})\,\mathrm{e}^{-2\pi\mathrm{i}\boldsymbol{k}\cdot(\boldsymbol{x}-\boldsymbol{y})}\,\mathrm{d}\boldsymbol{x}\,\mathrm{d}\boldsymbol{y}. \tag{7.2}$$

In particular,

$$\mathrm{err}^{\mathrm{sh}}_{N,d}(\mathcal{H}(K),\mathcal{P}) = \mathrm{err}_{N,d}(\mathcal{H}(K^{\mathrm{sh}}),\mathcal{P}). \tag{7.3}$$

Proof We apply the worst-case error formula in Theorem 1.27 to $\mathcal{P}_{\boldsymbol{\Delta}}$ and average over all $\boldsymbol{\Delta}$ in $[0,1]^d$. In this way we obtain

$$\begin{aligned}[\mathrm{err}^{\mathrm{sh}}_{N,d}(\mathcal{H}(K),\mathcal{P})]^2 &= \int_{[0,1]^d}\int_{[0,1]^d}\int_{[0,1]^d} K(\boldsymbol{x},\boldsymbol{y})\,\mathrm{d}\boldsymbol{x}\,\mathrm{d}\boldsymbol{y}\,\mathrm{d}\boldsymbol{\Delta}\\ &\quad -\frac{2}{N}\sum_{k=0}^{N-1}\int_{[0,1]^d}\int_{[0,1]^d} K(\{\boldsymbol{x}_k+\boldsymbol{\Delta}\},\boldsymbol{y})\,\mathrm{d}\boldsymbol{y}\,\mathrm{d}\boldsymbol{\Delta}\\ &\quad +\frac{1}{N^2}\sum_{k,\ell=0}^{N-1}\int_{[0,1]^d} K(\{\boldsymbol{x}_k+\boldsymbol{\Delta}\},\{\boldsymbol{x}_\ell+\boldsymbol{\Delta}\})\,\mathrm{d}\boldsymbol{\Delta}.\end{aligned}$$

Using the change of variables $\boldsymbol{x} = \{\boldsymbol{x}_k + \boldsymbol{\Delta}\}$ yields

$$\begin{aligned}&-\frac{2}{N}\sum_{k=0}^{N-1}\int_{[0,1]^d}\int_{[0,1]^d} K(\{\boldsymbol{x}_k+\boldsymbol{\Delta}\},\boldsymbol{y})\,\mathrm{d}\boldsymbol{y}\,\mathrm{d}\boldsymbol{\Delta}\\ &= -\frac{2}{N}\sum_{k=0}^{N-1}\int_{[0,1]^d}\int_{[0,1]^d} K(\{\boldsymbol{x}_k+\boldsymbol{\Delta}\},\boldsymbol{y})\,\mathrm{d}\boldsymbol{\Delta}\,\mathrm{d}\boldsymbol{y}\\ &= -\frac{2}{N}\sum_{k=0}^{N-1}\int_{[0,1]^d}\int_{[0,1]^d} K(\boldsymbol{x},\boldsymbol{y})\,\mathrm{d}\boldsymbol{x}\,\mathrm{d}\boldsymbol{y}\\ &= -2\int_{[0,1]^d}\int_{[0,1]^d} K(\boldsymbol{x},\boldsymbol{y})\,\mathrm{d}\boldsymbol{x}\,\mathrm{d}\boldsymbol{y},\end{aligned}$$

which, together with the definition of K^{sh}, implies

$$[\mathrm{err}^{\mathrm{sh}}_{N,d}(\mathcal{H}(K),\mathcal{P})]^2 = -\int_{[0,1]^d}\int_{[0,1]^d} K(\boldsymbol{x},\boldsymbol{y})\,\mathrm{d}\boldsymbol{x}\,\mathrm{d}\boldsymbol{y} + \frac{1}{N^2}\sum_{k,\ell=0}^{N-1} K^{\mathrm{sh}}(\boldsymbol{x}_k,\boldsymbol{x}_\ell), \tag{7.4}$$

as desired. From this, in combination with Theorem 1.27, where we use that

$$\int_{[0,1]^d}\int_{[0,1]^d} K^{\mathrm{sh}}(\boldsymbol{x},\boldsymbol{y})\,\mathrm{d}\boldsymbol{x}\,\mathrm{d}\boldsymbol{y} - \frac{2}{N}\sum_{k=0}^{N-1}\int_{[0,1]^d} K^{\mathrm{sh}}(\boldsymbol{x}_k,\boldsymbol{y})\,\mathrm{d}\boldsymbol{y}$$
$$= -\int_{[0,1]^d}\int_{[0,1]^d} K(\boldsymbol{x},\boldsymbol{y})\,\mathrm{d}\boldsymbol{x}\,\mathrm{d}\boldsymbol{y},$$

we immediately obtain (7.3).

Plugging (7.2) with $\boldsymbol{k} = \boldsymbol{0}$, and (7.1) into (7.4) yields

$$\begin{aligned}
[\mathrm{err}^{\mathrm{sh}}_{N,d}(\mathcal{H}(K),\mathcal{P})]^2 &= -\int_{[0,1]^d}\int_{[0,1]^d} K(\boldsymbol{x},\boldsymbol{y})\,\mathrm{d}\boldsymbol{x}\,\mathrm{d}\boldsymbol{y} + \frac{1}{N^2}\sum_{k,\ell=0}^{N-1} K^{\mathrm{sh}}(\boldsymbol{x}_k,\boldsymbol{x}_\ell) \\
&= -\widehat{K}(\boldsymbol{0},\boldsymbol{0}) + \frac{1}{N^2}\sum_{k,\ell=0}^{N-1}\sum_{\boldsymbol{k}\in\mathbb{Z}^d}\widehat{K}(\boldsymbol{k},\boldsymbol{k})\,\mathrm{e}^{2\pi\mathrm{i}\boldsymbol{k}\cdot(\boldsymbol{x}_k-\boldsymbol{x}_\ell)} \\
&= \sum_{\boldsymbol{k}\in\mathbb{Z}^d\setminus\{\boldsymbol{0}\}}\widehat{K}(\boldsymbol{k},\boldsymbol{k})\left|\frac{1}{N}\sum_{k=0}^{N-1}\mathrm{e}^{2\pi\mathrm{i}\boldsymbol{k}\cdot\boldsymbol{x}_k}\right|^2,
\end{aligned}$$

which shows the remaining claim. □

We now apply these findings to two important reference spaces for QMC rules, namely the weighted unanchored Sobolev space of smoothness one and the anchored Sobolev space of smoothness one.

The weighted unanchored Sobolev space of smoothness one

For $d \in \mathbb{N}$, $\mathfrak{u} \subseteq [d]$, and a function f on $[0,1]^d$ we write

$$f^{(\mathfrak{u})} = \frac{\partial^{|\mathfrak{u}|}}{\partial \boldsymbol{x}_{\mathfrak{u}}} f = \prod_{j\in\mathfrak{u}}\frac{\partial}{\partial x_j} f \quad \text{with } f^{(\emptyset)} = f.$$

We consider arbitrary weights $\boldsymbol{\gamma} = \{\gamma_{\mathfrak{u}}\}_{\mathfrak{u}\subseteq[d]}$. Let $\mathfrak{U}^+ := \{\mathfrak{u}\subseteq[d] \,:\, \gamma_{\mathfrak{u}} > 0\}$.

We define a reproducing kernel $K_{\mathrm{sob},d,\boldsymbol{\gamma}} : [0,1]^d \times [0,1]^d \to \mathbb{R}$ by

$$K_{\mathrm{sob},d,\boldsymbol{\gamma}}(\boldsymbol{x},\boldsymbol{y}) := \sum_{\mathfrak{u}\subseteq[d]}\gamma_{\mathfrak{u}}\prod_{j\in\mathfrak{u}}\eta(x_j,y_j), \tag{7.5}$$

where $\boldsymbol{x} = (x_1,\ldots,x_d)$ and $\boldsymbol{y} = (y_1,\ldots,y_d)$ are in $[0,1]^d$, and where η is given by

$$\eta(x, y) := \left(x - \frac{1}{2}\right)\left(y - \frac{1}{2}\right) + \frac{B_2(|x-y|)}{2}.$$

Here B_2 is the second Bernoulli polynomial (see also Chapters 2, 3, and 6), which is defined as $B_2(x) := x^2 - x + 1/6$. Observe also that the first Bernoulli polynomial B_1 is given by $B_1(x) = x - 1/2$.

Definition 7.6 The *weighted unanchored Sobolev space* $\mathcal{H}_{\text{sob},d,\boldsymbol{\gamma}}$ of smoothness one is the reproducing kernel Hilbert space with kernel $K_{\text{sob},d,\boldsymbol{\gamma}}$, i.e., $\mathcal{H}_{\text{sob},d,\boldsymbol{\gamma}} = \mathcal{H}(K_{\text{sob},d,\boldsymbol{\gamma}})$, and the corresponding inner product

$$\begin{aligned}&\langle f, g\rangle_{\text{sob},d,\boldsymbol{\gamma}}\\ &= \sum_{\mathfrak{u}\in\mathfrak{U}^+} \frac{1}{\gamma_{\mathfrak{u}}} \int_{[0,1]^{|\mathfrak{u}|}} \left(\int_{[0,1]^{d-|\mathfrak{u}|}} f^{(\mathfrak{u})}(\boldsymbol{x})\,\mathrm{d}\boldsymbol{x}_{[d]\setminus\mathfrak{u}}\right)\left(\int_{[0,1]^{d-|\mathfrak{u}|}} g^{(\mathfrak{u})}(\boldsymbol{x})\,\mathrm{d}\boldsymbol{x}_{[d]\setminus\mathfrak{u}}\right)\mathrm{d}\boldsymbol{x}_{\mathfrak{u}}.\end{aligned}$$

The norm is therefore given by

$$\|f\|_{\text{sob},d,\boldsymbol{\gamma}} = \left(\sum_{\mathfrak{u}\in\mathfrak{U}^+} \frac{1}{\gamma_{\mathfrak{u}}} \int_{[0,1]^{|\mathfrak{u}|}} \left(\int_{[0,1]^{d-|\mathfrak{u}|}} f^{(\mathfrak{u})}(\boldsymbol{x})\,\mathrm{d}\boldsymbol{x}_{[d]\setminus\mathfrak{u}}\right)^2 \mathrm{d}\boldsymbol{x}_{\mathfrak{u}}\right)^{1/2}.$$

The space $\mathcal{H}_{\text{sob},d,\boldsymbol{\gamma}}$ is also referred to as the *ANOVA Sobolev space*.

Remark 7.7 The weighted unanchored Sobolev space $\mathcal{H}_{\text{sob},d,\boldsymbol{\gamma}}$ contains all functions on $[0, 1]^d$ whose mixed first partial derivatives are square integrable. In particular, in the univariate case $d = 1$, the norm reduces to

$$\|f\|^2_{\text{sob},1,\gamma} = \left(\int_0^1 f(x)\,\mathrm{d}x\right)^2 + \frac{1}{\gamma}\int_0^1 (f'(x))^2\,\mathrm{d}x. \tag{7.6}$$

This formula should also be compared with the norm in the univariate weighted Korobov space with smoothness $\alpha = 1$ given in (2.20).

The following proposition shows that the integration problem in $\mathcal{H}_{\text{sob},d,\boldsymbol{\gamma}}$ is normalized.

Proposition 7.8 *The initial error of integration in* $\mathcal{H}_{\text{sob},d,\boldsymbol{\gamma}}$ *equals 1.*

Proof According to Remark 1.41 we have

$$\begin{aligned}[e(0,d)]^2 &= \int_{[0,1]^d}\int_{[0,1]^d} K_{\text{sob},d,\boldsymbol{\gamma}}(\boldsymbol{x},\boldsymbol{y})\,\mathrm{d}\boldsymbol{x}\,\mathrm{d}\boldsymbol{y}\\ &= \sum_{\mathfrak{u}\subseteq[d]} \gamma_{\mathfrak{u}} \prod_{j\in\mathfrak{u}} \int_0^1\int_0^1 \eta(x_j, y_j)\,\mathrm{d}x_j\,\mathrm{d}y_j. \end{aligned}\tag{7.7}$$

Note that

$$\int_0^1 B_2(|x-y|)\,\mathrm{d}y = \int_0^1 B_2(y)\,\mathrm{d}y = 0,$$

and hence

$$\int_0^1 \int_0^1 \eta(x, y)\, \mathrm{d}x\, \mathrm{d}y = \left(\int_0^1 \left(x - \frac{1}{2} \right) \mathrm{d}x \right)^2 = 0.$$

This implies that the only nonzero summand in (7.7) is the one corresponding to $\mathfrak{u} = \emptyset$. This yields

$$[e(0, d)]^2 = \gamma_\emptyset = 1,$$

according to our standing assumption that $\gamma_\emptyset = 1$. □

Remark 7.9 Proposition 7.8 implies that the squared shift-averaged worst-case error in $\mathcal{H}_{\mathrm{sob},d,\boldsymbol{\gamma}}$ with respect to a node set $\mathcal{P} = \{\boldsymbol{x}_0, \boldsymbol{x}_1, \ldots, \boldsymbol{x}_{N-1}\}$ equals

$$[\mathrm{err}^{\mathrm{sh}}_{N,d}(\mathcal{H}_{\mathrm{sob},d,\boldsymbol{\gamma}}, \mathcal{P})]^2 = -1 + \frac{1}{N^2} \sum_{k,\ell=0}^{N-1} K^{\mathrm{sh}}_{\mathrm{sob},d,\boldsymbol{\gamma}}(\boldsymbol{x}_k, \boldsymbol{x}_\ell).$$

In the following theorem we give the explicit form of the shift-invariant kernel for $\mathcal{H}_{\mathrm{sob},d,\boldsymbol{\gamma}}$.

Theorem 7.10 *Let $K_{\mathrm{sob},d,\boldsymbol{\gamma}}$ be given by* (7.5). *Then the corresponding shift-invariant kernel is given by*

$$K^{\mathrm{sh}}_{\mathrm{sob},d,\boldsymbol{\gamma}}(\boldsymbol{x}, \boldsymbol{y}) = \sum_{\mathfrak{u} \subseteq [d]} \gamma_{\mathfrak{u}} \prod_{j \in \mathfrak{u}} B_2(|x_j - y_j|).$$

Proof Equation (7.5) and Definition 7.4 imply

$$K^{\mathrm{sh}}_{\mathrm{sob},d,\boldsymbol{\gamma}}(\boldsymbol{x}, \boldsymbol{y}) = \sum_{\mathfrak{u} \subseteq [d]} \gamma_{\mathfrak{u}} \prod_{j \in \mathfrak{u}} \int_0^1 \eta(\{x_j + \Delta_j\}, \{y_j + \Delta_j\})\, \mathrm{d}\Delta_j.$$

From the definition of η and from the symmetry $B_2(x) = B_2(1 - x)$ for $x \in [0, 1]$ we obtain

$$\int_0^1 \eta(\{x + \Delta\}, \{y + \Delta\})\, \mathrm{d}\Delta = \int_0^1 \left(\{x + \Delta\} - \frac{1}{2} \right) \left(\{y + \Delta\} - \frac{1}{2} \right) \mathrm{d}\Delta + \frac{B_2(|x - y|)}{2}.$$

Therefore it remains to be shown that

$$I(x, y) := \int_0^1 \left(\{x + \Delta\} - \frac{1}{2} \right) \left(\{y + \Delta\} - \frac{1}{2} \right) \mathrm{d}\Delta = \frac{B_2(|x - y|)}{2}.$$

Obviously, $I(x, y) = I(y, x)$ and hence we may assume that $x \le y$ from now on. For x, y, Δ in $[0, 1)$ we have either

- $\Delta < 1 - y$, which implies $x + \Delta \le y + \Delta < 1$, or
- $1 - y \le \Delta < 1 - x$, which implies $x + \Delta < 1 \le y + \Delta$, or
- $1 - x \le \Delta$, which implies $1 \le x + \Delta \le y + \Delta$ and hence $1 - x \le \Delta$.

Thus we have

$$\begin{aligned} I(x,y) &= \int_0^{1-y} \left(x+\Delta-\frac{1}{2}\right)\left(y+\Delta-\frac{1}{2}\right) \mathrm{d}\Delta \\ &+ \int_{1-y}^{1-x} \left(x+\Delta-\frac{1}{2}\right)\left(y+\Delta-1-\frac{1}{2}\right) \mathrm{d}\Delta \\ &+ \int_{1-x}^{1} \left(x+\Delta-1-\frac{1}{2}\right)\left(y+\Delta-1-\frac{1}{2}\right) \mathrm{d}\Delta \\ &= \frac{(y-x)^2}{2} - \frac{y-x}{2} + \frac{1}{12} \\ &= \frac{B_2(|x-y|)}{2}. \end{aligned}$$

This implies the desired result. □

We now have the following corollary.

Corollary 7.11 *The squared shift-averaged worst-case error of a rank-1 lattice rule with node set* $\mathcal{P}(\boldsymbol{g},N)$ *in* $\mathcal{H}_{\mathrm{sob},d,\boldsymbol{\gamma}}$ *equals*

$$[\mathrm{err}^{\mathrm{sh}}_{N,d}(\mathcal{H}_{\mathrm{sob},d,\boldsymbol{\gamma}}, \mathcal{P}(\boldsymbol{g},N))]^2 = \sum_{\emptyset \neq \mathfrak{u} \subseteq [d]} \gamma_{\mathfrak{u}} \frac{1}{N} \sum_{k=0}^{N-1} \prod_{j\in\mathfrak{u}} B_2\left(\left\{\frac{k g_j}{N}\right\}\right).$$

Proof Using Remark 7.9 and Theorem 7.10 with $\boldsymbol{x}_k = \{(k/N)\boldsymbol{g}\}$, for $k \in \{0,1,\ldots,N-1\}$, and again the symmetry $B_2(x) = B_2(1-x)$ for $x \in [0,1]$, yields

$$\begin{aligned} [\mathrm{err}^{\mathrm{sh}}_{N,d}(\mathcal{H}_{\mathrm{sob},d,\boldsymbol{\gamma}}, \mathcal{P}(\boldsymbol{g},N))]^2 &= -1 + \sum_{\mathfrak{u} \subseteq [d]} \gamma_{\mathfrak{u}} \frac{1}{N^2} \sum_{k,\ell=0}^{N-1} \prod_{j\in\mathfrak{u}} B_2\left(\left|\left\{\frac{k g_j}{N}\right\} - \left\{\frac{\ell g_j}{N}\right\}\right|\right) \\ &= -1 + \sum_{\mathfrak{u} \subseteq [d]} \gamma_{\mathfrak{u}} \frac{1}{N^2} \sum_{k,\ell=0}^{N-1} \prod_{j\in\mathfrak{u}} B_2\left(\left\{\frac{(k-\ell) g_j}{N}\right\}\right). \end{aligned}$$

As k and ℓ range from 0 to $N-1$, $(k-\ell) \pmod N$ takes on the values $0,1,\ldots,N-1$ in some order, with each value occurring exactly N times. With this observation we can rewrite the double sum in the previous expression as a single sum, which yields

$$[\mathrm{err}^{\mathrm{sh}}_{N,d}(\mathcal{H}_{\mathrm{sob},d,\boldsymbol{\gamma}}, \mathcal{P}(\boldsymbol{g},N))]^2 = -1 + \sum_{\mathfrak{u} \subseteq [d]} \gamma_{\mathfrak{u}} \frac{1}{N} \sum_{k=0}^{N-1} \prod_{j\in\mathfrak{u}} B_2\left(\left\{\frac{k g_j}{N}\right\}\right),$$

and this gives the desired result, since, in the latter expression, the summand corresponding to $\mathfrak{u} = \emptyset$ equals 1. □

We remind the reader that, according to Remark 2.9, Equation (2.8), the Bernoulli polynomial B_2 has the Fourier expansion

$$B_2(x) = \frac{1}{2\pi^2} \sum_{h \in \mathbb{Z}\setminus\{0\}} \frac{\mathrm{e}^{2\pi \mathrm{i} hx}}{|h|^2} \quad \text{for all } x \in [0,1).$$

This expansion implies

$$\begin{aligned}
&[\mathrm{err}^{\mathrm{sh}}_{N,d}(\mathcal{H}_{\mathrm{sob},d,\boldsymbol{\gamma}}, \mathcal{P}(\boldsymbol{g},N))]^2 \\
&= \sum_{\emptyset \neq \mathfrak{u} \subseteq [d]} \frac{\gamma_{\mathfrak{u}}}{(2\pi^2)^{|\mathfrak{u}|}} \frac{1}{N} \sum_{k=0}^{N-1} \prod_{j \in \mathfrak{u}} \sum_{h_j \in \mathbb{Z}\setminus\{0\}} \frac{\mathrm{e}^{2\pi \mathrm{i} h_j k g_j/N}}{|h_j|^2} \\
&= \sum_{\emptyset \neq \mathfrak{u} \subseteq [d]} \frac{\gamma_{\mathfrak{u}}}{(2\pi^2)^{|\mathfrak{u}|}} \frac{1}{N} \sum_{k=0}^{N-1} \sum_{\boldsymbol{h}_{\mathfrak{u}} \in (\mathbb{Z}\setminus\{0\})^{|\mathfrak{u}|}} \left(\prod_{j \in \mathfrak{u}} \frac{1}{|h_j|^2} \right) \mathrm{e}^{2\pi \mathrm{i} k (\boldsymbol{h}_{\mathfrak{u}} \cdot \boldsymbol{g}_{\mathfrak{u}})/N} \\
&= \sum_{\boldsymbol{h} \in \mathbb{Z}^d \setminus\{\boldsymbol{0}\}} \frac{\gamma_{\mathfrak{u}(\boldsymbol{h})}}{(2\pi^2)^{|\mathfrak{u}(\boldsymbol{h})|}} \left(\prod_{j \in \mathfrak{u}(\boldsymbol{h})} \frac{1}{|h_j|^2} \right) \frac{1}{N} \sum_{k=0}^{N-1} \mathrm{e}^{2\pi \mathrm{i} k (\boldsymbol{h} \cdot \boldsymbol{g})/N} \\
&= \sum_{\substack{\boldsymbol{h} \in \mathbb{Z}^d \setminus\{\boldsymbol{0}\} \\ \boldsymbol{h}\cdot\boldsymbol{g} \equiv 0 \pmod N}} \frac{1}{r_{2,\widetilde{\boldsymbol{\gamma}}}(\boldsymbol{h})},
\end{aligned}$$

where $\widetilde{\boldsymbol{\gamma}} = \{\widetilde{\gamma}_{\mathfrak{u}}\}_{\mathfrak{u} \subseteq [d]}$ with

$$\widetilde{\gamma}_{\mathfrak{u}} := \frac{\gamma_{\mathfrak{u}}}{(2\pi^2)^{|\mathfrak{u}|}} \quad \text{for } \mathfrak{u} \subseteq [d]. \tag{7.8}$$

Comparing this result to Corollary 2.21, we observe that the shift-averaged worst-case error of a lattice rule in $\mathcal{H}_{\mathrm{sob},d,\boldsymbol{\gamma}}$ equals the worst-case error of a lattice rule in the Korobov space $\mathcal{H}_{\mathrm{kor},d,\alpha,\widetilde{\boldsymbol{\gamma}}}$ with the modified weights $\widetilde{\boldsymbol{\gamma}}$ and with smoothness parameter $\alpha = 1$. Putting it differently,

$$\mathrm{err}^{\mathrm{sh}}_{N,d}(\mathcal{H}_{\mathrm{sob},d,\boldsymbol{\gamma}}, \mathcal{P}(\boldsymbol{g},N)) = \mathrm{err}_{N,d,1,\widetilde{\boldsymbol{\gamma}}}(\boldsymbol{g}).$$

This allows us to rewrite all results previously shown for the worst-case error in Korobov spaces such that they apply to the shift-averaged worst-case error in the weighted unanchored Sobolev space (equipped with the corresponding weights).

In the vein of Theorems 2.24, 3.7, and 3.9 we obtain the following result, which we formulate for prime N for simplicity; note, though, that a similar result would also hold true for arbitrary $N \in \mathbb{N}$.

Theorem 7.12 *For* $\tau \in [1/2, 1)$ *set*

$$a_\tau := \frac{2^{1-1/(2\tau)}}{\pi^{1/\tau}} \zeta\left(\frac{1}{\tau}\right). \tag{7.9}$$

Then, for any prime number N and any dimension d, the following statements hold.

1. *There exists a generating vector* $\boldsymbol{g} \in G_d(N)$ *such that*

$$\mathrm{err}^{\mathrm{sh}}_{N,d}(\mathcal{H}_{\mathrm{sob},d,\boldsymbol{\gamma}}, \mathcal{P}(\boldsymbol{g}, N)) \leq \frac{1}{N^{\tau}} \left(\sum_{\emptyset \neq \mathfrak{u} \subseteq [d]} a_{\tau}^{|\mathfrak{u}|} \gamma_{\mathfrak{u}}^{1/(2\tau)} \right)^{\tau}$$

for all $\tau \in [1/2, 1)$. *This vector can be found using Algorithm 3.6 (by means of the weights* $\widetilde{\boldsymbol{\gamma}}$ *in* (7.8)).

2. *In the case of product weights the bound in Item 1 can be simplified to*

$$\mathrm{err}^{\mathrm{sh}}_{N,d}(\mathcal{H}_{\mathrm{sob},d,\boldsymbol{\gamma}}, \mathcal{P}(\boldsymbol{g}, N)) \leq \frac{1}{N^{\tau}} \left(-1 + \prod_{j=1}^{d} \left(1 + a_{\tau} \gamma_j^{1/(2\tau)} \right) \right)^{\tau}.$$

According to Remark 7.3, Theorem 7.12 can also be seen as an existence result for shifted lattice rules for integration, which we formulate as follows.

Corollary 7.13 *For any prime number N and any dimension d there exists a generating vector* $\boldsymbol{g} \in G_d(N)$ *and a shift* $\boldsymbol{\Delta} \in [0, 1]^d$ *such that*

$$\mathrm{err}_{N,d}(\mathcal{H}_{\mathrm{sob},d,\boldsymbol{\gamma}}, \mathcal{P}_{\boldsymbol{\Delta}}(\boldsymbol{g}, N)) \leq \frac{1}{N^{\tau}} \left(\sum_{\emptyset \neq \mathfrak{u} \subseteq [d]} a_{\tau}^{|\mathfrak{u}|} \gamma_{\mathfrak{u}}^{1/(2\tau)} \right)^{\tau}$$

for all $\tau \in [1/2, 1)$. *The vector* $\boldsymbol{g}$ *can be found using Algorithm 3.6 (by means of the weights* $\widetilde{\boldsymbol{\gamma}}$ *in* (7.8)).

There have been some attempts to construct also a good shift $\boldsymbol{\Delta}$ in the sense of Corollary 7.13 using a CBC algorithm, see [232]. The problem with this construction is, however, that the required number of elementary operations is of order $O(d\,N^3)$, which is too high for most practical applications. Furthermore, the error bounds can only be guaranteed to be of the suboptimal order $O(N^{-1/2})$. Nevertheless, Corollary 7.13 implies that the results on tractability of the integration problem in the weighted Korobov space, in particular Corollaries 2.27 and 2.28, can be carried over to the weighted unanchored Sobolev space.

The weighted anchored Sobolev space of smoothness one

A variant of the unanchored Sobolev space is the anchored Sobolev space which we briefly discuss here. We introduce a weighted generalization of the anchored Sobolev space of smoothness one with anchor **1** from Section 1.6.

For a so-called *anchor* $\boldsymbol{c} = (c_1, \ldots, c_d) \in [0, 1]^d$, we define a reproducing kernel $K^{\pitchfork}_{\mathrm{sob},d,\boldsymbol{\gamma},\boldsymbol{c}} : [0, 1]^d \times [0, 1]^d \to \mathbb{R}$ by

$$K^{\pitchfork}_{\mathrm{sob},d,\boldsymbol{\gamma},\boldsymbol{c}}(\boldsymbol{x}, \boldsymbol{y}) := \sum_{\mathfrak{u} \subseteq [d]} \gamma_{\mathfrak{u}} \prod_{j \in \mathfrak{u}} \eta^{\pitchfork}_{c_j}(x_j, y_j), \tag{7.10}$$

where, for $c \in [0,1]$, $\eta_c^{\pitchfork}$ is given by

$$\eta_c^{\pitchfork}(x,y) := \begin{cases} \min(x,y) - c & \text{if } x, y > c, \\ c - \max(x,y) & \text{if } x, y < c, \\ 0 & \text{otherwise.} \end{cases}$$

Definition 7.14 The *weighted anchored Sobolev space* $\mathcal{H}^{\pitchfork}_{\mathrm{sob},d,\boldsymbol{\gamma},\boldsymbol{c}}$ of smoothness one and with anchor $\boldsymbol{c} \in [0,1]^d$ is the reproducing kernel Hilbert space with kernel $K^{\pitchfork}_{\mathrm{sob},d,\boldsymbol{\gamma},\boldsymbol{c}}$, i.e., $\mathcal{H}^{\pitchfork}_{\mathrm{sob},d,\boldsymbol{\gamma},\boldsymbol{c}} = \mathcal{H}(K^{\pitchfork}_{\mathrm{sob},d,\boldsymbol{\gamma},\boldsymbol{c}})$, and with the corresponding inner product

$$\langle f, g\rangle_{\mathrm{sob},d,\boldsymbol{\gamma},\boldsymbol{c}} = \sum_{\mathfrak{u} \in \mathfrak{U}^+} \frac{1}{\gamma_{\mathfrak{u}}} \int_{[0,1]^{|\mathfrak{u}|}} f^{(\mathfrak{u})}(\boldsymbol{x}_{\mathfrak{u}}, \boldsymbol{c}_{[d]\setminus\mathfrak{u}}) g^{(\mathfrak{u})}(\boldsymbol{x}_{\mathfrak{u}}, \boldsymbol{c}_{[d]\setminus\mathfrak{u}}) \,\mathrm{d}\boldsymbol{x}_{\mathfrak{u}},$$

where we recall that $\mathfrak{U}^+ := \{\mathfrak{u} \subseteq [d] \;:\; \gamma_{\mathfrak{u}} > 0\}$. The norm is therefore given by

$$\|f\|_{\mathrm{sob},d,\boldsymbol{\gamma},\boldsymbol{c}_{[d]\setminus\mathfrak{u}}} = \left(\sum_{\mathfrak{u} \in \mathfrak{U}^+} \frac{1}{\gamma_{\mathfrak{u}}} \int_{[0,1]^{|\mathfrak{u}|}} \left(f^{(\mathfrak{u})}(\boldsymbol{x}_{\mathfrak{u}}, \boldsymbol{c}_{[d]\setminus\mathfrak{u}}) \right)^2 \mathrm{d}\boldsymbol{x}_{\mathfrak{u}} \right)^{1/2}.$$

In the univariate case $d = 1$ the norm reduces to

$$\|f\|^2_{\mathrm{sob},1,\gamma,c} = (f(c))^2 + \frac{1}{\gamma} \int_0^1 (f'(x))^2 \,\mathrm{d}x. \tag{7.11}$$

This formula should be compared with the corresponding formula (7.6) for the unanchored Sobolev space of smoothness one.

Remark 7.15 The role of the anchor $\boldsymbol{c}$ in the anchored Sobolev space is that the components in $[d] \setminus \mathfrak{u}$ are fixed in the summands corresponding to $\mathfrak{u} \in \mathfrak{U}^+$ in the definition of the inner product. In the unanchored Sobolev space, these components are not fixed by the anchor, but are integrated out.

Remark 7.16 If all weights are chosen to be one, i.e., $\gamma_{\mathfrak{u}} = 1$ for all $\mathfrak{u} \subseteq [d]$, and if the anchor $\boldsymbol{c}$ is chosen to be $\mathbf{1}$, we obtain the anchored Sobolev space of smoothness one with anchor $\mathbf{1}$ in Section 1.6.

Similar to the unanchored case, it is easy to identify the initial error of QMC integration.

Proposition 7.17 *The squared initial error of integration in* $\mathcal{H}^{\pitchfork}_{\mathrm{sob},d,\boldsymbol{\gamma},\boldsymbol{c}}$ *equals*

$$[e(0,d)]^2 = \sum_{\mathfrak{u} \subseteq [d]} \gamma_{\mathfrak{u}} \prod_{j \in \mathfrak{u}} \beta_{c_j},$$

where we put $\beta_c := c^2 - c + 1/3$ *for* $c \in [0,1]$.

Proof It is easily checked that

$$\int_0^1 \int_0^1 \eta_c^{\pitchfork}(x, y) \, \mathrm{d}x \, \mathrm{d}y = c^2 - c + \frac{1}{3} = \beta_c,$$

and hence, according to Remark 1.41,

$$[e(0, d)]^2 = \sum_{\mathfrak{u} \subseteq [d]} \gamma_{\mathfrak{u}} \prod_{j \in \mathfrak{u}} \int_0^1 \int_0^1 \eta_{c_j}^{\pitchfork}(x_j, y_j) \, \mathrm{d}x_j \, \mathrm{d}y_j = \sum_{\mathfrak{u} \subseteq [d]} \gamma_{\mathfrak{u}} \prod_{j \in \mathfrak{u}} \beta_{c_j}. \qquad \square$$

The following theorem is analogous to Theorem 7.10.

Theorem 7.18 *Let $K_{\mathrm{sob},d,\boldsymbol{\gamma},\boldsymbol{c}}^{\pitchfork}$ be given by (7.10). Then the corresponding shift-invariant kernel is given by*

$$K_{\mathrm{sob},d,\boldsymbol{\gamma},\boldsymbol{c}}^{\pitchfork,\mathrm{sh}}(\boldsymbol{x}, \boldsymbol{y}) = \sum_{\mathfrak{u} \subseteq [d]} \gamma_{\mathfrak{u}} \prod_{j \in \mathfrak{u}} (B_2(|x_j - y_j|) + \beta_{c_j}).$$

Proof Definition 7.4 and Equation (7.10) imply

$$K_{\mathrm{sob},d,\boldsymbol{\gamma}}^{\mathrm{sh}}(\boldsymbol{x}, \boldsymbol{y}) = \sum_{\mathfrak{u} \subseteq [d]} \gamma_{\mathfrak{u}} \prod_{j \in \mathfrak{u}} \int_0^1 \eta_{c_j}^{\pitchfork}(\{x_j + \Delta_j\}, \{y_j + \Delta_j\}) \, \mathrm{d}\Delta_j.$$

It is easily checked that $\eta_c^{\pitchfork}(x, y)$ can be represented in the form

$$\eta_c^{\pitchfork}(x, y) = \frac{B_2(|x - y|)}{2} + \left(x - \frac{1}{2}\right)\left(y - \frac{1}{2}\right) + \alpha_c(x) + \alpha_c(y) + \beta_c$$

for $x, y \in [0, 1]$, where

$$\alpha_c(x) := \max(x, c) - \frac{x^2}{2} - \frac{c^2}{2} - \frac{1}{3} \quad \text{and} \quad \beta_c := c^2 - c + \frac{1}{3}.$$

Thus,

$$\eta_c^{\pitchfork}(x, y) = \eta(x, y) + \alpha_c(x) + \alpha_c(y) + \beta_c.$$

Therefore we obtain, similarly to the proof of Theorem 7.10,

$$\begin{aligned} \int_0^1 \eta_c^{\pitchfork}(\{x + \Delta\}, \{y + \Delta\}) \, \mathrm{d}\Delta &= \int_0^1 \eta(\{x + \Delta\}, \{y + \Delta\}) \, \mathrm{d}\Delta \\ &\quad + \int_0^1 \alpha_c(\{x + \Delta\}) \, \mathrm{d}\Delta + \int_0^1 \alpha_c(\{y + \Delta\}) \, \mathrm{d}\Delta + \beta_c \\ &= B_2(|x - y|) + \beta_c, \end{aligned}$$

since $\int_0^1 \alpha_c(x) \, \mathrm{d}x = 0$. This implies the desired result. $\square$

Theorem 7.18 yields the following corollary.

Corollary 7.19 *The squared shift-averaged worst-case error of a rank-1 lattice rule with node set $\mathcal{P}(\boldsymbol{g}, N)$ in $\mathcal{H}^{\pitchfork}_{\mathrm{sob},d,\boldsymbol{\gamma},\boldsymbol{c}}$ equals*

$$[\mathrm{err}^{\mathrm{sh}}_{N,d}(\mathcal{H}^{\pitchfork}_{\mathrm{sob},d,\boldsymbol{\gamma},\boldsymbol{c}}, \mathcal{P}(\boldsymbol{g}, N))]^2 = \sum_{\emptyset \neq \mathfrak{u} \subseteq [d]} \gamma'_{\mathfrak{u}} \frac{1}{N} \sum_{k=0}^{N-1} \prod_{j \in \mathfrak{u}} B_2\left(\left\{\frac{k g_j}{N}\right\}\right),$$

where

$$\gamma'_{\mathfrak{u}} := \left(\prod_{j \in \mathfrak{u}} \frac{1}{\beta_{c_j}}\right) \sum_{\mathfrak{u} \subseteq \mathfrak{v} \subseteq [d]} \gamma_{\mathfrak{v}} \prod_{j \in \mathfrak{v}} \beta_{c_j},$$

and where we note that $\beta_{c_j} \neq 0$ independently of c_j.

Proof Using Theorem 7.18 with $\boldsymbol{x}_k = \{(k/N)\boldsymbol{g}\}$ for $k \in \{0, 1, \ldots, N-1\}$, and the symmetry $B_2(x) = B_2(1-x)$ for $x \in [0, 1]$, yields

$$\begin{aligned}
&[\mathrm{err}^{\mathrm{sh}}_{N,d}(\mathcal{H}^{\pitchfork}_{\mathrm{sob},d,\boldsymbol{\gamma},\boldsymbol{c}}, \mathcal{P}(\boldsymbol{g}, N))]^2 \\
&= -\sum_{\mathfrak{v} \subseteq [d]} \gamma_{\mathfrak{v}} \prod_{j \in \mathfrak{v}} \beta_{c_j} + \sum_{\mathfrak{v} \subseteq [d]} \gamma_{\mathfrak{v}} \frac{1}{N^2} \sum_{k,\ell=0}^{N-1} \prod_{j \in \mathfrak{v}} \left(B_2\left(\left|\left\{\frac{k g_j}{N}\right\} - \left\{\frac{\ell g_j}{N}\right\}\right|\right) + \beta_{c_j}\right) \\
&= -\sum_{\mathfrak{v} \subseteq [d]} \gamma_{\mathfrak{v}} \prod_{j \in \mathfrak{v}} \beta_{c_j} + \sum_{\mathfrak{v} \subseteq [d]} \gamma_{\mathfrak{v}} \frac{1}{N^2} \sum_{k,\ell=0}^{N-1} \prod_{j \in \mathfrak{v}} \left(B_2\left(\left\{\frac{(k-\ell) g_j}{N}\right\}\right) + \beta_{c_j}\right).
\end{aligned}$$

As k and ℓ range from 0 to $N-1$, $(k-\ell) \pmod N$ takes on the values $0, 1, \ldots, N-1$ in some order, with each value occurring exactly N times. With this observation we can rewrite the double sum in the latter expression as a single sum, which yields

$$\begin{aligned}
&[\mathrm{err}^{\mathrm{sh}}_{N,d}(\mathcal{H}^{\pitchfork}_{\mathrm{sob},d,\boldsymbol{\gamma},\boldsymbol{c}}, \mathcal{P}(\boldsymbol{g}, N))]^2 \\
&= -\sum_{\mathfrak{v} \subseteq [d]} \gamma_{\mathfrak{v}} \prod_{j \in \mathfrak{v}} \beta_{c_j} + \sum_{\mathfrak{v} \subseteq [d]} \gamma_{\mathfrak{v}} \frac{1}{N} \sum_{k=0}^{N-1} \prod_{j \in \mathfrak{v}} \left(B_2\left(\left\{\frac{k g_j}{N}\right\}\right) + \beta_{c_j}\right) \\
&= \sum_{\emptyset \neq \mathfrak{v} \subseteq [d]} \gamma_{\mathfrak{v}} \left(\frac{1}{N} \sum_{k=0}^{N-1} \prod_{j \in \mathfrak{v}} \left(B_2\left(\left\{\frac{k g_j}{N}\right\}\right) + \beta_{c_j}\right) - \prod_{j \in \mathfrak{v}} \beta_{c_j}\right).
\end{aligned}$$

We now rewrite the product over all $j \in \mathfrak{v}$ as a sum and obtain in this way that

$$\begin{aligned}
&[\mathrm{err}^{\mathrm{sh}}_{N,d}(\mathcal{H}^{\pitchfork}_{\mathrm{sob},d,\boldsymbol{\gamma},\boldsymbol{c}}, \mathcal{P}(\boldsymbol{g}, N))]^2 \\
&= \sum_{\emptyset \neq \mathfrak{v} \subseteq [d]} \gamma_{\mathfrak{v}} \frac{1}{N} \sum_{k=0}^{N-1} \sum_{\emptyset \neq \mathfrak{u} \subseteq \mathfrak{v}} \prod_{j \in \mathfrak{v} \setminus \mathfrak{u}} \beta_{c_j} \prod_{j \in \mathfrak{u}} B_2\left(\left\{\frac{k g_j}{N}\right\}\right) \\
&= \sum_{\emptyset \neq \mathfrak{u} \subseteq [d]} \sum_{\mathfrak{u} \subseteq \mathfrak{v} \subseteq [d]} \gamma_{\mathfrak{v}} \left(\prod_{j \in \mathfrak{v} \setminus \mathfrak{u}} \beta_{c_j}\right) \frac{1}{N} \sum_{k=0}^{N-1} \prod_{j \in \mathfrak{u}} B_2\left(\left\{\frac{k g_j}{N}\right\}\right).
\end{aligned}$$

Since

$$\sum_{\mathfrak{u}\subseteq\mathfrak{v}\subseteq[d]} \gamma_{\mathfrak{v}} \prod_{j\in\mathfrak{v}\setminus\mathfrak{u}} \beta_{c_j} = \left(\prod_{j\in\mathfrak{u}} \frac{1}{\beta_{c_j}}\right) \sum_{\mathfrak{u}\subseteq\mathfrak{v}\subseteq[d]} \gamma_{\mathfrak{v}} \prod_{j\in\mathfrak{v}} \beta_{c_j} = \gamma'_{\mathfrak{u}},$$

we obtain the result as claimed. □

The shift-invariant kernel of the anchored Sobolev space is of the same form as that of the unanchored Sobolev space with the only difference that the weights have to be modified to $\boldsymbol{\gamma}' = \{\gamma'_{\mathfrak{u}}\}_{\mathfrak{u}\subseteq[d]}$, where

$$\gamma'_{\mathfrak{u}} = \left(\prod_{j\in\mathfrak{u}} \frac{1}{\beta_{c_j}}\right) \sum_{\mathfrak{u}\subseteq\mathfrak{v}\subseteq[d]} \gamma_{\mathfrak{v}} \prod_{j\in\mathfrak{v}} \beta_{c_j} \quad \text{for } \mathfrak{u} \subseteq [d].$$

This means that, starting from Corollary 7.19, we can proceed as before and apply the Fourier series expansion of the Bernoulli polynomial B_2 in (2.8). This finally implies

$$[\mathrm{err}^{\mathrm{sh}}_{N,d}(\mathcal{H}^{⫛}_{\mathrm{sob},d,\boldsymbol{\gamma},\boldsymbol{c}}, \mathcal{P}(\boldsymbol{g},N))]^2 = \sum_{\substack{\boldsymbol{h}\in\mathbb{Z}^d\setminus\{\boldsymbol{0}\} \\ \boldsymbol{h}\cdot\boldsymbol{g}\equiv 0 \pmod N}} \frac{1}{r_{2,\widehat{\boldsymbol{\gamma}}}(\boldsymbol{h})}, \tag{7.12}$$

where now $\widehat{\boldsymbol{\gamma}} = \{\widehat{\gamma}_{\mathfrak{u}}\}_{\mathfrak{u}\subseteq[d]}$ with

$$\widehat{\gamma}_{\mathfrak{u}} := \frac{\gamma'_{\mathfrak{u}}}{(2\pi^2)^{|\mathfrak{u}|}} = \left(\prod_{j\in\mathfrak{u}} \frac{1}{2\pi^2\beta_{c_j}}\right) \sum_{\mathfrak{u}\subseteq\mathfrak{v}\subseteq[d]} \gamma_{\mathfrak{v}} \prod_{j\in\mathfrak{v}} \beta_{c_j} \quad \text{for } \mathfrak{u} \subseteq [d]. \tag{7.13}$$

Example 7.20 For product weights $\gamma_{\mathfrak{u}} = \prod_{j\in\mathfrak{u}} \gamma_j$ for $\mathfrak{u} \subseteq [d]$, we have

$$\begin{aligned}
\widehat{\gamma}_{\mathfrak{u}} &= \left(\prod_{j\in\mathfrak{u}} \frac{1}{2\pi^2\beta_{c_j}}\right) \sum_{\mathfrak{u}\subseteq\mathfrak{v}\subseteq[d]} \prod_{j\in\mathfrak{v}} \gamma_j\beta_{c_j} \\
&= \left(\prod_{j\in\mathfrak{u}} \frac{1}{2\pi^2\beta_{c_j}}\right)\left(\prod_{j\in\mathfrak{u}} \gamma_j\beta_{c_j}\right) \sum_{\mathfrak{u}\subseteq\mathfrak{v}\subseteq[d]} \prod_{j\in\mathfrak{v}\setminus\mathfrak{u}} \gamma_j\beta_{c_j} \\
&= \left(\prod_{j\in\mathfrak{u}} \frac{\gamma_j}{2\pi^2}\right) \sum_{\mathfrak{w}\subseteq[d]\setminus\mathfrak{u}} \prod_{j\in\mathfrak{w}} \gamma_j\beta_{c_j} \\
&= \left(\prod_{j\in\mathfrak{u}} \frac{\gamma_j}{2\pi^2}\right) \prod_{j\in[d]\setminus\mathfrak{u}} (1 + \gamma_j\beta_{c_j}).
\end{aligned}$$

Comparing the formula (7.12) with Corollary 2.21, we observe that the shift-averaged worst-case error of a lattice rule in $\mathcal{H}^{⫛}_{\mathrm{sob},d,\boldsymbol{\gamma},\boldsymbol{c}}$ equals the worst-case error of a lattice rule in the Korobov space $\mathcal{H}_{\mathrm{kor},d,\alpha,\widehat{\boldsymbol{\gamma}}}$ with the modified weights $\widehat{\boldsymbol{\gamma}}$ in (7.13) and with smoothness parameter $\alpha = 1$. I.e., we have

$$\mathrm{err}^{\mathrm{sh}}_{N,d}(\mathcal{H}^{\pitchfork}_{\mathrm{sob},d,\boldsymbol{\gamma},\boldsymbol{c}}, \mathcal{P}(\boldsymbol{g}, N)) = \mathrm{err}_{N,d,1,\widehat{\boldsymbol{\gamma}}}(\boldsymbol{g}).$$

This allows us to also rewrite all previously shown results for the worst-case error in Korobov spaces such that they apply to the shift-averaged worst-case error in the weighted anchored Sobolev space (equipped with the corresponding weights).

Similarly to Theorems 2.24, 3.7, and 3.9 we obtain the following result, which we again state only for prime N for the sake of simplicity.

Theorem 7.21 *For any prime number N and any dimension d the following statements hold.*

1. *There exists a generating vector $\boldsymbol{g} \in G_d(N)$ such that*

$$\mathrm{err}^{\mathrm{sh}}_{N,d}(\mathcal{H}^{\pitchfork}_{\mathrm{sob},d,\boldsymbol{\gamma},\boldsymbol{c}}, \mathcal{P}(\boldsymbol{g}, N)) \leq \frac{1}{N^{\tau}} \left(\sum_{\emptyset \neq \mathfrak{u} \subseteq [d]} \widehat{\gamma}_{\mathfrak{u}}^{1/(2\tau)} \left(2\zeta\left(\frac{1}{\tau}\right)\right)^{|\mathfrak{u}|} \right)^{\tau}$$

for all $\tau \in [1/2, 1)$. This vector can be found using Algorithm 3.6 (by means of the weights $\widehat{\boldsymbol{\gamma}}$ in (7.13)).

2. *In the case of product weights the bound in Item 1 can be simplified to*

$$\mathrm{err}^{\mathrm{sh}}_{N,d}(\mathcal{H}^{\pitchfork}_{\mathrm{sob},d,\boldsymbol{\gamma},\boldsymbol{c}}, \mathcal{P}(\boldsymbol{g}, N))$$
$$\leq \frac{1}{N^{\tau}} \left(-1 + \prod_{j=1}^{d} \left(1 + \gamma_j^{1/(2\tau)} a_{\tau} \prod_{\substack{i=1 \\ i \neq j}}^{d} (1 + \gamma_i \beta_{c_i})^{1/(2\tau)} \right) \right)^{\tau},$$

where a_{τ} is as in (7.9).

Remark 7.22 Assume that $\boldsymbol{c} = (c, c, \ldots, c)$ and that we are concerned with product weights. Assume furthermore that $\tau \in [1/2, 1)$ is such that $\Gamma_{\tau} := \sum_{j=1}^{\infty} \gamma_j^{1/(2\tau)} < \infty$. Then we have, using standard methods,

$$\prod_{\substack{i=1 \\ i \neq j}}^{d} (1 + \gamma_i \beta_{c_i})^{1/(2\tau)} \leq \mathrm{e}^{(\beta_c/(2\tau)) \sum_{i=1, i \neq j}^{d} \gamma_i} \leq \mathrm{e}^{(\beta_c/(2\tau))\Gamma_{1/2}},$$

and so

$$\prod_{j=1}^{d} \left(1 + \gamma_j^{1/(2\tau)} a_{\tau} \prod_{\substack{i=1 \\ i \neq j}}^{d} (1 + \gamma_i \beta_{c_i})^{1/(2\tau)} \right) \leq \prod_{j=1}^{d} \left(1 + \gamma_j^{1/(2\tau)} a_{\tau} \mathrm{e}^{(\beta_c/(2\tau))\Gamma_{1/2}} \right)$$
$$\leq \mathrm{e}^{a_{\tau} \mathrm{e}^{(\beta_c/(2\tau))\Gamma_{1/2}} \Gamma_{\tau}},$$

where the latter bound is independent of the dimension d. Consequently, we obtain

$$\mathrm{err}^{\mathrm{sh}}_{N,d}(\mathcal{H}^{\pitchfork}_{\mathrm{sob},d,\boldsymbol{\gamma},\boldsymbol{c}}, \mathcal{P}(\boldsymbol{g}, N)) \le \frac{1}{N^{\tau}} \, \mathrm{e}^{\tau a_{\tau} \mathrm{e}^{(\beta_c/(2\tau))\Gamma_{1/2}}\Gamma_{\tau}}.$$

The upper bound is independent of the dimension d. This implies that $\sum_{j=1}^{\infty} \gamma_j < \infty$ is a sufficient condition for strong polynomial tractability of the integration problem in $\mathcal{H}^{\pitchfork}_{\mathrm{sob},d,\boldsymbol{\gamma},\boldsymbol{c}}$.

According to Remark 7.3, Theorem 7.21 can also be interpreted as an existence result for good shifted lattice rules for integration.

Corollary 7.23 *For any prime number N and any dimension d there exists a generating vector $\boldsymbol{g} \in G_d(N)$ and a shift $\boldsymbol{\Delta} \in [0,1]^d$ such that*

$$\mathrm{err}_{N,d}(\mathcal{H}^{\pitchfork}_{\mathrm{sob},d,\boldsymbol{\gamma},\boldsymbol{c}}, \mathcal{P}_{\boldsymbol{\Delta}}(\boldsymbol{g}, N)) \le \frac{1}{N^{\tau}} \left(\sum_{\emptyset \neq \mathfrak{u} \subseteq [d]} \widetilde{\gamma}_{\mathfrak{u}}^{1/(2\tau)} \left(2\zeta\left(\frac{1}{\tau}\right)\right)^{|\mathfrak{u}|} \right)^{\tau}$$

for all $\tau \in [1/2, 1)$. The vector $\boldsymbol{g}$ can be found using Algorithm 3.6 (by means of the weights $\widehat{\boldsymbol{\gamma}}$ in (7.13)).

Remark 7.24 We will briefly discuss embeddings between the unanchored and the anchored Sobolev spaces of smoothness one, which allow us to transfer results on the worst-case error in one space to the other and vice versa, at the end of Section 7.2.

A relation to geometric discrepancy

We briefly discuss an important relation to the theory of geometric discrepancy. If we choose the anchor $\boldsymbol{c}$ to be $\mathbf{1}$, we have $\eta_1^{\pitchfork}(x, y) = 1 - \max(x, y)$ for $x, y \in [0, 1)$, and hence

$$\int_0^1 \int_0^1 \eta_1^{\pitchfork}(x, y) \,\mathrm{d}x \,\mathrm{d}y = \frac{1}{3} \quad \text{and} \quad \int_0^1 \eta_1^{\pitchfork}(x, y) \,\mathrm{d}y = \frac{1 - x^2}{2}.$$

According to Theorem 1.27, we therefore obtain

$$\begin{aligned}
&[\mathrm{err}_{N,d}(\mathcal{H}^{\pitchfork}_{\mathrm{sob},d,\boldsymbol{\gamma},\mathbf{1}}, \mathcal{P})]^2 \\
&= \int_{[0,1]^d} \int_{[0,1]^d} K^{\pitchfork}_{\mathrm{sob},d,\boldsymbol{\gamma},\mathbf{1}}(\boldsymbol{x}, \boldsymbol{y}) \,\mathrm{d}\boldsymbol{x} \,\mathrm{d}\boldsymbol{y} \\
&\quad - \frac{2}{N} \sum_{k=0}^{N-1} \int_{[0,1]^d} K^{\pitchfork}_{\mathrm{sob},d,\boldsymbol{\gamma},\mathbf{1}}(\boldsymbol{x}_k, \boldsymbol{y}) \,\mathrm{d}\boldsymbol{y} + \frac{1}{N^2} \sum_{k,\ell=0}^{N-1} K^{\pitchfork}_{\mathrm{sob},d,\boldsymbol{\gamma},\mathbf{1}}(\boldsymbol{x}_k, \boldsymbol{x}_\ell) \\
&= \sum_{\mathfrak{u} \subseteq [d]} \gamma_{\mathfrak{u}} \left(\frac{1}{3^{|\mathfrak{u}|}} - \frac{2}{N} \sum_{k=0}^{N-1} \prod_{j \in \mathfrak{u}} \left(\frac{1 - x_{k,j}^2}{2} \right) + \frac{1}{N^2} \sum_{k,\ell=0}^{N-1} \prod_{j \in \mathfrak{u}} (1 - \max(x_{k,j}, x_{\ell,j})) \right).
\end{aligned}$$

The term in the outer parentheses in the latter expression is exactly the squared L_2-discrepancy of the projected point set $\mathcal{P}_{\mathfrak{u}}$, i.e.,

$$\frac{1}{3^{|\mathfrak{u}|}} - \frac{2}{N}\sum_{k=0}^{N-1}\prod_{j\in\mathfrak{u}}\left(\frac{1-x_{k,j}^2}{2}\right) + \frac{1}{N^2}\sum_{k,\ell=0}^{N-1}\prod_{j\in\mathfrak{u}}(1-\max(x_{k,j},x_{\ell,j})) = \left(L_{2,N}(\mathcal{P}_{\mathfrak{u}})\right)^2. \tag{7.14}$$

Formula (7.14) is well known (see, e.g., [52, Proposition 2.15]) and can be obtained easily through a direct evaluation of the integral in the definition of the L_2-discrepancy. Obviously, the squared L_2-discrepancy of $\mathcal{P}_{\mathfrak{u}}$ can be rewritten in the form

$$\left(L_{2,N}(\mathcal{P}_{\mathfrak{u}})\right)^2 = \int_{[0,1]^{|\mathfrak{u}|}} |\Delta_{\mathcal{P}}(\boldsymbol{t}_{\mathfrak{u}},\mathbf{1})|^2 \,\mathrm{d}\boldsymbol{t}_{\mathfrak{u}},$$

where $\Delta_{\mathcal{P}}$ is the local discrepancy function from Definition 1.29. Therefore we obtain

$$[\mathrm{err}_{N,d}(\mathcal{H}^{\pitchfork}_{\mathrm{sob},d,\boldsymbol{\gamma},\mathbf{1}},\mathcal{P})]^2 = \sum_{\emptyset\neq\mathfrak{u}\subseteq[d]} \gamma_{\mathfrak{u}} \left(\int_{[0,1]^{|\mathfrak{u}|}} |\Delta_{\mathcal{P}}(\boldsymbol{t}_{\mathfrak{u}},\mathbf{1})|^2 \,\mathrm{d}\boldsymbol{t}_{\mathfrak{u}}\right).$$

Hence, the worst-case error of a QMC rule based on a point set $\mathcal{P}$ in the weighted anchored Sobolev space with smoothness one and anchor $\mathbf{1}$ is the weighted version of the combined L_2-discrepancy of $\mathcal{P}$ that we know from Definition 1.34.

Definition 7.25 For an N-element point set $\mathcal{P}$ in $[0,1)^d$ and for general weights $\boldsymbol{\gamma} = \{\gamma_{\mathfrak{u}}\}_{\mathfrak{u}\subseteq[d]}$ the *weighted L_2-discrepancy* of $\mathcal{P}$ is given by

$$L_{2,N,\boldsymbol{\gamma}}(\mathcal{P}) = \left(\sum_{\emptyset\neq\mathfrak{u}\subseteq[d]} \gamma_{\mathfrak{u}} \int_{[0,1]^{|\mathfrak{u}|}} |\Delta_{\mathcal{P}}(\boldsymbol{t}_{\mathfrak{u}},\mathbf{1})|^2 \,\mathrm{d}\boldsymbol{t}_{\mathfrak{u}}\right)^{1/2}.$$

The weighted L_2-discrepancy is an L_2-version of the weighted star-discrepancy from Section 5.3. It is obvious how to generalize this notion of discrepancy to the notion of a weighted L_p-discrepancy for arbitrary $p \in [1,\infty)$, see also [179, 211, 239].

We summarize our findings in the following theorem.

Theorem 7.26 *Let $\mathcal{P}$ be an N-element point set in $[0,1)^d$. Then we have*

$$\mathrm{err}_{N,d}(\mathcal{H}^{\pitchfork}_{\mathrm{sob},d,\boldsymbol{\gamma},\mathbf{1}},\mathcal{P}) = L_{2,N,\boldsymbol{\gamma}}(\mathcal{P}).$$

Above, we have considered the restriction of the anchor to the choice $\boldsymbol{c} = \mathbf{1}$ since this is the most important case. For a general anchor $\boldsymbol{c}$ one can obtain an analogous relation to the so-called weighted anchored L_2-discrepancy. We refer to the book of Novak and Woźniakowski [211, Section 9.5] for further information.

7.2 Sobolev Spaces of Higher Smoothness and Cosine Spaces

In this section we introduce further spaces of not necessarily periodic but smooth functions, and we discuss how they are related to the Korobov space of smooth periodic functions.

The weighted unanchored Sobolev space of smoothness α

The weighted unanchored Sobolev space $\mathcal{H}_{\mathrm{sob},d,\gamma}$ of smoothness one from the previous section can be generalized to higher order smoothness. For $\alpha \in \mathbb{N}$, consider the reproducing kernel $K_{\mathrm{sob},d,\alpha,\gamma} : [0,1]^d \times [0,1]^d \to \mathbb{R}$ given by

$$K_{\mathrm{sob},d,\alpha,\gamma}(\boldsymbol{x},\boldsymbol{y}) := \sum_{\mathfrak{u}\subseteq[d]} \gamma_{\mathfrak{u}} \prod_{j\in\mathfrak{u}} \left(\sum_{\tau=1}^{\alpha} \frac{B_\tau(x_j)B_\tau(y_j)}{(\tau!)^2} + (-1)^{\alpha+1} \frac{B_{2\alpha}(|x_j-y_j|)}{(2\alpha)!} \right),$$

where B_τ is the Bernoulli polynomial of order τ. For $\alpha = 1$ this definition matches the definition of $K_{\mathrm{sob},d,\gamma}$ in (7.5).

Definition 7.27 The *weighted unanchored Sobolev space* $\mathcal{H}_{\mathrm{sob},d,\alpha,\gamma}$ of smoothness α is the reproducing kernel Hilbert space with kernel $K_{\mathrm{sob},d,\alpha,\gamma}$, i.e., $\mathcal{H}_{\mathrm{sob},d,\alpha,\gamma} = \mathcal{H}(K_{\mathrm{sob},d,\alpha,\gamma})$, and the corresponding inner product

$$\langle f,g\rangle_{\mathrm{sob},d,\alpha,\gamma} = \sum_{\mathfrak{u}\in\mathfrak{U}^+} \frac{1}{\gamma_{\mathfrak{u}}} \sum_{\mathfrak{v}\subseteq\mathfrak{u}} \sum_{\boldsymbol{\tau}_{\mathfrak{u}\setminus\mathfrak{v}}\in[\alpha-1]^{|\mathfrak{u}\setminus\mathfrak{v}|}} \int_{[0,1]^{|\mathfrak{v}|}} \left(\int_{[0,1]^{d-|\mathfrak{v}|}} f^{(\boldsymbol{\tau}_{\mathfrak{u}\setminus\mathfrak{v}},\boldsymbol{\alpha}_{\mathfrak{v}},\boldsymbol{0})}(\boldsymbol{x})\,\mathrm{d}\boldsymbol{x}_{[d]\setminus\mathfrak{v}} \right)$$
$$\times \left(\int_{[0,1]^{d-|\mathfrak{v}|}} g^{(\boldsymbol{\tau}_{\mathfrak{u}\setminus\mathfrak{v}},\boldsymbol{\alpha}_{\mathfrak{v}},\boldsymbol{0})}(\boldsymbol{x})\,\mathrm{d}\boldsymbol{x}_{[d]\setminus\mathfrak{v}} \right) \mathrm{d}\boldsymbol{x}_{\mathfrak{v}}, \tag{7.15}$$

where $(\boldsymbol{\tau}_{\mathfrak{u}\setminus\mathfrak{v}},\boldsymbol{\alpha}_{\mathfrak{v}},\boldsymbol{0})$ denotes the d-dimensional vector whose j-th component is τ_j if $j \in \mathfrak{u}\setminus\mathfrak{v}$, α if $j \in \mathfrak{v}$, and 0 otherwise, and where $f^{(\boldsymbol{\tau}_{\mathfrak{u}\setminus\mathfrak{v}},\boldsymbol{\alpha}_{\mathfrak{v}},\boldsymbol{0})}$ denotes the $(\boldsymbol{\tau}_{\mathfrak{u}\setminus\mathfrak{v}},\boldsymbol{\alpha}_{\mathfrak{v}},\boldsymbol{0})$-th partial mixed derivative of f, i.e.,

$$f^{(\boldsymbol{\tau}_{\mathfrak{u}\setminus\mathfrak{v}},\boldsymbol{\alpha}_{\mathfrak{v}},\boldsymbol{0})} = \frac{\partial^{\sum_{i\in\mathfrak{u}\setminus\mathfrak{v}}\tau_i+|\mathfrak{v}|\alpha}}{\prod_{i\in\mathfrak{u}\setminus\mathfrak{v}}\partial x_i^{\tau_i}\prod_{i\in\mathfrak{v}}\partial x_i^{\alpha}} f.$$

If $\alpha = 1$, we interpret the definition of the inner product such that $[\alpha-1]$ is the empty set, and hence all summands of the second sum for $\mathfrak{v} \subsetneq \mathfrak{u}$ disappear. For the case $\mathfrak{v} = \mathfrak{u}$, $\boldsymbol{\tau}_{\mathfrak{u}\setminus\mathfrak{v}}$ in (7.15) is the "empty" vector with no components such that the corresponding summand in the second sum in (7.15) remains. Hence the definition indeed matches Definition 7.6 for the space with smoothness one.

The corresponding norm is therefore given by

$$\|f\|_{\mathrm{sob},d,\alpha,\boldsymbol{\gamma}} = \left(\sum_{\mathfrak{u}\in\mathfrak{U}^+} \frac{1}{\gamma_{\mathfrak{u}}} \sum_{\mathfrak{v}\subseteq\mathfrak{u}} \sum_{\boldsymbol{\tau}_{\mathfrak{u}\setminus\mathfrak{v}}\in[\alpha-1]^{|\mathfrak{u}\setminus\mathfrak{v}|}} \int_{[0,1]^{|\mathfrak{v}|}} \left(\int_{[0,1]^{d-|\mathfrak{v}|}} f^{(\boldsymbol{\tau}_{\mathfrak{u}\setminus\mathfrak{v}},\boldsymbol{\alpha}_{\mathfrak{v}},\boldsymbol{0})}(\boldsymbol{x}) \,\mathrm{d}\boldsymbol{x}_{[d]\setminus\mathfrak{v}} \right)^2 \mathrm{d}\boldsymbol{x}_{\mathfrak{v}} \right)^{1/2},$$

with the analogous modifications for $\alpha = 1$ as for the inner product.

Remark 7.28 The weighted unanchored Sobolev space $\mathcal{H}_{\mathrm{sob},d,\alpha,\boldsymbol{\gamma}}$ contains all functions on $[0,1]^d$ whose mixed partial derivatives up to order α in each coordinate are square integrable. In particular, in the univariate case $d = 1$, the norm reduces to

$$\|f\|_{\mathrm{sob},1,\alpha,\gamma}^2 = \left(\int_0^1 f(x)\,\mathrm{d}x\right)^2 + \frac{1}{\gamma}\left[\sum_{\tau=1}^{\alpha-1}\left(\int_0^1 f^{(\tau)}(x)\,\mathrm{d}x\right)^2 + \int_0^1 (f^{(\alpha)}(x))^2\,\mathrm{d}x\right].$$

The space $\mathcal{H}_{\mathrm{sob},1,\alpha,\gamma}$ is equivalent to the set of all functions $f : [0,1] \to \mathbb{R}$ such that $f^{(\tau)}$ is absolutely continuous for $0 \le \tau < \alpha$ and $f^{(\alpha)} \in L_2([0,1])$.

For the initial error we have the following result.

Proposition 7.29 *The initial error of integration in $\mathcal{H}_{\mathrm{sob},d,\alpha,\boldsymbol{\gamma}}$ equals 1.*

Proof The result follows in the same way as for the special instance $\alpha = 1$ (see Proposition 7.8) bearing in mind that $\int_0^1 B_\tau(x)\,\mathrm{d}x = 0$ for all $\tau \in \mathbb{N}$. □

The half-period cosine space

We begin the introduction of this space by recalling the well-known fact that the functions

$$1, \sqrt{2}\cos(\pi x), \sqrt{2}\cos(2\pi x), \sqrt{2}\cos(3\pi x), \ldots$$

on $[0,1]$ are an orthonormal basis of $L_2([0,1])$. In particular, for $m, n \in \mathbb{N}$ we have

$$\int_0^1 \sqrt{2}\cos(\pi m x)\sqrt{2}\cos(\pi n x)\,\mathrm{d}x = \begin{cases} 1 & \text{if } m = n, \\ 0 & \text{if } m \neq n, \end{cases}$$

and, for $n \in \mathbb{N}$,

$$\int_0^1 \sqrt{2}\cos(\pi n x)\,\mathrm{d}x = 0.$$

Any function $f \in L_2([0,1])$ has the cosine series expansion

$$\widetilde{f}(0) + \sum_{k=1}^{\infty} \widetilde{f}(k)\sqrt{2}\cos(\pi k x),$$

with cosine coefficients

$$\widetilde{f}(k) := \begin{cases} \int_0^1 f(x)\,\mathrm{d}x & \text{for } k = 0, \\ \int_0^1 f(x)\sqrt{2}\cos(\pi k x)\,\mathrm{d}x & \text{for } k \in \mathbb{N}. \end{cases}$$

The advantage of cosine series over Fourier series when expanding nonperiodic functions is that they overcome the well-known Gibbs phenomenon.

Next, we consider the multivariate case. For a function $f : [0,1]^d \to \mathbb{R}$ and for a vector $\boldsymbol{k} \in \mathbb{N}_0^d$ the $\boldsymbol{k}$-th cosine coefficient of f is defined as

$$\widetilde{f}(\boldsymbol{k}) := \int_{[0,1]^d} f(\boldsymbol{x}) \prod_{j=1}^{d} (2 - \delta_{0,k_j})^{1/2} \cos(\pi k_j x_j)\,\mathrm{d}\boldsymbol{x},$$

where δ denotes the Kronecker delta function, i.e., δ_{0,k_j} equals one if $k_j = 0$ and zero otherwise.

Definition 7.30 Let $\alpha > 1/2$. The *half-period cosine space of smoothness* α is the reproducing kernel Hilbert space $\mathcal{H}_{\cos,d,\alpha,\boldsymbol{\gamma}} := \mathcal{H}(K_{\cos,d,\alpha,\boldsymbol{\gamma}})$ with reproducing kernel $K_{\cos,d,\alpha,\boldsymbol{\gamma}} : [0,1]^d \times [0,1]^d \to \mathbb{R}$ given by

$$K_{\cos,d,\alpha,\boldsymbol{\gamma}}(\boldsymbol{x},\boldsymbol{y}) := \sum_{\mathfrak{u} \subseteq [d]} \gamma_{\mathfrak{u}} \sum_{\boldsymbol{k}_{\mathfrak{u}} \in \mathbb{N}^{|\mathfrak{u}|}} \prod_{j \in \mathfrak{u}} \frac{2\cos(\pi k_j x_j)\cos(\pi k_j y_j)}{k_j^{2\alpha}},$$

and with inner product

$$\langle f, g \rangle_{\cos,d,\alpha,\boldsymbol{\gamma}} := \sum_{\boldsymbol{k} \in \mathbb{N}_0^d} r_{2\alpha,\boldsymbol{\gamma}}(\boldsymbol{k})\, \widetilde{f}(\boldsymbol{k}) \widetilde{g}(\boldsymbol{k}).$$

The corresponding norm is then given by

$$\|f\|_{\cos,d,\alpha,\boldsymbol{\gamma}} := \left(\sum_{\boldsymbol{k} \in \mathbb{N}_0^d} r_{2\alpha,\boldsymbol{\gamma}}(\boldsymbol{k})\, |\widetilde{f}(\boldsymbol{k})|^2 \right)^{1/2}.$$

The reproducing kernel can also be written in the form

$$\begin{aligned} K_{\cos,d,\alpha,\boldsymbol{\gamma}}(\boldsymbol{x},\boldsymbol{y}) &= \sum_{\boldsymbol{k} \in \mathbb{N}_0^d} \gamma_{\mathfrak{u}(\boldsymbol{k})} \prod_{j \in \mathfrak{u}(\boldsymbol{k})} \frac{2\cos(\pi k_j x_j)\cos(\pi k_j y_j)}{k_j^{2\alpha}} \\ &= \sum_{\boldsymbol{k} \in \mathbb{N}_0^d} \frac{1}{r_{2\alpha,\boldsymbol{\gamma}}(\boldsymbol{k})} \prod_{j \in \mathfrak{u}(\boldsymbol{k})} \left(2\cos(\pi k_j x_j)\cos(\pi k_j y_j)\right), \end{aligned}$$

and the norm can be rewritten in the form

$$\|f\|_{\cos,d,\alpha,\gamma} = \left(\sum_{\boldsymbol{h} \in \mathbb{Z}^d} r_{2\alpha,\gamma}(\boldsymbol{h}) \left(\prod_{j=1}^{d} \frac{1}{2 - \delta_{0,h_j}} \right) |\widetilde{f}(|\boldsymbol{h}|)|^2 \right)^{1/2},$$

where $|\boldsymbol{h}| = (|h_1|, \ldots, |h_d|)$.

Usually, cosine series are considered over the whole period $[-1, 1)$. Here we only consider functions on $[0, 1]$, i.e., the "half" of the period and the term "half-period" we use refers to this fact.

The following result is analogous to Propositions 7.8, 7.17, and 7.29.

Proposition 7.31 *The initial error of integration in $\mathcal{H}_{\cos,d,\alpha,\gamma}$ equals 1.*

Proof The result follows easily from the fact that $\int_0^1 \cos(\pi k x)\,\mathrm{d}x$ equals one if $k = 0$ and zero if $k \in \mathbb{N}$. Hence, according to Remark 1.41,

$$[e(0, d)]^2 = \int_{[0,1]^d} \int_{[0,1]^d} K_{\cos,d,\alpha,\gamma}(\boldsymbol{x}, \boldsymbol{y})\,\mathrm{d}\boldsymbol{x}\,\mathrm{d}\boldsymbol{y} = 1. \qquad \square$$

Embeddings

With the Korobov space $\mathcal{H}_{\mathrm{kor},d,\alpha,\gamma}$, the unanchored Sobolev space $\mathcal{H}_{\mathrm{sob},d,\alpha,\gamma}$, and the half-period cosine space $\mathcal{H}_{\cos,d,\alpha,\gamma}$ we have introduced three reference spaces of higher smoothness for QMC integration and, in particular, for integration using lattice rules. We now discuss how these spaces are related to each other.

For two normed spaces $(\mathcal{H}_1, \|\cdot\|_1)$ and $(\mathcal{H}_2, \|\cdot\|_2)$ we say that $\mathcal{H}_1$ is *continuously embedded* in $\mathcal{H}_2$ if

$$\mathcal{H}_1 \subseteq \mathcal{H}_2,$$

and if

$$\|f\|_2 \le C\|f\|_1 \quad \text{for all } f \in \mathcal{H}_1$$

for some constant $C > 0$ that is independent of f. We write

$$\mathcal{H}_1 \hookrightarrow \mathcal{H}_2$$

in this case. If $\mathcal{H}_1 \hookrightarrow \mathcal{H}_2$ and $\mathcal{H}_2 \hookrightarrow \mathcal{H}_1$ we write

$$\mathcal{H}_1 \leftrightarrows \mathcal{H}_2.$$

On the other hand, it is possible that $\mathcal{H}_1$ is not even a subset of $\mathcal{H}_2$, i.e., there is a function in $\mathcal{H}_1$ which is not in $\mathcal{H}_2$. In this case we write $\mathcal{H}_1 \not\subset \mathcal{H}_2$.

The following proposition illustrates the effect of embeddings between function spaces on the relation between the respective worst-case integration errors.

Proposition 7.32 *Let $(\mathcal{H}_1, \|\cdot\|_1)$ and $(\mathcal{H}_2, \|\cdot\|_2)$ be two normed spaces of continuous, integrable functions on $[0, 1]^d$ such that $\mathcal{H}_1 \hookrightarrow \mathcal{H}_2$ and let $\iota_{\mathcal{H}_1,\mathcal{H}_2} : \mathcal{H}_1 \to \mathcal{H}_2$, $\iota_{\mathcal{H}_1,\mathcal{H}_2}(f) = f$ be the inclusion map with finite norm $\|\iota_{\mathcal{H}_1,\mathcal{H}_2}\|$. Then for any QMC rule (or linear integration rule) based on a point set $\mathcal{P}$ we have*

$$\mathrm{err}_{N,d}(\mathcal{H}_1, \mathcal{P}) \le \|\iota_{\mathcal{H}_1,\mathcal{H}_2}\| \, \mathrm{err}_{N,d}(\mathcal{H}_2, \mathcal{P}).$$

Proof We have

$$C := \|\iota_{\mathcal{H}_1,\mathcal{H}_2}\| = \sup_{\substack{f \in \mathcal{H}_1 \\ \|f\|_1 \le 1}} \|f\|_2.$$

Hence $\|f\|_1 \le 1/C$ implies $\|f\|_2 \le 1$. We therefore have

$$\begin{aligned}
\mathrm{err}_{N,d}(\mathcal{H}_1, \mathcal{P}) &= \sup_{\substack{f \in \mathcal{H}_1 \\ \|f\|_1 \le 1}} |\mathrm{err}_{N,d}(f, \mathcal{P})| \\
&= \sup_{\substack{f \in \mathcal{H}_1 \\ \|Cf\|_1 \le 1}} |\mathrm{err}_{N,d}(Cf, \mathcal{P})| \\
&= C \sup_{\substack{f \in \mathcal{H}_1 \\ \|f\|_1 \le 1/C}} |\mathrm{err}_{N,d}(f, \mathcal{P})| \\
&\le C \sup_{\substack{f \in \mathcal{H}_2 \\ \|f\|_2 \le 1}} |\mathrm{err}_{N,d}(f, \mathcal{P})| \\
&= C \, \mathrm{err}_{N,d}(\mathcal{H}_2, \mathcal{P}),
\end{aligned}$$

as claimed. □

We now consider the relations between the Korobov space, the unanchored Sobolev space, and the half-period cosine space.

Theorem 7.33 *The following relations hold.*

1. *For all* $\alpha \in \mathbb{N}$ *we have*

$$\mathcal{H}_{\mathrm{kor},d,\alpha,\gamma} \hookrightarrow \mathcal{H}_{\mathrm{sob},d,\alpha,\gamma} \quad \textit{but} \quad \mathcal{H}_{\mathrm{sob},d,\alpha,\gamma} \not\subset \mathcal{H}_{\mathrm{kor},d,\alpha,\gamma}.$$

2. *For* $\alpha = 1$ *we have*

$$\mathcal{H}_{\mathrm{sob},d,1,\gamma} \leftrightharpoons \mathcal{H}_{\mathrm{cos},d,1,\gamma}.$$

3. *For* $\alpha \in \mathbb{N}$ *with* $\alpha \ge 2$ *we have*

$$\mathcal{H}_{\mathrm{cos},d,\alpha,\gamma} \hookrightarrow \mathcal{H}_{\mathrm{sob},d,\alpha,\gamma} \quad \textit{but} \quad \mathcal{H}_{\mathrm{sob},d,\alpha,\gamma} \not\subset \mathcal{H}_{\mathrm{cos},d,\alpha,\gamma}.$$

4. *For real* $\alpha > 1/2$ *we have*

$$\begin{aligned}
&\mathcal{H}_{\mathrm{cos},d,\alpha,\gamma} \not\subset \mathcal{H}_{\mathrm{kor},d,\alpha,\gamma}, \\
\textit{and} \quad &\mathcal{H}_{\mathrm{kor},d,\alpha,\gamma} \not\subset \mathcal{H}_{\mathrm{cos},d,\alpha,\gamma} \quad \textit{for } \alpha \ge \frac{3}{2}.
\end{aligned}$$

Proof Since all spaces involved are tensor products of the corresponding univariate spaces, it suffices to show the result for $d = 1$ and a generic weight $\gamma > 0$.

Regarding Item 1 in the theorem, let $\alpha \in \mathbb{N}$. We first show that $\mathcal{H}_{\mathrm{kor},1,\alpha,\gamma} \hookrightarrow \mathcal{H}_{\mathrm{sob},1,\alpha,\gamma}$. To this end let $\widetilde{\mathcal{H}}_{\mathrm{sob},1,\alpha,\gamma}$ be the subspace of $\mathcal{H}_{\mathrm{sob},1,\alpha,\gamma}$ consisting of periodic functions,

$$\widetilde{\mathcal{H}}_{\text{sob},1,\alpha,\gamma} = \{f \in \mathcal{H}_{\text{sob},1,\alpha,\gamma} \ : \ f^{(\tau)}(0) = f^{(\tau)}(1) \ \text{ for all } \ \tau \in \{0, 1, \ldots, \alpha - 1\}\}.$$

We show that $\widetilde{\mathcal{H}}_{\text{sob},1,\alpha,\gamma}$ is a reproducing kernel Hilbert space with the same inner product as $\mathcal{H}_{\text{sob},1,\alpha,\gamma}$ and with the reproducing kernel

$$\widetilde{K}_{\text{sob},1,\alpha,\gamma}(x, y) = 1 + \gamma(-1)^{\alpha+1} \frac{B_{2\alpha}(|x - y|)}{(2\alpha)!}.$$

Since the kernel $\widetilde{K}_{\text{sob},1,\alpha,\gamma}$ differs from $K_{\text{sob},1,\alpha,\gamma}$ only by an additive term, which is a polynomial, and polynomials and $K_{\text{sob},1,\alpha,\gamma}(\cdot, y)$ for $y \in [0, 1]$ are in $\mathcal{H}_{\text{sob},1,\alpha,\gamma}$, it follows that $\widetilde{K}_{\text{sob},1,\alpha,\gamma}(\cdot, y) \in \mathcal{H}_{\text{sob},1,\alpha,\gamma}$ for $y \in [0, 1]$. To show that also the periodicity condition in $\widetilde{\mathcal{H}}_{\text{sob},1,\alpha,\gamma}$ is satisfied, we use the well-known formula $B'_\tau(x) = \tau B_{\tau-1}(x)$ for $\tau \in \mathbb{N}$ for Bernoulli polynomials. Then we obtain, for any $y \in [0, 1]$, that

$$\begin{aligned}
&\frac{\partial^\tau}{\partial x^\tau} \widetilde{K}_{\text{sob},1,\alpha,\gamma}(1, y) - \frac{\partial^\tau}{\partial x^\tau} \widetilde{K}_{\text{sob},1,\alpha,\gamma}(0, y) \\
&= \frac{\gamma(-1)^{\alpha+1}}{(2\alpha)!} \left(\frac{\partial^\tau}{\partial x^\tau} B_{2\alpha}(x - y)\Big|_{x=1} - \frac{\partial^\tau}{\partial x^\tau} B_{2\alpha}(y - x)\Big|_{x=0} \right) \\
&= \frac{\gamma(-1)^{\alpha+1}}{(2\alpha)!} \frac{(2\alpha)!}{(2\alpha - \tau)!} (B_{2\alpha-\tau}(1 - y) - (-1)^\tau B_{2\alpha-\tau}(y)) \\
&= \frac{\gamma(-1)^{\alpha+1}}{(2\alpha - \tau)!} ((-1)^\tau B_{2\alpha-\tau}(y) - (-1)^\tau B_{2\alpha-\tau}(y)) \\
&= 0,
\end{aligned}$$

as $B_{2\alpha-\tau}(1 - y) = (-1)^{2\alpha-\tau} B_{2\alpha-\tau}(y) = (-1)^\tau B_{2\alpha-\tau}(y)$ according to the symmetry $B_\tau(1 - x) = (-1)^\tau B_\tau(x)$ for $\tau \in \mathbb{N}_0$ for Bernoulli polynomials. Hence $(\partial^\tau/\partial x^\tau)\widetilde{K}_{\text{sob},1,\alpha,\gamma}(x, y)$ is periodic in x and therefore $\widetilde{K}_{\text{sob},1,\alpha,\gamma}(\cdot, y)$ belongs to $\widetilde{\mathcal{H}}_{\text{sob},1,\alpha,\gamma}$ for all $y \in [0, 1]$.

Next, we prove the reproducing property of $\widetilde{K}_{\text{sob},1,\alpha,\gamma}$. For $f \in \widetilde{\mathcal{H}}_{\text{sob},1,\alpha,\gamma}$ and for $y \in [0, 1]$ we have

$$\begin{aligned}
\langle f, \widetilde{K}_{\text{sob},1,\alpha,\gamma}(\cdot, y)\rangle_{\text{sob},1,\alpha,\gamma} = &\int_0^1 f(x) \,\mathrm{d}x \int_0^1 \widetilde{K}_{\text{sob},1,\alpha,\gamma}(x, y) \,\mathrm{d}x \\
&+ \frac{1}{\gamma} \sum_{\tau=1}^{\alpha-1} \int_0^1 f^{(\tau)}(x) \,\mathrm{d}x \int_0^1 \frac{\partial^\tau}{\partial x^\tau} \widetilde{K}_{\text{sob},1,\alpha,\gamma}(x, y) \,\mathrm{d}x \\
&+ \frac{1}{\gamma} \int_0^1 f^{(\alpha)}(x) \frac{\partial^\alpha}{\partial x^\alpha} \widetilde{K}_{\text{sob},1,\alpha,\gamma}(x, y) \,\mathrm{d}x.
\end{aligned}$$

We treat the integrals in the above formula separately. First we have

$$\int_0^1 \widetilde{K}_{\text{sob},1,\alpha,\gamma}(x, y) \,\mathrm{d}x = 1 + \frac{\gamma(-1)^{\alpha+1}}{(2\alpha)!} \int_0^1 B_{2\alpha}(|x - y|) \,\mathrm{d}x = 1,$$

because

$$\begin{aligned}\int_0^1 B_{2\alpha}(|x-y|)\,\mathrm{d}x &= \int_0^y B_{2\alpha}(y-x)\,\mathrm{d}x + \int_y^1 B_{2\alpha}(x-y)\,\mathrm{d}x\\ &= \int_0^y (-1)^{2\alpha} B_{2\alpha}(1-y+x)\,\mathrm{d}x + \int_y^1 B_{2\alpha}(x-y)\,\mathrm{d}x\\ &= \int_{1-y}^1 B_{2\alpha}(t)\,\mathrm{d}t + \int_0^{1-y} B_{2\alpha}(t)\,\mathrm{d}t\\ &= \int_0^1 B_{2\alpha}(t)\,\mathrm{d}t\\ &= 0.\end{aligned}$$

In the next step, we have

$$\int_0^1 \frac{\partial^\tau}{\partial x^\tau}\widetilde{K}_{\mathrm{sob},1,\alpha,\gamma}(x,y)\,\mathrm{d}x = \frac{\gamma(-1)^{\alpha+1}}{(2\alpha)!}\int_0^1 \frac{\partial^\tau}{\partial x^\tau} B_{2\alpha}(|x-y|)\,\mathrm{d}x = 0,$$

since

$$\begin{aligned}&\int_0^1 \frac{\partial^\tau}{\partial x^\tau} B_{2\alpha}(|x-y|)\,\mathrm{d}x\\ &= \int_0^y \frac{\partial^\tau}{\partial x^\tau} B_{2\alpha}(y-x)\,\mathrm{d}x + \int_y^1 \frac{\partial^\tau}{\partial x^\tau} B_{2\alpha}(x-y)\,\mathrm{d}x\\ &= \frac{(2\alpha)!}{(2\alpha-\tau)!}\left(\int_0^y (-1)^\tau B_{2\alpha-\tau}(y-x)\,\mathrm{d}x + \int_y^1 B_{2\alpha-\tau}(x-y)\,\mathrm{d}x\right)\\ &= \frac{(2\alpha)!}{(2\alpha-\tau)!}\left(\int_0^y B_{2\alpha-\tau}(1-y+x)\,\mathrm{d}x + \int_y^1 B_{2\alpha-\tau}(x-y)\,\mathrm{d}x\right)\\ &= \frac{(2\alpha)!}{(2\alpha-\tau)!}\int_0^1 B_{2\alpha-\tau}(t)\,\mathrm{d}t\\ &= 0.\end{aligned}$$

It remains to consider

$$\begin{aligned}&\int_0^1 f^{(\alpha)}(x)\frac{\partial^\alpha}{\partial x^\alpha}\widetilde{K}_{\mathrm{sob},1,\alpha,\gamma}(x,y)\,\mathrm{d}x\\ &= \frac{\gamma(-1)^{\alpha+1}}{(2\alpha)!}\left(\int_0^y f^{(\alpha)}(x)\frac{\partial^\alpha}{\partial x^\alpha} B_{2\alpha}(y-x)\,\mathrm{d}x + \int_y^1 f^{(\alpha)}(x)\frac{\partial^\alpha}{\partial x^\alpha} B_{2\alpha}(x-y)\,\mathrm{d}x\right).\end{aligned}$$

The partial derivatives of the Bernoulli polynomials involved can be computed as

$$\frac{\partial^\alpha}{\partial x^\alpha} B_{2\alpha}(y-x) = (-1)^\alpha \frac{(2\alpha)!}{\alpha!} B_\alpha(y-x) = \frac{(2\alpha)!}{\alpha!} B_\alpha(1-y+x),$$

and likewise

$$\frac{\partial^\alpha}{\partial x^\alpha} B_{2\alpha}(x-y) = \frac{(2\alpha)!}{\alpha!} B_\alpha(x-y).$$

Hence

$$\begin{aligned}
&\int_0^1 f^{(\alpha)}(x) \frac{\partial^\alpha}{\partial x^\alpha} \widetilde{K}_{\mathrm{sob},1,\alpha,\gamma}(x,y)\,\mathrm{d}x \\
&= \frac{\gamma(-1)^{\alpha+1}}{\alpha!} \left(\int_0^y f^{(\alpha)}(x) B_\alpha(1-y+x)\,\mathrm{d}x + \int_y^1 f^{(\alpha)}(x) B_\alpha(x-y)\,\mathrm{d}x \right).
\end{aligned}$$

We now use integration by parts. We have

$$\begin{aligned}
&\int_0^y f^{(\alpha)}(x) B_\alpha(1-y+x)\,\mathrm{d}x + \int_y^1 f^{(\alpha)}(x) B_\alpha(x-y)\,\mathrm{d}x \\
&= f^{(\alpha-1)}(x) B_\alpha(1-y+x)\Big|_0^y - \alpha \int_0^y f^{(\alpha-1)}(x) B_{\alpha-1}(1-y+x)\,\mathrm{d}x \\
&\quad + f^{(\alpha-1)}(x) B_\alpha(x-y)\Big|_y^1 - \alpha \int_y^1 f^{(\alpha-1)}(x) B_{\alpha-1}(x-y)\,\mathrm{d}x \\
&= -\alpha \left(\int_0^y f^{(\alpha-1)}(x) B_{\alpha-1}(1-y+x)\,\mathrm{d}x + \int_y^1 f^{(\alpha-1)}(x) B_{\alpha-1}(x-y)\,\mathrm{d}x \right),
\end{aligned}$$

where we used that $f^{(\alpha-1)}(0) = f^{(\alpha-1)}(1)$ and $B_\alpha(0) = B_\alpha(1)$ for $\alpha \geq 2$. Repeating this argument inductively we arrive at

$$\begin{aligned}
&\int_0^y f^{(\alpha)}(x) B_\alpha(1-y+x)\,\mathrm{d}x + \int_y^1 f^{(\alpha)}(x) B_\alpha(x-y)\,\mathrm{d}x \\
&= (-1)^{\alpha-1}\alpha! \left(\int_0^y f'(x) B_1(1-y+x)\,\mathrm{d}x + \int_y^1 f'(x) B_1(x-y)\,\mathrm{d}x \right).
\end{aligned}$$

Applying integration by parts one more time yields

$$\begin{aligned}
&\int_0^y f^{(\alpha)}(x) B_\alpha(1-y+x)\,\mathrm{d}x + \int_y^1 f^{(\alpha)}(x) B_\alpha(x-y)\,\mathrm{d}x \\
&\qquad = (-1)^{\alpha-1}\alpha! \left(f(y) - \int_0^1 f(x)\,\mathrm{d}x \right).
\end{aligned}$$

Putting everything together we finally obtain

$$\begin{aligned}
&\langle f, \widetilde{K}_{\mathrm{sob},1,\alpha,\gamma}(\cdot,y)\rangle_{\mathrm{sob},1,\alpha,\gamma} \\
&= \int_0^1 f(x)\,\mathrm{d}x + \frac{1}{\gamma}\frac{\gamma(-1)^{\alpha+1}}{\alpha!}(-1)^{\alpha-1}\alpha! \left(f(y) - \int_0^1 f(x)\,\mathrm{d}x \right) \\
&= f(y).
\end{aligned}$$

This proves the reproducing property of $\widetilde{K}_{\mathrm{sob},1,\alpha,\gamma}$ and therefore $\widetilde{K}_{\mathrm{sob},1,\alpha,\gamma}$ is the reproducing kernel of $\widetilde{\mathcal{H}}_{\mathrm{sob},1,\alpha,\gamma}$.

From the Fourier series expansion of the Bernoulli polynomials (see (2.8)), we obtain

$$B_{2\alpha}(|x-y|) = B_{2\alpha}(\{x-y\}) = \frac{(-1)^{\alpha+1}(2\alpha)!}{(2\pi)^{2\alpha}} \sum_{h\in\mathbb{Z}\setminus\{0\}} \frac{\mathrm{e}^{2\pi \mathrm{i} h(x-y)}}{|h|^{2\alpha}}, \tag{7.16}$$

and hence

$$\widetilde{K}_{\mathrm{sob},1,\alpha,\gamma}(x,y) = 1 + \frac{\gamma}{(2\pi)^{2\alpha}} \sum_{h\in\mathbb{Z}\setminus\{0\}} \frac{\mathrm{e}^{2\pi \mathrm{i} h(x-y)}}{|h|^{2\alpha}} = K_{\mathrm{kor},1,\alpha,\gamma(2\pi)^{-2\alpha}}(x,y),$$

where the latter term is the reproducing kernel of the Korobov space with smoothness α and weight $\gamma(2\pi)^{-2\alpha}$. This implies that

$$\widetilde{\mathcal{H}}_{\mathrm{sob},1,\alpha,\gamma} = \mathcal{H}_{\mathrm{kor},1,\alpha,\gamma(2\pi)^{-2\alpha}}$$

and

$$\|f\|_{\mathrm{sob},1,\alpha,\gamma} = \|f\|_{\mathrm{kor},1,\alpha,\gamma(2\pi)^{-2\alpha}} \quad \text{for } f \in \widetilde{\mathcal{H}}_{\mathrm{sob},1,\alpha,\gamma},$$

and thus also

$$\mathcal{H}_{\mathrm{kor},1,\alpha,\gamma} \hookrightarrow \mathcal{H}_{\mathrm{sob},1,\alpha,\gamma}.$$

To show the second claim in Item 1, we assume general $d \in \mathbb{N}$ for a moment. For instance, the function $g : [0,1]^d \to \mathbb{R}$, $g(\boldsymbol{x}) = x_1$ is in $\mathcal{H}_{\mathrm{sob},d,\alpha,\gamma}$ for all $\alpha \in \mathbb{N}$, but not in $\mathcal{H}_{\mathrm{kor},d,\alpha,\gamma}$, since g is not periodic. Thus

$$\mathcal{H}_{\mathrm{sob},d,\alpha,\gamma} \not\subset \mathcal{H}_{\mathrm{kor},d,\alpha,\gamma}.$$

In order to show the second item of the theorem, it is again sufficient to consider $d = 1$ only. We consider the case $\alpha = 1$ and show in a first step that

$$K_{\mathrm{sob},1,1,\gamma}(x,y) = 1 + \frac{\gamma}{\pi^2} \sum_{m=1}^{\infty} \frac{2}{m^2} \cos(\pi m x)\cos(\pi m y) = K_{\cos,1,1,\gamma\pi^{-2}}(x,y). \tag{7.17}$$

To this end we compute the cosine coefficients of $K_{\mathrm{sob},1,1,\gamma}$. It is easy to check that

$$\int_0^1 \int_0^1 K_{\mathrm{sob},1,1,\gamma}(x,y)\cos(\pi n x)\cos(\pi m y)\,\mathrm{d}x\,\mathrm{d}y = 0$$

if $n = 0$ and $m \geq 1$ or vice versa. Furthermore, if $n = m = 0$ we have

$$\int_0^1 \int_0^1 K_{\mathrm{sob},1,1,\gamma}(x,y)\,\mathrm{d}x\,\mathrm{d}y = 1$$

according to Proposition 7.8.

It remains to consider the case where both n and m are nonzero. From (7.16) we get

$$\begin{aligned} B_2(|x-y|) =& \frac{1}{2\pi^2} \sum_{k \in \mathbb{Z}\setminus\{0\}} \frac{\mathrm{e}^{2\pi \mathrm{i} k(x-y)}}{k^2} \\ =& \sum_{k=1}^{\infty} \frac{\cos(2\pi kx)\cos(2\pi ky)}{\pi^2 k^2} + \sum_{k=1}^{\infty} \frac{\sin(2\pi kx)\sin(2\pi ky)}{\pi^2 k^2}. \end{aligned}$$

Now we use that the cosine series expansion of the map $x \mapsto x - 1/2$ is given by

$$-\frac{4}{\pi^2} \sum_{\substack{k=1 \\ k \text{ odd}}}^{\infty} \frac{1}{k^2} \cos(\pi kx) \tag{7.18}$$

in order to obtain

$$\begin{aligned} &\left(x-\frac{1}{2}\right)\left(y-\frac{1}{2}\right) + \frac{B_2(|x-y|)}{2} \\ &= \sum_{k,\ell=1}^{\infty} \frac{16}{\pi^4 (2k-1)^2 (2\ell-1)^2} \cos(\pi(2k-1)x)\cos(\pi(2\ell-1)y) \\ &\quad + \sum_{k=1}^{\infty} \frac{\cos(2\pi kx)\cos(2\pi ky)}{2\pi^2 k^2} + \sum_{k=1}^{\infty} \frac{\sin(2\pi kx)\sin(2\pi ky)}{2\pi^2 k^2}. \end{aligned}$$

This immediately implies that

$$\int_0^1 \int_0^1 K_{\mathrm{sob},1,1,\gamma}(x,y)\sqrt{2}\cos(\pi mx)\sqrt{2}\cos(\pi ny)\,\mathrm{d}x\,\mathrm{d}y = 0$$

if m is even and n is odd or vice versa, or if m, n are even with $m \neq n$. If $m = n = k$ for even $k \geq 2$, we obtain

$$\int_0^1 \int_0^1 K_{\mathrm{sob},1,1,\gamma}(x,y)\sqrt{2}\cos(\pi kx)\sqrt{2}\cos(\pi ky)\,\mathrm{d}x\,\mathrm{d}y = \frac{\gamma}{\pi^2 k^2}.$$

In the next step, let $m, n \geq 1$ be odd. We have

$$\int_0^1 \sin(2\pi kx)\cos(\pi \ell x)\,\mathrm{d}x = \frac{2k(1-(-1)^{\ell})}{\pi(4k^2-\ell^2)} \quad \text{for } k, \ell \in \mathbb{Z},$$

and therefore

$$\int_0^1 \int_0^1 K_{\mathrm{sob},1,1,\gamma}(x,y)\sqrt{2}\cos(\pi m x)\sqrt{2}\cos(\pi n y)\,\mathrm{d}x\,\mathrm{d}y$$
$$= \gamma\left(\frac{8}{\pi^4 m^2 n^2} + \frac{16}{\pi^4}\sum_{k=1}^{\infty}\frac{1}{(4k^2-m^2)(4k^2-n^2)}\right).$$

For $m \neq n$ we have

$$\sum_{k=1}^{\infty}\frac{1}{(4k^2-m^2)(4k^2-n^2)} = -\frac{1}{2m^2n^2},$$

and we further observe that

$$\sum_{k=1}^{\infty}\frac{1}{(4k^2-m^2)^2} = \frac{\pi^2 m^2 - 8}{16 m^4}.$$

Thus we obtain

$$\int_0^1 \int_0^1 K_{\mathrm{sob},1,1,\gamma}(x,y)\sqrt{2}\cos(\pi m x)\sqrt{2}\cos(\pi n y)\,\mathrm{d}x\,\mathrm{d}y = \begin{cases} 0 & \text{if } m \neq n, \\ \gamma/(\pi m)^2 & \text{if } m = n. \end{cases}$$

In summary, we have, for $m, n \in \mathbb{N}_0$,

$$\widehat{K}_{\mathrm{sob},1,1,\gamma}(m,n) = \begin{cases} 1 & \text{if } m = n = 0, \\ \gamma/(\pi m)^2 & \text{if } m = n \geq 1, \\ 0 & \text{if } m \neq n. \end{cases}$$

Thus, the cosine series expansion of $K_{\mathrm{sob},1,1,\gamma}$ is given by

$$K_{\mathrm{sob},1,1,\gamma}(x,y) = 1 + \frac{\gamma}{\pi^2}\sum_{m=1}^{\infty}\frac{2}{m^2}\cos(\pi m x)\cos(\pi m y).$$

Note that the cosine series of $K_{\mathrm{sob},1,1,\gamma}$ converges absolutely. Since the function $K_{\mathrm{sob},1,1,\gamma}$ is continuous, the cosine series converges to the function pointwise. This completes the proof of the identity (7.17).

From (7.17) we find

$$\mathcal{H}_{\mathrm{sob},1,1,\gamma} = \mathcal{H}_{\cos,1,1,\gamma\pi^{-2}}, \text{ and } \|f\|_{\mathrm{sob},1,1,\gamma} = \|f\|_{\cos,1,1,\gamma\pi^{-2}} \text{ for } f \in \mathcal{H}_{\mathrm{sob},1,1,\gamma}. \tag{7.19}$$

This in turn implies that

$$\mathcal{H}_{\mathrm{sob},1,1,\gamma} \leftrightharpoons \mathcal{H}_{\cos,1,1,\gamma}.$$

Next, we show Item 3. Let $\alpha \in \mathbb{N}$, $\alpha \geq 2$. The function $x \mapsto x$ belongs to $\mathcal{H}_{\mathrm{sob},1,\alpha,\gamma}$ for all $\alpha \in \mathbb{N}$. On the other hand we have the expansion

$$x = \frac{1}{2} - \frac{4}{\pi^2} \sum_{\substack{k=1 \\ k \text{ odd}}}^{\infty} \frac{\cos(\pi k x)}{k^2}. \tag{7.20}$$

Hence the function $x \mapsto x$ is not in $\mathcal{H}_{\cos,1,\alpha,\gamma}$ for $\alpha \geq 3/2$ and therefore

$$\mathcal{H}_{\mathrm{sob},1,\alpha,\gamma} \not\subset \mathcal{H}_{\cos,1,\alpha,\gamma} \quad \text{for } \alpha \in \mathbb{N}, \alpha \geq 2.$$

Conversely, let $f \in \mathcal{H}_{\cos,1,\alpha,\gamma}$ be given by the cosine expansion

$$f(x) = \widetilde{f}(0) + \sum_{k=1}^{\infty} \widetilde{f}(k) \sqrt{2} \cos(\pi k x)$$

with

$$\|f\|^2_{\cos,1,\alpha,\gamma} = |\widetilde{f}(0)|^2 + \frac{1}{\gamma} \sum_{k=1}^{\infty} |\widetilde{f}(k)|^2 |k|^{2\alpha} < \infty.$$

Then for $\tau \in [\alpha]$ we have

$$f^{(\tau)}(x) = \sum_{k=1}^{\infty} \widetilde{f}(k)(-1)^{\lceil \tau/2 \rceil} (k\pi)^{\tau} \sqrt{2}\, \phi_{\tau}(\pi k x),$$

where $\phi_\tau(z) = \cos(z)$ for even τ and $\phi_\tau(z) = \sin(z)$ for odd τ, and where, for a real number x, $\lceil x \rceil$ denotes the smallest integer greater than or equal to x. This yields

$$\begin{aligned} \frac{1}{\gamma} \left| \int_0^1 f^{(\tau)}(x)\, \mathrm{d}x \right|^2 &\leq \frac{1}{\gamma} \int_0^1 |f^{(\tau)}(x)|^2\, \mathrm{d}x \\ &= \frac{\pi^{2\tau}}{\gamma} \sum_{k=1}^{\infty} |\widetilde{f}(k)|^2 k^{2\tau} \\ &\leq \pi^{2\tau} \|f\|^2_{\cos,1,\alpha,\gamma}. \end{aligned}$$

Thus we have

$$\|f\|_{\mathrm{sob},1,\alpha,\gamma} \leq \left(\sum_{\tau=0}^{\alpha} \pi^{2\tau} \right)^{1/2} \|f\|_{\cos,1,\alpha,\gamma} = \left(\frac{\pi^{2(\alpha+1)} - 1}{\pi^2 - 1} \right)^{1/2} \|f\|_{\cos,1,\alpha,\gamma},$$

and hence

$$\mathcal{H}_{\cos,1,\alpha,\gamma} \hookrightarrow \mathcal{H}_{\mathrm{sob},1,\alpha,\gamma}.$$

Finally, we come to the proof of Item 4. Let $\alpha > 1/2$. The function f with

$$f(x) = \sin(2\pi x) \quad \text{for } x \in [0, 1]$$

belongs to $\mathcal{H}_{\mathrm{kor},1,\alpha,\gamma}$ for all $\alpha > 1/2$. On the other hand we have, for $k \in \mathbb{N}$,

$$\widetilde{f}(k) = \int_0^1 \sin(2\pi x)\sqrt{2}\cos(\pi k x)\,\mathrm{d}x = \begin{cases} 4\sqrt{2}/(\pi(4-k^2)) & \text{if } k \text{ is odd,} \\ 0 & \text{otherwise.} \end{cases}$$

Consequently,

$$\|f\|^2_{\cos,1,\alpha,\gamma} = \frac{1}{\gamma}\sum_{\substack{k=1 \\ k \text{ odd}}}^{\infty} |\widetilde{f}(k)|^2 k^{2\alpha} = \frac{1}{\gamma}\frac{32}{\pi^2}\sum_{k=1}^{\infty}\frac{(2k-1)^{2\alpha}}{((2k-1)^2-4)^2}.$$

This implies that $\|f\|_{\cos,1,\alpha,\gamma} = \infty$ for $\alpha \geq 3/2$, and hence

$$\mathcal{H}_{\mathrm{kor},1,\alpha,\gamma} \not\subset \mathcal{H}_{\cos,1,\alpha,\gamma} \quad \text{for } \alpha \geq 3/2.$$

Conversely, let now

$$f(x) = \cos(\pi x) \quad \text{for } x \in [0,1].$$

Then $f \in \mathcal{H}_{\cos,1,\alpha,\gamma}$ for all $\alpha > 1/2$. However, for $h \in \mathbb{Z}$ the Fourier coefficients of f are given by

$$\widehat{f}(h) = \int_0^1 \cos(\pi x)\,\mathrm{e}^{-2\pi \mathrm{i} h x}\,\mathrm{d}x = \frac{4\mathrm{i}h}{\pi(1-4h^2)}$$

and hence

$$\|f\|^2_{\mathrm{kor},1,\alpha,\gamma} = \frac{1}{\gamma}\frac{16}{\pi^2}\sum_{h\in\mathbb{Z}\setminus\{0\}} |h|^{2\alpha}\frac{h^2}{(4h^2-1)^2} = \infty,$$

since $\alpha > 1/2$. This implies that $f \notin \mathcal{H}_{\mathrm{kor},1,\alpha,\gamma}$ and therefore we have

$$\mathcal{H}_{\cos,1,\alpha,\gamma} \not\subset \mathcal{H}_{\mathrm{kor},1,\alpha,\gamma}.$$

This concludes the proof. □

Relations between the unanchored Sobolev space of smoothness one and the anchored Sobolev space of smoothness one

We close this section with a brief discussion of the relation between the unanchored Sobolev space of smoothness one and the anchored Sobolev space of smoothness one as introduced in Section 7.1. The following result, which we state without a proof here, is the special L_2-case of [95, Proposition 13].

Proposition 7.34 *For general weights $\boldsymbol{\gamma} = \{\gamma_{\mathfrak{u}}\}_{\mathfrak{u}\subseteq[d]}$ the following holds. The $\boldsymbol{\gamma}$-weighted unanchored Sobolev space $\mathcal{H}_{\mathrm{sob},d,\boldsymbol{\gamma}}$ and the $\boldsymbol{\gamma}$-weighted anchored Sobolev space $\mathcal{H}^{\pitchfork}_{\mathrm{sob},d,\boldsymbol{\gamma},\mathbf{0}}$ are equal (as sets of functions) if and only if*

$$\gamma_{\mathfrak{u}} > 0 \text{ implies } \gamma_{\mathfrak{v}} > 0 \text{ for all } \mathfrak{v} \subseteq \mathfrak{u} \subseteq [d]. \tag{7.21}$$

Moreover, if (7.21) *does not hold, then*

$$\mathcal{H}^{\pitchfork}_{\mathrm{sob},d,\boldsymbol{\gamma},\mathbf{0}} \not\subset \mathcal{H}_{\mathrm{sob},d,\boldsymbol{\gamma}} \text{ and } \mathcal{H}_{\mathrm{sob},d,\boldsymbol{\gamma}} \not\subset \mathcal{H}^{\pitchfork}_{\mathrm{sob},d,\boldsymbol{\gamma},\mathbf{0}}.$$

Condition (7.21) is certainly satisfied for product weights $\boldsymbol{\gamma} = (\gamma_j)_{j\ge 1}$ with $\gamma_j \in (0, 1]$ for all $j \in [d]$. We will restrict ourselves to this case now. According to Proposition 7.32 it is desirable to have information on the norm of the embedding operator $\imath_{d,\boldsymbol{\gamma}} : \mathcal{H}^{\pitchfork}_{\mathrm{sob},d,\boldsymbol{\gamma},\mathbf{0}} \to \mathcal{H}_{\mathrm{sob},d,\boldsymbol{\gamma}}$ and its inverse $\imath^{-1}_{d,\boldsymbol{\gamma}} : \mathcal{H}_{\mathrm{sob},d,\boldsymbol{\gamma}} \to \mathcal{H}^{\pitchfork}_{\mathrm{sob},d,\boldsymbol{\gamma},\mathbf{0}}$ since these norms equal the factors involved when the worst-case errors for the respective Sobolev spaces are estimated against each other. The following result is taken from [145].

Theorem 7.35 *Consider product weights* $\boldsymbol{\gamma} = (\gamma_j)_{j\ge 1}$ *with* $\gamma_j \in (0, 1]$ *for all* $j \in [d]$. *Then we have*

$$\|\imath_{d,\boldsymbol{\gamma}}\| = \|\imath^{-1}_{d,\boldsymbol{\gamma}}\| = \prod_{j=1}^{d} \left(1 + \sqrt{\frac{\gamma_j}{3}} \left(\sqrt{1 + \frac{\gamma_j}{12}} + \sqrt{\frac{\gamma_j}{12}}\right)\right)^{1/2}. \tag{7.22}$$

Moreover, for any $j \in [d]$

$$1 + \sqrt{\frac{\gamma_j}{3}} + \frac{\gamma_j}{6} \le 1 + \sqrt{\frac{\gamma_j}{3}} \left(\sqrt{1 + \frac{\gamma_j}{12}} + \sqrt{\frac{\gamma_j}{12}}\right) \le 1 + \sqrt{\frac{\gamma_j}{3}} + \frac{\gamma_j}{6} + \frac{\gamma_j^{3/2}}{24\sqrt{3}}. \tag{7.23}$$

Proof Since the spaces $\mathcal{H}_{\mathrm{sob},d,\boldsymbol{\gamma}}$ and $\mathcal{H}^{\pitchfork}_{\mathrm{sob},d,\boldsymbol{\gamma},\mathbf{0}}$ with product weights are tensor products of the corresponding spaces of univariate functions, it is enough to prove (7.22) for $d = 1$ and a generic weight $\gamma \in (0, 1]$. Moreover we will only consider $\|f\|_{\mathrm{sob},1,\gamma,0}/\|f\|_{\mathrm{sob},1,\gamma}$ since the proof for $\|f\|_{\mathrm{sob},1,\gamma}/\|f\|_{\mathrm{sob},1,\gamma,0}$ is very similar. The formulas for the respective univariate norms are to be found in Remark 7.7 and in Equation (7.11) (with the choice $c = 0$).

Note that for constant $f \equiv c$, $\|f\|_{\mathrm{sob},1,\gamma,0}/\|f\|_{\mathrm{sob},1,\gamma} = 1$. Hence it is enough to consider f with

$$f(x) = \frac{c}{\sqrt{\gamma}} + \int_0^1 h(t)\, 1_{[0,x)}(t)\, \mathrm{d}t$$

for some $c \ge 0$ and h such that $\|h\|_{L_2} = 1$. Then

$$\|f\|^2_{\mathrm{sob},1,\gamma} = \frac{1 + c^2}{\gamma}.$$

Furthermore,

$$\begin{aligned}\int_0^1 f(x)\,\mathrm{d}x &= \frac{c}{\sqrt{\gamma}} + \int_0^1 h(t)\,(1 - t)\,\mathrm{d}t \\ &\le \frac{1}{\sqrt{\gamma}}\left(c + \frac{\sqrt{\gamma}\,\|h\|_{L_2}}{\sqrt{3}}\right)\end{aligned} \tag{7.24}$$

$$= \frac{1}{\sqrt{\gamma}} \left(c + \sqrt{\frac{\gamma}{3}} \right)$$

and, therefore,

$$\|f\|^2_{\text{sob},1,\gamma,0} \le \frac{1}{\gamma} \left(\left(c + \sqrt{\frac{\gamma}{3}} \right)^2 + 1 \right).$$

So,

$$\frac{\|f\|^2_{\text{sob},1,\gamma,0}}{\|f\|^2_{\text{sob},1,\gamma}} \le \frac{(c+\sqrt{\gamma/3})^2+1}{c^2+1} = \frac{c^2+1+2c\sqrt{\gamma/3}+\gamma/3}{c^2+1} = 1 + \sqrt{\frac{\gamma}{3}}\,\rho(c,\gamma),$$

where

$$\rho(c,\gamma) = \frac{2c+\sqrt{\gamma/3}}{c^2+1}.$$

It is easy to verify that

$$\max_{c\ge 0} \rho(c,\gamma) = \rho(c^*_\gamma,\gamma), \quad \text{where} \quad c^*_\gamma = \sqrt{1+\frac{\gamma}{12}} - \sqrt{\frac{\gamma}{12}},$$

which yields

$$\frac{\|f\|^2_{\text{sob},1,\gamma,0}}{\|f\|^2_{\text{sob},1,\gamma}} \le 1 + \sqrt{\frac{\gamma}{3}}\,\rho(c^*_\gamma,\gamma).$$

This shows that $\|\imath_{1,\gamma}\| \le 1 + \sqrt{\gamma/3}\,\rho(c^*_\gamma,\gamma)$. To prove equality it is enough to note that for $h(t) = \sqrt{3}\,(1-t)$ we have equality in (7.24), i.e.,

$$\int_0^1 h(t)\,(1-t)\,\mathrm{d}t = \frac{1}{\sqrt{3}}.$$

This proves that

$$\|\imath_{1,\gamma}\|^2 = 1 + \sqrt{\frac{\gamma}{3}}\,\rho(c^*_\gamma,\gamma). \tag{7.25}$$

It is easy to verify that

$$\rho(c^*_\gamma,\gamma) = \frac{\sqrt{1+\gamma/12}}{1+\gamma/12-\sqrt{\gamma/12+(\gamma/12)^2}} = \frac{1}{\sqrt{1+\gamma/12}-\sqrt{\gamma/12}}.$$

Therefore, applying the conjugate to the last fraction we get

$$\rho(c^*_\gamma,\gamma) = \sqrt{1+\frac{\gamma}{12}} + \sqrt{\frac{\gamma}{12}}.$$

This proves (7.22).

It remains to show (7.23). The first inequality is trivial. Clearly,

$$\sqrt{\frac{\gamma}{3}}\left(\sqrt{1+\frac{\gamma}{12}}+\sqrt{\frac{\gamma}{12}}\right)=\sqrt{\frac{\gamma}{3}}+\sqrt{\frac{\gamma}{3}}\left(\sqrt{\frac{\gamma}{12}}+\sqrt{1+\frac{\gamma}{12}}-1\right)$$
$$=\sqrt{\frac{\gamma}{3}}+\frac{\gamma}{6}+E(\gamma),$$

where

$$E(\gamma):=\sqrt{\frac{\gamma}{3}}\left(\sqrt{1+\frac{\gamma}{12}}-1\right).$$

Obviously, the term $E(\gamma)$ is nonnegative, and can be upper-bounded as

$$E(\gamma)=\sqrt{\frac{\gamma}{3}}\frac{\gamma/12}{\sqrt{1+\gamma/12}+1}=\frac{\gamma^{3/2}}{24\sqrt{3}}\frac{2}{\sqrt{1+\gamma/12}+1}\le\frac{\gamma^{3/2}}{24\sqrt{3}}.$$

Consequently,

$$1+\sqrt{\frac{\gamma}{3}}\left(\sqrt{1+\frac{\gamma}{12}}+\sqrt{\frac{\gamma}{12}}\right)\le 1+\sqrt{\frac{\gamma}{3}}+\frac{\gamma}{6}+\frac{\gamma^{3/2}}{24\sqrt{3}},$$

which completes the proof. □

Theorem 7.35 implies that the embedding norms can be bounded independently of the dimension d if $(\sqrt{\gamma_j})_{j\ge1}$ decays to zero fast enough that $\sum_{j=1}^{\infty}\sqrt{\gamma_j}<\infty$. Hence, even tractability results can be transferred from the unanchored Sobolev space to the anchored Sobolev space and vice versa, when this weight condition is satisfied.

Remark 7.36 Proposition 7.32 together with Condition (7.21) make it possible to bound the worst-case errors in the spaces $\mathcal{H}_{\mathrm{sob},d,\boldsymbol{\gamma}}$ and $\mathcal{H}^{\pitchfork}_{\mathrm{sob},d,\boldsymbol{\gamma},\boldsymbol{0}}$ in terms of each other. What, however, if Condition (7.21) is not met? In this case, one can use methods as in [147] to show that we can bound the worst-case error of a QMC rule directly in terms of the weighted L_2-discrepancy of the point set underlying the integration rule. To be more precise, we have the estimate

$$\mathrm{err}_{N,d}(\mathcal{H}_{\mathrm{sob},d,\boldsymbol{\gamma}},\mathcal{P})\le L_{2,N,\widetilde{\boldsymbol{\gamma}}}(\mathcal{P})$$

for any N-point set $\mathcal{P}$ in $[0,1)^d$, where the weights $\widetilde{\boldsymbol{\gamma}}$ are given by $\widetilde{\boldsymbol{\gamma}}=(\widetilde{\gamma}_{\mathfrak{u}})_{\mathfrak{u}\subseteq[d]}$ with

$$\widetilde{\gamma}_{\mathfrak{u}}:=\left(3^{|\mathfrak{u}|}\sum_{\substack{\mathfrak{u}\subseteq\mathfrak{v}\subseteq[d]\\ \gamma_{\mathfrak{v}}\ne0}}\gamma_{\mathfrak{v}}^4\left(\frac{2}{3}\right)^{|\mathfrak{v}|}\right)^{1/2}\quad\text{for } \mathfrak{u}\subseteq[d].$$

7.3 Folded Lattice Rules

A further method to obtain higher order convergence for smooth functions is an application of the tent transformation.

The *tent transformation* $\phi : [0, 1] \to [0, 1]$,

$$\phi(x) := 1 - |2x - 1|,$$

is a Lebesgue measure preserving function (see Figure 7.2). We apply the tent transformation coordinate-wise to points $\boldsymbol{x} = (x_1, \dots, x_d)$ in the d-dimensional unit cube $[0, 1]^d$, i.e.,

$$\phi(\boldsymbol{x}) := (\phi(x_1), \dots, \phi(x_d)).$$

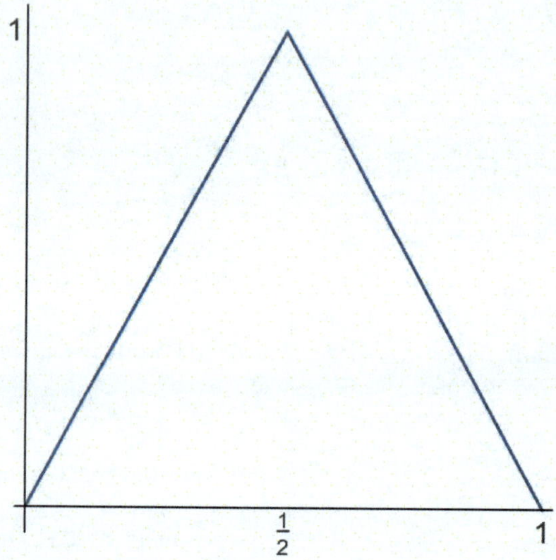

Fig. 7.2: The tent transformation $\phi : [0, 1] \to [0, 1]$.

Definition 7.37 Let $\mathcal{P} = \{\boldsymbol{x}_0, \boldsymbol{x}_1, \dots, \boldsymbol{x}_{N-1}\}$ be a point set in the unit cube $[0, 1]^d$. Then we call

$$\mathcal{P}_\phi := \{\phi(\boldsymbol{x}_0), \phi(\boldsymbol{x}_1), \dots, \phi(\boldsymbol{x}_{N-1})\}$$

the corresponding *folded point set* or the *tent-transformed point set*. If $\mathcal{P} = \mathcal{P}(\boldsymbol{g}, N)$ is the node set of a rank-1 lattice rule, then we write $\mathcal{P}_\phi(\boldsymbol{g}, N)$ for the corresponding folded version.

A QMC rule that is based on the folded version of a rank-1 lattice point set is called a *folded lattice rule* or a *tent-transformed lattice rule*.

Examples of tent-transformed lattice point sets are shown in Figure 7.3.

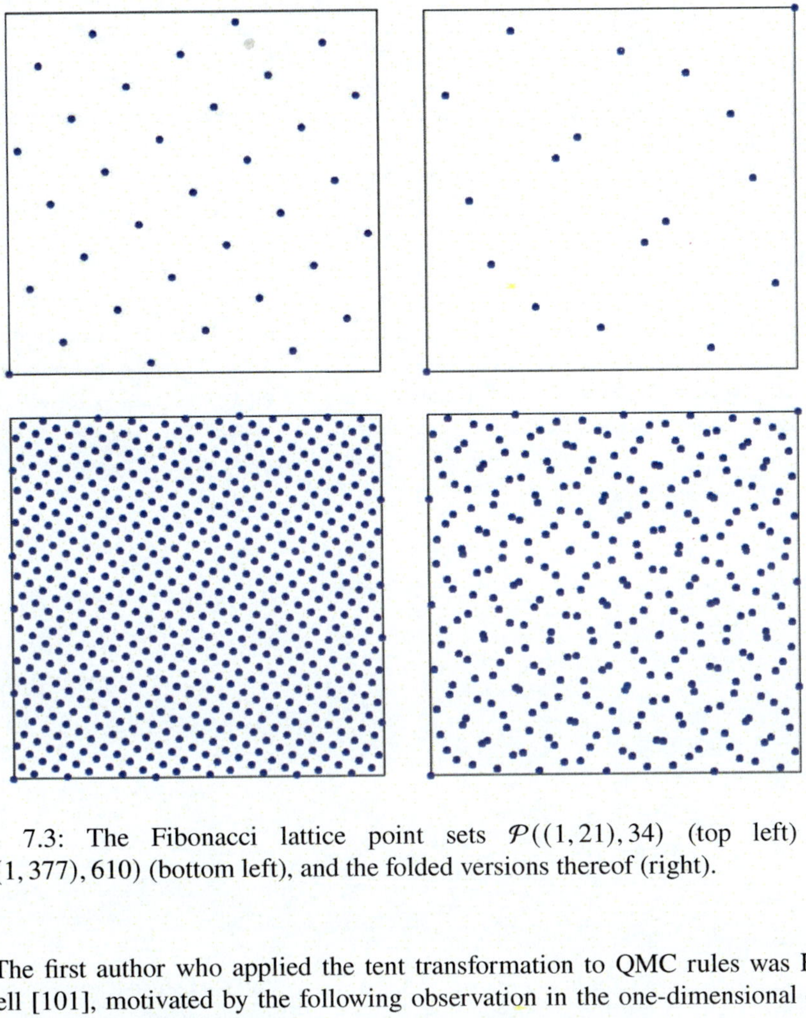

Fig. 7.3: The Fibonacci lattice point sets $\mathcal{P}((1,21),34)$ (top left) and $\mathcal{P}((1,377),610)$ (bottom left), and the folded versions thereof (right).

The first author who applied the tent transformation to QMC rules was Hickernell [101], motivated by the following observation in the one-dimensional case. Consider the left-rectangle rule

$$\mathcal{R}_N(f) = \frac{1}{N}\sum_{k=0}^{N-1} f\left(\frac{k}{N}\right) \quad \text{for } f : [0,1] \to \mathbb{R},$$

where we assume that N is even. Applying the tent transformation to the N nodes $0, 1/N, 2/N, \ldots, (N-1)/N$ of this rule leads to the sequence

$$0, \frac{2}{N}, \frac{4}{N}, \ldots, 1, \frac{N-2}{N}, \frac{N-4}{N}, \ldots, \frac{2}{N},$$

i.e., to the quadrature rule

$$\mathcal{T}_N(f) = \frac{1}{N}f(0) + \frac{2}{N}\sum_{k=1}^{(N-2)/2} f\left(\frac{2k}{N}\right) + \frac{1}{N}f(1).$$

The latter, however, is a trapezoidal rule. While the rectangle rule is known to have an error rate of order $O(N^{-1})$, the trapezoidal rule has a quadrature error of order $O(N^{-2})$ if applied to sufficiently smooth integrands.

Randomized folded lattice rules

Hickernell [101] studied integration using folded QMC rules in the (unweighted) unanchored Sobolev space of smoothness $\alpha = 2$. However, this study still requires a random shift in addition to folding. In particular, starting from a rank-1 lattice rule, a quadrature rule is a so-called *shifted and folded lattice rule* if it is of the form

$$Q_{N,d}^{\phi}(f, \boldsymbol{\Delta}) := \frac{1}{N}\sum_{k=0}^{N-1} f\left(\phi\left(\left\{\frac{k}{N}\boldsymbol{g} + \boldsymbol{\Delta}\right\}\right)\right).$$

An example of a shifted and folded lattice point set is shown in Figure 7.4.

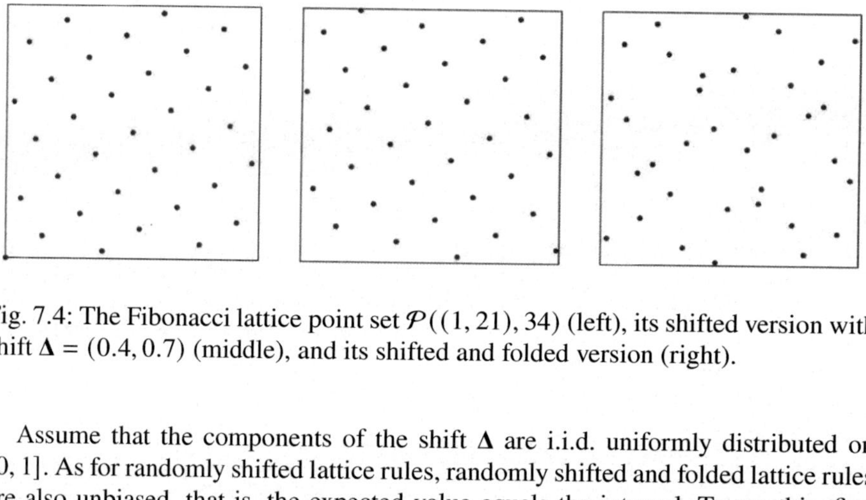

Fig. 7.4: The Fibonacci lattice point set $\mathcal{P}((1, 21), 34)$ (left), its shifted version with shift $\boldsymbol{\Delta} = (0.4, 0.7)$ (middle), and its shifted and folded version (right).

Assume that the components of the shift $\boldsymbol{\Delta}$ are i.i.d. uniformly distributed on $[0, 1]$. As for randomly shifted lattice rules, randomly shifted and folded lattice rules are also unbiased, that is, the expected value equals the integral. To see this, first note that, for any integrable function $g : [0, 1] \to \mathbb{R}$ and any $a \in [0, 1/2]$, we have

$$\int_0^1 g(\phi(\{a+x\}))\,\mathrm{d}x = \int_0^{1/2-a} g(2(a+x))\,\mathrm{d}x + \int_{1/2-a}^{1-a} g(2-2(a+x))\,\mathrm{d}x + \int_{1-a}^{1} g(2(a+x-1))\,\mathrm{d}x$$

$$= \int_a^{1/2} g(2x)\,\mathrm{d}x + \int_{1/2}^1 g(2-2x)\,\mathrm{d}x + \int_0^a g(2x)\,\mathrm{d}x$$
$$= \int_0^{1/2} g(2x)\,\mathrm{d}x + \int_{1/2}^1 g(2-2x)\,\mathrm{d}x$$
$$= \int_0^1 g(x)\,\mathrm{d}x.$$

The same property holds for $1/2 < a \leq 1$. Using this property component-wise, we obtain

$$\mathbb{E}\left[Q_{N,d}^{\phi}(f,\boldsymbol{\Delta})\right] = \int_{[0,1]^d} \frac{1}{N}\sum_{k=0}^{N-1} f\left(\phi\left(\left\{\frac{k}{N}\boldsymbol{g} + \boldsymbol{\Delta}\right\}\right)\right)\mathrm{d}\boldsymbol{\Delta}$$
$$= \frac{1}{N}\sum_{k=0}^{N-1}\int_{[0,1]^d} f\left(\phi\left(\left\{\frac{k}{N}\boldsymbol{g} + \boldsymbol{\Delta}\right\}\right)\right)\mathrm{d}\boldsymbol{\Delta}$$
$$= \int_{[0,1]^d} f(\boldsymbol{x})\,\mathrm{d}\boldsymbol{x}.$$

Hence the expected value of our estimator equals the integral which we want to approximate. Therefore, in order to get an estimate of the deviation of the random variable $Q_{N,d}^{\phi}(f,\boldsymbol{\Delta})$ from its expected value, we consider the root mean square worst-case error in the following. (A different kind of variance measure will be considered in Chapter 11.)

Definition 7.38 Assume that the components of the shift $\boldsymbol{\Delta}$ are i.i.d. uniformly distributed on $[0,1]$. Then the root mean square worst-case error of a shifted and folded QMC rule based on the point set $\mathcal{P}$ applied to functions in a Hilbert space $\mathcal{H}$,

$$\mathrm{err}_{N,d}^{\mathrm{sh},\phi}(\mathcal{H},\mathcal{P}) := \sqrt{\mathbb{E}\left[(\mathrm{err}_{N,d}(\mathcal{H},\mathcal{P}_{\boldsymbol{\Delta},\phi}))^2\right]},$$

is referred to as the *shift-averaged and folded worst-case error*. Here, expectation is considered with respect to the choice of $\boldsymbol{\Delta}$.

Theorem 7.5 can now easily be adapted to the following.

Theorem 7.39 *The shift-averaged and folded worst-case error of a QMC rule based on an N-element point set $\mathcal{P} = \{\boldsymbol{x}_0, \boldsymbol{x}_1, \ldots, \boldsymbol{x}_{N-1}\}$ in $[0,1)^d$ in a reproducing kernel Hilbert space $\mathcal{H}(K)$ is given as*

$$[\mathrm{err}_{N,d}^{\mathrm{sh},\phi}(\mathcal{H}(K),\mathcal{P})]^2 = -\int_{[0,1]^d}\int_{[0,1]^d} K(\boldsymbol{x},\boldsymbol{y})\,\mathrm{d}\boldsymbol{x}\,\mathrm{d}\boldsymbol{y} + \frac{1}{N^2}\sum_{k,\ell=0}^{N-1} K^{\mathrm{sh},\phi}(\boldsymbol{x}_k,\boldsymbol{x}_\ell),$$

where

$$K^{\mathrm{sh},\phi}(\boldsymbol{x}_k,\boldsymbol{x}_\ell) := \int_{[0,1]^d} K(\phi(\{\boldsymbol{x}+\boldsymbol{\Delta}\}),\phi(\{\boldsymbol{y}+\boldsymbol{\Delta}\}))\,\mathrm{d}\boldsymbol{\Delta}.$$

The effect of shifting and folding has been successfully used for integration in $\mathcal{H}_{\text{sob},d,2,\mathbf{1}}$, which is the unweighted version of the unanchored Sobolev space of smoothness $\alpha = 2$, by means of lattice rules. Based on the Fourier series expansion of Bernoulli polynomials one can compute $K^{\text{sh},\phi}_{\text{sob},d,2,\gamma}$, which reveals a direct relation to the Korobov kernel $K_{\text{kor},d,2,\gamma}$ of smoothness $\alpha = 2$. Using the known results for lattice rule integration in Korobov spaces, this implies that $\text{err}^{\text{sh},\phi}_{N,d}(\mathcal{H}_{\text{sob},d,2,\gamma}, \mathcal{P}(\boldsymbol{g}, N))$ is of order $O(N^{-2+\varepsilon})$ for every $\varepsilon > 0$. For details we refer to [101]. Below (see Corollary 7.46) we will more precisely explain a deterministic version of these findings.

Shifted and folded lattice rules for integration in half-period cosine spaces have been studied by Cools, Kuo, Nuyens, and Suryanarayana [30]. Also for this case, one can obtain almost optimal convergence rates, and also for this case we will give a deterministic version of these findings in the following.

Deterministic folded lattice rules

In the above considerations we require a random shift in addition to the tent transformation in order to obtain higher order convergence rates. There are cases where this random element is not needed. Indeed, in [50] folded lattice rules are applied to integrands in the half-period cosine space. The main observation is that the worst-case integration error of a folded lattice rule in $\mathcal{H}_{\cos,d,\alpha,\gamma}$ is bounded by the worst-case error of the nonfolded version of the same lattice rule in the Korobov space $\mathcal{H}_{\text{kor},d,\alpha,\gamma}$ with the same smoothness parameter α and the same weights $\boldsymbol{\gamma}$. The following result was first shown in [50].

Theorem 7.40 *The worst-case error of a folded lattice rule in the half-period cosine space $\mathcal{H}_{\cos,d,\alpha,\gamma}$ satisfies*

$$\text{err}_{N,d}(\mathcal{H}_{\cos,d,\alpha,\gamma}, \mathcal{P}_\phi(\boldsymbol{g}, N)) \le \text{err}_{N,d}(\mathcal{H}_{\text{kor},d,\alpha,\gamma}, \mathcal{P}(\boldsymbol{g}, N)).$$

Proof Using the worst-case error formula for QMC rules in reproducing kernel Hilbert spaces in Theorem 1.27 we obtain

$$[\text{err}_{N,d}(\mathcal{H}_{\cos,d,\alpha,\gamma}, \mathcal{P}_\phi(\boldsymbol{g}, N))]^2 = \int_{[0,1]^d}\int_{[0,1]^d} K_{\cos,d,\alpha,\gamma}(\boldsymbol{x},\boldsymbol{y})\,\mathrm{d}\boldsymbol{x}\,\mathrm{d}\boldsymbol{y}$$
$$-\frac{2}{N}\sum_{k=0}^{N-1}\int_{[0,1]^d} K_{\cos,d,\alpha,\gamma}(\phi(\boldsymbol{x}_k),\boldsymbol{y})\,\mathrm{d}\boldsymbol{y} + \frac{1}{N^2}\sum_{k,\ell=0}^{N-1} K_{\cos,d,\alpha,\gamma}(\phi(\boldsymbol{x}_k),\phi(\boldsymbol{x}_\ell)),$$

where $\mathcal{P}(\boldsymbol{g}, N) = \{\boldsymbol{x}_0, \boldsymbol{x}_1, \ldots, \boldsymbol{x}_{N-1}\}$. From Proposition 7.31 we know that

$$\int_{[0,1]^d}\int_{[0,1]^d} K_{\cos,d,\alpha,\gamma}(\boldsymbol{x},\boldsymbol{y})\,\mathrm{d}\boldsymbol{x}\,\mathrm{d}\boldsymbol{y} = 1,$$

and by the same argument as used in the proof of Proposition 7.31 we also obtain

$$\int_{[0,1]^d} K_{\cos,d,\alpha,\boldsymbol{\gamma}}(\phi(\boldsymbol{x}_k),\boldsymbol{y})\,\mathrm{d}\boldsymbol{y} = 1 \qquad \text{for all } k \in \{0,1,\ldots,N-1\}.$$

This yields

$$[\mathrm{err}_{N,d}(\mathcal{H}_{\cos,d,\alpha,\boldsymbol{\gamma}},\mathcal{P}_\phi(\boldsymbol{g},N))]^2 = -1 + \frac{1}{N^2}\sum_{k,\ell=0}^{N-1} K_{\cos,d,\alpha,\boldsymbol{\gamma}}(\phi(\boldsymbol{x}_k),\phi(\boldsymbol{x}_\ell)).$$

We now consider the kernel function $K_{\cos,d,\alpha,\boldsymbol{\gamma}}$. For $k,\ell \in \{0,1,\ldots,N-1\}$ we have

$$K_{\cos,d,\alpha,\boldsymbol{\gamma}}(\phi(\boldsymbol{x}_k),\phi(\boldsymbol{x}_\ell)) = \sum_{\mathfrak{u}\subseteq[d]}\gamma_{\mathfrak{u}}\sum_{\boldsymbol{h}_{\mathfrak{u}}\in\mathbb{N}^{|\mathfrak{u}|}}\prod_{j\in\mathfrak{u}}\frac{2\cos(\pi h_j\phi(x_{k,j}))\cos(\pi h_j\phi(x_{\ell,j}))}{h_j^{2\alpha}}.$$

Furthermore, for any $h \in \mathbb{N}_0$ and $x \in [0,1]$, it is true that

$$\cos(\pi h\phi(x)) = \cos(2\pi hx) = \frac{\mathrm{e}^{2\pi\mathrm{i}hx} + \mathrm{e}^{-2\pi\mathrm{i}hx}}{2}.$$

Therefore, for a vector $\boldsymbol{h}_{\mathfrak{u}} \in \mathbb{N}^{|\mathfrak{u}|}$ and for $\boldsymbol{x}_{\mathfrak{u}} \in [0,1]^{|\mathfrak{u}|}$, we have

$$\begin{aligned}\prod_{j\in\mathfrak{u}}\cos(\pi h_j\phi(x_j)) &= \prod_{j\in\mathfrak{u}}\frac{\mathrm{e}^{2\pi\mathrm{i}h_jx_j} + \mathrm{e}^{-2\pi\mathrm{i}h_jx_j}}{2}\\ &= \frac{1}{2^{|\mathfrak{u}|}}\sum_{\boldsymbol{\sigma}_{\mathfrak{u}}\in\{-1,1\}^{|\mathfrak{u}|}}\mathrm{e}^{2\pi\mathrm{i}(\boldsymbol{\sigma}_{\mathfrak{u}}*\boldsymbol{h}_{\mathfrak{u}})\cdot\boldsymbol{x}_{\mathfrak{u}}},\end{aligned} \tag{7.26}$$

where $\boldsymbol{\sigma}_{\mathfrak{u}} * \boldsymbol{h}_{\mathfrak{u}} := (\sigma_j h_j)_{j\in\mathfrak{u}}$ means component-wise multiplication. Inserting into the formula for the kernel yields

$$\begin{aligned}K_{\cos,d,\alpha,\boldsymbol{\gamma}}(\phi(\boldsymbol{x}_k),\phi(\boldsymbol{x}_\ell)) &= \sum_{\mathfrak{u}\subseteq[d]}\gamma_{\mathfrak{u}}2^{|\mathfrak{u}|}\sum_{\boldsymbol{h}_{\mathfrak{u}}\in\mathbb{N}^{|\mathfrak{u}|}}\prod_{j\in\mathfrak{u}}\frac{1}{h_j^{2\alpha}}\\ \times\left(\frac{1}{2^{|\mathfrak{u}|}}\sum_{\boldsymbol{\sigma}_{\mathfrak{u}}\in\{-1,1\}^{|\mathfrak{u}|}}\mathrm{e}^{2\pi\mathrm{i}(\boldsymbol{\sigma}_{\mathfrak{u}}*\boldsymbol{h}_{\mathfrak{u}})\cdot\boldsymbol{x}_{k,\mathfrak{u}}}\right)&\left(\frac{1}{2^{|\mathfrak{u}|}}\sum_{\widetilde{\boldsymbol{\sigma}}_{\mathfrak{u}}\in\{-1,1\}^{|\mathfrak{u}|}}\mathrm{e}^{2\pi\mathrm{i}(\widetilde{\boldsymbol{\sigma}}_{\mathfrak{u}}*\boldsymbol{h}_{\mathfrak{u}})\cdot\boldsymbol{x}_{\ell,\mathfrak{u}}}\right).\end{aligned}$$

Averaging over all $k,\ell \in \{0,1,\ldots,N-1\}$ gives

$$\begin{aligned}\frac{1}{N^2}\sum_{k,\ell=0}^{N-1}K_{\cos,d,\alpha,\boldsymbol{\gamma}}(\phi(\boldsymbol{x}_k),\phi(\boldsymbol{x}_\ell)) &= \sum_{\mathfrak{u}\subseteq[d]}\gamma_{\mathfrak{u}}2^{|\mathfrak{u}|}\sum_{\boldsymbol{h}_{\mathfrak{u}}\in\mathbb{N}^{|\mathfrak{u}|}}\prod_{j\in\mathfrak{u}}\frac{1}{h_j^{2\alpha}}\\ \times\left(\frac{1}{2^{|\mathfrak{u}|}}\sum_{\boldsymbol{\sigma}_{\mathfrak{u}}\in\{-1,1\}^{|\mathfrak{u}|}}\frac{1}{N}\sum_{k=0}^{N-1}\mathrm{e}^{2\pi\mathrm{i}(\boldsymbol{\sigma}_{\mathfrak{u}}*\boldsymbol{h}_{\mathfrak{u}})\cdot\boldsymbol{x}_{k,\mathfrak{u}}}\right)&\end{aligned}$$

$$\times \left(\frac{1}{2^{|\mathfrak{u}|}} \sum_{\widetilde{\boldsymbol{\sigma}}_{\mathfrak{u}} \in \{-1,1\}^{|\mathfrak{u}|}} \frac{1}{N} \sum_{\ell=0}^{N-1} \mathrm{e}^{2\pi \mathrm{i} (\widetilde{\boldsymbol{\sigma}}_{\mathfrak{u}} * \boldsymbol{h}_{\mathfrak{u}}) \cdot \boldsymbol{x}_{\ell,\mathfrak{u}}} \right)$$

$$= \sum_{\mathfrak{u} \subseteq [d]} \gamma_{\mathfrak{u}} 2^{|\mathfrak{u}|} \sum_{\boldsymbol{h}_{\mathfrak{u}} \in \mathbb{N}^{|\mathfrak{u}|}} \prod_{j \in \mathfrak{u}} \frac{1}{h_j^{2\alpha}}$$

$$\times \left(\frac{1}{2^{|\mathfrak{u}|}} \sum_{\substack{\boldsymbol{\sigma}_{\mathfrak{u}} \in \{-1,1\}^{|\mathfrak{u}|} \\ (\boldsymbol{\sigma}_{\mathfrak{u}} * \boldsymbol{h}_{\mathfrak{u}}) \cdot \boldsymbol{g}_{\mathfrak{u}} \equiv 0 \ (\mathrm{mod}\ N)}} 1 \right) \left(\frac{1}{2^{|\mathfrak{u}|}} \sum_{\substack{\widetilde{\boldsymbol{\sigma}}_{\mathfrak{u}} \in \{-1,1\}^{|\mathfrak{u}|} \\ (\widetilde{\boldsymbol{\sigma}}_{\mathfrak{u}} * \boldsymbol{h}_{\mathfrak{u}}) \cdot \boldsymbol{g}_{\mathfrak{u}} \equiv 0 \ (\mathrm{mod}\ N)}} 1 \right).$$

We trivially estimate

$$\frac{1}{2^{|\mathfrak{u}|}} \sum_{\substack{\widetilde{\boldsymbol{\sigma}}_{\mathfrak{u}} \in \{-1,1\}^{|\mathfrak{u}|} \\ (\widetilde{\boldsymbol{\sigma}}_{\mathfrak{u}} * \boldsymbol{h}_{\mathfrak{u}}) \cdot \boldsymbol{g}_{\mathfrak{u}} \equiv 0 \ (\mathrm{mod}\ N)}} 1 \leq 1,$$

and hence

$$\frac{1}{N^2} \sum_{k,\ell=0}^{N-1} K_{\cos,d,\alpha,\boldsymbol{\gamma}}(\phi(\boldsymbol{x}_k), \phi(\boldsymbol{x}_\ell)) \leq \sum_{\mathfrak{u} \subseteq [d]} \gamma_{\mathfrak{u}} \sum_{\boldsymbol{h}_{\mathfrak{u}} \in \mathbb{N}^{|\mathfrak{u}|}} \prod_{j \in \mathfrak{u}} \frac{1}{h_j^{2\alpha}} \sum_{\substack{\boldsymbol{\sigma}_{\mathfrak{u}} \in \{-1,1\}^{|\mathfrak{u}|} \\ (\boldsymbol{\sigma}_{\mathfrak{u}} * \boldsymbol{h}_{\mathfrak{u}}) \cdot \boldsymbol{g}_{\mathfrak{u}} \equiv 0 \ (\mathrm{mod}\ N)}} 1$$

$$= \sum_{\mathfrak{u} \subseteq [d]} \gamma_{\mathfrak{u}} \sum_{\substack{\boldsymbol{h}_{\mathfrak{u}} \in (\mathbb{Z} \setminus \{0\})^{|\mathfrak{u}|} \\ \boldsymbol{h}_{\mathfrak{u}} \cdot \boldsymbol{g}_{\mathfrak{u}} \equiv 0 \ (\mathrm{mod}\ N)}} \prod_{j \in \mathfrak{u}} \frac{1}{|h_j|^{2\alpha}}.$$

This implies that

$$[\mathrm{err}_{N,d}(\mathcal{H}_{\cos,d,\alpha,\boldsymbol{\gamma}}, \mathcal{P}_\phi(\boldsymbol{g}, N))]^2 \leq \sum_{\emptyset \neq \mathfrak{u} \subseteq [d]} \gamma_{\mathfrak{u}} \sum_{\substack{\boldsymbol{h}_{\mathfrak{u}} \in (\mathbb{Z} \setminus \{0\})^{|\mathfrak{u}|} \\ \boldsymbol{h}_{\mathfrak{u}} \cdot \boldsymbol{g}_{\mathfrak{u}} \equiv 0 \ (\mathrm{mod}\ N)}} \prod_{j \in \mathfrak{u}} \frac{1}{|h_j|^{2\alpha}}.$$

From Remark 2.22 we find that the latter expression is exactly the formula for the squared worst-case integration error of the lattice rule with node set $\mathcal{P}(\boldsymbol{g}, N)$ in $\mathcal{H}_{\mathrm{kor},d,\alpha,\boldsymbol{\gamma}}$. This completes the proof. □

Theorem 7.40 implies that all upper bounds from Chapters 2, 3, and 4 on the worst-case error in the weighted Korobov space using lattice rules also hold for the worst-case error in the half-period cosine space using folded lattice rules. For example, we can use Theorem 3.9 to obtain the following result.

Corollary 7.41 *Let $N \geq 2$ be an arbitrary integer, let $\boldsymbol{\gamma} = \{\gamma_{\mathfrak{u}}\}_{\mathfrak{u} \subseteq [d]}$ be general weights, and assume that $\boldsymbol{g}$ has been found by Algorithm 3.6. Then for arbitrary $\tau \in [1/2, \alpha)$ and for any $s \in [d]$ it is true that*

$$\mathrm{err}_{N,s}(\mathcal{H}_{\cos,s,\alpha,\boldsymbol{\gamma}}, \mathcal{P}_\phi((g_1, \ldots, g_s), N)) \leq \left(\frac{1}{\varphi(N)} \sum_{\emptyset \neq \mathfrak{u} \subseteq [s]} \gamma_{\mathfrak{u}}^{1/(2\tau)} \left(2\zeta\left(\frac{\alpha}{\tau}\right)\right)^{|\mathfrak{u}|} \right)^{\tau}.$$

Remark 7.42 The convergence rate obtained in Corollary 7.41 is essentially best possible for general linear rules (see Remark 1.28) employed in cosine spaces of smoothness α. It was shown in [50, Theorem 4] that for arbitrary N-element point sets $\mathcal{P}$ in $[0,1]^d$ and integration weights $\boldsymbol{w} \in \mathbb{R}^N$ we have

$$\mathrm{err}_{N,d}(\mathcal{H}_{\cos,d,\alpha,\boldsymbol{\gamma}}, \mathcal{P}, \boldsymbol{w}) \geq C(\alpha, \boldsymbol{\gamma}, d) \frac{(\log N)^{(d-1)/2}}{N^\alpha},$$

where we use notation as in Remark 1.28, and where $C(\alpha, \boldsymbol{\gamma}, d) > 0$ depends only on α, $\boldsymbol{\gamma}$, and d, but not on N and $\boldsymbol{w}$. The proof is based on the same method as the proof of Theorem 2.10. We refer to the appendix of [50] for details.

For the case of smoothness $\alpha = 1$ we can apply the above findings also to the weighted unanchored Sobolev space of smoothness one. From (7.19) we know that the unanchored Sobolev space $\mathcal{H}_{\mathrm{sob},d,1,\boldsymbol{\gamma}}$ and the cosine space $\mathcal{H}_{\cos,d,1,\widetilde{\boldsymbol{\gamma}}}$ with $\widetilde{\gamma}_{\mathfrak{u}} = \gamma_{\mathfrak{u}} \pi^{-2|\mathfrak{u}|}$ for $\mathfrak{u} \subseteq [d]$ coincide. Therefore we also get the following result.

Corollary 7.43 *Let $N \geq 2$ be an arbitrary integer, let $\boldsymbol{\gamma} = \{\gamma_{\mathfrak{u}}\}_{\mathfrak{u} \subseteq [d]}$ be general weights, and assume that $\boldsymbol{g}$ has been found by Algorithm 3.6. Then for arbitrary $\tau \in [1/2, 1)$ and for any $s \in [d]$ we have*

$$\mathrm{err}_{N,s}(\mathcal{H}_{\mathrm{sob},s,1,\boldsymbol{\gamma}}, \mathcal{P}_\phi((g_1, \ldots, g_s), N)) \leq \left(\frac{1}{\varphi(N)} \sum_{\emptyset \neq \mathfrak{u} \subseteq [s]} \left(\frac{\gamma_{\mathfrak{u}}}{\pi^{2|\mathfrak{u}|}} \right)^{1/(2\tau)} \left(2\zeta\left(\frac{1}{\tau}\right) \right)^{|\mathfrak{u}|} \right)^\tau .$$

Corollary 7.43 is a deterministic version of Theorem 7.12, where a random shift was required for an analogous bound to hold.

The results above can also be extended to the case of smoothness $\alpha = 2$, as shown by Goda, Suzuki, and Yoshiki [86]. The following theorem relates the error of a folded lattice rule in $\mathcal{H}_{\mathrm{sob},d,2,\boldsymbol{\gamma}}$ to the error of the same but nonfolded lattice rule in the Korobov space $\mathcal{H}_{\mathrm{kor},d,1,\boldsymbol{\gamma}}$ of smoothness $\alpha = 1$.

Theorem 7.44 *The worst-case error of folded lattice rules in the unanchored Sobolev space $\mathcal{H}_{\mathrm{sob},d,2,\boldsymbol{\gamma}}$ of smoothness two satisfies*

$$\mathrm{err}_{N,d}(\mathcal{H}_{\mathrm{sob},d,2,\boldsymbol{\gamma}}, \mathcal{P}_\phi(\boldsymbol{g}, N)) \leq [\mathrm{err}_{N,d}(\mathcal{H}_{\mathrm{kor},d,1,\widetilde{\boldsymbol{\gamma}}^{1/2}}, \mathcal{P}(\boldsymbol{g}, N))]^2,$$

where the weights $\widetilde{\boldsymbol{\gamma}}^{1/2} = \{\widetilde{\gamma}_{\mathfrak{u}}^{1/2}\}_{\mathfrak{u} \subseteq [d]}$ are defined as $\widetilde{\gamma}_\emptyset := 1$ and

$$\widetilde{\gamma}_{\mathfrak{u}} := \gamma_{\mathfrak{u}} \left(\frac{29}{6\pi^4} \right)^{|\mathfrak{u}|} \quad \text{for } \emptyset \neq \mathfrak{u} \subseteq [d]. \tag{7.27}$$

For the proof of Theorem 7.44 we require the following lemma.

Lemma 7.45 *For any* $x, y \in [0, 1]$ *we have*

$$\sum_{\tau=1}^{2} \frac{B_\tau(\phi(x))B_\tau(\phi(y))}{(\tau!)^2} - \frac{B_4(|\phi(x) - \phi(y)|)}{24}$$
$$= \frac{1}{\pi^4} \sum_{h,m=1}^{\infty} c(h, m) \cos(2\pi hx) \cos(2\pi my),$$

where

$$c(h, m) = \begin{cases} \frac{1}{h^4}\left(\frac{58}{3} - \frac{32}{\pi^2 h^2}\right) & \text{if } h \text{ and } m \text{ are odd and } h = m, \\ \frac{1}{h^2 m^2}\left(\frac{52}{3} - \frac{16}{\pi^2 h^2} - \frac{16}{\pi^2 m^2}\right) & \text{if } h \text{ and } m \text{ are odd and } h \neq m, \\ \frac{6}{h^4} & \text{if } h \text{ and } m \text{ are even and } h = m, \\ \frac{4}{h^2 m^2} & \text{if } h \text{ and } m \text{ are even and } h \neq m, \\ 0 & \text{otherwise.} \end{cases}$$

In particular, we have for all $h, m \in \mathbb{N}$,

$$0 \leq c(h, m) \leq \frac{58}{3} \frac{1}{h^2 m^2}.$$

Proof Let $k \in \mathbb{N}$ and $\ell \in \mathbb{N}_0$. Using the definition of the tent transformation ϕ it is easily checked that

$$\int_0^1 \sin(2\pi k\phi(x)) \sin(2\pi \ell x) \, dx = 0,$$

and

$$\int_0^1 \sin(2\pi k\phi(x)) \cos(2\pi \ell x) \, dx = \int_0^1 \sin(2\pi kx) \cos(\pi \ell x) \, dx$$
$$= \begin{cases} \frac{4}{\pi} \frac{k}{4k^2 - \ell^2} & \text{if } \ell \text{ is odd,} \\ 0 & \text{if } \ell \text{ is even.} \end{cases}$$

Therefore, the Fourier series expansion of $\sin(2\pi k\phi(\cdot))$ is given by

$$\sin(2\pi k\phi(x)) = 2 \sum_{\ell=1}^{\infty} \cos(2\pi \ell x) \int_0^1 \sin(2\pi k\phi(y)) \cos(2\pi \ell y) \, dy$$

$$= \frac{8}{\pi} \sum_{\substack{\ell=1 \\ \ell \text{ odd}}}^{\infty} \frac{k}{4k^2 - \ell^2} \cos(2\pi \ell x). \tag{7.28}$$

This preliminary result will be used below in the proof.

Let us write $b_\tau := B_\tau/\tau!$ for $\tau \in \mathbb{N}$. Furthermore, let $\widetilde{B}_\tau$ be the periodic extension of $B_\tau|_{[0,1)}$ to $\mathbb{R}$ and let $\widetilde{b}_\tau = \widetilde{B}_\tau/\tau!$. We now need to consider

$$b_1(\phi(x))b_1(\phi(y)) + b_2(\phi(x))b_2(\phi(y)) - \widetilde{b}_4(\phi(x) - \phi(y)). \tag{7.29}$$

From the expansion (7.20) we obtain

$$b_1(\phi(x)) = \phi(x) - \frac{1}{2} = -\frac{4}{\pi^2} \sum_{\substack{k=1 \\ k \text{ odd}}}^{\infty} \frac{\cos(2\pi k x)}{k^2}.$$

Therefore the Fourier series expansion of the first term in (7.29) is given by

$$b_1(\phi(x))b_1(\phi(y)) = \frac{16}{\pi^4} \sum_{\substack{k,\ell=1 \\ k,\ell \text{ odd}}}^{\infty} \frac{1}{k^2 \ell^2} \cos(2\pi k x)\cos(2\pi \ell y). \tag{7.30}$$

From (2.8) it follows that

$$b_2(x) = \frac{1}{4\pi^2} \sum_{k \in \mathbb{Z}\setminus\{0\}} \frac{\mathrm{e}^{2\pi \mathrm{i} k x}}{k^2}.$$

Since, for $k \in \mathbb{N}_0$ and $x \in [0, 1]$, we have $\cos(2\pi k \phi(x)) = \cos(4\pi k x)$, we obtain

$$b_2(\phi(x)) = \frac{1}{2\pi^2} \sum_{k=1}^{\infty} \frac{\cos(2\pi k \phi(x))}{k^2} = \frac{2}{\pi^2} \sum_{\substack{k=1 \\ k \text{ even}}}^{\infty} \frac{\cos(2\pi k x)}{k^2}.$$

This implies that the Fourier series expansion of the second term in (7.29) is given by

$$b_2(\phi(x))b_2(\phi(y)) = \frac{4}{\pi^4} \sum_{\substack{k,\ell=1 \\ k,\ell \text{ even}}}^{\infty} \frac{1}{k^2 \ell^2} \cos(2\pi k x)\cos(2\pi \ell x). \tag{7.31}$$

It remains to consider the third term in (7.29). From (2.8) we obtain the Fourier series expansion of $\widetilde{b}_4$, which is

$$\widetilde{b}_4(x) = -\frac{1}{(2\pi)^4} \sum_{h \in \mathbb{Z}\setminus\{0\}} \frac{\mathrm{e}^{2\pi \mathrm{i} h x}}{|h|^4}.$$

Therefore, and since $\cos(2\pi k \phi(x)) = \cos(4\pi k x)$ for $k \in \mathbb{N}_0$, we obtain

$$
\begin{aligned}
&\widetilde{b}_4(\phi(x)-\phi(y)) \\
&= -\frac{2}{(2\pi)^4}\sum_{k=1}^{\infty}\frac{\cos(2\pi k(\phi(x)-\phi(y)))}{k^4} \\
&= -\frac{2}{\pi^4}\sum_{\substack{k=1\\ k \text{ even}}}^{\infty}\frac{\cos(2\pi kx)\cos(2\pi ky)}{k^4} - \frac{1}{8\pi^4}\sum_{k=1}^{\infty}\frac{\sin(2\pi k\phi(x))\sin(2\pi k\phi(y))}{k^4}.
\end{aligned}
$$

Now we use the preliminary result in (7.28) and obtain, for the second term in the latter expression,

$$
\begin{aligned}
&-\frac{1}{8\pi^4}\sum_{k=1}^{\infty}\frac{\sin(2\pi k\phi(x))\sin(2\pi k\phi(y))}{k^4} \\
&= -\frac{8}{\pi^6}\sum_{k=1}^{\infty}\frac{1}{k^2}\sum_{\substack{\ell,m=1\\ \ell,m \text{ odd}}}^{\infty}\frac{1}{(4k^2-\ell^2)(4k^2-m^2)}\cos(2\pi\ell x)\cos(2\pi m y) \\
&= -\frac{8}{\pi^6}\sum_{\substack{\ell,m=1\\ \ell,m \text{ odd}}}^{\infty}\cos(2\pi\ell x)\cos(2\pi m y)\sum_{k=1}^{\infty}\frac{1}{k^2}\frac{1}{(4k^2-\ell^2)(4k^2-m^2)}.
\end{aligned} \tag{7.32}
$$

Next, we use the identities

$$
\sum_{k=1}^{\infty}\frac{1}{4k^2-\ell^2}=\frac{1}{2\ell^2} \quad \text{and} \quad \sum_{k=1}^{\infty}\frac{1}{(4k^2-\ell^2)^2}=\frac{1}{4\ell^2}\left(\frac{\pi^2}{4}-\frac{2}{\ell^2}\right),
$$

which hold for any positive odd integer ℓ. Thus, if $\ell = m$ in (7.32), we have

$$
\begin{aligned}
&\sum_{k=1}^{\infty}\frac{1}{k^2}\frac{1}{(4k^2-\ell^2)(4k^2-m^2)} \\
&= \sum_{k=1}^{\infty}\frac{1}{k^2}\frac{1}{(4k^2-\ell^2)^2} \\
&= \frac{1}{\ell^2}\sum_{k=1}^{\infty}\left(\frac{1}{\ell^2}\left(\frac{1}{k^2}-\frac{4}{4k^2-\ell^2}\right)+\frac{4}{(4k^2-\ell^2)^2}\right) \\
&= \frac{1}{\ell^2}\left(\frac{1}{\ell^2}\left(\frac{\pi^2}{6}-\frac{2}{\ell^2}\right)+\frac{1}{\ell^2}\left(\frac{\pi^2}{4}-\frac{2}{\ell^2}\right)\right) \\
&= \frac{1}{\ell^4}\left(\frac{5\pi^2}{12}-\frac{4}{\ell^2}\right).
\end{aligned}
$$

If $\ell \neq m$ in (7.32), we have

$$\begin{aligned}
&\sum_{k=1}^{\infty} \frac{1}{k^2} \frac{1}{(4k^2-\ell^2)(4k^2-m^2)} \\
&= \frac{1}{\ell^2 m^2} \sum_{k=1}^{\infty} \left(\frac{1}{k^2} + \frac{4}{\ell^2 - m^2} \left(\frac{m^2}{4k^2-\ell^2} - \frac{\ell^2}{4k^2-m^2} \right) \right) \\
&= \frac{1}{\ell^2 m^2} \left(\frac{\pi^2}{6} + \frac{4}{\ell^2 - m^2} \left(\frac{m^2}{2\ell^2} - \frac{\ell^2}{2m^2} \right) \right) \\
&= \frac{1}{\ell^2 m^2} \left(\frac{\pi^2}{6} - \frac{2}{\ell^2} - \frac{2}{m^2} \right).
\end{aligned}$$

This implies that

$$\begin{aligned}
&-\frac{1}{8\pi^4} \sum_{k=1}^{\infty} \frac{\sin(2\pi k \phi(x)) \sin(2\pi k \phi(y))}{k^4} \\
&= -\frac{8}{\pi^6} \sum_{\substack{\ell,m=1 \\ \ell,\, m \text{ odd} \\ \ell \neq m}}^{\infty} \cos(2\pi \ell x) \cos(2\pi m y) \frac{1}{\ell^2 m^2} \left(\frac{\pi^2}{6} - \frac{2}{\ell^2} - \frac{2}{m^2} \right) \\
&\quad - \frac{8}{\pi^6} \sum_{\substack{\ell=1 \\ \ell \text{ odd}}}^{\infty} \cos(2\pi \ell x) \cos(2\pi \ell y) \frac{1}{\ell^4} \left(\frac{5\pi^2}{12} - \frac{4}{\ell^2} \right).
\end{aligned}$$

Hence

$$\begin{aligned}
\widetilde{b}_4(\phi(x) - \phi(y)) = &-\frac{2}{\pi^4} \sum_{\substack{k=1 \\ k \text{ even}}}^{\infty} \frac{\cos(2\pi k x) \cos(2\pi k y)}{k^4} \\
&- \frac{8}{\pi^6} \sum_{\substack{\ell,m=1 \\ \ell,\, m \text{ odd} \\ \ell \neq m}}^{\infty} \cos(2\pi \ell x) \cos(2\pi m y) \frac{1}{\ell^2 m^2} \left(\frac{\pi^2}{6} - \frac{2}{\ell^2} - \frac{2}{m^2} \right) \\
&- \frac{8}{\pi^6} \sum_{\substack{\ell=1 \\ \ell \text{ odd}}}^{\infty} \cos(2\pi \ell x) \cos(2\pi \ell y) \frac{1}{\ell^4} \left(\frac{5\pi^2}{12} - \frac{4}{\ell^2} \right). \qquad (7.33)
\end{aligned}$$

Now, inserting (7.30), (7.31), and (7.33) into (7.29) we obtain the desired result. □

We are ready to present the proof of Theorem 7.44.

Proof of Theorem 7.44 It again follows from the worst-case error formula in Theorem 1.27 that

$$[\mathrm{err}_{N,d}(\mathcal{H}_{\mathrm{sob},d,2,\boldsymbol{\gamma}}, \mathcal{P}_\phi(\boldsymbol{g}, N))]^2 = -1 + \frac{1}{N^2} \sum_{k,\ell=0}^{N-1} K_{\mathrm{sob},d,2,\boldsymbol{\gamma}}(\phi(\boldsymbol{x}_k), \phi(\boldsymbol{x}_\ell)).$$

Next, we consider the kernel function. For $k, \ell \in \{0, 1, \ldots, N-1\}$ we have

$$\begin{aligned}
&K_{\mathrm{sob},d,2,\boldsymbol{\gamma}}(\phi(\boldsymbol{x}_k), \phi(\boldsymbol{x}_\ell)) \\
&= \sum_{\mathfrak{u} \subseteq [d]} \gamma_{\mathfrak{u}} \prod_{j \in \mathfrak{u}} \left(\sum_{\tau=1}^{2} \frac{B_\tau(\phi(x_{k,j})) B_\tau(\phi(x_{\ell,j}))}{(\tau!)^2} - \frac{B_4(|\phi(x_{k,j}) - \phi(x_{\ell,j})|)}{24} \right).
\end{aligned}$$

Using the expansion in Lemma 7.45 we obtain

$$\begin{aligned}
&K_{\mathrm{sob},d,2,\boldsymbol{\gamma}}(\phi(\boldsymbol{x}_k), \phi(\boldsymbol{x}_\ell)) \\
&= \sum_{\mathfrak{u} \subseteq [d]} \gamma_{\mathfrak{u}} \prod_{j \in \mathfrak{u}} \left(\frac{1}{\pi^4} \sum_{h,m=1}^{\infty} c(h,m) \cos(2\pi h x_{k,j}) \cos(2\pi m x_{\ell,j}) \right) \\
&= \sum_{\mathfrak{u} \subseteq [d]} \frac{\gamma_{\mathfrak{u}}}{\pi^{4|\mathfrak{u}|}} \sum_{\boldsymbol{h}_{\mathfrak{u}}, \boldsymbol{m}_{\mathfrak{u}} \in \mathbb{N}^{|\mathfrak{u}|}} \prod_{j \in \mathfrak{u}} c(h_j, m_j) \cos(2\pi h x_{k,j}) \cos(2\pi m x_{\ell,j}).
\end{aligned}$$

Employing (7.26) we obtain

$$\begin{aligned}
K_{\mathrm{sob},d,2,\boldsymbol{\gamma}}(\phi(\boldsymbol{x}_k), \phi(\boldsymbol{x}_\ell)) &= \sum_{\mathfrak{u} \subseteq [d]} \frac{\gamma_{\mathfrak{u}}}{\pi^{4|\mathfrak{u}|}} \sum_{\boldsymbol{h}_{\mathfrak{u}}, \boldsymbol{m}_{\mathfrak{u}} \in \mathbb{N}^{|\mathfrak{u}|}} \prod_{j \in \mathfrak{u}} c(h_j, m_j) \\
&\times \left(\frac{1}{2^{|\mathfrak{u}|}} \sum_{\boldsymbol{\sigma}_{\mathfrak{u}} \in \{-1,1\}^{|\mathfrak{u}|}} \mathrm{e}^{2\pi \mathrm{i} (\boldsymbol{\sigma}_{\mathfrak{u}} * \boldsymbol{h}_{\mathfrak{u}}) \cdot \boldsymbol{x}_{k,\mathfrak{u}}} \right) \left(\frac{1}{2^{|\mathfrak{u}|}} \sum_{\widetilde{\boldsymbol{\sigma}}_{\mathfrak{u}} \in \{-1,1\}^{|\mathfrak{u}|}} \mathrm{e}^{2\pi \mathrm{i} (\widetilde{\boldsymbol{\sigma}}_{\mathfrak{u}} * \boldsymbol{m}_{\mathfrak{u}}) \cdot \boldsymbol{x}_{\ell,\mathfrak{u}}} \right).
\end{aligned}$$

Averaging over $k, \ell \in \{0, 1, \ldots, N-1\}$ yields

$$\begin{aligned}
&\frac{1}{N^2} \sum_{k,\ell=0}^{N-1} K_{\mathrm{sob},d,2,\boldsymbol{\gamma}}(\phi(\boldsymbol{x}_k), \phi(\boldsymbol{x}_\ell)) \\
&= \sum_{\mathfrak{u} \subseteq [d]} \frac{\gamma_{\mathfrak{u}}}{\pi^{4|\mathfrak{u}|}} \sum_{\boldsymbol{h}_{\mathfrak{u}}, \boldsymbol{m}_{\mathfrak{u}} \in \mathbb{N}^{|\mathfrak{u}|}} \prod_{j \in \mathfrak{u}} c(h_j, m_j) \\
&\quad \times \left(\frac{1}{2^{|\mathfrak{u}|}} \sum_{\boldsymbol{\sigma}_{\mathfrak{u}} \in \{-1,1\}^{|\mathfrak{u}|}} \frac{1}{N} \sum_{k=0}^{N-1} \mathrm{e}^{2\pi \mathrm{i} (\boldsymbol{\sigma}_{\mathfrak{u}} * \boldsymbol{h}_{\mathfrak{u}}) \cdot \boldsymbol{x}_{k,\mathfrak{u}}} \right) \\
&\quad \times \left(\frac{1}{2^{|\mathfrak{u}|}} \sum_{\widetilde{\boldsymbol{\sigma}}_{\mathfrak{u}} \in \{-1,1\}^{|\mathfrak{u}|}} \frac{1}{N} \sum_{\ell=0}^{N-1} \mathrm{e}^{2\pi \mathrm{i} (\widetilde{\boldsymbol{\sigma}}_{\mathfrak{u}} * \boldsymbol{m}_{\mathfrak{u}}) \cdot \boldsymbol{x}_{\ell,\mathfrak{u}}} \right) \\
&= \sum_{\mathfrak{u} \subseteq [d]} \frac{\gamma_{\mathfrak{u}}}{\pi^{4|\mathfrak{u}|}} \sum_{\boldsymbol{h}_{\mathfrak{u}}, \boldsymbol{m}_{\mathfrak{u}} \in \mathbb{N}^{|\mathfrak{u}|}} \prod_{j \in \mathfrak{u}} c(h_j, m_j) \\
&\quad \times \left(\frac{1}{2^{|\mathfrak{u}|}} \sum_{\substack{\boldsymbol{\sigma}_{\mathfrak{u}} \in \{-1,1\}^{|\mathfrak{u}|} \\ (\boldsymbol{\sigma}_{\mathfrak{u}} * \boldsymbol{h}_{\mathfrak{u}}) \cdot \boldsymbol{g}_{\mathfrak{u}} \equiv 0 \pmod{N}}} 1 \right) \left(\frac{1}{2^{|\mathfrak{u}|}} \sum_{\substack{\widetilde{\boldsymbol{\sigma}}_{\mathfrak{u}} \in \{-1,1\}^{|\mathfrak{u}|} \\ (\widetilde{\boldsymbol{\sigma}}_{\mathfrak{u}} * \boldsymbol{m}_{\mathfrak{u}}) \cdot \boldsymbol{g}_{\mathfrak{u}} \equiv 0 \pmod{N}}} 1 \right)
\end{aligned}$$

$$= \sum_{\mathfrak{u}\subseteq[d]} \frac{\gamma_{\mathfrak{u}}}{(4\pi^4)^{|\mathfrak{u}|}} \sum_{\substack{\boldsymbol{h}_{\mathfrak{u}}\in(\mathbb{Z}\setminus\{0\})^{|\mathfrak{u}|} \\ \boldsymbol{h}_{\mathfrak{u}}\cdot\boldsymbol{g}_{\mathfrak{u}}\equiv 0 \pmod{N}}} \sum_{\substack{\boldsymbol{m}_{\mathfrak{u}}\in(\mathbb{Z}\setminus\{0\})^{|\mathfrak{u}|} \\ \boldsymbol{m}_{\mathfrak{u}}\cdot\boldsymbol{g}_{\mathfrak{u}}\equiv 0 \pmod{N}}} \prod_{j\in\mathfrak{u}} c(|h_j|,|m_j|).$$

Using the upper bound on the coefficients $c(h,m)$ in Lemma 7.45 we obtain

$$\begin{aligned}
&\frac{1}{N^2}\sum_{k,\ell=0}^{N-1} K_{\mathrm{sob},d,2,\boldsymbol{\gamma}}(\phi(\boldsymbol{x}_k),\phi(\boldsymbol{x}_\ell)) \\
&\le \sum_{\mathfrak{u}\subseteq[d]} \gamma_{\mathfrak{u}} \left(\frac{58}{12\pi^4}\right)^{|\mathfrak{u}|} \sum_{\substack{\boldsymbol{h}_{\mathfrak{u}}\in(\mathbb{Z}\setminus\{0\})^{|\mathfrak{u}|} \\ \boldsymbol{h}_{\mathfrak{u}}\cdot\boldsymbol{g}_{\mathfrak{u}}\equiv 0 \pmod{N}}} \sum_{\substack{\boldsymbol{m}_{\mathfrak{u}}\in(\mathbb{Z}\setminus\{0\})^{|\mathfrak{u}|} \\ \boldsymbol{m}_{\mathfrak{u}}\cdot\boldsymbol{g}_{\mathfrak{u}}\equiv 0 \pmod{N}}} \prod_{j\in\mathfrak{u}} \frac{1}{|h_j|^2|m_j|^2} \\
&= \sum_{\mathfrak{u}\subseteq[d]} \gamma_{\mathfrak{u}} \left(\frac{29}{6\pi^4}\right)^{|\mathfrak{u}|} \left(\sum_{\substack{\boldsymbol{h}_{\mathfrak{u}}\in(\mathbb{Z}\setminus\{0\})^{|\mathfrak{u}|} \\ \boldsymbol{h}_{\mathfrak{u}}\cdot\boldsymbol{g}_{\mathfrak{u}}\equiv 0 \pmod{N}}} \frac{1}{|h_j|^2} \right)^2 .
\end{aligned}$$

Therefore, and using Jensen's inequality (Lemma 2.25), we get

$$\begin{aligned}
[\mathrm{err}_{N,d}(\mathcal{H}_{\mathrm{sob},d,2,\boldsymbol{\gamma}}, \mathcal{P}_\phi(\boldsymbol{g},N))]^2 &\le \sum_{\emptyset\neq\mathfrak{u}\subseteq[d]} \widetilde{\gamma}_{\mathfrak{u}} \left(\sum_{\substack{\boldsymbol{h}_{\mathfrak{u}}\in(\mathbb{Z}\setminus\{0\})^{|\mathfrak{u}|} \\ \boldsymbol{h}_{\mathfrak{u}}\cdot\boldsymbol{g}_{\mathfrak{u}}\equiv 0 \pmod{N}}} \frac{1}{|h_j|^2} \right)^2 \\
&\le \left(\sum_{\emptyset\neq\mathfrak{u}\subseteq[d]} \widetilde{\gamma}_{\mathfrak{u}}^{1/2} \sum_{\substack{\boldsymbol{h}_{\mathfrak{u}}\in(\mathbb{Z}\setminus\{0\})^{|\mathfrak{u}|} \\ \boldsymbol{h}_{\mathfrak{u}}\cdot\boldsymbol{g}_{\mathfrak{u}}\equiv 0 \pmod{N}}} \frac{1}{|h_j|^2} \right)^2 \\
&= [\mathrm{err}_{N,d}(\mathcal{H}_{\mathrm{kor},d,1,\widetilde{\boldsymbol{\gamma}}^{1/2}}, \mathcal{P}(\boldsymbol{g},N))]^4,
\end{aligned}$$

where the final identity is due to Remark 2.22. □

Again, all upper bounds from Chapters 2, 3, and 4 on the worst-case error in the weighted Korobov space of smoothness $\alpha = 1$ using lattice rules also hold for the worst-case error in the unanchored Sobolev space of smoothness $\alpha = 2$ using folded lattice rules. For example, we can use Theorem 3.9 with $\alpha = 1$ to obtain the following.

Corollary 7.46 *Let $N \ge 2$ be an arbitrary integer, let $\boldsymbol{\gamma} = \{\gamma_{\mathfrak{u}}\}_{\mathfrak{u}\subseteq[d]}$ be general weights, and assume that $\boldsymbol{g}$ has been found by Algorithm 3.6 (by means of weights $\widetilde{\boldsymbol{\gamma}}^{1/2}$, where the $\widetilde{\gamma}_{\mathfrak{u}}$ are given by (7.27), and smoothness parameter $\alpha = 1$). Then for arbitrary $\tau \in [1,2)$ and for any $s \in [d]$ we have*

$$\mathrm{err}_{N,s}(\mathcal{H}_{\mathrm{sob},s,2,\boldsymbol{\gamma}}, \mathcal{P}_\phi((g_1,\ldots,g_s),N)) \le \left(\frac{1}{\varphi(N)} \sum_{\emptyset\neq\mathfrak{u}\subseteq[s]} \gamma_{\mathfrak{u}}^{1/(2\tau)} c_\tau^{|\mathfrak{u}|} \right)^\tau, \quad (7.34)$$

where $c_\tau := \left(29/(6\pi^4)\right)^{1/(2\tau)} 2\zeta(2/\tau)$.

Proof Combining Theorem 7.44, and Theorem 3.9 with $\alpha = 1$, we obtain for every $\widetilde{\tau} \in [1/2, 1)$ and any $s \in [d]$ that

$$\mathrm{err}_{N,s}(\mathcal{H}_{\mathrm{sob},s,2,\gamma}, \mathcal{P}_\phi((g_1,\ldots,g_s),N)) \le \left(\frac{1}{\varphi(N)} \sum_{\emptyset \ne \mathfrak{u} \subseteq [s]} \widetilde{\gamma}_{\mathfrak{u}}^{1/(4\widetilde{\tau})} \left(2\zeta\left(\frac{1}{\widetilde{\tau}}\right)\right)^{|\mathfrak{u}|}\right)^{2\widetilde{\tau}}.$$

Writing $\tau := 2\widetilde{\tau}$, i.e., $\tau \in [1, 2)$, and using $\widetilde{\gamma}_{\mathfrak{u}} = \gamma_{\mathfrak{u}} \left(29/(6\pi^4)\right)^{|\mathfrak{u}|}$ gives the desired result. □

We remark that Corollary 7.46 is a deterministic version of the findings of Hickernell [101], which we outlined on p. 304.

7.4 Symmetrized Lattice Rules

We close this chapter with a brief discussion of a further method that allows us to obtain higher order convergence rates for certain subspaces of Sobolev spaces. The underlying idea is to apply the transformation $x \mapsto 1 - x$ to each possible set of coordinates separately in order to obtain a "symmetrization" of a given point set. To be more precise, let $\boldsymbol{x} = (x_1, \ldots, x_d) \in [0, 1]^d$ and let $\mathfrak{u} \subseteq [d]$. Then let $\mathrm{sym}_{\mathfrak{u}}(\boldsymbol{x})$ denote the vector whose j-th coordinate is $1 - x_j$ if $j \in \mathfrak{u}$ and x_j if $j \notin \mathfrak{u}$, i.e., $\mathrm{sym}_{\mathfrak{u}}(\boldsymbol{x}) = (y_1, \ldots, y_d)$ with

$$y_j = \begin{cases} 1 - x_j & \text{if } j \in \mathfrak{u}, \\ x_j & \text{if } j \notin \mathfrak{u}. \end{cases}$$

Definition 7.47 For $N \in \mathbb{N}$, $N \ge 2$, and for $\boldsymbol{g} \in \mathbb{Z}^d$ define

$$\mathcal{P}_{\mathrm{sym}}(\boldsymbol{g}, N) := \left\{\mathrm{sym}_{\mathfrak{u}}\left(\left\{\frac{k}{N}\boldsymbol{g}\right\}\right) \,:\, k \in \{0, 1, \ldots, N-1\},\ \mathfrak{u} \subseteq [d]\right\}$$

to be a *symmetrized (rank-1) lattice point set*. A QMC rule that is based on the nodes $\mathcal{P}_{\mathrm{sym}}(\boldsymbol{g}, N)$ is called a *symmetrized (rank-1) lattice rule*.

Examples of symmetrized lattice point sets are shown in Figure 7.5.

A consequence of the symmetrization procedure is that all functions of the form

$$\sum_{\substack{k_1,\ldots,k_d \in \mathbb{N} \\ k_1,\ldots,k_d \text{ odd}}} b_{k_1,\ldots,k_d} \prod_{j=1}^{d} \cos(\pi k_j x_j) \qquad \text{with coefficients } b_{k_1,\ldots,k_d} \in \mathbb{R}$$

are integrated exactly. Likewise, all polynomials of the form

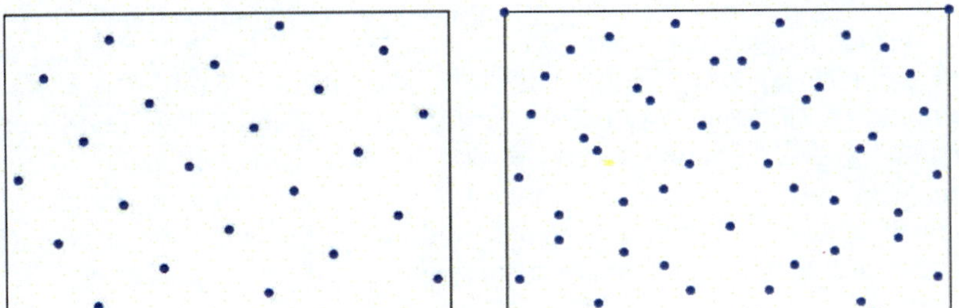

Fig. 7.5: The Fibonacci lattice point sets $\mathcal{P}((1,21),34)$ (top left) and $\mathcal{P}((1,377),610)$ (bottom left), and their symmetrized versions (right).

$$\sum_{\substack{k_1,\ldots,k_d \in \mathbb{N} \\ k_1,\ldots,k_d \text{ odd}}} a_{k_1,\ldots,k_d} \prod_{j=1}^{d} \left(x_j - \frac{1}{2}\right)^{k_j} \quad \text{with coefficients } a_{k_1,\ldots,k_d} \in \mathbb{R}$$

are integrated exactly. This is because all the odd frequencies in a cosine series are integrated exactly. Specifically, for the half-period cosine space, we can state the following result which is [50, Corollary 4].

Theorem 7.48 *The squared worst-case error of a symmetrized rank-1 lattice rule in the half-period cosine space $\mathcal{H}_{\cos,d,\alpha,\gamma}$ is given by*

$$[\mathrm{err}_{N,d}(\mathcal{H}_{\cos,d,\alpha,\gamma}, \mathcal{P}_{\mathrm{sym}}(\boldsymbol{g}, N))]^2 = \sum_{\substack{\boldsymbol{h} \in \mathbb{Z}^d \setminus \{\boldsymbol{0}\} \\ \boldsymbol{h} \cdot \boldsymbol{g} \equiv 0 \ (\mathrm{mod}\ N)}} \frac{1}{r_{2\alpha,\gamma}(2\boldsymbol{h})}.$$

Theorem 7.48 implies that symmetrized lattice rules applied to elements of cosine spaces of smoothness α can yield a convergence rate of order $O(N^{-\alpha+\varepsilon})$ for every $\varepsilon > 0$. Here $N \asymp_d |\mathcal{P}_{\mathrm{sym}}(\boldsymbol{g}, N)|$ but the implied factors highly depend on the dimension d (see Lemma 7.49 below), where, for two number theoretic functions $f, g : D \subseteq \mathbb{N} \to \mathbb{R}^+$, we write $f(N) \asymp g(N)$ if there exist positive reals c and C such that $c\, g(N) \le f(N) \le C\, g(N)$ for all $N \in D$. The notation $\asymp_d$ indicates that the implied factors c or C may depend on d. Indeed, the following lemma shows how the number of function evaluations used in symmetrized lattice rules grows with the dimension.

Lemma 7.49 *Let $d \in \mathbb{N}$. The number of function evaluations employed in the symmetrized lattice rule based on $\mathcal{P}_{\mathrm{sym}}(\boldsymbol{g}, N)$ is $2^{d-1}(N+1)$ if N is odd and $2^{d-1}N + 1$ if N is even.*

Proof We have the symmetry

$$k g_j \equiv N - (N-k) g_j \pmod{N}, \qquad \text{for all } k \in \{0, 1, \ldots, N-1\} \text{ and for } j \in [d],$$

which exactly corresponds to the symmetry $x_{k,j} = 1 - x_{N-k,j}$. This means that we only have to compute and symmetrize the points for $0 \le k \le N/2$. We make the following observations.

- For $0 < k < N/2$ we have $2^d(N-1)/2$ points if N is odd and $2^d(N/2-1)$ if N is even.
- For $k = 0$ symmetrization returns all 2^d corners of $[0,1]^d$.
- If $2 < N$ and N is even, then for $k = N/2$ we have $\boldsymbol{x}_{N/2} = (1/2, \ldots, 1/2)$ and thus no symmetrization is needed.

Counting the number of function evaluations yields the desired result. □

This shows that symmetrized lattice rules may work well for very small dimension d, but show unfavorable dependence on the dimension when d is large.

Notes and Remarks

For more thorough descriptions of classical periodization methods we refer to the monographs of Hua and Wang [115, Chapter 6], Niederreiter [199, p. 107], and Sloan and Joe [230], or to the survey articles of Niederreiter [195, pp. 983–984] and [48, Section 5.10] and the references therein.

A detailed discussion of weighted anchored and unanchored Sobolev spaces can also be found in [210, Appendix A]. Results on tractability of integration in weighted Sobolev spaces can be found in [211, Section 16.9].

The formula (7.14) for the L_2-discrepancy is sometimes credited to Warnock [260], which is historically not entirely correct, since it was already provided by Koksma [135] in 1942 but only for $d = 1$. Later, Warnock used the same proof method for arbitrary dimension (see also [192]). An extension of the formula (7.14) to the weighted L_2-discrepancy was given in [239] (see also [52, Proposition 3.60] or [53, Chapter 1]) and to weighted L_p-discrepancy for even p in [179].

The half-period cosine space was introduced and studied as a reference space for numerical integration in [50]. Further subspaces of Sobolev spaces and the relations between them have been presented in [50] and by Goda, Suzuki, and Yoshiki in [86].

Embedding results for various function spaces are an important tool in the theory of numerical integration. In Section 7.2 we have discussed embeddings between the anchored Sobolev space of smoothness one and the unanchored Sobolev space of smoothness one. For further embedding results in this context we refer to [80, 81, 82, 94, 95, 109, 145, 146].

The proof of Theorem 7.40 follows the outline in [30].

The same result as in Theorem 7.48 holds true for subspaces of the unanchored Sobolev space comprising a greater class of functions, see [50, 86] for further information.

Symmetrization is a well-known concept from discrepancy theory in order to construct point sets and sequences with optimal order of discrepancy. For example, the first explicit construction of a point set in dimension two with L_2-discrepancy of order $O(\sqrt{\log N}/N)$, which is optimal with respect to Roth's general lower bound (1.20), according to Davenport [34] uses the symmetrization technique. Since then many authors have used this method (see, e.g., [15, 84, 151, 152, 153, 175, 176]). From this point of view, and because of the close relation of discrepancy to numerical integration, it is not a big surprise that symmetrization can also yield improvements for integration errors of QMC rules.

Chapter 8
Integration With Respect to Probability Measures

In many applications (see for instance Appendix A) one needs integration rules over domains D^d, $D \subseteq \mathbb{R}$, other than the unit cube $[0,1]^d$. Of particular importance is integration over the Euclidean space $\mathbb{R}^d$ with respect to a normal distribution. In this chapter we consider numerical integration of functions with respect to some probability density ϕ,

$$\int_{D^d} f(\boldsymbol{x})\phi(\boldsymbol{x})\,\mathrm{d}\boldsymbol{x}. \tag{8.1}$$

In general, one can apply a transformation to obtain an integral over the unit cube. However, this changes the integrand, and often certain smoothness assumptions are not satisfied anymore by the transformed integrand. The approach which we pursue in this chapter is to transform the points to D^d, where the transformation depends on the probability density ϕ. Using this approach one can obtain bounds on the integration error for a number of important choices of probability densities.

8.1 Transforming the Points Versus Transforming the Integrand

Consider a domain $D = \overline{(a,b)}$, the closure of the interval (a,b), where $a,b \in \mathbb{R} \cup \{-\infty,\infty\}$, i.e., we also allow cases such as $(-\infty,b]$, $[a,b]$, $[a,\infty)$, and $\mathbb{R} = (-\infty,\infty)$. Consider a probability density function ϕ of product form defined on D^d,

$$\phi(\boldsymbol{x}) = \prod_{j=1}^{d} \phi_j(x_j) \quad \text{for } \boldsymbol{x} = (x_1,\ldots,x_d) \in D^d$$

for Lebesgue measurable univariate probability densities ϕ_j defined on D. The cumulative distribution function Φ_j is given by

$$\Phi_j(y) = \int_{-\infty}^{y} \phi_j(x)\,\mathrm{d}x \quad \text{for } j \in [d].$$

J. Dick et al., *Lattice Rules*, Springer Series in Computational Mathematics 58,
https://doi.org/10.1007/978-3-031-09951-9_8

In this chapter we assume that its inverse always exists and is well-defined, and denote it by Φ_j^{-1}, and for vectors by $\Phi^{-1} = (\Phi_1^{-1}, \ldots, \Phi_d^{-1})$, where $\Phi^{-1}(\boldsymbol{x}) = (\Phi_1^{-1}(x_1), \ldots, \Phi_d^{-1}(x_d))$ for $\boldsymbol{x} = (x_1, \ldots, x_d)$ in $[0, 1]^d$. (Technically, $\boldsymbol{x}$ may be in $[0, 1]^d$, $(0, 1]^d$, $[0, 1)^d$, or $(0, 1)^d$, depending on whether D is a closed interval, half line or the whole real line. To simplify the notation, we will ignore this technicality in the following.) The integral (8.1) can then be mapped to the unit cube $[0, 1]^d$,

$$\int_{D^d} f(\boldsymbol{x})\phi(\boldsymbol{x})\, \mathrm{d}\boldsymbol{x} = \int_{[0,1]^d} f(\Phi^{-1}(\boldsymbol{x}))\, \mathrm{d}\boldsymbol{x}.$$

In the case of a standard Gaussian density

$$\phi(\boldsymbol{x}) = \prod_{j=1}^{d} \frac{1}{\sqrt{2\pi}}\, \mathrm{e}^{-x_j^2/2} = \frac{1}{(2\pi)^{d/2}}\, \mathrm{e}^{-\|\boldsymbol{x}\|_2^2/2},$$

the integrand becomes

$$\int_{\mathbb{R}^d} f(\boldsymbol{x}) \prod_{j=1}^{d} \frac{1}{\sqrt{2\pi}}\, \mathrm{e}^{-x_j^2/2}\, \mathrm{d}\boldsymbol{x} = \int_{[0,1]^d} f(\Phi^{-1}(\boldsymbol{x}))\, \mathrm{d}\boldsymbol{x},$$

where, in this instance, Φ^{-1} is the inverse of the standard normal cumulative distribution function.

Hence to approximate such integrals one can use a quadrature rule over the unit cube

$$\int_{D^d} f(\boldsymbol{x})\phi(\boldsymbol{x})\, \mathrm{d}\boldsymbol{x} \approx \frac{1}{N} \sum_{k=0}^{N-1} f(\Phi^{-1}(\boldsymbol{x}_k)).$$

The right-hand side can be interpreted in at least two ways. One is to consider it as a quadrature rule with quadrature points $\Phi^{-1}(\boldsymbol{x}_k)$, $k \in \{0, 1, \ldots, N-1\}$, where the $\boldsymbol{x}_k$, $k \in \{0, 1, \ldots, N-1\}$, are i.i.d. points uniformly distributed on $[0, 1]^d$. The transformed points $\boldsymbol{z}_k = \Phi^{-1}(\boldsymbol{x}_k)$, $k \in \{0, 1, \ldots, N-1\}$ are then i.i.d. random variables with law ϕ. In this interpretation, the idea is to generate points $\boldsymbol{z}_k$ with the correct distribution and then approximate the expectation $\mathbb{E}[f] = \int_{D^d} f(\boldsymbol{x})\phi(\boldsymbol{x})\, \mathrm{d}\boldsymbol{x}$ with the Monte Carlo estimator $(1/N)\sum_{k=0}^{N-1} f(\boldsymbol{z}_k)$. This is often the standard way of thinking in Monte Carlo methods. A similar approach also works for probability density functions not of product form, using for instance the Rosenblatt transform to map the points from the unit cube to the target domain.

The second interpretation is where we consider (8.1) as an integration problem over the unit cube $[0, 1]^d$. In this case the quadrature points are $\boldsymbol{x}_k$, $k \in \{0, 1, \ldots, N-1\}$, and the integrand is given by the composition $f(\Phi^{-1}) := f \circ \Phi^{-1}$. This interpretation is usually adopted in QMC, as the quadrature rules are naturally defined over $[0, 1]^d$. The difficulty is now that, due to the transformation, the integrand $f(\Phi^{-1})$ often has particular features and the standard function space setting (as, e.g., in the Korobov or the Sobolev space) usually does not apply. This is in particular true when the probability density is a standard multivariate normal density. It is therefore necessary to consider a new function space setting. We describe such an approach in the following.

8.2 Function Space Setting

Let again $D = \overline{(a, b)}$, and let $\phi : D \to \mathbb{R}$ be a probability density function with $\phi \geq 0$ and $\int_D \phi(x)\,\mathrm{d}x = 1$. The cumulative distribution function is $\Phi(x) = \int_a^x \phi(y)\,\mathrm{d}y$ for $x \in D$ and its inverse is denoted by $\Phi^{-1} : [0, 1] \to D$.

Let $f : D^d \to \mathbb{R}$ denote the integrand function. We are interested in approximating the integral

$$I_{d,\phi}(f) = \int_{D^d} f(\boldsymbol{x}) \prod_{j=1}^{d} \phi(x_j)\,\mathrm{d}\boldsymbol{x}.$$

We assume that f belongs to a reproducing kernel Hilbert space $\mathcal{F}$ with reproducing kernel $K_{\mathcal{F}} : D^d \times D^d \to \mathbb{R}$, inner product $\langle\cdot,\cdot\rangle_{\mathcal{F}}$, and corresponding norm $\|\cdot\|_{\mathcal{F}}$ (see Definition 1.17 and Proposition 1.18). The analysis of the worst-case error of numerical integration of functions in $\mathcal{F}$ is in many ways analogous to the integration problem over $[0, 1]^d$.

We assume that

$$\int_{D^d} \sqrt{K_{\mathcal{F}}(\boldsymbol{x}, \boldsymbol{x})} \prod_{j=1}^{d} \phi(x_j)\,\mathrm{d}\boldsymbol{x} < \infty,$$

which implies that the integration functional is bounded, since for any $f \in \mathcal{F}$ with $\|f\|_{\mathcal{F}} < \infty$ we have

$$\begin{aligned} |I_{d,\phi}(f)| &\leq \int_{D^d} \|f\|_{\mathcal{F}} \|K_{\mathcal{F}}(\cdot, \boldsymbol{x})\|_{\mathcal{F}} \prod_{j=1}^{d} \phi(x_j)\,\mathrm{d}\boldsymbol{x} \\ &= \|f\|_{\mathcal{F}} \int_{D^d} \sqrt{K_{\mathcal{F}}(\boldsymbol{x}, \boldsymbol{x})} \prod_{j=1}^{d} \phi(x_j)\,\mathrm{d}\boldsymbol{x} \\ &< \infty. \end{aligned}$$

For $f \in \mathcal{F}$ we can use the reproducing property $f(\boldsymbol{x}) = \langle f, K_{\mathcal{F}}(\cdot, \boldsymbol{x})\rangle_{\mathcal{F}}$ for $\boldsymbol{x} \in D^d$ to obtain

$$\begin{aligned} I_{d,\phi}(f) &= \int_{D^d} \langle f, K_{\mathcal{F}}(\cdot, \boldsymbol{x})\rangle_{\mathcal{F}} \prod_{j=1}^{d} \phi(x_j)\,\mathrm{d}\boldsymbol{x} \\ &= \left\langle f, \int_{D^d} K_{\mathcal{F}}(\cdot, \boldsymbol{x}) \prod_{j=1}^{d} \phi(x_j)\,\mathrm{d}\boldsymbol{x} \right\rangle_{\mathcal{F}} = \langle f, h\rangle_{\mathcal{F}}, \end{aligned}$$

where

$$h(\boldsymbol{y}) := \int_{D^d} K_{\mathcal{F}}(\boldsymbol{y},\boldsymbol{x}) \prod_{j=1}^{d} \phi(x_j)\,\mathrm{d}\boldsymbol{x} \in \mathcal{F}$$

is the representer of the integration functional (the existence of such a function is guaranteed by the Fréchet–Riesz representation theorem since integration is a bounded linear functional; see Section 1.5).

The initial error of integration is

$$\begin{aligned}
\mathrm{err}_{0,d}(\mathcal{F}) &:= \sup_{\substack{f\in\mathcal{F}\\ \|f\|_{\mathcal{F}}\le 1}} |I_{d,\phi}(f)| \\
&= \sqrt{\langle h, h\rangle_{\mathcal{F}}} \\
&= \left(\int_{D^d}\int_{D^d} K_{\mathcal{F}}(\boldsymbol{x},\boldsymbol{y}) \prod_{j=1}^{d} (\phi(x_j)\phi(y_j))\,\mathrm{d}\boldsymbol{x}\,\mathrm{d}\boldsymbol{y}\right)^{1/2} \\
&< \infty,
\end{aligned}$$

where the boundedness follows from the fact that $h \in \mathcal{F}$. The worst-case error can be written as (see Theorem 1.27)

$$\begin{aligned}
[\mathrm{err}_{N,d}(\mathcal{F},\mathcal{P})]^2 &= \int_{D^d}\int_{D^d} K_{\mathcal{F}}(\boldsymbol{x},\boldsymbol{y}) \prod_{j=1}^{d} (\phi(x_j)\phi(y_j))\,\mathrm{d}\boldsymbol{x}\,\mathrm{d}\boldsymbol{y} \\
&\quad -\frac{2}{N}\sum_{k=0}^{N-1}\int_{D^d} K_{\mathcal{F}}(\boldsymbol{z}_k,\boldsymbol{x}) \prod_{j=1}^{d} \phi(x_j)\,\mathrm{d}\boldsymbol{x} + \frac{1}{N^2}\sum_{k,\ell=0}^{N-1} K_{\mathcal{F}}(\boldsymbol{z}_k,\boldsymbol{z}_\ell),
\end{aligned}$$

using the quadrature points $\mathcal{P} = \{\boldsymbol{z}_0, \boldsymbol{z}_1, \ldots, \boldsymbol{z}_{N-1}\}$ in D^d.

As before, we use the inverse cumulative distribution function Φ^{-1} to transform points from the unit cube to D^d (we will consider concrete examples below; for now we just assume that the inverse cumulative distribution function exists and is well-defined). We define the space of functions $\mathcal{G}$ given by

$$\mathcal{G} = \{g : [0,1]^d \to \mathbb{R} : g = f(\Phi^{-1}), f \in \mathcal{F}\},$$

and an inner product on $\mathcal{G}$ in the following way: for $g_1, g_2 \in \mathcal{G}$ let

$$\langle g_1, g_2\rangle_{\mathcal{G}} = \langle g_1(\Phi), g_2(\Phi)\rangle_{\mathcal{F}}.$$

Then for any function $g \in \mathcal{G}$ we have

$$\begin{aligned}
g(\boldsymbol{x}) &= f(\Phi^{-1}(\boldsymbol{x})) \\
&= \langle f, K_{\mathcal{F}}(\cdot, \Phi^{-1}(\boldsymbol{x}))\rangle_{\mathcal{F}} \\
&= \langle f(\Phi^{-1}), K_{\mathcal{F}}(\Phi^{-1}(\cdot), \Phi^{-1}(\boldsymbol{x}))\rangle_{\mathcal{G}} \\
&= \langle g, K_{\mathcal{G}}(\cdot, \boldsymbol{x})\rangle_{\mathcal{G}},
\end{aligned}$$

where

$$K_{\mathcal{G}}(\boldsymbol{x},\boldsymbol{y}) := K_{\mathcal{F}}(\Phi^{-1}(\boldsymbol{x}),\Phi^{-1}(\boldsymbol{y})) \quad \text{for } \boldsymbol{x},\boldsymbol{y} \in [0,1]^d.$$

The transformed kernel $K_{\mathcal{G}}$ is a reproducing kernel and is in fact the reproducing kernel for the space $\mathcal{G}$. The initial error in $\mathcal{G}$ is

$$\mathrm{err}_{0,d}(\mathcal{G}) = \sup_{\substack{g\in\mathcal{G}\\ \|g\|_{\mathcal{G}}\le 1}} |I(g)| = \sup_{\substack{f\in\mathcal{F}\\ \|f\|_{\mathcal{F}}\le 1}} |I_{d,\phi}(f)| = \mathrm{err}_{0,d}(\mathcal{F}),$$

where the norm $\|\cdot\|_{\mathcal{G}}$ is the norm induced by $\langle\cdot,\cdot\rangle_{\mathcal{G}}$, and where $I(g) = \int_{[0,1]^d} g(\boldsymbol{x})\,\mathrm{d}\boldsymbol{x}$. For $\mathcal{P}_{\mathcal{G}} = \{\boldsymbol{x}_0,\boldsymbol{x}_1,\ldots,\boldsymbol{x}_{N-1}\}$ in $[0,1]^d$ let $z_k = \Phi^{-1}(\boldsymbol{x}_k)$ for $k \in \{0,1,\ldots,N-1\}$, and let $\mathcal{P}_{\mathcal{F}} = \{z_0,z_1,\ldots,z_{N-1}\}$ in D^d. Then the squared worst-case error of the QMC rule using $\mathcal{P}_{\mathcal{G}}$ in $\mathcal{G}$ is

$$\begin{aligned}
[\mathrm{err}_{N,d}(\mathcal{G},\mathcal{P}_{\mathcal{G}})]^2 &= \int_{[0,1]^d}\int_{[0,1]^d} K_{\mathcal{G}}(\boldsymbol{x},\boldsymbol{y})\,\mathrm{d}\boldsymbol{x}\,\mathrm{d}\boldsymbol{y}\\
&\quad -\frac{2}{N}\sum_{k=0}^{N-1}\int_{[0,1]^d} K_{\mathcal{G}}(\boldsymbol{x}_k,\boldsymbol{x})\,\mathrm{d}\boldsymbol{x} + \frac{1}{N^2}\sum_{k,\ell=0}^{N-1} K_{\mathcal{G}}(\boldsymbol{x}_k,\boldsymbol{x}_\ell)\\
&= \int_{D^d}\int_{D^d} K_{\mathcal{F}}(\boldsymbol{x},\boldsymbol{y})\prod_{j=1}^d(\phi(x_j)\phi(y_j))\,\mathrm{d}\boldsymbol{x}\,\mathrm{d}\boldsymbol{y}\\
&\quad -\frac{2}{N}\sum_{k=0}^{N-1}\int_{D^d} K_{\mathcal{F}}(z_k,\boldsymbol{x})\prod_{j=1}^d\phi(x_j)\,\mathrm{d}\boldsymbol{x} + \frac{1}{N^2}\sum_{k,\ell=0}^{N-1} K_{\mathcal{F}}(z_k,z_\ell)\\
&= [\mathrm{err}_{N,d}(\mathcal{F},\mathcal{P}_{\mathcal{F}})]^2.
\end{aligned}$$

The transformation therefore allows us to switch between numerical integration in $\mathcal{F}$ and $\mathcal{G}$. In particular, we have, for any $f \in \mathcal{F}$,

$$\begin{aligned}
&\left|\int_{D^d} f(\boldsymbol{x})\prod_{j=1}^d\phi(x_j)\,\mathrm{d}\boldsymbol{x} - \frac{1}{N}\sum_{k=0}^{N-1} f(z_k)\right|\\
&= \left|\int_{[0,1]^d} f(\Phi^{-1}(\boldsymbol{x}))\,\mathrm{d}\boldsymbol{x} - \frac{1}{N}\sum_{k=0}^{N-1} f(\Phi^{-1}(\boldsymbol{x}_k))\right|\\
&\le \|f\|_{\mathcal{F}}\,\mathrm{err}_{N,d}(\mathcal{F},\mathcal{P}_{\mathcal{F}})\\
&= \|f\|_{\mathcal{F}}\,\mathrm{err}_{N,d}(\mathcal{G},\mathcal{P}_{\mathcal{G}}).
\end{aligned}$$

The latter expression is convenient in the sense that it suffices to study the integration error $\mathrm{err}_{N,d}(\mathcal{G},\mathcal{P}_{\mathcal{G}})$, while the overall error is then still bounded in terms of the norm $\|\cdot\|_{\mathcal{F}}$ of the original space $\mathcal{F}$.

We now have developed the integration theory for generic reproducing kernel Hilbert spaces. In the next section we introduce a particular class of reproducing kernel Hilbert spaces.

8.3 Unanchored Spaces

Since we will consider tensor product spaces later, we first develop the theory for the univariate case only, to simplify the notation. The following result was shown in [191].

Lemma 8.1 *Let the domain $D = \overline{(a,b)}$ be fixed. Let ϕ be a probability density function, let Φ be the corresponding cumulative distribution function, and let $\psi : D \to (0,\infty)$ be a given weight function defined on D. Assume that Φ^{-1} exists and is well-defined and that*

$$\int_a^b \frac{\Phi(t)(1-\Phi(t))}{(\psi(t))^2}\,\mathrm{d}t < \infty. \tag{8.2}$$

Let $\mathcal{F}$ be the space of absolutely continuous functions $f : D \to \mathbb{R}$ such that

$$\int_a^b f(x)\phi(x)\,\mathrm{d}x < \infty \quad \text{and} \quad \int_a^b (f'(x)\psi(x))^2\,\mathrm{d}x < \infty.$$

For functions $f_1, f_2 \in \mathcal{F}$ we define the inner product

$$\langle f_1, f_2\rangle_{\mathcal{F}} := \int_a^b f_1(x)\phi(x)\,\mathrm{d}x \int_a^b f_2(x)\phi(x)\,\mathrm{d}x + \frac{1}{\gamma}\int_a^b f_1'(x) f_2'(x)\psi^2(x)\,\mathrm{d}x$$

for some generic weight $\gamma > 0$.

Then $\mathcal{F}$ is a reproducing kernel Hilbert space with reproducing kernel

$$K_{\mathcal{F},\gamma}(x,y)$$
$$= 1+\gamma\left(\int_a^{\min(x,y)} \frac{\Phi(t)}{(\psi(t))^2}\,\mathrm{d}t + \int_{\max(x,y)}^b \frac{1-\Phi(t)}{(\psi(t))^2}\,\mathrm{d}t - \int_a^b \frac{\Phi(t)(1-\Phi(t))}{(\psi(t))^2}\,\mathrm{d}t\right)$$

for $x, y \in D$.

Proof Assumption (8.2) ensures that $K_{\mathcal{F},\gamma}$ is well-defined.

We check that the reproducing property is satisfied. Indeed we have, for any $x, y \in D$,

$$\begin{aligned}\int_a^b \int_a^{\min(x,y)} \frac{\Phi(t)}{(\psi(t))^2}\,\mathrm{d}t\phi(x)\,\mathrm{d}x &= \int_a^b \frac{\Phi(t)}{(\psi(t))^2}\int_a^b \chi_{[a,\min(x,y))}(t)\phi(x)\,\mathrm{d}x\,\mathrm{d}t \\ &= \int_a^y \frac{\Phi(t)}{(\psi(t))^2}(1-\Phi(t))\,\mathrm{d}t, \end{aligned} \tag{8.3}$$

$$\begin{aligned}\int_a^b \int_{\max(x,y)}^b \frac{1-\Phi(t)}{(\psi(t))^2}\,\mathrm{d}t\phi(x)\,\mathrm{d}x &= \int_a^b \frac{1-\Phi(t)}{(\psi(t))^2}\int_a^b \chi_{(\max(x,y),b]}(t)\phi(x)\,\mathrm{d}x\,\mathrm{d}t \\ &= \int_y^b \frac{1-\Phi(t)}{(\psi(t))^2}\Phi(t)\,\mathrm{d}t, \end{aligned} \tag{8.4}$$

where we remind the reader that χ_A denotes the indicator function of a set A, which implies that

$$\int_a^b K_{\mathcal{F},\gamma}(x,y)\phi(x)\,\mathrm{d}x = 1.$$

Hence the representer of the integration functional is 1.

Further, for $x, y \in D$ satisfying $x < y$ we have

$$\frac{\partial}{\partial x}K_{\mathcal{F},\gamma}(x,y) = \gamma\frac{\Phi(x)}{(\psi(x))^2},$$

and for $x, y \in D$ satisfying $x > y$ we have

$$\frac{\partial}{\partial x}K_{\mathcal{F},\gamma}(x,y) = -\gamma\frac{1-\Phi(x)}{(\psi(x))^2}.$$

Thus, for any $f \in \mathcal{F}$ and any $y \in D$,

$$\begin{aligned}
&\frac{1}{\gamma}\int_a^b f'(x)\frac{\partial}{\partial x}K_{\mathcal{F},\gamma}(x,y)(\psi(x))^2\,\mathrm{d}x \\
&= \int_a^y f'(x)\Phi(x)\,\mathrm{d}x - \int_y^b f'(x)(1-\Phi(x))\,\mathrm{d}x \\
&= \int_a^b f'(x)\Phi(x)\,\mathrm{d}x - \int_y^b f'(x)\,\mathrm{d}x \\
&= f(b) - \int_a^b f(x)\phi(x)\,\mathrm{d}x - f(b) + f(y) \\
&= f(y) - \int_a^b f(x)\phi(x)\,\mathrm{d}x.
\end{aligned}$$

Combining these results we obtain the reproducing property

$$\langle f, K_{\mathcal{F},\gamma}(\cdot,y)\rangle_{\mathcal{F}} = f(y).$$

In particular, we have for $y, z \in D$ with $y < z$, that

$$\begin{aligned}
&\langle K_{\mathcal{F},\gamma}(\cdot,z), K_{\mathcal{F},\gamma}(\cdot,y)\rangle_{\mathcal{F}} \\
&= 1+\gamma\left(\int_a^y \frac{(\Phi(t))^2}{(\psi(t))^2}\,\mathrm{d}t - \int_y^z \frac{\Phi(t)(1-\Phi(t))}{(\psi(t))^2}\,\mathrm{d}t + \int_z^b \frac{(1-\Phi(t))^2}{(\psi(t))^2}\,\mathrm{d}t\right) \\
&= 1+\gamma\left(\int_a^y \frac{(\Phi(t))^2+\Phi(t)(1-\Phi(t))}{(\psi(t))^2}\,\mathrm{d}t + \int_z^b \frac{(1-\Phi(t))^2+\Phi(t)(1-\Phi(t))}{(\psi(t))^2}\,\mathrm{d}t\right. \\
&\quad \left. - \int_a^b \frac{\Phi(t)(1-\Phi(t))}{(\psi(t))^2}\,\mathrm{d}t\right) = K_{\mathcal{F},\gamma}(z,y).
\end{aligned}$$

The same result follows if $z < y$. Hence $K_{\mathcal{F},\gamma}(\cdot, y) \in \mathcal{F}$. □

Some important properties

We now derive some important properties of the unanchored space discussed previously. The representer of integration is given by

$$\int_a^b K_{\mathcal{F},\gamma}(x, y)\phi(x)\,\mathrm{d}x = 1 \quad \text{for any } y \in D,$$

and thus we also have

$$\int_a^b \int_a^b K_{\mathcal{F},\gamma}(x, y)\phi(x)\phi(y)\,\mathrm{d}x\,\mathrm{d}y = 1. \tag{8.5}$$

This implies that, according to Remark 1.41, the initial error is $\mathrm{err}_{0,d}(\mathcal{F}) = 1$.

Further, we require that $\int_a^b \sqrt{K_{\mathcal{F},\gamma}(x,x)}\,\phi(x)\,\mathrm{d}x < \infty$ to ensure that $I_{1,\phi}$ is a bounded linear functional (see Condition (1.17) in Section 1.5). Note that for every $f \in \mathcal{F}$ we have $\|f\|_{\mathcal{F}} < \infty$, which implies $|I_{1,\phi}(f)| < \infty$. We can also show this property directly. Using (8.3) with $y = b$ and (8.4) with $y = a$, we obtain

$$\begin{aligned}\int_a^b \sqrt{K_{\mathcal{F},\gamma}(x,x)}\,\phi(x)\,\mathrm{d}x &\le \left(\int_a^b K_{\mathcal{F},\gamma}(x,x)\,\phi(x)\,\mathrm{d}x\right)^{1/2} \\ &= \left(1 + \gamma \int_a^b \frac{\Phi(t)(1-\Phi(t))}{(\psi(t))^2}\,\mathrm{d}t\right)^{1/2} < \infty. \end{aligned} \tag{8.6}$$

This implies that numerical integration is a bounded linear functional.

We can express the transformed kernel $K_{\mathcal{G},\gamma}$ as defined in Section 8.2, as

$$\begin{aligned} K_{\mathcal{G},\gamma}(x, y) &= K_{\mathcal{F},\gamma}(\Phi^{-1}(x), \Phi^{-1}(y)) \\ &= 1 + \gamma\left(\int_a^{\min(\Phi^{-1}(x),\Phi^{-1}(y))} \frac{\Phi(t)}{(\psi(t))^2}\,\mathrm{d}t \right. \\ &\quad \left. + \int_{\max(\Phi^{-1}(x),\Phi^{-1}(y))}^b \frac{1-\Phi(t)}{(\psi(t))^2}\,\mathrm{d}t - \int_a^b \frac{\Phi(t)(1-\Phi(t))}{(\psi(t))^2}\,\mathrm{d}t\right) \end{aligned}$$

for $x, y \in [0, 1]$.

The kernel $K_{\mathcal{G},\gamma}$ is in general difficult to handle, since, depending on the choices of ϕ and ψ, it may have singularities. One way to proceed is to consider the shift-invariant kernel associated with $K_{\mathcal{G},\gamma}$ to obtain a space based on Fourier series (see Definition 7.4).

8.4 The Shift-Invariant Kernel

The shift-invariant kernel (see Section 7.1) associated with $K_{\mathcal{G},\gamma}$ is given by

$$K^{\mathrm{sh}}_{\mathcal{G},\gamma}(x,y) = \int_0^1 K_{\mathcal{G},\gamma}(\{x+\Delta\},\{y+\Delta\})\,\mathrm{d}\Delta = 1+\gamma\,\Theta(\{x-y\}),$$

for $x, y \in [0,1]$, where, for $z \in [0,1)$,

$$\begin{aligned}\Theta(z) = &\int_0^1 \int_a^{\min(\Phi^{-1}(\{z+\Delta\}),\Phi^{-1}(\Delta))} \frac{\Phi(t)}{(\psi(t))^2}\,\mathrm{d}t\,\mathrm{d}\Delta \\ &+\int_0^1 \int_{\max(\Phi^{-1}(\{z+\Delta\}),\Phi^{-1}(\Delta))}^{b} \frac{1-\Phi(t)}{(\psi(t))^2}\,\mathrm{d}t\,\mathrm{d}\Delta \\ &-\int_a^b \frac{\Phi(t)(1-\Phi(t))}{(\psi(t))^2}\,\mathrm{d}t. \end{aligned} \tag{8.7}$$

We treat each of the terms in (8.7) separately. For the first term in (8.7) we have

$$\begin{aligned}&\int_0^1 \int_a^{\min(\Phi^{-1}(\{z+\Delta\}),\Phi^{-1}(\Delta))} \frac{\Phi(t)}{(\psi(t))^2}\,\mathrm{d}t\,\mathrm{d}\Delta \\ &= \int_a^b \frac{\Phi(t)}{(\psi(t))^2} \int_0^1 \mathcal{X}_{[a,\min(\Phi^{-1}(\{z+\Delta\}),\Phi^{-1}(\Delta)))}(t)\,\mathrm{d}\Delta\,\mathrm{d}t.\end{aligned}$$

Since Φ^{-1} is monotonically nondecreasing, we have

$$\min(\Phi^{-1}(\{z+\Delta\}),\Phi^{-1}(\Delta)) = \Phi^{-1}(\min(\{z+\Delta\},\Delta)),$$

and therefore $t \in [a,\min(\Phi^{-1}(\{z+\Delta\}),\Phi^{-1}(\Delta)))$ is equivalent to $\Phi(t) \in [0,\min(\{z+\Delta\},\Delta))$. Consequently,

$$\begin{aligned}\int_0^1 \mathcal{X}_{[a,\min(\Phi^{-1}(\{z+\Delta\}),\Phi^{-1}(\Delta)))}(t)\,\mathrm{d}\Delta &= \int_0^1 \mathcal{X}_{[0,\min(\{z+\Delta\},\Delta))}(\Phi(t))\,\mathrm{d}\Delta \\ &= \lambda(\{\Delta\in[0,1] : \Phi(t) < \min(\{z+\Delta\},\Delta)\}),\end{aligned}$$

and this expression is the length (i.e., the Lebesgue measure λ) of the set $\{\Delta \in [0,1] : \Phi(t) < \min(\{z+\Delta\},\Delta)\}$, which is a union of disjoint intervals. When $\Delta \in [0,\Phi(t)]$ the inequality $\Phi(t) < \min(\{z+\Delta\},\Delta)$ cannot be satisfied. When $\Delta \in (\Phi(t), 1-z]$, the inequality is satisfied (provided that $\Phi(t) < 1-z$). When $\Delta \in (1-z, 1-z+\Phi(t)]$, the inequality is not satisfied. When $\Delta \in (1-z+\Phi(t), 1]$, the inequality is again satisfied (provided that $1-z+\Phi(t) < 1$). Thus,

$$\begin{aligned}&\int_0^1 \mathcal{X}_{[a,\min(\Phi^{-1}(\{z+\Delta\}),\Phi^{-1}(\Delta)))}(t)\,\mathrm{d}\Delta \\ &= \max(0, 1-z-\Phi(t)) + \max(0, 1-1+z-\Phi(t))\end{aligned}$$

$$= \max(0, 1 - z - \Phi(t)) + \max(0, z - \Phi(t)).$$

Therefore

$$\int_0^1 \int_a^{\min(\Phi^{-1}(\{z+\Delta\}),\Phi^{-1}(\Delta))} \frac{\Phi(t)}{(\psi(t))^2} \, \mathrm{d}t \, \mathrm{d}\Delta$$
$$= \int_a^b \frac{\Phi(t)}{(\psi(t))^2} \left(\max(0, 1 - z - \Phi(t)) + \max(0, z - \Phi(t))\right) \mathrm{d}t. \tag{8.8}$$

We can proceed in a similar manner for the second term in (8.7),

$$\int_0^1 \int_{\max(\Phi^{-1}(\{z+\Delta\}),\Phi^{-1}(\Delta))}^b \frac{1 - \Phi(t)}{(\psi(t))^2} \, \mathrm{d}t \, \mathrm{d}\Delta$$
$$= \int_a^b \frac{1 - \Phi(t)}{(\psi(t))^2} \int_0^1 \chi_{(\max(\Phi^{-1}(\{z+\Delta\},\Delta)),b]}(t) \, \mathrm{d}\Delta \, \mathrm{d}t$$
$$= \int_a^b \frac{1 - \Phi(t)}{(\psi(t))^2} \int_0^1 \chi_{(\max(\{z+\Delta\},\Delta),1]}(\Phi(t)) \, \mathrm{d}\Delta \, \mathrm{d}t$$
$$= \int_a^b \frac{1 - \Phi(t)}{(\psi(t))^2} \left(\max(0, \Phi(t) - z) + \max(0, z - (1 - \Phi(t)))\right) \mathrm{d}t. \tag{8.9}$$

Therefore, combining (8.8) and (8.9),

$$\Theta(z) = \int_a^b \frac{\Phi(t)}{(\psi(t))^2} \left(\max(0, 1 - z - \Phi(t)) + \max(0, z - \Phi(t))\right) \mathrm{d}t$$
$$+ \int_a^b \frac{1 - \Phi(t)}{(\psi(t))^2} \left(\max(0, \Phi(t) - z) + \max(0, z - (1 - \Phi(t)))\right) \mathrm{d}t$$
$$- \int_a^b \frac{\Phi(t)(1 - \Phi(t))}{(\psi(t))^2} \, \mathrm{d}t. \tag{8.10}$$

Fourier coefficients

In the next step, we investigate under which assumptions the function Θ can be expressed as a Fourier series

$$\Theta(z) = \sum_{h \in \mathbb{Z}} \widehat{\Theta}(h) \, \mathrm{e}^{2\pi \mathrm{i} h z},$$

with Fourier coefficients

$$\widehat{\Theta}(h) = \int_0^1 \Theta(z) \, \mathrm{e}^{-2\pi \mathrm{i} h z} \, \mathrm{d}z.$$

In this case then, since the kernel $K^{\mathrm{sh}}_{\mathcal{G},\gamma}$ with $K^{\mathrm{sh}}_{\mathcal{G},\gamma}(x, y) = 1 + \gamma\,\Theta(\{x - y\})$ for $x, y \in [0, 1]$ is a reproducing kernel with a similar structure as that of a Korobov space, the convergence rate of numerical integration then comes down solely to the decay rate of the Fourier coefficients $\widehat{\Theta}(h)$.

Lemma 8.2 *Let the domain $D = \overline{(a, b)}$ be fixed. Let $\phi : D \to \mathbb{R}$ be a probability density function and let $\Phi : D \to \mathbb{R}$ be the associated cumulative probability density function. Assume that the inverse $\Phi^{-1} : [0, 1] \to D$ exists and is well-defined. Let $\psi : D \to (0, \infty)$. Assume that*

$$\int_a^b \frac{\Phi(t)(1 - \Phi(t))}{(\psi(t))^2}\,\mathrm{d}t < \infty. \tag{8.11}$$

Then the Fourier coefficients of Θ, the function given by (8.10), *are well-defined and given by*

$$\widehat{\Theta}(0) = 0,$$

and, for $h \in \mathbb{N}$,

$$\begin{aligned}\widehat{\Theta}(h) = \widehat{\Theta}(-h) &= \frac{1}{\pi^2 h^2} \int_a^b \frac{\sin^2(\pi h \Phi(t))}{(\psi(t))^2}\,\mathrm{d}t \\ &= \frac{1}{\pi^2 h^2} \int_0^1 \frac{\sin^2(\pi h u)}{\phi(\Phi^{-1}(u))(\psi(\Phi^{-1}(u)))^2}\,\mathrm{d}u.\end{aligned} \tag{8.12}$$

Proof If condition (8.11) is satisfied, then (8.10) implies that Θ is well-defined. For $\widehat{\Theta}(0)$ we have

$$\begin{aligned}\widehat{\Theta}(0) &= \int_0^1 \Theta(z)\,\mathrm{d}z \\ &= \int_a^b \frac{\Phi(t)}{(\psi(t))^2} \int_0^1 (\max(0, 1 - z - \Phi(t)) + \max(0, z - \Phi(t)))\,\mathrm{d}z\,\mathrm{d}t \\ &\quad + \int_a^b \frac{1 - \Phi(t)}{(\psi(t))^2} \int_0^1 (\max(0, \Phi(t) - z) + \max(0, z - (1 - \Phi(t))))\,\mathrm{d}z\,\mathrm{d}t \\ &\quad - \int_a^b \frac{\Phi(t)(1 - \Phi(t))}{(\psi(t))^2}\,\mathrm{d}t \\ &= \int_a^b \frac{\Phi(t)(1 - \Phi(t))^2}{(\psi(t))^2}\,\mathrm{d}t + \int_a^b \frac{(1 - \Phi(t))(\Phi(t))^2}{(\psi(t))^2}\,\mathrm{d}t \\ &\quad - \int_a^b \frac{\Phi(t)(1 - \Phi(t))}{(\psi(t))^2}\,\mathrm{d}t \\ &= 0.\end{aligned}$$

We now consider $h \in \mathbb{Z} \setminus \{0\}$. The Fourier coefficients $\widehat{\Theta}(h)$ are given by

$$\begin{aligned}
&\widehat{\Theta}(h) \\
&= \int_a^b \frac{\Phi(t)}{(\psi(t))^2} \int_0^1 \mathrm{e}^{-2\pi \mathrm{i} hz} \left(\max(0, 1 - z - \Phi(t)) + \max(0, z - \Phi(t))\right) \mathrm{d}z \, \mathrm{d}t \\
&+ \int_a^b \frac{1 - \Phi(t)}{(\psi(t))^2} \int_0^1 \mathrm{e}^{-2\pi \mathrm{i} hz} \left(\max(0, \Phi(t) - z) + \max(0, \Phi(t) - (1 - z))\right) \mathrm{d}z \, \mathrm{d}t \\
&- \int_0^1 \mathrm{e}^{-2\pi \mathrm{i} hz} \, \mathrm{d}z \int_a^b \frac{\Phi(t)(1 - \Phi(t))}{(\psi(t))^2} \, \mathrm{d}t. \qquad (8.13)
\end{aligned}$$

Since $h \in \mathbb{Z} \setminus \{0\}$ we have $\int_0^1 \mathrm{e}^{-2\pi \mathrm{i} hz} \, \mathrm{d}z = 0$, so the third term in (8.13) vanishes. We consider the remaining terms appearing in (8.13) separately. The inner integral in the first term can be simplified using integration by parts,

$$\begin{aligned}
&\int_0^1 \mathrm{e}^{-2\pi \mathrm{i} hz} \left(\max(0, 1 - z - \Phi(t)) + \max(0, z - \Phi(t))\right) \mathrm{d}z \\
&= \int_0^{1-\Phi(t)} \mathrm{e}^{-2\pi \mathrm{i} hz} (1 - z - \Phi(t)) \, \mathrm{d}z + \int_{\Phi(t)}^1 \mathrm{e}^{-2\pi \mathrm{i} hz} (z - \Phi(t)) \, \mathrm{d}z \\
&= \frac{1 - \Phi(t)}{2\pi \mathrm{i} h} - \frac{\mathrm{e}^{2\pi \mathrm{i} h\Phi(t)} - 1}{(2\pi h)^2} - \frac{1 - \Phi(t)}{2\pi \mathrm{i} h} - \frac{\mathrm{e}^{-2\pi \mathrm{i} h\Phi(t)} - 1}{(2\pi h)^2} \\
&= \frac{\sin^2(\pi h \Phi(t))}{(\pi h)^2}.
\end{aligned}$$

The inner integral of the second term can be simplified in a similar manner using integration by parts,

$$\begin{aligned}
&\int_0^1 \mathrm{e}^{-2\pi \mathrm{i} hz} \left(\max(0, \Phi(t) - z) + \max(0, z - (1 - \Phi(t)))\right) \mathrm{d}z \\
&= \int_0^{\Phi(t)} \mathrm{e}^{-2\pi \mathrm{i} hz} (\Phi(t) - z) \, \mathrm{d}z + \int_{1-\Phi(t)}^1 \mathrm{e}^{-2\pi \mathrm{i} hz} (z - 1 + \Phi(t)) \, \mathrm{d}z \\
&= \frac{\Phi(t)}{2\pi \mathrm{i} h} - \frac{\mathrm{e}^{-2\pi \mathrm{i} h\Phi(t)} - 1}{(2\pi h)^2} - \frac{\Phi(t)}{2\pi \mathrm{i} h} - \frac{\mathrm{e}^{2\pi \mathrm{i} h\Phi(t)} - 1}{(2\pi h)^2} \\
&= \frac{\sin^2(\pi h \Phi(t))}{(\pi h)^2}.
\end{aligned}$$

Substituting these results into (8.13), and using the transformation $u = \Phi(t)$ yields the result for $h \in \mathbb{Z} \setminus \{0\}$. □

The decay rate of the Fourier coefficients

We have seen that (8.11) implies that the Fourier coefficients of the function Θ are well-defined. In order for Θ to be represented pointwise by a Fourier series, we need to ensure that the Fourier coefficients $\widehat{\Theta}(h)$ decay fast enough. This depends on the particular choices of Φ and ψ. We discuss several combinations of Φ and ψ for which Θ can indeed be represented pointwise by a Fourier series. This is the case for instance if the function Θ is continuous and the Fourier coefficients satisfy $|\widehat{\Theta}(h)| < C_\delta |h|^{-1-\delta}$ for some constant $C_\delta > 0$ independent of h and $\delta > 0$.

When Θ can be represented pointwise by a Fourier series, we obtain that $\mathcal{G}$ is a space of Fourier series, similar to a Korobov space.

The following lemma will be useful to obtain bounds on the decay rate of the Fourier coefficients $\widehat{\Theta}$.

Lemma 8.3 *For $h \in \mathbb{N}$ and $a \in (0, 1)$, we have*

$$\int_0^{1/2} \frac{\sin^2(\pi h u)}{u^{1+a}} \, \mathrm{d}u \le h^a \frac{2\pi^a}{(2-a)a}.$$

Proof Using the substitution $t = hu$, replacing the upper limit $h/2$ by ∞, and employing the estimates $\sin^2(\pi t) \le (\pi t)^2$ for $0 \le t \le 1/\pi$ and $\sin^2(\pi t) \le 1$ for $t \ge 1/\pi$, we obtain

$$\begin{aligned}
\int_0^{1/2} \frac{\sin^2(\pi h u)}{u^{1+a}} \, \mathrm{d}u &= h^a \int_0^{h/2} \frac{\sin^2(\pi t)}{t^{1+a}} \, \mathrm{d}t \\
&\le h^a \left(\pi^2 \int_0^{1/\pi} t^{1-a} \, \mathrm{d}t + \int_{1/\pi}^{\infty} t^{-1-a} \, \mathrm{d}t \right) \\
&= h^a \frac{2\pi^a}{(2-a)a}.
\end{aligned}$$

□

As we will consider some examples involving the normal distribution, we will collect some useful inequalities regarding this distribution first.

Useful inequalities regarding the normal distribution

The probability density function of a normal distribution with mean 0 and variance σ^2 is given by

$$\phi_\sigma(t) = \frac{1}{\sigma\sqrt{2\pi}} \, \mathrm{e}^{-t^2/(2\sigma^2)} \quad \text{for } t \in \mathbb{R}, \tag{8.14}$$

and the cumulative distribution function is given by

$$\Phi_\sigma(t) = \int_{-\infty}^{t} \phi_\sigma(x) \, \mathrm{d}x \quad \text{for } t \in \mathbb{R}.$$

The main difficulty in estimating the Fourier coefficients $\widehat{\Theta}(h)$ in (8.12) arises from the term $1/\phi(\Phi^{-1}(u))$. This term tends to infinity as $u \to 0^+$.

For the cumulative normal distribution function, for $t < 0$, we have the tail estimate

$$\begin{aligned}\Phi_\sigma(t) &= \frac{1}{\sigma\sqrt{2\pi}} \int_{-\infty}^{t} \mathrm{e}^{-x^2/(2\sigma^2)}\,\mathrm{d}x \\ &< \frac{1}{\sigma\sqrt{2\pi}} \int_{-\infty}^{t} \frac{x}{t}\,\mathrm{e}^{-x^2/(2\sigma^2)}\,\mathrm{d}x \\ &= -\frac{\sigma}{t\sqrt{2\pi}}\,\mathrm{e}^{-t^2/(2\sigma^2)}. \end{aligned} \tag{8.15}$$

Similarly, for $t > 0$, we also have

$$1 - \Phi_\sigma(t) < \frac{\sigma}{t\sqrt{2\pi}}\,\mathrm{e}^{-t^2/(2\sigma^2)}. \tag{8.16}$$

We claim that for all $u \in (0, 1/2)$,

$$\mathrm{e}^{(\Phi_\sigma^{-1}(u))^2/(2\sigma^2)} \le \frac{1}{u}.$$

This claim is equivalent to

$$-\Phi_\sigma^{-1}(u) \le \sqrt{-2\sigma^2 \log u}\,, \tag{8.17}$$

which in turn is equivalent to

$$u \ge \Phi_\sigma\left(-\sqrt{-2\sigma^2 \log u}\right),$$

where we used that $\Phi_\sigma^{-1}(u) < 0$ for $u \in (0, 1/2)$. This inequality, however, holds true since

$$\Phi_\sigma\left(-\sqrt{-2\sigma^2 \log u}\right) \le \frac{\sigma\,\mathrm{e}^{\log u}}{\sqrt{-4\pi\sigma^2 \log u}} \le \frac{u}{\sqrt{4\pi \log(1/u)}} \le u,$$

for all $u \in (0, 1/2)$, where we used (8.15).

In particular, (8.17) implies that for $u \in (0, 1/2)$ we have

$$\mathrm{e}^{-\Phi_\sigma^{-1}(u)} \le \mathrm{e}^{\sqrt{-2\sigma^2 \log u}}.$$

For $\delta > 0$ we have $\mathrm{e}^{\sqrt{-2\sigma^2 \log u}} \le C_\delta u^{-\delta}$ for $u \in (0, 1/2)$, where

$$C_\delta = \sup_{u \in (0,1/2)} \mathrm{e}^{\delta \log u + \sqrt{-2\sigma^2 \log u}}$$

$$
\begin{aligned}
&= \mathrm{e}^{\sup_{u\in(0,1/2)}\left(\delta \log u+\sqrt{-2\sigma^2 \log u}\right)} \\
&\le \mathrm{e}^{\sigma^2/(2\delta)}.
\end{aligned}
$$

Therefore, for any $\delta > 0$ and $u \in (0, 1/2)$, we have

$$
\mathrm{e}^{-\Phi_\sigma^{-1}(u)} \le \mathrm{e}^{\sigma^2/(2\delta)} u^{-\delta}.
$$

Normal distribution and exponential decay

We consider the case where $\phi = \phi_\sigma$ is the density function of a normal distribution as in (8.14) and ψ is given by

$$
\psi_\beta(t) = \mathrm{e}^{-|t|/(2\beta)} \quad \text{for } t \in \mathbb{R},
$$

where β is a positive real. Using (8.15), we obtain that the condition (8.11) is satisfied for any $\beta, \sigma > 0$, as

$$
\int_{-\infty}^{\infty} \Phi_\sigma(t)(1 - \Phi_\sigma(t))\, \mathrm{e}^{|t|/\beta}\, \mathrm{d}t < \infty.
$$

Using Lemmas 8.2 and 8.3 we obtain for $h \in \mathbb{Z} \setminus \{0\}$ and $\delta \in (0, 1)$ that

$$
\begin{aligned}
\pi^2 h^2\, \widehat{\Theta}(h) &= 2 \int_0^{1/2} \frac{\sin^2(\pi h u)}{\mathrm{e}^{\Phi_\sigma^{-1}(u)/\beta} \phi_\sigma(\Phi_\sigma^{-1}(u))}\, \mathrm{d}u \\
&= 2\sigma\sqrt{2\pi} \int_0^{1/2} \sin^2(\pi h u)\, \mathrm{e}^{(\Phi_\sigma^{-1}(u))^2/(2\sigma^2) - \Phi_\sigma^{-1}(u)/\beta}\, \mathrm{d}u \\
&\le 2\sigma\sqrt{2\pi}\, \mathrm{e}^{\sigma^2/(2\delta\beta)} \int_0^{1/2} \frac{\sin^2(\pi h u)}{u^{1+\delta/\beta}}\, \mathrm{d}u \\
&\le 2\sigma\sqrt{2\pi}\, \mathrm{e}^{\sigma^2/(2\delta\beta)} |h|^{\delta/\beta} \frac{2\beta\pi^{\delta/\beta}}{(2 - \delta/\beta)\delta}.
\end{aligned}
$$

Hence we obtain that for any $h \in \mathbb{Z} \setminus \{0\}$ and any $\delta \in (0, 1)$ we have

$$
|\widehat{\Theta}(h)| \le \frac{C_{\beta,\delta,\sigma}}{|h|^{2-\delta/\beta}},
$$

where $C_{\beta,\delta,\sigma} := (2/\pi)^{3/2}\sigma \mathrm{e}^{\sigma^2/(2\delta\beta)} 2\beta\pi^{\delta/\beta}/((2 - \delta/\beta)\delta)$.

Thus, for instance, for $\beta \in (0, 2)$ and $\delta = \beta/2$ we obtain that $|\widehat{\Theta}(h)| \le C_{\beta,\beta/2,\sigma}|h|^{-3/2}$, which shows that for this combination of ϕ and ψ, Θ can be represented pointwise by a Fourier series.

Two normal distributions

Consider now the case where $\phi = \phi_\sigma$ is a normal distribution as in (8.14) and where

$$\psi(t) = \mathrm{e}^{-t^2/(2\beta^2)} \quad \text{for } t \in \mathbb{R},$$

where we assume that $0 < \sigma < \beta/\sqrt{2}$. Then the inequalities (8.15) and (8.16) imply that the condition (8.11) is satisfied.

Using Lemmas 8.2 and 8.3 we obtain for $h \in \mathbb{Z} \setminus \{0\}$ and $\delta \in (0, 1)$ that

$$\begin{aligned}
\pi^2 h^2 \widehat{\Theta}(h) &= 2 \int_0^{1/2} \frac{\sin^2(\pi h u)}{\mathrm{e}^{-(\Phi_\sigma^{-1}(u))^2/\beta^2} \phi_\sigma(\Phi_\sigma^{-1}(u))} \, \mathrm{d}u \\
&= 2\sigma\sqrt{2\pi} \int_0^{1/2} \sin^2(\pi h u)\, \mathrm{e}^{(\Phi_\sigma^{-1}(u))^2/(2\sigma^2) + (\Phi_\sigma^{-1}(u))^2/\beta^2} \, \mathrm{d}u \\
&= 2\sigma\sqrt{2\pi} \int_0^{1/2} \sin^2(\pi h u)\, \mathrm{e}^{\left((\Phi_\sigma^{-1}(u))^2/(2\sigma^2)\right)\left(1 + (2\sigma^2)/\beta^2\right)} \, \mathrm{d}u \\
&\leq 2\sigma\sqrt{2\pi} \int_0^{1/2} \frac{\sin^2(\pi h u)}{u^{1+2\sigma^2/\beta^2}} \, \mathrm{d}u \\
&\leq 2\sigma\sqrt{2\pi}\, |h|^{2\sigma^2/\beta^2} \frac{\pi^{2\sigma^2/\beta^2}}{2(1 - \sigma^2/\beta^2)\sigma^2/\beta^2}.
\end{aligned}$$

Hence we obtain that for any $h \in \mathbb{Z} \setminus \{0\}$

$$|\widehat{\Theta}(h)| \leq \frac{C_{\beta,\sigma}}{|h|^{2(1-\sigma^2/\beta^2)}},$$

where $C_{\beta,\sigma} := (2/\pi)^{3/2} \pi^{2\sigma^2/\beta^2} \beta^2 / (2(1 - \sigma^2/\beta^2)\sigma)$.

Since $\sigma < \beta/\sqrt{2}$ ensures that $2(1 - \sigma^2/\beta^2) > 1$, we obtain that Θ can be represented pointwise by a Fourier series.

Logistic distribution and exponential function

We now consider the logistic distribution with density

$$\phi_\nu(t) = \frac{\mathrm{e}^{t/\nu}}{\nu(1 + \mathrm{e}^{t/\nu})^2} \quad \text{for } t \in \mathbb{R}$$

with parameter $\nu > 0$, and put $\phi = \phi_\nu$, and the exponential function

$$\psi(t) = \mathrm{e}^{-|t|/\beta} \quad \text{for } t \in \mathbb{R},$$

with parameter $\beta > 0$. We assume that $\beta > 2\nu$.

The cumulative distribution function of the logistic distribution is given by

$$\Phi_\nu(t) = \frac{\mathrm{e}^{t/\nu}}{1+\mathrm{e}^{t/\nu}},$$

and its inverse is given by

$$\Phi_\nu^{-1}(t) = \nu \log\left(\frac{t}{1-t}\right).$$

Combining these, we obtain for $u \in (0, 1/2)$ that

$$\phi_\nu(\Phi_\nu^{-1}(u)) = \frac{u(1-u)}{\nu} \quad \text{and} \quad \psi(\Phi_\nu^{-1}(u)) = \left(\frac{u}{1-u}\right)^{\nu/\beta}.$$

Using Lemma 8.2, we obtain for $h \in \mathbb{Z} \setminus \{0\}$ that

$$\begin{aligned}\pi^2 h^2\, \widehat{\Theta}(h) &= 2\int_0^{1/2} \frac{\sin^2(\pi h u)}{(\psi(\Phi_\nu^{-1}(u)))^2 \phi_\nu(\Phi_\nu^{-1}(u))}\,\mathrm{d}u \\ &= 2\nu \int_0^{1/2} u^{-1-2\nu/\beta}(1-u)^{-1+2\nu/\beta} \sin^2(\pi h u)\,\mathrm{d}u.\end{aligned}$$

Using the fact that $1 - u \geq 1/2$ for $u \in (0, 1/2)$ and Lemma 8.3, we obtain for $h \in \mathbb{Z} \setminus \{0\}$,

$$|\widehat{\Theta}(h)| \leq \frac{2^{2-2\nu/\beta}}{\pi^2 h^2} \int_0^{1/2} \frac{\sin^2(\pi h u)}{u^{1+2\nu/\beta}}\,\mathrm{d}u \leq \frac{C_{\nu,\beta}}{|h|^{2(1-\nu/\beta)}},$$

where $C_{\nu,\beta} := (2/\pi)^{2(1-\nu/\beta)}/((1-\nu/\beta)\nu/\beta)$.

Again, $\beta > 2\nu$ implies that $2(1-\nu/\beta) > 1$ and hence for this combination of ϕ and ν, Θ can be represented pointwise by a Fourier series.

8.5 Integration Error

Next, we consider weighted tensor product spaces based on the univariate unanchored Sobolev space introduced in Section 8.3. Let $\boldsymbol{\gamma} = (\gamma_j)_{j\geq 1}$ be a sequence of product weights. Let $D = \overline{(a, b)}$, as above. The kernel of the weighted tensor product space $\mathcal{F}_d$ is of the form

$$K_{\mathcal{F}_d,\boldsymbol{\gamma}}(\boldsymbol{x}, \boldsymbol{y}) = \prod_{j=1}^{d} K_{\mathcal{F},\gamma_j}(x_j, y_j) \quad \text{for } \boldsymbol{x}, \boldsymbol{y} \in D^d.$$

As for the univariate case, using (8.5), we obtain that the initial error is given by

$$\mathrm{err}_{0,d}(\mathcal{F}_d) = 1.$$

To approximate $I_{d,\phi}(f)$ for $f \in \mathcal{F}_d$ we use the quadrature rule

$$\frac{1}{N}\sum_{k=0}^{N-1} f(\Phi^{-1}(\{\boldsymbol{x}_k + \boldsymbol{\Delta}\})),$$

where $\boldsymbol{\Delta}$ is a random shift whose components are i.i.d. uniformly distributed on $[0, 1]$, and where $\mathcal{P} = \{\boldsymbol{x}_0, \boldsymbol{x}_1, \ldots, \boldsymbol{x}_{N-1}\}$ is an N-element point set in $[0, 1]^d$. The root mean square worst-case error of the randomly shifted point set $\mathcal{P}$ is given by

$$[\mathrm{err}^{\mathrm{sh}}_{N,d}(\mathcal{F}_d, \Phi^{-1}(\mathcal{P}))]^2 = [\mathrm{err}^{\mathrm{sh}}_{N,d}(\mathcal{G}_d, \mathcal{P})]^2 = -1 + \frac{1}{N^2}\sum_{k,\ell=0}^{N-1} K^{\mathrm{sh}}_{\mathcal{G}_d,\boldsymbol{\gamma}}(\boldsymbol{x}_k, \boldsymbol{x}_\ell).$$

If $\mathcal{P} = \mathcal{P}(\boldsymbol{g}, N)$ is a lattice point set, then the latter formula simplifies to

$$[\mathrm{err}^{\mathrm{sh}}_{N,d}(\mathcal{F}_d, \mathcal{P})]^2 = -1 + \frac{1}{N}\sum_{k=0}^{N-1} K^{\mathrm{sh}}_{\mathcal{G}_d,\boldsymbol{\gamma}}(\boldsymbol{x}_k, \boldsymbol{0}).$$

Using the character property of lattice point sets (see Lemma 1.9) we obtain

$$[\mathrm{err}^{\mathrm{sh}}_{N,d}(\mathcal{F}_d, \mathcal{P})]^2 = \sum_{\boldsymbol{h} \in \mathcal{L}^{\perp}(\boldsymbol{g},N)\setminus\{\boldsymbol{0}\}} \prod_{j \in \mathfrak{u}(\boldsymbol{h})} \gamma_j \widehat{\Theta}(h_j),$$

where Θ is the univariate function defined as in (8.10), and where, as in Section 2.4, $\mathfrak{u}(\boldsymbol{h}) := \{j \in [d] \,:\, h_j \neq 0\}$.

As we have seen before, the value of $\widehat{\Theta}(h_j)$, $j \in [d]$, depends on the functions ϕ and ψ. In most instances we do not easily obtain explicit values of the Fourier coefficients. However, they can be estimated numerically (using a one-dimensional integration rule). Using such estimates, one can numerically estimate the values of $\Theta(k/N)$ for $k \in \{0, 1, \ldots, N-1\}$ and use those to get approximations of $K^{\mathrm{sh}}_{\mathcal{F}_d,\boldsymbol{\gamma}}(\boldsymbol{x}_k, \boldsymbol{0})$, for a lattice point set $\{\boldsymbol{x}_0, \boldsymbol{x}_1, \ldots, \boldsymbol{x}_{N-1}\}$. These can then be used in a component-by-component algorithm (see, Sections 3.3 and the following) to obtain good generating vectors for lattice rules.

A bound on the convergence rate can be obtained as in Theorem 7.21. The convergence rate will depend on the decay rate of the coefficients $\widehat{\Theta}(h)$ as $|h| \to \infty$. The proof of the following theorem follows along the same lines as the proof of Theorem 3.7, where the values $|h|^{-2\alpha}$ are replaced by $|\widehat{\Theta}(h)|$ (or an upper bound on $|\widehat{\Theta}(h)|$).

Theorem 8.4 *Let $d \in \mathbb{N}$, and let $D = \overline{(a, b)}$ for some $a, b \in \mathbb{R} \cup \{-\infty, \infty\}$. Let $\phi : D \to \mathbb{R}$ be a probability density function and let $\psi : D \to (0, \infty)$. Assume that (8.11) is satisfied for ϕ and ψ.*

Let $\alpha > 1/2$ be such that

$$|\widehat{\Theta}(h)| \le \frac{C}{|h|^{2\alpha}} \quad \textit{for all } h \in \mathbb{Z} \setminus \{0\}.$$

Let N be a prime number and assume that the generating vector $\boldsymbol{g} \in G_d(N)$ has been constructed using Algorithm 3.6 based on $\mathrm{err}^{\mathrm{sh}}_{N,d}(\mathcal{F}_d, \mathcal{P})$. *Then the root mean square error satisfies, for any $\tau \in [1/2, \alpha)$,*

$$\mathrm{err}^{\mathrm{sh}}_{N,d}(\mathcal{F}_d, \mathcal{P}(\boldsymbol{g}, N)) \le \frac{2^\tau}{N^\tau} \prod_{j=1}^{d} \left(1 + 2\gamma_j^{1/(2\tau)} c\right)^\tau,$$

where $c > 0$ is a constant chosen in a way such that $\sum_{h \in \mathbb{Z}\setminus\{0\}} |\widehat{\Theta}(h)|^{1/(2\tau)} < c$.

Notes and Remarks

Integration over domains other than the unit cube comes up naturally in applications as for instance in PDEs with random coefficients (see Appendix A and the references there), in option pricing in financial mathematics (see, e.g., [78, 173, 257, 258]), or in applications in statistics (see [157]).

Regarding the inner product and norm in the space $\mathcal{G}$ as introduced in Section 8.2, note that the norm in $\mathcal{G}$ is in general given by $\|g\|_{\mathcal{G}} = \inf\{\|f\|_{\mathcal{F}} : f = g(\Phi)\}$. Since we assume Φ to be invertible and Φ^{-1} to be well-defined, it holds that there is only one function $f \in \mathcal{F}$ for which $f = g(\Phi)$, which implies $\|g\|_{\mathcal{G}} = \|f\|_{\mathcal{F}}$ for that f. This property can then also be transferred to the inner product.

The theory of transformed randomly shifted lattice rules was first introduced in [164] and further developed in [161] and [191]. The main part of the theory presented in this chapter stems from those three papers.

In Lemma 8.1 we verified that the function $K_{\mathcal{F},\gamma}$ is the reproducing kernel of the given space $\mathcal{F}$. The approach in [191] is different, as the authors there show how one may derive this kernel from scratch.

As we have seen in Section 8.4, the decay rate of the Fourier coefficients of the shift-invariant kernel depends on Φ and ψ. We have obtained bounds on the decay rate for some common examples. Further examples are given in [191].

Note that it would be no problem to assume in Theorem 8.4 and the other observations in Section 8.5 that there are potentially different functions ϕ_j and ψ_j for each $j \in [d]$, instead of assuming the same ϕ and ψ in each component. This would then lead to potentially different Θ_j for $j \in [d]$ instead of Θ for every component. Since all considerations in this chapter have been outlined for tensor product spaces, such an assumption would cause no difficulties. For the sake of simplicity, we have outlined the results for the simpler case, where all components are based on the same ϕ and ψ.

We also refer to [144] regarding QMC for weighted integration over unbounded domains. Other quasi-Monte Carlo methods for integration over other domains have been discussed in [42, 56, 116, 118, 119].

Chapter 9
Integration of Analytic Functions

In Chapter 2 we introduced the Korobov spaces $\mathcal{H}_{\mathrm{kor},d,\alpha,\gamma}$ and studied numerical integration using lattice rules for these spaces. The parameter α is related to the smoothness of the elements of $\mathcal{H}_{\mathrm{kor},d,\alpha,\gamma}$ (see Propositions 2.2 and 2.4) and is called the smoothness parameter of these spaces. For an optimal integration rule the rate of convergence of the worst-case error is essentially of order $O(N^{-\alpha})$ and therefore reflects the smoothness of the reference space. In the cases considered so far, the smoothness parameter α has always been finite and hence we have observed convergence rates of polynomial order.

In this section we consider numerical integration of functions with infinite smoothness and introduce a Korobov space of periodic functions that are analytic. In this case we can obtain exponential convergence rates, which are optimal. We are also interested in the dependence of the error bounds on the dimension d. In order to investigate this question systematically, we will reinterpret the classical notions of tractability which have been designed for polynomial error convergence rates, and introduce several so-called exponential tractability notions.

9.1 General Korobov Spaces and Korobov Spaces of Analytic Functions

We introduce a general Korobov space as a reproducing kernel Hilbert space with kernel function

$$K_{\mathrm{kor},\rho}(\boldsymbol{x},\boldsymbol{y}) = \sum_{\boldsymbol{h}\in\mathbb{Z}^d} \rho(\boldsymbol{h})\, \mathrm{e}^{2\pi \mathrm{i}\boldsymbol{h}\cdot(\boldsymbol{x}-\boldsymbol{y})} \quad \text{for } \boldsymbol{x},\boldsymbol{y} \in [0,1]^d. \tag{9.1}$$

The nonnegative $\rho(\boldsymbol{h})$ for $\boldsymbol{h} \in \mathbb{Z}^d$, which may also depend on d and on other parameters, are assumed to satisfy the condition

J. Dick et al., *Lattice Rules*, Springer Series in Computational Mathematics 58,
https://doi.org/10.1007/978-3-031-09951-9_9

$$\sum_{\boldsymbol{h}\in\mathbb{Z}^d} \rho(\boldsymbol{h}) < \infty.$$

This assumption implies

$$|K_{\mathrm{kor},\rho}(\boldsymbol{x},\boldsymbol{y})| \le K_{\mathrm{kor},\rho}(\boldsymbol{x},\boldsymbol{x}) = \sum_{\boldsymbol{h}\in\mathbb{Z}^d} \rho(\boldsymbol{h}) < \infty.$$

Obviously, $K_{\mathrm{kor},\rho}$ is conjugate symmetric and positive semi-definite, since for any choice of $n \in \mathbb{N}$, $a_1, a_2, \ldots, a_n \in \mathbb{C}$, and $\boldsymbol{x}_1, \boldsymbol{x}_2, \ldots, \boldsymbol{x}_n \in [0,1]^d$ we have

$$\begin{aligned}
\sum_{i,j=1}^{n} \overline{a}_i a_j K_{\mathrm{kor},\rho}(\boldsymbol{x}_i, \boldsymbol{x}_j) &= \sum_{i,j=1}^{n} \overline{a}_i a_j \sum_{\boldsymbol{h}\in\mathbb{Z}^d} \rho(\boldsymbol{h})\, \mathrm{e}^{2\pi \mathrm{i}\boldsymbol{h}\cdot(\boldsymbol{x}_i-\boldsymbol{x}_j)} \\
&= \sum_{\boldsymbol{h}\in\mathbb{Z}^d} \rho(\boldsymbol{h}) \left(\sum_{i=1}^{n} \overline{a}_i\, \mathrm{e}^{2\pi\mathrm{i}\boldsymbol{h}\cdot\boldsymbol{x}_i}\right)\left(\sum_{j=1}^{n} a_j\, \mathrm{e}^{-2\pi\mathrm{i}\boldsymbol{h}\cdot\boldsymbol{x}_j}\right) \\
&= \sum_{\boldsymbol{h}\in\mathbb{Z}^d} \rho(\boldsymbol{h}) \left|\sum_{j=1}^{n} a_j\, \mathrm{e}^{-2\pi\mathrm{i}\boldsymbol{h}\cdot\boldsymbol{x}_j}\right|^2 \\
&\ge 0.
\end{aligned}$$

Therefore, $K_{\mathrm{kor},\rho}$ is indeed a reproducing kernel.

Definition 9.1 The reproducing kernel Hilbert space $\mathcal{H}_{\mathrm{kor},\rho} = \mathcal{H}(K_{\mathrm{kor},\rho})$ is called a *general Korobov space*. The corresponding inner product is given by

$$\langle f, g\rangle_{\mathrm{kor},\rho} := \sum_{\boldsymbol{h}\in\mathbb{Z}^d} \frac{1}{\rho(\boldsymbol{h})} \widehat{f}(\boldsymbol{h})\overline{\widehat{g}(\boldsymbol{h})},$$

and the norm is

$$\|f\|_{\mathrm{kor},\rho} := \left(\sum_{\boldsymbol{h}\in\mathbb{Z}^d} \frac{1}{\rho(\boldsymbol{h})}\, |\widehat{f}(\boldsymbol{h})|^2\right)^{1/2}.$$

The smoothness of the functions in $\mathcal{H}_{\mathrm{kor},\rho}$ is determined by the decay of $\rho(\boldsymbol{h})$ as $\|\boldsymbol{h}\|_\infty := \max_{j\in[d]} |h_j|$ grows. In the classical case introduced in Chapter 2 we had (see Equation (2.4))

$$\rho(\boldsymbol{h}) = \frac{1}{r_{2\alpha}(\boldsymbol{h})},$$

or, in the weighted case (see Equation (2.19)),

$$\rho(\boldsymbol{h}) = \frac{1}{r_{2\alpha,\gamma}(\boldsymbol{h})}.$$

In these cases $\rho(\boldsymbol{h})$ decays like $\prod_{j=1}^{d} \max(1, |h_j|)^{-2\alpha}$, and this rate implies that the functions in the corresponding Korobov space have finite smoothness of order α (see Section 2.1).

Infinite smoothness

If we would like to study functions of infinite smoothness we need to demand that $\rho(\boldsymbol{h})$ decays exponentially as $\|\boldsymbol{h}\|_\infty$ grows. Also then, we are interested in the possibly different degrees of influence of different components and use, to model this, two weight sequences

$$\boldsymbol{a} = (a_j)_{j \ge 1} \quad \text{and} \quad \boldsymbol{b} = (b_j)_{j \ge 1}.$$

In order to guarantee that the kernel that we are going to study is well-defined, we demand that $a_j > 0$ and $b_j > 0$ for all $j \in \mathbb{N}$. In fact, we assume a little more throughout this chapter, namely that with a proper ordering of variables we have

$$0 < a_1 \le a_2 \le \cdots \quad \text{and} \quad b_* := \inf b_j > 0. \tag{9.2}$$

We also write $a_* := \inf a_j = a_1$.

We now fix $\omega \in (0, 1)$ and put

$$\rho(\boldsymbol{h}) = \rho_{\boldsymbol{a},\boldsymbol{b}}(\boldsymbol{h}) := \omega^{\sum_{j=1}^d a_j |h_j|^{b_j}} \quad \text{for } \boldsymbol{h} = (h_1, \ldots, h_d) \in \mathbb{Z}^d.$$

With this choice we denote the kernel in (9.1) by $K_{\text{kor},d,\boldsymbol{a},\boldsymbol{b}}$. We suppress the dependence on ω in the notation since ω is assumed to be fixed, but $\boldsymbol{a}, \boldsymbol{b}$ can be varied. It is easily seen that $K_{\text{kor},d,\boldsymbol{a},\boldsymbol{b}}$ is well-defined, since

$$\sum_{\boldsymbol{h} \in \mathbb{Z}^d} \rho_{\boldsymbol{a},\boldsymbol{b}}(\boldsymbol{h}) = \prod_{j=1}^d \left(1 + 2 \sum_{h=1}^\infty \omega^{a_j h^{b_j}}\right) \le \left(1 + 2 \sum_{h=1}^\infty \omega^{a_* h^{b_*}}\right)^d < \infty,$$

due to the fact that $a_* > 0$ and $b_* > 0$.

We denote the corresponding reproducing kernel Hilbert space by $\mathcal{H}_{\text{kor},d,\boldsymbol{a},\boldsymbol{b}} = \mathcal{H}(K_{\text{kor},d,\boldsymbol{a},\boldsymbol{b}})$. For functions $f \in \mathcal{H}_{\text{kor},d,\boldsymbol{a},\boldsymbol{b}}$ we have

$$f(\boldsymbol{x}) = \sum_{\boldsymbol{h} \in \mathbb{Z}^d} \widehat{f}(\boldsymbol{h}) \, \mathrm{e}^{2\pi \mathrm{i} \boldsymbol{h} \cdot \boldsymbol{x}} \quad \text{for all } \boldsymbol{x} \in [0, 1]^d,$$

where again $\widehat{f}(\boldsymbol{h})$ is the $\boldsymbol{h}$-th Fourier coefficient of f and the norm of $f \in \mathcal{H}_{\text{kor},d,\boldsymbol{a},\boldsymbol{b}}$ is

$$\|f\|_{\text{kor},d,\boldsymbol{a},\boldsymbol{b}} = \left(\sum_{\boldsymbol{h} \in \mathbb{Z}^d} \omega^{-\sum_{j=1}^d a_j |h_j|^{b_j}} |\widehat{f}(\boldsymbol{h})|^2\right)^{1/2} < \infty.$$

We now show that the functions in $\mathcal{H}_{\text{kor},d,\boldsymbol{a},\boldsymbol{b}}$ are infinitely many times differentiable, and even more, they are analytic.

Theorem 9.2 *The elements of* $\mathcal{H}_{\text{kor},d,\boldsymbol{a},\boldsymbol{b}}$ *are analytic.*

Proof Let $\boldsymbol{a}_*$ and $\boldsymbol{b}_*$ denote the constant sequences $(a_*)_{j \ge 1}$ and $(b_*)_{j \ge 1}$, respectively. Since $\mathcal{H}_{\text{kor},d,\boldsymbol{a},\boldsymbol{b}} \subseteq \mathcal{H}_{\text{kor},d,\boldsymbol{a}_*,\boldsymbol{b}_*}$, it suffices to show the assertion for $f \in \mathcal{H}_{\text{kor},d,\boldsymbol{a}_*,\boldsymbol{b}_*}$.

Let $\boldsymbol{\eta} = (\eta_1, \eta_2, \ldots, \eta_d) \in \mathbb{N}_0^d$ and recall that $\|\boldsymbol{\eta}\|_1 = \eta_1 + \cdots + \eta_d$. For $f \in \mathcal{H}_{\mathrm{kor},d,\boldsymbol{a}_*,\boldsymbol{b}_*}$, consider the operator of partial differentiation,

$$D^{\boldsymbol{\eta}} f = \frac{\partial^{\eta_1+\cdots+\eta_d}}{\partial x_1^{\eta_1}\, \partial x_2^{\eta_2} \ldots \partial x_d^{\eta_d}} f.$$

Then

$$D^{\boldsymbol{\eta}} f(\boldsymbol{x}) = \sum_{\boldsymbol{h}\in\mathbb{Z}^d} \left(\widehat{f}(\boldsymbol{h})\, (2\pi \mathrm{i})^{\|\boldsymbol{\eta}\|_1} \prod_{j=1}^d h_j^{\eta_j} \right) \mathrm{e}^{2\pi \mathrm{i} \boldsymbol{h}\cdot\boldsymbol{x}},$$

where, by convention, we take $0^0 := 1$.

Fix $\omega_1 \in (\omega, 1)$ and $q = \omega/\omega_1 < 1$. For any $\eta \in \mathbb{N}$ consider $g(x) = x^{2\eta} q^x$ for $x \geq 0$. Then $g'(x) = 0$ if $x = 2\eta/(\log q^{-1})$ and

$$g''\left(\frac{2\eta}{\log q^{-1}}\right) = \frac{1}{2}\left(\frac{2}{\mathrm{e}}\right)^{2\eta} \left(\frac{\eta}{\log q^{-1}}\right)^{2\eta-1} \log q < 0.$$

Hence,

$$g(x) \leq g\left(\frac{2\eta}{\log q^{-1}}\right) = \left(\frac{2\eta}{\mathrm{e}\log q^{-1}}\right)^{2\eta}.$$

Since

$$\eta^{2\eta} = \left(\eta! \frac{\eta^\eta}{\eta!}\right)^2 \leq \mathrm{e}^{2\eta} (\eta!)^2,$$

we obtain

$$g(x) \leq \left(\frac{2}{\log q^{-1}}\right)^{2\eta} (\eta!)^2.$$

Therefore, we have

$$x^{2\eta}\omega^x \leq \left(\frac{2}{\log\omega_1 - \log\omega}\right)^{2\eta} (\eta!)^2 \omega_1^x = C^{2\eta} (\eta!)^2 \omega_1^x,$$

where $C := 2/(\log\omega_1 - \log\omega) = 2/\log(\omega_1/\omega) > 0$ depends only on ω and ω_1.

Now consider the absolute value of the partial derivative $D^{\boldsymbol{\eta}} f$ in $\boldsymbol{x} \in [0, 1]^d$. Recall that $\rho_{\boldsymbol{a}_*,\boldsymbol{b}_*}(\boldsymbol{h}) = \omega^{a_*|h_1|^{b_*}+\cdots+a_*|h_d|^{b_*}}$. Dividing and multiplying each term of the series by the corresponding value of $\sqrt{\rho_{\boldsymbol{a}_*,\boldsymbol{b}_*}(\boldsymbol{h})}$, and using the Cauchy–Schwarz inequality yields

$$\begin{aligned} |D^{\boldsymbol{\eta}} f(\boldsymbol{x})| &= \left| \sum_{\boldsymbol{h}\in\mathbb{Z}^d} \left(\frac{\widehat{f}(\boldsymbol{h})}{\sqrt{\rho_{\boldsymbol{a}_*,\boldsymbol{b}_*}(\boldsymbol{h})}}\right) \left(\sqrt{\rho_{\boldsymbol{a}_*,\boldsymbol{b}_*}(\boldsymbol{h})} (2\pi\mathrm{i})^{\|\boldsymbol{\eta}\|_1} \prod_{j=1}^d h_j^{\eta_j} \right) \mathrm{e}^{2\pi\mathrm{i}\boldsymbol{h}\cdot\boldsymbol{x}} \right| \\ &\leq \|f\|_{\mathrm{kor},d,\boldsymbol{a}_*,\boldsymbol{b}_*} \left(\sum_{\boldsymbol{h}\in\mathbb{Z}^d} (2\pi)^{2\|\boldsymbol{\eta}\|_1} \prod_{j=1}^d |h_j|^{2\eta_j} \omega^{a_*|h_j|^{b_*}} \right)^{1/2} \end{aligned}$$

$$
\begin{aligned}
&\leq \|f\|_{\mathrm{kor},d,\boldsymbol{a}_*,\boldsymbol{b}_*} \left(\sum_{\boldsymbol{h}\in\mathbb{Z}^d} (2\pi)^{2\|\boldsymbol{\eta}\|_1} \prod_{j=1}^{d} \left(C^{2\eta_j} (\eta_j!)^2 \right) \omega_1^{a_*|h_j|^{b_*}} \right)^{1/2} \\
&= \|f\|_{\mathrm{kor},d,\boldsymbol{a}_*,\boldsymbol{b}_*} (2\pi C)^{\|\boldsymbol{\eta}\|_1} \prod_{j=1}^{d} (\eta_j!) \left(\sum_{\boldsymbol{h}\in\mathbb{Z}^d} \prod_{j=1}^{d} \omega_1^{a_*|h_j|^{b_*}} \right)^{1/2} \\
&= \|f\|_{\mathrm{kor},d,\boldsymbol{a}_*,\boldsymbol{b}_*} (2\pi C)^{\|\boldsymbol{\eta}\|_1} \prod_{j=1}^{d} (\eta_j!) \left(1 + 2\sum_{h=1}^{\infty} \omega_1^{a_* h^{b_*}} \right)^{d/2} \\
&= C_1 \cdot C_2^{\|\boldsymbol{\eta}\|_1} \prod_{j=1}^{d} (\eta_j!),
\end{aligned}
$$

where $C_1 := \|f\|_{\mathrm{kor},d,\boldsymbol{a}_*,\boldsymbol{b}_*}(1 + 2\sum_{h=1}^{\infty} \omega_1^{a_* h^{b_*}})^{d/2} \geq 0$ and $C_2 := 2\pi C > 0$.

Then for any $\boldsymbol{\zeta} = (\zeta_1, \ldots, \zeta_d)$ and $\boldsymbol{x} = (x_1, \ldots, x_d)$ with $\|\boldsymbol{x} - \boldsymbol{\zeta}\|_\infty < C_2^{-1}$ we have

$$
\begin{aligned}
\left| \sum_{\boldsymbol{\eta}\in\mathbb{N}_0^d} \frac{D^{\boldsymbol{\eta}} f(\boldsymbol{\zeta})}{(\eta_1!)\cdots(\eta_d!)} \prod_{j=1}^{d} (x_j - \zeta_j)^{\eta_j} \right| &\leq C_1 \sum_{\boldsymbol{\eta}\in\mathbb{N}_0^d} \prod_{j=1}^{d} (C_2|x_j - \zeta_j|)^{\eta_j} \\
&\leq C_1 \left(\sum_{\eta=0}^{\infty} (C_2\|\boldsymbol{x} - \boldsymbol{\zeta}\|_\infty)^{\eta} \right)^d \\
&= C_1 \left(\frac{1}{1 - C_2\|\boldsymbol{x} - \boldsymbol{\zeta}\|_\infty} \right)^d < \infty.
\end{aligned}
$$

It remains to show that the remainder R_n of the Taylor polynomial, given by

$$
R_n := \sum_{\|\boldsymbol{\eta}\|_1=n+1} \frac{n+1}{(\eta_1!)\cdots(\eta_d!)} \left(\prod_{j=1}^{d} (x_j - \zeta_j)^{\eta_j} \right) \int_0^1 (1-t)^n D^{\boldsymbol{\eta}} f(\boldsymbol{\zeta} + t(\boldsymbol{x} - \boldsymbol{\zeta}))\, \mathrm{d}t
$$

for $n \in \mathbb{N}_0$, vanishes if n goes to infinity, whenever $\boldsymbol{x}$ and $\boldsymbol{\zeta}$ are close enough to each other. Let again $\|\boldsymbol{x} - \boldsymbol{\zeta}\|_\infty < C_2^{-1}$. Then we have

$$
\begin{aligned}
|R_n| &\leq \sum_{\|\boldsymbol{\eta}\|_1=n+1} \frac{n+1}{(\eta_1!)\cdots(\eta_d!)} \left(\prod_{j=1}^{d} |x_j - \zeta_j|^{\eta_j} \right) \int_0^1 |1-t|^n \, |D^{\boldsymbol{\eta}} f(\boldsymbol{\zeta} + t(\boldsymbol{x} - \boldsymbol{\zeta}))|\, \mathrm{d}t \\
&\leq C_1 \cdot C_2^{n+1} \sum_{\|\boldsymbol{\eta}\|_1=n+1} (n+1) \left(\prod_{j=1}^{d} \|\boldsymbol{x} - \boldsymbol{\zeta}\|_\infty^{\eta_j} \right) \int_0^1 |1-t|^n \, \mathrm{d}t \\
&= C_1 \cdot (C_2\|\boldsymbol{x} - \boldsymbol{\zeta}\|_\infty)^{n+1} \binom{n+d}{d+1},
\end{aligned}
$$

because $\int_0^1 |1-t|^n \,\mathrm{d}t = 1/(n+1)$ and $\sum_{\|\boldsymbol{\eta}\|_1=n+1} 1 = \binom{n+d}{d+1}$. Since $C_2\|\boldsymbol{x}-\boldsymbol{\zeta}\|_\infty < 1$ and $\binom{n+d}{d+1} = O(n^{d-1})$, we get that $\lim_{n\to\infty} R_n = 0$. Thus, f is analytic. □

9.2 Integration in Korobov Spaces of Analytic Functions

In this section we study numerical integration of functions from the Korobov space $\mathcal{H}_{\mathrm{kor},d,\boldsymbol{a},\boldsymbol{b}}$. We consider the worst-case error of QMC rules or linear rules. From Remark 1.28 and Theorem 1.27, respectively, we find that the worst-case error of a linear rule based on the N-element point set $\mathcal{P} = \{\boldsymbol{x}_0, \boldsymbol{x}_1, \ldots, \boldsymbol{x}_{N-1}\}$ in $[0,1)^d$ and integration weights $\boldsymbol{w} = (w_0, w_1, \ldots, w_{N-1}) \in \mathbb{C}^N$ is equal to

$$\mathrm{err}_{N,d}(\mathcal{H}_{\mathrm{kor},d,\boldsymbol{a},\boldsymbol{b}}, \mathcal{P}, \boldsymbol{w}) = \left(1 - 2\sum_{k=0}^{N-1} \mathrm{Re}(w_k) + \sum_{k,\ell=0}^{N-1} \overline{w_k} w_\ell K_{\mathrm{kor},d,\boldsymbol{a},\boldsymbol{b}}(\boldsymbol{x}_k, \boldsymbol{x}_\ell)\right)^{1/2}.$$

For QMC rules this reduces to

$$\mathrm{err}_{N,d}(\mathcal{H}_{\mathrm{kor},d,\boldsymbol{a},\boldsymbol{b}}, \mathcal{P}) = \left(-1 + \frac{1}{N^2}\sum_{k,\ell=0}^{N-1} K_{\mathrm{kor},d,\boldsymbol{a},\boldsymbol{b}}(\boldsymbol{x}_k, \boldsymbol{x}_\ell)\right)^{1/2}. \tag{9.3}$$

Let again

$$e(N,d) = \inf_{\mathcal{P},\boldsymbol{w}} \mathrm{err}_{N,d}(\mathcal{H}_{\mathrm{kor},d,\boldsymbol{a},\boldsymbol{b}}, \mathcal{P}, \boldsymbol{w})$$

be the N-th minimal worst-case error of the integration problem in $\mathcal{H}_{\mathrm{kor},d,\boldsymbol{a},\boldsymbol{b}}$. In the same manner as for the classical Korobov space of finite smoothness we see that the initial error of integration in $\mathcal{H}_{\mathrm{kor},d,\boldsymbol{a},\boldsymbol{b}}$ equals one, i.e., $e(0,d) = 1$. Hence, the integration problem is normalized for all d.

Since with $\mathcal{H}_{\mathrm{kor},d,\boldsymbol{a},\boldsymbol{b}}$ we are concerned with analytic functions it is natural to expect even exponential convergence rates for optimal integration rules.

Definition 9.3 We say that we have *exponential convergence* of $e(N,d)$ if there exist a number $q \in (0,1)$ and functions $p, C, M : \mathbb{N} \to (0,\infty)$ such that

$$e(N,d) \le C(d)\, q^{(N/M(d))^{p(d)}} \quad \text{for all } d, N \in \mathbb{N}. \tag{9.4}$$

If (9.4) holds, then it is favorable to find $p(d)$ as large as possible. We call the largest possible rate

$$p^*(d) = \sup\{p(d) \ : \ p(d) \text{ satisfies (9.4)}\}$$

the *exponent of exponential convergence*.

Uniform exponential convergence of $e(N,d)$ holds if we can take $p(d) = p > 0$ for all $d \in \mathbb{N}$ in (9.4). The supremum of such p is called the *exponent of uniform exponential convergence* and is denoted by p^*.

Note that if (9.4) holds, then the information complexity (see Definition 1.42) is bounded according to

$$N(\varepsilon, d) \le \left\lceil M(d) \left(\frac{\log C(d) + \log \varepsilon^{-1}}{\log q^{-1}} \right)^{1/p(d)} \right\rceil \quad \text{for all } d \in \mathbb{N} \text{ and all } \varepsilon \in (0,1). \tag{9.5}$$

Furthermore, if (9.5) holds then

$$e(N+1, d) \le C(d)\, q^{(N/M(d))^{p(d)}} \quad \text{for all } d, N \in \mathbb{N}.$$

Indeed, let $d, N \in \mathbb{N}$ and let $\varepsilon := e(N+1, d)$. Then (9.5) implies

$$N + 1 = \min\{L \,:\, e(L,d) \le \varepsilon\} \le \left\lceil M(d) \left(\frac{\log C(d) + \log \varepsilon^{-1}}{\log q^{-1}} \right)^{1/p(d)} \right\rceil,$$

and hence

$$N \le M(d) \left(\frac{\log C(d) + \log \varepsilon^{-1}}{\log q^{-1}} \right)^{1/p(d)}.$$

This yields, after some elementary algebra,

$$e(N+1, d) = \varepsilon \le C(d)\, q^{(N/M(d))^{p(d)}}.$$

This observation means that (9.4) and (9.5) are practically equivalent.

In the following we study lower and upper bounds on the N-th minimal error of integration in $\mathcal{H}_{\mathrm{kor},d,\boldsymbol{a},\boldsymbol{b}}$. These results lead to necessary and sufficient conditions for (uniform) exponential convergence rates. We follow [148] in our outline.

Lower bounds

We present a lower bound on the N-th minimal worst-case error of integration based on [49, Theorem 1], whose proof is adopted from [238].

Lemma 9.4 *For any choice of nonnegative integers $t_1, t_2, \ldots, t_d$ we have*

$$e(N, d) \ge \omega^{2^{-1} \sum_{j=1}^d a_j t_j^{b_j}} \prod_{j=1}^d (1 + t_j)^{-1/2} \quad \textit{for all } N < \textstyle\prod_{j=1}^d (1 + t_j).$$

Proof We first show that for any finite subset $\mathcal{A}_d$ of $\mathbb{Z}^d$ we have

$$e(N, d) \ge \left(\max_{\boldsymbol{h}_* \in \mathcal{A}_d} \sum_{\boldsymbol{h} \in \mathcal{A}_d} \frac{1}{\rho_{\boldsymbol{a},\boldsymbol{b}}(\boldsymbol{h} - \boldsymbol{h}_*)} \right)^{-1/2} \quad \text{for all} \quad N < |\mathcal{A}_d|. \tag{9.6}$$

Consider an arbitrary linear algorithm $A_{N,d}(f) = \sum_{k=0}^{N-1} w_k f(\boldsymbol{x}_k)$, and define

$$g(\boldsymbol{x}) = \sum_{\boldsymbol{h} \in \mathcal{A}_d} b_{\boldsymbol{h}} \, \mathrm{e}^{2\pi \mathrm{i} \boldsymbol{h} \cdot \boldsymbol{x}} \quad \text{for all } \boldsymbol{x} \in [0,1]^d$$

such that $g(\boldsymbol{x}_k) = 0$ for all $k \in \{0, 1, \ldots, N-1\}$. Since we have N homogeneous linear equations and $|\mathcal{A}_d| > N$ unknowns $b_{\boldsymbol{h}}$, we can choose the $b_{\boldsymbol{h}}$, $\boldsymbol{h} \in \mathcal{A}_d$, such that at least one $b_{\boldsymbol{h}}$ is nonzero, and we can normalize the $b_{\boldsymbol{h}}$ by assuming that

$$\max_{\boldsymbol{h} \in \mathcal{A}_d} |b_{\boldsymbol{h}}| = b_{\boldsymbol{h}_*} = 1 \quad \text{for some } \boldsymbol{h}_* \in \mathcal{A}_d.$$

Define the function f by

$$f(\boldsymbol{x}) = c \, \mathrm{e}^{-2\pi \mathrm{i} \boldsymbol{h}_* \cdot \boldsymbol{x}} g(\boldsymbol{x}) = c \sum_{\boldsymbol{h} \in \mathcal{A}_d} b_{\boldsymbol{h}} \, \mathrm{e}^{2\pi \mathrm{i} (\boldsymbol{h} - \boldsymbol{h}_*) \cdot \boldsymbol{x}} \quad \text{for } \boldsymbol{x} \in [0,1]^d,$$

where the positive constant c is chosen such that $\|f\|_{\mathrm{kor},d,\boldsymbol{a},\boldsymbol{b}} \leq 1$. More precisely, we have

$$\|f\|^2_{\mathrm{kor},d,\boldsymbol{a},\boldsymbol{b}} = c^2 \sum_{\boldsymbol{h} \in \mathcal{A}_d} |b_{\boldsymbol{h}}|^2 \frac{1}{\rho_{\boldsymbol{a},\boldsymbol{b}}(\boldsymbol{h} - \boldsymbol{h}_*)} \leq c^2 \max_{\boldsymbol{h}_* \in \mathcal{A}_d} \sum_{\boldsymbol{h} \in \mathcal{A}_d} \frac{1}{\rho_{\boldsymbol{a},\boldsymbol{b}}(\boldsymbol{h} - \boldsymbol{h}_*)}.$$

Hence we can take

$$c = \left(\max_{\boldsymbol{h}_* \in \mathcal{A}_d} \sum_{\boldsymbol{h} \in \mathcal{A}_d} \frac{1}{\rho_{\boldsymbol{a},\boldsymbol{b}}(\boldsymbol{h} - \boldsymbol{h}_*)} \right)^{-1/2}.$$

Note that $f(\boldsymbol{x}_k) = 0$ for all $k \in \{0, 1, \ldots, N-1\}$, which implies that $A_{N,d}(f) = 0$. Furthermore,

$$\int_{[0,1]^d} f(\boldsymbol{x}) \, \mathrm{d}\boldsymbol{x} = c \, b_{\boldsymbol{h}_*} = c.$$

As a consequence,

$$\mathrm{err}_{N,d}(\mathcal{H}_{\mathrm{kor},d,\boldsymbol{a},\boldsymbol{b}}, \mathcal{P}, \boldsymbol{w}) \geq \left| \int_{[0,1]^d} f(\boldsymbol{x}) \, \mathrm{d}\boldsymbol{x} - A_{N,d}(f) \right| = \int_{[0,1]^d} f(\boldsymbol{x}) \, \mathrm{d}\boldsymbol{x} = c.$$

Since this holds for all $\mathcal{P}$ and $\boldsymbol{w}$, we conclude that $e(N,d) \geq c$ and this proves (9.6).

Now, for $\boldsymbol{t} = (t_1, \ldots, t_d)$ with nonnegative integers t_j for $j \in [d]$, we consider the set

$$\mathcal{A}_d = \mathcal{A}_{d,\boldsymbol{t}} = \{\boldsymbol{h} \in \mathbb{Z}^d \ : \ h_j \in \{0, 1, \ldots, t_j\} \quad \text{for all } j \in [d]\}.$$

Clearly, $|\mathcal{A}_{d,\boldsymbol{t}}| = \prod_{j=1}^d (1 + t_j)$. For $\boldsymbol{h}, \boldsymbol{h}_* \in \mathcal{A}_{d,\boldsymbol{t}}$ we have

$$\frac{1}{\rho_{\boldsymbol{a},\boldsymbol{b}}(\boldsymbol{h} - \boldsymbol{h}_*)} = \omega^{-\sum_{j=1}^d a_j |h_j - h_j^*|^{b_j}} \leq \omega^{-\sum_{j=1}^d a_j t_j^{b_j}}.$$

Therefore

$$\max_{\boldsymbol{h}_* \in \mathcal{A}_{d,\boldsymbol{t}}} \sum_{\boldsymbol{h} \in \mathcal{A}_{d,\boldsymbol{t}}} \frac{1}{\rho_{\boldsymbol{a},\boldsymbol{b}}(\boldsymbol{h}-\boldsymbol{h}_*)} \le \omega^{-\sum_{j=1}^d a_j t_j^{b_j}} \, |\mathcal{A}_{d,\boldsymbol{t}}|.$$

Inserting this estimate into (9.6) gives the desired result. □

We are ready to prove necessary conditions for uniform exponential convergence.

Theorem 9.5 *Assume that we have uniform exponential convergence, i.e.,* (9.4) *holds with* $p(d) = p$. *Then we have*

$$B := \sum_{j=1}^{\infty} \frac{1}{b_j} < \infty \quad \textit{and} \quad p \le \frac{1}{B},$$

independently of the choice of the positive $\boldsymbol{a}$ *and the choice of* ω.

Proof First we show that uniform exponential convergence with $p(d) = p$ implies, for any fixed $d \in \mathbb{N}$,

$$\liminf_{\|\boldsymbol{t}\|_\infty \to \infty} \frac{\sum_{j=1}^d a_j t_j^{b_j}}{\prod_{j=1}^d (1+t_j)^p} \ge \frac{2}{(M(d))^p} \frac{\log q^{-1}}{\log \omega^{-1}} > 0, \tag{9.7}$$

where for the limit inferior we consider the vectors $\boldsymbol{t} = (t_1, \dots, t_d) \in \mathbb{N}_0^d$ with $\|\boldsymbol{t}\|_\infty := \max_{j \in [d]} t_j$ tending to infinity.

From (9.4) with $p(d) = p$ and Lemma 9.4 with $N = -1 + \prod_{j=1}^d (1+t_j)$ we have

$$\log C(d) \ge -\frac{1}{2} \sum_{j=1}^d a_j t_j^{b_j} \log \frac{1}{\omega} - \frac{1}{2} \sum_{j=1}^d \log(1+t_j) + \left(\frac{N}{M(d)}\right)^p \log \frac{1}{q}.$$

This implies that

$$\begin{aligned} &\frac{\sum_{j=1}^d a_j t_j^{b_j}}{\prod_{j=1}^d (1+t_j)^p} + \frac{2\log C(d) + \sum_{j=1}^d \log(1+t_j)}{(\log \omega^{-1}) \prod_{j=1}^d (1+t_j)^p} \\ &\ge \left(1 - \frac{1}{\prod_{j=1}^d (1+t_j)}\right)^p \frac{2}{(M(d))^p} \frac{\log q^{-1}}{\log \omega^{-1}}. \end{aligned} \tag{9.8}$$

For fixed d, when $\|\boldsymbol{t}\|_\infty$ goes to infinity, the second term on the left-hand side of (9.8) tends to zero, and the term on the right-hand side tends to

$$\frac{2}{(M(d))^p} \frac{\log q^{-1}}{\log \omega^{-1}},$$

which is strictly positive. Thus, the necessary condition (9.7) holds.

For a positive t take now

$$t_j = \lceil t^{1/b_j} \rceil \quad \text{for all } j \in [d].$$

Clearly, $\lim_{t\to\infty} \lceil t^{1/b_j} \rceil / t^{1/b_j} = 1$. Then for t tending to infinity we have

$$\begin{aligned}\frac{\sum_{j=1}^d a_j t_j^{b_j}}{\prod_{j=1}^d (1+t_j)^p} &= t^{1-p\sum_{j=1}^d b_j^{-1}} \frac{\sum_{j=1}^d a_j (\lceil t^{1/b_j} \rceil / t^{1/b_j})^{b_j}}{\prod_{j=1}^d \left(\lceil t^{1/b_j} \rceil / t^{1/b_j} + t^{-1/b_j}\right)^p} \\ &= t^{1-p\sum_{j=1}^d b_j^{-1}} (1+o(1)) \sum_{j=1}^d a_j.\end{aligned}$$

Since this expression is positive when t tends to infinity, we must have $p \sum_{j=1}^d \frac{1}{b_j} \leq 1$. This holds for all d. Hence for d tending to infinity we conclude that

$$p \sum_{j=1}^\infty \frac{1}{b_j} = pB \leq 1,$$

which completes the proof. □

Remark 9.6 It follows from the same arguments as used in the proof of Theorem 9.5 that if we have exponential convergence, then the exponent $p^*(d)$ of exponential convergence can be at most $1/B(d)$, where $B(d) := \sum_{j=1}^d b_j^{-1}$.

Upper bounds

Upper bounds on the N-th minimal error in $\mathcal{H}_{\mathrm{kor},d,\boldsymbol{a},\boldsymbol{b}}$ can be obtained using regular grids with different mesh sizes. This may be surprising, since for numerical integration of functions with partial derivatives up to order α in each variable regular grids are known to yield very poor integration rules even for moderate values of d.

Definition 9.7 For $d \in \mathbb{N}$, a *regular grid with mesh sizes* $m_1, m_2, \ldots, m_d \in \mathbb{N}$ is defined as the point set $\mathcal{G}_{N,d}$ with points

$$\left(\frac{k_1}{m_1}, \frac{k_2}{m_2}, \ldots, \frac{k_d}{m_d} \right)$$

for $k_j \in \{0, 1, \ldots, m_j - 1\}$ and $j \in [d]$, where $N := \prod_{j=1}^d m_j$ is the cardinality of $\mathcal{G}_{N,d}$.

A regular grid with mesh sizes $m_1, m_2, \ldots, m_d$ obviously is a lattice point set that corresponds to the lattice

$$\mathcal{L} = \left\{ \sum_{j=1}^d \frac{k_j}{m_j} \boldsymbol{e}_j \ : \ k_j \in \mathbb{Z} \text{ for } j \in [d] \right\},$$

where $e_1, e_2, \ldots, e_d$ are the canonical basis vectors in $\mathbb{R}^d$. To be more precise,

$$\mathcal{G}_{N,d} = \mathcal{L} \cap [0,1)^d = \mathcal{P}(\mathcal{L}).$$

The dual lattice is, in this case, accordingly given by

$$\mathcal{L}^\perp = \{\boldsymbol{h} \in \mathbb{Z}^d \ : \ h_j \equiv 0 \pmod{m_j} \text{ for all } j \in [d]\}. \tag{9.9}$$

The QMC rule based on $\mathcal{G}_{N,d}$ then is a lattice rule of the form

$$\begin{aligned} Q_{N,d}(f) &= \frac{1}{N} \sum_{\boldsymbol{x} \in \mathcal{G}_{N,d}} f(\boldsymbol{x}) \\ &= \frac{1}{m_1 \cdots m_d} \sum_{k_1=0}^{m_1-1} \cdots \sum_{k_d=0}^{m_d-1} f\left(\sum_{j=1}^{d} \frac{k_j}{m_j} \boldsymbol{e}_j\right) \\ &= \frac{1}{m_1 \cdots m_d} \sum_{k_1=0}^{m_1-1} \cdots \sum_{k_d=0}^{m_d-1} f\left(\frac{k_1}{m_1}, \frac{k_2}{m_2}, \ldots, \frac{k_d}{m_d}\right). \end{aligned} \tag{9.10}$$

Note that for dimension $d = 1$ this is the trapezoidal rule applied to a one-periodic function f. For the trapezoidal rule it is well known that exponential convergence rates can be obtained for periodic integrands; see, for example, [18, Chapter 8] or [255]. Obviously, (9.10) may be interpreted as the d-fold product of trapezoidal rules of different mesh sizes $m_1, m_2, \ldots, m_d$, and we will see that this again leads to exponential convergence in $\mathcal{H}_{\mathrm{kor},d,\boldsymbol{a},\boldsymbol{b}}$, and even to uniform exponential convergence for suitably chosen parameters $\boldsymbol{b}$.

We first present an explicit formula for the worst-case error of the lattice rule that uses a regular grid $\mathcal{G}_{N,d}$ with arbitrary mesh sizes $m_j \in \mathbb{N}$, $j \in [d]$.

Lemma 9.8 *Let $m_1, m_2, \ldots, m_d \in \mathbb{N}$ and let $N = m_1 m_2 \cdots m_d$. Then*

$$\mathrm{err}_{N,d}(\mathcal{H}_{\mathrm{kor},d,\boldsymbol{a},\boldsymbol{b}}, \mathcal{G}_{N,d}) = \left(-1 + \prod_{j=1}^{d}\left(1 + 2\sum_{h=1}^{\infty} \omega^{a_j (m_j h)^{b_j}}\right)\right)^{1/2}.$$

Proof From (9.3) and the definition of the kernel $K_{\mathrm{kor},d,\boldsymbol{a},\boldsymbol{b}}$ we obtain

$$\begin{aligned} [\mathrm{err}_{N,d}(\mathcal{H}_{\mathrm{kor},d,\boldsymbol{a},\boldsymbol{b}}, \mathcal{G}_{N,d})]^2 &= -1 + \frac{1}{N^2} \sum_{\boldsymbol{x},\boldsymbol{y} \in \mathcal{G}_{N,d}} K_{\mathrm{kor},d,\boldsymbol{a},\boldsymbol{b}}(\boldsymbol{x},\boldsymbol{y}) \\ &= \frac{1}{N^2} \sum_{\boldsymbol{x},\boldsymbol{y} \in \mathcal{G}_{N,d}} \sum_{\boldsymbol{h} \in \mathbb{Z}^d \setminus \{\boldsymbol{0}\}} \omega^{\sum_{j=1}^d a_j |h_j|^{b_j}} \mathrm{e}^{2\pi \mathrm{i} \boldsymbol{h} \cdot (\boldsymbol{x}-\boldsymbol{y})} \\ &= \sum_{\boldsymbol{h} \in \mathbb{Z}^d \setminus \{\boldsymbol{0}\}} \omega^{\sum_{j=1}^d a_j |h_j|^{b_j}} \left| \frac{1}{N} \sum_{\boldsymbol{x} \in \mathcal{G}_{N,d}} \mathrm{e}^{2\pi \mathrm{i} \boldsymbol{h} \cdot \boldsymbol{x}} \right|^2 \end{aligned}$$

$$= \sum_{\boldsymbol{h} \in \mathcal{L}^{\perp} \setminus \{\boldsymbol{0}\}} \omega^{\sum_{j=1}^{d} a_j |h_j|^{b_j}}.$$

Recalling that the dual lattice $\mathcal{L}^{\perp}$ is given by (9.9) we obtain

$$[\mathrm{err}_{N,d}(\mathcal{H}_{\mathrm{kor},d,\boldsymbol{a},\boldsymbol{b}}, \mathcal{G}_{N,d})]^2 = \sum_{\boldsymbol{h} \in \mathbb{Z}^d \setminus \{\boldsymbol{0}\}} \omega^{\sum_{j=1}^{d} a_j (m_j |h_j|)^{b_j}}$$

$$= -1 + \prod_{j=1}^{d} \left(1 + 2 \sum_{h=1}^{\infty} \omega^{a_j (m_j h)^{b_j}} \right).$$

Taking the square root gives the desired result. □

We now show that with an appropriate choice of the m_j the corresponding QMC algorithm has a worst-case error of at most ε and yields exponential convergence.

Theorem 9.9 *For $d \in \mathbb{N}$ and $\varepsilon \in (0, 1)$ define*

$$m := \max_{j \in [d]} \left\lceil \left(\frac{2^{b_j}}{a_j} \frac{\log\left(1 + (2\, d\, F)/\log(1 + \varepsilon^2)\right)}{\log \omega^{-1}} \right)^{B(d)} \right\rceil,$$

where

$$B(d) := \sum_{j=1}^{d} \frac{1}{b_j},$$

and

$$F := \max_{j \in [d]} \sum_{h=1}^{\infty} \omega^{m^{1/B(d)} a_j 2^{-b_j} (h^{b_j} - 1)} < \infty.$$

Let $\mathcal{G}_{N,d}$ be a regular grid with mesh sizes $m_1, m_2, \ldots, m_d$ given by

$$m_j := \left\lfloor m^{1/(B(d) \cdot b_j)} \right\rfloor \quad \text{for } j \in [d] \text{ and } N = \prod_{j=1}^{d} m_j.$$

Then

$$\mathrm{err}_{N,d}(\mathcal{H}_{\mathrm{kor},d,\boldsymbol{a},\boldsymbol{b}}, \mathcal{G}_{N,d}) \le \varepsilon \quad \text{and} \quad N(\varepsilon, d) \le N = O\left(\left(\log\left(1 + \frac{1}{\varepsilon} \right) \right)^{B(d)} \right),$$

with the factor in the O-notation independent of ε but dependent on d. This means that we have exponential convergence.

Proof Note first that

$$N = \prod_{j=1}^{d} m_j = \prod_{j=1}^{d} \left\lfloor m^{1/(B(d) \cdot b_j)} \right\rfloor$$

$$\leq m^{(1/B(d))\sum_{j=1}^{d} 1/b_j} = m = O\left(\left(\log\left(1+\frac{1}{\varepsilon}\right)\right)^{B(d)}\right).$$

Since $\lfloor x \rfloor \geq x/2$ for all $x \geq 1$, we have

$$a_j |m_j h_j|^{b_j} \geq a_j (|h_j|/2)^{b_j}\, m^{1/B(d)}$$

for every $j \in [d]$. From Lemma 9.8 we obtain

$$[\mathrm{err}_{N,d}(\mathcal{H}_{\mathrm{kor},d,\boldsymbol{a},\boldsymbol{b}}, \mathcal{G}_{N,d})]^2 \leq -1 + \prod_{j=1}^{d}\left(1 + 2\sum_{h=1}^{\infty} \omega^{m^{1/B(d)} a_j 2^{-b_j} h^{b_j}}\right).$$

We further estimate

$$\begin{aligned}\sum_{h=1}^{\infty} \omega^{m^{1/B(d)} a_j 2^{-b_j} h^{b_j}} &= \omega^{m^{1/B(d)} a_j 2^{-b_j}} \sum_{h=1}^{\infty} \omega^{m^{1/B(d)} a_j 2^{-b_j} (h^{b_j}-1)} \\ &\leq \omega^{m^{1/B(d)} a_j 2^{-b_j}} F.\end{aligned}$$

From the definition of m, we have for all $j \in [d]$,

$$\omega^{m^{1/B(d)} a_j 2^{-b_j}} F \leq \frac{\log(1+\varepsilon^2)}{2d}.$$

This proves

$$\begin{aligned}\mathrm{err}_{N,d}(\mathcal{H}_{\mathrm{kor},d,\boldsymbol{a},\boldsymbol{b}}, \mathcal{G}_{N,d}) &\leq \left(-1 + \left(1 + \frac{\log(1+\varepsilon^2)}{d}\right)^d\right)^{1/2} \\ &\leq \left(-1 + \mathrm{e}^{\log(1+\varepsilon^2)}\right)^{1/2} \\ &= \varepsilon,\end{aligned}$$

and completes the proof. □

An exact characterization of exponential convergence

We now combine the previously shown lower and upper bounds in order to obtain an exact characterization of weight sequences $\boldsymbol{a}$ and $\boldsymbol{b}$ for which (uniform) exponential convergence can be obtained, and we present the exact values of the respective exponents $p^*(d)$ and p^*.

Corollary 9.10 *Consider integration in the Korobov space $\mathcal{H}_{\mathrm{kor},d,\boldsymbol{a},\boldsymbol{b}}$ with weight sequences $\boldsymbol{a}$ and $\boldsymbol{b}$ satisfying* (9.2)*. The following results hold.*

1. *Exponential convergence can be obtained for all $\boldsymbol{a}$ and $\boldsymbol{b}$ considered, with the exponent of exponential convergence satisfying $p^*(d) = 1/B(d)$.*

2. *Uniform exponential convergence can be obtained if and only if* $\boldsymbol{b}$ *is such that*

$$B := \sum_{j=1}^{\infty} \frac{1}{b_j} < \infty.$$

If so, then $p^* = 1/B$.

Proof Regarding Item 1, we know from Theorem 9.9 that for all $\boldsymbol{a}$ and $\boldsymbol{b}$ we have

$$N(\varepsilon, d) = O\left(\left(\log\left(1 + \frac{1}{\varepsilon}\right)\right)^{B(d)}\right).$$

This implies that we have exponential convergence and, furthermore, $p^*(d) \geq 1/B(d)$. On the other hand, Remark 9.6 implies $p^*(d) \leq 1/B(d)$ and hence the first result is proven.

Regarding Item 2, suppose first that $\boldsymbol{a}$ is an arbitrary sequence and that $\boldsymbol{b}$ is such that

$$B = \sum_{j=1}^{\infty} \frac{1}{b_j} < \infty.$$

Then we can replace $B(d)$ by B in Theorem 9.9, and we obtain

$$N(\varepsilon, d) = O\left(\log^B\left(1 + \frac{1}{\varepsilon}\right)\right).$$

This means that we have uniform exponential convergence with exponent $p^* \geq 1/B$.

On the other hand, if we have uniform exponential convergence, then Theorem 9.5 implies that $B < \infty$ and $p^* \leq 1/B$. This completes the proof. □

Corollary 9.10 demonstrates that exponential convergence can always be reached with a suitable choice of a regular grid with different mesh sizes $m_1, m_2, \ldots, m_d$ (see also Theorem 9.9). However, a necessary and sufficient condition for uniform exponential convergence is that the weight sequence $\boldsymbol{b} = (b_j)_{j \geq 1}$ tends to infinity so fast that $B = \sum_{j=1}^{\infty} b_j^{-1} < \infty$, with no extra condition on $\boldsymbol{a}$ and ω. If B is small, then the exponent p^* is large. For instance, for $b_j = j^2$ for all $j \in \mathbb{N}$ we have $B = \pi^2/6$ and $p^* = 6/\pi^2 = 0.6079\ldots$.

Note that the factors hidden in the O-notation in the above results still depend on the dimension d. Making assertions also about tractability properties requires a more detailed analysis, which will be discussed in the next section.

9.3 Exponential Tractability

Exponential error convergence implies that asymptotically, with respect to ε tending to zero, we require only $O((\log \varepsilon^{-1})^{1/p(d)})$ function evaluations to compute an ε-approximation to the exact value of the integral. However, the implied factors in the O-notation can depend on the dimension d, and therefore it is not clear how long one has to wait to actually see this excellent asymptotic behavior, especially for large dimension d. This is again the subject of tractability theory.

In the context of exponential convergence it is natural to adapt the concepts of tractability in Definition 1.44 by replacing ε^{-1} by $\log \varepsilon^{-1}$, or, to avoid technical issues if $\log \varepsilon^{-1} \leq 1$, by $1 + \log \varepsilon^{-1}$. This adaption yields new versions of weak, polynomial, and strong polynomial tractability, with the prefix "EXP", which indicates "exponential-convergence". In the following definition we use the same notation as introduced in Section 1.7.

Definition 9.11 The sequence of integration problems $(I : \mathcal{F}_d \to \mathbb{R})_{d \geq 1}$ is called

- *EXP-weakly tractable* if

$$\lim_{d+\log \varepsilon^{-1} \to \infty} \frac{\log N(\varepsilon, d)}{d + \log \varepsilon^{-1}} = 0;$$

  We remark that if $N(\varepsilon, d) = 0$, which can only happen in trivial cases, we interpret $\log N(\varepsilon, d)$ as 0 in this context;
- *EXP-polynomially tractable* if there exist constants $C, \sigma > 0$ and $\tau \geq 0$ such that

$$N(\varepsilon, d) \leq C\, d^{\tau}\, (1 + \log \varepsilon^{-1})^{\sigma} \quad \text{for all } \varepsilon \in (0, 1) \text{ and all } d \in \mathbb{N};$$

- *EXP-strongly polynomially tractable* if there exist constants $C > 0$ and $\sigma > 0$ such that

$$N(\varepsilon, d) \leq C\, (1 + \log \varepsilon^{-1})^{\sigma} \quad \text{for all } \varepsilon \in (0, 1) \text{ and all } d \in \mathbb{N}. \tag{9.11}$$

  The infimum of all $\sigma > 0$ such that a bound of the form (9.11) holds is called the *ε-exponent* of EXP-strong polynomial tractability.

Further, and more refined, notions of EXP-tractability have been introduced and studied in [119]. The use of the prefix "EXP" to indicate "exponential convergence" in the notation is motivated by the following result.

Proposition 9.12 *The following assertions hold true.*

1. *EXP-polynomial tractability (and therefore also EXP-strong polynomial tractability) implies uniform exponential convergence.*
2. *EXP-weak tractability implies that $e(N, d)$ converges to zero faster than any power of N^{-1} as N tends to infinity, i.e., for any $\alpha > 0$ we have*

$$\lim_{N \to \infty} N^{\alpha} e(N, d) = 0. \tag{9.12}$$

Proof For the proof of Item 1, EXP-polynomial tractability means that $N(\varepsilon, d) \le C\, d^{\tau}\, (1 + \log \varepsilon^{-1})^{\sigma}$ for all $\varepsilon \in (0, 1)$ and all $d \in \mathbb{N}$, with constants $C, \sigma > 0$ and $\tau \ge 0$. For (sufficiently large) N, we choose $\varepsilon \in (0, 1)$ such that

$$N = \left\lfloor C d^{\tau} (1 + \log \varepsilon^{-1})^{\sigma} \right\rfloor.$$

As $N(\varepsilon, d) \le C\, d^{\tau}\, (1 + \log \varepsilon^{-1})^{\sigma}$ and $N(\varepsilon, d) \in \mathbb{N}$, we obtain $N(\varepsilon, d) \le N$, and hence $e(N, d) \le \varepsilon$. On the other hand, we have

$$N \le C d^{\tau} (1 + \log \varepsilon^{-1})^{\sigma},$$

and therefore

$$\varepsilon \le \mathrm{e}^{1-(N/(Cd^{\tau}))^{1/\sigma}},$$

and this implies that

$$e(N, d) \le \mathrm{e}^{1-(N/(Cd^{\tau}))^{1/\sigma}}. \tag{9.13}$$

So, uniform exponential convergence holds with $p = 1/\sigma$.

With respect to Item 2, choose $\delta \in (0, 1/\alpha)$ for given $\alpha > 0$. For a fixed dimension d, EXP-weak tractability implies the existence of a number $K = K(\delta) > 0$ such that for all $\varepsilon \in (0, 1)$ with $\log \varepsilon^{-1} > K$ we have

$$\frac{\log N(\varepsilon, d)}{\log \varepsilon^{-1}} < \delta.$$

This, however, is equivalent to $N(\varepsilon, d) < \varepsilon^{-\delta}$. The latter implies that for sufficiently large $N \in \mathbb{N}$ we have $e(N, d) < N^{-1/\delta}$ and hence we have

$$N^{\alpha} e(N, d) < N^{\alpha - 1/\delta} \to 0 \quad \text{as} \quad N \to \infty.$$ □

Next, we characterize EXP-tractability of the integration problem in $\mathcal{H}_{\mathrm{kor},d,\boldsymbol{a},\boldsymbol{b}}$ by means of the weight sequences $\boldsymbol{a}$ and $\boldsymbol{b}$.

Theorem 9.13 *Consider integration in the Korobov space $\mathcal{H}_{\mathrm{kor},d,\boldsymbol{a},\boldsymbol{b}}$ with weight sequences $\boldsymbol{a}$ and $\boldsymbol{b}$ satisfying* (9.2)*. The following results hold.*

1. *EXP-weak tractability holds if and only if*

$$\lim_{j \to \infty} a_j = \infty.$$

2. *EXP-polynomial and EXP-strong polynomial tractability are equivalent.*
3. *EXP-strong polynomial tractability holds if and only if*

$$B := \sum_{j=1}^{\infty} \frac{1}{b_j} < \infty \quad \textit{and} \quad \alpha_* := \liminf_{j \to \infty} \frac{\log a_j}{j} > 0.$$

Then the exponent σ_* *of EXP-strong polynomial tractability satisfies*

$$\sigma_* \in \left[B, B + \frac{\log 2}{\alpha_*}\right].$$

In particular, if $\alpha_* = \infty$, *then* $\sigma_* = B$.

Before giving the proof of Theorem 9.13, we would like to make some remarks.

Remark 9.14 EXP-weak tractability holds if and only if the weight sequence $\boldsymbol{a} = (a_j)_{j\geq 1}$ tends to infinity. This holds independently of the weights $\boldsymbol{b}$ and independently of the rate of convergence of $\boldsymbol{a}$ to infinity. Proposition 9.12 then implies that (9.12) holds. If we additionally have $B < \infty$, then we have uniform exponential convergence and EXP-weak tractability. On the other hand, if $\lim_{j\to\infty} a_j = \infty$ and $B = \infty$, then EXP-weak tractability holds without uniform exponential convergence.

EXP-strong polynomial tractability holds if and only if $B < \infty$ and the elements a_j of the weight sequence $\boldsymbol{a}$ are exponentially large in j for large $j \in \mathbb{N}$.

We now give the proof of the theorem.

Proof of Theorem 9.13. Regarding Item 1, assume first that the sequence of the a_j is bounded, say $a_j \leq A < \infty$ for all $j \in \mathbb{N}$. From Lemma 9.4 with $t_j = 1$ for all $j \in [d]$ it follows that for all $N < 2^d$ we have

$$e(N, d) \geq 2^{-d/2}\omega^{2^{-1}\sum_{j=1}^{d} a_j} \geq 2^{-d/2}\omega^{dA/2} = \eta^d,$$

where $\eta := \sqrt{\omega^A/2} \in (0, 1)$. As a consequence, for $\varepsilon = \eta^d/2$ we have $e(N, d) > \varepsilon$ for all $N < 2^d$. This implies that $N(\varepsilon, d) \geq 2^d$ and

$$\frac{\log N(\varepsilon, d)}{d + \log \varepsilon^{-1}} \geq \frac{d \log 2}{d + \log 2 + d \log \eta^{-1}},$$

with the right-hand side of the latter inequality tending to

$$\frac{\log 2}{1 + \log \eta^{-1}} > 0$$

as $d \to \infty$. Thus we do not have EXP-weak tractability. This reasoning shows that EXP-weak tractability implies $\lim_{j\to\infty} a_j = \infty$.

Assume now that $\lim_{j\to\infty} a_j = \infty$. For the next part of the proof we use rank-1 lattice rules. It follows from (9.3) that the squared worst-case error of a rank-1 lattice rule in $\mathcal{H}_{\mathrm{kor},d,\boldsymbol{a},\boldsymbol{b}}$ is given by

$$\begin{aligned}
[\mathrm{err}_{N,d}(\mathcal{H}_{\mathrm{kor},d,\boldsymbol{a},\boldsymbol{b}}, \mathcal{P}(\boldsymbol{g}, N))]^2 &= \sum_{\boldsymbol{h}\in\mathbb{Z}^d\setminus\{\boldsymbol{0}\}} \rho_{\boldsymbol{a},\boldsymbol{b}}(\boldsymbol{h}) \left|\frac{1}{N}\sum_{k=0}^{N-1} \mathrm{e}^{2\pi \mathrm{i} \boldsymbol{h}\cdot\boldsymbol{g}/N}\right|^2 \\
&= \sum_{\substack{\boldsymbol{h}\in\mathbb{Z}^d\setminus\{\boldsymbol{0}\} \\ \boldsymbol{h}\cdot\boldsymbol{g}\equiv 0 \pmod{N}}} \omega^{\sum_{j=1}^{d} a_j |h_j|^{b_j}}.
\end{aligned}$$

By Jensen's inequality (Lemma 2.25) we obtain for arbitrary $\lambda \in (0,1]$,

$$[\mathrm{err}_{N,d}(\mathcal{H}_{\mathrm{kor},d,\boldsymbol{a},\boldsymbol{b}}, \mathcal{P}(\boldsymbol{g},N))]^{2\lambda} \le \sum_{\substack{\boldsymbol{h}\in\mathbb{Z}^d\setminus\{\boldsymbol{0}\} \\ \boldsymbol{h}\cdot\boldsymbol{g}\equiv 0 \ (\mathrm{mod}\ N)}} \omega^{\sum_{j=1}^d \lambda a_j |h_j|^{b_j}}.$$

Let now N be a prime number and average $[\mathrm{err}_{N,d}(\mathcal{H}_{\mathrm{kor},d,\boldsymbol{a},\boldsymbol{b}}, \mathcal{P}(\boldsymbol{g},N))]^{2\lambda}$ over all possible values $\boldsymbol{g} \in G_d(N)$. In the same way as in the proof of (2.25) we find that there exists a $\boldsymbol{g} \in G_d(N)$ such that

$$\begin{aligned}[\mathrm{err}_{N,d}(\mathcal{H}_{\mathrm{kor},d,\boldsymbol{a},\boldsymbol{b}}, \mathcal{P}(\boldsymbol{g},N))]^{2\lambda} &\le \frac{1}{N} \sum_{\boldsymbol{h}\in\mathbb{Z}^d\setminus\{\boldsymbol{0}\}} \omega^{\sum_{j=1}^d \lambda a_j |h_j|^{b_j}} \\ &\le \frac{1}{N} \prod_{j=1}^d \left(\sum_{h\in\mathbb{Z}} \omega^{\lambda a_j |h|^{b_j}} \right).\end{aligned}$$

We have

$$\sum_{h\in\mathbb{Z}} \omega^{\lambda a_j |h|^{b_j}} \le 1 + 2\omega^{\lambda a_j} \sum_{h=1}^{\infty} \omega^{\lambda a_j (h^{b_j}-1)} \le 1 + 2A_\lambda \omega^{\lambda a_j},$$

where $A_\lambda := \sum_{h=1}^\infty \omega^{\lambda a_* (h^{b_*}-1)} < \infty$. Hence,

$$\mathrm{err}_{N,d}(\mathcal{H}_{\mathrm{kor},d,\boldsymbol{a},\boldsymbol{b}}, \mathcal{P}(\boldsymbol{g},N)) \le \frac{1}{N^{1/(2\lambda)}} \prod_{j=1}^d (1 + 2A_\lambda \omega^{\lambda a_j})^{1/(2\lambda)}. \tag{9.14}$$

Note that, since $G_d(N)$ is a finite set, there even must exist a $\boldsymbol{g} \in G_d(N)$ that satisfies (9.14) for all $\lambda \in (0,1]$ simultaneously. Indeed, we can deduce from (9.14) that

$$\min_{\boldsymbol{g}\in G_d(N)} \mathrm{err}_{N,d}(\mathcal{H}_{\mathrm{kor},d,\boldsymbol{a},\boldsymbol{b}}, \mathcal{P}(\boldsymbol{g},N)) \le \frac{1}{N^{1/(2\lambda)}} \prod_{j=1}^d (1 + 2A_\lambda \omega^{\lambda a_j})^{1/(2\lambda)},$$

and this holds for any choice of $\lambda \in (0,1]$. However, the minimum over all $\boldsymbol{g} \in G_d(N)$ is independent of λ, so the corresponding minimizer yields an error that satisfies (9.14) for all $\lambda \in (0,1]$ simultaneously.

Let $\varepsilon > 0$, let $\lambda \in (0,1]$, and let N be the smallest prime number greater than or equal to $\lceil \varepsilon^{-2\lambda} \prod_{j=1}^d (1 + 2A_\lambda \omega^{\lambda a_j}) \rceil =: M$. Then there exists $\boldsymbol{g} \in G_d(N)$ such that

$$\mathrm{err}_{N,d}(\mathcal{H}_{\mathrm{kor},d,\boldsymbol{a},\boldsymbol{b}}, \mathcal{P}(\boldsymbol{g},N)) \le \varepsilon$$

and hence

$$N(\varepsilon,d) \le N \le 2M \le 4\varepsilon^{-2\lambda} \prod_{j=1}^d (1 + 2A_\lambda \omega^{\lambda a_j}).$$

Consequently,

$$\frac{\log N(\varepsilon,d)}{d+\log\varepsilon^{-1}} \le \frac{\log 4+2\lambda\log\varepsilon^{-1}+\sum_{j=1}^{d}\log(1+2A_\lambda\omega^{\lambda a_j})}{d+\log\varepsilon^{-1}}$$
$$\le \frac{\log 4}{d+\log\varepsilon^{-1}}+\frac{2\lambda\log\varepsilon^{-1}}{d+\log\varepsilon^{-1}}+\frac{2A_\lambda\sum_{j=1}^{d}\omega^{\lambda a_j}}{d+\log\varepsilon^{-1}}.$$

Note that $\lim_{j\to\infty} a_j=\infty$ implies $\lim_{j\to\infty}\omega^{\lambda a_j}=0$, and so

$$\lim_{d\to\infty}\frac{1}{d}\sum_{j=1}^{d}\omega^{\lambda a_j}=0.$$

This implies that

$$\limsup_{d+\varepsilon^{-1}\to\infty}\frac{\log N(\varepsilon,d)}{d+\log\varepsilon^{-1}}\le 2\lambda.$$

Since $\lambda\in(0,1]$ can be arbitrarily close to zero we obtain

$$\lim_{d+\varepsilon^{-1}\to\infty}\frac{\log N(\varepsilon,d)}{d+\log\varepsilon^{-1}}=0,$$

and this proves EXP-weak tractability.

With respect to Item 2 of the theorem, in order to prove the equivalence of EXP-polynomial and EXP-strong polynomial tractability, it suffices to show that EXP-polynomial tractability implies EXP-strong polynomial tractability.

We already know from Proposition 9.12 that EXP-polynomial tractability implies uniform exponential convergence, which in turn implies, by Corollary 9.10, $B<\infty$.

Furthermore, EXP-polynomial tractability implies (9.13). From Lemma 9.4 with $t_j=1$ for all $j\in[d]$ and $N=2^d-1$ we therefore obtain

$$\mathrm{e}^{1-(N/(Cd^{\tau}))^{1/\sigma}}\ge\omega^{2^{-1}\sum_{j=1}^{d}a_j}\,2^{-d/2}.$$

Taking the logarithm and reordering we get

$$\frac{d^{\tau/\sigma}\sum_{j=1}^{d}a_j}{2^{d/\sigma}}+\frac{2d^{\tau/\sigma}+d^{\tau/\sigma+1}\log 2}{2^{d/\sigma}\log\omega^{-1}}\ge\frac{2}{C^{1/\sigma}}\left(1-\frac{1}{2^d}\right)^{1/\sigma}\frac{1}{\log\omega^{-1}}.$$

This implies that

$$\liminf_{d\to\infty}\frac{d^{\tau/\sigma}\sum_{j=1}^{d}a_j}{2^{d/\sigma}}\ge\frac{2}{C^{1/\sigma}\log\omega^{-1}}>0.$$

Consequently, there exists a positive number β such that we have, for all sufficiently large d,

$$\sum_{j=1}^{d} a_j \geq \beta \frac{2^{d/\sigma}}{d^{\tau/\sigma}}.$$

Since the a_j are ordered we obtain $da_d \geq \sum_{j=1}^{d} a_j$, and so it holds for all sufficiently large d that

$$a_d \geq \beta \frac{2^{d/\sigma}}{d^{1+\tau/\sigma}}.$$

Taking the logarithm and dividing by d implies

$$\frac{\log a_d}{d} \geq \frac{\log 2}{\sigma} + \frac{\log \beta - (1+\tau/\sigma)\log d}{d}$$

for d large enough. Thus,

$$\alpha_* = \liminf_{d\to\infty} \frac{\log a_d}{d} \geq \frac{\log 2}{\sigma} > 0.$$

We have now shown that EXP-polynomial tractability implies $B < \infty$ and $\alpha_* > 0$. In the following step we will prove that these assumptions imply EXP-strong polynomial tractability.

For Item 3, it remains to prove that $B < \infty$ and $\alpha_* > 0$ imply EXP-strong polynomial tractability. We will show even more and explain how EXP-strong polynomial tractability and uniform exponential convergence can be obtained. To this end we again use a regular grid $\mathcal{G}_{N,d}$ of different mesh sizes $m_1, m_2, \ldots, m_d$ with

$$m_j = \left\lceil \left(\frac{\log M}{a_j^{\beta} \log \omega^{-1}} \right)^{1/b_j} \right\rceil \qquad \text{for } j \in [d],$$

where $M \geq 2$ and $\beta \in (0,1)$. Note that $m_j \geq 1$ for all $j \in [d]$ and $m_j = 1$ if $a_j \geq ((\log M)/(\log \omega^{-1}))^{1/\beta}$. By the assumption $\alpha_* > 0$ we obtain that for all $\delta \in (0, \alpha_*)$ there exists a positive integer j_δ^* such that

$$a_j \geq \mathrm{e}^{\delta j} \quad \text{for all } j \geq j_\delta^*.$$

We conclude that

$$j \geq j_{\beta,\delta}^* := \max\left(j_\delta^*, \frac{\log(((\log M)/(\log \omega^{-1}))^{1/\beta})}{\delta} \right) \qquad \text{implies } m_j = 1.$$

From Lemma 9.8 we have

$$[\mathrm{err}_{N,d}(\mathcal{H}_{\mathrm{kor},d,\boldsymbol{a},\boldsymbol{b}}, \mathcal{G}_{N,d})]^2 = -1 + \prod_{j=1}^{d} \left(1 + 2 \sum_{h=1}^{\infty} \omega^{a_j (m_j h)^{b_j}} \right).$$

The sum in the latter expression can be estimated in the form

$$\sum_{h=1}^{\infty} \omega^{a_j (m_j h)^{b_j}} \le \omega^{a_j m_j^{b_j}} \sum_{h=1}^{\infty} \omega^{a_j (h^{b_j}-1)} \le A\, \omega^{a_j m_j^{b_j}},$$

for $j \in [d]$, where $A := A_1 = \sum_{h=1}^{\infty} \omega^{a_*(h^{b_*}-1)} < \infty$. Hence

$$[\mathrm{err}_{N,d}(\mathcal{H}_{\mathrm{kor},d,\boldsymbol{a},\boldsymbol{b}}, \mathcal{G}_{N,d})]^2 \le -1 + \prod_{j=1}^{d} \left(1 + 2A\, \omega^{a_j m_j^{b_j}}\right).$$

From the definition of m_j we obtain $a_j m_j^{b_j} \ge a_j^{1-\beta} (\log M)/(\log \omega^{-1})$ and so

$$\omega^{a_j m_j^{b_j}} \le \omega^{a_j^{1-\beta} (\log M)/(\log \omega^{-1})} = \left(\frac{1}{M}\right)^{a_j^{1-\beta}}.$$

Therefore,

$$\begin{aligned}
\log\left(\prod_{j=1}^{d} \left(1 + 2A\, \omega^{a_j m_j^{b_j}}\right)\right) &= \sum_{j=1}^{d} \log\left(1 + 2A\, \omega^{a_j m_j^{b_j}}\right) \\
&\le 2A \sum_{j=1}^{d} \omega^{a_j m_j^{b_j}} \\
&\le 2A \sum_{j=1}^{d} \left(\frac{1}{M}\right)^{a_j^{1-\beta}}.
\end{aligned}$$

Using $a_j \ge a_*$ for all $j \le j^*_{\beta,\delta} - 1$ and $a_j \ge \mathrm{e}^{\delta j}$ for all $j \ge j^*_{\beta,\delta}$ we further obtain

$$\begin{aligned}
\log\left(\prod_{j=1}^{d} \left(1 + 2A\, \omega^{a_j m_j^{b_j}}\right)\right) &\le 2A \left(\frac{j^*_{\beta,\delta} - 1}{M^{a_*^{1-\beta}}} + \sum_{j=j^*_{\beta,\delta}}^{d} \left(\frac{1}{M}\right)^{\mathrm{e}^{(1-\beta)\delta j}}\right) \\
&\le \frac{C_{\beta,\delta}}{M^{\min(a_*^{1-\beta},1)}},
\end{aligned}$$

with

$$C_{\beta,\delta} := 2A \left(j^*_{\beta,\delta} - 1 + \sum_{j=j^*_{\beta,\delta}}^{\infty} \left(\frac{1}{2}\right)^{\mathrm{e}^{(1-\beta)\delta j} - 1}\right) < \infty,$$

where we used the assumption $M \ge 2$. In particular, this shows that $M \ge C_{\beta,\delta}^{1/\min(a_*^{1-\beta},1)}$ implies

$$\log\left(\prod_{j=1}^{d} \left(1 + 2A\, \omega^{a_j m_j^{b_j}}\right)\right) \le 1.$$

Using convexity it can be checked that $-1 + \mathrm{e}^x \le (\mathrm{e}-1)x$ for all $x \in [0,1]$. We choose

$$M = \max\left(\left(\varepsilon^{-2} C_{\beta,\delta}(\mathrm{e}-1)\right)^{1/\min(a_*^{1-\beta},1)}, \omega^{-a_*^\beta}, \omega^{-\mathrm{e}^{\beta\delta j^*_\delta}}, 2\right).$$

As $M \ge \left(\varepsilon^{-2} C_{\beta,\delta}(\mathrm{e}-1)\right)^{1/\min(a_*^{1-\beta},1)}$, we obtain

$$\begin{aligned}
\left[\mathrm{err}_{N,d}(\mathcal{H}_{\mathrm{kor},d,\boldsymbol{a},\boldsymbol{b}}, \mathcal{G}_{N,d})\right]^2 &\le -1 + \exp\left(\log\left(\prod_{j=1}^d \left(1 + 2A\,\omega^{a_j m_j^{b_j}}\right)\right)\right)\\
&\le -1 + \mathrm{e}^{C_{\beta,\delta}/M^{\min(a_*^{1-\beta},1)}}\\
&\le \frac{(\mathrm{e}-1)\, C_{\beta,\delta}}{M^{\min(a_*^{1-\beta},1)}}\\
&\le \varepsilon^2.
\end{aligned}$$

This shows that $\mathrm{err}_{N,d}(\mathcal{H}_{\mathrm{kor},d,\boldsymbol{a},\boldsymbol{b}}, \mathcal{G}_{N,d}) \le \varepsilon$. Hence, as $m_j = 1$ for $j \ge j^*_{\beta,\delta}$ and $M \ge \omega^{-a_*^\beta}$, which implies $(\log M)/(a_*^\beta \log \omega^{-1}) \ge 1$,

$$\begin{aligned}
N(\varepsilon, d) \le N &= \prod_{j=1}^{\min(d, j^*_{\beta,\delta})} m_j\\
&\le \prod_{j=1}^{\min(d, j^*_{\beta,\delta})} \left\lceil \left(\frac{\log M}{a_*^\beta \log \omega^{-1}}\right)^{1/b_j} \right\rceil\\
&\le 2^{\min(d, j^*_{\beta,\delta})} a_*^{-\beta B} \left(\frac{\log M}{\log \omega^{-1}}\right)^B,
\end{aligned}$$

where we also used $B < \infty$. By definition of $j^*_{\beta,\delta}$ and since $M \ge \omega^{-\mathrm{e}^{\beta\delta j^*_\delta}}$ we have

$$2^{j^*_{\beta,\delta}} = \left(\frac{\log M}{\log \omega^{-1}}\right)^{(\log 2)/(\beta\delta)}.$$

Therefore,

$$N(\varepsilon, d) \le a_*^{-\beta B} \left(\frac{\log M}{\log \omega^{-1}}\right)^{B + (\log 2)/(\beta\delta)} = \mathcal{O}\left(\left(1 + \log \varepsilon^{-1}\right)^{B + (\log 2)/(\beta\delta)}\right),$$

with the factor in the $\mathcal{O}$-notation depending only on $\beta, \delta, B, a_*, b_*$, and ω, but not on ε and d. This implies EXP-strong polynomial tractability with

$$\sigma = B + \frac{\log 2}{\beta\delta}.$$

Since β can be arbitrarily close to one, and δ can be arbitrarily close to α_*, the exponent σ_* of EXP-strong polynomial tractability is at most

$$B + \frac{\log 2}{\alpha_*}.$$

Furthermore, from Theorem 9.5, combined with the proof of the first point of Proposition 9.12, it follows that $\sigma_* \geq B$. For $\alpha_* = \infty$ we have $(\log 2)/\alpha_* = 0$.

This completes the proof of Theorem 9.13. □

Notes and Remarks

In this chapter we have discussed numerical integration in Korobov spaces of analytic functions. The study of this problem was initiated in the paper [49], however in a slightly different setting. The function space setting considered here was first considered in [148]. The present outline follows this paper. The notions of EXP-tractability were introduced in [149] (see also [46, 49, 148]). Theorem 9.13 on EXP-tractability of the integration problem in the Korobov space of analytic functions summarizes results from [46, 148, 219].

Methods similar to those used in this chapter can be applied to and also work for numerical integration in cosine spaces of analytic functions (see [117]) and in Hermite spaces of analytic functions over the whole $\mathbb{R}^d$ (see [116]).

Furthermore, also L_p-approximation has been studied in the context of analytic functions. We refer to [46, 117, 118, 119, 149, 150, 184, 256, 265] for information on this subject. For L_p-approximation for $p \in \{2, \infty\}$ in Korobov spaces of finite smoothness we refer to Chapters 13–15.

Chapter 10
Korobov's p-Sets

The so-called p-sets go back to definitions due to Korobov in the 1950s and Hua and Wang in the 1970s. Since then, these sets have been largely ignored because a number of other constructions have been discovered which yield a better convergence rate in terms of the cardinality of the point sets in integration rules. However, it was discovered later in [37] that p-sets perform very well with respect to the dependence on the dimension d. In this chapter we study the weighted star-discrepancy of p-sets and numerical integration of a sub-class of the Wiener algebra that consists of absolutely convergent Fourier series satisfying a Hölder condition. In the latter example no weights are necessary in order to get polynomial tractability of the integration problem.

10.1 The Construction of Korobov's p-Sets

In this section we introduce three types of point sets which are often collectively called Korobov's p-sets. Throughout this chapter let p be a prime number.

Definition 10.1 Let p be a prime number. We consider the following point sets in $[0,1)^d$.

- Define $\mathcal{P}_{p,d} = \{\boldsymbol{x}_0, \boldsymbol{x}_1, \ldots, \boldsymbol{x}_{p-1}\}$ with

$$\boldsymbol{x}_k = \left(\left\{ \frac{k}{p} \right\}, \left\{ \frac{k^2}{p} \right\}, \ldots, \left\{ \frac{k^d}{p} \right\} \right) \quad \text{for } k \in \{0, 1, \ldots, p-1\}.$$

- Define $\mathcal{Q}_{p^2,d} = \{\boldsymbol{x}_0, \boldsymbol{x}_1, \ldots, \boldsymbol{x}_{p^2-1}\}$ with

$$\boldsymbol{x}_k = \left(\left\{ \frac{k}{p^2} \right\}, \left\{ \frac{k^2}{p^2} \right\}, \ldots, \left\{ \frac{k^d}{p^2} \right\} \right) \quad \text{for } k \in \{0, 1, \ldots, p^2-1\}.$$

J. Dick et al., *Lattice Rules*, Springer Series in Computational Mathematics 58,
https://doi.org/10.1007/978-3-031-09951-9_10

- Define $\mathcal{R}_{p^2,d} = \{\boldsymbol{x}_{g,k} \ : \ g,k \in \{0,1,\ldots,p-1\}\}$ with

$$\boldsymbol{x}_{g,k} = \left(\left\{\frac{k}{p}\right\}, \left\{\frac{gk}{p}\right\}, \ldots, \left\{\frac{g^{d-1}k}{p}\right\}\right) \quad \text{for } g,k \in \{0,1,\ldots,p-1\}.$$

Note that we have

$$|\mathcal{P}_{p,d}| = p \ \text{ and } \ |Q_{p^2,d}| = |\mathcal{R}_{p^2,d}| = p^2.$$

The point set $\mathcal{P}_{p,d}$ was introduced by Korobov in [140], the set $Q_{p^2,d}$ also by Korobov in [137], and the set $\mathcal{R}_{p^2,d}$ by Hua and Wang in [115]. The term "p-sets" for the point sets $\mathcal{P}_{p,d}$, $Q_{p^2,d}$, and $\mathcal{R}_{p^2,d}$ was introduced by Hua and Wang, see [115, Section 4.3]. Obviously, Korobov's p-sets are not lattice point sets as defined in Chapter 1 and as studied in this book so far. However, in some sense their construction is very similar to that of rank-1 lattice point sets and they can be analyzed using similar tools, in particular by means of exponential sums.

Remark 10.2 A particular relation to lattice point sets can be observed for the p-set $\mathcal{R}_{p^2,d}$, which is the multi-set union of all Korobov lattice point sets $\mathcal{P}(\boldsymbol{g}_d(g), p)$ with modulus p as given in Definition 3.1, i.e.,

$$\mathcal{R}_{p^2,d} = \bigcup_{g=0}^{p-1} \mathcal{P}(\boldsymbol{g}_d(g), p).$$

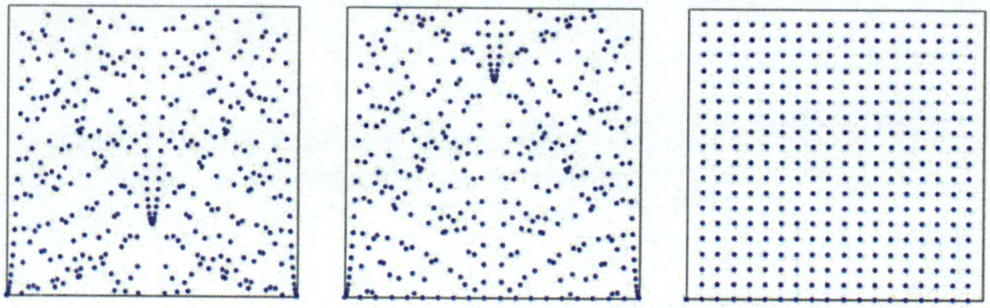

Fig. 10.1: The two-dimensional p-sets $\mathcal{P}_{359,2}$, $Q_{361,2}$, and $\mathcal{R}_{361,2}$ (left to right). We remark that the regular grid structure of the set $\mathcal{R}_{361,2}$ is a phenomenon that usually only applies to the two-dimensional case. For higher dimensions, this regular structure does not occur anymore, see also Figure 10.2.

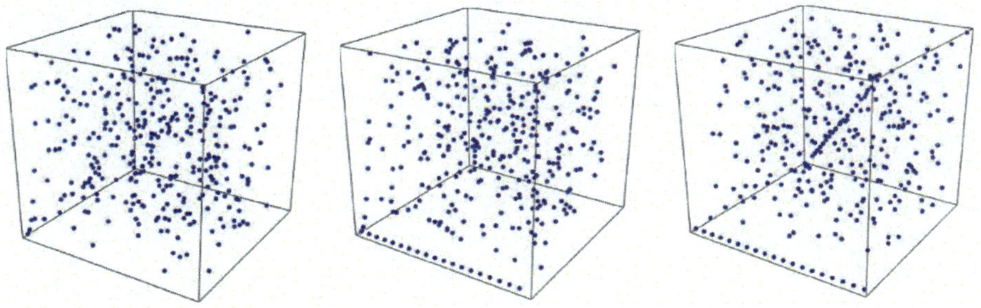

Fig. 10.2: The three-dimensional p-sets $\mathcal{P}_{359,3}$, $\mathcal{Q}_{361,3}$, and $\mathcal{R}_{361,3}$ (left to right).

10.2 The Weighted Star-Discrepancy of the p-Sets

The classical, unweighted star-discrepancy of the p-sets was studied in [115, Theorems 4.7–4.9]. There it was shown that for $\mathcal{P} \in \{\mathcal{P}_{p,d}, \mathcal{Q}_{p^2,d}, \mathcal{R}_{p^2,d}\}$ the star-discrepancy satisfies

$$D_N^*(\mathcal{P}) = O\left(\frac{(\log N)^d}{\sqrt{N}}\right), \quad \text{where } N = |\mathcal{P}|.$$

Here we consider the weighted star-discrepancy. The following result is due to [55].

Theorem 10.3 *Let p be a prime number. For arbitrary weights $\boldsymbol{\gamma} = \{\gamma_{\mathfrak{u}}\}_{\mathfrak{u}\subseteq[d]}$ we have*

$$D_{p,\boldsymbol{\gamma}}^*(\mathcal{P}_{p,d}) \le \frac{2}{\sqrt{p}} \max_{\emptyset\neq\mathfrak{u}\subseteq[d]} \gamma_{\mathfrak{u}}(\max \mathfrak{u})\,(4\log p)^{|\mathfrak{u}|},$$

$$D_{p^2,\boldsymbol{\gamma}}^*(\mathcal{Q}_{p^2,d}) \le \frac{3}{p} \max_{\emptyset\neq\mathfrak{u}\subseteq[d]} \gamma_{\mathfrak{u}}(\max \mathfrak{u})\,(6\log p)^{|\mathfrak{u}|}, \quad \textit{and}$$

$$D_{p^2,\boldsymbol{\gamma}}^*(\mathcal{R}_{p^2,d}) \le \frac{2}{p} \max_{\emptyset\neq\mathfrak{u}\subseteq[d]} \gamma_{\mathfrak{u}}(\max \mathfrak{u})\,(4\log p)^{|\mathfrak{u}|}.$$

Remark 10.4 Note that the point sets $\mathcal{P}_{p,d}, \mathcal{Q}_{p^2,d}$, and $\mathcal{R}_{p^2,d}$ in Theorem 10.3 are independent of the choice of the weights. In this sense Korobov's p-sets are universal. This is an advantage over the common CBC construction, where the resulting lattice point set depends on the given weights $\boldsymbol{\gamma}$. Note, however, that there has been recent progress on CBC constructions which work for different choices of weights simultaneously. We outline these results in Chapter 12 of this book.

For the proof of Theorem 10.3 we need a weighted version of Theorem 5.1, which we state in the following proposition.

Proposition 10.5 *For $M \in \mathbb{N}$, $M \geq 2$, and for $\boldsymbol{y}_0, \boldsymbol{y}_1, \ldots, \boldsymbol{y}_{N-1} \in \mathbb{Z}^d$, let $\mathcal{P} = \{\boldsymbol{x}_0, \boldsymbol{x}_1, \ldots, \boldsymbol{x}_{N-1}\}$ be the point set consisting of the fractional parts $\boldsymbol{x}_k = \{\boldsymbol{y}_k/M\}$ for $k \in \{0, 1, \ldots, N-1\}$. Then we have*

$$D^*_{N,\gamma}(\mathcal{P}) \leq \max_{\emptyset \neq \mathfrak{u} \subseteq [d]} \gamma_{\mathfrak{u}} \frac{|\mathfrak{u}|}{M} + \max_{\emptyset \neq \mathfrak{u} \subseteq [d]} \gamma_{\mathfrak{u}} \sum_{\boldsymbol{h} \in C^*_{|\mathfrak{u}|}(M)} \frac{1}{r_1(\boldsymbol{h})} \left| \frac{1}{N} \sum_{k=0}^{N-1} \mathrm{e}^{2\pi \mathrm{i} \boldsymbol{h} \cdot \boldsymbol{y}_{k,\mathfrak{u}}/M} \right|,$$

where $\boldsymbol{y}_{k,\mathfrak{u}} \in [0,1)^{|\mathfrak{u}|}$ is the projection of $\boldsymbol{y}_k$ onto the coordinates given by $\mathfrak{u}$.

Proof The result follows from a combination of Formula (5.5), Theorem 5.1, and Lemma 5.12. □

We shall also use the following estimates for exponential sums, which were already stated in the book of Hua and Wang [115].

Lemma 10.6 *Let p be a prime number and let $d \in \mathbb{N}$. Then for all $h_1, h_2, \ldots, h_d \in \mathbb{Z}$ such that $p \nmid h_j$ for at least one $j \in [d]$ we have*

$$\left| \sum_{k=0}^{p-1} \mathrm{e}^{2\pi \mathrm{i}(h_1 k + h_2 k^2 + \cdots + h_d k^d)/p} \right| \leq (d-1)\sqrt{p}, \tag{10.1}$$

$$\left| \sum_{k=0}^{p^2-1} \mathrm{e}^{2\pi \mathrm{i}(h_1 k + h_2 k^2 + \cdots + h_d k^d)/p^2} \right| \leq (d-1)p, \quad \text{and} \tag{10.2}$$

$$\left| \sum_{g=0}^{p-1} \sum_{k=0}^{p-1} \mathrm{e}^{2\pi \mathrm{i} k(h_1 + h_2 g + \cdots + h_d g^{d-1})/p} \right| \leq (d-1)p. \tag{10.3}$$

Proof Equation (10.1) follows from a bound due to Weil [262] on exponential sums, which is widely known as the *Weil bound* (see also [183]). For details we refer to [37]. For a proof of Equation (10.2) we refer to [115, Lemma 4.6]. It remains to show Equation (10.3). Under the assumption that $p \nmid \gcd(h_1, h_2, \ldots, h_d)$, the number of solutions of the congruence $h_1 + h_2 x + \cdots + h_d x^{d-1} \equiv 0 \pmod{p}$ in $\{0, 1, \ldots, p-1\}$ is at most $d-1$. Thus,

$$\left| \sum_{g=0}^{p-1} \sum_{k=0}^{p-1} \mathrm{e}^{2\pi \mathrm{i} k(h_1 + h_2 g + \cdots + h_d g^{d-1})/p} \right| = \left| \sum_{\substack{g=0 \\ h_1 + h_2 g + \cdots + h_d g^{d-1} \equiv 0 \pmod{p}}}^{p-1} p \right| \leq (d-1)p,$$

as claimed. □

We can now give the proof of Theorem 10.3.

Proof of Theorem 10.3 We estimate the exponential sum that appears in the upper bound in Proposition 10.5 with the help of Lemma 10.6. We do this separately for each of the three p-sets.

First, we consider $\mathcal{P}_{p,d}$. Here $\boldsymbol{x}_k$ is of the form $\boldsymbol{x}_k = \{\boldsymbol{y}_k/M\}$, where $\boldsymbol{y}_k = (k, k^2, \ldots, k^d) \in \mathbb{Z}^d$, for $k \in \{0, 1, \ldots, p-1\}$, and $M = p$.

For $\emptyset \neq \mathfrak{u} \subseteq [d]$ we obtain from (10.1) in Lemma 10.6,

$$\sum_{\boldsymbol{h} \in C^*_{|\mathfrak{u}|}(p)} \frac{1}{r_1(\boldsymbol{h})} \left| \frac{1}{p} \sum_{k=0}^{p-1} \mathrm{e}^{2\pi \mathrm{i} \boldsymbol{h} \cdot \boldsymbol{y}_{k,\mathfrak{u}}/p} \right| \leq \frac{1}{p} \sum_{\boldsymbol{h} \in C^*_{|\mathfrak{u}|}(p)} \frac{(\max \mathfrak{u})\sqrt{p}}{r_1(\boldsymbol{h})} \leq \frac{\max \mathfrak{u}}{\sqrt{p}} \left(1 + S_p\right)^{|\mathfrak{u}|},$$

where $S_p = \sum_{h \in C_1^*(p)} |h|^{-1}$. From (2.16) we obtain

$$S_p \leq 2\left(1 + \log\left(\frac{p}{2}\right)\right). \tag{10.4}$$

Therefore we get

$$\sum_{\boldsymbol{h} \in C^*_{|\mathfrak{u}|}(p)} \frac{1}{r_1(\boldsymbol{h})} \left| \frac{1}{p} \sum_{k=0}^{p-1} \mathrm{e}^{2\pi \mathrm{i} \boldsymbol{h} \cdot \boldsymbol{y}_{k,\mathfrak{u}}/p} \right| \leq \frac{\max \mathfrak{u}}{\sqrt{p}} \left(2 + 2\log\left(\frac{p}{2}\right)\right)^{|\mathfrak{u}|}.$$

Inserting this estimate into Proposition 10.5 gives

$$\begin{aligned} D^*_{p,\boldsymbol{\gamma}}(\mathcal{P}_{p,d}) &\leq \max_{\emptyset \neq \mathfrak{u} \subseteq [d]} \gamma_{\mathfrak{u}} \frac{|\mathfrak{u}|}{p} + \max_{\emptyset \neq \mathfrak{u} \subseteq [d]} \gamma_{\mathfrak{u}} \frac{\max \mathfrak{u}}{\sqrt{p}} \left(2 + 2\log\left(\frac{p}{2}\right)\right)^{|\mathfrak{u}|} \\ &\leq \frac{2}{\sqrt{p}} \max_{\emptyset \neq \mathfrak{u} \subseteq [d]} \gamma_{\mathfrak{u}} (\max \mathfrak{u}) \left(4 \log p\right)^{|\mathfrak{u}|}. \end{aligned}$$

Next, we consider $\mathcal{Q}_{p^2,d}$. Here $\boldsymbol{x}_k$ is of the form $\boldsymbol{x}_k = \{\boldsymbol{y}_k/M\}$, where $\boldsymbol{y}_k = (k, k^2, \ldots, k^d) \in \mathbb{Z}^d$, for $k \in \{0, 1, \ldots, p-1\}$, and $M = p^2$. For $\boldsymbol{h} = (h_j)_{j \in \mathfrak{u}} \in \mathbb{Z}^{|\mathfrak{u}|}$ we write $p | \boldsymbol{h}$ if $p | h_j$ for all $j \in \mathfrak{u}$ and $p \nmid \boldsymbol{h}$ if this is not the case.

For $\emptyset \neq \mathfrak{u} \subseteq [d]$ we obtain from (10.2) in Lemma 10.6,

$$\begin{aligned} &\sum_{\boldsymbol{h} \in C^*_{|\mathfrak{u}|}(p^2)} \frac{1}{r_1(\boldsymbol{h})} \left| \frac{1}{p^2} \sum_{k=0}^{p^2-1} \mathrm{e}^{2\pi \mathrm{i} \boldsymbol{h} \cdot \boldsymbol{y}_{k,\mathfrak{u}}/p^2} \right| \\ &\leq \sum_{\substack{\boldsymbol{h} \in C^*_{|\mathfrak{u}|}(p^2) \\ p | \boldsymbol{h}}} \frac{1}{r_1(\boldsymbol{h})} \left| \frac{1}{p^2} \sum_{k=0}^{p^2-1} \mathrm{e}^{2\pi \mathrm{i} \boldsymbol{h} \cdot \boldsymbol{y}_{k,\mathfrak{u}}/p^2} \right| + \frac{1}{p^2} \sum_{\substack{\boldsymbol{h} \in C^*_{|\mathfrak{u}|}(p^2) \\ p \nmid \boldsymbol{h}}} \frac{(\max \mathfrak{u}) p}{r_1(\boldsymbol{h})}. \end{aligned}$$

We consider the first sum in the latter expression, where $p | \boldsymbol{h}$. Let $\boldsymbol{m} = \boldsymbol{h}/p \in \mathbb{Z}^{\mathfrak{u}}$. Since $\boldsymbol{y}_{k,\mathfrak{u}} = (k^j)_{j \in \mathfrak{u}}$, we have from (10.1) in Lemma 10.6 that

$$\begin{aligned}\left|\frac{1}{p^2}\sum_{k=0}^{p^2-1} \mathrm{e}^{2\pi \mathrm{i}\boldsymbol{h}\cdot \boldsymbol{y}_{k,\mathfrak{u}}/p^2}\right| &= \left|\frac{1}{p^2}\sum_{\ell=0}^{p-1}\sum_{k=\ell p}^{\ell p+p-1} \mathrm{e}^{2\pi \mathrm{i}\boldsymbol{m}\cdot \boldsymbol{y}_{k,\mathfrak{u}}/p}\right|\\ &= \left|\frac{1}{p}\sum_{k=0}^{p-1} \mathrm{e}^{2\pi \mathrm{i}\boldsymbol{m}\cdot \boldsymbol{y}_{k,\mathfrak{u}}/p}\right|\\ &\le \frac{\max \mathfrak{u}}{\sqrt{p}}.\end{aligned}$$

Furthermore,

$$\sum_{\substack{\boldsymbol{h}\in C^*_{|\mathfrak{u}|}(p^2)\\ p|\boldsymbol{h}}} \frac{1}{r_1(\boldsymbol{h})} \le \frac{1}{p}\sum_{\boldsymbol{h}\in C^*_{|\mathfrak{u}|}(p^2)} \frac{1}{r_1(\boldsymbol{h})} \le \frac{1}{p}(1+S_{p^2})^{|\mathfrak{u}|},$$

where $S_{p^2} = \sum_{h\in C_1^*(p^2)} |h|^{-1}$. From (10.4) we get

$$S_{p^2} \le 2\left(1+\log\left(\frac{p^2}{2}\right)\right).$$

Thus we have

$$\sum_{\boldsymbol{h}\in C^*_{|\mathfrak{u}|}(p^2)} \frac{1}{r_1(\boldsymbol{h})}\left|\frac{1}{p^2}\sum_{k=0}^{p^2-1} \mathrm{e}^{2\pi \mathrm{i}\boldsymbol{h}\cdot \boldsymbol{y}_{k,\mathfrak{u}}/p^2}\right| \le \frac{2\max \mathfrak{u}}{p}\left(2+2\log\left(\frac{p^2}{2}\right)\right)^{|\mathfrak{u}|}.$$

Inserting this into Proposition 10.5 gives

$$\begin{aligned}D^*_{p^2,\boldsymbol{\gamma}}(\mathcal{Q}_{p^2,d}) &\le \max_{\emptyset\ne\mathfrak{u}\subseteq[d]} \gamma_{\mathfrak{u}}\frac{|\mathfrak{u}|}{p^2} + \max_{\emptyset\ne\mathfrak{u}\subseteq[d]} \gamma_{\mathfrak{u}}\frac{2(\max \mathfrak{u})}{p}\left(2+2\log\left(\frac{p^2}{2}\right)\right)^{|\mathfrak{u}|}\\ &\le \frac{3}{p}\max_{\emptyset\ne\mathfrak{u}\subseteq[d]} \gamma_{\mathfrak{u}}(\max \mathfrak{u})\,(6\log p)^{|\mathfrak{u}|}.\end{aligned}$$

Finally, let us consider $\mathcal{R}_{p^2,d}$. Here $\boldsymbol{x}_{g,k}$ is of the form $\boldsymbol{x}_{g,k} = \{\boldsymbol{y}_{g,k}/M\}$, where $\boldsymbol{y}_{g,k} = (gk, g^2k, \ldots, g^{d-1}k) \in \mathbb{Z}^d$, for $g, k \in \{0, 1, \ldots, p-1\}$, and $M = p$.

For $\emptyset \ne \mathfrak{u} \subseteq [d]$ we obtain from (10.3) in Lemma 10.6,

$$\begin{aligned}\sum_{\boldsymbol{h}\in C^*_{|\mathfrak{u}|}(p)} \frac{1}{r_1(\boldsymbol{h})}\left|\frac{1}{p^2}\sum_{g=0}^{p-1}\sum_{k=0}^{p-1} \mathrm{e}^{2\pi \mathrm{i}\boldsymbol{h}\cdot \boldsymbol{y}_{g,k,\mathfrak{u}}/p}\right| &\le \frac{1}{p^2}\sum_{\boldsymbol{h}\in C^*_{|\mathfrak{u}|}(p)} \frac{(\max \mathfrak{u})p}{r_1(\boldsymbol{h})}\\ &\le \frac{\max \mathfrak{u}}{p}\left(2+2\log\left(\frac{p}{2}\right)\right)^{|\mathfrak{u}|},\end{aligned}$$

where we used (10.4). Inserting this into Proposition 10.5 gives

$$D^*_{p^2,\boldsymbol{\gamma}}(\mathcal{R}_{p^2,d}) \le \max_{\emptyset \ne \mathfrak{u} \subseteq [d]} \gamma_{\mathfrak{u}} \frac{|\mathfrak{u}|}{p} + \max_{\emptyset \ne \mathfrak{u} \subseteq [d]} \gamma_{\mathfrak{u}} \frac{(\max \mathfrak{u})}{p} \left(2 + 2\log\left(\frac{p}{2}\right)\right)^{|\mathfrak{u}|}$$
$$\le \frac{2}{p} \max_{\emptyset \ne \mathfrak{u} \subseteq [d]} \gamma_{\mathfrak{u}} (\max \mathfrak{u}) \, (4 \log p)^{|\mathfrak{u}|}.$$

This concludes the proof of all three discrepancy estimates. □

The weighted star-discrepancy of the p-sets for product weights

We now consider product weights $\boldsymbol{\gamma} = (\gamma_j)_{j \ge 1}$ and study tractability properties. We again use the reasonable assumption that the sequence $(\gamma_j)_{j \ge 1}$ is nonincreasing, i.e., $\gamma_1 \ge \gamma_2 \ge \gamma_3 \cdots > 0$. The following result gives upper bounds on the discrepancies of the p-sets for this case.

Theorem 10.7 *Assume that the weights $\boldsymbol{\gamma} = (\gamma_j)_{j \ge 1}$ are nonincreasing such that*

$$A := \limsup_{j \to \infty} j\gamma_j < \infty. \tag{10.5}$$

Then for every $\delta > 0$ there exist positive reals $C_1(\boldsymbol{\gamma}, \delta)$, $C_2(\boldsymbol{\gamma}, \delta)$, and $C_3(\boldsymbol{\gamma}, \delta)$, which are independent of d and p, such that

$$D^*_{p,\boldsymbol{\gamma}}(\mathcal{P}_{p,d}) \le \frac{C_1(\boldsymbol{\gamma}, \delta)}{p^{1/2 - 4A - \delta}},$$
$$D^*_{p^2,\boldsymbol{\gamma}}(\mathcal{Q}_{p^2,d}) \le \frac{C_2(\boldsymbol{\gamma}, \delta)}{p^{1 - 6A - \delta}}, \quad \textit{and}$$
$$D^*_{p^2,\boldsymbol{\gamma}}(\mathcal{R}_{p^2,d}) \le \frac{C_3(\boldsymbol{\gamma}, \delta)}{p^{1 - 4A - \delta}}.$$

Proof We show the result for $\mathcal{P}_{p,d}$ only. The remaining results follow by the same arguments with possibly different constants. First we find from Condition (10.5) that there exists a $\Gamma_0 > 0$ such that $j\gamma_j \le \Gamma_0$ for all $j \in \mathbb{N}$. Thus for any finite set $\mathfrak{u} \subseteq \mathbb{N}$ we have

$$\gamma_{\max \mathfrak{u}} (\max \mathfrak{u}) \le \Gamma_0 < \infty.$$

Therefore,

$$D^*_{p,\boldsymbol{\gamma}}(\mathcal{P}_{p,d}) \le \frac{2}{\sqrt{p}} \max_{\emptyset \ne \mathfrak{u} \subseteq [d]} (\max \mathfrak{u}) \prod_{j \in \mathfrak{u}} (4\gamma_j \log p)$$
$$\le \frac{8\,\Gamma_0 \log p}{\sqrt{p}} \max_{\emptyset \ne \mathfrak{u} \subseteq [d]} \prod_{j \in \mathfrak{u} \setminus \{\max \mathfrak{u}\}} (4\gamma_j \log p)$$
$$\le \frac{8\,\Gamma_0 \log p}{\sqrt{p}} \max_{1 \le r \le d-1} \prod_{j=1}^{r} (4\gamma_j \log p),$$

where in the third inequality we used that the weights $\boldsymbol{\gamma}$ are nonincreasing.

Let $\ell = \ell(\boldsymbol{\gamma}, p)$ be the largest integer such that $4\gamma_\ell \log p > 1$. Then we have

$$D^*_{p,\boldsymbol{\gamma}}(\mathcal{P}_{p,d}) \le \frac{8\,\Gamma_0 \log p}{\sqrt{p}} \prod_{j=1}^{\ell} (4\gamma_j \log p).$$

Condition (10.5) implies that for any $\delta > 0$ there exists an $L = L(\boldsymbol{\gamma}, \delta) > 0$ such that $j\gamma_j < A + \delta$ for all $j \ge L$. We consider two cases.

If $\ell \ge L$, by the definition of ℓ we have

$$1 < 4\gamma_\ell \log p < \frac{4(A+\delta)\log p}{\ell},$$

and hence $\ell < 4(A+\delta)\log p$, or $\ell \le \lfloor 4(A+\delta)\log p \rfloor$, since ℓ is an integer. Therefore we obtain

$$\begin{aligned}\prod_{j=1}^{\ell}(4\gamma_j \log p) &= \prod_{j=1}^{L-1}(4\gamma_j \log p) \prod_{j=L}^{\ell}(4\gamma_j \log p)\\ &\le \frac{(4\Gamma_0 \log p)^{L-1}}{(L-1)!} \prod_{j=L}^{\lfloor 4(A+\delta)\log p \rfloor} \frac{4(A+\delta)\log p}{j}.\end{aligned}$$

Let $x := 4(A+\delta)\log p$. Then

$$\begin{aligned}\prod_{j=1}^{\ell}(4\gamma_j \log p) &\le \left(\frac{4\,\Gamma_0 \log p}{x}\right)^{L-1} \frac{x^{\lfloor x \rfloor}}{\lfloor x \rfloor !}\\ &\le \left(\frac{\Gamma_0}{A+\delta}\right)^{L-1} \mathrm{e}^x\\ &= \left(\frac{\Gamma_0}{A+\delta}\right)^{L-1} p^{4(A+\delta)}.\end{aligned}$$

Note that L only depends on $\boldsymbol{\gamma}$ and δ.

If $\ell < L$, we have

$$\begin{aligned}\prod_{j=1}^{\ell}(4\gamma_j \log p) &\le \frac{(4\Gamma_0 \log p)^{\ell}}{\ell!}\\ &= \left(\frac{4\,\Gamma_0 \log p}{x}\right)^{\ell} \frac{x^{\ell}}{\ell!}\\ &= \left(\frac{\Gamma_0}{A+\delta}\right)^{\ell} \mathrm{e}^x\\ &= \left(\frac{\Gamma_0}{A+\delta}\right)^{\ell} p^{4(A+\delta)}\end{aligned}$$

$$\leq \max\left(1, \left(\frac{\Gamma_0}{A+\delta}\right)^{L-1}\right) p^{4(A+\delta)}.$$

This gives, in any of the two cases,

$$\begin{aligned} D^*_{p,\boldsymbol{\gamma}}(\mathcal{P}_{p,d}) &\leq \frac{8\,\Gamma_0 \log p}{\sqrt{p}}\; p^{4(A+\delta)} \;\max\left(1, \left(\frac{\Gamma_0}{A+\delta}\right)^{L-1}\right) \\ &\leq \frac{C(\boldsymbol{\gamma},\delta)}{p^{1/2-4A-5\delta}}, \end{aligned}$$

with a suitably chosen positive real $C(\boldsymbol{\gamma},\delta)$. This observation then immediately yields the desired result. □

Strong polynomial tractability

If $\limsup_{j\to\infty} j\gamma_j$ is small enough we can deduce results on strong polynomial tractability for the weighted star-discrepancy from Theorem 10.7. For example, consider the p-set $\mathcal{P}_{p,d}$. Assume that $A := \limsup_{j\to\infty} j\gamma_j < 1/8$. Fix $\delta > 0$ such that

$$t := \frac{1}{2} - 4A - \delta > 0,$$

and choose $C_1(\boldsymbol{\gamma},\delta)$ accordingly. Let $\varepsilon > 0$ and let p be the smallest prime number that is greater than or equal to $\lceil (C_1(\boldsymbol{\gamma},\delta)\varepsilon^{-1})^{1/t}\rceil =: M$. Then we have

$$D^*_{p,\boldsymbol{\gamma}}(\mathcal{P}_{p,d}) \leq \varepsilon$$

and hence

$$N^*_{\boldsymbol{\gamma}}(\varepsilon,d) \leq p < 2M = 2\left\lceil (C_1(\boldsymbol{\gamma},\delta)\varepsilon^{-1})^{1/t}\right\rceil,$$

where we used Bertrand's postulate which implies that $M \leq p < 2M$. Consequently, the weighted star-discrepancy is strongly polynomially tractable with an ε-exponent of at most $2/(1-8A) > 0$. If, in particular, $A = 0$, then the ε-exponent is at most 2.

Remark 10.8 Note that $\Gamma := \sum_{j=1}^{\infty} \gamma_j < \infty$ and the monotonicity of the weights $\boldsymbol{\gamma}$ imply

$$A = \lim_{j\to\infty} j\gamma_j = 0. \tag{10.6}$$

Indeed, from $\sum_{j=1}^{\infty} \gamma_j < \infty$ it follows by the Cauchy condensation test that also $\sum_{k=0}^{\infty} 2^k \gamma_{2^k} < \infty$. In particular, $\lim_{k\to\infty} 2^k \gamma_{2^k} = 0$. This means that for any $\varepsilon > 0$ we have $\gamma_{2^k} \leq \varepsilon/2^{k+1}$ for sufficiently large k. Thus, for large enough j with $2^k \leq j < 2^{k+1}$ we obtain, using the monotonicity of $\boldsymbol{\gamma}$, that

$$\gamma_j \leq \gamma_{2^k} \leq \frac{\varepsilon}{2^{k+1}} < \frac{\varepsilon}{j}.$$

In particular, for sufficiently large j, we have $j\gamma_j < \varepsilon$. Since $\varepsilon > 0$ can be chosen arbitrarily close to zero this implies (10.6).

Obviously, the converse is not true in general. As a counterexample serves the sequence $(\gamma_j)_{j\geq 1}$ with $\gamma_1 = \gamma_2 = 1$ and $\gamma_j = 1/(j \log j)$ for $j \geq 3$.

10.3 Integration of Hölder Continuous Fourier Series

In the previous section we have seen that for suitably chosen weights strong polynomial tractability can be obtained for the weighted star-discrepancy by means of p-sets. It was surprising when it was discovered in [37] that p-sets may even yield some notion of tractability for unweighted problems. In this context we study integration of one-periodic functions with absolutely convergent Fourier series that satisfy a Hölder condition.

For $f \in L_2([0,1]^d)$ we can associate with f its Fourier series,

$$f(\boldsymbol{x}) \sim \sum_{\boldsymbol{h}\in\mathbb{Z}^d} \widehat{f}(\boldsymbol{h})\, \mathrm{e}^{2\pi \mathrm{i}\boldsymbol{h}\cdot\boldsymbol{x}}, \tag{10.7}$$

where the $\boldsymbol{h}$-th Fourier coefficient is given by

$$\widehat{f}(\boldsymbol{h}) = \int_{[0,1]^d} f(\boldsymbol{x})\, \mathrm{e}^{-2\pi \mathrm{i}\boldsymbol{h}\cdot\boldsymbol{x}}\, \mathrm{d}\boldsymbol{x} \quad \text{for } \boldsymbol{h} \in \mathbb{Z}^d.$$

For $\alpha \in (0,1]$ and $t \in [1,\infty]$ consider the norm

$$\|f\|_{K_{\alpha,t}} := \|f\|_{H_{\alpha,t}} + \sum_{\boldsymbol{h}\in\mathbb{Z}^d} |\widehat{f}(\boldsymbol{h})|,$$

where

$$\|f\|_{H_{\alpha,t}} := \sup_{\substack{\boldsymbol{x},\boldsymbol{h}\in[0,1]^d\\ \boldsymbol{x}+\boldsymbol{h}\in[0,1]^d}} \frac{|f(\boldsymbol{x}+\boldsymbol{h}) - f(\boldsymbol{x})|}{\|\boldsymbol{h}\|_t^{\alpha}}$$

is the *Hölder seminorm*, with $\|\cdot\|_t$ denoting the ℓ_t-norm.

We consider the following subclass of the Wiener algebra.

Definition 10.9 For $\alpha \in (0,1]$ and $t \in [1,\infty]$ define the subclass $K_{\alpha,t}$ of the *Wiener algebra* as

$$K_{\alpha,t} := \{f \in L_2([0,1]^d) \ :\ f \text{ is one-periodic and } \|f\|_{K_{\alpha,t}} < \infty\}.$$

Remark 10.10 Some brief remarks on the class $K_{\alpha,t}$ are in order.

1. For $f \in K_{\alpha,t}$ we also have $\|f\|_{H_{\alpha,t}} < \infty$. This means that f satisfies a Hölder condition and therefore is also continuous.

2. For $f \in K_{\alpha,t}$ the Fourier series (10.7) of f converges to f at every point $\boldsymbol{x} \in [0,1]^d$. This follows directly from [244, Corollary 1.8, p. 249], using that f is continuous.

Further information on $K_{\alpha,t}$ can be found in [37].

We study QMC integration in $K_{\alpha,t}$ by means of QMC rules based on the p-set $\mathcal{P}_{p,d}$. As the error criterion we consider the worst-case error as in Definition 1.26 extended over the unit ball of the class $K_{\alpha,t}$. The choice of t in the definition of the function class $K_{\alpha,t}$ influences the dependence on the dimension d of the upper bound on the worst-case error. The following theorem is according to [37].

Theorem 10.11 *Let $\mathcal{P}_{p,d}$ be as in Definition 10.1. Then, for $\alpha \in (0,1]$ and $t \in [1,\infty]$, we have*

$$\mathrm{err}_{p,d}(K_{\alpha,t},\mathcal{P}_{p,d}) \le \max\left(\frac{d-1}{\sqrt{p}}, \frac{d^{\alpha/t}}{p^\alpha}\right).$$

For the proof of Theorem 10.11 we require the following lemma.

Lemma 10.12 *For any $\alpha \in (0,1]$, $t \in [1,\infty]$, $f \in K_{\alpha,t}$, and $L \in \mathbb{N}$ we have*

$$\left|\sum_{\boldsymbol{h}\in\mathbb{Z}^d\setminus\{\boldsymbol{0}\}} \widehat{f}(L\boldsymbol{h})\right| \le \frac{d^{\alpha/t}}{L^\alpha}\,\|f\|_{H_{\alpha,t}}.$$

Proof Using the Fourier series expansion of f we obtain

$$\begin{aligned}\frac{1}{L^d}\sum_{\boldsymbol{k}\in\{0,1,\ldots,L-1\}^d} f\left(\frac{\boldsymbol{k}}{L}\right) &= \sum_{\boldsymbol{h}\in\mathbb{Z}^d} \widehat{f}(\boldsymbol{h})\,\frac{1}{L^d}\sum_{\boldsymbol{k}\in\{0,1,\ldots,L-1\}^d} \mathrm{e}^{2\pi\mathrm{i}\boldsymbol{h}\cdot\boldsymbol{k}/L}\\ &= \sum_{\boldsymbol{h}\in\mathbb{Z}^d}\widehat{f}(L\boldsymbol{h}),\end{aligned}$$

where the latter equality follows since

$$\frac{1}{L^d}\sum_{\boldsymbol{k}\in\{0,1,\ldots,L-1\}^d} \mathrm{e}^{2\pi\mathrm{i}\boldsymbol{h}\cdot\boldsymbol{k}/L} = \prod_{j=1}^d \frac{1}{L}\sum_{k=0}^{L-1}\mathrm{e}^{2\pi\mathrm{i}h_j k/L} = \begin{cases}1 & \text{if } L|\boldsymbol{h},\\ 0 & \text{otherwise,}\end{cases}$$

and where we write $L|\boldsymbol{h}$ if all coordinates of $\boldsymbol{h}$ are divisible by L (and $L\nmid\boldsymbol{h}$ otherwise). Therefore we get

$$\begin{aligned}\left|\sum_{\boldsymbol{h}\in\mathbb{Z}^d\setminus\{\boldsymbol{0}\}}\widehat{f}(L\boldsymbol{h})\right| &= \left|\frac{1}{L^d}\sum_{\boldsymbol{k}\in\{0,1,\ldots,L-1\}^d} f\left(\frac{\boldsymbol{k}}{L}\right) - \int_{[0,1]^d} f(\boldsymbol{x})\,\mathrm{d}\boldsymbol{x}\right|\\ &= \left|\frac{1}{L^d}\sum_{\boldsymbol{k}\in\{0,1,\ldots,L-1\}^d}\left(f\left(\frac{\boldsymbol{k}}{L}\right) - L^d\int_{I_L(\boldsymbol{k})} f(\boldsymbol{x})\,\mathrm{d}\boldsymbol{x}\right)\right|,\end{aligned}$$

where

$$\mathcal{I}_L(\boldsymbol{k}) := \left[\frac{\boldsymbol{k}}{L}, \frac{\boldsymbol{k}+\boldsymbol{1}}{L}\right) = \bigotimes_{j=1}^{d} \left[\frac{k_j}{L}, \frac{k_j+1}{L}\right) \quad \text{for } \boldsymbol{k} = (k_1, \ldots, k_d).$$

Since f is continuous, as it satisfies a Hölder condition, for every $\boldsymbol{k} \in \{0, 1, \ldots, L-1\}^d$ there is a $\boldsymbol{y}_{\boldsymbol{k}} \in \mathcal{I}_L(\boldsymbol{k})$ such that

$$L^d \int_{\mathcal{I}_L(\boldsymbol{k})} f(\boldsymbol{x}) \, \mathrm{d}\boldsymbol{x} = f(\boldsymbol{y}_{\boldsymbol{k}}).$$

Obviously, the distance $\|\boldsymbol{y}_{\boldsymbol{k}} - \boldsymbol{k}/L\|_t$ is not larger than the diameter of the box $\mathcal{I}_L(\boldsymbol{k})$ measured in the ℓ_t-norm, which is $d^{1/t}/L$. Therefore

$$\begin{aligned}
\left| \sum_{\boldsymbol{h} \in \mathbb{Z}^d \setminus \{\boldsymbol{0}\}} \widehat{f}(L\boldsymbol{h}) \right| &= \left| \frac{1}{L^d} \sum_{\boldsymbol{k} \in \{0,1,\ldots,L-1\}^d} \left(f\left(\frac{\boldsymbol{k}}{L}\right) - f(\boldsymbol{y}_{\boldsymbol{k}}) \right) \right| \\
&\le \frac{1}{L^d} \sum_{\boldsymbol{k} \in \{0,1,\ldots,L-1\}^d} \left| f\left(\frac{\boldsymbol{k}}{L}\right) - f(\boldsymbol{y}_{\boldsymbol{k}}) \right| \\
&\le \frac{1}{L^d} \sum_{\boldsymbol{k} \in \{0,1,\ldots,L-1\}^d} \|\boldsymbol{y}_{\boldsymbol{k}} - \boldsymbol{k}/L\|_t^{\alpha} \, \|f\|_{H_{\alpha,t}} \\
&\le \frac{d^{\alpha/t}}{L^{\alpha}} \|f\|_{H_{\alpha,t}},
\end{aligned}$$

as desired. □

We now present the proof of Theorem 10.11.

Proof of Theorem 10.11 Let $\mathcal{P}_{p,d} = \{\boldsymbol{x}_0, \boldsymbol{x}_1, \ldots, \boldsymbol{x}_{p-1}\}$. For $f \in K_{\alpha,t}$ we have

$$\begin{aligned}
&\left| \frac{1}{p} \sum_{k=0}^{p-1} f(\boldsymbol{x}_k) - \int_{[0,1]^d} f(\boldsymbol{x}) \, \mathrm{d}\boldsymbol{x} \right| \\
&= \left| \sum_{\boldsymbol{h} \in \mathbb{Z}^d \setminus \{\boldsymbol{0}\}} \widehat{f}(\boldsymbol{h}) \frac{1}{p} \sum_{k=0}^{p-1} \mathrm{e}^{2\pi \mathrm{i} \boldsymbol{h} \cdot \boldsymbol{x}_k} \right| \\
&\le \frac{1}{p} \sum_{\substack{\boldsymbol{h} \in \mathbb{Z}^d \\ p \nmid \boldsymbol{h}}} |\widehat{f}(\boldsymbol{h})| \left| \sum_{k=0}^{p-1} \mathrm{e}^{(2\pi \mathrm{i}/p)\, \boldsymbol{h} \cdot (k, k^2, \ldots, k^d)} \right| + \sum_{\substack{\boldsymbol{h} \in \mathbb{Z}^d \setminus \{\boldsymbol{0}\} \\ p \mid \boldsymbol{h}}} |\widehat{f}(\boldsymbol{h})|.
\end{aligned}$$

Next, we apply (10.1) in Lemma 10.6 to the first sum and Lemma 10.12 to the second sum in the latter term and obtain

$$\left| \frac{1}{p} \sum_{k=0}^{p-1} f(\boldsymbol{x}_k) - \int_{[0,1]^d} f(\boldsymbol{x}) \, \mathrm{d}\boldsymbol{x} \right| \le \frac{d-1}{\sqrt{p}} \sum_{\boldsymbol{h} \in \mathbb{Z}^d} |\widehat{f}(\boldsymbol{h})| + \frac{d^{\alpha/t}}{p^{\alpha}} \|f\|_{H_{\alpha,t}}$$

$$\leq \max\left(\frac{d-1}{\sqrt{p}}, \frac{d^{\alpha/t}}{p^{\alpha}}\right) \|f\|_{K_{\alpha,t}}.$$

From here the desired result follows immediately. □

Theorem 10.11 yields the following corollary.

Corollary 10.13 *Integration in $K_{\alpha,t}$ is polynomially tractable and polynomial tractability can be obtained by means of QMC rules based on the p-set $\mathcal{P}_{p,d}$.*

Proof Let $\varepsilon > 0$, and let p be the smallest prime number that is greater than or equal to

$$\left\lceil \max\left(d^2\varepsilon^{-2}, d^{1/t}\varepsilon^{-1/\alpha}\right)\right\rceil =: M.$$

Then Theorem 10.11 implies

$$\mathrm{err}_{p,d}(K_{\alpha,t}, \mathcal{P}_{p,d}) \leq \varepsilon,$$

and hence the information complexity of the integration problem in $K_{\alpha,t}$ satisfies

$$N(\varepsilon, d) \leq p < 2M = 2\left\lceil \max\left(d^2\varepsilon^{-2}, d^{1/t}\varepsilon^{-1/\alpha}\right)\right\rceil \leq 2(d^{\max(2,1/t)}\varepsilon^{-\max(2,1/\alpha)} + 1),$$

where we again used Bertrand's postulate. Therefore, integration in $K_{\alpha,t}$ is polynomially tractable. □

Notes and Remarks

Korobov's p-sets were introduced by Korobov in [137, 140] and by Hua and Wang in their book [115]. This book also provides a detailed analysis of the (unweighted) star-discrepancy of all three types of p-sets.

A construction of point sets with very similar properties as those of p-sets, and which is analyzed in the same vein as in the present chapter, is given in [40]. This construction is based on pseudo-random vectors, in particular, digital inverse vectors.

The p-sets, as well as the point sets based on digital inverse vectors, can also be efficiently employed for the numerical integration of Hölder-continuous, absolutely convergent cosine and Walsh series. See [37] and [40] for further information. A discussion of Korobov's p-sets in the context of the inverse of the star-discrepancy (see Example 1.45) can be found in [54].

Chapter 11
Lattice Rules in the Randomized Setting

In this chapter, we discuss a randomization method for rank-1 lattice rules that is an adaption of a technique introduced by Bakhvalov in [7] in 1961. We present a randomized algorithm $A^{\text{ran}}_{N,d}$ for numerical integration of elements of the weighted Korobov space $\mathcal{H}_{\text{kor},d,\alpha,\gamma}$ that uses at most N integration nodes and that is based on rank-1 lattice rules as building blocks.

We have already discussed randomized lattice rules above, as for example in Section 7.1, where we considered the situation where one fixed rank-1 lattice rule is transformed by adding a random shift. However, the route taken in this chapter is different, in the sense that the randomization considered here is due to a random selection of one rank-1 lattice rule out of several with varying sizes. Indeed, the random algorithm $A^{\text{ran}}_{N,d}$, for given $N \in \mathbb{N}$, will be defined as a rank-1 lattice rule with randomly chosen prime modulus $p \in \{\lceil N/2 \rceil + 1, \dots, N\}$ and then randomly chosen generating vector $\boldsymbol{g}$ from a certain set $\mathcal{Z}_p \subseteq G_d(p)$ of "good" generating vectors (see Algorithm 11.2 below).

Choosing the number of points p randomly is a necessary step in this setting. Indeed, consider the integration error in dimension one. Let us fix a prime number p, then the integration error of integrating an absolutely convergent Fourier series f by a rank-1 lattice rule is

$$\int_0^1 f(x)\,\mathrm{d}x - \frac{1}{p}\sum_{k=0}^{p-1} f\left(\left\{\frac{kg}{p}\right\}\right) = \sum_{k\in\mathbb{Z}\setminus\{0\}} \widehat{f}(kp),$$

which follows by Proposition 1.12. Thus the integration error on the right-hand side does not depend on g, and randomly choosing g does not provide any randomization. Furthermore, for fixed p, the Fourier coefficients whose frequencies are multiples of p always appear in the error term. As in Monte Carlo, we want to introduce a randomization which, at least with some nonzero probability, can integrate any Fourier coefficient. One way to achieve this is by choosing the number of points p, as well as the generating vector $\boldsymbol{g}$, randomly.

J. Dick et al., *Lattice Rules*, Springer Series in Computational Mathematics 58,
https://doi.org/10.1007/978-3-031-09951-9_11

An algorithm of such a form was first analyzed by Bakhvalov in [7] for generating vectors of Korobov type (see Section 3.2), i.e., for $\boldsymbol{g} = (1, g, \ldots, g^{d-1})$ for some $g \in \mathbb{Z} \setminus \{0\}$. Bakhvalov proved that this algorithm has almost the optimal order of convergence in a Sobolev space with dominating mixed smoothness, but the error bound depends on the dimension d. In [143], Bakhvalov's original idea was adapted such that the generating vectors under consideration are no longer restricted to the Korobov type. This has the significant advantage that the randomized lattice algorithm $A_{N,d}^{\text{ran}}$ can yield almost the optimal order of convergence in the Korobov space, with the error bound independent of d under the usual conditions on the weights $\boldsymbol{\gamma}$. However, we need to give some explanation of what we mean by "optimal order of convergence in the Korobov space" in this context. This is because we do not study the worst-case error of integration as in the previous chapters, but the randomized (worst-case) error of the algorithm $A_{N,d}^{\text{ran}}$ in the unit ball of $\mathcal{H}_{\text{kor},d,\alpha,\boldsymbol{\gamma}}$, defined by

$$\text{err}_{d,\alpha,\boldsymbol{\gamma}}^{\text{ran}}(A_{N,d}^{\text{ran}}) := \sup_{\substack{f \in \mathcal{H}_{\text{kor},d,\alpha,\boldsymbol{\gamma}} \\ \|f\|_{\text{kor},d,\alpha,\boldsymbol{\gamma}} \leq 1}} \mathbb{E}\left[|A_{N,d}^{\text{ran}}(f) - I_d(f)|\right]. \tag{11.1}$$

The details of the expectation in (11.1) will be made clear when we formally specify the randomized lattice algorithm in the following section.

The randomized lattice algorithm achieving an upper bound independent of the dimension was first discussed in [143]. However, this algorithm is difficult to implement in practice. In [39], a variation of this method was studied which can be implemented, but requires knowledge of the smoothness parameter α in advance. We follow the presentation in the latter paper in this chapter.

11.1 The Randomized Algorithm for Korobov Spaces

In this section, we present the algorithm $A_{N,d}^{\text{ran}}$ and show how it can be applied to elements of the weighted Korobov space $\mathcal{H}_{\text{kor},d,\alpha,\boldsymbol{\gamma}}$ (see Definition 2.16). To this end we introduce a weighted version of the quality criterion P_α from Definition 1.14.

Definition 11.1 For a given $d \in \mathbb{N}$, a prime number p, a generating vector $\boldsymbol{g} \in G_d(p)$, and given $\tau > 1$, define the quantity

$$P_{\tau,\boldsymbol{\gamma}}(\boldsymbol{g}, p) := \sum_{\boldsymbol{h} \in \mathcal{L}^\perp(\boldsymbol{g},p) \setminus \{\boldsymbol{0}\}} \frac{1}{r_{\tau,\boldsymbol{\gamma}}(\boldsymbol{h})},$$

where $\mathcal{L}^\perp(\boldsymbol{g}, p)$ is the dual lattice corresponding to $\mathcal{P}(\boldsymbol{g}, p)$ as introduced in Equation (1.7).

The randomized algorithm

Now, we will outline how the randomized algorithm works in detail. Let $N \in \mathbb{N}$, $N \geq 2$, and define the set

$$P_N := \{p \ : \ p \text{ is prime and } \lceil N/2 \rceil + 1 \leq p \leq N\},$$

which is nonempty due to Bertrand's postulate.

We are ready to formally define the randomized algorithm $A^{\mathrm{ran}}_{N,d}$.

Algorithm 11.2 (Randomized algorithm $A^{\mathrm{ran}}_{N,d}$) Let $N \in \mathbb{N}$, $N \geq 2$, and let $d \in \mathbb{N}$, $\alpha > 1/2$, and weights $\boldsymbol{\gamma} = \{\gamma_{\mathfrak{u}}\}_{\mathfrak{u} \subseteq [d]}$ be given. Choose $\eta \in (0, 1)$.

(1) Choose an element $p \in P_N$ randomly and uniformly.
(2) Set $g_1 = 1$.
(3) For s from 1 to $d - 1$:
Assume that $g_1, \ldots, g_s$ have already been found, and consider them as fixed.

(3a) Compute

$$P_{2\alpha,\boldsymbol{\gamma}}((g_1, \ldots, g_s, g), p)$$

for all values of $g \in \{1, 2, \ldots, p - 1\}$.
(3b) Construct a $\lceil \eta(p - 1) \rceil$-element set $Z_{s+1} \subseteq \{1, 2, \ldots, p - 1\}$ such that

$$P_{2\alpha,\boldsymbol{\gamma}}((g_1, \ldots, g_s, g), p) \leq P_{2\alpha,\boldsymbol{\gamma}}((g_1, \ldots, g_s, a), p)$$

for all $g \in Z_{s+1}$ and $a \in \{1, 2, \ldots, p - 1\} \setminus Z_{s+1}$. Randomly and uniformly pick g_{s+1} from Z_{s+1}.

End for.
(4) Set $A^{\mathrm{ran}}_{N,d}$ to be the p-element lattice rule with generating vector $\boldsymbol{g} := (g_1, \ldots, g_d)$.

It should be noted that the output of Algorithm 11.2 is a random generating vector chosen from a set of possible generating vectors. Based on the choice $g_1 = 1$, several choices are possible for g_2. Based on the choice of g_1 and g_2, several choices are possible for g_3, etc. The structure of the algorithm is such that one possible choice $(g_1, \ldots, g_d)$ is selected at random among several choices. Indeed, for a given p, we denote the set of possible generating vectors $\boldsymbol{g}$ generated by Algorithm 11.2 in Steps (2) and (3) by $\mathcal{Z}_{p,d,\alpha,\boldsymbol{\gamma},\eta}$. The size of this set is

$$|\mathcal{Z}_{p,d,\alpha,\boldsymbol{\gamma},\eta}| = 1 \times (\lceil \eta(p - 1) \rceil)^{d-1} \geq \eta^{d-1}(p - 1)^{d-1}.$$

The worst-case error

In the first step of our analysis of Algorithm 11.2, we prove a bound on the worst-case error, which is needed later to prove a bound on the randomized error. Algorithm 11.2 is a variation of the classical CBC construction principle in Algorithm 3.5, where in Step (3b) we do not choose the component which yields the smallest worst-case error, but we randomly select an element from the set of the "best" $\lceil \eta(p-1) \rceil$ choices (with respect to $P_{2\alpha,\gamma}$). Markov's inequality states that

$$\mathbb{P}[X \geq c] \leq \frac{\mathbb{E}[X]}{c}$$

for any nonnegative random variable X and any positive real c. This inequality can then be used to show that any choice $g_{s+1} \in Z_{s+1}$ in Step (3) of Algorithm 11.2 yields a generating vector which gives almost the optimal rate of convergence (the parameter $\eta \in (0, 1)$ only affects the constant).

The following result, which applies to Algorithm 11.2, is an extension of Theorem 2.24 in Section 2.6.

Theorem 11.3 *Let $d, N \in \mathbb{N}$, let $\alpha > 1/2$, let $\boldsymbol{\gamma} = \{\gamma_{\mathfrak{u}}\}_{\mathfrak{u} \subseteq [d]}$ be general weights and let $\eta \in (0, 1)$ be given. For any $p \in P_N$ and any generating vector $\mathbf{g} = (g_1, \ldots, g_d)$ found by Algorithm 11.2 we have, for every $s \in [d]$ and every $\tau \in [1/2, \alpha)$,*

$$\mathrm{err}_{p,s,\alpha,\gamma}((g_1, \ldots, g_s)) \leq \left(\frac{2}{(1-\eta)p} \sum_{\emptyset \neq \mathfrak{u} \subseteq [s]} \gamma_{\mathfrak{u}}^{1/(2\tau)} \left(2\zeta\left(\frac{\alpha}{\tau}\right)\right)^{|\mathfrak{u}|} \right)^{\tau}.$$

Proof According to Theorem 2.19 we have

$$\mathrm{err}_{p,s,\alpha,\gamma}((g_1, \ldots, g_s)) = (P_{2\alpha,\gamma}((g_1, \ldots, g_s), p))^{1/2}.$$

Hence it suffices to show that $(P_{2\alpha,\gamma}((g_1, \ldots, g_s), p))^{1/2}$ satisfies the desired bound. First we prove by induction on s that, for any $\lambda \in (1/(2\alpha), 1]$,

$$P_{2\alpha,\gamma}((g_1, \ldots, g_s), p) \leq \left(\frac{2}{(1-\eta)p} \sum_{\emptyset \neq \mathfrak{u} \subseteq [s]} \gamma_{\mathfrak{u}}^{\lambda} (2\zeta(2\alpha\lambda))^{|\mathfrak{u}|} \right)^{1/\lambda}. \tag{11.2}$$

For $s = 1$ we have $g_1 = 1$, and so

$$\begin{aligned} P_{2\alpha,\gamma}(g_1, p) &= \sum_{h \in \mathbb{Z} \setminus \{0\}} \frac{1}{r_{2\alpha,\gamma}(ph)} \\ &= 2\gamma_{\{1\}} \sum_{h=1}^{\infty} \frac{1}{(ph)^{2\alpha}} \\ &= \frac{2\gamma_{\{1\}}\zeta(2\alpha)}{p^{2\alpha}} \\ &\leq \left(\frac{2}{(1-\eta)p} \gamma_{\{1\}}^{\lambda} 2\zeta(2\alpha\lambda) \right)^{1/\lambda} \end{aligned}$$

for every $\lambda \in (1/(2\alpha), 1]$, where we used Jensen's inequality (see Lemma 2.25).

Assume that the result holds for dimension s for some fixed $s \in [d-1]$, i.e.,

$$P_{2\alpha,\gamma}((g_1,\ldots,g_s),p) \le \left(\frac{2}{(1-\eta)p}\sum_{\emptyset\neq\mathfrak{u}\subseteq[s]}\gamma_{\mathfrak{u}}^{\lambda}(2\zeta(2\alpha\lambda))^{|\mathfrak{u}|}\right)^{1/\lambda},$$

for any $\lambda \in (1/(2\alpha),1]$.

We write, again with some abuse of notation, $\boldsymbol{g}^{(s)} := (g_1,\ldots,g_s)$ and $(\boldsymbol{g}^{(s)},g_{s+1}) := (g_1,\ldots,g_s,g_{s+1})$. Recall from (4.5) the definition of $\mathcal{L}_{\mathfrak{u}}^{\perp}$ which is given by

$$\mathcal{L}_{\mathfrak{u}}^{\perp}(\boldsymbol{g},p) := \{\boldsymbol{h}_{\mathfrak{u}} \in (\mathbb{Z}\setminus\{0\})^{|\mathfrak{u}|} \,:\, \boldsymbol{h}_{\mathfrak{u}}\cdot\boldsymbol{g}_{\mathfrak{u}} \equiv 0 \pmod{p}\}.$$

From the definition of $P_{2\alpha,\gamma}((\boldsymbol{g}^{(s)},g_{s+1}),p)$ we have

$$\begin{aligned}
P_{2\alpha,\gamma}((\boldsymbol{g}^{(s)},g_{s+1}),p) &= \sum_{\boldsymbol{h}\in\mathcal{L}^{\perp}((\boldsymbol{g}^{(s)},g_{s+1}),p)\setminus\{\boldsymbol{0}\}} \frac{1}{r_{2\alpha,\gamma}(\boldsymbol{h})}\\
&= \sum_{\emptyset\neq\mathfrak{u}\subseteq[s+1]}\ \sum_{\boldsymbol{h}_{\mathfrak{u}}\in\mathcal{L}_{\mathfrak{u}}^{\perp}((\boldsymbol{g}^{(s)},g_{s+1}),p)} \frac{1}{r_{2\alpha,\gamma}(\boldsymbol{h}_{\mathfrak{u}})}\\
&= \sum_{\emptyset\neq\mathfrak{u}\subseteq[s]}\ \sum_{\boldsymbol{h}_{\mathfrak{u}}\in\mathcal{L}_{\mathfrak{u}}^{\perp}(\boldsymbol{g}^{(s)},p)} \frac{1}{r_{2\alpha,\gamma}(\boldsymbol{h}_{\mathfrak{u}})}\\
&\quad + \sum_{\substack{\mathfrak{u}\subseteq[s+1]\\ s+1\in\mathfrak{u}}}\ \sum_{\boldsymbol{h}_{\mathfrak{u}}\in\mathcal{L}_{\mathfrak{u}}^{\perp}((\boldsymbol{g}^{(s)},g_{s+1}),p)} \frac{1}{r_{2\alpha,\gamma}(\boldsymbol{h}_{\mathfrak{u}})}\\
&= P_{2\alpha,\gamma}(\boldsymbol{g}^{(s)},p) + \Theta_{\alpha,\gamma}(\boldsymbol{g}^{(s)},g_{s+1}),
\end{aligned}$$

where

$$\Theta_{\alpha,\gamma}(\boldsymbol{g}^{(s)},g_{s+1}) := \sum_{\substack{\mathfrak{u}\subseteq[s+1]\\ s+1\in\mathfrak{u}}}\ \sum_{\boldsymbol{h}_{\mathfrak{u}}\in\mathcal{L}_{\mathfrak{u}}^{\perp}((\boldsymbol{g}^{(s)},g_{s+1}),p)} \frac{1}{r_{2\alpha,\gamma}(\boldsymbol{h}_{\mathfrak{u}})}.$$

Next we need a bound on the average of $\Theta_{\alpha,\gamma}(\boldsymbol{g}^{(s)},g)$ over all possible choices of g. Such a bound was shown in the proof of Theorem 3.9, which gives

$$\frac{1}{p-1}\sum_{g=1}^{p-1}(\Theta_{\alpha,\gamma}(\boldsymbol{g}^{(s)},g_{s+1}))^{\lambda} \le \frac{1}{p-1}\sum_{\substack{\mathfrak{u}\subseteq[s+1]\\ s+1\in\mathfrak{u}}}\gamma_{\mathfrak{u}}^{\lambda}(2\zeta(2\alpha\lambda))^{|\mathfrak{u}|}.$$

Markov's inequality implies that there are at least $\lceil\eta(p-1)\rceil$ possible choices for the $(s+1)$-st component $g_{s+1} \in \{1,2,\ldots,p-1\}$ of the generating vector $\boldsymbol{g}$ such that

$$(\Theta_{\alpha,\gamma}(\boldsymbol{g}^{(s)},g_{s+1}))^{\lambda} \le \frac{1}{(1-\eta)(p-1)}\sum_{g=1}^{p-1}\Theta_{\alpha,\gamma}(\boldsymbol{g}^{(s)},g)^{\lambda}$$

$$\leq \frac{1}{(1-\eta)(p-1)} \sum_{\substack{\mathfrak{u} \subseteq [s+1] \\ s+1 \in \mathfrak{u}}} \gamma_{\mathfrak{u}}^{\lambda} (2\zeta(2\alpha\lambda))^{|\mathfrak{u}|},$$

or equivalently

$$\Theta_{\alpha,\boldsymbol{\gamma}}(\boldsymbol{g}^{(s)}, g_{s+1}) \leq \left(\frac{1}{(1-\eta)(p-1)} \sum_{\substack{\mathfrak{u} \subseteq [s+1] \\ s+1 \in \mathfrak{u}}} \gamma_{\mathfrak{u}}^{\lambda} (2\zeta(2\alpha\lambda))^{|\mathfrak{u}|} \right)^{1/\lambda}. \tag{11.3}$$

This means that the $\lceil \eta(p-1) \rceil$ choices of $g_{s+1} \in \{1, 2, \ldots, p-1\}$ with the smallest values of $\Theta_{\alpha,\boldsymbol{\gamma}}(\boldsymbol{g}^{(s)}, g_{s+1})$ satisfy the bound (11.3). The induction assumption and Jensen's inequality now lead to

$$\begin{aligned}
(P_{2\alpha,\boldsymbol{\gamma}}((\boldsymbol{g}^{(s)}, g_{s+1}), p))^{\lambda} &\leq (P_{2\alpha,\boldsymbol{\gamma}}(\boldsymbol{g}^{(s)}, p))^{\lambda} + (\Theta_{\alpha,\boldsymbol{\gamma}}(\boldsymbol{g}^{(s)}, g_{s+1}))^{\lambda} \\
&\leq \frac{2}{(1-\eta)p} \sum_{\emptyset \neq \mathfrak{u} \subseteq [s]} \gamma_{\mathfrak{u}}^{\lambda} (2\zeta(2\alpha\lambda))^{|\mathfrak{u}|} \\
&\quad + \frac{1}{(1-\eta)(p-1)} \sum_{\substack{\mathfrak{u} \subseteq [s+1] \\ s+1 \in \mathfrak{u}}} \gamma_{\mathfrak{u}}^{\lambda} (2\zeta(2\alpha\lambda))^{|\mathfrak{u}|} \\
&\leq \frac{2}{(1-\eta)p} \sum_{\emptyset \neq \mathfrak{u} \subseteq [s+1]} \gamma_{\mathfrak{u}}^{\lambda} (2\zeta(2\alpha\lambda))^{|\mathfrak{u}|}.
\end{aligned}$$

This concludes the proof of (11.2). Putting $\tau := 1/(2\lambda)$, which implies that $\tau \in [1/2, \alpha)$, gives the result in the theorem. □

The randomized error

The randomized (worst-case) error of $A_{N,d}^{\mathrm{ran}}$ is given by

$$\begin{aligned}
\mathrm{err}_{d,\alpha,\boldsymbol{\gamma}}^{\mathrm{ran}}(A_{N,d}^{\mathrm{ran}}) &:= \sup_{\substack{f \in \mathcal{H}_{\mathrm{kor},d,\alpha,\boldsymbol{\gamma}} \\ \|f\|_{\mathrm{kor},d,\alpha,\boldsymbol{\gamma}} \leq 1}} \mathbb{E}\left[|A_{N,d}^{\mathrm{ran}}(f) - I_d(f)| \right] \\
&= \sup_{\substack{f \in \mathcal{H}_{\mathrm{kor},d,\alpha,\boldsymbol{\gamma}} \\ \|f\|_{\mathrm{kor},d,\alpha,\boldsymbol{\gamma}} \leq 1}} \left(\frac{1}{|P_N|} \sum_{p \in P_N} \frac{1}{|\mathcal{Z}_{p,\eta}|} \sum_{\boldsymbol{g} \in \mathcal{Z}_{p,\eta}} \left| Q_{(\boldsymbol{g},p),d}(f) - I_d(f) \right| \right),
\end{aligned} \tag{11.4}$$

where $Q_{(\boldsymbol{g},p),d}$ denotes the p-element lattice rule with generating vector $\boldsymbol{g}$ obtained by Algorithm 11.2, and where we write $\mathcal{Z}_{p,\eta} = \mathcal{Z}_{p,d,\alpha,\boldsymbol{\gamma},\eta}$ for short. It is essential to stress here that the randomized error defined in (11.4) is not the same as the average of the worst-case errors of a set of deterministic rank-1 lattice rules, because the averaging occurs inside the supremum rather than outside. This is in contrast to the shift-averaged worst-case error as given in Definition 7.2.

The following theorem, which is the first main result in [39], shows that for sufficiently large N the convergence order of the randomized error of $A^{\mathrm{ran}}_{N,d}$ can be arbitrarily close to the optimal order of magnitude.

Theorem 11.4 *Let $d \in \mathbb{N}$, let $\alpha > 1/2$, let $\eta \in (0,1)$, and let $\boldsymbol{\gamma} = \{\gamma_{\mathfrak{u}}\}_{\mathfrak{u}\subseteq[d]}$ be weights with $\gamma_{\mathfrak{u}} \in [0,1]$ for $\mathfrak{u} \subseteq [d]$. For $N \in \mathbb{N}$ such that*

$$N \geq \inf_{1/2<\tau<\alpha} \frac{4}{1-\eta} \sum_{\emptyset\neq\mathfrak{u}\subseteq[d]} \gamma_{\mathfrak{u}}^{1/(2\tau)} \left(2\zeta\left(\frac{\alpha}{\tau}\right)\right)^{|\mathfrak{u}|} \tag{11.5}$$

it is true that the randomized error (11.4) of $A^{\mathrm{ran}}_{N,d}$ as in Algorithm 11.2 satisfies

$$\mathrm{err}^{\mathrm{ran}}_{d,\alpha,\boldsymbol{\gamma}}(A^{\mathrm{ran}}_{N,d}) \leq \frac{C_{\alpha,\tau,\delta,\eta}}{N^{\tau+1/2-\delta}} \left(\sum_{\emptyset\neq\mathfrak{u}\subseteq[d]} \gamma_{\mathfrak{u}}^{1/(2\tau)} \left(2\zeta\left(\frac{\alpha}{\tau}\right)\right)^{|\mathfrak{u}|} \right)^{\tau-\delta}$$

for any $\tau \in (1/2,\alpha)$ and $\delta \in (0, \min(\tau-1/2, 2\tau/\alpha))$, where the positive factor $C_{\alpha,\tau,\delta,\eta}$ depends only on α, τ, δ, and η.

Proof Considering the Fourier series of an individual function f in the unit ball of $\mathcal{H}_{\mathrm{kor},d,\alpha,\boldsymbol{\gamma}}$ and applying the character property of lattice rules (see Lemma 1.9 and Remark 1.10) and the triangle inequality we obtain

$$\begin{aligned}
&\mathrm{err}^{\mathrm{ran}}_{d,\alpha,\boldsymbol{\gamma}}(A^{\mathrm{ran}}_{N,d}) \\
&= \sup_{\substack{f\in\mathcal{H}_{\mathrm{kor},d,\alpha,\boldsymbol{\gamma}}\\ \|f\|_{\mathrm{kor},d,\alpha,\boldsymbol{\gamma}}\leq 1}} \frac{1}{|P_N|}\sum_{p\in P_N} \frac{1}{|\mathcal{Z}_{p,\eta}|}\sum_{\boldsymbol{g}\in\mathcal{Z}_{p,\eta}} \left| \sum_{\boldsymbol{h}\in\mathbb{Z}^d\setminus\{\boldsymbol{0}\}} \widehat{f}(\boldsymbol{h}) \frac{1}{p}\sum_{k=0}^{p-1} \mathrm{e}^{2\pi\mathrm{i}k\boldsymbol{h}\cdot\boldsymbol{g}/p} \right| \\
&\leq \sup_{\substack{f\in\mathcal{H}_{\mathrm{kor},d,\alpha,\boldsymbol{\gamma}}\\ \|f\|_{\mathrm{kor},d,\alpha,\boldsymbol{\gamma}}\leq 1}} \frac{1}{|P_N|}\sum_{p\in P_N} \frac{1}{|\mathcal{Z}_{p,\eta}|}\sum_{\boldsymbol{g}\in\mathcal{Z}_{p,\eta}} \sum_{\boldsymbol{h}\in\mathcal{L}^\perp(\boldsymbol{g},p)\setminus\{\boldsymbol{0}\}} |\widehat{f}(\boldsymbol{h})| \\
&= \sup_{\substack{f\in\mathcal{H}_{\mathrm{kor},d,\alpha,\boldsymbol{\gamma}}\\ \|f\|_{\mathrm{kor},d,\alpha,\boldsymbol{\gamma}}\leq 1}} \sum_{\boldsymbol{h}\in\mathbb{Z}^d\setminus\{\boldsymbol{0}\}} \sqrt{r_{2\alpha,\boldsymbol{\gamma}}(\boldsymbol{h})}\, |\widehat{f}(\boldsymbol{h})| \frac{\omega(\boldsymbol{h})}{\sqrt{r_{2\alpha,\boldsymbol{\gamma}}(\boldsymbol{h})}},
\end{aligned}$$

where for $\boldsymbol{h} \in \mathbb{Z}^d \setminus \{\boldsymbol{0}\}$ we write

$$\omega(\boldsymbol{h}) := \frac{1}{|P_N|}\sum_{p\in P_N} \frac{1}{|\mathcal{Z}_{p,\eta}|} \sum_{\substack{\boldsymbol{g}\in\mathcal{Z}_{p,\eta}\\ \boldsymbol{h}\in\mathcal{L}^\perp(\boldsymbol{g},p)}} 1.$$

If for a given $\boldsymbol{h} \in \mathbb{Z}^d \setminus \{\boldsymbol{0}\}$ there is no p and $\boldsymbol{g}$ such that $\boldsymbol{h} \in \mathcal{L}^\perp(\boldsymbol{g},p)$, then we set $\omega(\boldsymbol{h}) := 0$. Now, applying the Cauchy–Schwarz inequality, we get

$$\begin{aligned}
&\mathrm{err}^{\mathrm{ran}}_{d,\alpha,\boldsymbol{\gamma}}(A^{\mathrm{ran}}_{N,d}) \\
&\le \sup_{\substack{f\in\mathcal{H}_{\mathrm{kor},d,\alpha,\boldsymbol{\gamma}}\\ \|f\|_{\mathrm{kor},d,\alpha,\boldsymbol{\gamma}}\le 1}} \left(\sum_{\boldsymbol{h}\in\mathbb{Z}^d\setminus\{\boldsymbol{0}\}} r_{2\alpha,\boldsymbol{\gamma}}(\boldsymbol{h})|\widehat{f}(\boldsymbol{h})|^2\right)^{1/2} \left(\sum_{\boldsymbol{h}\in\mathbb{Z}^d\setminus\{\boldsymbol{0}\}} \frac{(\omega(\boldsymbol{h}))^2}{r_{2\alpha,\boldsymbol{\gamma}}(\boldsymbol{h})}\right)^{1/2} \\
&= \sup_{\substack{f\in\mathcal{H}_{\mathrm{kor},d,\alpha,\boldsymbol{\gamma}}\\ \|f\|_{\mathrm{kor},d,\alpha,\boldsymbol{\gamma}}\le 1}} \|f\|_{\mathrm{kor},d,\alpha,\boldsymbol{\gamma}} \left(\sum_{\boldsymbol{h}\in\mathbb{Z}^d\setminus\{\boldsymbol{0}\}} \frac{(\omega(\boldsymbol{h}))^2}{r_{2\alpha,\boldsymbol{\gamma}}(\boldsymbol{h})}\right)^{1/2} \\
&= \left(\sum_{\boldsymbol{h}\in\mathbb{Z}^d\setminus\{\boldsymbol{0}\}} \frac{(\omega(\boldsymbol{h}))^2}{r_{2\alpha,\boldsymbol{\gamma}}(\boldsymbol{h})}\right)^{1/2}.
\end{aligned}$$

Thus it suffices to give an upper bound on

$$B_N := \left(\sum_{\boldsymbol{h}\in\mathbb{Z}^d\setminus\{\boldsymbol{0}\}} \frac{(\omega(\boldsymbol{h}))^2}{r_{2\alpha,\boldsymbol{\gamma}}(\boldsymbol{h})}\right)^{1/2}.$$

Let us define

$$H_N := \inf_{1/2<\tau<\alpha} \left(\frac{4}{(1-\eta)N} \sum_{\emptyset\neq\mathfrak{u}\subseteq[d]} \gamma_{\mathfrak{u}}^{1/(2\tau)} \left(2\zeta\left(\frac{\alpha}{\tau}\right)\right)^{|\mathfrak{u}|}\right)^{2\tau}. \tag{11.6}$$

We note here that $H_N \le 1$ by the assumption (11.5). It follows from (11.2) that we have

$$\sum_{\boldsymbol{h}\in\mathcal{L}^{\perp}(\boldsymbol{g},p)\setminus\{\boldsymbol{0}\}} \frac{1}{r_{2\alpha,\boldsymbol{\gamma}}(\boldsymbol{h})} = P_{2\alpha,\boldsymbol{\gamma}}(\boldsymbol{g},p) \le H_N$$

for any $p \in P_N$ and any $\boldsymbol{g} \in \mathcal{Z}_{p,\eta}$. This means that for $\boldsymbol{h} \in \mathbb{Z}^d \setminus \{\boldsymbol{0}\}$ with

$$\frac{1}{r_{2\alpha,\boldsymbol{\gamma}}(\boldsymbol{h})} > H_N$$

it holds that $\boldsymbol{h} \notin \mathcal{L}^{\perp}(\boldsymbol{g},p)$ for all $p \in P_N$ and all $\boldsymbol{g} \in \mathcal{Z}_{p,\eta}$, such that

$$\omega(\boldsymbol{h}) = 0.$$

If $1/r_{2\alpha,\boldsymbol{\gamma}}(\boldsymbol{h}) \le H_N$, the following observation holds. If $p|\boldsymbol{h}$, i.e., every component of $\boldsymbol{h}$ is divisible by p, then such $\boldsymbol{h}$ is always included in the dual lattice $\mathcal{L}^{\perp}(\boldsymbol{g},p)$ for any choice of $\boldsymbol{g}$, since $\boldsymbol{h}\cdot\boldsymbol{g} \equiv 0 \pmod p$ holds.

Now we focus on the case $p \nmid \boldsymbol{h}$. Let $\mathfrak{u}$ be the nonempty set consisting of all $j \in [d]$ such that $p \nmid h_j$. Then the condition $\boldsymbol{h} \in \mathcal{L}^{\perp}(\boldsymbol{g},p)$ is equivalent to $\boldsymbol{h}_{\mathfrak{u}}\cdot\boldsymbol{g}_{\mathfrak{u}} \equiv 0 \pmod p$. Note that we can first exclude the case $\mathfrak{u} = \{1\}$, since it cannot occur due to the fact that $g_1 = 1$. For the remaining sets $\mathfrak{u}$, defining $\ell := \max \mathfrak{u}$, such that $\ell \ge 2$, the condition can be further rewritten as

$$h_\ell g_\ell \equiv -\boldsymbol{h}_{\mathfrak{u}\setminus\{\ell\}} \cdot \boldsymbol{g}_{\mathfrak{u}\setminus\{\ell\}} \equiv -\boldsymbol{h}_{[\ell-1]} \cdot \boldsymbol{g}_{[\ell-1]} \pmod{p}. \tag{11.7}$$

If $p|\boldsymbol{h}_{[\ell-1]} \cdot \boldsymbol{g}_{[\ell-1]}$, no $g_\ell \in \{1, 2, \dots, p-1\}$ satisfies (11.7). If this is not the case, as p is a prime, there is exactly one g_ℓ which satisfies (11.7), although such a solution may not be in the set $\mathcal{Z}_{p,\eta}$. It follows from the structure of $\mathcal{Z}_{p,\eta}$ that we need to consider at most $1 \times (\lceil \eta(p-1) \rceil)^{\ell-2}$ patterns of $\boldsymbol{g}_{[\ell-1]}$ occurring in elements of $\mathcal{Z}_{p,\eta}$, for each of which there is at most one possible choice of g_ℓ. Furthermore, the number of possible patterns for the remaining components $\boldsymbol{g}_{\{\ell+1,\dots,d\}}$ for each $\boldsymbol{g}_{[\ell]}$ is exactly $(\lceil \eta(p-1) \rceil)^{d-\ell}$, so that the total number of $\boldsymbol{g} \in \mathcal{Z}_{p,\eta}$ such that $\boldsymbol{h} \in \mathcal{L}^\perp(\boldsymbol{g}, p)$ is at most $(\lceil \eta(p-1) \rceil)^{\ell-2} \times (\lceil \eta(p-1) \rceil)^{d-\ell} = (\lceil \eta(p-1) \rceil)^{d-2}$.

It follows from the above argument that

$$\frac{1}{|\mathcal{Z}_{p,\eta}|} \sum_{\substack{\boldsymbol{g} \in \mathcal{Z}_{p,\eta} \\ \boldsymbol{h} \in \mathcal{L}^\perp(\boldsymbol{g},p)}} 1 \le \begin{cases} 1 & \text{if } p|\boldsymbol{h}, \\ \dfrac{(\lceil \eta(p-1) \rceil)^{d-2}}{|\mathcal{Z}_{p,\eta}|} = \dfrac{1}{\lceil \eta(p-1) \rceil} \le \dfrac{2}{\eta N} & \text{if } p \nmid \boldsymbol{h}. \end{cases}$$

Therefore we have

$$\omega(\boldsymbol{h}) \le \frac{1}{|P_N|} \left[\sum_{\substack{p \in P_N \\ p|\boldsymbol{h}}} 1 + \frac{2}{\eta N} \sum_{\substack{p \in P_N \\ p \nmid \boldsymbol{h}}} 1 \right] \le \frac{1}{|P_N|} \sum_{\substack{p \in P_N \\ p|\boldsymbol{h}}} 1 + \frac{2}{\eta N}.$$

Note that for given $M \in \mathbb{N}$ any number $h \in \mathbb{N}$ has at most $\log_M h$ prime divisors greater than M. So, for $\boldsymbol{h} \neq \boldsymbol{0}$, the number of primes $p \ge \lceil N/2 \rceil + 1$ for which $\boldsymbol{h} \equiv \boldsymbol{0} \pmod{p}$ is at most $\log_{\lceil N/2 \rceil + 1}(\|\boldsymbol{h}\|_\infty)$, i.e.,

$$\sum_{\substack{p \in P_N \\ p|\boldsymbol{h}}} 1 \le \log_{\lceil N/2 \rceil + 1}(\|\boldsymbol{h}\|_\infty) = \frac{\log(\|\boldsymbol{h}\|_\infty)}{\log(\lceil N/2 \rceil + 1)} \le \frac{2\log(\|\boldsymbol{h}\|_\infty)}{\log N}$$

for all $N \ge 2$. It is well known from number theory that $|P_N| > c'N/\log N$ for some absolute constant $c' > 0$, which can be deduced from the prime number theorem. Using explicit lower and upper bounds on the prime-counting function, see, for example, [221, Corollary 3], the constant c' can be made concrete.) We conclude that

$$\omega(\boldsymbol{h}) \le \frac{2\log(\|\boldsymbol{h}\|_\infty)}{c' N} + \frac{2}{\eta N} \le c \, \frac{\log(1 + \|\boldsymbol{h}\|_\infty)}{\eta N} \tag{11.8}$$

for some real $c > 0$ that is independent of d, N, and η.

For what follows, recall that $\mathfrak{u}(\boldsymbol{h}) := \{j \in [d] \ : \ h_j \neq 0\}$ for $\boldsymbol{h} = (h_1, \dots, h_d) \in \mathbb{Z}^d$. Substituting the bound (11.8) on $\omega(\boldsymbol{h})$ into B_N first, and then using the elementary inequality $\log(1 + x) \le x^\beta/\beta$, which holds for any $\beta \in (0, 1)$ and $x > 0$, we obtain

$$
\begin{aligned}
B_N &= \left(\sum_{\boldsymbol{h} \in \mathbb{Z}^d \setminus \{\boldsymbol{0}\}} \frac{(\omega(\boldsymbol{k}))^2}{r_{2\alpha,\boldsymbol{\gamma}}(\boldsymbol{h})} \right)^{1/2} \\
&\le \frac{c}{\eta N} \left(\sum_{\substack{\boldsymbol{h} \in \mathbb{Z}^d \setminus \{\boldsymbol{0}\} \\ 1/r_{2\alpha,\boldsymbol{\gamma}}(\boldsymbol{h}) \le H_N}} (\log(1 + \|\boldsymbol{h}\|_\infty))^2 \; \gamma_{\mathfrak{u}(\boldsymbol{h})} \prod_{j \in \mathfrak{u}(\boldsymbol{h})} \frac{1}{|h_j|^{2\alpha}} \right)^{1/2} \\
&\le \frac{c}{\beta \eta N} \left(\sum_{\substack{\boldsymbol{h} \in \mathbb{Z}^d \setminus \{\boldsymbol{0}\} \\ 1/r_{2\alpha,\boldsymbol{\gamma}}(\boldsymbol{h}) \le H_N}} \|\boldsymbol{h}\|_\infty^{2\beta} \; \gamma_{\mathfrak{u}(\boldsymbol{h})} \prod_{j \in \mathfrak{u}(\boldsymbol{h})} \frac{1}{|h_j|^{2\alpha}} \right)^{1/2} \\
&\le \frac{c}{\beta \eta N} \left(\sum_{\substack{\boldsymbol{h} \in \mathbb{Z}^d \setminus \{\boldsymbol{0}\} \\ 1/r_{2\alpha,\boldsymbol{\gamma}}(\boldsymbol{h}) \le H_N}} \gamma_{\mathfrak{u}(\boldsymbol{h})} \prod_{j \in \mathfrak{u}(\boldsymbol{h})} \frac{1}{|h_j|^{2(\alpha-\beta)}} \right)^{1/2} \\
&\le \frac{c}{\beta \eta N} \left(\sum_{\substack{\boldsymbol{h} \in \mathbb{Z}^d \setminus \{\boldsymbol{0}\} \\ 1/r_{2\alpha,\boldsymbol{\gamma}}(\boldsymbol{h}) \le H_N}} \left(\gamma_{\mathfrak{u}(\boldsymbol{h})} \prod_{j \in \mathfrak{u}(\boldsymbol{h})} \frac{1}{|h_j|^{2\alpha}} \right)^{1-\beta/\alpha} \right)^{1/2} \\
&= \frac{c}{\beta \eta N} \left(\sum_{\substack{\boldsymbol{h} \in \mathbb{Z}^d \setminus \{\boldsymbol{0}\} \\ 1/r_{2\alpha,\boldsymbol{\gamma}}(\boldsymbol{h}) \le H_N}} \frac{1}{(r_{2\alpha,\boldsymbol{\gamma}}(\boldsymbol{h}))^{1-\beta/\alpha}} \right)^{1/2},
\end{aligned}
$$

for any $\beta \in (0, \min(\alpha - 1/2, 1))$. We shall restrict the range of β slightly further below in order to obtain a suitable upper bound on the latter expression.

For $\ell \in \mathbb{N}$, let us define

$$
A_{\alpha,\boldsymbol{\gamma}}(\ell) := \sum_{\substack{\boldsymbol{h} \in \mathbb{Z}^d \setminus \{\boldsymbol{0}\} \\ r_{2\alpha,\boldsymbol{\gamma}}(\boldsymbol{h}) < \ell}} 1.
$$

As it holds that $H_N \le 1$ by the assumption (11.5), we have $\lfloor 1/H_N \rfloor \ge 1$. With this notation, we obtain

$$
\sum_{\substack{\boldsymbol{h} \in \mathbb{Z}^d \setminus \{\boldsymbol{0}\} \\ 1/r_{2\alpha,\boldsymbol{\gamma}}(\boldsymbol{h}) \le H_N}} \frac{1}{(r_{2\alpha,\boldsymbol{\gamma}}(\boldsymbol{h}))^{1-\beta/\alpha}} \le \sum_{\ell = \lfloor 1/H_N \rfloor}^{\infty} \sum_{\substack{\boldsymbol{h} \in \mathbb{Z}^d \setminus \{\boldsymbol{0}\} \\ \ell \le r_{2\alpha,\boldsymbol{\gamma}}(\boldsymbol{h}) < \ell+1}} \frac{1}{(r_{2\alpha,\boldsymbol{\gamma}}(\boldsymbol{h}))^{1-\beta/\alpha}}
$$

$$
\begin{aligned}
&\le \sum_{\ell=\lfloor 1/H_N \rfloor}^{\infty} \frac{1}{\ell^{1-\beta/\alpha}} \sum_{\substack{\boldsymbol{h}\in\mathbb{Z}^d\setminus\{\boldsymbol{0}\} \\ \ell \le r_{2\alpha,\boldsymbol{\gamma}}(\boldsymbol{h}) < \ell+1}} 1 \\
&= \sum_{\ell=\lfloor 1/H_N \rfloor}^{\infty} \frac{1}{\ell^{1-\beta/\alpha}} \left(A_{\alpha,\boldsymbol{\gamma}}(\ell+1) - A_{\alpha,\boldsymbol{\gamma}}(\ell)\right) \\
&\le \sum_{\ell=\lfloor 1/H_N \rfloor}^{\infty} A_{\alpha,\boldsymbol{\gamma}}(\ell+1) \left(\frac{1}{\ell^{1-\beta/\alpha}} - \frac{1}{(\ell+1)^{1-\beta/\alpha}}\right) \\
&\le \sum_{\ell=\lfloor 1/H_N \rfloor}^{\infty} \frac{A_{\alpha,\boldsymbol{\gamma}}(\ell+1)}{\ell^{1-\beta/\alpha}(\ell+1)^{1-\beta/\alpha}}.
\end{aligned}
$$

For any $\tau \in (1/2, \alpha)$ it follows from the definition of $A_{\alpha,\boldsymbol{\gamma}}$ that

$$
\begin{aligned}
\frac{A_{\alpha,\boldsymbol{\gamma}}(\ell+1)}{(\ell+1)^{1/(2\tau)}} &= \sum_{\substack{\boldsymbol{h}\in\mathbb{Z}^d\setminus\{\boldsymbol{0}\} \\ r_{2\alpha,\boldsymbol{\gamma}}(\boldsymbol{h}) < \ell+1}} \frac{1}{(\ell+1)^{1/(2\tau)}} \\
&\le \sum_{\boldsymbol{h}\in\mathbb{Z}^d\setminus\{\boldsymbol{0}\}} \frac{1}{(r_{2\alpha,\boldsymbol{\gamma}}(\boldsymbol{h}))^{1/(2\tau)}} \\
&= \sum_{\emptyset\neq\mathfrak{u}\subseteq[d]} \gamma_{\mathfrak{u}}^{1/(2\tau)} \left(2\zeta\left(\frac{\alpha}{\tau}\right)\right)^{|\mathfrak{u}|},
\end{aligned}
$$

which leads to

$$
\begin{aligned}
&\sum_{\substack{\boldsymbol{h}\in\mathbb{Z}^d\setminus\{\boldsymbol{0}\} \\ 1/r_{2\alpha,\boldsymbol{\gamma}}(\boldsymbol{h}) \le H_N}} \frac{1}{(r_{2\alpha,\boldsymbol{\gamma}}(\boldsymbol{h}))^{1-\beta/\alpha}} \\
&\le \sum_{\ell=\lfloor 1/H_N \rfloor}^{\infty} \frac{A_{\alpha,\boldsymbol{\gamma}}(\ell+1)}{\ell^{1-\beta/\alpha}(\ell+1)^{1-\beta/\alpha}} \\
&\le \sum_{\emptyset\neq\mathfrak{u}\subseteq[d]} \gamma_{\mathfrak{u}}^{1/(2\tau)} \left(2\zeta\left(\frac{\alpha}{\tau}\right)\right)^{|\mathfrak{u}|} \sum_{\ell=\lfloor 1/H_N \rfloor}^{\infty} \frac{1}{\ell^{1-\beta/\alpha}(\ell+1)^{1-\beta/\alpha-1/(2\tau)}} \\
&\le \sum_{\emptyset\neq\mathfrak{u}\subseteq[d]} \gamma_{\mathfrak{u}}^{1/(2\tau)} \left(2\zeta\left(\frac{\alpha}{\tau}\right)\right)^{|\mathfrak{u}|} \sum_{\ell=\lfloor 1/H_N \rfloor}^{\infty} \frac{1}{\ell^{2(1-\beta/\alpha)-1/(2\tau)}},
\end{aligned}
$$

for $\beta \in (0, \min(\alpha(1-1/(2\tau))/2, 1))$ (note that for every $\tau \in (1/2, \alpha)$ it holds that $\alpha(1-1/(2\tau))/2 < \alpha - 1/2$). The innermost sum over ℓ in the latter expression is bounded by

$$\begin{aligned}
&\sum_{\ell=\lfloor 1/H_N \rfloor}^{\infty} \frac{1}{\ell^{2(1-\beta/\alpha)-1/(2\tau)}} \\
&\leq \frac{1}{(\lfloor 1/H_N \rfloor)^{2(1-\beta/\alpha)-1/(2\tau)}} + \int_{\lfloor 1/H_N \rfloor}^{\infty} \frac{1}{x^{2(1-\beta/\alpha)-1/(2\tau)}} \,\mathrm{d}x \\
&= \frac{1}{(\lfloor 1/H_N \rfloor)^{2(1-\beta/\alpha)-1/(2\tau)}} + \frac{1}{2(1-\beta/\alpha)-1/(2\tau)-1} \frac{1}{(\lfloor 1/H_N \rfloor)^{2(1-\beta/\alpha)-1/(2\tau)-1}} \\
&\leq \left(1 + \frac{1}{1-1/(2\tau)-2\beta/\alpha}\right) (2H_N)^{1-1/(2\tau)-2\beta/\alpha}.
\end{aligned}$$

Therefore, for any $\tau \in (1/2, \alpha)$ and $\beta \in (0, \min(\alpha(1-1/(2\tau))/2, 1))$, we get

$$\begin{aligned}
B_N &\leq \frac{c}{\beta\eta N} \left(\sum_{\substack{\boldsymbol{h} \in \mathbb{Z}^d \setminus \{\boldsymbol{0}\} \\ 1/r_{2\alpha,\boldsymbol{\gamma}}(\boldsymbol{h}) \leq H_N}} \frac{1}{(r_{2\alpha,\boldsymbol{\gamma}}(\boldsymbol{h}))^{1-\beta/\alpha}} \right)^{1/2} \\
&\leq \frac{c}{\beta\eta N} \left(1 + \frac{1}{1-1/(2\tau)-2\beta/\alpha}\right)^{1/2} (2H_N)^{(1-1/(2\tau))/2-\beta/\alpha} \\
&\quad \times \left(\sum_{\emptyset \neq \mathfrak{u} \subseteq [d]} \gamma_{\mathfrak{u}}^{1/(2\tau)} \left(2\zeta\left(\frac{\alpha}{\tau}\right)\right)^{|\mathfrak{u}|} \right)^{1/2} \\
&\leq \frac{C_{\alpha,\tau,\beta,\eta}}{N^{\tau+1/2-2\tau\beta/\alpha}} \left(\sum_{\emptyset \neq \mathfrak{u} \subseteq [d]} \gamma_{\mathfrak{u}}^{1/(2\tau)} \left(2\zeta\left(\frac{\alpha}{\tau}\right)\right)^{|\mathfrak{u}|} \right)^{\tau-2\tau\beta/\alpha},
\end{aligned}$$

where the last inequality follows from the fact that

$$H_N \leq \left(\frac{4}{(1-\eta)N} \sum_{\emptyset \neq \mathfrak{u} \subseteq [d]} \gamma_{\mathfrak{u}}^{1/(2\tau)} \left(2\zeta\left(\frac{\alpha}{\tau}\right)\right)^{|\mathfrak{u}|} \right)^{2\tau},$$

which holds for $\tau \in (1/2, \alpha)$, according to (11.6).

Finally, putting $\delta := 2\tau\beta/\alpha$ leads to the desired result. This completes the proof of Theorem 11.4. □

Observe that the upper bound in Theorem 11.4 can be bounded uniformly in d if the weights $\gamma_{\mathfrak{u}}$ satisfy certain summability conditions. For example, for product weights this is true if $\sum_{j=1}^{\infty} \gamma_j^{1/(2\tau)} < \infty$ is satisfied. This observation is important in the discussion of tractability results, which we return to in Section 11.3.

We have proven in Theorem 11.4 that the algorithm $A^{\mathrm{ran}}_{N,d}$ can yield an order of convergence that is arbitrarily close to $N^{-\alpha-1/2}$, which is exactly the optimal rate according to [253]. However, we will show in Theorem 11.5 below that with the present method we cannot match the optimal rate exactly, since one cannot get rid of logarithmic terms by means of Algorithm 11.2.

Theorem 11.5 *Let $d \in \mathbb{N}$, let $\alpha > 1/2$, and let $\boldsymbol{\gamma} = \{\gamma_{\mathfrak{u}}\}_{\mathfrak{u}\subseteq[d]}$ be positive weights. Then the randomized error of the algorithm $A^{\mathrm{ran}}_{N,d}$ for any $N \in \mathbb{N}$, $N \geq 2$, is bounded from below,*

$$\mathrm{err}^{\mathrm{ran}}_{d,\alpha,\boldsymbol{\gamma}}(A^{\mathrm{ran}}_{N,d}) \geq \frac{\sqrt{\gamma_{\{1\}}}\,\log N}{2\,N^{\alpha+1/2}}.$$

Proof To prove the lower bound on $\mathrm{err}^{\mathrm{ran}}_{d,\alpha,\boldsymbol{\gamma}}(A^{\mathrm{ran}}_{N,d})$ it is enough to construct, for each $N \in \mathbb{N}$, $N \geq 2$, a function, say f_N, for which the absolute integration error satisfies the lower bound. To this end, we define f_N via its Fourier coefficients,

$$\widehat{f_N}(\boldsymbol{h}) = \begin{cases} (r_{2\alpha,\boldsymbol{\gamma}}(\boldsymbol{h})\,|P_N|)^{-1/2} & \text{if } h_1 \in P_N \text{ and } h_2 = \cdots = h_d = 0, \\ 0 & \text{otherwise.} \end{cases}$$

Clearly, $\|f_N\|_{\mathrm{kor},d,\alpha,\boldsymbol{\gamma}} = 1$ and f_N integrates to 0 over $[0,1]^d$ because $I_d(f_N) = \widehat{f_N}(\boldsymbol{0}) = 0$. Moreover, for $p \in P_N$, we have

$$\begin{aligned} |Q_{(\boldsymbol{g},p),d}(f_N) - I(f_N)| &= \left| \sum_{\substack{\boldsymbol{h}\in\mathbb{Z}^d\setminus\{\boldsymbol{0}\} \\ \boldsymbol{h}\cdot\boldsymbol{g}\equiv 0 \ (\mathrm{mod}\ p)}} \widehat{f_N}(\boldsymbol{h}) \right| \\ &= (r_{2\alpha,\boldsymbol{\gamma}}((p,0,\ldots,0))\,|P_N|)^{-1/2} \\ &= \frac{\sqrt{\gamma_{\{1\}}}}{p^{\alpha}\sqrt{|P_N|}} \\ &\geq \frac{\sqrt{\gamma_{\{1\}}}\,\log N}{2\,N^{\alpha+1/2}}, \end{aligned}$$

where we used $|P_N| \leq \frac{2N}{\log N}$ from [221]. This proves

$$\begin{aligned} \mathrm{err}^{\mathrm{ran}}_{d,\alpha,\boldsymbol{\gamma}}(A^{\mathrm{ran}}_{N,d}) &\geq \frac{1}{|P_N|} \sum_{p\in P_N} \frac{1}{|\mathcal{Z}_{p,\eta}|} \sum_{\boldsymbol{g}\in\mathcal{Z}_{p,\eta}} |Q_{(\boldsymbol{g},p),d}(f_N) - I_d(f_N)| \\ &\geq \frac{\sqrt{\gamma_{\{1\}}}\,\log N}{2\,N^{\alpha+1/2}}, \end{aligned}$$

as claimed. □

11.2 Randomized Folded Lattice Rules

Using folded rank-1 lattice rules (see Section 7.3), the result from the previous section can be extended to the weighted half-period cosine space, as defined in Section 7.2.

From Theorem 7.40 it follows that Theorem 11.3 also applies to the half-period cosine space. Therefore any generating vector $\boldsymbol{g}$ and any $p \in P_N$ selected by Algorithm 11.2 (without any modification) satisfy the worst-case error bound

$$\operatorname{err}_{p,d}(\mathcal{H}_{\cos,d,\alpha,\boldsymbol{\gamma}}, \mathcal{P}_\phi(\boldsymbol{g}, p)) \le \left(\frac{2}{(1-\eta)p} \sum_{\emptyset \neq \mathfrak{u} \subseteq [d]} \gamma_{\mathfrak{u}}^{1/(2\tau)} \left(2\zeta\left(\frac{\alpha}{\tau}\right)\right)^{|\mathfrak{u}|} \right)^{\tau},$$

for any $\tau \in [1/2, \alpha)$, where $\mathcal{P}_\phi(\boldsymbol{g}, p)$ is as in Chapter 7.

We now consider the randomized error for the weighted half-period cosine space. Let $A_{N,d}^{\mathrm{ran},\phi}$ be the folded p-element lattice rule with generating vector $\boldsymbol{g}$ obtained by Algorithm 11.2. We apply this randomized algorithm to functions from $\mathcal{H}_{\cos,d,\alpha,\boldsymbol{\gamma}}$ and study the randomized error

$$\begin{aligned} &\operatorname{err}_{d,\alpha,\boldsymbol{\gamma}}^{\mathrm{ran}}(A_{N,d}^{\mathrm{ran},\phi}) \\ &:= \sup_{\substack{f \in \mathcal{H}_{\cos,d,\alpha,\boldsymbol{\gamma}} \\ \|f\|_{\cos,d,\alpha,\boldsymbol{\gamma}} \le 1}} \frac{1}{|P_N|} \sum_{p \in P_N} \frac{1}{|\mathcal{Z}_{p,\eta}|} \sum_{\boldsymbol{g} \in \mathcal{Z}_{p,\eta}} |Q_{(\boldsymbol{g},p),d}^{\phi}(f) - I_d(f)|, \end{aligned} \tag{11.9}$$

where $Q_{(\boldsymbol{g},p),d}^{\phi}$ denotes the folded p-element lattice rule with generating vector $\boldsymbol{g}$.

Theorem 11.6 *Let $d, N \in \mathbb{N}$, let $\alpha > 1/2$, let $\boldsymbol{\gamma} = \{\gamma_{\mathfrak{u}}\}_{\mathfrak{u} \subseteq [d]}$ be general weights with $\gamma_{\mathfrak{u}} \in [0,1]$ for $\mathfrak{u} \subseteq [d]$, and let $\eta \in (0,1)$ be given. Assume that* (11.5) *holds. For the randomized and folded algorithm $A_{N,d}^{\mathrm{ran},\phi}$ obtained by Algorithm 11.2, the randomized error* (11.9) *is bounded from above by*

$$\operatorname{err}_{d,\alpha,\boldsymbol{\gamma}}^{\mathrm{ran}}(A_{N,d}^{\mathrm{ran},\phi}) \le \frac{C_{\alpha,\tau,\delta,\eta}}{N^{\tau+1/2-\delta}} \left(\sum_{\emptyset \neq \mathfrak{u} \subseteq [d]} \gamma_{\mathfrak{u}}^{1/(2\tau)} \left(2\zeta\left(\frac{\alpha}{\tau}\right)\right)^{|\mathfrak{u}|} \right)^{\tau-\delta},$$

for any $\tau \in (1/2, \alpha)$ and $\delta \in (0, \min(\tau - 1/2, 2\tau/\alpha))$, where the positive factor $C_{\alpha,\tau,\delta,\eta}$ depends only on α, τ, δ, and η.

Proof For $f \in \mathcal{H}_{\cos,d,\alpha,\boldsymbol{\gamma}}$ we have

$$\begin{aligned} f(\phi(\boldsymbol{x})) &= \sum_{\mathfrak{u} \subseteq [d]} 2^{|\mathfrak{u}|/2} \sum_{\boldsymbol{k}_{\mathfrak{u}} \in \mathbb{N}_0^{|\mathfrak{u}|}} \widetilde{f}((\boldsymbol{k}_{\mathfrak{u}}, \boldsymbol{0})) \prod_{j \in \mathfrak{u}} \cos(\pi k_j \phi(x_j)) \\ &= \sum_{\mathfrak{u} \subseteq [d]} 2^{-|\mathfrak{u}|/2} \sum_{\boldsymbol{k}_{\mathfrak{u}} \in \mathbb{N}_0^{|\mathfrak{u}|}} \widetilde{f}((\boldsymbol{k}_{\mathfrak{u}}, \boldsymbol{0})) \sum_{\boldsymbol{\sigma}_{\mathfrak{u}} \in \{-1,1\}^{|\mathfrak{u}|}} \mathrm{e}^{2\pi \mathrm{i} (\boldsymbol{\sigma}_{\mathfrak{u}} * \boldsymbol{k}_{\mathfrak{u}}) \cdot \boldsymbol{x}_{\mathfrak{u}}} \\ &= \sum_{\mathfrak{u} \subseteq [d]} 2^{-|\mathfrak{u}|/2} \sum_{\boldsymbol{k}_{\mathfrak{u}} \in \mathbb{Z}^{|\mathfrak{u}|}} \widetilde{f}((|\boldsymbol{k}_{\mathfrak{u}}|, \boldsymbol{0}))\, \mathrm{e}^{2\pi \mathrm{i} \boldsymbol{k}_{\mathfrak{u}} \cdot \boldsymbol{x}_{\mathfrak{u}}}, \end{aligned}$$

where we used (7.26) to obtain the second equality, which implies that

$$\begin{aligned}
\|f\|^2_{\cos,d,\alpha,\gamma} &= \sum_{\mathfrak{u}\subseteq[d]} \sum_{\boldsymbol{k}_{\mathfrak{u}}\in\mathbb{N}^{|\mathfrak{u}|}} r_{2\alpha,\gamma}(\boldsymbol{k}_{\mathfrak{u}})\,|\widetilde{f}((\boldsymbol{k}_{\mathfrak{u}},\boldsymbol{0}))|^2 \\
&= \sum_{\mathfrak{u}\subseteq[d]} \sum_{\boldsymbol{k}_{\mathfrak{u}}\in(\mathbb{Z}\setminus\{0\})^{|\mathfrak{u}|}} r_{2\alpha,\gamma}(\boldsymbol{k}_{\mathfrak{u}})\,2^{-|\mathfrak{u}|}\,|\widetilde{f}((|\boldsymbol{k}_{\mathfrak{u}}|,\boldsymbol{0}))|^2 \\
&= \|f\circ\phi\|^2_{\mathrm{kor},d,\alpha,\gamma}.
\end{aligned}$$

Thus, for any function $f \in \mathcal{H}_{\cos,d,\alpha,\gamma}$, we have

$$f\circ\phi \in \mathcal{H}_{\mathrm{kor},d,\alpha,\gamma} \quad \text{and} \quad \|f\circ\phi\|_{\mathrm{kor},d,\alpha,\gamma} = \|f\|_{\cos,d,\alpha,\gamma}.$$

Since $\int_{[0,1]^d} f(\boldsymbol{x})\,\mathrm{d}\boldsymbol{x} = \int_{[0,1]^d} f(\phi(\boldsymbol{x}))\,\mathrm{d}\boldsymbol{x}$, we obtain that

$$\begin{aligned}
&\mathrm{err}^{\mathrm{ran}}_{d,\alpha,\gamma}(A^{\mathrm{ran},\phi}_{N,d}) \\
&= \sup_{\substack{f\in\mathcal{H}_{\cos,d,\alpha,\gamma}\\ \|f\|_{\cos,d,\alpha,\gamma}\le 1}} \frac{1}{|P_N|}\sum_{p\in P_N}\frac{1}{|\mathcal{Z}_{p,\eta}|}\sum_{\boldsymbol{g}\in\mathcal{Z}_{p,\eta}} |Q^{\phi}_{(\boldsymbol{g},p),d}(f) - I_d(f)| \\
&= \sup_{\substack{f\in\mathcal{H}_{\cos,d,\alpha,\gamma}\\ \|f\|_{\cos,d,\alpha,\gamma}\le 1}} \frac{1}{|P_N|}\sum_{p\in P_N}\frac{1}{|\mathcal{Z}_{p,\eta}|}\sum_{\boldsymbol{g}\in\mathcal{Z}_{p,\eta}} \left|Q_{(p,\boldsymbol{g}),d}(f\circ\phi) - I_d(f\circ\phi)\right| \\
&\le \sup_{\substack{g\in\mathcal{H}_{\mathrm{kor},d,\alpha,\gamma}\\ \|g\|_{\mathrm{kor},d,\alpha,\gamma}\le 1}} \frac{1}{|P_N|}\sum_{p\in P_N}\frac{1}{|\mathcal{Z}_{p,\eta}|}\sum_{\boldsymbol{g}\in\mathcal{Z}_{p,\eta}} \left|Q_{(p,\boldsymbol{g}),d}(g) - I_d(g)\right| \\
&= \mathrm{err}^{\mathrm{ran}}_{d,\alpha,\gamma}(A^{\mathrm{ran}}_{N,d}).
\end{aligned}$$

Thus the bound on $\mathrm{err}^{\mathrm{ran}}_{d,\alpha,\gamma}(A^{\mathrm{ran}}_{N,d})$ shown in Theorem 11.4 directly applies. □

11.3 A Brief Discussion of Tractability

We would like to give a few comments on tractability in this section. Suppose that for fixed $d \in \mathbb{N}$ and $\varepsilon \in (0,1)$, we would like to achieve $\mathrm{err}^{\mathrm{ran}}_{d,\alpha,\gamma}(A^{\mathrm{ran}}_{N,d}) \le \varepsilon$. From Theorem 11.4 we conclude that for $\mathcal{H}_{\mathrm{kor},d,\alpha,\gamma}$ with $\alpha > 1/2$, provided that N satisfies (11.5), it is sufficient to choose N such that

$$N \ge \frac{C_{\alpha,\tau,\delta,\eta,\gamma,d}}{\varepsilon^{1/(\tau+1/2-\delta)}},$$

where

$$C_{\alpha,\tau,\delta,\eta,\gamma,d} = C_{\alpha,\tau,\delta,\eta}^{1/(\tau+1/2-\delta)} \left(\sum_{\emptyset \neq \mathfrak{u} \subseteq [d]} \gamma_{\mathfrak{u}}^{1/(2\tau)} \left(2\zeta\left(\frac{\alpha}{\tau}\right)\right)^{|\mathfrak{u}|} \right)^{(\tau-\delta)/(\tau+1/2-\delta)},$$

and where $C_{\alpha,\tau,\delta,\eta}$ is defined as in Theorem 11.4.

Thus, the information complexity $N^{\mathrm{ran}}(\varepsilon, d)$ in the randomized setting, i.e., the minimal number of function evaluations that are required by any kind of random algorithm to obtain a randomized error not exceeding the threshold ε, satisfies

$$N^{\mathrm{ran}}(\varepsilon, d) \leq \max \left(\frac{C_{\alpha,\tau,\delta,\eta,\gamma,d}}{\varepsilon^{1/(\tau+1/2-\delta)}}, \inf_{1/2<\tau<\alpha} \frac{4}{1-\eta} \sum_{\emptyset \neq \mathfrak{u} \subseteq [d]} \gamma_{\mathfrak{u}}^{1/(2\tau)} \left(2\zeta\left(\frac{\alpha}{\tau}\right)\right)^{|\mathfrak{u}|} \right),$$

where the second term in the maximum is due to Condition (11.5).

For the sake of simplicity, we consider product weights $\gamma_{\mathfrak{u}} = \prod_{j \in \mathfrak{u}} \gamma_j$ for $\mathfrak{u} \subseteq [d]$ in the following. Then

$$\sum_{\mathfrak{u} \subseteq [d]} \gamma_{\mathfrak{u}}^{1/(2\tau)} \left(2\zeta\left(\frac{\alpha}{\tau}\right)\right)^{|\mathfrak{u}|} = \prod_{j=1}^{d} \left(1 + 2\gamma_j^{1/(2\tau)} \zeta\left(\frac{\alpha}{\tau}\right)\right)$$

and the latter expression is uniformly bounded in d if $\sum_{j=1}^{\infty} \gamma_j^{1/(2\tau)} < \infty$. This implies the following result.

Theorem 11.7 *Let $\alpha > 1/2$, let $\boldsymbol{\gamma} = (\gamma_j)_{j \geq 1}$ be product weights with $\gamma_j \in (0, 1]$ for $j \in \mathbb{N}$. Let $\tau_0 \geq 0$ be the supremum of the numbers τ such that*

$$\sum_{j=1}^{\infty} \gamma_j^{1/(2\tau)} < \infty.$$

Then the integration problems in the space $\mathcal{H}_{\mathrm{kor},d,\alpha,\boldsymbol{\gamma}}$ and in the space $\mathcal{H}_{\cos,d,\alpha,\boldsymbol{\gamma}}$, respectively, are strongly polynomially tractable in the randomized setting, with the respective exponents of strong polynomial tractability lying in the interval

$$\left[\frac{1}{\alpha + 1/2}, \frac{1}{\min(\tau_0, \alpha) + 1/2} \right].$$

In particular, if $\tau_0 \geq \alpha$, then the exponent of strong polynomial tractability is $1/(\alpha + 1/2)$.

Proof The upper bound on the exponent of strong polynomial tractability follows from Theorem 11.4, Theorem 11.6, and the discussion preceding Theorem 11.7, noting that we have $\tau < \min(\tau_0, \alpha)$.

For the lower bound we use a result stating that $N^{\mathrm{ran}}(\varepsilon, d) \geq N^{\mathrm{ran}}(\varepsilon, 1) \geq c \cdot \varepsilon^{-1/(\alpha+1/2)}$ for some $c > 0$, see, e.g., [205].

The same arguments can also be used for the half-period cosine space and hence the same result applies. □

Notes and Remarks

In this section we have studied the randomized setting instead of the worst-case setting like in the remaining parts of this book. In the randomized setting one considers randomized algorithms, and, correspondingly, their randomized errors. General information on the randomized setting can be found in [210, Sections 3.2.7–3.2.8 and Chapter 7] and in [211, Chapter 17]. The most prominent kind of randomized algorithms are Monte Carlo methods (see Section 1.1), which is sometimes even used as a synonym for any kind of randomization. The randomized error of Monte Carlo integration for the weighted Korobov space is studied in [240, Section 5]. In particular, in [240, Theorem 8] matching necessary and sufficient conditions on product weights for (strong) polynomial tractability of MC integration in the weighted Korobov space are provided.

Rather than the MC method, we presented here a kind of randomization that is especially suited for lattice rules. This randomization technique has first been introduced and analyzed by Bakhvalov in [7]. We also refer to Bakhvalov's papers [6, 9] for general results on randomized algorithms. However, the presentation in this chapter follows the papers [39, 143].

As in all other parts of this book, the results presented in Section 11.1 are valid for weighted Korobov spaces with smoothness parameter $\alpha > 1/2$. In [143] also the more general case of $\alpha \in (0, 1/2]$ is considered. This case corresponds to spaces of one-periodic functions whose Fourier series converge almost everywhere but not necessarily pointwise. The algorithms considered require an additional random shift, uniformly distributed on $[0,1]^d$, like in Chapter 7. For $d, N \in \mathbb{N}$ with $N \geq 2$, and $f \in \mathcal{H}_{\mathrm{kor},d,\alpha,\boldsymbol{\gamma}}$ with $\alpha > 0$, and product weights $\boldsymbol{\gamma} = (\gamma_j)_{j \geq 1}$ with $\gamma_j \in (0,1]$ for $j \in \mathbb{N}$, define

$$\widetilde{A}_{N,d}^{\mathrm{ran}}(f) := A_{N,d}^{\mathrm{ran}}(f(\{\cdot + \boldsymbol{\Delta}\})).$$

Then [143, Theorem 11] guarantees the existence of a prime number p and a generating vector $\boldsymbol{g}$ such that the root mean square randomized error, where the "mean" in "root mean square" is considered with respect to all random shifts $\boldsymbol{\Delta} \in [0,1]^d$, satisfies for any $\tau \in (0, \alpha)$ and any $N \in \mathbb{N}$, $N \geq 2$, the upper bound

$$\mathrm{err}_{d,\alpha,\boldsymbol{\gamma}}^{\mathrm{rms}}(\widetilde{A}_{N,d}^{\mathrm{ran}}) \leq \frac{C_{\alpha,\delta,\tau}}{N^{\tau+1/2-\delta}} \prod_{j=1}^{d} \left(1 + 2\gamma_j^{1/(2\tau)} \zeta\left(\frac{\alpha}{\tau}\right)\right)^{\tau}$$

for arbitrary $\delta \in (0, \tau \min(1/(2\alpha), 1))$, where the factor $C_{\alpha,\delta,\tau}$ only depends on α, δ, and τ. For details see [143, Section 3.2].

In the same context, but using randomly shifted and dilated lattice point sets, Ullrich [253] was able to prove optimal convergence rates. The algorithm employed was introduced by Krieg and Novak in [142] and is based on the deterministic cubature rule of Frolov [75] (see also [252]).

In Theorem 11.7 we have presented a characterization of strong polynomial tractability of integration in the Korobov space in the randomized setting. With the usual methods, it is possible to also analyze other tractability notions, under modified summability assumptions on the weights.

The paper [85] by Goda and L'Ecuyer uses a new approach based on the median of the estimator of several lattice rules to approximate the integral of a function in a Korobov space or Sobolev space.

Chapter 12
Stability of Lattice Rules

We have seen in Chapters 3 and 4 that the component-by-component construction can identify generating vectors of lattice rules which yield almost the optimal rate of convergence of the worst-case error for a given weighted function space with given weights. One drawback of this approach is that the constructed generating vector is specific to the given function space parameters like smoothness and weights. Hence one question which arises naturally is: What happens if we construct a generating vector with respect to a given set of parameters, but then use it for a space with different parameters? Do we still get a fast rate of convergence? In other words, we ask whether lattice rules are stable with respect to a change of parameters. In the following we provide some results in this direction.

12.1 A Stability Result

Theorem 3.9 implies that one can use a CBC construction, for $d, N \in \mathbb{N}$, to find a generating vector $\boldsymbol{g} \in G_d^{\varphi}(N)$ such that we have

$$\mathrm{err}_{N,d,\alpha,\boldsymbol{\gamma}}(\boldsymbol{g}) \leq \left(\frac{1}{\varphi(N)} \sum_{\emptyset \neq \mathfrak{u} \subseteq [d]} \gamma_{\mathfrak{u}}^{1/(2\tau)} \left(2\zeta\left(\frac{\alpha}{\tau}\right)\right)^{|\mathfrak{u}|} \right)^{\tau}$$

for any $\tau \in [1/2, \alpha)$.

Since the CBC construction uses a quality criterion which depends on the parameters α and $\boldsymbol{\gamma}$, it follows that this bound holds only for these specific α and $\boldsymbol{\gamma}$. But what if we change α and/or $\boldsymbol{\gamma}$ but keep the same generating vector? Does a similar result still hold?

One way to obtain a result in this direction is by using Jensen's inequality, stated in Lemma 2.25, which implies that, for any $\delta \in (0, 1]$, we have

J. Dick et al., *Lattice Rules*, Springer Series in Computational Mathematics 58,
https://doi.org/10.1007/978-3-031-09951-9_12

$$
\begin{aligned}
[\mathrm{err}_{N,d,\alpha/\delta,\boldsymbol{\gamma}^{1/\delta}}(\boldsymbol{g})]^{2\delta} &= \left(\sum_{\boldsymbol{h}\in\mathcal{L}^\perp(\boldsymbol{g},N)\setminus\{\boldsymbol{0}\}} \frac{1}{r_{2\alpha/\delta,\boldsymbol{\gamma}^{1/\delta}}(\boldsymbol{h})} \right)^{\delta} \\
&\le \sum_{\boldsymbol{h}\in\mathcal{L}^\perp(\boldsymbol{g},N)\setminus\{\boldsymbol{0}\}} \frac{1}{(r_{2\alpha/\delta,\boldsymbol{\gamma}^{1/\delta}}(\boldsymbol{h}))^{\delta}} \\
&= \sum_{\boldsymbol{h}\in\mathcal{L}^\perp(\boldsymbol{g},N)\setminus\{\boldsymbol{0}\}} \frac{1}{r_{2\alpha,\boldsymbol{\gamma}}(\boldsymbol{h})} \\
&= [\mathrm{err}_{N,d,\alpha,\boldsymbol{\gamma}}(\boldsymbol{g})]^2.
\end{aligned}
$$

Thus a generating vector $\boldsymbol{g}$ that is constructed based on the criterion $\mathrm{err}_{N,d,\alpha,\boldsymbol{\gamma}}$, i.e., the worst-case error in the weighted Korobov space $\mathcal{H}_{\mathrm{kor},d,\alpha,\boldsymbol{\gamma}}$, satisfies, for any $\delta \in (0,1]$, the following bound on the worst-case error in $\mathcal{H}_{\mathrm{kor},d,\alpha/\delta,\boldsymbol{\gamma}^{1/\delta}}$,

$$
\mathrm{err}_{N,d,\alpha/\delta,\boldsymbol{\gamma}^{1/\delta}}(\boldsymbol{g}) \le \left(\frac{1}{\varphi(N)} \sum_{\emptyset\ne\mathfrak{u}\subseteq[d]} \gamma_{\mathfrak{u}}^{1/(2\tau)} \left(2\zeta\left(\frac{\alpha}{\tau}\right)\right)^{|\mathfrak{u}|} \right)^{\tau/\delta}
$$

for $\tau \in [1/2, \alpha)$. From this result it follows that if we construct a generating vector $\boldsymbol{g}$ for a given smoothness α, then we can obtain the almost optimal convergence rate of the integration error for numerical integration using a lattice rule with the same generating vector $\boldsymbol{g}$ in a Korobov space for any smoothness $\beta \ge \alpha$. This approach does not yield any result for Korobov spaces with smoothness $\beta < \alpha$. Also the set of weights is restricted to the case $\boldsymbol{\gamma}^{1/\delta}$. It is, however, possible to achieve a much more general result, stated in the following theorem.

Theorem 12.1 *Let $d, N \in \mathbb{N}$ and $\boldsymbol{g} \in G_d^{\varphi}(N)$. For any $\alpha, \alpha' > 1/2$ and sets of positive weights $\boldsymbol{\gamma} = \{\gamma_{\mathfrak{u}}\}_{\mathfrak{u}\subseteq[d]}$, $\boldsymbol{\gamma}' = \{\gamma'_{\mathfrak{u}}\}_{\mathfrak{u}\subseteq[d]}$ such that $\gamma_{\mathfrak{v}} \ge \gamma_{\mathfrak{u}} > 0$ whenever $\mathfrak{v} \subseteq \mathfrak{u}$, we have*

$$
\begin{aligned}
&\mathrm{err}_{N,d,\alpha',\boldsymbol{\gamma}'}(\boldsymbol{g}) \\
&\le [\mathrm{err}_{N,d,\alpha,\boldsymbol{\gamma}}(\boldsymbol{g})]^{\alpha'/\alpha} \left(c_{\alpha'} \sum_{\emptyset\ne\mathfrak{u}\subseteq[d]} \frac{\gamma'_{\mathfrak{u}}}{\gamma_{\mathfrak{u}}^{\alpha'/\alpha}} \left(\frac{2^{2\alpha'+1}}{2^{2\alpha'-1}-1} \right)^{|\mathfrak{u}|} (\log_2 N)^{|\mathfrak{u}|-1} \right)^{1/2},
\end{aligned}
$$

where $\log_2$ denotes the logarithm in base 2, and where

$$
c_{\alpha'} := 1 + \zeta(2\alpha') + (2^{2\alpha'} + \zeta(2\alpha'))\, \frac{2^{2\alpha'-1}-1}{2^{4\alpha'}}. \tag{12.1}
$$

Before we give the proof of Theorem 12.1, we outline some of its consequences.

Let $\boldsymbol{g}$ be the generating vector constructed by the CBC algorithm based on the criterion $\mathrm{err}_{N,d,\alpha,\boldsymbol{\gamma}}$ for given $\alpha > 1/2$ and given $\boldsymbol{\gamma}$ such that $\gamma_{\mathfrak{v}} \ge \gamma_{\mathfrak{u}} > 0$ whenever $\mathfrak{v} \subseteq \mathfrak{u}$. Applying Theorems 3.9 and 12.1 shows that, for any $\alpha' > 1/2$ and any positive weights $\boldsymbol{\gamma}'$, we have

$$[\mathrm{err}_{N,d,\alpha',\boldsymbol{\gamma}'}(\boldsymbol{g})]^2 \le c_{\alpha'}\left(\frac{1}{\varphi(N)}\sum_{\emptyset\neq\mathfrak{u}\subseteq[d]}\gamma_{\mathfrak{u}}^{1/(2\tau)}\left(2\zeta\left(\frac{\alpha}{\tau}\right)\right)^{|\mathfrak{u}|}\right)^{\tau\alpha'/\alpha}$$
$$\times\sum_{\emptyset\neq\mathfrak{u}\subseteq[d]}\frac{\gamma'_{\mathfrak{u}}}{\gamma_{\mathfrak{u}}^{\alpha'/\alpha}}\left(\frac{2^{2\alpha'+1}}{2^{2\alpha'-1}-1}\right)^{|\mathfrak{u}|}(\log_2 N)^{|\mathfrak{u}|-1} \tag{12.2}$$

for any $\tau \in [1/2, \alpha)$. As τ can be arbitrarily close to α, the exponent $\tau\alpha'/\alpha$ can be arbitrarily close to α'. Furthermore, for arbitrarily small but positive δ, the term $(\log_2 N)^{|\mathfrak{u}|}$ can be bounded by $C_\delta N^\delta$, for some factor $C_\delta > 0$ depending only on δ and $|\mathfrak{u}|$. Consequently, the overall convergence rate can be arbitrarily close to α', which is almost best possible.

Under suitable conditions on the weights $\boldsymbol{\gamma}$ and $\boldsymbol{\gamma}'$ we can show that the worst-case error $\mathrm{err}_{N,d,\alpha',\boldsymbol{\gamma}'}(\boldsymbol{g})$ depends only polynomially on the dimension d, or even that the bound is completely independent of the dimension.

Corollary 12.2 *Let $d, N \in \mathbb{N}$, let $\alpha, \alpha' > 1/2$, and let $\boldsymbol{\gamma} = \{\gamma_{\mathfrak{u}}\}_{\mathfrak{u}\subseteq[d]}$, $\boldsymbol{\gamma}' = \{\gamma'_{\mathfrak{u}}\}_{\mathfrak{u}\subseteq[d]}$ be positive weights such that $\gamma_{\mathfrak{v}} \ge \gamma_{\mathfrak{u}} > 0$ whenever $\mathfrak{v} \subseteq \mathfrak{u}$. Assume that $\boldsymbol{g} \in G_d^{\varphi}(N)$ has been constructed by Algorithm 3.6 based on the criterion $\mathrm{err}_{N,d,\alpha,\boldsymbol{\gamma}}$. Then the following assertions hold true.*

1. *For general weights $\boldsymbol{\gamma}$ and $\boldsymbol{\gamma}'$ satisfying the above condition, assume that there exist $\tau, \delta, q, q' \ge 0$ such that for $\tau \in [1/2, \alpha)$ and $\delta \in (0, \tau\alpha'/(2\alpha))$,*

$$\sup_{d\in\mathbb{N}}\frac{1}{d^{2q}}\sum_{\emptyset\neq\mathfrak{u}\subseteq[d]}\gamma_{\mathfrak{u}}^{1/(2\tau)}\left(2\zeta\left(\frac{\alpha}{\tau}\right)\right)^{|\mathfrak{u}|} < \infty,$$

and

$$\sup_{d,N\in\mathbb{N}}\frac{1}{d^{2q'}N^{2\delta}}\sum_{\emptyset\neq\mathfrak{u}\subseteq[d]}\frac{\gamma'_{\mathfrak{u}}}{\gamma_{\mathfrak{u}}^{\alpha'/\alpha}}\left(\frac{2^{2\alpha'+1}}{2^{2\alpha'-1}-1}\right)^{|\mathfrak{u}|}(\log_2 N)^{|\mathfrak{u}|-1} < \infty.$$

Then the worst-case error $\mathrm{err}_{N,d,\alpha',\boldsymbol{\gamma}'}(\boldsymbol{g})$ depends at most polynomially on d and is bounded by

$$\mathrm{err}_{N,d,\alpha',\boldsymbol{\gamma}'}(\boldsymbol{g}) \le C_{\boldsymbol{\gamma},\boldsymbol{\gamma}',\alpha,\alpha',q,q',\tau,\delta}\, d^{\tau q\alpha'/\alpha+q'}\frac{N^\delta}{\varphi(N)^{\tau\alpha'/(2\alpha)}},$$

for some quantity $C_{\boldsymbol{\gamma},\boldsymbol{\gamma}',\alpha,\alpha',q,q',\tau,\delta} > 0$, which depends on the parameters $\boldsymbol{\gamma}, \boldsymbol{\gamma}', \alpha, \alpha', q, q', \tau$, and δ, but is independent of d and N. If the above conditions hold for $q = q' = 0$, the worst-case error $\mathrm{err}_{N,d,\alpha',\boldsymbol{\gamma}'}(\boldsymbol{g})$ is bounded independently of d.

2. *In particular, in the case of product weights $\boldsymbol{\gamma}$ and $\boldsymbol{\gamma}'$, assume the existence of $\tau \in [1/2, \alpha)$ such that*

$$\sum_{j=1}^{\infty}\gamma_j^{1/(2\tau)} < \infty \quad \text{and} \quad \sum_{j=1}^{\infty}\frac{\gamma'_j}{\gamma_j^{\alpha'/\alpha}} < \infty. \tag{12.3}$$

Then the worst-case error $\mathrm{err}_{N,d,\alpha',\gamma'}(\boldsymbol{g})$ *can be bounded independently of d by*

$$\mathrm{err}_{N,d,\alpha',\gamma'}(\boldsymbol{g}) \leq C_{\gamma,\gamma',\alpha,\alpha',\tau,\delta} \frac{N^{\delta}}{(\varphi(N))^{\tau\alpha'/(2\alpha)}},$$

for arbitrarily small $\delta > 0$, *where the factor* $C_{\gamma,\gamma',\alpha,\alpha',\tau,\delta} > 0$ *depends on* $\boldsymbol{\gamma}, \boldsymbol{\gamma}', \alpha, \alpha', \tau$, *and* δ, *but is independent of d and N.*

Proof The first item immediately follows from the bound (12.2). The second item can be shown using the bound (12.2) and Lemma 3.20. □

One of the most important consequences of the first item of Corollary 12.2 is that by choosing $\boldsymbol{\gamma}$ of product form such that the required conditions are satisfied, the fast implementation of the CBC algorithm (see Algorithm 3.6 and Section 3.4) returns a lattice rule in $O(dN \log N)$ arithmetic operations with $O(N)$ memory, which yields the almost optimal rate of convergence in $\mathcal{H}_{\mathrm{kor},d,\alpha',\gamma'}$ with good tractability properties even for general weights $\boldsymbol{\gamma}'$.

We illustrate the conditions for tractability for product weights in the following example.

Example 12.3 Assume that both $\boldsymbol{\gamma}$ and $\boldsymbol{\gamma}'$ are product weights, and that $\gamma_j = j^{-r}$ for some $r > 1$, for $j \in \mathbb{N}$. Then

$$\sum_{j=1}^{\infty} \gamma_j^{1/(2\tau)} = \sum_{j=1}^{\infty} \frac{1}{j^{r/(2\tau)}} < \infty$$

holds for any $\tau \in [1/2, r/2)$. Hence we can choose $\tau \in [1/2, \min(\alpha, r/2))$ to obtain a bound on $\mathrm{err}_{N,d,\alpha,\gamma}(\boldsymbol{g})$ which is independent of the dimension. Considering the special case $r = 2\alpha$ the second necessary condition on the weights in the second item of Corollary 12.2 reads

$$\sum_{j=1}^{\infty} \frac{\gamma_j'}{\gamma_j^{\alpha'/\alpha}} = \sum_{j=1}^{\infty} j^{r\alpha'/\alpha} \gamma_j' = \sum_{j=1}^{\infty} j^{2\alpha'} \gamma_j' < \infty.$$

This is guaranteed, e.g., with $\gamma_j' = j^{-2\alpha'-1-\varepsilon}$ for some $\varepsilon > 0$ for $j \in \mathbb{N}$. In these instances we obtain an error bound for $\mathrm{err}_{N,d,\alpha,\gamma'}(\boldsymbol{g})$ that is independent of the dimension. On the other hand, if we were to directly construct a lattice rule using $\mathrm{err}_{N,d,\alpha,\gamma'}(\boldsymbol{g})$, then $\sum_{j=1}^{\infty} \gamma_j'^{1/(2\alpha')} < \infty$ would be sufficient. This condition is weaker as, for instance, $\gamma_j' = j^{-2\alpha'-\varepsilon}$ for some $\varepsilon > 0$, for $j \in \mathbb{N}$, is sufficient to guarantee a bound independent of the dimension.

The proof of Theorem 12.1

In the proof of Theorem 12.1, we shall use the following elementary inequality.

Lemma 12.4 *For any real $b > 1$ and any $k, t_0 \in \mathbb{N}$, we have*

$$\sum_{t=t_0}^{\infty} b^{-t}\binom{t+k-1}{k-1} \le b^{-t_0}\binom{t_0+k-1}{k-1}\left(1-\frac{1}{b}\right)^{-k}.$$

Proof By the binomial theorem we have

$$\sum_{t=t_0}^{\infty} b^{-t}\binom{t-t_0+k-1}{k-1} = b^{-t_0}\left(1-\frac{1}{b}\right)^{-k}.$$

The result then follows from the inequality

$$\binom{t+k-1}{k-1}\Big/\binom{t-t_0+k-1}{k-1} = \frac{(t+k-1)(t+k-2)\cdots(t-t_0+k)}{t(t-1)\cdots(t-t_0+1)} \le \binom{t_0+k-1}{k-1}.$$

□

We are now ready to give the proof of Theorem 12.1.

Proof of Theorem 12.1 We prove the stability result using a weighted higher order version of the Zaremba index that was in its original form introduced in Definition 1.51. Set

$$\rho_{\alpha,\gamma}(\boldsymbol{g}, N) := \min_{\boldsymbol{h}\in\mathcal{L}^{\perp}(\boldsymbol{g},N)\setminus\{\boldsymbol{0}\}} r_{2\alpha,\gamma}(\boldsymbol{h}).$$

Note that $1/\rho_{\alpha,\gamma}(\boldsymbol{g}, N)$ is the largest term in the worst-case error expression

$$[\mathrm{err}_{N,d,\alpha,\gamma}(\boldsymbol{g})]^2 = \sum_{\boldsymbol{h}\in\mathcal{L}^{\perp}(\boldsymbol{g},N)\setminus\{\boldsymbol{0}\}} \frac{1}{r_{2\alpha,\gamma}(\boldsymbol{h})},$$

which implies

$$\frac{1}{\rho_{\alpha,\gamma}(\boldsymbol{g}, N)} \le [\mathrm{err}_{N,d,\alpha,\gamma}(\boldsymbol{g})]^2. \tag{12.4}$$

Moreover, we define

$$\widetilde{\rho}_{\mathfrak{u}}(\boldsymbol{g}, N) := \min_{\boldsymbol{h}_{\mathfrak{u}}\in\mathcal{L}_{\mathfrak{u}}^{\perp}(\boldsymbol{g},N)} \prod_{j\in\mathfrak{u}} |h_j|,$$

where we recall the notation

$$\mathcal{L}_{\mathfrak{u}}^{\perp}(\boldsymbol{g}, N) := \{\boldsymbol{h}_{\mathfrak{u}} \in (\mathbb{Z}\setminus\{0\})^{|\mathfrak{u}|} \ : \ \boldsymbol{h}_{\mathfrak{u}}\cdot\boldsymbol{g}_{\mathfrak{u}} \equiv 0 \pmod{N}\}$$

from (4.5). Then we can represent $\rho_{\alpha,\gamma}(\boldsymbol{g}, N)$ in terms of $\widetilde{\rho}_{\mathfrak{u}}(\boldsymbol{g}, N)$ in the following way,

$$\rho_{\alpha,\boldsymbol{\gamma}}(\boldsymbol{g},N) = \min_{\emptyset \neq \mathfrak{u} \subseteq [d]} \frac{1}{\gamma_{\mathfrak{u}}} \min_{\boldsymbol{h}_{\mathfrak{u}} \in \mathcal{L}_{\mathfrak{u}}^{\perp}(\boldsymbol{g},N)} \prod_{j \in \mathfrak{u}} |h_j|^{2\alpha} = \min_{\emptyset \neq \mathfrak{u} \subseteq [d]} \frac{1}{\gamma_{\mathfrak{u}}} (\widetilde{\rho}_{\mathfrak{u}}(\boldsymbol{g},N))^{2\alpha}.$$

Therefore, it holds for any nonempty subset $\mathfrak{u}$ of $[d]$ that

$$\widetilde{\rho}_{\mathfrak{u}}(\boldsymbol{g},N) \geq \left(\gamma_{\mathfrak{u}}\, \rho_{\alpha,\boldsymbol{\gamma}}(\boldsymbol{g},N)\right)^{1/(2\alpha)}.$$

Now, recall the definition of the unweighted Zaremba index from Definition 1.51, which is, for $\boldsymbol{g}_{\mathfrak{u}}$, given by

$$\rho(\boldsymbol{g}_{\mathfrak{u}},N) = \min_{\substack{\boldsymbol{h}_{\mathfrak{u}} \in \mathbb{Z}^{|\mathfrak{u}|} \setminus \{\boldsymbol{0}\} \\ \boldsymbol{h}_{\mathfrak{u}} \cdot \boldsymbol{g}_{\mathfrak{u}} \equiv 0 \pmod{N}}} \prod_{j \in \mathfrak{u}} \max(1,|h_j|).$$

As mentioned in Section 1.8 it is a classical result that $\rho(\boldsymbol{g}_{\mathfrak{u}},N) \leq N/2$ (see [199, Lemma 5.8]). Then, by assuming $\gamma_{\mathfrak{v}} \geq \gamma_{\mathfrak{u}}$ whenever $\mathfrak{v} \subseteq \mathfrak{u}$, we obtain a lower bound,

$$\begin{aligned} \rho(\boldsymbol{g}_{\mathfrak{u}},N) &= \min_{\emptyset \neq \mathfrak{v} \subseteq \mathfrak{u}} \widetilde{\rho}_{\mathfrak{v}}(\boldsymbol{g},N) \\ &\geq \min_{\emptyset \neq \mathfrak{v} \subseteq \mathfrak{u}} \left(\gamma_{\mathfrak{v}}\, \rho_{\alpha,\boldsymbol{\gamma}}(\boldsymbol{g},N)\right)^{1/(2\alpha)} \\ &= \left(\gamma_{\mathfrak{u}}\, \rho_{\alpha,\boldsymbol{\gamma}}(\boldsymbol{g},N)\right)^{1/(2\alpha)}. \end{aligned} \tag{12.5}$$

Next, we consider the worst-case error in $\mathcal{H}_{\mathrm{kor},d,\alpha',\boldsymbol{\gamma}'}$. According to Theorem 2.19 we have

$$[\mathrm{err}_{N,d,\alpha',\boldsymbol{\gamma}'}(\boldsymbol{g})]^2 = \sum_{\emptyset \neq \mathfrak{u} \subseteq [d]} \gamma'_{\mathfrak{u}} \sum_{\boldsymbol{h}_{\mathfrak{u}} \in \mathcal{L}_{\mathfrak{u}}^{\perp}(\boldsymbol{g},N)} \prod_{j \in \mathfrak{u}} \frac{1}{|h_j|^{2\alpha'}}. \tag{12.6}$$

For a nonempty subset $\mathfrak{u}$ of $[d]$, we denote by $\mu_{\mathfrak{u}}$ the largest nonnegative integer such that $2^{\mu_{\mathfrak{u}}} < \rho(\boldsymbol{g}_{\mathfrak{u}},N)$ holds. It follows from the proof of [199, Theorem 5.34] (see the bound on $S_1(\mathbf{d})$ in Equation (5.46) on p. 134 and the bound on $S_2(\mathbf{d})$ on p. 135 in that reference) that the inner sum in (12.6) for a given $\mathfrak{u}$ with $|\mathfrak{u}| \geq 2$ is bounded from above by

$$\begin{aligned} \sum_{\boldsymbol{h}_{\mathfrak{u}} \in \mathcal{L}_{\mathfrak{u}}^{\perp}(\boldsymbol{g},N)} \prod_{j \in \mathfrak{u}} \frac{1}{|h_j|^{2\alpha'}} &\leq \sum_{\substack{\boldsymbol{h}_{\mathfrak{u}} \in \mathbb{Z}^{|\mathfrak{u}|} \setminus \{\boldsymbol{0}\} \\ \boldsymbol{h}_{\mathfrak{u}} \cdot \boldsymbol{g}_{\mathfrak{u}} \equiv 0 \pmod{N}}} \prod_{j \in \mathfrak{u}} \frac{1}{\max(1,|h_j|)^{2\alpha'}} \\ &\leq \frac{2^{|\mathfrak{u}|}}{(\rho(\boldsymbol{g}_{\mathfrak{u}},N))^{2\alpha'}} \left((1+\zeta(2\alpha')) \binom{\mu_{\mathfrak{u}} + |\mathfrak{u}| - 1}{|\mathfrak{u}| - 1} \right. \\ &\quad \left. + (2^{2\alpha'} + \zeta(2\alpha')) \sum_{k=1}^{\infty} 2^{(1-2\alpha')k} \binom{k + \mu_{\mathfrak{u}} + |\mathfrak{u}| - 2}{|\mathfrak{u}| - 2} \right). \end{aligned} \tag{12.7}$$

For the first term in the parentheses on the right-hand side of (12.7), we have

$$\binom{\mu_{\mathfrak{u}}+|\mathfrak{u}|-1}{|\mathfrak{u}|-1}=\prod_{i=1}^{|\mathfrak{u}|-1}\frac{\mu_{\mathfrak{u}}+i}{i}\leq(\mu_{\mathfrak{u}}+1)^{|\mathfrak{u}|-1}.$$

For the second term in the parentheses on the right-hand side of (12.7), Lemma 12.4 with $t_0=\mu_{\mathfrak{u}}+1$, $k=|\mathfrak{u}|-1$, and $b=2^{2\alpha'-1}$ gives

$$\begin{aligned}\sum_{k=1}^{\infty}2^{(1-2\alpha')k}\binom{k+\mu_{\mathfrak{u}}+|\mathfrak{u}|-2}{|\mathfrak{u}|-2}&=2^{(2\alpha'-1)\mu_{\mathfrak{u}}}\sum_{k=\mu_{\mathfrak{u}}+1}^{\infty}2^{-(2\alpha'-1)k}\binom{k+|\mathfrak{u}|-2}{|\mathfrak{u}|-2}\\&\leq 2^{-(2\alpha'-1)}\binom{\mu_{\mathfrak{u}}+|\mathfrak{u}|-1}{|\mathfrak{u}|-2}\left(\frac{2^{2\alpha'-1}}{2^{2\alpha'-1}-1}\right)^{|\mathfrak{u}|-1}\\&=\frac{2^{2\alpha'-1}-1}{2^{4\alpha'-2}}\left(\frac{2^{2\alpha'-1}}{2^{2\alpha'-1}-1}\right)^{|\mathfrak{u}|}\prod_{i=1}^{|\mathfrak{u}|-2}\frac{\mu_{\mathfrak{u}}+i+1}{i}\\&\leq\frac{2^{2\alpha'-1}-1}{2^{4\alpha'-2}}\left(\frac{2^{2\alpha'-1}}{2^{2\alpha'-1}-1}\right)^{|\mathfrak{u}|}(\mu_{\mathfrak{u}}+2)^{|\mathfrak{u}|-2}\\&\leq\frac{2^{2\alpha'-1}-1}{2^{4\alpha'}}\left(\frac{2^{2\alpha'}}{2^{2\alpha'-1}-1}\right)^{|\mathfrak{u}|}(\mu_{\mathfrak{u}}+1)^{|\mathfrak{u}|-2}.\end{aligned}$$

Using these bounds, we obtain

$$\begin{aligned}&\sum_{\boldsymbol{h}_{\mathfrak{u}}\in\mathcal{L}_{\mathfrak{u}}^{\perp}(\boldsymbol{g},N)}\prod_{j\in\mathfrak{u}}\frac{1}{|h_j|^{2\alpha'}}\\&\leq\frac{2^{|\mathfrak{u}|}}{(\rho(\boldsymbol{g}_{\mathfrak{u}},N))^{2\alpha'}}\Bigg((1+\zeta(2\alpha'))(\mu_{\mathfrak{u}}+1)^{|\mathfrak{u}|-1}\\&\quad+(2^{2\alpha'}+\zeta(2\alpha'))\frac{2^{2\alpha'-1}-1}{2^{4\alpha'}}\left(\frac{2^{2\alpha'}}{2^{2\alpha'-1}-1}\right)^{|\mathfrak{u}|}(\mu_{\mathfrak{u}}+1)^{|\mathfrak{u}|-2}\Bigg)\\&\leq c_{\alpha'}\left(\frac{2^{2\alpha'+1}}{2^{2\alpha'-1}-1}\right)^{|\mathfrak{u}|}\frac{(\mu_{\mathfrak{u}}+1)^{|\mathfrak{u}|-1}}{(\rho(\boldsymbol{g}_{\mathfrak{u}},N))^{2\alpha'}},\end{aligned}$$

with $c_{\alpha'}$ given in (12.1). Note that this bound on the inner sum in the formula (12.6) for $\mathrm{err}_{N,d,\alpha',\gamma'}(\boldsymbol{g})$ also applies to the case $|\mathfrak{u}|=1$.

From $2^{\mu_{\mathfrak{u}}}<\rho(\boldsymbol{g}_{\mathfrak{u}},N)\leq N/2$ we directly obtain $\mu_{\mathfrak{u}}\leq\log_2 N-1$. Using this fact and the lower bound (12.5) we can bound the square of $\mathrm{err}_{N,d,\alpha',\gamma'}(\boldsymbol{g})$ as

$$\begin{aligned}&[\mathrm{err}_{N,d,\alpha',\gamma'}(\boldsymbol{g})]^2\\&\leq\frac{c_{\alpha'}}{(\rho_{\alpha,\gamma}(\boldsymbol{g},N))^{\alpha'/\alpha}}\sum_{\emptyset\neq\mathfrak{u}\subseteq[d]}\frac{\gamma'_{\mathfrak{u}}}{\gamma_{\mathfrak{u}}^{\alpha'/\alpha}}\left(\frac{2^{2\alpha'+1}}{2^{2\alpha'-1}-1}\right)^{|\mathfrak{u}|}(\log_2 N)^{|\mathfrak{u}|-1}.\end{aligned}$$

Applying (12.4) and taking the square root gives the desired result. □

An approach for fixed smoothness but different product weight sequences

We present some further considerations for product weights. Assume we have sets of product weights $\boldsymbol{\gamma}^{(i)} = (\gamma_j^{(i)})_{j\geq 1}$ for $i \in \mathcal{I}$, where $\mathcal{I}$ is a given index set, satisfying the usual condition $\gamma_1^{(i)} \geq \gamma_2^{(i)} \geq \cdots > 0$ for every $i \in \mathcal{I}$. Consider numerical integration in the weighted Korobov spaces $\mathcal{H}_{\mathrm{kor},d,\alpha,\boldsymbol{\gamma}^{(i)}}$ for $i \in \mathcal{I}$. This means that we consider various Korobov spaces with the same smoothness parameter α but with different weight sequences. If we aim at finding a universal rank-1 lattice rule which guarantees for each of these spaces an error bound that is independent of the dimension, it is obviously necessary to claim that

$$\sum_{j=1}^{\infty} \gamma_j^{(i)} < \infty \quad \text{for all } i \in \mathcal{I}.$$

Let now, for $j \in \mathbb{N}$,

$$\nu_j := \sup_{i \in \mathcal{I}} \gamma_j^{(i)}$$

and $\boldsymbol{\nu} = (\nu_j)_{j\geq 1}$. Now we assume that also the weights ν_j are summable, i.e., $\sum_{j=1}^{\infty} \nu_j < \infty$. Then construct, e.g., by means of Algorithm 3.6 with a prime N, a lattice point $\boldsymbol{g}$ with respect to the quality measure $\mathrm{err}_{N,d,\alpha,\boldsymbol{\nu}}$, i.e., the worst-case error in $\mathcal{H}_{\mathrm{kor},d,\alpha,\boldsymbol{\nu}}$. As a result we obtain from Theorem 3.7 that

$$\mathrm{err}_{N,d,\alpha,\boldsymbol{\gamma}^{(i)}}(\boldsymbol{g}) \leq \mathrm{err}_{N,d,\alpha,\boldsymbol{\nu}}(\boldsymbol{g}) \leq \frac{2^{\tau}}{N^{\tau}} \prod_{j=1}^{d} \left(1 + 2\nu_j^{1/(2\tau)} \zeta\left(\frac{\alpha}{\tau}\right)\right)^{\tau}$$

for any $\tau \in [1/2, \alpha)$ and any $i \in \mathcal{I}$. If $\sum_{j=1}^{\infty} \nu_j^{1/(2\tau)} < \infty$ for some $\tau \in [1/2, \alpha)$, then the above product with this τ can be bounded uniformly in d, i.e.,

$$\prod_{j=1}^{d} \left(1 + 2\nu_j^{1/(2\tau)} \zeta\left(\frac{\alpha}{\tau}\right)\right)^{\tau} \leq \prod_{j=1}^{\infty} \left(1 + 2\nu_j^{1/(2\tau)} \zeta\left(\frac{\alpha}{\tau}\right)\right)^{\tau} =: C_{\boldsymbol{\nu},\tau,\alpha} < \infty$$

and therefore

$$\mathrm{err}_{N,d,\alpha,\boldsymbol{\gamma}^{(i)}}(\boldsymbol{g}) \leq \frac{2^{\tau} C_{\boldsymbol{\nu},\tau,\alpha}}{N^{\tau}} \quad \text{for all } i \in \mathcal{I}.$$

As simple as this method is, it is applicable for specific weights that are not covered by Corollary 12.2.

Example 12.5 Assume we are given the two weight sequences $\boldsymbol{\gamma} = (j^{-2})_{j\geq 1}$ and $\boldsymbol{\gamma}' = (j^{-3})_{j\geq 1}$. Then it is clear that

$$\nu_j = \sup\{j^{-2}, j^{-3}\} = j^{-2} \quad \text{and} \quad \sum_{j=1}^{\infty} \nu_j < \infty$$

and, according to the above considerations we can find a generating vector $\boldsymbol{g}$ that yields a bound on the worst-case error in both weighted Korobov spaces with weights $\boldsymbol{\gamma}$ and $\boldsymbol{\gamma}'$, respectively, that is uniformly bounded in the dimension d. However, it is not possible to deduce this result from Corollary 12.2 since there the condition

$$\sum_{j=1}^{\infty} \frac{\gamma_j'}{\gamma_j} < \infty$$

in (12.3) is violated, as

$$\sum_{j=1}^{\infty} \frac{\gamma_j'}{\gamma_j} = \sum_{j=1}^{\infty} \frac{1}{j} = \infty.$$

(Interchanging the role of γ_j and γ_j' would make the situation even worse.)

12.2 The CBC Algorithm With Respect to More Than One Criterion

In this section, we would like to outline the principle ideas behind a CBC construction with more than one criterion, which are based on the following two observations, the second of which we already encountered in Remark 6.6.

(a) Consider Step (2) of the CBC algorithm (Algorithm 3.5). There we choose g_{s+1} which minimizes the error criterion. However, we will explain below that actually most choices of g_{s+1} are not too far off the best possible choice and give a reasonably good result.
(b) Let X be a finite set and let $A, B \subseteq X$. If $|A|, |B| > |X|/2$, then $A \cap B \neq \emptyset$, i.e., if we have two subsets of a finite set, each of which has more than $|X|/2$ elements, then the intersection of A and B is nonempty. This principle can of course be generalized, for instance, if $A, B, C \subseteq X$ with $|A|, |B|, |C| > 2|X|/3$, then $A \cap B \cap C \neq \emptyset$.

Compare (a) and (b) with the underlying idea for constructing extensible lattice rules as outlined in Remark 6.6. The main difference here is that we apply this idea to different sets of weights, rather than different numbers of points.

Using these considerations, we can, e.g., employ a CBC algorithm to find a generating vector $\boldsymbol{g}$ such that both $\mathrm{err}_{N,d,\alpha,\boldsymbol{\gamma}}(\boldsymbol{g})$ and $\mathrm{err}_{N,d,\alpha,\boldsymbol{\gamma}'}(\boldsymbol{g})$ are small for two sets of weights $\boldsymbol{\gamma}$ and $\boldsymbol{\gamma}'$, respectively. Thus, this generating vector works well with respect to $\boldsymbol{\gamma}$ and $\boldsymbol{\gamma}'$ simultaneously. In fact, we can say even a bit more. If there are, for $\tau \in [1/2, \alpha)$, B_τ and B_τ', such that

$$[\mathrm{err}_{N,d,\alpha,\gamma}(\boldsymbol{g})]^{1/\tau} \le \sum_{\emptyset \neq \mathfrak{u} \subseteq [d]} \gamma_{\mathfrak{u}}^{1/(2\tau)} \sum_{\boldsymbol{h}_{\mathfrak{u}} \in \mathcal{L}_{\mathfrak{u}}^{\perp}(\boldsymbol{g},N)} \prod_{j \in \mathfrak{u}} \frac{1}{|h_j|^{\alpha/\tau}} \le B_\tau,$$

$$[\mathrm{err}_{N,d,\alpha,\gamma'}(\boldsymbol{g})]^{1/\tau} \le \sum_{\emptyset \neq \mathfrak{u} \subseteq [d]} (\gamma'_{\mathfrak{u}})^{1/(2\tau)} \sum_{\boldsymbol{h}_{\mathfrak{u}} \in \mathcal{L}_{\mathfrak{u}}^{\perp}(\boldsymbol{g},N)} \prod_{j \in \mathfrak{u}} \frac{1}{|h_j|^{\alpha/\tau}} \le B'_\tau,$$

where the first inequalities in both of the previous lines follow from an application of Jensen's inequality (Lemma 2.25) to the respective error formulas in Remark 2.22, then for any $\kappa, \kappa' \ge 0$, not both of them zero, we have

$$\begin{aligned}
&[\mathrm{err}_{N,d,\alpha,\kappa\gamma+\kappa'\gamma'}(\boldsymbol{g})]^{1/\tau} \\
&\le \sum_{\emptyset \neq \mathfrak{u} \subseteq [d]} (\kappa\gamma_{\mathfrak{u}} + \kappa'\gamma'_{\mathfrak{u}})^{1/(2\tau)} \sum_{\boldsymbol{h}_{\mathfrak{u}} \in \mathcal{L}_{\mathfrak{u}}^{\perp}(\boldsymbol{g},N)} \prod_{j \in \mathfrak{u}} \frac{1}{|h_j|^{\alpha/\tau}} \\
&\le \sum_{\emptyset \neq \mathfrak{u} \subseteq [d]} ((\kappa\gamma_{\mathfrak{u}})^{1/(2\tau)} + (\kappa'\gamma'_{\mathfrak{u}})^{1/(2\tau)}) \sum_{\boldsymbol{h}_{\mathfrak{u}} \in \mathcal{L}_{\mathfrak{u}}^{\perp}(\boldsymbol{g},N)} \prod_{j \in \mathfrak{u}} \frac{1}{|h_j|^{\alpha/\tau}} \\
&\le \kappa^{1/(2\tau)} B_\tau + (\kappa')^{1/(2\tau)} B'_\tau. \qquad (12.8)
\end{aligned}$$

I.e., if we have a small worst-case error with respect to two sets of weights $\boldsymbol{\gamma}$ and $\boldsymbol{\gamma}'$, then the worst-case error with respect to certain linear combinations of those sets of weights is also small.

Item (a) of our preliminary observations above actually follows from the proof of Theorem 11.3. There it is shown that for an arbitrary prime number N and any $\eta \in (0,1)$, in each step corresponding to $s \in [d]$ of Algorithm 11.2, at least $\lceil \eta(N-1) \rceil$ choices of $g \in [N-1]$ satisfy the bound

$$\mathrm{err}_{N,s+1,\alpha,\gamma}((g_1, g_2, \ldots, g_s, g)) \le \left(\frac{2}{(1-\eta)N} \sum_{\emptyset \neq \mathfrak{u} \subseteq [s+1]} \gamma_{\mathfrak{u}}^{1/(2\tau)} \left(2\zeta\left(\frac{\alpha}{\tau}\right)\right)^{|\mathfrak{u}|} \right)^{\tau},$$

for any $\tau \in [1/2, \alpha)$, where each g_j, $j \in [s]$, has been chosen as one of the "best" $\lceil \eta(N-1) \rceil$ components in one of the previous steps. Actually, using the notation of Algorithm 11.2, we obtain this result by making the fixed choice $p = N$ in that algorithm, and not selecting $p \in P_N$ at random. In fact, we can set $g_1 = 1$, as we have also done in Algorithm 11.2. So, we can use the following CBC algorithm. For simplicity we again assume that N is a prime number, but a similar result could also be derived for composite N.

Algorithm 12.6 (CBC construction principle with two quality criteria) Let prime N, $d \in \mathbb{N}$, two sets of weights $\boldsymbol{\gamma} = \{\gamma_{\mathfrak{u}}\}_{\mathfrak{u} \subseteq [d]}$ and $\boldsymbol{\gamma}' = \{\gamma'_{\mathfrak{u}}\}_{\mathfrak{u} \subseteq [d]}$, and $\eta \in (1/2, 1)$ be given. Construct a generating vector $\boldsymbol{g} = (g_1, \ldots, g_d) \in G_d^{\varphi}(N)$ as follows.

(1) Set $g_1 = 1$.
(2) For s from 1 to $d-1$:
Assume that $g_1, \ldots, g_s \in [N-1]$ have already been found. Consider them as fixed and choose g_{s+1} from the set $\mathcal{G}_{\boldsymbol{\gamma}} \cap \mathcal{G}_{\boldsymbol{\gamma}'}$, where, for $\boldsymbol{\nu} \in \{\boldsymbol{\gamma}, \boldsymbol{\gamma}'\}$,

$$\mathcal{G}_\nu = \{\mathfrak{g}_1^{(\nu)}, \mathfrak{g}_2^{(\nu)}, \ldots, \mathfrak{g}_{\lceil \eta(N-1) \rceil}^{(\nu)}\} \subseteq [N-1],$$

such that

$$\begin{aligned}
&\mathrm{err}_{N,s+1,\alpha,\nu}(g_1, \ldots, g_s, \mathfrak{g}_1^{(\nu)}) \\
&\le \mathrm{err}_{N,s+1,\alpha,\nu}(g_1, \ldots, g_s, \mathfrak{g}_2^{(\nu)}) \\
&\vdots \\
&\le \mathrm{err}_{N,s+1,\alpha,\nu}(g_1, \ldots, g_s, \mathfrak{g}_{\lceil \eta(N-1) \rceil}^{(\nu)}) \\
&\le \mathrm{err}_{N,s+1,\alpha,\nu}(g_1, \ldots, g_s, g) \quad \text{for all } g \in [N-1] \setminus \{\mathfrak{g}_1^{(\nu)}, \ldots, \mathfrak{g}_{\lceil \eta(N-1) \rceil}^{(\nu)}\}.
\end{aligned}$$

If the error criterion has the same minimal value for several distinct elements in one step of the algorithm, it is allowed to choose any of them.
End for.

(3) Set $\boldsymbol{g} = (g_1, \ldots, g_d)$.

We have the following theorem.

Theorem 12.7 *Let N be a prime number, let $d \in \mathbb{N}$, and let $\boldsymbol{\gamma} = \{\gamma_{\mathfrak{u}}\}_{\mathfrak{u} \subseteq [d]}$ and $\boldsymbol{\gamma}' = \{\gamma'_{\mathfrak{u}}\}_{\mathfrak{u} \subseteq [d]}$ be weights. Assume that $\boldsymbol{g} = (g_1, g_2, \ldots, g_d)$ has been found by Algorithm 12.6 for some $\eta \in (1/2, 1)$. Then, for any $s \in [d]$ and any $\tau \in [1/2, \alpha)$, we have*

$$\mathrm{err}_{N,s,\alpha,\boldsymbol{\gamma}}((g_1, g_2, \ldots, g_s)) \le \left(\frac{2}{(1-\eta)N} \sum_{\emptyset \ne \mathfrak{u} \subseteq [s]} \gamma_{\mathfrak{u}}^{1/(2\tau)} \left(2\zeta\left(\frac{\alpha}{\tau}\right) \right)^{|\mathfrak{u}|} \right)^{\tau}$$

and

$$\mathrm{err}_{N,s,\alpha,\boldsymbol{\gamma}'}((g_1, g_2, \ldots, g_s)) \le \left(\frac{2}{(1-\eta)N} \sum_{\emptyset \ne \mathfrak{u} \subseteq [s]} (\gamma'_{\mathfrak{u}})^{1/(2\tau)} \left(2\zeta\left(\frac{\alpha}{\tau}\right) \right)^{|\mathfrak{u}|} \right)^{\tau}.$$

Algorithm 12.6 and Theorem 12.7 can be generalized to $k \ge 2$ criteria. In this case we need to ensure that $k(1-\eta) < 1$, that is, $1 - 1/k < \eta < 1$. More generally, we can choose a different value η_i instead of η for each criterion $i \in [k]$. In this case the condition $\sum_{i=1}^{k}(1-\eta_i) < 1$ ensures that there is a generating vector satisfying all k conditions simultaneously.

12.3 Random Weights

Another approach to obtaining results for sets of weights different from the set of weights used in the common CBC algorithm (Algorithm 3.5), is to assume that the weights $\{\gamma_{\mathfrak{u}}\}_{\mathfrak{u}\subseteq[d]}$ are random variables. Assume that the random variables $\gamma_{\mathfrak{u}}$ for $\mathfrak{u} \subseteq [d]$ (on some underlying probability space $(\Omega, \mathcal{F}, \mathbb{P})$) have the following properties.

- $\gamma_\emptyset = 1$, and $\gamma_{\mathfrak{u}} \geq 0$ for $\mathfrak{u} \neq \emptyset$;
- $\overline{\gamma}_{\mathfrak{u}} := \mathbb{E}[\gamma_{\mathfrak{u}}] > 0$;
- $\mathbb{E}[\gamma_{\mathfrak{u}}\gamma_{\mathfrak{v}}] = \mathbb{E}[\gamma_{\mathfrak{u}}]\,\mathbb{E}[\gamma_{\mathfrak{v}}]$ for $\mathfrak{u} \neq \mathfrak{v}$;
- $\sigma_{\mathfrak{u}}^2 := \mathrm{Var}[\gamma_{\mathfrak{u}}] < \infty$ for $\mathfrak{u} \neq \emptyset$.

In this setup, the squared worst-case error is now a random variable,

$$[\mathrm{err}_{N,d,\alpha,\boldsymbol{\gamma}}(\boldsymbol{g})]^2 = \sum_{\boldsymbol{h}\in\mathcal{L}^\perp(\boldsymbol{g},N)\setminus\{\boldsymbol{0}\}} \frac{1}{r_{2\alpha,\boldsymbol{\gamma}}(\boldsymbol{h})} = \sum_{\emptyset\neq\mathfrak{u}\subseteq[d]} \gamma_{\mathfrak{u}} \sum_{\boldsymbol{h}_{\mathfrak{u}}\in\mathcal{L}_{\mathfrak{u}}^\perp(\boldsymbol{g},N)} \prod_{j\in\mathfrak{u}} \frac{1}{|h_j|^{2\alpha}},$$

(see Remark 2.22 for the second identity), and the expected value of the squared worst-case error is

$$\begin{aligned}\mathbb{E}\left[[\mathrm{err}_{N,d,\alpha,\boldsymbol{\gamma}}(\boldsymbol{g})]^2\right] &= \sum_{\emptyset\neq\mathfrak{u}\subseteq[d]} \mathbb{E}[\gamma_{\mathfrak{u}}] \sum_{\boldsymbol{h}_{\mathfrak{u}}\in\mathcal{L}_{\mathfrak{u}}^\perp(\boldsymbol{g},N)} \prod_{j\in\mathfrak{u}} \frac{1}{|h_j|^{2\alpha}}\\ &= \sum_{\emptyset\neq\mathfrak{u}\subseteq[d]} \overline{\gamma}_{\mathfrak{u}} \sum_{\boldsymbol{h}_{\mathfrak{u}}\in\mathcal{L}_{\mathfrak{u}}^\perp(\boldsymbol{g},N)} \prod_{j\in\mathfrak{u}} \frac{1}{|h_j|^{2\alpha}} = [\mathrm{err}_{N,d,\alpha,\overline{\boldsymbol{\gamma}}}(\boldsymbol{g})]^2,\end{aligned}$$

where $\overline{\boldsymbol{\gamma}} := \{\overline{\gamma}_{\mathfrak{u}}\}_{\mathfrak{u}\subseteq[d]}$. One can use $[\mathrm{err}_{N,d,\alpha,\overline{\boldsymbol{\gamma}}}(\boldsymbol{g})]^2$ as a quality criterion in a CBC algorithm. Doing so only gives us information on the mean of the squared worst-case error with respect to the weights but not on how much the squared worst-case error changes as we move away from the mean. To incorporate such information in the CBC algorithm, one may look at the variance of the squared worst-case error,

$$\begin{aligned}\mathrm{Var}\left[[\mathrm{err}_{N,d,\alpha,\boldsymbol{\gamma}}(\boldsymbol{g})]^2\right] &= \mathbb{E}\left[[\mathrm{err}_{N,d,\alpha,\boldsymbol{\gamma}}(\boldsymbol{g})]^4\right] - \left(\mathbb{E}\left[[\mathrm{err}_{N,d,\alpha,\boldsymbol{\gamma}}(\boldsymbol{g})]^2\right]\right)^2\\ &= \sum_{\emptyset\neq\mathfrak{u}\subseteq[d]} \left(\sigma_{\mathfrak{u}} \sum_{\boldsymbol{h}_{\mathfrak{u}}\in\mathcal{L}_{\mathfrak{u}}^\perp(\boldsymbol{g},N)} \prod_{j\in\mathfrak{u}} \frac{1}{|h_j|^{2\alpha}}\right)^2.\end{aligned}$$

The standard deviation is given by

$$\begin{aligned}\sigma\left[[\mathrm{err}_{N,d,\alpha,\boldsymbol{\gamma}}(\boldsymbol{g})]^2\right] &= \left(\sum_{\emptyset\neq\mathfrak{u}\subseteq[d]} \left(\sigma_{\mathfrak{u}} \sum_{\boldsymbol{h}_{\mathfrak{u}}\in\mathcal{L}_{\mathfrak{u}}^\perp(\boldsymbol{g},N)} \prod_{j\in\mathfrak{u}} \frac{1}{|h_j|^{2\alpha}}\right)^2\right)^{1/2}\\ &\leq \sum_{\emptyset\neq\mathfrak{u}\subseteq[d]} \sigma_{\mathfrak{u}} \sum_{\boldsymbol{h}_{\mathfrak{u}}\in\mathcal{L}_{\mathfrak{u}}^\perp(\boldsymbol{g},N)} \prod_{j\in\mathfrak{u}} \frac{1}{|h_j|^{2\alpha}}\end{aligned}$$

$$= [\mathrm{err}_{N,d,\alpha,\boldsymbol{\sigma}}(\boldsymbol{g})]^2,$$

where $\boldsymbol{\sigma} := \{\sigma_{\mathfrak{u}}\}_{\mathfrak{u}\subseteq[d]}$, and where we formally set $\sigma_\emptyset := 1$. Using the standard deviation directly is difficult to do numerically, but we can use the upper bound given by $[\mathrm{err}_{N,d,\alpha,\boldsymbol{\sigma}}(\boldsymbol{g})]^2$ instead.

Hence we can now use Algorithm 12.6 with the two criteria $[\mathrm{err}_{N,d,\alpha,\overline{\boldsymbol{\gamma}}}(\boldsymbol{g})]^2$ and $[\mathrm{err}_{N,d,\alpha,\boldsymbol{\sigma}}(\boldsymbol{g})]^2$ to find a generating vector for which the expected value and the variance of $[\mathrm{err}_{N,d,\alpha,\boldsymbol{\gamma}}(\boldsymbol{g})]^2$ are both small. This allows us to obtain a more precise probabilistic statement on the squared worst-case error of the obtained lattice rule by using the Chebyshev inequality.

Indeed, we can use the one-sided Chebyshev inequality, which states that for a random variable X with standard deviation $\sigma[X]$ we have

$$\mathbb{P}[X - \mathbb{E}[X]] \geq c\,\sigma[X] \geq \frac{1}{1+c^2} \quad \text{for any } c > 0.$$

This implies the following result.

Theorem 12.8 *Let N be a prime number and let $d \in \mathbb{N}$. Let $(\Omega, \mathcal{F}, \mathbb{P})$ be a probability space and let $\boldsymbol{\gamma} = (\gamma_{\mathfrak{u}})_{\mathfrak{u}\subseteq[d]}$ be a random vector on $(\Omega, \mathcal{F}, \mathbb{P})$ which satisfies*

- $\gamma_\emptyset = 1$, *and* $\gamma_{\mathfrak{u}} \geq 0$ *for* $\mathfrak{u} \neq \emptyset$*;*
- $\overline{\gamma}_{\mathfrak{u}} := \mathbb{E}[\gamma_{\mathfrak{u}}] > 0$*;*
- $\mathbb{E}[\gamma_{\mathfrak{u}}\gamma_{\mathfrak{v}}] = \mathbb{E}[\gamma_{\mathfrak{u}}]\,\mathbb{E}[\gamma_{\mathfrak{v}}]$ *for* $\mathfrak{u} \neq \mathfrak{v}$*;*
- $\sigma_{\mathfrak{u}}^2 := \mathrm{Var}[\gamma_{\mathfrak{u}}] < \infty$ *for* $\mathfrak{u} \neq \emptyset$ *(formally also set* $\sigma_\emptyset := 1$*).*

Put $\overline{\boldsymbol{\gamma}} := \{\overline{\gamma}_{\mathfrak{u}}\}_{\mathfrak{u}\subseteq[d]}$ and $\boldsymbol{\sigma} := \{\sigma_{\mathfrak{u}}\}_{\mathfrak{u}\subseteq[d]}$. If $\boldsymbol{g}$ is constructed by Algorithm 12.6 based on the criteria $[\mathrm{err}_{N,d,\alpha,\overline{\boldsymbol{\gamma}}}(\boldsymbol{g})]^2$ and $[\mathrm{err}_{N,d,\alpha,\boldsymbol{\sigma}}(\boldsymbol{g})]^2$, then for any $c > 0$ we have

$$\mathbb{P}\left[[\mathrm{err}_{N,d,\alpha,\boldsymbol{\gamma}}(\boldsymbol{g})]^2 \leq [\mathrm{err}_{N,d,\alpha,\overline{\boldsymbol{\gamma}}}(\boldsymbol{g})]^2 + c[\mathrm{err}_{N,d,\alpha,\boldsymbol{\sigma}}(\boldsymbol{g})]^2\right] \geq \frac{c^2}{1+c^2}.$$

For instance, if $c = 10$, then $c^2/(1+c^2) = 0.99009\ldots$, and hence

$$\mathbb{P}\left[[\mathrm{err}_{N,d,\alpha,\boldsymbol{\gamma}}(\boldsymbol{g})]^2 \leq [\mathrm{err}_{N,d,\alpha,\overline{\boldsymbol{\gamma}}}(\boldsymbol{g})]^2 + 10[\mathrm{err}_{N,d,\alpha,\boldsymbol{\sigma}}(\boldsymbol{g})]^2\right] \geq 0.99.$$

Notes and Remarks

The results in Section 12.1 were first shown in [38]. This paper also discusses further examples, in particular the combinations of product weights with general weights and POD weights with POD weights. The method can also be applied to other criteria, for instance, [38] also includes results for the weighted discrepancy.

The approach for fixed smoothness but different product weight sequences on p. 402 was suggested by Larcher (personal communication).

Random weights and the CBC algorithm with more than one quality criterion were discussed in [36]. This paper studies the general case of r criteria in the CBC algorithm and considers geometric aspects related to (12.8).

Chapter 13
L_2-Approximation Using Lattice Rules

In the preceding chapters, we have seen how (rank-1) lattice rules can be used for numerical integration, in particular in Korobov and Sobolev spaces. Moreover, we have outlined how we can efficiently construct the generating vectors of such (good) lattice rules, e.g., by the CBC algorithm.

However, this is not the end of the story of lattice rules. As it has been outlined repeatedly in the literature, we can also employ lattice rules in suitably designed algorithms for approximating functions. In this chapter, we will outline what role lattice rules can play in function approximation with respect to the L_2-norm. We shall start by describing a quite natural approach to this question.

13.1 L_2-Approximation of Functions in Korobov Spaces

Let us again consider the weighted Korobov space $\mathcal{H}_{\mathrm{kor},d,\alpha,\boldsymbol{\gamma}}$ of smoothness α with weights $\boldsymbol{\gamma}$, as defined in Section 2.4. For the sake of simplicity, we assume product weights $\boldsymbol{\gamma} = (\gamma_j)_{j\geq 1}$ with $\gamma_j \in (0, 1]$ for $j \in \mathbb{N}$ throughout this section. The cases of general weights and of POD weights will be discussed in Section 13.4. As usual, we assume without loss of generality that the product weights are ordered in a nonincreasing fashion. For $f \in \mathcal{H}_{\mathrm{kor},d,\alpha,\boldsymbol{\gamma}}$ we know that we can represent f pointwise by its Fourier series,

$$f(\boldsymbol{x}) = \sum_{\boldsymbol{h}\in\mathbb{Z}^d} \widehat{f}(\boldsymbol{h})\, \mathrm{e}^{2\pi \mathrm{i}\boldsymbol{h}\cdot\boldsymbol{x}} \qquad \text{for } \boldsymbol{x} \in [0, 1)^d.$$

We now study the problem of approximating the function f itself in the L_2-norm instead of approximating the integral of f. To formalize this problem, we define the embedding operator $\mathrm{EMB}_d : \mathcal{H}_{\mathrm{kor},d,\alpha,\boldsymbol{\gamma}} \to L_2([0, 1]^d)$ by $\mathrm{EMB}_d(f) = f$. We would like to approximate $\mathrm{EMB}_d(f)$ for $f \in \mathcal{H}_{\mathrm{kor},d,\alpha,\boldsymbol{\gamma}}$ instead of approximating the integral $\int_{[0,1]^d} f(\boldsymbol{x})\, \mathrm{d}\boldsymbol{x}$.

J. Dick et al., *Lattice Rules*, Springer Series in Computational Mathematics 58,
https://doi.org/10.1007/978-3-031-09951-9_13

Naturally, this problem has been very well studied in the literature, and a common approach in the classical literature on approximation theory is to approximate the function $f \in \mathcal{H}_{\mathrm{kor},d,\alpha,\boldsymbol{\gamma}}$ by a truncated Fourier series,

$$\sum_{\boldsymbol{h} \in \mathcal{A}_d} \widehat{f}(\boldsymbol{h})\, \mathrm{e}^{2\pi \mathrm{i} \boldsymbol{h} \cdot \boldsymbol{x}},$$

where $\mathcal{A}_d$ is some suitably chosen finite subset of $\mathbb{Z}^d$.

In many practical problems one will not have direct access to the Fourier coefficients $\widehat{f}(\boldsymbol{h})$ for $\boldsymbol{h} \in \mathcal{A}_d$, so it makes sense to consider approximating also these. This is the point where we can make use of lattice rules; to be more precise, we can approximate the integral

$$\widehat{f}(\boldsymbol{h}) = \int_{[0,1]^d} f(\boldsymbol{x})\, \mathrm{e}^{-2\pi \mathrm{i} \boldsymbol{h} \cdot \boldsymbol{x}}\, \mathrm{d}\boldsymbol{x} \quad \text{for } \boldsymbol{h} \in \mathcal{A}_d,$$

by a QMC rule using a rank-1 lattice point set.

First of all, it is necessary to state precisely what choice of $\mathcal{A}_d$ should be made. Naturally, we want to choose the size of the set as small as possible to keep the computational cost low, but also such that it contains as many indices that correspond to "large" Fourier coefficients as possible. To this end, define, for a real $M \geq 0$, $\mathcal{A}_d = \mathcal{A}_{d,M}$ as

$$\mathcal{A}_{d,M} := \left\{\boldsymbol{h} \in \mathbb{Z}^d \;:\; r_{2\alpha,\boldsymbol{\gamma}}(\boldsymbol{h}) \leq M\right\}. \tag{13.1}$$

Then, we will approximate $f \in \mathcal{H}_{\mathrm{kor},d,\alpha,\boldsymbol{\gamma}}$ by an algorithm of the form

$$A_{N,M,d,\alpha,\boldsymbol{\gamma}}(\boldsymbol{g})(f)(\boldsymbol{x}) := \sum_{\boldsymbol{h} \in \mathcal{A}_{d,M}} \left(\frac{1}{N} \sum_{k=0}^{N-1} f\left(\left\{\frac{k\boldsymbol{g}}{N}\right\}\right) \mathrm{e}^{-2\pi \mathrm{i} k \boldsymbol{h} \cdot \boldsymbol{g}/N} \right) \mathrm{e}^{2\pi \mathrm{i} \boldsymbol{h} \cdot \boldsymbol{x}} \tag{13.2}$$

for $\boldsymbol{x} \in [0,1)^d$, where $\boldsymbol{g}$ is the generating vector of a rank-1 lattice point set $\mathcal{P}(\boldsymbol{g}, N)$.

Similarly to the worst-case error of integration, we can then define the worst-case error of the approximation algorithm $A_{N,M,d,\alpha,\boldsymbol{\gamma}}(\boldsymbol{g})$ in $\mathcal{H}_{\mathrm{kor},d,\alpha,\boldsymbol{\gamma}}$ by

$$\mathrm{err}^{L_2-\mathrm{app}}(\mathcal{H}_{\mathrm{kor},d,\alpha,\boldsymbol{\gamma}}, A_{N,M,d,\alpha,\boldsymbol{\gamma}}(\boldsymbol{g})) := \sup_{\substack{f \in \mathcal{H}_{\mathrm{kor},d,\alpha,\boldsymbol{\gamma}} \\ \|f\|_{\mathrm{kor},d,\alpha,\boldsymbol{\gamma}} \leq 1}} \|f - A_{N,M,d,\alpha,\boldsymbol{\gamma}}(\boldsymbol{g})(f)\|_{L_2}. \tag{13.3}$$

In order to analyze this worst-case error of approximation, we first need to collect some properties of the set $\mathcal{A}_{d,M}$, which we show in the following lemma, reminding the reader that we restrict ourselves to considering product weights not exceeding 1 in this section.

Lemma 13.1 *For $\alpha > 1/2$, $d \in \mathbb{N}$, a weight sequence $\boldsymbol{\gamma} = (\gamma_j)_{j \geq 1}$ with $\gamma_j \in (0,1]$ for $j \in \mathbb{N}$, and $M \geq 0$, let $\mathcal{A}_{d,M} := \left\{\boldsymbol{h} \in \mathbb{Z}^d \;:\; r_{2\alpha,\boldsymbol{\gamma}}(\boldsymbol{h}) \leq M\right\}$. Then the following assertions hold.*

1. *For* $\boldsymbol{h} = (h_1, \ldots, h_d) \in \mathcal{A}_{d,M}$ *it is true that* $|h_j| \le (\gamma_j M)^{1/(2\alpha)}$ *for* $j \in [d]$.
2. *For any* $d \in \mathbb{N}$ *we have*

$$\mathcal{A}_{d+1,M} = \{(\boldsymbol{h},0) \,:\, \boldsymbol{h} \in \mathcal{A}_{d,M}\} \cup \bigcup_{h_{d+1} \in \mathbb{Z}\setminus\{0\}} \{(\boldsymbol{h},h_{d+1}) \,:\, \boldsymbol{h} \in \mathcal{A}_{d,\widetilde{M}(h_{d+1})}\},$$

where $\widetilde{M}(h_{d+1}) := (\gamma_{d+1}M)/|h_{d+1}|^{2\alpha}$. *We remind the reader that for* $\boldsymbol{h} = (h_1, \ldots, h_d)$ *in* $\mathbb{Z}^d$ *and* $h_{d+1} \in \mathbb{Z}$ *we write, with some abuse of notation,* $(\boldsymbol{h}, h_{d+1})$ *for the vector* $(h_1, \ldots, h_d, h_{d+1})$ *in* $\mathbb{Z}^{d+1}$, *and in particular* $(\boldsymbol{h}, 0)$ *if* $h_{d+1} = 0$.
3. *For any* $d \in \mathbb{N}$ *it is true that*

$$|\mathcal{A}_{d+1,M}| = |\mathcal{A}_{d,M}| + 2 \sum_{h_{d+1}=1}^{\infty} |\mathcal{A}_{d,\widetilde{M}(h_{d+1})}|,$$

with $\widetilde{M}(h_{d+1})$ *defined as in Item 2.*
4. *For any* $\lambda > 1/(2\alpha)$ *we have*

$$|\mathcal{A}_{d,M}| \le M^{\lambda} \prod_{j=1}^{d} (1 + 2\gamma_j^{\lambda} \zeta(2\alpha\lambda)).$$

Proof Regarding Item 1, note that $\mathcal{A}_{d,M}$ is empty as long as $M < 1$. For $M \ge 1$, suppose that an index $\boldsymbol{h} \in \mathbb{Z}^d$ is contained in $\mathcal{A}_{d,M}$. For such an $\boldsymbol{h}$ we have

$$r_{2\alpha,\boldsymbol{\gamma}}(\boldsymbol{h}) = \prod_{j=1}^{d} r_{2\alpha,\gamma_j}(h_j) \le M.$$

Due to our assumption that no weight γ_j exceeds 1, we have $r_{2\alpha,\gamma_j}(h_j) \ge 1$ for each $j \in [d]$, and hence we must have $r_{2\alpha,\gamma_j}(h_j) \le M$ for each j. If $h_j = 0$, this condition and also the statement in Item 1 is trivially fulfilled. On the other hand, if $h_j \neq 0$, the inequality in Item 1 follows immediately by the definition of $r_{2\alpha,\gamma_j}(h_j)$ and the fact that $\gamma_j \le 1$.

For the proof of Item 2, note that again the case $M < 1$ is trivial, as all sets involved are then empty. So, let us assume $M \ge 1$. Suppose that $(\boldsymbol{h}, h_{d+1})$ with $\boldsymbol{h} \in \mathbb{Z}^d$ and $h_{d+1} \in \mathbb{Z}$ is an element of $\mathcal{A}_{d+1,M}$. If $h_{d+1} = 0$, we have $r_{2\alpha,\boldsymbol{\gamma}}((\boldsymbol{h}, h_{d+1})) = r_{2\alpha,\boldsymbol{\gamma}}(\boldsymbol{h})$. Then, we must have $r_{2\alpha,\boldsymbol{\gamma}}(\boldsymbol{h}) \le M$, which is the same as $\boldsymbol{h} \in \mathcal{A}_{d,M}$. If, however, $h_{d+1} \neq 0$, we must have $r_{2\alpha,\boldsymbol{\gamma}}((\boldsymbol{h}, h_{d+1})) = r_{2\alpha,\boldsymbol{\gamma}}(\boldsymbol{h}) r_{2\alpha,\gamma_{d+1}}(h_{d+1}) \le M$, which means

$$r_{2\alpha,\boldsymbol{\gamma}}(\boldsymbol{h}) \le \frac{\gamma_{d+1} M}{|h_{d+1}|^{2\alpha}} = \widetilde{M}(h_{d+1}),$$

and this is equivalent to $\boldsymbol{h} \in \mathcal{A}_{d,\widetilde{M}(h_{d+1})}$.

Item 3 immediately follows from Item 2.

The result in Item 4 can be proven by induction on d. We can again assume $M \geq 1$, as otherwise the set $\mathcal{A}_{d,M}$ is empty. For $d = 1$, it follows from the definition of $\mathcal{A}_{1,M}$ that

$$|\mathcal{A}_{1,M}| = 1 + 2\lfloor(\gamma_1 M)^{1/(2\alpha)}\rfloor.$$

If $\gamma_1 M < 1$, then obviously $0 = \lfloor(\gamma_1 M)^{1/(2\alpha)}\rfloor \leq \gamma_1^\lambda M^\lambda$. If $\gamma_1 M \geq 1$, then we have

$$\lfloor(\gamma_1 M)^{1/(2\alpha)}\rfloor \leq (\gamma_1 M)^{1/(2\alpha)} \leq (\gamma_1 M)^\lambda.$$

Combining these two cases with the fact that $\zeta(2\alpha\lambda) \geq 1$ immediately yields that

$$|\mathcal{A}_{1,M}| \leq M^\lambda(1 + 2\zeta(2\alpha\lambda)\gamma_1^\lambda),$$

which is the desired result for $d = 1$. Suppose now that the estimate in Item 4 holds for fixed $d \in \mathbb{N}$. Then we can use Item 3 and the induction assumption to see that

$$\begin{aligned}
|\mathcal{A}_{d+1,M}| &\leq M^\lambda \prod_{j=1}^{d}(1 + 2\gamma_j^\lambda\zeta(2\alpha\lambda)) + 2\sum_{h_{d+1}=1}^{\infty}(\widetilde{M}(h_{d+1}))^\lambda \prod_{j=1}^{d}(1 + 2\zeta(2\alpha\lambda)\gamma_j^\lambda) \\
&= (1 + 2\zeta(2\alpha\lambda)\gamma_{d+1}^\lambda)M^\lambda \prod_{j=1}^{d}(1 + 2\zeta(2\alpha\lambda)\gamma_j^\lambda) \\
&= M^\lambda \prod_{j=1}^{d+1}(1 + 2\zeta(2\alpha\lambda)\gamma_j^\lambda),
\end{aligned}$$

as claimed. □

In the next step, let us study the square of the error of $A_{N,M,d,\alpha,\boldsymbol{\gamma}}(\boldsymbol{g})(f)$ for a given generating vector $\boldsymbol{g}$ and a given function f in $\mathcal{H}_{\mathrm{kor},d,\alpha,\boldsymbol{\gamma}}$. Using Parseval's identity, we obtain

$$\begin{aligned}
\left(\|f - A_{N,M,d,\alpha,\boldsymbol{\gamma}}(\boldsymbol{g})(f)\|_{L_2}\right)^2 &= \sum_{\boldsymbol{h}\notin\mathcal{A}_{d,M}} |\widehat{f}(\boldsymbol{h})|^2 \\
&+ \sum_{\boldsymbol{h}\in\mathcal{A}_{d,M}} \left| \int_{[0,1]^d} f(\boldsymbol{x})\, \mathrm{e}^{2\pi \mathrm{i}\boldsymbol{h}\cdot\boldsymbol{x}}\,\mathrm{d}\boldsymbol{x} - \frac{1}{N}\sum_{k=0}^{N-1} f\left(\left\{\frac{k\boldsymbol{g}}{N}\right\}\right)\mathrm{e}^{-2\pi \mathrm{i}k\boldsymbol{h}\cdot\boldsymbol{g}/N}\right|^2.
\end{aligned} \tag{13.4}$$

Obviously, we have

$$\sum_{\boldsymbol{h}\notin\mathcal{A}_{d,M}} |\widehat{f}(\boldsymbol{h})|^2 = \sum_{\boldsymbol{h}\notin\mathcal{A}_{d,M}} |\widehat{f}(\boldsymbol{h})|^2\, r_{2\alpha,\boldsymbol{\gamma}}(\boldsymbol{h})\frac{1}{r_{2\alpha,\boldsymbol{\gamma}}(\boldsymbol{h})} \leq \frac{1}{M}\|f\|^2_{\mathrm{kor},d,\alpha,\boldsymbol{\gamma}}.$$

The next, and more involved, step is to analyze the second term in (13.4). To this end, let $\boldsymbol{h} \in \mathcal{A}_{d,M}$ be fixed, and consider

$$\left| \int_{[0,1]^d} f(\boldsymbol{x}) \, \mathrm{e}^{-2\pi \mathrm{i} \boldsymbol{h} \cdot \boldsymbol{x}} \, \mathrm{d}\boldsymbol{x} - \frac{1}{N} \sum_{k=0}^{N-1} f\left(\left\{ \frac{k\boldsymbol{g}}{N} \right\} \right) \mathrm{e}^{-2\pi \mathrm{i} k \boldsymbol{h} \cdot \boldsymbol{g}/N} \right|^2 . \tag{13.5}$$

We now discuss two possibilities to deal with the term in (13.5).

A direct approach

The first way to deal with (13.5) is to view it as the squared integration error of the involved lattice rule applied to the function $f_{\boldsymbol{h}}$ with

$$f_{\boldsymbol{h}}(\boldsymbol{x}) := f(\boldsymbol{x}) \, \mathrm{e}^{-2\pi \mathrm{i} \boldsymbol{h} \cdot \boldsymbol{x}} \quad \text{for} \quad \boldsymbol{x} \in [0,1]^d.$$

It is easy to see that $f_{\boldsymbol{h}}$ is also contained in $\mathcal{H}_{\mathrm{kor},d,\alpha,\boldsymbol{\gamma}}$ (this is a property that could be called an "algebra property" of $\mathcal{H}_{\mathrm{kor},d,\alpha,\boldsymbol{\gamma}}$, and we remark that an analogous property need not be satisfied by other function spaces with norms based on series expansions), and that $\widehat{f_{\boldsymbol{h}}}(\boldsymbol{\ell}) = \widehat{f}(\boldsymbol{h} + \boldsymbol{\ell})$ for $\boldsymbol{\ell} \in \mathbb{Z}^d$. It follows that

$$\begin{aligned} &\left| \int_{[0,1]^d} f(\boldsymbol{x}) \, \mathrm{e}^{-2\pi \mathrm{i} \boldsymbol{h} \cdot \boldsymbol{x}} \, \mathrm{d}\boldsymbol{x} - \frac{1}{N} \sum_{k=0}^{N-1} f\left(\left\{ \frac{k\boldsymbol{g}}{N} \right\} \right) \mathrm{e}^{-2\pi \mathrm{i} k \boldsymbol{h} \cdot \boldsymbol{g}/N} \right|^2 \\ &\qquad\qquad \leq [\mathrm{err}_{N,d,\alpha,\boldsymbol{\gamma}}(\boldsymbol{g})]^2 \, \|f_{\boldsymbol{h}}\|^2_{\mathrm{kor},d,\alpha,\boldsymbol{\gamma}} , \end{aligned}$$

where, as usual, $\mathrm{err}_{N,d,\alpha,\boldsymbol{\gamma}}(\boldsymbol{g})$ denotes the worst-case integration error of the lattice rule generated by $\boldsymbol{g}$. It remains to treat the (squared) norm of $f_{\boldsymbol{h}}$, which can be done as follows. By definition, we have

$$\|f_{\boldsymbol{h}}\|^2_{\mathrm{kor},d,\alpha,\boldsymbol{\gamma}} = \sum_{\boldsymbol{\ell} \in \mathbb{Z}^d} |\widehat{f_{\boldsymbol{h}}}(\boldsymbol{\ell})|^2 r_{2\alpha,\boldsymbol{\gamma}}(\boldsymbol{\ell}) = \sum_{\boldsymbol{\ell} \in \mathbb{Z}^d} |\widehat{f}(\boldsymbol{h}+\boldsymbol{\ell})|^2 r_{2\alpha,\boldsymbol{\gamma}}(\boldsymbol{h}+\boldsymbol{\ell}) \frac{r_{2\alpha,\boldsymbol{\gamma}}(\boldsymbol{\ell})}{r_{2\alpha,\boldsymbol{\gamma}}(\boldsymbol{h}+\boldsymbol{\ell})}. \tag{13.6}$$

For any choice of the indices $\boldsymbol{h}, \boldsymbol{\ell} \in \mathbb{Z}^d$, it is true that

$$\frac{r_{2\alpha,\boldsymbol{\gamma}}(\boldsymbol{\ell})}{r_{2\alpha,\boldsymbol{\gamma}}(\boldsymbol{h}+\boldsymbol{\ell})} \leq r_{2\alpha,\boldsymbol{\gamma}}(\boldsymbol{h}) \prod_{j=1}^{d} \max(1, 2^{2\alpha}\gamma_j). \tag{13.7}$$

It suffices to prove this estimate for the univariate case by simply checking the four cases in which $(\ell, h+\ell)$ in $\{(0,0)\}$, in $\{0\} \times \mathbb{Z}^*$, in $\mathbb{Z}^* \times \{0\}$, or in $\mathbb{Z}^* \times \mathbb{Z}^*$, respectively, where in the latter case one has to consider the two sub-cases $h = 0$ and $h \neq 0$, and where, as usual, $\mathbb{Z}^* := \mathbb{Z} \setminus \{0\}$; going through these cases separately, it is then easy to see that (13.7) holds. Consequently, it follows from (13.6) that

$$\|f_{\boldsymbol{h}}\|^2_{\mathrm{kor},d,\alpha,\boldsymbol{\gamma}} \leq \|f\|^2_{\mathrm{kor},d,\alpha,\boldsymbol{\gamma}} \, r_{2\alpha,\boldsymbol{\gamma}}(\boldsymbol{h}) \prod_{j=1}^{d} \max(1, 2^{2\alpha}\gamma_j).$$

This enables us to make use of the usual construction methods of a rank-1 lattice generating vector $\boldsymbol{g}$ that makes the worst-case integration error small, in order to obtain an effective bound on the approximation error. Indeed, let us assume that $\boldsymbol{g}$ has been constructed according to, say, the usual (fast) CBC construction outlined in Algorithm 3.6. Then, for prime N, Theorem 3.7 implies for any $\tau \in [1/2, \alpha)$ that

$$\begin{aligned}
&\sum_{\boldsymbol{h}\in\mathcal{A}_{d,M}} \left| \int_{[0,1]^d} f(\boldsymbol{x})\, \mathrm{e}^{-2\pi \mathrm{i}\boldsymbol{h}\cdot\boldsymbol{x}}\, \mathrm{d}\boldsymbol{x} - \frac{1}{N}\sum_{k=0}^{N-1} f\left(\left\{\frac{k\boldsymbol{g}}{N}\right\}\right) \mathrm{e}^{-2\pi \mathrm{i} k\boldsymbol{h}\cdot\boldsymbol{g}/N} \right|^2 \\
&\le \|f\|^2_{\mathrm{kor},d,\alpha,\boldsymbol{\gamma}} \frac{2^{2\tau}}{N^{2\tau}} \left(\prod_{j=1}^d \left(1 + 2\gamma_j^{1/(2\tau)}\zeta\left(\frac{\alpha}{\tau}\right)\right)^{2\tau}\right)\left(\prod_{j=1}^d \max(1, 2^{2\alpha}\gamma_j)\right) \\
&\quad \times \sum_{\boldsymbol{h}\in\mathcal{A}_{d,M}} r_{2\alpha,\boldsymbol{\gamma}}(\boldsymbol{h}) \\
&\le \|f\|^2_{\mathrm{kor},d,\alpha,\boldsymbol{\gamma}} \frac{2^{2\tau}}{N^{2\tau}} \left(\prod_{j=1}^d \left(1 + 2\gamma_j^{1/(2\tau)}\zeta\left(\frac{\alpha}{\tau}\right)\right)^{2\tau}\right)\left(\prod_{j=1}^d \max(1, 2^{2\alpha}\gamma_j)\right) \sum_{\boldsymbol{h}\in\mathcal{A}_{d,M}} M \\
&= \|f\|^2_{\mathrm{kor},d,\alpha,\boldsymbol{\gamma}} \frac{2^{2\tau} M|\mathcal{A}_{d,M}|}{N^{2\tau}} \left(\prod_{j=1}^d \left(1 + 2\gamma_j^{1/(2\tau)}\zeta\left(\frac{\alpha}{\tau}\right)\right)^{2\tau}\right)\prod_{j=1}^d \max(1, 2^{2\alpha}\gamma_j).
\end{aligned}$$

Consequently,

$$\begin{aligned}
&[\mathrm{err}^{L_2-\mathrm{app}}(\mathcal{H}_{\mathrm{kor},d,\alpha,\boldsymbol{\gamma}}, A_{N,M,d,\alpha,\boldsymbol{\gamma}}(\boldsymbol{g}))]^2 \\
&\quad \le \frac{1}{M} + \frac{2^{2\tau} M|\mathcal{A}_{d,M}|}{N^{2\tau}} \left(\prod_{j=1}^d \left(1 + 2\gamma_j^{1/(2\tau)}\zeta\left(\frac{\alpha}{\tau}\right)\right)^{2\tau}\right)\prod_{j=1}^d \max(1, 2^{2\alpha}\gamma_j).
\end{aligned}$$

Finally we employ Lemma 13.1, Item 4, with $\lambda = 1/(2\tau)$ and $\tau \in [1/2, \alpha)$ in order to obtain

$$\begin{aligned}
&[\mathrm{err}^{L_2-\mathrm{app}}(\mathcal{H}_{\mathrm{kor},d,\alpha,\boldsymbol{\gamma}}, A_{N,M,d,\alpha,\boldsymbol{\gamma}}(\boldsymbol{g}))]^2 \qquad (13.8)\\
&\quad \le \frac{1}{M} + \frac{2^{2\tau} M^{1+1/(2\tau)}}{N^{2\tau}} \left(\prod_{j=1}^d \left(1 + 2\gamma_j^{1/(2\tau)}\zeta\left(\frac{\alpha}{\tau}\right)\right)^{2\tau+1}\right)\prod_{j=1}^d \max(1, 2^{2\alpha}\gamma_j).
\end{aligned}$$

For fixed $\alpha > 1/2$ and a fixed nonincreasing weight sequence $\boldsymbol{\gamma}$, we define $d_0(\alpha, \boldsymbol{\gamma}) = d_0 \in \mathbb{N}_0$ to be the minimal index j for which it is true that $\gamma_{j+1} \le 2^{-2\alpha} < \gamma_j$. If $\gamma_1 \le 2^{-2\alpha}$, we set $d_0 := 0$. If no such index exists, we set $d_0 := \infty$. Using this notation, we can easily formulate the following result.

Proposition 13.2 *Let $\alpha > 1/2$, let N be a prime number, and let $\boldsymbol{\gamma} = (\gamma_j)_{j\ge 1}$ be positive and nonincreasing product weights with $\gamma_j \in (0, 1]$ for $j \in \mathbb{N}$. Assume that $\boldsymbol{g}$ has been found by Algorithm 3.6. Then, for any $\tau \in [1/2, \alpha)$ it is true that*

$$\mathrm{err}^{L_2\text{-app}}(\mathcal{H}_{\mathrm{kor},d,\alpha,\gamma}, A_{N,M,d,\alpha,\gamma}(\boldsymbol{g})) \le \frac{C_{d,\tau,\alpha,\gamma}}{N^{2\tau^2/(4\tau+1)}},$$

where $M = N^{4\tau^2/(4\tau+1)}$ *and where*

$$C_{d,\tau,\alpha,\gamma} = \left(1 + 2^{2\tau} 2^{2\alpha \min(d_0,d)} \left(\prod_{j=1}^{\min(d_0,d)} \gamma_j\right) \prod_{j=1}^{d} \left(1 + 2\gamma_j^{1/(2\tau)} \zeta\left(\frac{\alpha}{\tau}\right)\right)^{2\tau+1}\right)^{1/2}.$$

Proof From (13.8) together with

$$\prod_{j=1}^{d} \max(1, 2^{2\alpha}\gamma_j) = 2^{2\alpha \min(d_0,d)} \prod_{j=1}^{\min(d_0,d)} \gamma_j$$

we immediately obtain

$$[\mathrm{err}^{L_2\text{-app}}(\mathcal{H}_{\mathrm{kor},d,\alpha,\gamma}, A_{N,M,d,\alpha,\gamma}(\boldsymbol{g}))]^2 \le \frac{1}{M} + \frac{\widetilde{C}_{d,\tau,\alpha,\gamma} M^{1+1/(2\tau)}}{N^{2\tau}}, \tag{13.9}$$

where

$$\widetilde{C}_{d,\tau,\alpha,\gamma} = 2^{2\tau} 2^{2\alpha \min(d_0,d)} \left(\prod_{j=1}^{\min(d_0,d)} \gamma_j\right) \prod_{j=1}^{d} \left(1 + 2\gamma_j^{1/(2\tau)} \zeta\left(\frac{\alpha}{\tau}\right)\right)^{2\tau+1}.$$

In order to balance the two summands in (13.9) choose M, depending on N, in such a way that

$$\frac{1}{M} = \frac{M^{1+1/(2\tau)}}{N^{2\tau}}.$$

This is achieved by the choice $M = N^{4\tau^2/(4\tau+1)}$, and so we obtain

$$[\mathrm{err}^{L_2\text{-app}}(\mathcal{H}_{\mathrm{kor},d,\alpha,\gamma}, A_{N,M,d,\alpha,\gamma}(\boldsymbol{g}))]^2 \le \frac{1 + \widetilde{C}_{d,\tau,\alpha,\gamma}}{N^{4\tau^2/(4\tau+1)}}.$$

Now the result follows by taking the square root and the fact that $C_{d,\tau,\alpha,\gamma} = (1 + \widetilde{C}_{d,\tau,\alpha,\gamma})^{1/2}$. □

Note that Proposition 13.2 provides an error convergence rate that is arbitrarily close to $O(N^{-(\alpha/2)(1-1/(4\alpha+1))})$. This is slightly weaker than $O(N^{-\alpha/2})$.

Obviously, it is also possible to derive tractability results for the approximation problem, using Proposition 13.2, as outlined in [162]. However, we will not include this discussion at this point, but first outline an alternative approach to obtaining an upper bound on the worst-case integration error, yielding a slightly different result and, more importantly, an improved convergence rate.

Improving the convergence rate

Let us once more go back to Equation (13.5), which is the squared integration error of a lattice rule with generating vector $\boldsymbol{g}$ when integrating the function $f_{\boldsymbol{h}}$ for $\boldsymbol{h} \in \mathcal{A}_{d,M}$. However, due to Proposition 1.12, this error can be expressed as

$$\left|\int_{[0,1]^d} f(\boldsymbol{x})\mathrm{e}^{-2\pi\mathrm{i}\boldsymbol{h}\cdot\boldsymbol{x}}\,\mathrm{d}\boldsymbol{x} - \frac{1}{N}\sum_{k=0}^{N-1} f\left(\left\{\frac{k\boldsymbol{g}}{N}\right\}\right)\mathrm{e}^{-2\pi\mathrm{i}k\boldsymbol{h}\cdot\boldsymbol{g}/N}\right| = \left|\sum_{\substack{\boldsymbol{\ell}\in\mathbb{Z}^d\setminus\{\boldsymbol{0}\}\\ \boldsymbol{\ell}\cdot\boldsymbol{g}\equiv 0 \ (\mathrm{mod}\ N)}} \widehat{f_{\boldsymbol{h}}}(\boldsymbol{\ell})\right|.$$

On the other hand,

$$\sum_{\substack{\boldsymbol{\ell}\in\mathbb{Z}^d\setminus\{\boldsymbol{0}\}\\ \boldsymbol{\ell}\cdot\boldsymbol{g}\equiv 0 \ (\mathrm{mod}\ N)}} \widehat{f_{\boldsymbol{h}}}(\boldsymbol{\ell}) = \sum_{\substack{\boldsymbol{\ell}\in\mathbb{Z}^d\setminus\{\boldsymbol{0}\}\\ \boldsymbol{\ell}\cdot\boldsymbol{g}\equiv 0 \ (\mathrm{mod}\ N)}} \widehat{f}(\boldsymbol{h}+\boldsymbol{\ell}).$$

Hence we obtain from (13.4) that

$$\begin{aligned}
&\left(\|f - A_{N,M,d,\alpha,\boldsymbol{\gamma}}(\boldsymbol{g})(f)\|_{L_2}\right)^2 \\
&= \sum_{\boldsymbol{h}\notin\mathcal{A}_{d,M}} |\widehat{f}(\boldsymbol{h})|^2 + \sum_{\boldsymbol{h}\in\mathcal{A}_{d,M}} \left|\sum_{\substack{\boldsymbol{\ell}\in\mathbb{Z}^d\setminus\{\boldsymbol{0}\}\\ \boldsymbol{\ell}\cdot\boldsymbol{g}\equiv 0 \ (\mathrm{mod}\ N)}} \widehat{f}(\boldsymbol{h}+\boldsymbol{\ell})\right|^2 \\
&\le \frac{1}{M}\|f\|_{\mathrm{kor},d,\alpha,\boldsymbol{\gamma}}^2 + \sum_{\boldsymbol{h}\in\mathcal{A}_{d,M}} \left|\sum_{\substack{\boldsymbol{\ell}\in\mathbb{Z}^d\setminus\{\boldsymbol{0}\}\\ \boldsymbol{\ell}\cdot\boldsymbol{g}\equiv 0 \ (\mathrm{mod}\ N)}} \widehat{f}(\boldsymbol{h}+\boldsymbol{\ell})\right|^2.
\end{aligned}$$

Dividing and multiplying the terms in the latter sum by the square roots of the corresponding values of $r_{2\alpha,\boldsymbol{\gamma}}$, we obtain, using the Cauchy–Schwarz inequality,

$$\begin{aligned}
\left|\sum_{\substack{\boldsymbol{\ell}\in\mathbb{Z}^d\setminus\{\boldsymbol{0}\}\\ \boldsymbol{\ell}\cdot\boldsymbol{g}\equiv 0 \ (\mathrm{mod}\ N)}} \widehat{f}(\boldsymbol{h}+\boldsymbol{\ell})\right|^2 &\le \left(\sum_{\substack{\boldsymbol{\ell}\in\mathbb{Z}^d\setminus\{\boldsymbol{0}\}\\ \boldsymbol{\ell}\cdot\boldsymbol{g}\equiv 0 \ (\mathrm{mod}\ N)}} |\widehat{f}(\boldsymbol{h}+\boldsymbol{\ell})|^2 r_{2\alpha,\boldsymbol{\gamma}}(\boldsymbol{h}+\boldsymbol{\ell})\right) \\
&\quad\times \sum_{\substack{\boldsymbol{\ell}\in\mathbb{Z}^d\setminus\{\boldsymbol{0}\}\\ \boldsymbol{\ell}\cdot\boldsymbol{g}\equiv 0 \ (\mathrm{mod}\ N)}} \frac{1}{r_{2\alpha,\boldsymbol{\gamma}}(\boldsymbol{h}+\boldsymbol{\ell})} \\
&\le \|f\|_{\mathrm{kor},d,\alpha,\boldsymbol{\gamma}}^2 \sum_{\substack{\boldsymbol{\ell}\in\mathbb{Z}^d\setminus\{\boldsymbol{0}\}\\ \boldsymbol{\ell}\cdot\boldsymbol{g}\equiv 0 \ (\mathrm{mod}\ N)}} \frac{1}{r_{2\alpha,\boldsymbol{\gamma}}(\boldsymbol{h}+\boldsymbol{\ell})}.
\end{aligned}$$

This implies, for the squared worst-case error of L_2-approximation,

$$[\mathrm{err}^{L_2-\mathrm{app}}(\mathcal{H}_{\mathrm{kor},d,\alpha,\boldsymbol{\gamma}}, A_{N,M,d,\alpha,\boldsymbol{\gamma}}(\boldsymbol{g}))]^2 \le \frac{1}{M} + \sum_{\boldsymbol{h}\in\mathcal{A}_{d,M}} \sum_{\substack{\boldsymbol{\ell}\in\mathbb{Z}^d\setminus\{\boldsymbol{0}\} \\ \boldsymbol{\ell}\cdot\boldsymbol{g}\equiv 0 \pmod N}} \frac{1}{r_{2\alpha,\boldsymbol{\gamma}}(\boldsymbol{h}+\boldsymbol{\ell})}. \tag{13.10}$$

Let us now study the second term in the right-hand side of (13.10). For analyzing this term, we follow what is outlined in [44]. Using that $r_{2\alpha,\boldsymbol{\gamma}}(\boldsymbol{h}) \le M$ for $\boldsymbol{h} \in \mathcal{A}_{d,M}$, we obtain

$$\begin{aligned}
&\sum_{\boldsymbol{h}\in\mathcal{A}_{d,M}} \sum_{\substack{\boldsymbol{\ell}\in\mathbb{Z}^d\setminus\{\boldsymbol{0}\} \\ \boldsymbol{\ell}\cdot\boldsymbol{g}\equiv 0 \pmod N}} \frac{1}{r_{2\alpha,\boldsymbol{\gamma}}(\boldsymbol{h}+\boldsymbol{\ell})} \\
&\le \sum_{\boldsymbol{h}\in\mathcal{A}_{d,M}} \frac{M}{r_{2\alpha,\boldsymbol{\gamma}}(\boldsymbol{h})} \sum_{\substack{\boldsymbol{\ell}\in\mathbb{Z}^d\setminus\{\boldsymbol{0}\} \\ \boldsymbol{\ell}\cdot\boldsymbol{g}\equiv 0 \pmod N}} \frac{1}{r_{2\alpha,\boldsymbol{\gamma}}(\boldsymbol{h}+\boldsymbol{\ell})} \\
&\le M \sum_{\boldsymbol{h}\in\mathbb{Z}^d} \frac{1}{r_{2\alpha,\boldsymbol{\gamma}}(\boldsymbol{h})} \sum_{\substack{\boldsymbol{\ell}\in\mathbb{Z}^d\setminus\{\boldsymbol{0}\} \\ \boldsymbol{\ell}\cdot\boldsymbol{g}\equiv 0 \pmod N}} \frac{1}{r_{2\alpha,\boldsymbol{\gamma}}(\boldsymbol{h}+\boldsymbol{\ell})}.
\end{aligned}$$

For short, we write

$$\begin{aligned}
S_{N,d,\alpha,\boldsymbol{\gamma}}(\boldsymbol{g}) &:= \left(\sum_{\boldsymbol{h}\in\mathbb{Z}^d} \frac{1}{r_{2\alpha,\boldsymbol{\gamma}}(\boldsymbol{h})} \sum_{\substack{\boldsymbol{\ell}\in\mathbb{Z}^d\setminus\{\boldsymbol{0}\} \\ \boldsymbol{\ell}\cdot\boldsymbol{g}\equiv 0 \pmod N}} \frac{1}{r_{2\alpha,\boldsymbol{\gamma}}(\boldsymbol{h}+\boldsymbol{\ell})} \right)^{1/2} \\
&= \left(\sum_{\substack{\boldsymbol{\ell}\in\mathbb{Z}^d\setminus\{\boldsymbol{0}\} \\ \boldsymbol{\ell}\cdot\boldsymbol{g}\equiv 0 \pmod N}} \sum_{\boldsymbol{h}\in\mathbb{Z}^d} \frac{1}{r_{2\alpha,\boldsymbol{\gamma}}(\boldsymbol{h})\, r_{2\alpha,\boldsymbol{\gamma}}(\boldsymbol{h}+\boldsymbol{\ell})} \right)^{1/2},
\end{aligned} \tag{13.11}$$

and therefore obtain

$$[\mathrm{err}^{L_2-\mathrm{app}}(\mathcal{H}_{\mathrm{kor},d,\alpha,\boldsymbol{\gamma}}, A_{N,M,d,\alpha,\boldsymbol{\gamma}}(\boldsymbol{g}))]^2 \le \frac{1}{M} + M[S_{N,d,\alpha,\boldsymbol{\gamma}}(\boldsymbol{g})]^2. \tag{13.12}$$

This suggests using the quantity $S_{N,d,\alpha,\boldsymbol{\gamma}}$ as a new figure of merit for lattice rules for approximation. In order to obtain a small bound on the worst-case approximation error, it is sufficient to find a generating vector $\boldsymbol{g}$ such that $S_{N,d,\alpha,\boldsymbol{\gamma}}(\boldsymbol{g})$ is small. This again can be achieved by a component-wise approach, which we outline now. First, we formulate the following algorithm. For the sake of simplicity, let us assume that N is a prime number.

Algorithm 13.3 (CBC construction for L_2-approximation in the weighted Korobov space) Let prime N and $d \in \mathbb{N}$ be given. Let $\boldsymbol{\gamma} = (\gamma_j)_{j\ge 1}$ be product weights with $\gamma_j \in (0,1]$ for $j \in \mathbb{N}$. Construct a generating vector $\boldsymbol{g} = (g_1,\dots,g_d) \in G_d^{\varphi}(N)$ as follows.

(1) Choose $g_1 = 1$.
(2) For s from 1 to $d-1$:
Assume that $g_1,\dots,g_s \in G_1^{\varphi}(N)$ have already been found. Choose $g_{s+1} \in G_1^{\varphi}(N)$ as

$$g_{s+1} := \underset{g\in G_1^{\varphi}(N)}{\operatorname{argmin}}\; S_{N,s+1,\alpha,\boldsymbol{\gamma}}((g_1,\dots,g_s,g)).$$

End for.
(3) Set $\boldsymbol{g} = (g_1,\dots,g_d)$.

Remark 13.4 As pointed out in [44], Algorithm 13.3 can be practically implemented easily, as we have, for a given generating vector $\boldsymbol{g} = (g_1,\dots,g_d)$,

$$\begin{aligned}
[S_{N,d,\alpha,\boldsymbol{\gamma}}(\boldsymbol{g})]^2 &= -\sum_{\boldsymbol{h}\in\mathbb{Z}^d} \frac{1}{(r_{2\alpha,\boldsymbol{\gamma}}(\boldsymbol{h}))^2} + \sum_{\substack{\boldsymbol{\ell}\in\mathbb{Z}^d\\ \boldsymbol{\ell}\cdot\boldsymbol{g}\equiv 0 \pmod N}} \sum_{\boldsymbol{h}\in\mathbb{Z}^d} \frac{1}{r_{2\alpha,\boldsymbol{\gamma}}(\boldsymbol{h})\, r_{2\alpha,\boldsymbol{\gamma}}(\boldsymbol{h}+\boldsymbol{\ell})}\\
&= -\prod_{j=1}^{d}(1+2\zeta(4\alpha)\gamma_j^2) + \sum_{\substack{\boldsymbol{\ell}\in\mathbb{Z}^d\\ \boldsymbol{\ell}\cdot\boldsymbol{g}\equiv 0 \pmod N}} \sum_{\boldsymbol{h}\in\mathbb{Z}^d} \frac{1}{r_{2\alpha,\boldsymbol{\gamma}}(\boldsymbol{h})\, r_{2\alpha,\boldsymbol{\gamma}}(\boldsymbol{h}+\boldsymbol{\ell})}.
\end{aligned}$$

However,

$$\begin{aligned}
&\sum_{\substack{\boldsymbol{\ell}\in\mathbb{Z}^d\\ \boldsymbol{\ell}\cdot\boldsymbol{g}\equiv 0 \pmod N}} \sum_{\boldsymbol{h}\in\mathbb{Z}^d} \frac{1}{r_{2\alpha,\boldsymbol{\gamma}}(\boldsymbol{h})\, r_{2\alpha,\boldsymbol{\gamma}}(\boldsymbol{h}+\boldsymbol{\ell})}\\
&= \sum_{\boldsymbol{\ell}\in\mathbb{Z}^d} \sum_{\boldsymbol{h}\in\mathbb{Z}^d} \frac{1}{r_{2\alpha,\boldsymbol{\gamma}}(\boldsymbol{h})\, r_{2\alpha,\boldsymbol{\gamma}}(\boldsymbol{h}+\boldsymbol{\ell})} \frac{1}{N}\sum_{k=0}^{N-1} \mathrm{e}^{2\pi \mathrm{i} k\boldsymbol{g}\cdot\boldsymbol{\ell}/N}\\
&= \frac{1}{N}\sum_{k=0}^{N-1} \sum_{\boldsymbol{\ell}\in\mathbb{Z}^d} \sum_{\boldsymbol{h}\in\mathbb{Z}^d} \frac{\mathrm{e}^{2\pi \mathrm{i} k\boldsymbol{g}\cdot\boldsymbol{\ell}/N}}{r_{2\alpha,\boldsymbol{\gamma}}(\boldsymbol{h})\, r_{2\alpha,\boldsymbol{\gamma}}(\boldsymbol{h}+\boldsymbol{\ell})}\\
&= \frac{1}{N}\sum_{k=0}^{N-1} \sum_{\boldsymbol{\ell}\in\mathbb{Z}^d} \sum_{\boldsymbol{h}\in\mathbb{Z}^d} \frac{\mathrm{e}^{-2\pi \mathrm{i} k\boldsymbol{g}\cdot\boldsymbol{h}/N}\, \mathrm{e}^{2\pi \mathrm{i} k\boldsymbol{g}\cdot(\boldsymbol{h}+\boldsymbol{\ell})/N}}{r_{2\alpha,\boldsymbol{\gamma}}(\boldsymbol{h})\, r_{2\alpha,\boldsymbol{\gamma}}(\boldsymbol{h}+\boldsymbol{\ell})}\\
&= \frac{1}{N}\sum_{k=0}^{N-1} \sum_{\boldsymbol{h}\in\mathbb{Z}^d} \frac{\mathrm{e}^{-2\pi \mathrm{i} k\boldsymbol{g}\cdot\boldsymbol{h}/N}}{r_{2\alpha,\boldsymbol{\gamma}}(\boldsymbol{h})} \sum_{\boldsymbol{\ell}\in\mathbb{Z}^d} \frac{\mathrm{e}^{2\pi \mathrm{i} k\boldsymbol{g}\cdot\boldsymbol{\ell}/N}}{r_{2\alpha,\boldsymbol{\gamma}}(\boldsymbol{\ell})}\\
&= \frac{1}{N}\sum_{k=0}^{N-1} \left(\sum_{\boldsymbol{h}\in\mathbb{Z}^d} \frac{\mathrm{e}^{2\pi \mathrm{i} k\boldsymbol{g}\cdot\boldsymbol{h}/N}}{r_{2\alpha,\boldsymbol{\gamma}}(\boldsymbol{h})}\right)^2
\end{aligned}$$

$$= \frac{1}{N} \sum_{k=0}^{N-1} \prod_{j=1}^{d} \left(1 + \gamma_j \sum_{h \in \mathbb{Z}^d \setminus \{0\}} \frac{\mathrm{e}^{2\pi \mathrm{i} k g_j h/N}}{|h|^{2\alpha}} \right)^2 .$$

This shows that $S_{N,d,\alpha,\boldsymbol{\gamma}}(\boldsymbol{g})$ is very similar to the worst-case integration error of the rank-1 lattice rule generated by $\boldsymbol{g}$. Hence, one can practically use the same machinery as in Section 3.4 and obtain a construction cost of $O(d\,N \log N)$ operations.

We then have the following theorem, which shows that the generating vectors constructed by Algorithm 13.3 indeed yield a low value of $S_{N,d,\alpha,\boldsymbol{\gamma}}$.

Theorem 13.5 *Let $N \geq 3$ be a prime number, let $\boldsymbol{\gamma} = (\gamma_j)_{j\geq 1}$ be product weights with $\gamma_j \in (0,1]$ for $j \in \mathbb{N}$, and assume that $\mathbf{g}$ has been found by Algorithm 13.3. Then for arbitrary $\tau \in [1/2, \alpha)$ and for any $s \in [d]$ we have*

$$S_{N,s,\alpha,\boldsymbol{\gamma}}((g_1,\ldots,g_s)) \leq \frac{1}{N^\tau} \prod_{j=1}^{s} \left(1 + 2^{4\alpha+1} \zeta\left(\frac{\alpha}{\tau}\right) \gamma_j^{1/(2\tau)} \right)^{2\tau} .$$

The proof of Theorem 13.5 uses standard ideas that have been employed for similar results in the context of CBC constructions already earlier in this book. However, the details are rather technical and tedious. Busy readers may proceed directly to the statement of Theorem 13.6.

Proof of Theorem 13.5 We prove the result by induction on s. For $s = 1$, we have, as $g_1 = 1$,

$$[S_{N,1,\alpha,\gamma_1}(1)]^2 = \sum_{h \in \mathbb{Z}} \sum_{\substack{\ell \in \mathbb{Z}\setminus\{0\} \\ \ell \equiv 0 \pmod{N}}} \frac{1}{r_{2\alpha,\gamma_1}(h)} \frac{1}{r_{2\alpha,\gamma_1}(h+\ell)}$$

and, using Jensen's inequality (see Lemma 2.25), for $\lambda \in (1/(2\alpha), 1]$,

$$\begin{aligned}
[S_{N,1,\alpha,\gamma_1}(1)]^{2\lambda} &\leq \sum_{h \in \mathbb{Z}} \sum_{\substack{\ell \in \mathbb{Z}\setminus\{0\} \\ \ell \equiv 0 \pmod{N}}} \left(\frac{1}{r_{2\alpha,\gamma_1}(h)} \frac{1}{r_{2\alpha,\gamma_1}(h+\ell)} \right)^\lambda \\
&= \sum_{\substack{h \in \mathbb{Z} \\ h \equiv 0 \pmod{N}}} \frac{1}{r_{2\alpha\lambda,\gamma_1^\lambda}(h)} \sum_{\substack{\ell \in \mathbb{Z}\setminus\{0\} \\ \ell \equiv 0 \pmod{N}}} \frac{1}{r_{2\alpha\lambda,\gamma_1^\lambda}(h+\ell)} \\
&\quad + \sum_{\substack{h \in \mathbb{Z} \\ h \not\equiv 0 \pmod{N}}} \frac{1}{r_{2\alpha\lambda,\gamma_1^\lambda}(h)} \sum_{\substack{\ell \in \mathbb{Z}\setminus\{0\} \\ \ell \equiv 0 \pmod{N}}} \frac{1}{r_{2\alpha\lambda,\gamma_1^\lambda}(h+\ell)} .
\end{aligned}$$

We have

$$\begin{aligned}
&\sum_{\substack{h\in\mathbb{Z}\\ h\equiv 0 \pmod N}} \frac{1}{r_{2\alpha\lambda,\gamma_1^\lambda}(h)} \sum_{\substack{\ell\in\mathbb{Z}\setminus\{0\}\\ \ell\equiv 0 \pmod N}} \frac{1}{r_{2\alpha\lambda,\gamma_1^\lambda}(h+\ell)}\\
&= \sum_{h\in\mathbb{Z}} \frac{1}{r_{2\alpha\lambda,\gamma_1^\lambda}(hN)} \sum_{\ell\in\mathbb{Z}\setminus\{0\}} \frac{1}{r_{2\alpha\lambda,\gamma_1^\lambda}(hN+\ell N)}\\
&= \sum_{h\in\mathbb{Z}} \frac{1}{r_{2\alpha\lambda,\gamma_1^\lambda}(hN)} \sum_{\ell\in\mathbb{Z}} \frac{1}{r_{2\alpha\lambda,\gamma_1^\lambda}(hN+\ell N)} - \sum_{h\in\mathbb{Z}} \frac{1}{(r_{2\alpha\lambda,\gamma_1^\lambda}(hN))^2}\\
&\le \left(1+\frac{2\zeta(2\alpha\lambda)\gamma_1^\lambda}{N^{2\alpha\lambda}}\right)^2 - 1\\
&\le \frac{4\zeta(2\alpha\lambda)\gamma_1^\lambda}{N} + \frac{4(\zeta(2\alpha\lambda))^2\gamma_1^{2\lambda}}{N}.
\end{aligned}$$

Furthermore, using that $N \ge 3$ is prime and therefore odd,

$$\begin{aligned}
&\sum_{\substack{h\in\mathbb{Z}\\ h\not\equiv 0 \pmod N}} \frac{1}{r_{2\alpha\lambda,\gamma_1^\lambda}(h)} \sum_{\substack{\ell\in\mathbb{Z}\setminus\{0\}\\ \ell\equiv 0 \pmod N}} \frac{1}{r_{2\alpha\lambda,\gamma_1^\lambda}(h+\ell)}\\
&= \sum_{\substack{h\in\mathbb{Z}\\ h\not\equiv 0 \pmod N}} \frac{1}{r_{2\alpha\lambda,\gamma_1^\lambda}(h)} \sum_{\ell\in\mathbb{Z}\setminus\{0\}} \frac{1}{r_{2\alpha\lambda,\gamma_1^\lambda}(h+\ell N)}\\
&= \sum_{\substack{h\in\mathbb{Z}\\ h\not\equiv 0 \pmod N}} \frac{1}{r_{2\alpha\lambda,\gamma_1^\lambda}(h)} \sum_{\ell\in\mathbb{Z}} \frac{1}{r_{2\alpha\lambda,\gamma_1^\lambda}(h+\ell N)} - \sum_{\substack{h\in\mathbb{Z}\\ h\not\equiv 0 \pmod N}} \left(\frac{1}{r_{2\alpha\lambda,\gamma_1^\lambda}(h)}\right)^2\\
&= \sum_{\substack{k=-(N-1)/2\\ k\neq 0}}^{(N-1)/2} \sum_{h\in\mathbb{Z}} \frac{1}{r_{2\alpha\lambda,\gamma_1^\lambda}(hN+k)} \sum_{\ell\in\mathbb{Z}} \frac{1}{r_{2\alpha\lambda,\gamma_1^\lambda}(hN+k+\ell N)}\\
&\quad - \sum_{\substack{k=-(N-1)/2\\ k\neq 0}}^{(N-1)/2} \sum_{h\in\mathbb{Z}} \left(\frac{1}{r_{2\alpha\lambda,\gamma_1^\lambda}(hN+k)}\right)^2\\
&= \sum_{\substack{k=-(N-1)/2\\ k\neq 0}}^{(N-1)/2} \left(\sum_{\ell\in\mathbb{Z}} \frac{1}{r_{2\alpha\lambda,\gamma_1^\lambda}(\ell N+k)}\right)^2 - \sum_{\substack{k=-(N-1)/2\\ k\neq 0}}^{(N-1)/2} \sum_{h\in\mathbb{Z}} \left(\frac{1}{r_{2\alpha\lambda,\gamma_1^\lambda}(hN+k)}\right)^2\\
&= \gamma_1^{2\lambda} \sum_{\substack{k=-(N-1)/2\\ k\neq 0}}^{(N-1)/2} \left(\left(\sum_{\ell\in\mathbb{Z}} \frac{1}{|\ell N+k|^{2\alpha\lambda}}\right)^2 - \sum_{h\in\mathbb{Z}} \frac{1}{|hN+k|^{4\alpha\lambda}}\right)
\end{aligned}$$

$$\leq \gamma_1^{2\lambda} \sum_{\substack{k=-(N-1)/2 \\ k\neq 0}}^{(N-1)/2} \left(\left(\frac{1}{k^{2\alpha\lambda}} + \sum_{\ell \in \mathbb{Z}\setminus\{0\}} \frac{1}{|\ell N|^{2\alpha\lambda} |1 + k/(\ell N)|^{2\alpha\lambda}} \right)^2 - \frac{1}{k^{4\alpha\lambda}} \right)$$

$$\leq \gamma_1^{2\lambda} \sum_{\substack{k=-(N-1)/2 \\ k\neq 0}}^{(N-1)/2} \left(\left(\frac{1}{k^{2\alpha\lambda}} + \sum_{\ell \in \mathbb{Z}\setminus\{0\}} \frac{1}{|\ell N|^{2\alpha\lambda} (1/2)^{2\alpha\lambda}} \right)^2 - \frac{1}{k^{4\alpha\lambda}} \right)$$

$$\leq \gamma_1^{2\lambda} \sum_{\substack{k=-(N-1)/2 \\ k\neq 0}}^{(N-1)/2} \left(\frac{1}{k^{2\alpha\lambda}} \frac{2^{2\alpha\lambda+2}\zeta(2\alpha\lambda)}{N^{2\alpha\lambda}} + \frac{2^{4\alpha\lambda+2}(\zeta(2\alpha\lambda))^2}{N^{4\alpha\lambda}} \right)$$

$$\leq \frac{2^{4\alpha\lambda+3}(\zeta(2\alpha\lambda))^2\gamma_1^{2\lambda}}{N}.$$

In summary, we get

$$\begin{aligned}
[S_{N,1,\alpha,\gamma_1}(1)]^{2\lambda} &\leq \frac{4\zeta(2\alpha\lambda)\gamma_1^{\lambda}}{N} + \frac{4(\zeta(2\alpha\lambda))^2\gamma_1^{2\lambda}}{N} + \frac{2^{4\alpha\lambda+3}(\zeta(2\alpha\lambda))^2\gamma_1^{2\lambda}}{N} \\
&= \frac{1}{N}\left(2^2\zeta(2\alpha\lambda)\gamma_1^{\lambda} + 2^2(1 + 2^{4\alpha\lambda+1})(\zeta(2\alpha\lambda))^2\gamma_1^{2\lambda}\right) \\
&\leq \frac{1}{N}\left(2\,(2\zeta(2\alpha\lambda)\gamma_1^{\lambda}) + (1 + 2^{6\alpha\lambda})(2\zeta(2\alpha\lambda)\gamma_1^{\lambda})^2\right) \\
&\leq \frac{1}{N}\left(1 + 2\sqrt{1 + 2^{6\alpha\lambda}}\,\zeta(2\alpha\lambda)\gamma_1^{\lambda}\right)^2 \\
&\leq \frac{1}{N}\left(1 + 2^{4\alpha+1}\zeta(2\alpha\lambda)\gamma_1^{\lambda}\right)^2,
\end{aligned}$$

where we used that $2\sqrt{1 + 2^{6\alpha\lambda}} \leq 2^{4\alpha+1}$. We therefore obtain

$$S_{N,1,\alpha,\gamma_1}(1) \leq \frac{1}{N^{1/(2\lambda)}} \left(1 + 2^{4\alpha+1}\zeta(2\alpha\lambda)\gamma_1^{\lambda}\right)^{1/\lambda}.$$

By setting $\tau = 1/(2\lambda)$, we get the result for $s = 1$.

For the induction step, let $\boldsymbol{g}^{(s)} = (g_1, \ldots, g_s) \in G_s^{\varphi}(N)$ denote the s-dimensional generating vector that has been found in the first s steps of the algorithm, and suppose that the claimed error bound holds for the lattice rule generated by $\boldsymbol{g}^{(s)}$. Again, we write $(\boldsymbol{g}^{(s)}, g)$ for the vector $(g_1, \ldots, g_s, g) \in G_{s+1}^{\varphi}(N)$.

By definition,

$$\begin{aligned}
&[S_{N,s+1,\alpha,\boldsymbol{\gamma}}((\boldsymbol{g}^{(s)}, g))]^2 \\
&= \sum_{\substack{\boldsymbol{\ell} \in \mathbb{Z}^{s+1}\setminus\{\boldsymbol{0}\} \\ \boldsymbol{\ell}\cdot(\boldsymbol{g}^{(s)}, g) \equiv 0 \pmod{N}}} \sum_{\boldsymbol{h} \in \mathbb{Z}^{s+1}} \frac{1}{r_{2\alpha,\boldsymbol{\gamma}}(\boldsymbol{h})\, r_{2\alpha,\boldsymbol{\gamma}}(\boldsymbol{h} + \boldsymbol{\ell})}
\end{aligned}$$

$$
\begin{aligned}
&= \sum_{h\in\mathbb{Z}} \frac{1}{(r_{2\alpha,\gamma_{s+1}}(h))^2} \sum_{\substack{\boldsymbol{\ell}\in\mathbb{Z}^s\setminus\{\boldsymbol{0}\} \\ \boldsymbol{\ell}\cdot\boldsymbol{g}^{(s)}\equiv 0 \ (\mathrm{mod}\ N)}} \sum_{\boldsymbol{h}\in\mathbb{Z}^s} \frac{1}{r_{2\alpha,\boldsymbol{\gamma}}(\boldsymbol{h})\, r_{2\alpha,\boldsymbol{\gamma}}(\boldsymbol{h}+\boldsymbol{\ell})} \\
&\quad + \sum_{h\in\mathbb{Z}} \sum_{\ell_{s+1}\in\mathbb{Z}\setminus\{0\}} \frac{1}{r_{2\alpha,\gamma_{s+1}}(h)\, r_{2\alpha,\gamma_{s+1}}(h+\ell_{s+1})} \\
&\qquad \times \sum_{\boldsymbol{h}\in\mathbb{Z}^s} \sum_{\substack{\boldsymbol{\ell}\in\mathbb{Z}^s \\ \boldsymbol{\ell}\cdot\boldsymbol{g}^{(s)}\equiv -\ell_{s+1}g \ (\mathrm{mod}\ N)}} \frac{1}{r_{2\alpha,\boldsymbol{\gamma}}(\boldsymbol{h})\, r_{2\alpha,\boldsymbol{\gamma}}(\boldsymbol{h}+\boldsymbol{\ell})} \\
&= (1+2\zeta(4\alpha)\gamma_{s+1}^2)[S_{N,s,\alpha,\boldsymbol{\gamma}}(\boldsymbol{g}^{(s)})]^2 + \theta_{N,s,\alpha,\boldsymbol{\gamma}}(\boldsymbol{g}^{(s)}, g),
\end{aligned}
\tag{13.13}
$$

where

$$
\begin{aligned}
\theta_{N,s,\alpha,\boldsymbol{\gamma}}(\boldsymbol{g}^{(s)}, g) &:= \sum_{h\in\mathbb{Z}} \sum_{\ell_{s+1}\in\mathbb{Z}\setminus\{0\}} \frac{1}{r_{2\alpha,\gamma_{s+1}}(h)\, r_{2\alpha,\gamma_{s+1}}(h+\ell_{s+1})} \\
&\qquad \times \sum_{\boldsymbol{h}\in\mathbb{Z}^s} \sum_{\substack{\boldsymbol{\ell}\in\mathbb{Z}^s \\ \boldsymbol{\ell}\cdot\boldsymbol{g}^{(s)}\equiv -\ell_{s+1}g \ (\mathrm{mod}\ N)}} \frac{1}{r_{2\alpha,\boldsymbol{\gamma}}(\boldsymbol{h})\, r_{2\alpha,\boldsymbol{\gamma}}(\boldsymbol{h}+\boldsymbol{\ell})}.
\end{aligned}
$$

We now use the standard averaging argument (see Remark 2.14) and consider the average of $\theta_{N,s,\alpha,\boldsymbol{\gamma}}(\boldsymbol{g}^{(s)}, g)$ over all possible values of $g \in G_1^{\varphi}(N) = \{1, 2, \ldots, N-1\}$. For the optimal choice of g_{s+1} we have, for every $\lambda \in (1/(2\alpha), 1]$,

$$
(\theta_{N,s,\alpha,\boldsymbol{\gamma}}(\boldsymbol{g}^{(s)}, g_{s+1}))^{\lambda} \le \frac{1}{N-1}\sum_{g=1}^{N-1} (\theta_{N,s,\alpha,\boldsymbol{\gamma}}(\boldsymbol{g}^{(s)}, g))^{\lambda}.
$$

By Jensen's inequality, and by the properties of $r_{2\alpha,\boldsymbol{\gamma}}$,

$$
\begin{aligned}
&\frac{1}{N-1}\sum_{g=1}^{N-1} (\theta_{N,s,\alpha,\boldsymbol{\gamma}}(\boldsymbol{g}^{(s)}, g))^{\lambda} \\
&\le \frac{1}{N-1}\sum_{g=1}^{N-1} \sum_{h\in\mathbb{Z}} \sum_{\ell_{s+1}\in\mathbb{Z}\setminus\{0\}} \frac{1}{r_{2\alpha\lambda,\gamma_{s+1}^{\lambda}}(h)\, r_{2\alpha\lambda,\gamma_{s+1}^{\lambda}}(h+\ell_{s+1})} \\
&\qquad \times \sum_{\boldsymbol{h}\in\mathbb{Z}^s} \sum_{\substack{\boldsymbol{\ell}\in\mathbb{Z}^s \\ \boldsymbol{\ell}\cdot\boldsymbol{g}^{(s)}\equiv -\ell_{s+1}g \ (\mathrm{mod}\ N)}} \frac{1}{r_{2\alpha\lambda,\boldsymbol{\gamma}^{\lambda}}(\boldsymbol{h})\, r_{2\alpha\lambda,\boldsymbol{\gamma}^{\lambda}}(\boldsymbol{h}+\boldsymbol{\ell})} \\
&=: \Theta_{N,s,\alpha,\boldsymbol{\gamma},\lambda}(\boldsymbol{g}^{(s)}).
\end{aligned}
$$

By separating the cases where $\ell_{s+1} \equiv 0 \pmod N$ and where $\ell_{s+1} \not\equiv 0 \pmod N$, we obtain

$$
\begin{aligned}
&\Theta_{N,s,\alpha,\boldsymbol{\gamma},\lambda}(\boldsymbol{g}^{(s)}) \\
&= \sum_{h\in\mathbb{Z}} \sum_{\substack{\ell_{s+1}\in\mathbb{Z}\setminus\{0\}\\ \ell_{s+1}\equiv 0 \pmod N}} \frac{1}{r_{2\alpha\lambda,\gamma_{s+1}^{\lambda}}(h)\, r_{2\alpha\lambda,\gamma_{s+1}^{\lambda}}(h+\ell_{s+1})} \\
&\qquad \times \sum_{\boldsymbol{h}\in\mathbb{Z}^s} \sum_{\substack{\boldsymbol{\ell}\in\mathbb{Z}^s\\ \boldsymbol{\ell}\cdot\boldsymbol{g}^{(s)}\equiv 0 \pmod N}} \frac{1}{r_{2\alpha\lambda,\boldsymbol{\gamma}^{\lambda}}(\boldsymbol{h})\, r_{2\alpha\lambda,\boldsymbol{\gamma}^{\lambda}}(\boldsymbol{h}+\boldsymbol{\ell})} \\
&\quad + \frac{1}{N-1} \sum_{g=1}^{N-1} \sum_{h\in\mathbb{Z}} \sum_{\substack{\ell_{s+1}\in\mathbb{Z}\setminus\{0\}\\ \ell_{s+1}\not\equiv 0 \pmod N}} \frac{1}{r_{2\alpha\lambda,\gamma_{s+1}^{\lambda}}(h)\, r_{2\alpha\lambda,\gamma_{s+1}^{\lambda}}(h+\ell_{s+1})} \\
&\qquad \times \sum_{\boldsymbol{h}\in\mathbb{Z}^s} \sum_{\substack{\boldsymbol{\ell}\in\mathbb{Z}^s\\ \boldsymbol{\ell}\cdot\boldsymbol{g}^{(s)}\equiv -\ell_{s+1} g \pmod N}} \frac{1}{r_{2\alpha\lambda,\boldsymbol{\gamma}^{\lambda}}(\boldsymbol{h})\, r_{2\alpha\lambda,\boldsymbol{\gamma}^{\lambda}}(\boldsymbol{h}+\boldsymbol{\ell})} \\
&= \sum_{h\in\mathbb{Z}} \sum_{\substack{\ell_{s+1}\in\mathbb{Z}\setminus\{0\}\\ \ell_{s+1}\equiv 0 \pmod N}} \frac{1}{r_{2\alpha\lambda,\gamma_{s+1}^{\lambda}}(h)\, r_{2\alpha\lambda,\gamma_{s+1}^{\lambda}}(h+\ell_{s+1})} \\
&\qquad \times \sum_{\boldsymbol{h}\in\mathbb{Z}^s} \sum_{\substack{\boldsymbol{\ell}\in\mathbb{Z}^s\\ \boldsymbol{\ell}\cdot\boldsymbol{g}^{(s)}\equiv 0 \pmod N}} \frac{1}{r_{2\alpha\lambda,\boldsymbol{\gamma}^{\lambda}}(\boldsymbol{h})\, r_{2\alpha\lambda,\boldsymbol{\gamma}^{\lambda}}(\boldsymbol{h}+\boldsymbol{\ell})} \\
&\quad + \frac{1}{N-1} \sum_{h\in\mathbb{Z}} \sum_{\substack{\ell_{s+1}\in\mathbb{Z}\setminus\{0\}\\ \ell_{s+1}\not\equiv 0 \pmod N}} \frac{1}{r_{2\alpha\lambda,\gamma_{s+1}^{\lambda}}(h)\, r_{2\alpha\lambda,\gamma_{s+1}^{\lambda}}(h+\ell_{s+1})} \\
&\qquad \times \sum_{\boldsymbol{h}\in\mathbb{Z}^s} \sum_{\substack{\boldsymbol{\ell}\in\mathbb{Z}^s\\ \boldsymbol{\ell}\cdot\boldsymbol{g}^{(s)}\not\equiv 0 \pmod N}} \frac{1}{r_{2\alpha\lambda,\boldsymbol{\gamma}^{\lambda}}(\boldsymbol{h})\, r_{2\alpha\lambda,\boldsymbol{\gamma}^{\lambda}}(\boldsymbol{h}+\boldsymbol{\ell})},
\end{aligned}
$$

where we used

$$
\begin{aligned}
&\sum_{g=1}^{N-1} \sum_{\substack{\boldsymbol{\ell}\in\mathbb{Z}^s\\ \boldsymbol{\ell}\cdot\boldsymbol{g}^{(s)}\equiv -\ell_{s+1} g \pmod N}} \frac{1}{r_{2\alpha\lambda,\boldsymbol{\gamma}^{\lambda}}(\boldsymbol{h})\, r_{2\alpha\lambda,\boldsymbol{\gamma}^{\lambda}}(\boldsymbol{h}+\boldsymbol{\ell})} \\
&\qquad\qquad = \sum_{\substack{\boldsymbol{\ell}\in\mathbb{Z}^s\\ \boldsymbol{\ell}\cdot\boldsymbol{g}^{(s)}\not\equiv 0 \pmod N}} \frac{1}{r_{2\alpha\lambda,\boldsymbol{\gamma}^{\lambda}}(\boldsymbol{h})\, r_{2\alpha\lambda,\boldsymbol{\gamma}^{\lambda}}(\boldsymbol{h}+\boldsymbol{\ell})}
\end{aligned}
$$

for $\ell_{s+1} \in \mathbb{Z}\setminus\{0\}$, $\ell_{s+1} \not\equiv 0 \pmod N$. We now write

$$
\Xi(\boldsymbol{g}^{(s)}) := \sum_{\boldsymbol{h}\in\mathbb{Z}^s} \sum_{\substack{\boldsymbol{\ell}\in\mathbb{Z}^s\\ \boldsymbol{\ell}\cdot\boldsymbol{g}^{(s)}\equiv 0 \pmod N}} \frac{1}{r_{2\alpha\lambda,\boldsymbol{\gamma}^{\lambda}}(\boldsymbol{h})\, r_{2\alpha\lambda,\boldsymbol{\gamma}^{\lambda}}(\boldsymbol{h}+\boldsymbol{\ell})},
$$

and

$$\widetilde{\Xi} := \sum_{\boldsymbol{h}\in\mathbb{Z}^s}\sum_{\boldsymbol{\ell}\in\mathbb{Z}^s} \frac{1}{r_{2\alpha\lambda,\boldsymbol{\gamma}^\lambda}(\boldsymbol{h})\, r_{2\alpha\lambda,\boldsymbol{\gamma}^\lambda}(\boldsymbol{h}+\boldsymbol{\ell})} = \prod_{j=1}^{s}(1+2\zeta(2\alpha\lambda)\gamma_j^\lambda)^2.$$

Then we obviously have $\Xi(\boldsymbol{g}^{(s)}) \le \widetilde{\Xi}$, and

$$\sum_{\boldsymbol{h}\in\mathbb{Z}^s}\sum_{\substack{\boldsymbol{\ell}\in\mathbb{Z}^s\\ \boldsymbol{\ell}\cdot\boldsymbol{g}^{(s)}\not\equiv 0 \pmod N}} \frac{1}{r_{2\alpha\lambda,\boldsymbol{\gamma}^\lambda}(\boldsymbol{h})\, r_{2\alpha\lambda,\boldsymbol{\gamma}^\lambda}(\boldsymbol{h}+\boldsymbol{\ell})} = \widetilde{\Xi} - \Xi(\boldsymbol{g}^{(s)}).$$

This yields

$$\begin{aligned}
\Theta_{N,s,\alpha,\boldsymbol{\gamma},\lambda}(\boldsymbol{g}^{(s)}) &= \Xi(\boldsymbol{g}^{(s)}) \sum_{h\in\mathbb{Z}}\sum_{\substack{\ell_{s+1}\in\mathbb{Z}\setminus\{0\}\\ \ell_{s+1}\equiv 0 \pmod N}} \frac{1}{r_{2\alpha\lambda,\gamma_{s+1}^\lambda}(h)\, r_{2\alpha\lambda,\gamma_{s+1}^\lambda}(h+\ell_{s+1})}\\
&\quad+\frac{\widetilde{\Xi}-\Xi(\boldsymbol{g}^{(s)})}{N-1} \sum_{h\in\mathbb{Z}}\sum_{\substack{\ell_{s+1}\in\mathbb{Z}\setminus\{0\}\\ \ell_{s+1}\not\equiv 0 \pmod N}} \frac{1}{r_{2\alpha\lambda,\gamma_{s+1}^\lambda}(h)\, r_{2\alpha\lambda,\gamma_{s+1}^\lambda}(h+\ell_{s+1})}\\
&= \Xi(\boldsymbol{g}^{(s)}) \sum_{h\in\mathbb{Z}}\sum_{\substack{\ell_{s+1}\in\mathbb{Z}\setminus\{0\}\\ \ell_{s+1}\equiv 0 \pmod N}} \frac{1}{r_{2\alpha\lambda,\gamma_{s+1}^\lambda}(h)\, r_{2\alpha\lambda,\gamma_{s+1}^\lambda}(h+\ell_{s+1})}\\
&\quad+\frac{\widetilde{\Xi}-\Xi(\boldsymbol{g}^{(s)})}{N-1} \sum_{h\in\mathbb{Z}}\sum_{\ell_{s+1}\in\mathbb{Z}\setminus\{0\}} \frac{1}{r_{2\alpha\lambda,\gamma_{s+1}^\lambda}(h)\, r_{2\alpha\lambda,\gamma_{s+1}^\lambda}(h+\ell_{s+1})}\\
&\quad-\frac{\widetilde{\Xi}-\Xi(\boldsymbol{g}^{(s)})}{N-1} \sum_{h\in\mathbb{Z}}\sum_{\substack{\ell_{s+1}\in\mathbb{Z}\setminus\{0\}\\ \ell_{s+1}\equiv 0 \pmod N}} \frac{1}{r_{2\alpha\lambda,\gamma_{s+1}^\lambda}(h)\, r_{2\alpha\lambda,\gamma_{s+1}^\lambda}(h+\ell_{s+1})}\\
&= \frac{\widetilde{\Xi}-\Xi(\boldsymbol{g}^{(s)})}{N-1} \sum_{h\in\mathbb{Z}}\sum_{\ell_{s+1}\in\mathbb{Z}\setminus\{0\}} \frac{1}{r_{2\alpha\lambda,\gamma_{s+1}^\lambda}(h)\, r_{2\alpha\lambda,\gamma_{s+1}^\lambda}(h+\ell_{s+1})}\\
&\quad+\frac{N\Xi(\boldsymbol{g}^{(s)})-\widetilde{\Xi}}{N-1} \sum_{h\in\mathbb{Z}}\sum_{\substack{\ell_{s+1}\in\mathbb{Z}\setminus\{0\}\\ \ell_{s+1}\equiv 0 \pmod N}} \frac{1}{r_{2\alpha\lambda,\gamma_{s+1}^\lambda}(h)\, r_{2\alpha\lambda,\gamma_{s+1}^\lambda}(h+\ell_{s+1})}.
\end{aligned} \tag{13.14}$$

Regarding the first term in (13.14), we have

$$\begin{aligned}
&\sum_{h\in\mathbb{Z}}\sum_{\ell_{s+1}\in\mathbb{Z}\setminus\{0\}} \frac{1}{r_{2\alpha\lambda,\gamma_{s+1}^\lambda}(h)\, r_{2\alpha\lambda,\gamma_{s+1}^\lambda}(h+\ell_{s+1})}\\
&= \sum_{h\in\mathbb{Z}}\sum_{\ell_{s+1}\in\mathbb{Z}} \frac{1}{r_{2\alpha\lambda,\gamma_{s+1}^\lambda}(h)\, r_{2\alpha\lambda,\gamma_{s+1}^\lambda}(h+\ell_{s+1})} - \sum_{h\in\mathbb{Z}}\frac{1}{(r_{2\alpha\lambda,\gamma_{s+1}^\lambda}(h))^2}\\
&= (1+2\zeta(2\alpha\lambda)\gamma_{s+1}^\lambda)^2 - (1+2\zeta(4\alpha\lambda)\gamma_{s+1}^{2\lambda})\\
&\le 2^2\zeta(2\alpha\lambda)\gamma_{s+1}^\lambda + 2^2\zeta(2\alpha\lambda)^2\gamma_{s+1}^{2\lambda}.
\end{aligned}$$

Regarding the second term in (13.14), we see in exactly the same way as in the case $s = 1$ that

$$\sum_{h\in\mathbb{Z}} \sum_{\substack{\ell_{s+1}\in\mathbb{Z}\setminus\{0\}\\ \ell_{s+1}\equiv 0 \pmod N}} \frac{1}{r_{2\alpha\lambda,\gamma_{s+1}^{\lambda}}(h)\, r_{2\alpha\lambda,\gamma_{s+1}^{\lambda}}(h+\ell_{s+1})}$$
$$\leq \frac{2^2\zeta(2\alpha\lambda)\gamma_{s+1}^{\lambda} + 2^2(1+2^{4\alpha\lambda+1})(\zeta(2\alpha\lambda))^2\gamma_{s+1}^{2\lambda}}{N}.$$

This gives

$$\begin{aligned}
\Theta_{N,s,\alpha,\boldsymbol{\gamma},\lambda}(\boldsymbol{g}^{(s)}) &\leq \frac{\widetilde{\Xi} - \Xi(\boldsymbol{g}^{(s)})}{N-1}\left(2^2\zeta(2\alpha\lambda)\gamma_{s+1}^{\lambda} + 2^2(\zeta(2\alpha\lambda))^2\gamma_{s+1}^{2\lambda}\right)\\
&+\frac{N\Xi(\boldsymbol{g}^{(s)}) - \widetilde{\Xi}}{N-1}\,\frac{2^2\zeta(2\alpha\lambda)\gamma_{s+1}^{\lambda} + 2^2(1+2^{4\alpha\lambda+1})(\zeta(2\alpha\lambda))^2\gamma_{s+1}^{2\lambda}}{N}\\
&= \frac{\widetilde{\Xi}}{N}\left(2^2\zeta(2\alpha\lambda)\gamma_{s+1}^{\lambda} + 2^2(\zeta(2\alpha\lambda))^2\gamma_{s+1}^{2\lambda}\right) + \frac{N\Xi(\boldsymbol{g}^{(s)}) - \widetilde{\Xi}}{N-1}\,\frac{2^{4\alpha\lambda+3}(\zeta(2\alpha\lambda))^2\gamma_{s+1}^{2\lambda}}{N}\\
&\leq \frac{\widetilde{\Xi}}{N}\left(2^2\zeta(2\alpha\lambda)\gamma_{s+1}^{\lambda} + 2^2(\zeta(2\alpha\lambda))^2\gamma_{s+1}^{2\lambda}\right) + \widetilde{\Xi}\,\frac{2^{4\alpha\lambda+3}(\zeta(2\alpha\lambda))^2\gamma_{s+1}^{2\lambda}}{N}\\
&= \frac{2^2\zeta(2\alpha\lambda)\gamma_{s+1}^{\lambda} + 2^2(1+2^{4\alpha\lambda+1})(\zeta(2\alpha\lambda))^2\gamma_{s+1}^{2\lambda}}{N}\prod_{j=1}^{s}(1+2\zeta(2\alpha\lambda)\gamma_j^{\lambda})^2.
\end{aligned}$$

Using (13.13) and the induction assumption, we obtain

$$\begin{aligned}
&[S_{N,s+1,\alpha,\boldsymbol{\gamma}}((\boldsymbol{g}^{(s)}, g_{s+1}))]^2\\
&\leq (1+2\zeta(4\alpha)\gamma_{s+1}^2)\frac{1}{N^{2\tau}}\prod_{j=1}^{s}\left(1+2^{4\alpha+1}\zeta\left(\frac{\alpha}{\tau}\right)\gamma_j^{1/(2\tau)}\right)^{4\tau}\\
&\quad+\frac{1}{N^{1/\lambda}}\left(2^2\zeta(2\alpha\lambda)\gamma_{s+1}^{\lambda} + 2^2(1+2^{4\alpha\lambda+1})(\zeta(2\alpha\lambda))^2\gamma_{s+1}^{2\lambda}\right)^{1/\lambda}\\
&\quad\quad\times\prod_{j=1}^{s}(1+2\zeta(2\alpha\lambda)\gamma_j^{\lambda})^{2/\lambda}.
\end{aligned}$$

We partly replace τ by $1/(2\lambda)$ in the previous inequality, which yields

$$\begin{aligned}
&[S_{N,s+1,\alpha,\boldsymbol{\gamma}}((\boldsymbol{g}^{(s)}, g_{s+1}))]^2\\
&\leq (1+2\zeta(4\alpha)\gamma_{s+1}^2)\frac{1}{N^{2\tau}}\prod_{j=1}^{s}\left(1+2^{4\alpha+1}\zeta\left(\frac{\alpha}{\tau}\right)\gamma_j^{1/(2\tau)}\right)^{4\tau}\\
&\quad+\frac{1}{N^{2\tau}}\left(2^2\zeta(2\alpha\lambda)\gamma_{s+1}^{\lambda} + 2^2(1+2^{4\alpha\lambda+1})(\zeta(2\alpha\lambda))^2\gamma_{s+1}^{2\lambda}\right)^{1/\lambda}\\
&\quad\quad\times\prod_{j=1}^{s}\left(1+2\zeta\left(\frac{\alpha}{\tau}\right)\gamma_j^{1/(2\tau)}\right)^{4\tau}
\end{aligned}$$

$$\le \frac{1}{N^{2\tau}} \prod_{j=1}^{s} \left(1 + 2^{4\alpha+1} \zeta\left(\frac{\alpha}{\tau}\right) \gamma_j^{1/(2\tau)}\right)^{4\tau}$$
$$\times \left((1 + 2\zeta(4\alpha)\gamma_{s+1}^2) + \left(2^2\zeta(2\alpha\lambda)\gamma_{s+1}^{\lambda} + 2^2(1 + 2^{4\alpha\lambda+1})(\zeta(2\alpha\lambda))^2\gamma_{s+1}^{2\lambda}\right)^{1/\lambda}\right).$$

However, using Jensen's inequality,

$$(1 + 2\zeta(4\alpha)\gamma_{s+1}^2) + \left(2^2\zeta(2\alpha\lambda)\gamma_{s+1}^{\lambda} + 2^2(1 + 2^{4\alpha\lambda+1})(\zeta(2\alpha\lambda))^2\gamma_{s+1}^{2\lambda}\right)^{1/\lambda}$$
$$\le \left(1 + 2^{\lambda}(\zeta(4\alpha))^{\lambda}\gamma_{s+1}^{2\lambda} + 2^2\zeta(2\alpha\lambda)\gamma_{s+1}^{\lambda} + 2^2(1 + 2^{4\alpha\lambda+1})(\zeta(2\alpha\lambda))^2\gamma_{s+1}^{2\lambda}\right)^{1/\lambda}$$
$$\le \left(1 + 6\zeta(2\alpha\lambda)\gamma_{s+1}^{\lambda} + \left(2\sqrt{1 + 2^{4\alpha+1}}\zeta(2\alpha\lambda)\gamma_{s+1}^{\lambda}\right)^2\right)^{1/\lambda}$$
$$\le \left(1 + 2^{4\alpha+1}\zeta(2\alpha\lambda)\gamma_{s+1}^{\lambda}\right)^{2/\lambda},$$

where we used $3 \le 2\sqrt{1 + 2^{4\alpha+1}} \le 2^{4\alpha+1}$ for $\alpha > 1/2$. Replacing the remaining instances of λ by $1/(2\tau)$ yields

$$[S_{N,s+1,\alpha,\boldsymbol{\gamma}}((\boldsymbol{g}^{(s)}, g_{s+1}))]^2 \le \frac{1}{N^{2\tau}} \prod_{j=1}^{s+1} \left(1 + 2^{4\alpha+1} \zeta\left(\frac{\alpha}{\tau}\right) \gamma_j^{1/(2\tau)}\right)^{4\tau}.$$

Now the result is obtained by taking the square root. □

The following theorem should be compared with Proposition 13.2.

Theorem 13.6 *Let $N \ge 3$ be a prime number and let $\boldsymbol{\gamma} = (\gamma_j)_{j\ge1}$ be product weights with $\gamma_j \in (0, 1]$ for $j \in \mathbb{N}$. Assume that $\boldsymbol{g}$ has been found by Algorithm 13.3. Then, for arbitrary $\tau \in [1/2, \alpha)$, we have*

$$\mathrm{err}^{L_2\text{-app}}(\mathcal{H}_{\mathrm{kor},d,\alpha,\boldsymbol{\gamma}}, A_{N,M,d,\alpha,\boldsymbol{\gamma}}(\boldsymbol{g})) \le \frac{C_{d,\alpha,\boldsymbol{\gamma},\tau}}{N^{\tau/2}},$$

where $M = N^{\tau}$, and where

$$C_{d,\alpha,\boldsymbol{\gamma},\tau} = \left(1 + \prod_{j=1}^{d} \left(1 + 2^{4\alpha+1} \zeta\left(\frac{\alpha}{\tau}\right) \gamma_j^{1/(2\tau)}\right)^{4\tau}\right)^{1/2}.$$

Proof Plugging the result in Theorem 13.5 into Equation (13.12), we obtain

$$[\mathrm{err}^{L_2\text{-app}}(\mathcal{H}_{\mathrm{kor},d,\alpha,\boldsymbol{\gamma}}, A_{N,M,d,\alpha,\boldsymbol{\gamma}}(\boldsymbol{g}))]^2 \le \frac{1}{M} + \frac{\widetilde{C}_{d,\alpha,\boldsymbol{\gamma},\tau} M}{N^{2\tau}},$$

where

$$\widetilde{C}_{d,\alpha,\gamma,\tau} := \prod_{j=1}^{d} \left(1 + 2^{4\alpha+1} \zeta\left(\frac{\alpha}{\tau}\right) \gamma_j^{1/(2\tau)}\right)^{4\tau}.$$

In order to balance the two summands in the latter error bound choose $M = N^\tau$. Then we have

$$[\mathrm{err}^{L_2\text{-app}}(\mathcal{H}_{\mathrm{kor},d,\alpha,\gamma}, A_{N,M,d,\alpha,\gamma}(\boldsymbol{g}))]^2 \leq \frac{1 + \widetilde{C}_{d,\alpha,\gamma,\tau}}{N^\tau}.$$

The result now follows by taking the square root. □

Note that Theorem 13.6 provides an error convergence rate that is arbitrarily close to $O(N^{-\alpha/2})$. This is an improvement over the initial, more direct approach, with the error bound in Proposition 13.2.

13.2 Lower Error Bounds for L_2-Approximation in Korobov Spaces Using Lattice-Based Algorithms

The results in Section 13.1 (in particular, Theorem 13.6) show that by using an algorithm based on a suitable rank-1 lattice rule we can obtain a convergence order arbitrarily close to $O(N^{-\alpha/2})$. The question is to which extent this upper bound on the error is sharp. As it turns out, whenever one uses an algorithm based on function values along a rank-1 lattice point set, this convergence order cannot be improved. This result was shown, in a slightly more general fashion, in [24].

Before we proceed with the lower bound, we explain the idea of aliasing. For $\boldsymbol{h} \in \mathbb{Z}^d$, recall the approximation of the $\boldsymbol{h}$-th Fourier coefficient by

$$\widehat{f}(\boldsymbol{h}) = \int_{[0,1]^d} f(\boldsymbol{x})\, \mathrm{e}^{-2\pi \mathrm{i} \boldsymbol{h}\cdot\boldsymbol{x}}\, \mathrm{d}\boldsymbol{x} \approx \frac{1}{N} \sum_{k=0}^{N-1} f\left(\left\{\frac{k\boldsymbol{g}}{N}\right\}\right) \mathrm{e}^{-2\pi \mathrm{i} k \boldsymbol{h}\cdot\boldsymbol{g}/N}. \tag{13.15}$$

Consider now an $\boldsymbol{\ell} \in \mathbb{Z}^d$ in the dual lattice $\mathcal{L}^\perp(\boldsymbol{g}, N)$ of the lattice rule with generating vector $\boldsymbol{g}$, i.e., $\boldsymbol{\ell} \cdot \boldsymbol{g} \equiv 0 \pmod N$. If $\boldsymbol{\ell} \cdot \boldsymbol{g} = vN$ for some $v \in \mathbb{Z}$, then

$$\begin{aligned}
\frac{1}{N} \sum_{k=0}^{N-1} f\left(\left\{\frac{k\boldsymbol{g}}{N}\right\}\right) \mathrm{e}^{-2\pi \mathrm{i} k(\boldsymbol{h}+\boldsymbol{\ell})\cdot\boldsymbol{g}/N} &= \frac{1}{N} \sum_{k=0}^{N-1} f\left(\left\{\frac{k\boldsymbol{g}}{N}\right\}\right) \mathrm{e}^{-2\pi \mathrm{i} k \boldsymbol{h}\cdot\boldsymbol{g}/N} \mathrm{e}^{-2\pi \mathrm{i} k v} \\
&= \frac{1}{N} \sum_{k=0}^{N-1} f\left(\left\{\frac{k\boldsymbol{g}}{N}\right\}\right) \mathrm{e}^{-2\pi \mathrm{i} k \boldsymbol{h}\cdot\boldsymbol{g}/N}.
\end{aligned}$$

Thus the approximation of the Fourier coefficient $\widehat{f}(\boldsymbol{h})$ is the same as the approximation of the Fourier coefficient $\widehat{f}(\boldsymbol{h}+\boldsymbol{\ell})$. We say that these two Fourier coefficient approximations are aliased. For a rank-1 lattice rule, the dual lattice can be defined via the character property of the lattice point set, and hence the aliasing can be viewed as a consequence of the character property.

Aliasing has already come up in the approximation error of integrals. Using Proposition 1.12 for a rank-1 lattice rule shows

$$\frac{1}{N}\sum_{k=0}^{N-1} f\left(\left\{\frac{k\boldsymbol{g}}{N}\right\}\right) - \int_{[0,1]^d} f(\boldsymbol{x})\,\mathrm{d}\boldsymbol{x} = \sum_{\substack{\boldsymbol{h}\in\mathbb{Z}\setminus\{\boldsymbol{0}\} \\ \boldsymbol{h}\cdot\boldsymbol{g}\equiv \boldsymbol{0} \pmod N}} \widehat{f}(\boldsymbol{h}), \tag{13.16}$$

i.e., the integration error is just the sum over all Fourier coefficients which are aliased with $\widehat{f}(\boldsymbol{0})$. Roughly speaking, to make the error of numerical integration in a Korobov space small, we sought generating vectors $\boldsymbol{g}$ such that the closest frequency in the dual lattice is "far" from the origin $\boldsymbol{0}$.

We can use (13.16) for expressing the error of approximating the Fourier coefficient $\widehat{f}(\boldsymbol{h})$,

$$\frac{1}{N}\sum_{k=0}^{N-1} f\left(\left\{\frac{k\boldsymbol{g}}{N}\right\}\right) \mathrm{e}^{-2\pi \mathrm{i} k\boldsymbol{h}\cdot\boldsymbol{g}/N} - \int_{[0,1]^d} f(\boldsymbol{x})\,\mathrm{e}^{-2\pi\mathrm{i}\boldsymbol{h}\cdot\boldsymbol{x}}\,\mathrm{d}\boldsymbol{x} = \sum_{\substack{\boldsymbol{\ell}\in\mathbb{Z}^d\setminus\{\boldsymbol{0}\} \\ \boldsymbol{\ell}\cdot\boldsymbol{g}\equiv \boldsymbol{0} \pmod N}} \widehat{f}(\boldsymbol{h}+\boldsymbol{\ell}).$$

Now the error is the sum over all Fourier coefficients which are aliased with $\widehat{f}(\boldsymbol{h})$. Geometrically this means that we shift the dual lattice by $\boldsymbol{h}$ (the dual lattice is now an affine space), and hence the dual lattice contains now points much closer to the origin than the points in the "original" dual lattice (see Figure 13.1). For approximation in Korobov spaces in which the Fourier coefficients decay as one moves away from the origin (where we measure "distance" by $r_1(\boldsymbol{h}) = \prod_{j=1}^d \max(1,|h_j|)$), this means that it is natural to expect the approximation error to be larger than the integration error.

One can use these ideas to obtain a lower bound on the approximation error. It turns out that it is sufficient to consider the two-dimensional case. Say we are given a generating vector $\boldsymbol{g} = (1,g)$ and number of points N. In Lemma 13.8 below we show that there is a vector $(h_1,h_2) \in \mathcal{L}^{\perp}(\boldsymbol{g},N) \setminus \{(0,0)\}$ with $|h_1|,|h_2| \le \sqrt{N}$. If we use (13.16) to approximate the Fourier coefficients of a function f, then the Fourier coefficients $\widehat{f}(0,0)$ and $\widehat{f}(h_1,h_2)$ are aliased. This is not sufficient to obtain a good lower bound on the approximation error however, since the aliasing between the Fourier coefficient $\widehat{f}(0,0) = \int_0^1\int_0^1 f(\boldsymbol{x})\,\mathrm{d}\boldsymbol{x}$ and other Fourier coefficient only yields a lower bound on the integration error.

However, consider now the case of approximating the Fourier coefficient $\widehat{f}(h_1,0)$ using (13.15). Due to aliasing, the approximation of $\widehat{f}(h_1,0)$ is the same as the approximation of $\widehat{f}(0,-h_2)$. The approximation is given by the sum over all aliased Fourier coefficients

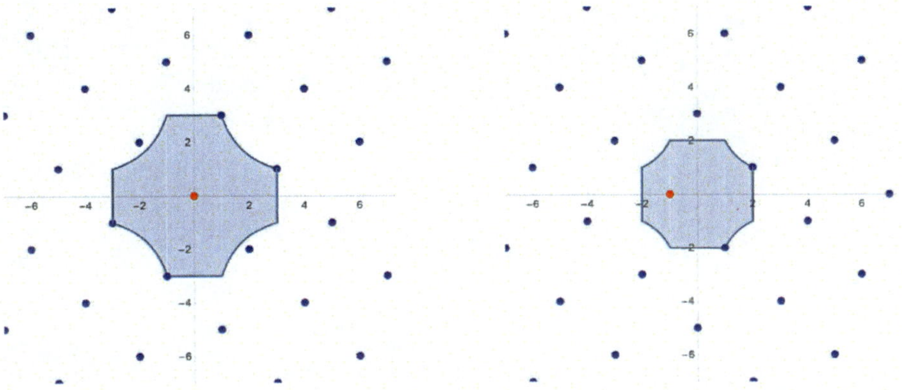

Fig. 13.1: We illustrate the approximation of the Fourier coefficients $\widehat{f}(0,0)$ (left) and $\widehat{f}(-1,0)$ (right) using a hyperbolic cross, see p. 16. The left-hand side shows the dual lattice of a lattice rule with generating vector $\boldsymbol{g} = (1,5)$ and $N = 8$ points. The blue-shaded region shows the hyperbolic cross $\{(h_1,h_2) \in \mathbb{R}^2 \ : \ \max(1,|h_1|)\max(1,|h_2|) \le 3\}$. The only point in the interior of the hyperbolic cross is the point $(0,0)$ (indicated by the red dot), which is exactly the frequency we want to approximate. If we measure "distance" from the origin by $\max(1,|h_1|)\max(1,|h_2|)$, then the points $(1,3)$, $(3,1)$, $(-1,-3)$ and $(-3,-1)$ are "closest" to the origin (i.e., they minimize r_1 over $\mathcal{L}^\perp(\boldsymbol{g},N)\setminus\{\boldsymbol{0}\}$). The figure on the right shows the dual lattice shifted by $(-1,0)$ (now indicated by the red dot), which corresponds to the case where we want to approximate $\widehat{f}(-1,0)$. Here the blue-shaded region shows the hyperbolic cross $\{(h_1,h_2) \in \mathbb{R}^2 \ : \ \max(1,|h_1|)\max(1,|h_2|) \le 2\}$. In this case the "closest" points are $(2,1)$ and $(1,-2)$. The hyperbolic cross on the right is smaller than the hyperbolic cross on the left, which illustrates that in general the error of approximating $\widehat{f}(-1,0)$ is larger than the error of approximating $\widehat{f}(0,0)$.

$$\frac{1}{N}\sum_{k=0}^{N-1} f\left(\left\{\frac{k\boldsymbol{g}}{N}\right\}\right) \mathrm{e}^{-2\pi \mathrm{i} k \boldsymbol{h}\cdot\boldsymbol{g}/N} = \sum_{\substack{\boldsymbol{\ell}\in\mathbb{Z}^2 \\ \boldsymbol{\ell}\cdot\boldsymbol{g}\equiv \boldsymbol{0} \pmod{N}}} \widehat{f}((h_1,0)+\boldsymbol{\ell}). \tag{13.17}$$

We can now construct a Fourier polynomial with only two terms which are aliased and which is difficult to approximate. It is convenient to set $\widehat{f}(h_1,0) = 1$ and $\widehat{f}(0,-h_2) = -1$, since then (13.17) becomes 0, which means that (13.2) approximates the Fourier polynomial

$$\mathrm{e}^{2\pi \mathrm{i}(h_1,0)\cdot\boldsymbol{x}} - \mathrm{e}^{2\pi \mathrm{i}(0,-h_2)\cdot\boldsymbol{x}} \tag{13.18}$$

by 0. This is convenient, because now we do not need to distinguish between the cases whether $(h_1,0)$, $(0,-h_2)$ are in the set $\mathcal{A}_{2,M}$ or not. The lower bound then comes from estimating the norm of the Fourier polynomial (13.18) (the details are given in the proof of Theorem 13.7 below).

The above approach can be generalised—the lower bound proof does not rely on the specific form of the approximation algorithm (it applies to (13.2) but also any other method which uses the same function values). The details are given in Theorem 13.7 and its proof.

Theorem 13.7 *Let $d, N \in \mathbb{N}$, let $\alpha > 1/2$, and let $\boldsymbol{\gamma} = (\gamma_j)_{j\geq 1}$ be product weights. Furthermore, let $A_{N,d}(\boldsymbol{g})$ be an arbitrary linear L_2-approximation algorithm using function evaluations at the points of a rank-1 lattice point set with generating vector $\boldsymbol{g}$ for $f \in \mathcal{H}_{\mathrm{kor},d,\alpha,\boldsymbol{\gamma}}$, i.e.,*

$$A_{N,d}(\boldsymbol{g})(f) = \sum_{k=0}^{N-1} f\left(\left\{\frac{k\boldsymbol{g}}{N}\right\}\right) a_k, \tag{13.19}$$

with some $a_k \in L_2([0,1]^d)$. Then it is true that

$$\mathrm{err}^{L_2\text{-app}}(\mathcal{H}_{\mathrm{kor},d,\alpha,\boldsymbol{\gamma}}, A_{N,d}(\boldsymbol{g})) \geq \frac{C_{d,\alpha,\boldsymbol{\gamma}}}{N^{\alpha/2}},$$

where $C_{d,\alpha,\boldsymbol{\gamma}}$ is a positive real that is independent of N.

For the proof of this result we require some preparation. For given $d, N \in \mathbb{N}$, let

$$\widetilde{X}_{d,\sqrt{N}} := \{-\lfloor\sqrt{N}\rfloor, \ldots, \lfloor\sqrt{N}\rfloor\} \times \{-\lfloor\sqrt{N}\rfloor, \ldots, \lfloor\sqrt{N}\rfloor\} \times \underbrace{\{0\} \times \cdots \times \{0\}}_{d-2 \text{ times}},$$

and $\widetilde{X}^{+}_{d,\sqrt{N}} := \widetilde{X}_{d,\sqrt{N}} \cap \mathbb{N}_0^d$. Furthermore, again for $d, N \in \mathbb{N}$, define

$$X_{d,\sqrt{N}} := \{\boldsymbol{h} \in \widetilde{X}_{d,\sqrt{N}} \ : \ \|\boldsymbol{h}\|_1 = \|\boldsymbol{h}\|_\infty\}.$$

Next, we show the following lemma.

Lemma 13.8 *Let $d, N \in \mathbb{N}$, and let $\{\boldsymbol{x}_0, \boldsymbol{x}_1, \ldots, \boldsymbol{x}_{N-1}\}$ be an arbitrary N-point set in $[0,1)^d$, such that*

$$\sum_{k=0}^{N-1} \mathrm{e}^{2\pi \mathrm{i} \boldsymbol{h}\cdot\boldsymbol{x}_k} \in \{0, N\}$$

for all $\boldsymbol{h} \in \widetilde{X}_{d,\sqrt{N}}$. Then, there exist at least two distinct indices $\boldsymbol{h}^{(1)}, \boldsymbol{h}^{(2)} \in X_{d,\sqrt{N}}$ such that

$$\mathrm{e}^{2\pi \mathrm{i} \boldsymbol{h}^{(1)}\cdot\boldsymbol{x}_k} = \mathrm{e}^{2\pi \mathrm{i} \boldsymbol{h}^{(2)}\cdot\boldsymbol{x}_k} \quad \textit{for all } k \in \{0, 1, \ldots, N-1\}.$$

Proof If $\boldsymbol{h} = \boldsymbol{0} \in \widetilde{X}_{d,\sqrt{N}}$, then we obviously have

$$\sum_{k=0}^{N-1} \mathrm{e}^{2\pi \mathrm{i} \boldsymbol{h}\cdot\boldsymbol{x}_k} = N.$$

Now assume that

$$\sum_{k=0}^{N-1} \mathrm{e}^{2\pi \mathrm{i} \boldsymbol{h}\cdot \boldsymbol{x}_k} = 0 \quad \text{for all } \boldsymbol{h} \in \widetilde{X}_{d,\sqrt{N}} \setminus \{\boldsymbol{0}\}.$$

For any two $\boldsymbol{h}^{(1)}, \boldsymbol{h}^{(2)} \in \widetilde{X}^+_{d,\sqrt{N}}$, we have $\boldsymbol{h}^{(1)} - \boldsymbol{h}^{(2)} \in \widetilde{X}_{d,\sqrt{N}}$, and furthermore

$$\sum_{k=0}^{N-1} \mathrm{e}^{2\pi \mathrm{i} (\boldsymbol{h}^{(2)}-\boldsymbol{h}^{(1)})\cdot \boldsymbol{x}_k} = \begin{cases} N & \text{if } \boldsymbol{h}^{(2)} - \boldsymbol{h}^{(1)} = \boldsymbol{0}, \\ 0 & \text{otherwise.} \end{cases}$$

We can write this relation in matrix-vector form as

$$A^* A = N U_{(\lfloor \sqrt{N} \rfloor + 1)^2},$$

where we remind the reader that U_k denotes the $k \times k$ identity matrix, and where

$$A = \left(\mathrm{e}^{2\pi \mathrm{i} \boldsymbol{h}\cdot \boldsymbol{x}_k} \right)_{k \in \{0,1,\ldots,N-1\}, \boldsymbol{h} \in \widetilde{X}^+_{d,\sqrt{N}}},$$

and where $A^* = \overline{A}^\top$ is the adjoint matrix of A. However, the relation $A^* A = N U_{(\lfloor \sqrt{N} \rfloor + 1)^2}$ means that A must have full column rank, which is impossible as $N < (\lfloor \sqrt{N} \rfloor + 1)^2$. Accordingly, this case cannot occur and therefore there must exist at least one $\boldsymbol{h}^{(0)} = (h_1^{(0)}, h_2^{(0)}, 0, \ldots, 0) \in \widetilde{X}_{d\sqrt{N}} \setminus \{\boldsymbol{0}\}$ such that

$$\sum_{k=0}^{N-1} \mathrm{e}^{2\pi \mathrm{i} \boldsymbol{h}^{(0)}\cdot \boldsymbol{x}_k} = N.$$

This implies that $\mathrm{e}^{2\pi \mathrm{i} \boldsymbol{h}^{(0)}\cdot \boldsymbol{x}_k} = 1$ for all $k \in \{0, 1, \ldots, N-1\}$. In this case, we can choose $\boldsymbol{h}^{(1)}, \boldsymbol{h}^{(2)} \in X_{d,\sqrt{N}}$ as

$$\boldsymbol{h}^{(1)} = (h_1^{(0)}, 0, 0, \ldots, 0) \quad \text{and} \quad \boldsymbol{h}^{(2)} = (0, -h_2^{(0)}, 0, \ldots, 0).$$

Then, we obviously have

$$\mathrm{e}^{2\pi \mathrm{i} \boldsymbol{h}^{(1)}\cdot \boldsymbol{x}_k} \mathrm{e}^{-2\pi \mathrm{i} \boldsymbol{h}^{(2)}\cdot \boldsymbol{x}_k} = 1 \quad \text{or, equivalently,} \quad \mathrm{e}^{2\pi \mathrm{i} \boldsymbol{h}^{(1)}\cdot \boldsymbol{x}_k} = \mathrm{e}^{2\pi \mathrm{i} \boldsymbol{h}^{(2)}\cdot \boldsymbol{x}_k}$$

for all $k \in \{0, 1, \ldots, N-1\}$. This yields the result in the lemma. □

Using Lemma 13.8, we can proceed to show Theorem 13.7.

Proof of Theorem 13.7 Due to Lemma 13.8, we know that there exist distinct elements $\boldsymbol{h}^{(1)}, \boldsymbol{h}^{(2)}$ of $X_{d,\sqrt{N}}$ such that

$$\mathrm{e}^{2\pi \mathrm{i} k \boldsymbol{h}^{(1)}\cdot \boldsymbol{g}/N} = \mathrm{e}^{2\pi \mathrm{i} k \boldsymbol{h}^{(2)}\cdot \boldsymbol{g}/N} \quad \text{for all } k \in \{0, 1, \ldots, N-1\}, \tag{13.20}$$

which implies $\boldsymbol{h}^{(1)} \cdot \boldsymbol{g} \equiv \boldsymbol{h}^{(2)} \cdot \boldsymbol{g} \pmod{N}$.

We now define a special function $\widetilde{f} \in \mathcal{H}_{\mathrm{kor},d,\alpha,\boldsymbol{\gamma}}$ by

$$\widetilde{f}(\boldsymbol{x}) := \mathrm{e}^{2\pi \mathrm{i} \boldsymbol{h}^{(1)} \cdot \boldsymbol{x}} - \mathrm{e}^{2\pi \mathrm{i} \boldsymbol{h}^{(2)} \cdot \boldsymbol{x}} \quad \text{for } \boldsymbol{x} \in [0,1]^d,$$

such that $\|\widetilde{f}\|_{\mathrm{kor},d,\alpha,\boldsymbol{\gamma}} = (r_{2\alpha,\boldsymbol{\gamma}}(\boldsymbol{h}^{(1)}) + r_{2\alpha,\boldsymbol{\gamma}}(\boldsymbol{h}^{(2)}))^{1/2}$. We then define $f_0 \in \mathcal{H}_{\mathrm{kor},d,\alpha,\boldsymbol{\gamma}}$ by

$$f_0(\boldsymbol{x}) := \frac{\widetilde{f}(\boldsymbol{x})}{(r_{2\alpha,\boldsymbol{\gamma}}(\boldsymbol{h}^{(1)}) + r_{2\alpha,\boldsymbol{\gamma}}(\boldsymbol{h}^{(2)}))^{1/2}}$$

such that $\|f_0\|_{\mathrm{kor},d,\alpha,\boldsymbol{\gamma}} = 1$. Since (13.20) holds, we obtain that $f_0(\{k\boldsymbol{g}/N\}) = 0$ for all $k \in \{0, 1, \ldots, N-1\}$. Furthermore, by Parseval's identity,

$$\|f_0\|_{L_2} = \left(\sum_{\boldsymbol{h} \in \mathbb{Z}^d} |\widehat{f_0}(\boldsymbol{h})|^2 \right)^{1/2} = \frac{\sqrt{2}}{(r_{2\alpha,\boldsymbol{\gamma}}(\boldsymbol{h}^{(1)}) + r_{2\alpha,\boldsymbol{\gamma}}(\boldsymbol{h}^{(2)}))^{1/2}}.$$

Since $\boldsymbol{h}^{(1)}, \boldsymbol{h}^{(2)} \in X_{d,\sqrt{N}}$, we have $\|\boldsymbol{h}^{(1)}\|_\infty = \|\boldsymbol{h}^{(1)}\|_1$ and $\|\boldsymbol{h}^{(2)}\|_\infty = \|\boldsymbol{h}^{(2)}\|_1$. Without loss of generality, we assume $\|\boldsymbol{h}^{(1)}\|_1 = \|\boldsymbol{h}^{(1)}\|_\infty \geq \|\boldsymbol{h}^{(2)}\|_\infty = \|\boldsymbol{h}^{(2)}\|_1$, which implies

$$r_{2\alpha,\boldsymbol{\gamma}}(\boldsymbol{h}^{(1)}) \geq r_{2\alpha,\boldsymbol{\gamma}}(\boldsymbol{h}^{(2)}).$$

Consequently,

$$\|f_0\|_{L_2} \geq \frac{1}{(r_{2\alpha,\boldsymbol{\gamma}}(\boldsymbol{h}^{(1)}))^{1/2}}.$$

On the other hand, since $\boldsymbol{h}^{(1)} \in X_{d,\sqrt{N}}$, we get

$$r_{2\alpha,\boldsymbol{\gamma}}(\boldsymbol{h}^{(1)}) \leq \frac{N^\alpha}{\gamma_{j_0}},$$

where either $j_0 = 1$ or $j_0 = 2$. This yields

$$\|f_0\|_{L_2} \geq \frac{\sqrt{\gamma_{j_0}}}{N^{\alpha/2}}.$$

Let now $A_{N,d}(\boldsymbol{g})$ be an arbitrary linear L_2-approximation algorithm using function evaluations at the lattice points $\{k\boldsymbol{g}/N\}$, $k \in \{0, 1, \ldots, N-1\}$, for $f \in \mathcal{H}_{\mathrm{kor},d,\alpha,\boldsymbol{\gamma}}$, as given in (13.19). We then obviously have $A_{N,d}(\boldsymbol{g})(f_0) = 0$ and therefore

$$\frac{\sqrt{\gamma_{j_0}}}{N^{\alpha/2}} \leq \|f_0\|_{L_2} = \|f_0 - A_{N,d}(\boldsymbol{g})(f_0)\|_{L_2} \leq \sup_{\substack{f \in \mathcal{H}_{\mathrm{kor},d,\alpha,\boldsymbol{\gamma}} \\ \|f\|_{\mathrm{kor},d,\alpha,\boldsymbol{\gamma}} \leq 1}} \|f - A_{N,d}(\boldsymbol{g})(f)\|_{L_2}.$$

This yields the result. □

13.3 Tractability of L_2-Approximation Using Lattice Rules

In Section 1.7, we introduced the notions of information complexity and tractability for the problem of numerical integration. Analogously, we can proceed for the problem of L_2-approximation in the Korobov space $\mathcal{H}_{\mathrm{kor},d,\alpha,\boldsymbol{\gamma}}$. However, in the case of function approximation, we have to distinguish regarding which class of information we have access to. To be more precise, we could consider general linear algorithms of the form

$$A_{N,d}(f) := \sum_{k=0}^{N-1} a_k \mathcal{L}_k(f),$$

where the coefficients a_k are elements of $L_2([0,1]^d)$, and the $\mathcal{L}_k$ are continuous linear functionals defined on $\mathcal{H}_{\mathrm{kor},d,\alpha,\boldsymbol{\gamma}}$ (in fact, it can be shown that it is no essential restriction to consider only linear algorithms like $A_{N,d}$ for this problem, see, e.g., [207] or [210, Chapter 4], and the references therein). If we allow arbitrary continuous linear functionals for the $\mathcal{L}_k$, we say that we use information from the class Λ^{all}. If, on the other hand, we only allow function evaluations for $\mathcal{L}_k$, i.e., $\mathcal{L}_k(f) = f(\boldsymbol{x}_k)$ for some $\boldsymbol{x}_k \in [0,1]^d$, $k \in \{0,1,\ldots,N-1\}$, we speak of standard information or information from the class Λ^{std}. With some abuse of notation we also write $A_{N,d} \in \Lambda$ for $\Lambda \in \{\Lambda^{\mathrm{all}}, \Lambda^{\mathrm{std}}\}$ when the algorithm $A_{N,d}$ uses information exclusively from the class Λ.

We can then define the worst-case error of a linear algorithm $A_{N,d}$ for the problem of L_2-approximation in the Korobov space $\mathcal{H}_{\mathrm{kor},d,\alpha,\boldsymbol{\gamma}}$ as

$$\mathrm{err}^{L_2\text{-app}}(\mathcal{H}_{\mathrm{kor},d,\alpha,\boldsymbol{\gamma}}, A_{N,d}) := \sup_{\substack{f\in\mathcal{H}_{\mathrm{kor},d,\alpha,\boldsymbol{\gamma}}\\ \|f\|_{\mathrm{kor},d,\alpha,\boldsymbol{\gamma}}\le 1}} \|f - A_{N,d}(f)\|_{L_2}. \tag{13.21}$$

The initial error, corresponding to using no information about the functions to be approximated, is in this case given by

$$\mathrm{err}^{L_2\text{-app}}(\mathcal{H}_{\mathrm{kor},d,\alpha,\boldsymbol{\gamma}}, 0) := \sup_{\substack{f\in\mathcal{H}_{\mathrm{kor},d,\alpha,\boldsymbol{\gamma}}\\ \|f\|_{\mathrm{kor},d,\alpha,\boldsymbol{\gamma}}\le 1}} \|f\|_{L_2}.$$

As we have $\|f\|_{L_2} \le \|f\|_{\mathrm{kor},d,\alpha,\boldsymbol{\gamma}}$ for all $f \in \mathcal{H}_{\mathrm{kor},d,\alpha,\boldsymbol{\gamma}}$, and since for $f \equiv 1$ we have $\|f\|_{L_2} = \|f\|_{\mathrm{kor},d,\alpha,\boldsymbol{\gamma}} = 1$, we know that the initial error in this problem is actually equal to one, which means that the specific problem of L_2-approximation in the Korobov space is already normalized.

Adapting the definition of the information complexity that was previously given in Definition 1.42 to the approximation problem, and to the different classes of information, we call, for given $\varepsilon \in (0,1)$ and $d \in \mathbb{N}$, and given $\Lambda \in \{\Lambda^{\mathrm{std}}, \Lambda^{\mathrm{all}}\}$,

$$\begin{aligned} &N^{L_2\text{-app}}(\varepsilon, d, \Lambda) \\ &:= \min\{N \in \mathbb{N} : \exists A_{N,d} \in \Lambda \text{ such that } \mathrm{err}^{L_2\text{-app}}(\mathcal{H}_{\mathrm{kor},d,\alpha,\boldsymbol{\gamma}}, A_{N,d}) \le \varepsilon\} \end{aligned}$$

the *information complexity* of L_2-approximation in $\mathcal{H}_{\mathrm{kor},d,\alpha,\boldsymbol{\gamma}}$ using information from Λ.

We can then define the various notions of tractability, depending on the information class Λ under consideration, in analogy to Definition 1.44. In particular, we speak of polynomial tractability for the class $\Lambda \in \{\Lambda^{\mathrm{std}}, \Lambda^{\mathrm{all}}\}$, if there are constants $C, \sigma > 0$ and $\eta \geq 0$ such that

$$N^{L_2\text{-app}}(\varepsilon, d, \Lambda) \leq C d^{\eta} \varepsilon^{-\sigma} \quad \text{for all } \varepsilon \in (0,1) \text{ and all } d \in \mathbb{N}. \tag{13.22}$$

If (13.22) holds with $\eta = 0$, we speak of strong polynomial tractability. If strong polynomial tractability holds, we call the infimum of those σ for which (13.22) holds with $\eta = 0$ the ε-exponent of strong polynomial tractability for the class Λ, denoted by $\sigma^*(\Lambda)$.

Next, we discuss tractability properties of the L_2-approximation problem for the weighted Korobov space $\mathcal{H}_{\mathrm{kor},d,\alpha,\boldsymbol{\gamma}}$.

L_2-approximation based on Λ^{all}

Conveniently, for the information class Λ^{all}, it is known how the optimal algorithm for approximation in the Korobov space is defined. Indeed, in Λ^{all}, we have access to Fourier coefficients, and the optimal algorithm $A_{N,d}$ to obtain an error of at most ε is given by choosing $N = |\mathcal{A}_{d,\varepsilon^{-2}}|$, and

$$A_{N,d}(f)(\boldsymbol{x}) := \sum_{\boldsymbol{h} \in \mathcal{A}_{d,\varepsilon^{-2}}} \widehat{f}(\boldsymbol{h})\, \mathrm{e}^{2\pi \mathrm{i} \boldsymbol{h}\cdot\boldsymbol{x}} \quad \text{for } f \in \mathcal{H}_{\mathrm{kor},d,\alpha,\boldsymbol{\gamma}},$$

where we recall the definition of $\mathcal{A}_{d,M}$ for $M \geq 0$ from (13.1). This is a special case of a well-known general result for Hilbert spaces (see, e.g., [210] for further details). From this, one can deduce precise conditions on the weights $\boldsymbol{\gamma}$ in order to achieve (strong) polynomial tractability. These conditions are stated in terms of the so-called sum-exponent of the weights $\boldsymbol{\gamma} = (\gamma_j)_{j\geq 1}$, which is given by

$$S_{\boldsymbol{\gamma}} := \inf \left\{ s > 0 \; : \; \sum_{j=1}^{\infty} \gamma_j^s < \infty \right\}.$$

We now state a characterization of (strong) polynomial tractability in terms of the sum-exponent of the weight sequence $\boldsymbol{\gamma}$.

Theorem 13.9 *For Λ^{all}, it is true that polynomial tractability and strong polynomial tractability of L_2-approximation for the weighted Korobov space are equivalent and they are equivalent to the condition $S_{\boldsymbol{\gamma}} < \infty$. If this is the case, the ε-exponent of strong polynomial tractability is given by*

$$\sigma^*(\Lambda^{\mathrm{all}}) = 2 \max(1/(2\alpha), S_{\boldsymbol{\gamma}}).$$

This result follows from very general findings due to Wasilkowski and Woźniakowski [261]. A direct proof for the present special instance of L_2-approximation for Korobov spaces can also be found in [72]. Since the proof is beyond the scope of this book, we omit it and refer to the existing literature instead. A complete overview of the weight conditions for all current standard tractability notions of this approximation problem can be found in [72, Theorem 1].

L_2-approximation based on Λ^{std}

For Λ^{std}, the situation is more involved than for Λ^{all}. The following result is due to Novak, Sloan, and Woźniakowski [209].

Theorem 13.10 *For Λ^{std}, strong polynomial tractability of L_2-approximation for the weighted Korobov space holds if and only if*

$$\sum_{j=1}^{\infty} \gamma_j < \infty,$$

which implies $S_\gamma \leq 1$ and $\sigma^*(\Lambda^{\text{all}}) \leq 2$. For the ε-exponent of strong polynomial tractability for standard information, it is then true that

$$\sigma^*(\Lambda^{\text{std}}) \in [\sigma^*(\Lambda^{\text{all}}), \sigma^*(\Lambda^{\text{all}}) + 2].$$

In particular, we have $\sigma^*(\Lambda^{\text{std}}) \leq 4$.

Furthermore, polynomial tractability holds for Λ^{std} if and only if

$$\limsup_{d \to \infty} \frac{1}{\log d} \sum_{j=1}^{d} \gamma_j < \infty.$$

Again, we omit the proof of these results and refer to [209] or to [162].

We now demonstrate how the error bound presented in Theorem 13.6 can be used to meet the results stated in Theorem 13.10. In this way we show how (strong) polynomial tractability can be achieved by means of algorithms that are based on lattice rules.

Theorem 13.11 *The L_2-approximation problem for the weighted Korobov space with respect to information from the class Λ^{std}, using function evaluations at the points of a rank-1 lattice point set, is strongly polynomially tractable if*

$$\sum_{j=1}^{\infty} \gamma_j < \infty.$$

If

$$\sum_{j=1}^{\infty} \gamma_j^{1/(2\tau)} < \infty \tag{13.23}$$

holds for some $\tau \in [1/2, \alpha)$*, then the* ε*-exponent is at most* $2/\tau$*. If* (13.23) *even holds for every* $\tau \in [1/2, \alpha)$*, then the* ε*-exponent is at most* $2/\alpha$*. Furthermore, polynomial tractability holds if*

$$\limsup_{d\to\infty} \frac{1}{\log d} \sum_{j=1}^{d} \gamma_j < \infty. \tag{13.24}$$

Proof For every odd prime number N one can construct an algorithm $A_{N,d}(\boldsymbol{g})$ based on a rank-1 lattice point set $\mathcal{P}(\boldsymbol{g}, N)$ such that according to Theorem 13.6 we have, for every $\tau \in [1/2, \alpha)$,

$$\mathrm{err}^{L_2-\mathrm{app}}(\mathcal{H}_{\mathrm{kor},d,\alpha,\boldsymbol{\gamma}}, A_{N,d}(\boldsymbol{g})) \le \frac{C_{d,\alpha,\tau,\boldsymbol{\gamma}}}{N^{\tau/2}}, \tag{13.25}$$

where

$$C_{d,\alpha,\tau,\boldsymbol{\gamma}} \le \sqrt{2} \prod_{j=1}^{d} \left(1 + 2^{4\alpha+1} \gamma_j^{1/(2\tau)} \zeta\left(\frac{\alpha}{\tau}\right)\right)^{2\tau}.$$

Assume that (13.23) holds for some $\tau \in [1/2, \alpha)$. Then

$$\prod_{j=1}^{d} \left(1 + 2^{4\alpha+1} \gamma_j^{1/(2\tau)} \zeta\left(\frac{\alpha}{\tau}\right)\right) \le \mathrm{e}^{2^{4\alpha+1}\zeta(\alpha/\tau)\sum_{j=1}^{d}\gamma_j^{1/(2\tau)}} \le \mathrm{e}^{2^{4\alpha+1}\zeta(\alpha/\tau)\sum_{j=1}^{\infty}\gamma_j^{1/(2\tau)}}$$

is bounded uniformly in d. This means that

$$\mathrm{err}^{L_2-\mathrm{app}}(\mathcal{H}_{\mathrm{kor},d,\alpha,\boldsymbol{\gamma}}, A_{N,d}(\boldsymbol{g})) \le \frac{C_{\alpha,\tau,\boldsymbol{\gamma}}}{N^{\tau/2}}$$

for some $C_{\alpha,\tau,\boldsymbol{\gamma}} < \infty$, which is independent of the dimension d. Let $Q := \lceil (C_{\alpha,\tau,\boldsymbol{\gamma}}\varepsilon^{-1})^{2/\tau} \rceil$ and let N be the smallest odd prime number greater than or equal to Q. Note that obviously $N \in [Q, 2Q)$. Now construct a generating vector $\boldsymbol{g}$ according to Algorithm 13.3. For this $\boldsymbol{g}$ we then obtain

$$\mathrm{err}^{L_2-\mathrm{app}}(\mathcal{H}_{\mathrm{kor},d,\alpha,\boldsymbol{\gamma}}, A_{N,d}(\boldsymbol{g})) \le \varepsilon,$$

and hence the information complexity satisfies

$$N^{L_2-\mathrm{app}}(\varepsilon, d, \Lambda^{\mathrm{std}}) \le N \le 2Q = 2\left\lceil \left(C_{\alpha,\tau,\boldsymbol{\gamma}}\varepsilon^{-1}\right)^{2/\tau} \right\rceil.$$

Thus, the problem is strongly polynomially tractable with an ε-exponent of at most $2/\tau$. In particular, for $\tau = 1/2$, we see that summability of the weights γ_j, $j \in \mathbb{N}$, is sufficient to obtain strong polynomial tractability. If (13.23) even holds for all $\tau \in [1/2, \alpha)$, then the ε-exponent is at most $2/\alpha$.

Finally, let us assume that (13.24) holds. Then we can choose $\tau = 1/2$ in (13.25) and see by standard arguments as in the proof of Corollary 2.28 (in particular, see (2.33)) that

$$\prod_{j=1}^{d}\left(1+2^{4\alpha+1}\gamma_j\,\zeta(2\alpha)\right)$$

is bounded from above by a term that depends only polynomially on d. In this way, we obtain polynomial tractability. □

13.4 Adaptions for General Weights

In this section, we will highlight some additional aspects of the theory developed in this chapter so far. Up to now, we have only allowed product weights for approximation by means of lattice rules. Here we would like to generalize these observations to the case of general (nonnegative) weights with a special focus on product-and-order-dependent (POD) weights (recall that product weights are a special case of POD weights). As for integration problems, the significance of POD weights is mostly due to applications such as the analysis of PDEs with random coefficients (see Appendix A).

Derivation of a CBC algorithm and an error bound

Analogously to what we did in Section 13.1, we define the set $\mathcal{A}_{d,M}$ as in (13.1), i.e.,

$$\mathcal{A}_{d,M} := \left\{\boldsymbol{h}\in\mathbb{Z}^d \;:\; r_{2\alpha,\boldsymbol{\gamma}}(\boldsymbol{h})\le M\right\},$$

but now for general (nonnegative) weights $\boldsymbol{\gamma}=\{\gamma_{\mathfrak{u}}\}_{\mathfrak{u}\subseteq[d]}$.

For given $N\in\mathbb{N}$ and $M\ge 0$, a given rank-1 lattice point set with generating vector $\boldsymbol{g}\in G_d^{\varphi}(N)$, and for the Korobov space $\mathcal{H}_{\mathrm{kor},d,\alpha,\boldsymbol{\gamma}}$ with weights $\boldsymbol{\gamma}=\{\gamma_{\mathfrak{u}}\}_{\mathfrak{u}\subseteq[d]}$, the algorithm $A_{N,M,d,\alpha,\boldsymbol{\gamma}}(\boldsymbol{g})$ and its error are defined exactly as in (13.2) and (13.3), respectively.

First, we need an estimate on the size of the set $\mathcal{A}_{d,M}$. The following lemma, which is a generalization of Item 4 of Lemma 13.1, gives such a bound. For a proof, we refer to [28].

Lemma 13.12 *For $\alpha>1/2$, $d\in\mathbb{N}$, general (nonnegative) weights $\boldsymbol{\gamma}=\{\gamma_{\mathfrak{u}}\}_{\mathfrak{u}\subseteq[d]}$, and $M\ge 0$, let $\mathcal{A}_{d,M}:=\left\{\boldsymbol{h}\in\mathbb{Z}^d \;:\; r_{2\alpha,\boldsymbol{\gamma}}(\boldsymbol{h})\le M\right\}$. Then, for any $\lambda>1/(2\alpha)$, we have*

$$|\mathcal{A}_{d,M}|\le M^{\lambda}\sum_{\mathfrak{u}\subseteq[d]}\gamma_{\mathfrak{u}}^{\lambda}(2\zeta(2\alpha\lambda))^{|\mathfrak{u}|}.$$

Regarding the error analysis of the algorithm $A_{N,M,d,\alpha,\boldsymbol{\gamma}}(\boldsymbol{g})$, it can be shown in exactly the same way as in the computations leading to (13.12) that

$$[\mathrm{err}^{L_2\text{-app}}(\mathcal{H}_{\mathrm{kor},d,\alpha,\boldsymbol{\gamma}},A_{N,M,d,\alpha,\boldsymbol{\gamma}}(\boldsymbol{g}))]^2\le\frac{1}{M}+M[S_{N,d,\alpha,\boldsymbol{\gamma}}(\boldsymbol{g})]^2, \tag{13.26}$$

where $S_{N,d,\alpha,\boldsymbol{\gamma}}(\boldsymbol{g})$ is defined as in (13.11), i.e.,

$$S_{N,d,\alpha,\boldsymbol{\gamma}}(\boldsymbol{g}) := \left(\sum_{\boldsymbol{h}\in\mathbb{Z}^d} \frac{1}{r_{2\alpha,\boldsymbol{\gamma}}(\boldsymbol{h})} \sum_{\substack{\boldsymbol{\ell}\in\mathbb{Z}^d\setminus\{\boldsymbol{0}\} \\ \boldsymbol{\ell}\cdot\boldsymbol{g}\equiv 0 \pmod N}} \frac{1}{r_{2\alpha,\boldsymbol{\gamma}}(\boldsymbol{h}+\boldsymbol{\ell})} \right)^{1/2}.$$

In Section 13.1, we then proceeded with a CBC construction of generating vectors $\boldsymbol{g}$ such that $S_{N,d,\alpha,\boldsymbol{\gamma}}(\boldsymbol{g})$ is small. This step, however, requires additional work for the case of general weights as compared to the special case of product weights.

First, we need the following lemma to "decompose" the square of the quantity $S_{N,d,\alpha,\boldsymbol{\gamma}}(\boldsymbol{g})$ component-wise. Before we state the lemma we remind the reader that for $d \in \mathbb{N}$ and $\boldsymbol{h} \in \mathbb{Z}^d$ we use the notation $\mathfrak{u}(\boldsymbol{h}) := \{j \in [d] \,:\, h_j \neq 0\}$.

Lemma 13.13 *Let $d, N \in \mathbb{N}$, let $\alpha > 1/2$, and let $\boldsymbol{\gamma} = \{\gamma_{\mathfrak{u}}\}_{\mathfrak{u}\subseteq[d]}$ be general (nonnegative) weights. Let $S_{N,d,\alpha,\boldsymbol{\gamma}}(\boldsymbol{g})$ be as in* (13.11)*, where $\boldsymbol{g} = (g_1, \ldots, g_d) \in G_d^{\varphi}(N)$. Then it is true that*

$$[S_{N,d,\alpha,\boldsymbol{\gamma}}(\boldsymbol{g})]^2 = \sum_{s=1}^{d} T_{N,d,s,\alpha,\boldsymbol{\gamma}}(g_1, \ldots, g_s), \tag{13.27}$$

where, for $s \in [d]$,

$$\begin{aligned} &T_{N,d,s,\alpha,\boldsymbol{\gamma}}(g_1, \ldots, g_s) \\ &:= \sum_{\mathfrak{w}\subseteq\{s+1,\ldots,d\}} (2\zeta(4\alpha))^{|\mathfrak{w}|}\, \widetilde{T}_{N,s,\alpha}(g_1, \ldots, g_s, \{\gamma_{\mathfrak{u}\cup\mathfrak{w}}\}_{\mathfrak{u}\subseteq[s]}), \end{aligned} \tag{13.28}$$

with

$$\widetilde{T}_{N,s,\alpha}(g_1, \ldots, g_s, \{\gamma_{\mathfrak{u}\cup\mathfrak{w}}\}_{\mathfrak{u}\subseteq[s]}) := \sum_{\boldsymbol{h}\in\mathbb{Z}^s} \sum_{\substack{\boldsymbol{\ell}\in\mathbb{Z}^s \\ \ell_s\neq 0 \\ \boldsymbol{\ell}\cdot(g_1,\ldots,g_s)\equiv 0 \pmod N}} \frac{\gamma_{\mathfrak{u}(\boldsymbol{h})\cup\mathfrak{w}}}{r_{2\alpha}(\boldsymbol{h})} \frac{\gamma_{\mathfrak{u}(\boldsymbol{h}+\boldsymbol{\ell})\cup\mathfrak{w}}}{r_{2\alpha}(\boldsymbol{h}+\boldsymbol{\ell})}. \tag{13.29}$$

The set $\{s+1, \ldots, d\}$ in (13.28) *is to be interpreted as $\emptyset$ if $s = d$, i.e., in that case, the summation is considered only over the empty set.*

Proof Formally, we put $S_{N,0,\alpha,\boldsymbol{\gamma}} := 0$. Note that we can write, for $d \in \mathbb{N}$

$$\begin{aligned} [S_{N,d,\alpha,\boldsymbol{\gamma}}(\boldsymbol{g})]^2 &= [S_{N,d,\alpha,\{\gamma_{\mathfrak{u}}\}_{\mathfrak{u}\subseteq[d]}}((g_1, \ldots, g_d))]^2 \\ &= \sum_{\boldsymbol{h}\in\mathbb{Z}^d} \sum_{\substack{\boldsymbol{\ell}\in\mathbb{Z}^d\setminus\{\boldsymbol{0}\} \\ \boldsymbol{\ell}\cdot\boldsymbol{g}\equiv 0 \pmod N}} \frac{\gamma_{\mathfrak{u}(\boldsymbol{h})}}{r_{2\alpha}(\boldsymbol{h})} \frac{\gamma_{\mathfrak{u}(\boldsymbol{h}+\boldsymbol{\ell})}}{r_{2\alpha}(\boldsymbol{h}+\boldsymbol{\ell})}. \end{aligned}$$

Now we can split up the double sum in the latter expression according to three cases depending on $\boldsymbol{h} = (h_1, \ldots, h_d)$ and $\boldsymbol{\ell} = (\ell_1, \ldots, \ell_d)$, namely when $h_d = \ell_d = 0$, when $\ell_d = 0$ but $h_d \neq 0$, and when $\ell_d \neq 0$. This yields

$$[S_{N,d,\alpha,\boldsymbol{\gamma}}(\boldsymbol{g})]^2 = \sum_{\boldsymbol{h}\in\mathbb{Z}^{d-1}} \sum_{\substack{\boldsymbol{\ell}\in\mathbb{Z}^{d-1}\setminus\{\boldsymbol{0}\} \\ \boldsymbol{\ell}\cdot(g_1,\ldots,g_{d-1})\equiv 0 \pmod N}} \frac{\gamma_{\mathrm{u}(\boldsymbol{h})}}{r_{2\alpha}(\boldsymbol{h})} \frac{\gamma_{\mathrm{u}(\boldsymbol{h}+\boldsymbol{\ell})}}{r_{2\alpha}(\boldsymbol{h}+\boldsymbol{\ell})}$$
$$+ \sum_{h_d\in\mathbb{Z}\setminus\{0\}} \frac{1}{|h_d|^{4\alpha}} \sum_{\boldsymbol{h}\in\mathbb{Z}^{d-1}} \sum_{\substack{\boldsymbol{\ell}\in\mathbb{Z}^{d-1}\setminus\{\boldsymbol{0}\} \\ \boldsymbol{\ell}\cdot(g_1,\ldots,g_{d-1})\equiv 0 \pmod N}} \frac{\gamma_{\mathrm{u}(\boldsymbol{h})\cup\{d\}}}{r_{2\alpha}(\boldsymbol{h})} \frac{\gamma_{\mathrm{u}(\boldsymbol{h}+\boldsymbol{\ell})\cup\{d\}}}{r_{2\alpha}(\boldsymbol{h}+\boldsymbol{\ell})}$$
$$+ \sum_{\boldsymbol{h}\in\mathbb{Z}^{d}} \sum_{\substack{\boldsymbol{\ell}\in\mathbb{Z}^{d} \\ \ell_d\neq 0 \\ \boldsymbol{\ell}\cdot(g_1,\ldots,g_{d})\equiv 0 \pmod N}} \frac{\gamma_{\mathrm{u}(\boldsymbol{h})}}{r_{2\alpha}(\boldsymbol{h})} \frac{\gamma_{\mathrm{u}(\boldsymbol{h}+\boldsymbol{\ell})}}{r_{2\alpha}(\boldsymbol{h}+\boldsymbol{\ell})}.$$

Adapting notation slightly, we can therefore write

$$\begin{aligned}[S_{N,d,\alpha,\{\gamma_{\mathrm{u}}\}_{\mathrm{u}\subseteq[d]}}(\boldsymbol{g})]^2 &= [S_{N,d-1,\alpha,\{\gamma_{\mathrm{u}}\}_{\mathrm{u}\subseteq[d-1]}}((g_1,\ldots,g_{d-1}))]^2 \\ &\quad +2\zeta(4\alpha)[S_{N,d-1,\alpha,\{\gamma_{\mathrm{u}\cup\{d\}}\}_{\mathrm{u}\subseteq[d-1]}}((g_1,\ldots,g_{d-1}))]^2 \\ &\quad +\widetilde{T}_{N,d,\alpha}(g_1,\ldots,g_d,\{\gamma_{\mathrm{u}}\}_{\mathrm{u}\subseteq[d]}).\end{aligned}$$

For a moment, we can omit the arguments $\boldsymbol{g}$ and $g_1,\ldots,g_d$, respectively, from the latter recursion for simplicity, and continue with

$$\begin{aligned}&[S_{N,d,\alpha,\{\gamma_{\mathrm{u}}\}_{\mathrm{u}\subseteq[d]}}]^2 \\ &= [S_{N,d-1,\alpha,\{\gamma_{\mathrm{u}}\}_{\mathrm{u}\subseteq[d-1]}}]^2 + 2\zeta(4\alpha)[S_{N,d-1,\alpha,\{\gamma_{\mathrm{u}\cup\{d\}}\}_{\mathrm{u}\subseteq[d-1]}}]^2 + \widetilde{T}_{N,d,\alpha}(\{\gamma_{\mathrm{u}}\}_{\mathrm{u}\subseteq[d]}) \\ &= [S_{N,d-2,\alpha,\{\gamma_{\mathrm{u}}\}_{\mathrm{u}\subseteq[d-2]}}]^2 + 2\zeta(4\alpha)[S_{N,d-2,\alpha,\{\gamma_{\mathrm{u}\cup\{d-1\}}\}_{\mathrm{u}\subseteq[d-2]}}]^2 \\ &\quad +\widetilde{T}_{N,d-1,\alpha}(\{\gamma_{\mathrm{u}}\}_{\mathrm{u}\subseteq[d-1]}) \\ &\quad +2\zeta(4\alpha)[S_{N,d-2,\alpha,\{\gamma_{\mathrm{u}\cup\{d\}}\}_{\mathrm{u}\subseteq[d-2]}}]^2 + (2\zeta(4\alpha))^2[S_{N,d-2,\alpha,\{\gamma_{\mathrm{u}\cup\{d-1\}\cup\{d\}}\}_{\mathrm{u}\subseteq[d-2]}}]^2 \\ &\quad +2\zeta(4\alpha)\widetilde{T}_{N,d-1,\alpha}(\{\gamma_{\mathrm{u}\cup\{d\}}\}_{\mathrm{u}\subseteq[d-1]}) + \widetilde{T}_{N,d,\alpha}(\{\gamma_{\mathrm{u}}\}_{\mathrm{u}\subseteq[d]}).\end{aligned}$$

We can continue this procedure repeatedly, until finally reaching the point where we can use $S_{N,0,\alpha,\boldsymbol{\gamma}} = 0$, which yields (13.27). □

We are ready to formulate the CBC construction for L_2-approximation in the weighted Korobov space with general weights.

Algorithm 13.14 (CBC construction for L_2-approximation in the weighted Korobov space with general weights) Let prime N and $d \in \mathbb{N}$ be given, let $\alpha > 1/2$, and let $\boldsymbol{\gamma} = \{\gamma_{\mathrm{u}}\}_{\mathrm{u}\subseteq[d]}$ be given (nonnegative) weights. Construct a generating vector $\boldsymbol{g} = (g_1,\ldots,g_d) \in G_d^{\varphi}(N)$ as follows.

(1) Choose g_1 as

$$g_1 := \underset{g\in G_1^{\varphi}(N)}{\operatorname{argmin}}\ T_{N,d,1,\alpha,\boldsymbol{\gamma}}(g_1).$$

(2) For s from 2 to d:
Assume that $g_1,\ldots,g_{s-1} \in G_1^{\varphi}(N)$ have already been found. Choose $g_s \in G_1^{\varphi}(N)$ as

$$g_s := \operatorname*{argmin}_{g \in G_1^{\varphi}(N)} T_{N,d,s,\alpha,\boldsymbol{\gamma}}(g_1, \ldots, g_{s-1}, g).$$

End for.

(3) Set $\boldsymbol{g} = (g_1, \ldots, g_d)$.

Remark 13.15 It should be noted that, even though Algorithm 13.14 can in principle be run for any $d \in \mathbb{N}$, it is as such not extensible in d. This is due to the fact that we optimize $T_{N,d,s,\alpha,\boldsymbol{\gamma}}$, and to define this quantity, d needs to be fixed beforehand.

Next, we show the following theorem which indicates that the generating vectors constructed by Algorithm 13.14 yield a low value of $S_{N,d,\alpha,\boldsymbol{\gamma}}$ for general weights $\boldsymbol{\gamma}$.

Theorem 13.16 *Let $N \geq 3$ be a prime number, let $\alpha > 1/2$, let $\boldsymbol{\gamma} = \{\gamma_{\mathfrak{u}}\}_{\mathfrak{u} \subseteq [d]}$ be given (nonnegative) weights, and assume that $\boldsymbol{g}$ has been found by Algorithm 13.14. Then for arbitrary $\tau \in [1/2, \alpha)$ we have*

$$S_{N,d,\alpha,\boldsymbol{\gamma}}(\boldsymbol{g}) \leq \frac{C_{\alpha,\tau}}{N^{\tau}} \left(\sum_{\emptyset \neq \mathfrak{v} \subseteq [d]} |\mathfrak{v}| \, \gamma_{\mathfrak{v}}^{1/(2\tau)} \left(2\zeta\left(\frac{\alpha}{\tau}\right)\right)^{|\mathfrak{v}|} \right)^{\tau} \times \left(\sum_{\mathfrak{u} \subseteq [d]} \gamma_{\mathfrak{u}}^{1/(2\tau)} \left(2\zeta\left(\frac{\alpha}{\tau}\right)\right)^{|\mathfrak{u}|} \right)^{\tau},$$

where $C_{\alpha,\tau} = \left(17/2 + 2^{2\alpha/\tau+1}\right)^{\tau}$.

Proof Suppose that $\boldsymbol{g} = (g_1, \ldots, g_d)$ has been constructed according to Algorithm 13.14. Then, Equation (13.27) implies that we can express $[S_{N,d,\alpha,\boldsymbol{\gamma}}(\boldsymbol{g})]^2$ in terms of the quantities $T_{N,d,s,\alpha,\boldsymbol{\gamma}}(g_1, \ldots, g_s)$ for $s \in [d]$. By the formulation of the algorithm, for $s \in [d]$, the s-th component g_s of $\boldsymbol{g}$ has been chosen in a greedy fashion to minimize $T_{N,d,s,\alpha,\boldsymbol{\gamma}}(g_1, \ldots, g_{s-1}, g)$ as a function of $g \in G_1^{\varphi}(N)$, with $T_{N,d,s,\alpha,\boldsymbol{\gamma}}$ defined as in (13.28).

Let $\lambda \in (1/(2\alpha), 1]$. We again use the standard averaging argument (see Remark 2.14) that the choice of g as the minimizer must yield a value that is at least as good as the average, and also use Jensen's inequality (see Lemma 2.25) to get

$$\begin{aligned}
&[T_{N,d,s,\alpha,\boldsymbol{\gamma}}(g_1, \ldots, g_{s-1}, g_s)]^{\lambda} \leq \frac{1}{N-1} \sum_{g=1}^{N-1} [T_{N,d,s,\alpha,\boldsymbol{\gamma}}(g_1, \ldots, g_{s-1}, g)]^{\lambda} \\
&= \frac{1}{N-1} \sum_{g=1}^{N-1} \left(\sum_{\mathfrak{w} \subseteq \{s+1,\ldots,d\}} (2\zeta(4\alpha))^{|\mathfrak{w}|} \, \widetilde{T}_{N,s,\alpha}(g_1, \ldots, g_{s-1}, g, \{\gamma_{\mathfrak{u} \cup \mathfrak{w}}\}_{\mathfrak{u} \subseteq [s]}) \right)^{\lambda} \\
&\leq \frac{1}{N-1} \sum_{g=1}^{N-1} \sum_{\mathfrak{w} \subseteq \{s+1,\ldots,d\}} (2\zeta(4\alpha))^{\lambda|\mathfrak{w}|} \left[\widetilde{T}_{N,s,\alpha}(g_1, \ldots, g_{s-1}, g, \{\gamma_{\mathfrak{u} \cup \mathfrak{w}}\}_{\mathfrak{u} \subseteq [s]}) \right]^{\lambda} \\
&= \sum_{\mathfrak{w} \subseteq \{s+1,\ldots,d\}} (2\zeta(4\alpha))^{\lambda|\mathfrak{w}|} \frac{1}{N-1} \sum_{g=1}^{N-1} \left[\widetilde{T}_{N,s,\alpha}(g_1, \ldots, g_{s-1}, g, \{\gamma_{\mathfrak{u} \cup \mathfrak{w}}\}_{\mathfrak{u} \subseteq [s]}) \right]^{\lambda}.
\end{aligned}$$

(13.30)

Consider now an arbitrarily chosen but fixed $\mathfrak{w} \subseteq \{s+1, \ldots, d\}$. Writing $\boldsymbol{g}^{(s-1)} := (g_1, \ldots, g_{s-1})$ and using Jensen's inequality again, we have

$$
\begin{aligned}
&\Theta_{N,s,\alpha}(\boldsymbol{g}^{(s-1)}) \\
&:= \frac{1}{N-1} \sum_{g=1}^{N-1} \left[\widetilde{T}_{N,s+1,\alpha}(g_1, \ldots, g_{s-1}, g, \{\gamma_{\mathfrak{u}\cup\mathfrak{w}}\}_{\mathfrak{u}\subseteq[s]}) \right]^{\lambda} \\
&\le \frac{1}{N-1} \sum_{g=1}^{N-1} \sum_{\boldsymbol{h}\in\mathbb{Z}^s} \sum_{\substack{\boldsymbol{\ell}\in\mathbb{Z}^s \\ \ell_s\neq 0 \\ (\ell_1,\ldots,\ell_{s-1})\cdot\boldsymbol{g}^{(s-1)} \equiv -\ell_s g \pmod N}} \left(\frac{\gamma_{\mathfrak{u}(\boldsymbol{h})\cup\mathfrak{w}}}{r_{2\alpha}(\boldsymbol{h})} \frac{\gamma_{\mathfrak{u}(\boldsymbol{h}+\boldsymbol{\ell})\cup\mathfrak{w}}}{r_{2\alpha}(\boldsymbol{h}+\boldsymbol{\ell})} \right)^{\lambda} \\
&= \frac{1}{N-1} \sum_{\boldsymbol{h}\in\mathbb{Z}^s} \sum_{\substack{\boldsymbol{\ell}\in\mathbb{Z}^s \\ \ell_s \not\equiv 0 \pmod N \\ (\ell_1,\ldots,\ell_{s-1})\cdot\boldsymbol{g}^{(s-1)} \not\equiv 0 \pmod N}} \left(\frac{\gamma_{\mathfrak{u}(\boldsymbol{h})\cup\mathfrak{w}}}{r_{2\alpha}(\boldsymbol{h})} \frac{\gamma_{\mathfrak{u}(\boldsymbol{h}+\boldsymbol{\ell})\cup\mathfrak{w}}}{r_{2\alpha}(\boldsymbol{h}+\boldsymbol{\ell})} \right)^{\lambda} \\
&\quad + \sum_{\boldsymbol{h}\in\mathbb{Z}^s} \sum_{\substack{\boldsymbol{\ell}\in\mathbb{Z}^s \\ \ell_s\neq 0,\ \ell_s\equiv 0 \pmod N \\ (\ell_1,\ldots,\ell_{s-1})\cdot\boldsymbol{g}^{(s-1)} \equiv 0 \pmod N}} \left(\frac{\gamma_{\mathfrak{u}(\boldsymbol{h})\cup\mathfrak{w}}}{r_{2\alpha}(\boldsymbol{h})} \frac{\gamma_{\mathfrak{u}(\boldsymbol{h}+\boldsymbol{\ell})\cup\mathfrak{w}}}{r_{2\alpha}(\boldsymbol{h}+\boldsymbol{\ell})} \right)^{\lambda} \\
&\le \frac{1}{N-1} \sum_{\boldsymbol{h}\in\mathbb{Z}^s} \sum_{\substack{\boldsymbol{\ell}\in\mathbb{Z}^s \\ \ell_s\neq 0}} \left(\frac{\gamma_{\mathfrak{u}(\boldsymbol{h})\cup\mathfrak{w}}}{r_{2\alpha}(\boldsymbol{h})} \frac{\gamma_{\mathfrak{u}(\boldsymbol{h}+\boldsymbol{\ell})\cup\mathfrak{w}}}{r_{2\alpha}(\boldsymbol{h}+\boldsymbol{\ell})} \right)^{\lambda} \\
&\quad + \sum_{\boldsymbol{h}\in\mathbb{Z}^s} \sum_{\substack{\boldsymbol{\ell}\in\mathbb{Z}^s \\ \ell_s\neq 0,\ \ell_s\equiv 0 \pmod N}} \left(\frac{\gamma_{\mathfrak{u}(\boldsymbol{h})\cup\mathfrak{w}}}{r_{2\alpha}(\boldsymbol{h})} \frac{\gamma_{\mathfrak{u}(\boldsymbol{h}+\boldsymbol{\ell})\cup\mathfrak{w}}}{r_{2\alpha}(\boldsymbol{h}+\boldsymbol{\ell})} \right)^{\lambda}.
\end{aligned}
$$

To obtain the penultimate relation, we distinguished the cases where ℓ_s is a multiple of N and where this is not the case, and used the fact that if $\ell_s \not\equiv 0 \pmod N$, then $\ell_s g \pmod N$ runs through the whole set $G_1^{\varphi}(N)$ when g does. To obtain the last inequality, we simply dropped some of the conditions on the indices.

For short, we write

$$
B(h_s, \ell_s) := \sum_{\boldsymbol{h}\in\mathbb{Z}^{s-1}} \sum_{\boldsymbol{\ell}\in\mathbb{Z}^{s-1}} \left(\frac{\gamma_{\mathfrak{u}((\boldsymbol{h},h_s))\cup\mathfrak{w}}}{r_{2\alpha}((\boldsymbol{h},h_s))} \frac{\gamma_{\mathfrak{u}((\boldsymbol{h},h_s)+(\boldsymbol{\ell},\ell_s))\cup\mathfrak{w}}}{r_{2\alpha}((\boldsymbol{h},h_s)+(\boldsymbol{\ell},\ell_s))} \right)^{\lambda}
$$

for $h_s, \ell_s \in \mathbb{Z}$. This yields

$$
\begin{aligned}
&\Theta_{N,s,\alpha}(\boldsymbol{g}^{(s-1)}) \\
&\le \frac{1}{N-1} \sum_{h_s\in\mathbb{Z}} \sum_{\ell_s\in\mathbb{Z}\setminus\{0\}} B(h_s,\ell_s) + \sum_{h_s\in\mathbb{Z}} \sum_{\substack{\ell_s\in\mathbb{Z}\setminus\{0\} \\ \ell_s\equiv 0 \pmod N}} B(h_s,\ell_s)
\end{aligned}
$$

$$= \frac{1}{N-1} \sum_{h_s \in \mathbb{Z}} \sum_{\ell_s \in \mathbb{Z}\setminus\{0\}} B(h_s, \ell_s) \\ + \sum_{\substack{h_s \in \mathbb{Z} \\ h_s \equiv 0 \pmod N}} \sum_{\substack{\ell_s \in \mathbb{Z}\setminus\{0\} \\ \ell_s \equiv 0 \pmod N}} B(h_s, \ell_s) + \sum_{\substack{h_s \in \mathbb{Z} \\ h_s \not\equiv 0 \pmod N}} \sum_{\substack{\ell_s \in \mathbb{Z}\setminus\{0\} \\ \ell_s \equiv 0 \pmod N}} B(h_s, \ell_s).$$

As a further abbreviation, we put

$$\Xi_1 := \sum_{h_s \in \mathbb{Z}} \sum_{\ell_s \in \mathbb{Z}\setminus\{0\}} B(h_s, \ell_s),$$

$$\Xi_2 := \sum_{\substack{h_s \in \mathbb{Z} \\ h_s \equiv 0 \pmod N}} \sum_{\substack{\ell_s \in \mathbb{Z}\setminus\{0\} \\ \ell_s \equiv 0 \pmod N}} B(h_s, \ell_s),$$

$$\Xi_3 := \sum_{\substack{h_s \in \mathbb{Z} \\ h_s \not\equiv 0 \pmod N}} \sum_{\substack{\ell_s \in \mathbb{Z}\setminus\{0\} \\ \ell_s \equiv 0 \pmod N}} B(h_s, \ell_s).$$

Putting $\boldsymbol{q} = \boldsymbol{h} + \boldsymbol{\ell}$ for $\boldsymbol{h}, \boldsymbol{\ell} \in \mathbb{Z}^{s-1}$, we can write

$$B(h_s, \ell_s) = \sum_{\boldsymbol{h} \in \mathbb{Z}^{s-1}} \frac{\gamma^{\lambda}_{\mathfrak{u}((\boldsymbol{h}, h_s)) \cup \mathfrak{w}}}{(r_{2\alpha}((\boldsymbol{h}, h_s)))^{\lambda}} \sum_{\boldsymbol{q} \in \mathbb{Z}^{s-1}} \frac{\gamma^{\lambda}_{\mathfrak{u}((\boldsymbol{q}, h_s + \ell_s)) \cup \mathfrak{w}}}{(r_{2\alpha}((\boldsymbol{q}, h_s + \ell_s)))^{\lambda}}.$$

Using this representation, we have, for $h_s = 0$ and $\ell_s \in \mathbb{Z} \setminus \{0\}$,

$$\begin{aligned} B(0, \ell_s) &= \sum_{\mathfrak{u} \subseteq [s-1]} \sum_{\boldsymbol{h} \in (\mathbb{Z}\setminus\{0\})^{|\mathfrak{u}|}} \frac{\gamma^{\lambda}_{\mathfrak{u} \cup \mathfrak{w}}}{(r_{2\alpha}(\boldsymbol{h}))^{\lambda}} \sum_{\mathfrak{v} \subseteq [s-1]} \sum_{\boldsymbol{q} \in (\mathbb{Z}\setminus\{0\})^{|\mathfrak{v}|}} \frac{\gamma^{\lambda}_{\mathfrak{v} \cup \{s\} \cup \mathfrak{w}}}{(r_{2\alpha}(\boldsymbol{q}))^{\lambda}} \frac{1}{|\ell_s|^{2\alpha\lambda}} \\ &= \frac{1}{|\ell_s|^{2\alpha\lambda}} \sum_{\mathfrak{u} \subseteq [s-1]} \gamma^{\lambda}_{\mathfrak{u} \cup \mathfrak{w}} (2\zeta(2\alpha\lambda))^{|\mathfrak{u}|} \sum_{\mathfrak{v} \subseteq [s-1]} \gamma^{\lambda}_{\mathfrak{v} \cup \{s\} \cup \mathfrak{w}} (2\zeta(2\alpha\lambda))^{|\mathfrak{v}|} \\ &= \frac{\mathcal{G} \cdot Q}{|\ell_s|^{2\alpha\lambda}}, \end{aligned}$$

where we use the abbreviations

$$\mathcal{G} := \sum_{\mathfrak{u} \subseteq [s-1]} \gamma^{\lambda}_{\mathfrak{u} \cup \mathfrak{w}} (2\zeta(2\alpha\lambda))^{|\mathfrak{u}|} \quad \text{and} \quad Q := \sum_{\mathfrak{v} \subseteq [s-1]} \gamma^{\lambda}_{\mathfrak{v} \cup \{s\} \cup \mathfrak{w}} (2\zeta(2\alpha\lambda))^{|\mathfrak{v}|}.$$

Analogously, we obtain

$$B(h_s,\ell_s)=\begin{cases}\dfrac{\mathcal{G}\cdot Q}{|h_s|^{2\alpha\lambda}} & \text{if } h_s\neq 0 \text{ and } \ell_s=-h_s,\\[2ex] \dfrac{Q^2}{|h_s|^{2\alpha\lambda}\,|h_s+\ell_s|^{2\alpha\lambda}} & \text{if } h_s\neq 0 \text{ and } \ell_s\neq -h_s.\end{cases}$$

Using these observations, we get

$$\begin{aligned}\Xi_1 &= \sum_{\ell_s\in\mathbb{Z}\setminus\{0\}}\frac{\mathcal{G}\cdot Q}{|\ell_s|^{2\alpha\lambda}}+\sum_{h_s\in\mathbb{Z}\setminus\{0\}}\frac{\mathcal{G}\cdot Q}{|h_s|^{2\alpha\lambda}}\\ &\quad+\sum_{h_s\in\mathbb{Z}\setminus\{0\}}\sum_{\ell_s\in\mathbb{Z}\setminus\{0,-h_s\}}\frac{Q^2}{|h_s|^{2\alpha\lambda}\,|h_s+\ell_s|^{2\alpha\lambda}}\\ &\le 4\zeta(2\alpha\lambda)\,\mathcal{G}\cdot Q+\sum_{h_s\in\mathbb{Z}\setminus\{0\}}\sum_{q\in\mathbb{Z}\setminus\{0\}}\frac{Q^2}{|h_s|^{2\alpha\lambda}\,|q|^{2\alpha\lambda}}\\ &= 4\zeta(2\alpha\lambda)\,\mathcal{G}\cdot Q+(2\zeta(2\alpha\lambda))^2Q^2.\end{aligned}$$

Similarly, we can derive

$$\Xi_2\le\frac{4\zeta(2\alpha\lambda)\,\mathcal{G}\cdot Q}{N^{2\alpha\lambda}}+\frac{(2\zeta(2\alpha\lambda))^2\,Q^2}{N^{4\alpha\lambda}}\le\frac{4\zeta(2\alpha\lambda)\,\mathcal{G}\cdot Q+2(\zeta(2\alpha\lambda))^2\,Q^2}{N},$$

where we used that $N^{-2\alpha\lambda}\le N^{-1}$ and $N^{-4\alpha\lambda}\le(2N)^{-1}$. Finally,

$$\begin{aligned}\Xi_3 &= \sum_{\substack{h_s\in\mathbb{Z}\\ h_s\not\equiv 0\ (\mathrm{mod}\ N)}}\ \sum_{\ell_s\in\mathbb{Z}\setminus\{0\}}\frac{Q^2}{|h_s|^{2\alpha\lambda}\,|h_s+\ell_sN|^{2\alpha\lambda}}\\ &= Q^2\sum_{\substack{h_s\in\mathbb{Z}\\ h_s\not\equiv 0\ (\mathrm{mod}\ N)}}\left(\frac{1}{|h_s|^{2\alpha\lambda}}\sum_{\ell_s\in\mathbb{Z}}\frac{1}{|h_s+\ell_sN|^{2\alpha\lambda}}-\frac{1}{|h_s|^{4\alpha\lambda}}\right).\end{aligned}$$

The rest of the analysis of Ξ_3 works in a way that is very analogous to the derivation of the bound on $[S_{N,1,\alpha,\gamma_1}(1)]^2$ in the proof of Theorem 13.5 (or see [28]), and eventually yields

$$\Xi_3\le\frac{2^{4\alpha\lambda+1}(2\zeta(2\alpha\lambda))^2\,Q^2}{N^{2\alpha\lambda}}\le\frac{2^{4\alpha\lambda+1}(2\zeta(2\alpha\lambda))^2\,Q^2}{N},$$

where we again used $N^{-2\alpha\lambda}\le N^{-1}$.

Combining the estimates on Ξ_1,Ξ_2, and Ξ_3, and noting that $1/(N-1)\le 2/N$ gives

$$\begin{aligned}\Theta_{N,s,\alpha}(\boldsymbol{g}^{(s-1)}) &\le \frac{8\zeta(2\alpha\lambda)\,\mathcal{G}\cdot Q+2(2\zeta(2\alpha\lambda))^2\,Q^2}{N}\\ &\quad+\frac{4\zeta(2\alpha\lambda)\,\mathcal{G}\cdot Q+2(\zeta(2\alpha\lambda))^2\,Q^2}{N}+\frac{2^{4\alpha\lambda+1}(2\zeta(2\alpha\lambda))^2\,Q^2}{N}\\ &=\frac{6(2\zeta(2\alpha\lambda))\,\mathcal{G}\cdot Q+(5/2+2^{4\alpha\lambda+1})(2\zeta(2\alpha\lambda))^2\,Q^2}{N}\\ &\le\frac{(17/2+2^{4\alpha\lambda+1})2\zeta(2\alpha\lambda)\,Q\,(\mathcal{G}+2\zeta(2\alpha\lambda)Q)}{N}.\end{aligned}$$

Inserting back for $\mathcal{G}$ and Q and writing $\widetilde{C}_{\alpha,\lambda} := 17/2+2^{4\alpha\lambda+1}$, we obtain

$$\begin{aligned}\Theta_{N,s,\alpha}(\boldsymbol{g}^{(s-1)}) &\le \frac{\widetilde{C}_{\alpha,\lambda}}{N}\left(2\zeta(2\alpha\lambda)\sum_{\mathfrak{v}\subseteq[s-1]}\gamma^{\lambda}_{\mathfrak{v}\cup\{s\}\cup\mathfrak{w}}(2\zeta(2\alpha\lambda))^{|\mathfrak{v}|}\right)\\ &\quad\times\left(\sum_{\mathfrak{u}\subseteq[s-1]}\gamma^{\lambda}_{\mathfrak{u}\cup\mathfrak{w}}(2\zeta(2\alpha\lambda))^{|\mathfrak{u}|}+2\zeta(2\alpha\lambda)\sum_{\mathfrak{v}\subseteq[s-1]}\gamma^{\lambda}_{\mathfrak{v}\cup\{s\}\cup\mathfrak{w}}(2\zeta(2\alpha\lambda))^{|\mathfrak{v}|}\right)\\ &=\frac{\widetilde{C}_{\alpha,\lambda}}{N}\left(\sum_{s\in\mathfrak{v}\subseteq[s]}\gamma^{\lambda}_{\mathfrak{v}\cup\mathfrak{w}}(2\zeta(2\alpha\lambda))^{|\mathfrak{v}|}\right)\\ &\quad\times\left(\sum_{\mathfrak{u}\subseteq[s-1]}\gamma^{\lambda}_{\mathfrak{u}\cup\mathfrak{w}}(2\zeta(2\alpha\lambda))^{|\mathfrak{u}|}+\sum_{s\in\mathfrak{v}\subseteq[s]}\gamma^{\lambda}_{\mathfrak{v}\cup\mathfrak{w}}(2\zeta(2\alpha\lambda))^{|\mathfrak{v}|}\right)\\ &\le\frac{\widetilde{C}_{\alpha,\lambda}}{N}\left(\sum_{s\in\mathfrak{v}\subseteq[s]}\gamma^{\lambda}_{\mathfrak{v}\cup\mathfrak{w}}(2\zeta(2\alpha\lambda))^{|\mathfrak{v}|}\right)\left(\sum_{\mathfrak{u}\subseteq[s]}\gamma^{\lambda}_{\mathfrak{u}\cup\mathfrak{w}}(2\zeta(2\alpha\lambda))^{|\mathfrak{u}|}\right).\end{aligned}$$

We can now insert this estimate into (13.30) to obtain

$$\begin{aligned}[T_{N,d,s,\alpha,\boldsymbol{\gamma}}(g_1,\ldots,g_{s-1},g_s)]^{\lambda} &\le \sum_{\mathfrak{w}\subseteq\{s+1,\ldots,d\}}(2\zeta(4\alpha))^{\lambda|\mathfrak{w}|}\\ &\quad\times\frac{\widetilde{C}_{\alpha,\lambda}}{N}\left(\sum_{s\in\mathfrak{v}\subseteq[s]}\gamma^{\lambda}_{\mathfrak{v}\cup\mathfrak{w}}(2\zeta(2\alpha\lambda))^{|\mathfrak{v}|}\right)\left(\sum_{\mathfrak{u}\subseteq[s]}\gamma^{\lambda}_{\mathfrak{u}\cup\mathfrak{w}}(2\zeta(2\alpha\lambda))^{|\mathfrak{u}|}\right)\\ &\le\frac{\widetilde{C}_{\alpha,\lambda}}{N}\sum_{\mathfrak{w}\subseteq\{s+1,\ldots,d\}}\left(\sum_{s\in\mathfrak{v}\subseteq[s]}\gamma^{\lambda}_{\mathfrak{v}\cup\mathfrak{w}}(2\zeta(2\alpha\lambda))^{|\mathfrak{v}\cup\mathfrak{w}|}\right)\\ &\quad\times\left(\sum_{\mathfrak{u}\subseteq[s]}\gamma^{\lambda}_{\mathfrak{u}\cup\mathfrak{w}}(2\zeta(2\alpha\lambda))^{|\mathfrak{u}\cup\mathfrak{w}|}\right)\end{aligned}$$

$$
\begin{aligned}
&\le \frac{\widetilde{C}_{\alpha,\lambda}}{N}\left(\max_{\mathfrak{w}\subseteq\{s+1,\dots,d\}}\sum_{s\in\mathfrak{v}\subseteq[s]}\gamma_{\mathfrak{v}\cup\mathfrak{w}}^{\lambda}(2\zeta(2\alpha\lambda))^{|\mathfrak{v}\cup\mathfrak{w}|}\right)\\
&\qquad\times\left(\sum_{\mathfrak{w}\subseteq\{s+1,\dots,d\}}\sum_{\mathfrak{u}\subseteq[s]}\gamma_{\mathfrak{u}\cup\mathfrak{w}}^{\lambda}(2\zeta(2\alpha\lambda))^{|\mathfrak{u}\cup\mathfrak{w}|}\right)\\
&= \frac{\widetilde{C}_{\alpha,\lambda}}{N}\left(\max_{\mathfrak{w}\subseteq\{s+1,\dots,d\}}\sum_{s\in\mathfrak{v}\subseteq[s]}\gamma_{\mathfrak{v}\cup\mathfrak{w}}^{\lambda}(2\zeta(2\alpha\lambda))^{|\mathfrak{v}\cup\mathfrak{w}|}\right)\left(\sum_{\mathfrak{u}\subseteq[d]}\gamma_{\mathfrak{u}}^{\lambda}(2\zeta(2\alpha\lambda))^{|\mathfrak{u}|}\right).
\end{aligned}
$$

Now, using (13.27) and applying the previous estimate to all $\mathfrak{w}\subseteq\{s+1,\dots,d\}$, we obtain

$$
\begin{aligned}
[S_{N,d,\alpha,\boldsymbol{\gamma}}(\boldsymbol{g})]^2 &\le \sum_{s=1}^{d}\left(\frac{\widetilde{C}_{\alpha,\lambda}}{N}\left(\max_{\mathfrak{w}\subseteq\{s+1,\dots,d\}}\sum_{s\in\mathfrak{v}\subseteq[s]}\gamma_{\mathfrak{v}\cup\mathfrak{w}}^{\lambda}(2\zeta(2\alpha\lambda))^{|\mathfrak{v}\cup\mathfrak{w}|}\right)\right.\\
&\qquad\left.\times\left(\sum_{\mathfrak{u}\subseteq[d]}\gamma_{\mathfrak{u}}^{\lambda}(2\zeta(2\alpha\lambda))^{|\mathfrak{u}|}\right)\right)^{1/\lambda}\\
&\le \left(\frac{\widetilde{C}_{\alpha,\lambda}}{N}\left(\sum_{s=1}^{d}\max_{\mathfrak{w}\subseteq\{s+1,\dots,d\}}\sum_{s\in\mathfrak{v}\subseteq[s]}\gamma_{\mathfrak{v}\cup\mathfrak{w}}^{\lambda}(2\zeta(2\alpha\lambda))^{|\mathfrak{v}\cup\mathfrak{w}|}\right)\right.\\
&\qquad\left.\times\left(\sum_{\mathfrak{u}\subseteq[d]}\gamma_{\mathfrak{u}}^{\lambda}(2\zeta(2\alpha\lambda))^{|\mathfrak{u}|}\right)\right)^{1/\lambda}\\
&\le \left(\frac{\widetilde{C}_{\alpha,\lambda}}{N}\left(\sum_{s=1}^{d}\sum_{\mathfrak{w}\subseteq\{s+1,\dots,d\}}\sum_{s\in\mathfrak{v}\subseteq[s]}\gamma_{\mathfrak{v}\cup\mathfrak{w}}^{\lambda}(2\zeta(2\alpha\lambda))^{|\mathfrak{v}\cup\mathfrak{w}|}\right)\right.\\
&\qquad\left.\times\left(\sum_{\mathfrak{u}\subseteq[d]}\gamma_{\mathfrak{u}}^{\lambda}(2\zeta(2\alpha\lambda))^{|\mathfrak{u}|}\right)\right)^{1/\lambda}\\
&= \left(\frac{\widetilde{C}_{\alpha,\lambda}}{N}\left(\sum_{\emptyset\ne\mathfrak{v}\subseteq[d]}|\mathfrak{v}|\,\gamma_{\mathfrak{v}}^{\lambda}(2\zeta(2\alpha\lambda))^{|\mathfrak{v}|}\right)\left(\sum_{\mathfrak{u}\subseteq[d]}\gamma_{\mathfrak{u}}^{\lambda}(2\zeta(2\alpha\lambda))^{|\mathfrak{u}|}\right)\right)^{1/\lambda},
\end{aligned}
$$

where we used that $\lambda\le 1$ in the second estimate.

The proof is concluded by taking the square root of the latter estimate and setting $\tau = 1/(2\lambda)$, which implies that $\tau\in[1/2,\alpha)$ and $\widetilde{C}_{\alpha,\lambda}^{1/(2\lambda)} = (17/2+2^{2\alpha/\tau+1})^{\tau} = C_{\alpha,\tau}$. □

We can now formulate the following error bound, which can be seen as a generalization of Theorem 13.6.

Theorem 13.17 *Let $N \geq 3$ be a prime number, let $\alpha > 1/2$, and let $\boldsymbol{\gamma} = \{\gamma_{\mathfrak{u}}\}_{\mathfrak{u} \subseteq [d]}$ be general (nonnegative) weights. Assume that $\boldsymbol{g}$ has been found by Algorithm 13.14. Then for arbitrary $\tau \in [1/2, \alpha)$ we have*

$$\mathrm{err}^{L_2\text{-app}}(\mathcal{H}_{\mathrm{kor},d,\alpha,\boldsymbol{\gamma}}, A_{N,M,d,\alpha,\boldsymbol{\gamma}}(\boldsymbol{g})) \leq \frac{C_{d,\alpha,\boldsymbol{\gamma},\tau}}{N^{\tau/2}},$$

where $M = N^{\tau}$, and where

$$C_{d,\alpha,\boldsymbol{\gamma},\tau} = \left(1 + \left(\frac{17}{2} + 2^{2\alpha/\tau+1}\right)^{2\tau}\right)^{1/2} \left(\sum_{\mathfrak{u} \subseteq [d]} (|\mathfrak{u}| + 1)\gamma_{\mathfrak{u}}^{1/(2\tau)} \left(2\zeta\left(\frac{\alpha}{\tau}\right)\right)^{|\mathfrak{u}|}\right)^{2\tau}.$$

Proof We combine the result in Theorem 13.16 with Equation (13.26) to obtain

$$\begin{aligned}
[\mathrm{err}^{L_2\text{-app}}(\mathcal{H}_{\mathrm{kor},d,\alpha,\boldsymbol{\gamma}}, A_{N,M,d,\alpha,\boldsymbol{\gamma}}(\boldsymbol{g}))]^2 &\leq \frac{1}{M} \\
&+ \frac{M\, C_{\alpha,\tau}^2}{N^{2\tau}} \left(\sum_{\emptyset \neq \mathfrak{v} \subseteq [d]} |\mathfrak{v}|\, \gamma_{\mathfrak{v}}^{1/(2\tau)} \left(2\zeta\left(\frac{\alpha}{\tau}\right)\right)^{|\mathfrak{v}|}\right)^{2\tau} \left(\sum_{\mathfrak{u} \subseteq [d]} \gamma_{\mathfrak{u}}^{1/(2\tau)} \left(2\zeta\left(\frac{\alpha}{\tau}\right)\right)^{|\mathfrak{u}|}\right)^{2\tau} \\
&\leq \frac{1}{M} + \frac{M\, C_{\alpha,\tau}^2}{N^{2\tau}} \left(\sum_{\mathfrak{u} \subseteq [d]} (|\mathfrak{u}| + 1)\gamma_{\mathfrak{u}}^{1/(2\tau)} \left(2\zeta\left(\frac{\alpha}{\tau}\right)\right)^{|\mathfrak{u}|}\right)^{4\tau},
\end{aligned}$$

where $C_{\alpha,\tau}$ is defined as in Theorem 13.16. In order to balance the latter bound with respect to M and N, we choose $M = N^{\tau}$, and get

$$\begin{aligned}
&[\mathrm{err}^{L_2\text{-app}}(\mathcal{H}_{\mathrm{kor},d,\alpha,\boldsymbol{\gamma}}, A_{N,M,d,\alpha,\boldsymbol{\gamma}}(\boldsymbol{g}))]^2 \\
&\qquad \leq \frac{1 + C_{\alpha,\tau}^2}{N^{\tau}} \left(\sum_{\mathfrak{u} \subseteq [d]} (|\mathfrak{u}| + 1)\gamma_{\mathfrak{u}}^{1/(2\tau)} \left(2\zeta\left(\frac{\alpha}{\tau}\right)\right)^{|\mathfrak{u}|}\right)^{4\tau}.
\end{aligned}$$

Taking the square root and inserting the expression for $C_{\alpha,\tau}$ yields the result as claimed. □

Remark 13.18 Note that by an appropriate choice of τ we can obtain a convergence rate arbitrarily close to $N^{-\alpha/2}$ in Theorem 13.17. This is in correspondence with the case of product weights in Theorem 13.6.

Tractability of approximation with general weights

From Theorem 13.17, we can deduce a sufficient condition for strong polynomial tractability in the case of general weights, as we demonstrate in the following theorem.

Theorem 13.19 *The L_2-approximation problem for the weighted Korobov space $\mathcal{H}_{\mathrm{kor},d,\alpha,\boldsymbol{\gamma}}$ with general (nonnegative) weights $\boldsymbol{\gamma}$, using information from the class Λ^{std}, is strongly polynomially tractable if*

$$\sum_{\substack{\mathfrak{u}\subseteq\mathbb{N}\\ |\mathfrak{u}|<\infty}} \gamma_{\mathfrak{u}}\,(4\zeta(2\alpha))^{|\mathfrak{u}|} < \infty, \tag{13.31}$$

where the summation is meant to be extended over all finite subsets $\mathfrak{u}$ of $\mathbb{N}$. If

$$\sum_{\substack{\mathfrak{u}\subseteq\mathbb{N}\\ |\mathfrak{u}|<\infty}} \gamma_{\mathfrak{u}}^{1/(2\tau)} \left(4\zeta\left(\frac{\alpha}{\tau}\right)\right)^{|\mathfrak{u}|} < \infty \tag{13.32}$$

for some $\tau\in[1/2,\alpha)$, then we get strong polynomial tractability with an ε-exponent of at most $2/\tau$.

Proof Theorem 13.17 yields the existence of an algorithm $A_{N,d}(\boldsymbol{g})$ such that

$$\mathrm{err}^{L_2-\mathrm{app}}(\mathcal{H}_{\mathrm{kor},d,\alpha,\boldsymbol{\gamma}}, A_{N,d}(\boldsymbol{g})) \le \frac{C_{d,\alpha,\boldsymbol{\gamma},\tau}}{N^{\tau/2}},$$

where

$$\begin{aligned} C_{d,\alpha,\boldsymbol{\gamma},\tau} &= \left(1+\left(\frac{17}{2}+2^{2\alpha/\tau+1}\right)^{2\tau}\right)^{1/2} \left(\sum_{\mathfrak{u}\subseteq[d]} (|\mathfrak{u}|+1)\gamma_{\mathfrak{u}}^{1/(2\tau)} \left(2\zeta\left(\frac{\alpha}{\tau}\right)\right)^{|\mathfrak{u}|}\right)^{4\tau} \\ &\le \left(1+\left(\frac{17}{2}+2^{2\alpha/\tau+1}\right)^{2\tau}\right)^{1/2} \left(\sum_{\mathfrak{u}\subseteq[d]} \gamma_{\mathfrak{u}}^{1/(2\tau)} \left(4\zeta\left(\frac{\alpha}{\tau}\right)\right)^{|\mathfrak{u}|}\right)^{4\tau} \\ &\le \left(1+\left(\frac{17}{2}+2^{2\alpha/\tau+1}\right)^{2\tau}\right)^{1/2} \left(\sum_{\substack{\mathfrak{u}\subseteq\mathbb{N}\\ |\mathfrak{u}|<\infty}} \gamma_{\mathfrak{u}}^{1/(2\tau)} \left(4\zeta\left(\frac{\alpha}{\tau}\right)\right)^{|\mathfrak{u}|}\right)^{4\tau}. \end{aligned}$$

Now, if Condition (13.31) holds, we can bound $C_{d,\alpha,\boldsymbol{\gamma},1/2}$ independently of d, and we get strong polynomial tractability. If Condition (13.32) holds for some $\tau\in[1/2,\alpha)$, $C_{d,\alpha,\boldsymbol{\gamma},\tau}$ can be bounded independently of d, and we get strong polynomial tractability with an ε-exponent of at most $2/\tau$. □

Remark 13.20 Similarly to Theorem 13.11, one could also show a sufficient condition for polynomial tractability, by modifying (13.31) accordingly.

Implementation of the CBC algorithm for POD weights

Finally, let us outline how Algorithm 13.14 can be implemented efficiently for POD weights. The first step in our analysis is to rewrite the error criterion that is used in Algorithm 13.14 in a suitable way. This is done in the following lemma.

Lemma 13.21 *Let $d, N \in \mathbb{N}$, let $\alpha > 1/2$, and let $\boldsymbol{\gamma} = \{\gamma_{\mathfrak{u}}\}_{\mathfrak{u}\subseteq[d]}$ be general (nonnegative) weights. Let $\boldsymbol{g} = (g_1, \ldots, g_d) \in G_d^{\varphi}(N)$. We write $\boldsymbol{g}^{(s)} := (g_1, \ldots, g_s)$ and let $T_{N,d,s,\alpha,\boldsymbol{\gamma}}(\boldsymbol{g}^{(s)})$ be defined as in (13.28) for $s \in [d]$. Then,*

$$
\begin{aligned}
T_{N,d,s,\alpha,\boldsymbol{\gamma}}(\boldsymbol{g}^{(s)}) = & \frac{1}{N}\sum_{t=0}^{N-1} \psi_\alpha\left(\frac{tg_s}{N}\right) V_{N,d,s,\alpha,\boldsymbol{\gamma}}(\boldsymbol{g}^{(s-1)}, t) \\
& + \frac{2}{N}\sum_{t=0}^{N-1} \varphi_\alpha\left(\frac{tg_s}{N}\right) W_{N,d,s,\alpha,\boldsymbol{\gamma}}(\boldsymbol{g}^{(s-1)}, t),
\end{aligned}
$$

where φ_α is defined in (3.7), i.e., for real x

$$
\varphi_\alpha(x) := \sum_{h\in\mathbb{Z}\setminus\{0\}} \frac{\mathrm{e}^{2\pi \mathrm{i} hx}}{|h|^{2\alpha}},
$$

and where

$$
\psi_\alpha(x) := (\varphi_\alpha(x))^2 - 2\zeta(4\alpha),
$$

as well as

$$
\begin{aligned}
& V_{N,d,s,\alpha,\boldsymbol{\gamma}}(\boldsymbol{g}^{(s-1)}, t) \\
& := \sum_{\mathfrak{w}\subseteq\{s+1,\ldots,d\}} (2\zeta(4\alpha))^{|\mathfrak{w}|} \left(\sum_{\mathfrak{u}\subseteq[s-1]} \gamma_{\mathfrak{u}\cup\{s\}\cup\mathfrak{w}} \prod_{j\in\mathfrak{u}} \varphi_\alpha\left(\frac{tg_j}{N}\right)\right)^2,
\end{aligned}
$$

and

$$
\begin{aligned}
& W_{N,d,s,\alpha,\boldsymbol{\gamma}}(\boldsymbol{g}^{(s-1)}, t) \\
& := \sum_{\mathfrak{w}\subseteq\{s+1,\ldots,d\}} (2\zeta(4\alpha))^{|\mathfrak{w}|} \left(\sum_{\mathfrak{u}\subseteq[s-1]} \gamma_{\mathfrak{u}\cup\{s\}\cup\mathfrak{w}} \prod_{j\in\mathfrak{u}} \varphi_\alpha\left(\frac{tg_j}{N}\right)\right) \\
& \qquad \times \left(\sum_{\mathfrak{u}\subseteq[s-1]} \gamma_{\mathfrak{u}\cup\mathfrak{w}} \prod_{j\in\mathfrak{u}} \varphi_\alpha\left(\frac{tg_j}{N}\right)\right).
\end{aligned}
$$

In the previous two definitions, the set $\{s+1, \ldots, d\}$ in (13.28) is again to be interpreted as $\emptyset$ if $s = d$, i.e., in that case, the summation is considered only over the empty set.

Proof We remind the reader of Equations (13.28) and (13.29), and use the notation $(q_1, \ldots, q_s) = \boldsymbol{q} = \boldsymbol{h} + \boldsymbol{\ell}$ for $\boldsymbol{h} = (h_1, \ldots, h_s)$ and $\boldsymbol{\ell} = (\ell_1, \ldots, \ell_s)$ in $\mathbb{Z}^s$ to write $\widetilde{T}_{N,s,\alpha}(\boldsymbol{g}^{(s)}, \{\gamma_{\mathfrak{u}\cup\mathfrak{w}}\}_{\mathfrak{u}\subseteq[s]})$, for fixed $\mathfrak{w} \subseteq \{s+1, \ldots, d\}$, as

$$\begin{aligned}\widetilde{T}_{N,s,\alpha}(\boldsymbol{g}^{(s)}, \{\gamma_{\mathfrak{u}\cup\mathfrak{w}}\}_{\mathfrak{u}\subseteq[s]}) &= \sum_{\boldsymbol{h}\in\mathbb{Z}^s} \sum_{\substack{\boldsymbol{q}\in\mathbb{Z}^s \\ q_s\neq h_s \\ (\boldsymbol{q}-\boldsymbol{h})\cdot\boldsymbol{g}^{(s)}\equiv 0 \ (\mathrm{mod}\ N)}} \frac{\gamma_{\mathfrak{u}(\boldsymbol{h})\cup\mathfrak{w}}}{r_{2\alpha}(\boldsymbol{h})} \frac{\gamma_{\mathfrak{u}(\boldsymbol{q})\cup\mathfrak{w}}}{r_{2\alpha}(\boldsymbol{q})} \\ &= \frac{1}{N} \sum_{t=0}^{N-1} \sum_{\boldsymbol{h}\in\mathbb{Z}^s} \sum_{\substack{\boldsymbol{q}\in\mathbb{Z}^s \\ q_s\neq h_s}} \frac{\gamma_{\mathfrak{u}(\boldsymbol{h})\cup\mathfrak{w}}}{r_{2\alpha}(\boldsymbol{h})} \frac{\gamma_{\mathfrak{u}(\boldsymbol{q})\cup\mathfrak{w}}}{r_{2\alpha}(\boldsymbol{q})} \mathrm{e}^{2\pi\mathrm{i}t(\boldsymbol{q}-\boldsymbol{h})\cdot\boldsymbol{g}^{(s)}/N}.\end{aligned}$$

For fixed $t \in \{0, 1, \ldots, N-1\}$, let us for a moment drop the condition $q_s \neq h_s$ in the inner sum of the latter expression, such that we get

$$\begin{aligned}\Sigma_1 &:= \sum_{\boldsymbol{h}\in\mathbb{Z}^s} \sum_{\boldsymbol{q}\in\mathbb{Z}^s} \frac{\gamma_{\mathfrak{u}(\boldsymbol{h})\cup\mathfrak{w}}}{r_{2\alpha}(\boldsymbol{h})} \frac{\gamma_{\mathfrak{u}(\boldsymbol{q})\cup\mathfrak{w}}}{r_{2\alpha}(\boldsymbol{q})} \mathrm{e}^{2\pi\mathrm{i}t(\boldsymbol{q}-\boldsymbol{h})\cdot\boldsymbol{g}^{(s)}/N} \\ &= \left(\sum_{\boldsymbol{h}\in\mathbb{Z}^s} \gamma_{\mathfrak{u}(\boldsymbol{h})\cup\mathfrak{w}} \frac{\mathrm{e}^{2\pi\mathrm{i}t\boldsymbol{h}\cdot\boldsymbol{g}^{(s)}/N}}{r_{2\alpha}(\boldsymbol{h})} \right)^2 \\ &= \left(\sum_{\mathfrak{v}\subseteq[s]} \sum_{\substack{\boldsymbol{h}\in\mathbb{Z}^s \\ \mathfrak{u}(\boldsymbol{h})=\mathfrak{v}}} \gamma_{\mathfrak{v}\cup\mathfrak{w}} \prod_{j\in\mathfrak{v}} \frac{\mathrm{e}^{2\pi\mathrm{i}th_jg_j/N}}{r_{2\alpha}(h_j)} \right)^2 \\ &= \left(\sum_{\mathfrak{u}\subseteq[s]} \gamma_{\mathfrak{u}\cup\mathfrak{w}} \prod_{j\in\mathfrak{u}} \varphi_\alpha\left(\frac{tg_j}{N}\right) \right)^2 \\ &= \left(\sum_{s\in\mathfrak{u}\subseteq[s]} \gamma_{\mathfrak{u}\cup\mathfrak{w}} \prod_{j\in\mathfrak{u}} \varphi_\alpha\left(\frac{tg_j}{N}\right) + \sum_{s\notin\mathfrak{u}\subseteq[s]} \gamma_{\mathfrak{u}\cup\mathfrak{w}} \prod_{j\in\mathfrak{u}} \varphi_\alpha\left(\frac{tg_j}{N}\right) \right)^2 \\ &= \left(\varphi_\alpha\left(\frac{tg_s}{N}\right) \sum_{\mathfrak{u}\subseteq[s-1]} \gamma_{\mathfrak{u}\cup\{s\}\cup\mathfrak{w}} \prod_{j\in\mathfrak{u}} \varphi_\alpha\left(\frac{tg_j}{N}\right) \right. \\ &\quad \left. + \sum_{\mathfrak{u}\subseteq[s-1]} \gamma_{\mathfrak{u}\cup\mathfrak{w}} \prod_{j\in\mathfrak{u}} \varphi_\alpha\left(\frac{tg_j}{N}\right) \right)^2.\end{aligned}$$

On the other hand, with similar reasoning as for Σ_1,

$$\Sigma_2 := \sum_{\substack{\boldsymbol{h}\in\mathbb{Z}^s \\ h_s=0}} \sum_{\substack{\boldsymbol{q}\in\mathbb{Z}^s \\ q_s=0}} \frac{\gamma_{\mathfrak{u}(\boldsymbol{h})\cup\mathfrak{w}}}{r_{2\alpha}(\boldsymbol{h})} \frac{\gamma_{\mathfrak{u}(\boldsymbol{q})\cup\mathfrak{w}}}{r_{2\alpha}(\boldsymbol{q})} \mathrm{e}^{2\pi\mathrm{i}t(\boldsymbol{q}-\boldsymbol{h})\cdot\boldsymbol{g}^{(s)}/N}$$

$$= \left(\sum_{\mathfrak{u} \subseteq [s-1]} \gamma_{\mathfrak{u} \cup \mathfrak{w}} \prod_{j \in \mathfrak{u}} \varphi_\alpha \left(\frac{t g_j}{N} \right) \right)^2 .$$

Furthermore, again with a similar argumentation,

$$\begin{aligned}
\Sigma_3 &:= \sum_{\substack{\boldsymbol{h} \in \mathbb{Z}^s \\ h_s \neq 0}} \sum_{\substack{\boldsymbol{q} \in \mathbb{Z}^s \\ q_s = h_s}} \frac{\gamma_{\mathfrak{u}(\boldsymbol{h}) \cup \mathfrak{w}}}{r_{2\alpha}(\boldsymbol{h})} \frac{\gamma_{\mathfrak{u}(\boldsymbol{q}) \cup \mathfrak{w}}}{r_{2\alpha}(\boldsymbol{q})} \mathrm{e}^{2\pi \mathrm{i} t (\boldsymbol{q} - \boldsymbol{h}) \cdot \boldsymbol{g}^{(s)}/N} \\
&= \sum_{h_s \in \mathbb{Z} \setminus \{0\}} \frac{1}{h_s^{4\alpha}} \sum_{\boldsymbol{h} \in \mathbb{Z}^{s-1}} \sum_{\boldsymbol{q} \in \mathbb{Z}^{s-1}} \frac{\gamma_{\mathfrak{u}(\boldsymbol{h}) \cup \{s\} \cup \mathfrak{w}}}{r_{2\alpha}(\boldsymbol{h})} \frac{\gamma_{\mathfrak{u}(\boldsymbol{q}) \cup \{s\} \cup \mathfrak{w}}}{r_{2\alpha}(\boldsymbol{q})} \mathrm{e}^{2\pi \mathrm{i} t (\boldsymbol{q} - \boldsymbol{h}) \cdot \boldsymbol{g}^{(s-1)}/N} \\
&= 2\zeta(4\alpha) \left(\sum_{\mathfrak{u} \subseteq [s-1]} \gamma_{\mathfrak{u} \cup \{s\} \cup \mathfrak{w}} \prod_{j \in \mathfrak{u}} \varphi_\alpha \left(\frac{t g_j}{N} \right) \right)^2 .
\end{aligned}$$

Putting these observations together, we obtain

$$\begin{aligned}
\widetilde{T}_{N,s,\alpha}(\boldsymbol{g}^{(s)}, \{\gamma_{\mathfrak{u} \cup \mathfrak{w}}\}_{\mathfrak{u} \subseteq [s]}) &= \frac{1}{N} \sum_{t=0}^{N-1} (\Sigma_1 - \Sigma_2 - \Sigma_3) \\
&= \frac{1}{N} \sum_{t=0}^{N-1} \left(\left(\varphi_\alpha \left(\frac{t g_s}{N} \right) \right)^2 - 2\zeta(4\alpha) \right) \left(\sum_{\mathfrak{u} \subseteq [s-1]} \gamma_{\mathfrak{u} \cup \{s\} \cup \mathfrak{w}} \prod_{j \in \mathfrak{u}} \varphi_\alpha \left(\frac{t g_j}{N} \right) \right)^2 \\
&\quad + \frac{2}{N} \sum_{t=0}^{N-1} \varphi_\alpha \left(\frac{t g_s}{N} \right) \left(\sum_{\mathfrak{u} \subseteq [s-1]} \gamma_{\mathfrak{u} \cup \{s\} \cup \mathfrak{w}} \prod_{j \in \mathfrak{u}} \varphi_\alpha \left(\frac{t g_j}{N} \right) \right) \\
&\quad \times \left(\sum_{\mathfrak{u} \subseteq [s-1]} \gamma_{\mathfrak{u} \cup \mathfrak{w}} \prod_{j \in \mathfrak{u}} \varphi_\alpha \left(\frac{t g_j}{N} \right) \right).
\end{aligned}$$

The result of the lemma now follows by inserting the latter expression into (13.28). □

From now on assume that we are given POD weights as in (3.25), i.e., weights $\boldsymbol{\gamma}$ of the form

$$\gamma_{\mathfrak{u}} = \Gamma_{|\mathfrak{u}|} \prod_{j \in \mathfrak{u}} \gamma_j, \tag{13.33}$$

for $\mathfrak{u} \subseteq [d]$, where $(\gamma_j)_{j \geq 1}$ is a given (positive) weight sequence, and where the $\Gamma_{|\mathfrak{u}|}$ may depend on $|\mathfrak{u}|$ but not on the elements of $\mathfrak{u}$. Furthermore, we set $\Gamma_0 := \gamma_\emptyset = 1$.

In order to implement Algorithm 13.14 efficiently for POD weights, it is necessary to compute, for $s \in [d]$, the term $T_{N,d,s,\alpha,\boldsymbol{\gamma}}(g_1, \ldots, g_{s-1}, g)$ in a fast manner for all possible values of $g \in G_1^{\varphi}(N)$. This can be done by making use of matrix-vector multiplications for which we define the following vectors and matrices. For fixed N, α, and $\boldsymbol{\gamma}$, let

$$\boldsymbol{v}_{d,s}(\boldsymbol{g}^{(s-1)}) := \left(V_{N,d,s,\alpha,\boldsymbol{\gamma}}(\boldsymbol{g}^{(s-1)},0),\ldots,V_{N,d,s,\alpha,\boldsymbol{\gamma}}(\boldsymbol{g}^{(s-1)},N-1)\right)^{\top},$$

$$\boldsymbol{w}_{d,s}(\boldsymbol{g}^{(s-1)}) := \left(W_{N,d,s,\alpha,\boldsymbol{\gamma}}(\boldsymbol{g}^{(s-1)},0),\ldots,W_{N,d,s,\alpha,\boldsymbol{\gamma}}(\boldsymbol{g}^{(s-1)},N-1)\right)^{\top},$$

so both $\boldsymbol{v}_{d,s}$ and $\boldsymbol{w}_{d,s}$ are column-vectors of length N. Moreover, we define two $(N-1)\times N$ matrices

$$\boldsymbol{\Omega}_{N,\alpha} := \left(\varphi_\alpha\left(\frac{tg}{N}\right)\right)_{g\in G_1^{\varphi}(N),t\in\{0,1,\ldots,N-1\}},$$

$$\boldsymbol{\Psi}_{N,\alpha} := \left(\psi_\alpha\left(\frac{tg}{N}\right)\right)_{g\in G_1^{\varphi}(N),t\in\{0,1,\ldots,N-1\}},$$

where φ_α and ψ_α are as in Lemma 13.21. Using this notation, computing the quantity $T_{N,d,s,\alpha,\boldsymbol{\gamma}}(g_1,\ldots,g_{s-1},g)$ for all $g\in G_1^{\varphi}(N)$ at once boils down to calculating

$$\frac{1}{N}\boldsymbol{\Psi}_{N,\alpha}\boldsymbol{v}_{d,s}(\boldsymbol{g}^{(s-1)})+\frac{2}{N}\boldsymbol{\Omega}_{N,\alpha}\boldsymbol{w}_{d,s}(\boldsymbol{g}^{(s-1)}),$$

i.e., to calculating two matrix-vector products. Obviously, here one can use the machinery in Chapter 3 to compute the two matrix-vector products by reordering the involved matrices to become circulant, and by employing the fast Fourier transform. This needs, by taking into account all d components, $O(d\,N\log N)$ operations, similar to what we showed in Chapter 3, and we do not repeat the details here. The crucial open point for the construction in the present section is the efficient computation (and storage) of the quantities $V_{N,d,s,\alpha,\boldsymbol{\gamma}}(\boldsymbol{g}^{(s-1)},t)$ and $W_{N,d,s,\alpha,\boldsymbol{\gamma}}(\boldsymbol{g}^{(s-1)},t)$, respectively, for $t\in\{0,1,\ldots,N-1\}$ and $s\in[d]$. To this end, we need the following lemma.

Lemma 13.22 *Let $d,N\in\mathbb{N}$, let $\alpha>1/2$, and let $\boldsymbol{\gamma}=\{\gamma_{\mathfrak{u}}\}_{\mathfrak{u}\subseteq[d]}$ be POD weights as in* (13.33). *Let $\boldsymbol{g}=(g_1,\ldots,g_d)\in G_d^{\varphi}(N)$, and $\boldsymbol{g}^{(s-1)}:=(g_1,\ldots,g_{s-1})$ for $s\in[d]$. Then, for $t\in\{0,1,\ldots,N-1\}$, the quantities $V_{N,d,s,\alpha,\boldsymbol{\gamma}}(\boldsymbol{g}^{(s-1)},t)$ and $W_{N,d,s,\alpha,\boldsymbol{\gamma}}(\boldsymbol{g}^{(s-1)},t)$ defined in Lemma 13.21 can be expressed as*

$$V_{N,d,s,\alpha,\boldsymbol{\gamma}}(\boldsymbol{g}^{(s-1)},t)=\gamma_s^2\sum_{m=0}^{d-s}Q_{d,s,m,\alpha,\boldsymbol{\gamma}}\left(\sum_{\ell=0}^{s-1}\Gamma_{\ell+m+1}P_{s-1,\ell,\alpha,\boldsymbol{\gamma}}(\boldsymbol{g}^{(s-1)},t)\right)^2,$$

and

$$W_{N,d,s,\alpha,\boldsymbol{\gamma}}(\boldsymbol{g}^{(s-1)},t)=\gamma_s\sum_{m=0}^{d-s}Q_{d,s,m,\alpha,\boldsymbol{\gamma}}\left(\sum_{\ell=0}^{s-1}\Gamma_{\ell+m+1}P_{s-1,\ell,\alpha,\boldsymbol{\gamma}}(\boldsymbol{g}^{(s-1)},t)\right)\times\left(\sum_{\ell=0}^{s-1}\Gamma_{\ell+m}P_{s-1,\ell,\alpha,\boldsymbol{\gamma}}(\boldsymbol{g}^{(s-1)},t)\right),$$

where

$$Q_{d,s,m,\alpha,\boldsymbol{\gamma}} := \sum_{\substack{\mathfrak{w} \subseteq \{s+1,\ldots,d\} \\ |\mathfrak{w}|=m}} \prod_{j \in \mathfrak{w}} \left(2\zeta(4\alpha)\gamma_j^2\right) \quad \textit{for } m \in \{0,1,\ldots,d-s\},$$

and

$$P_{s-1,\ell,\alpha,\boldsymbol{\gamma}}(\boldsymbol{g}^{(s-1)},t) := \sum_{\substack{\mathfrak{u} \subseteq [s-1] \\ |\mathfrak{u}|=\ell}} \prod_{j \in \mathfrak{u}} \gamma_j \varphi_\alpha\left(\frac{tg_j}{N}\right) \quad \textit{for } \ell \in \{0,1,\ldots,s-1\}.$$

Proof Suppose that $\mathfrak{u}, \mathfrak{w} \subseteq [d]$ with $\mathfrak{u} \cap \mathfrak{w} = \emptyset$. Then we have

$$\gamma_{\mathfrak{u}\cup\mathfrak{w}} = \Gamma_{|\mathfrak{u}|+|\mathfrak{w}|}\left(\prod_{j\in\mathfrak{u}} \gamma_j\right)\left(\prod_{j\in\mathfrak{w}} \gamma_j\right).$$

Using this equality for the quantity $V_{N,d,s,\alpha,\boldsymbol{\gamma}}(\boldsymbol{g}^{(s-1)},t)$ as defined in Lemma 13.21 yields

$$\begin{aligned}
V_{N,d,s,\alpha,\boldsymbol{\gamma}}(\boldsymbol{g}^{(s-1)},t) &= \sum_{\mathfrak{w} \subseteq \{s+1,\ldots,d\}} (2\zeta(4\alpha))^{|\mathfrak{w}|} \\
&\quad \times \left(\sum_{\mathfrak{u} \subseteq [s-1]} \Gamma_{|\mathfrak{u}|+1+|\mathfrak{w}|}\gamma_s \left(\prod_{j\in\mathfrak{u}} \gamma_j\right)\left(\prod_{j\in\mathfrak{w}} \gamma_j\right)\prod_{j\in\mathfrak{u}} \varphi_\alpha\left(\frac{tg_j}{N}\right)\right)^2 \\
&= \gamma_s^2 \sum_{\mathfrak{w} \subseteq \{s+1,\ldots,d\}} \left(\prod_{j\in\mathfrak{w}} \left(2\zeta(4\alpha)\gamma_j^2\right)\right)\left(\sum_{\mathfrak{u} \subseteq [s-1]} \Gamma_{|\mathfrak{u}|+1+|\mathfrak{w}|} \prod_{j\in\mathfrak{u}} \gamma_j \varphi_\alpha\left(\frac{tg_j}{N}\right)\right)^2 \\
&= \gamma_s^2 \sum_{m=0}^{d-s} \sum_{\substack{\mathfrak{w} \subseteq \{s+1,\ldots,d\} \\ |\mathfrak{w}|=m}} \left(\prod_{j\in\mathfrak{w}} \left(2\zeta(4\alpha)\gamma_j^2\right)\right)\left(\sum_{\ell=0}^{s-1} \Gamma_{\ell+m+1} \sum_{\substack{\mathfrak{u} \subseteq [s-1] \\ |\mathfrak{u}|=\ell}} \prod_{j\in\mathfrak{u}} \gamma_j \varphi_\alpha\left(\frac{tg_j}{N}\right)\right)^2.
\end{aligned}$$

This yields the result for $V_{N,d,s,\alpha,\boldsymbol{\gamma}}(\boldsymbol{g}^{(s-1)},t)$ as claimed. The result for the term $W_{N,d,s,\alpha,\boldsymbol{\gamma}}(\boldsymbol{g}^{(s-1)},t)$ follows analogously. □

We will use the result and the notation in Lemma 13.22 to express the quantities $V_{N,d,s,\alpha,\boldsymbol{\gamma}}(\boldsymbol{g}^{(s-1)},t)$ and $W_{N,d,s,\alpha,\boldsymbol{\gamma}}(\boldsymbol{g}^{(s-1)},t)$ in a way that makes it possible to compute them efficiently. To this end, we need some further notation. First, define, for $s \in [d]$,

$$\begin{aligned}
\boldsymbol{p}_{s-1,\alpha,\boldsymbol{\gamma}} &= \boldsymbol{p}_{s-1,\alpha,\boldsymbol{\gamma}}(\boldsymbol{g}^{(s-1)},t) \\
&:= \left(P_{s-1,0,\alpha,\boldsymbol{\gamma}}(\boldsymbol{g}^{(s-1)},t),\ldots,P_{s-1,s-1,\alpha,\boldsymbol{\gamma}}(\boldsymbol{g}^{(s-1)},t)\right)^\top,
\end{aligned}$$

which is a column-vector of length s. Moreover, we remind the reader of the notation $\mathrm{diag}(a_1,\ldots,a_m)$ for $m \in \mathbb{N}$, by which we mean an $m \times m$ diagonal matrix

$$(A_{i,j})_{i,j\in\{1,2,\dots,m\}} \quad \text{with} \quad A_{i,j} = \begin{cases} a_i & \text{if } i = j, \\ 0 & \text{otherwise.} \end{cases}$$

Using this notation, we put, for $s \in [d]$,

$$\boldsymbol{Q}_{d,s,\alpha,\gamma} := \operatorname{diag}\left(Q_{d,s,0,\alpha,\gamma}, \dots, Q_{d,s,d-s,\alpha,\gamma}\right).$$

Finally, again for $s \in [d]$, we define the two $(d - s + 1) \times s$ matrices

$$H^{(1)}_{d,s,\gamma} := \begin{pmatrix} \Gamma_1 & \Gamma_2 & \dots & \Gamma_s \\ \Gamma_2 & \Gamma_3 & \dots & \Gamma_{s+1} \\ \vdots & \vdots & \ddots & \vdots \\ \Gamma_{d-s+1} & \Gamma_{d-s+2} & \dots & \Gamma_d \end{pmatrix} \quad \text{and} \quad H^{(0)}_{d,s,\gamma} := \begin{pmatrix} \Gamma_0 & \Gamma_1 & \dots & \Gamma_{s-1} \\ \Gamma_1 & \Gamma_2 & \dots & \Gamma_s \\ \vdots & \vdots & \ddots & \vdots \\ \Gamma_{d-s} & \Gamma_{d-s+1} & \dots & \Gamma_{d-1} \end{pmatrix}.$$

We now formulate the following lemma.

Lemma 13.23 *Let $d, N \in \mathbb{N}$, let $\alpha > 1/2$, and let $\boldsymbol{\gamma} = \{\gamma_{\mathfrak{u}}\}_{\mathfrak{u}\subseteq[d]}$ be POD weights as in* (13.33). *Let $\boldsymbol{g} = (g_1, \dots, g_d) \in G_d^{\varphi}(N)$, and $\boldsymbol{g}^{(s-1)} := (g_1, \dots, g_{s-1})$ for $s \in [d]$. Then for $t \in \{0, 1, \dots, N-1\}$, the quantities $V_{N,d,s,\alpha,\gamma}(\boldsymbol{g}^{(s-1)}, t)$ and $W_{N,d,s,\alpha,\gamma}(\boldsymbol{g}^{(s-1)}, t)$ defined in Lemma 13.21 can be expressed, by using the notation introduced above, as*

$$V_{N,d,s,\alpha,\gamma}(\boldsymbol{g}^{(s-1)}, t) = \gamma_s^2 \left(H^{(1)}_{d,s,\gamma} \boldsymbol{p}_{s-1,\alpha,\gamma}\right)^{\top} \boldsymbol{Q}_{d,s,\alpha,\gamma} \left(H^{(1)}_{d,s,\gamma} \boldsymbol{p}_{s-1,\alpha,\gamma}\right)$$

and

$$W_{N,d,s,\alpha,\gamma}(\boldsymbol{g}^{(s-1)}, t) = \gamma_s \left(H^{(1)}_{d,s,\gamma} \boldsymbol{p}_{s-1,\alpha,\gamma}\right)^{\top} \boldsymbol{Q}_{d,s,\alpha,\gamma} \left(H^{(0)}_{d,s,\gamma} \boldsymbol{p}_{s-1,\alpha,\gamma}\right).$$

Proof Let us first derive the result for $V_{N,d,s,\alpha,\gamma}(\boldsymbol{g}^{(s-1)}, t)$. Considering the formula derived for this term in Lemma 13.22, we see that

$$\sum_{\ell=0}^{s-1} \Gamma_{\ell+m+1} P_{s-1,\ell,\alpha,\gamma}(\boldsymbol{g}^{(s-1)}, t)$$

is exactly the $(m+1)$-st component of $H^{(1)}_{d,s,\gamma} \boldsymbol{p}_{s-1,\alpha,\gamma}$ for $m \in \{0, 1, \dots, d-s\}$. Hence, calculating the product

$$\left(H^{(1)}_{d,s,\gamma} \boldsymbol{p}_{s-1,\alpha,\gamma}\right)^{\top} \boldsymbol{Q}_{d,s,\alpha,\gamma} \left(H^{(1)}_{d,s,\gamma} \boldsymbol{p}_{s-1,\alpha,\gamma}\right)$$

yields the sum over m in the expression for $V_{N,d,s,\alpha,\gamma}(\boldsymbol{g}^{(s-1)}, t)$ in Lemma 13.22, by using that $\boldsymbol{Q}_{d,s,\alpha,\gamma}$ is a diagonal matrix. This gives the result for $V_{N,d,s,\alpha,\gamma}(\boldsymbol{g}^{(s-1)}, t)$.

The result for $W_{N,d,s,\alpha,\gamma}(\boldsymbol{g}^{(s-1)}, t)$ follows analogously, by noting that

$$\sum_{\ell=0}^{s-1} \Gamma_{\ell+m} P_{s-1,\ell,\alpha,\gamma}(\boldsymbol{g}^{(s-1)}, t)$$

is the $(m+1)$-st entry of $H_{d,s,\gamma}^{(0)} \boldsymbol{p}_{s-1,\alpha,\gamma}$ for $m \in \{0, 1, \ldots, d-s\}$. □

We now have all the ingredients to outline how the terms $V_{N,d,s,\alpha,\gamma}(\boldsymbol{g}^{(s-1)}, t)$ and $W_{N,d,s,\alpha,\gamma}(\boldsymbol{g}^{(s-1)}, t)$ can be computed efficiently for POD weights, hence making the construction in Algorithm 13.14 fast. Indeed, it is necessary to compute these two quantities efficiently for all $t \in \{0, 1, \ldots, N-1\}$, step by step for every component.

For fixed $s \in [d]$ and fixed $t \in \{0, 1, \ldots, N-1\}$, let us first analyze the computational effort to compute $V_{N,d,s,\alpha,\gamma}(\boldsymbol{g}^{(s-1)}, t)$ and $W_{N,d,s,\alpha,\gamma}(\boldsymbol{g}^{(s-1)}, t)$. To this end, it should be noted that $H_{d,s,\gamma}^{(1)}$ is the left upper $(d-s+1) \times s$ submatrix of the Hankel matrix

$$H_{d,\gamma} = \begin{pmatrix} \Gamma_1 & \Gamma_2 & \ldots & \Gamma_{d-1} & \Gamma_d \\ \Gamma_2 & \Gamma_3 & \ldots & \Gamma_d & 0 \\ \vdots & \vdots & \iddots & \iddots & \vdots \\ \Gamma_d & 0 & \ldots & 0 & 0 \end{pmatrix}.$$

Therefore, $H_{d,s,\gamma}^{(1)}$ can be embedded in the Hankel matrix $H_{d,\gamma}$, which, in turn, can be embedded in a $2d \times 2d$ circulant matrix. By extending the input vector by zeros and then using only a suitably selected initial part of the output vector, we can in this way compute the matrix-vector products involving $H_{d,s,\gamma}^{(1)}$. A similar observation holds for $H_{d,s,\gamma}^{(0)}$ with a slightly modified Hankel matrix. Consequently the matrix-vector products in the result in Lemma 13.23 can be computed using $O(d \log d)$ operations. The matrix multiplications with the diagonal matrix $\boldsymbol{Q}_{d,s,\alpha,\gamma}$ obviously do not increase this order of magnitude. Thus, the computational cost of computing $V_{N,d,s,\alpha,\gamma}(\boldsymbol{g}^{(s-1)}, t)$ and $W_{N,d,s,\alpha,\gamma}(\boldsymbol{g}^{(s-1)}, t)$ for fixed $s \in [d]$ and fixed $t \in \{0, 1, \ldots, N-1\}$ is of order $O(d \log d)$. However, it is also necessary for these computations to have the quantities $Q_{d,s,m,\alpha,\gamma}$ and $P_{s-1,\ell,\alpha,\gamma}(\boldsymbol{g}^{(s-1)}, t)$ available for the various s, ℓ, and m. This can be achieved by updating them recursively.

Indeed, by setting $P_{r,0,\alpha,\gamma}(\boldsymbol{g}^{(r)}, t) := 1$ for any r, and $P_{r,\ell,\alpha,\gamma}(\boldsymbol{g}^{(r)}, t) := 0$ if $\ell > r$, we can use the recursion

$$P_{r,\ell,\alpha,\gamma}(\boldsymbol{g}^{(r)}, t) = P_{r-1,\ell,\alpha,\gamma}(\boldsymbol{g}^{(r-1)}, t) + \gamma_r \varphi_\alpha\left(\frac{t g_r}{N}\right) P_{r-1,\ell-1,\alpha,\gamma}(\boldsymbol{g}^{(r-1)}, t).$$

If r is increased by 1, one can compute the according values starting from $\ell = r$, and ending with $\ell = 1$, and overwrite the previous values.

Regarding $Q_{d,s,m,\alpha,\gamma}$, one can again proceed recursively. Putting $Q_{d,s,0,\alpha,\gamma} := 1$ for any s, and $Q_{d,s,m,\alpha,\gamma} := 0$ if $m > d-s$, one can use the recursion

$$Q_{d,s,m,\alpha,\gamma} = Q_{d,s+1,m,\alpha,\gamma} + 2\zeta(4\alpha)\gamma_{s+1}^2 Q_{d,s+1,m-1,\alpha,\gamma}.$$

Again, this recursion works "backwards", i.e., it is started from $s = d$, and ends with $s = 1$. The different values of $Q_{d,s,m,\alpha,\gamma}$ can be precomputed and stored in a triangular matrix (as for each s, one needs to calculate the values for m ranging from 0 to $d - s$) at a cost of $O(d^2)$ computations and $O(d^2)$ storage.

In summary, we can bound the order of magnitude of operations necessary to run Algorithm 13.14 as follows. The computation of the matrix-vector products involving $\boldsymbol{\Omega}_{N,\alpha}$ and $\boldsymbol{\Psi}_{N,\alpha}$ requires $O(d\,N \log N)$ operations. The computation of $V_{N,d,s,\alpha,\gamma}(\boldsymbol{g}^{(s-1)}, t)$ and $W_{N,d,s,\alpha,\gamma}(\boldsymbol{g}^{(s-1)}, t)$ and updating $Q_{d,s,m,\alpha,\gamma}$ and $P_{s-1,\ell,\alpha,\gamma}(\boldsymbol{g}^{(s-1)}, t)$ needs $O(d \log d)$ operations, but these calculations need to be done in each step of the algorithm and for each $t \in \{0, 1, \dots, N-1\}$. In total this yields that Algorithm 13.14 can be efficiently implemented such that it requires

$$O(d\,N \log N + N d^2 \log d)$$

operations. While the quadratic dependence on d is a slight drawback, this is what can reasonably be expected for the case of POD weights, and a further improvement of this order of magnitude does not seem to be within reach according to the current state of the art.

Notes and Remarks

The idea of using lattice rules for approximation in Korobov spaces goes back to Korobov himself (see [140], and also [115]). Approximation in unweighted Korobov spaces was studied by Temlyakov in the papers [247, 249]. Furthermore, a method similar to the one presented in Section 13.1 was also considered before in [182, 271]. Moreover, we would like to point out that a similar approach based on so-called reconstructing rank-1 lattices for trigonometric polynomials is due to Kämmerer and his co-authors, see, e.g. [125, 126, 129, 130]. This approach is also applicable to more general function spaces beyond Korobov spaces. We shall return to further methods by Kämmerer et al. in the context of multiple rank-1 lattice point sets in Chapter 15.

In the present chapter, however, we have followed Kuo, Sloan, and Woźniakowski [162] in our outline, as this was the first paper to show the result for the weighted setting, and to include tractability results.

The lower bound presented in Section 13.2 for approximation was shown in a more general setting in [24]. Here, we have restricted ourselves to the case of approximation in $\mathcal{H}_{\mathrm{kor},d,\alpha,\gamma}$, but the bound can also be shown for more general/slightly different function spaces with the same technique.

It is beyond the scope of this book to give an extensive overview of tractability of function approximation, so we have focused only on the most important aspects here. For further details on tractability of function approximation, we refer to the paper [209], and the trilogy [210]–[212], as well as the references therein. A complete overview of the weight conditions for all current standard notions of tractability beyond (strong) polynomial tractability of the L_2-approximation problem in weighted Korobov spaces can be found in [72].

In the outline of adaptions for general weights in Section 13.4 we have mostly followed the papers [28] and [29]. In [29], there is also an analysis of how to implement Algorithm 13.14 for order-dependent weights and SPOD weights.

Lattice rules for approximation in the average case setting are considered in the paper [163] by Kuo, Sloan, and Woźniakowski.

In this chapter we have only considered Korobov spaces of finite smoothness α. L_2-approximation for Korobov spaces of analytic functions as introduced in Chapter 9 is studied in the papers [46, 119] (see also [149] for a survey).

Chapter 14
L_∞-Approximation Using Lattice Rules

Besides L_2-approximation, which we outlined in Chapter 13, it is natural to consider whether one can also use lattice point sets in order to approximate functions in the Korobov space in the L_∞-norm. Obviously, L_∞-approximation is in general a much more difficult task than L_2-approximation, so it is, a priori, not clear whether lattice rules can help also in the more demanding L_∞-case. This is an interesting problem, and there are several results showing that lattice rules can be also effectively used in L_∞-approximation. These results (from, e.g., [165, 167, 270]) will be discussed in this chapter.

14.1 L_∞-Approximation of Functions in Korobov Spaces

The approach outlined in Section 13.1 can also be analyzed such that we can derive bounds for L_∞-approximation of elements of the Korobov space $\mathcal{H}_{\mathrm{kor},d,\alpha,\boldsymbol{\gamma}}$. Indeed, we will use the same algorithm (given in (13.2)) as we used for L_2-approximation for L_∞-approximation, and analyze its worst-case error,

$$\begin{aligned}\mathrm{err}^{L_\infty\text{-app}}(\mathcal{H}_{\mathrm{kor},d,\alpha,\boldsymbol{\gamma}}, A_{N,M,d,\alpha,\boldsymbol{\gamma}}(\boldsymbol{g}))\\ := \sup_{\substack{f\in\mathcal{H}_{\mathrm{kor},d,\alpha,\boldsymbol{\gamma}}\\ \|f\|_{\mathrm{kor},d,\alpha,\boldsymbol{\gamma}}\le 1}} \sup_{\boldsymbol{x}\in[0,1]^d} |f(\boldsymbol{x}) - A_{N,M,d,\alpha,\boldsymbol{\gamma}}(\boldsymbol{g})(f)(\boldsymbol{x})|.\end{aligned}$$

As in Section 13.1, we will restrict ourselves—for the sake of simplicity—to considering only product weights $\boldsymbol{\gamma} = (\gamma_j)_{j\ge 1}$, where we shall additionally often assume that $\gamma_j \in (0, 1]$ for all $j \in \mathbb{N}$.

Let again

$$\mathcal{A}_{d,M} := \left\{\boldsymbol{h} \in \mathbb{Z}^d \ : \ r_{2\alpha,\boldsymbol{\gamma}}(\boldsymbol{h}) \le M\right\}$$

J. Dick et al., *Lattice Rules*, Springer Series in Computational Mathematics 58,
https://doi.org/10.1007/978-3-031-09951-9_14

for a real $M > 0$ and $d \in \mathbb{N}$. Let N be a prime number, and assume that $\boldsymbol{g}$ is a generating vector of a rank-1 lattice rule with N points. Then, we can define the algorithm $A_{N,M,d,\alpha,\boldsymbol{\gamma}}(\boldsymbol{g})$ exactly as in (13.2) and use it for L_∞-approximation of $f \in \mathcal{H}_{\mathrm{kor},d,\alpha,\boldsymbol{\gamma}}$.

For fixed $\boldsymbol{h} \in \mathcal{A}_{d,M}$, define a function $\tau_{\boldsymbol{h}} \in \mathcal{H}_{\mathrm{kor},d,\alpha,\boldsymbol{\gamma}}$ as

$$\tau_{\boldsymbol{h}}(\boldsymbol{x}) := -\sum_{\substack{\boldsymbol{\ell}\in\mathbb{Z}^d\setminus\{\boldsymbol{h}\}\\ (\boldsymbol{h}-\boldsymbol{\ell})\cdot\boldsymbol{g}\equiv 0 \pmod N}} \frac{\mathrm{e}^{2\pi\mathrm{i}\boldsymbol{\ell}\cdot\boldsymbol{x}}}{r_{2\alpha,\boldsymbol{\gamma}}(\boldsymbol{\ell})}. \tag{14.1}$$

Note that $\tau_{\boldsymbol{h}}$ implicitly also depends on $\boldsymbol{g}$, N, and $\mathcal{A}_{d,M}$, but we suppress this dependence in the notation to keep it as simple as possible. Then, for $f \in \mathcal{H}_{\mathrm{kor},d,\alpha,\boldsymbol{\gamma}}$, we get

$$\langle f, \tau_{\boldsymbol{h}}\rangle_{\mathrm{kor},d,\alpha,\boldsymbol{\gamma}} = -\sum_{\substack{\boldsymbol{\ell}\in\mathbb{Z}^d\setminus\{\boldsymbol{h}\}\\ (\boldsymbol{h}-\boldsymbol{\ell})\cdot\boldsymbol{g}\equiv 0 \pmod N}} \widehat{f}(\boldsymbol{\ell}).$$

On the other hand, the integration error using $\mathcal{P}(\boldsymbol{g}, N)$ for the function defined by $f_{\boldsymbol{h}}(\boldsymbol{x}) := f(\boldsymbol{x})\,\mathrm{e}^{-2\pi\mathrm{i}\boldsymbol{h}\cdot\boldsymbol{x}}$ for $\boldsymbol{x} \in [0,1)^d$ is, by Proposition 1.12,

$$\begin{aligned}
\int_{[0,1]^d} f_{\boldsymbol{h}}(\boldsymbol{x})\,\mathrm{d}\boldsymbol{x} - \frac{1}{N}\sum_{k=0}^{N-1} f_{\boldsymbol{h}}\left(\left\{\frac{k\boldsymbol{g}}{N}\right\}\right) &= -\sum_{\substack{\boldsymbol{\ell}\in\mathbb{Z}^d\setminus\{\boldsymbol{0}\}\\ \boldsymbol{\ell}\cdot\boldsymbol{g}\equiv 0 \pmod N}} \widehat{f_{\boldsymbol{h}}}(\boldsymbol{\ell})\\
&= -\sum_{\substack{\boldsymbol{\ell}\in\mathbb{Z}^d\setminus\{\boldsymbol{0}\}\\ \boldsymbol{\ell}\cdot\boldsymbol{g}\equiv 0 \pmod N}} \widehat{f}(\boldsymbol{\ell}+\boldsymbol{h})\\
&= -\sum_{\substack{\boldsymbol{\ell}\in\mathbb{Z}^d\setminus\{\boldsymbol{h}\}\\ (\boldsymbol{\ell}-\boldsymbol{h})\cdot\boldsymbol{g}\equiv 0 \pmod N}} \widehat{f}(\boldsymbol{\ell})\\
&= -\sum_{\substack{\boldsymbol{\ell}\in\mathbb{Z}^d\setminus\{\boldsymbol{h}\}\\ (\boldsymbol{h}-\boldsymbol{\ell})\cdot\boldsymbol{g}\equiv 0 \pmod N}} \widehat{f}(\boldsymbol{\ell}).
\end{aligned}$$

This immediately yields that

$$\langle f, \tau_{\boldsymbol{h}}\rangle_{\mathrm{kor},d,\alpha,\boldsymbol{\gamma}} = \widehat{f}(\boldsymbol{h}) - \frac{1}{N}\sum_{k=0}^{N-1} f\left(\left\{\frac{k\boldsymbol{g}}{N}\right\}\right)\mathrm{e}^{-2\pi\mathrm{i}k\boldsymbol{h}\cdot\boldsymbol{g}/N}.$$

Furthermore, we get for $\boldsymbol{h}, \boldsymbol{p} \in \mathcal{A}_{d,M}$,

$$\langle \tau_{\boldsymbol{h}}, \tau_{\boldsymbol{p}}\rangle_{\mathrm{kor},d,\alpha,\boldsymbol{\gamma}} = \sum_{\substack{\boldsymbol{\ell}\in\mathbb{Z}^d\setminus\{\boldsymbol{h},\boldsymbol{p}\}\\ (\boldsymbol{h}-\boldsymbol{\ell})\cdot\boldsymbol{g}\equiv 0 \pmod N\\ (\boldsymbol{p}-\boldsymbol{\ell})\cdot\boldsymbol{g}\equiv 0 \pmod N}} \frac{1}{r_{2\alpha,\boldsymbol{\gamma}}(\boldsymbol{\ell})}$$

$$= \sum_{\substack{\boldsymbol{\ell}\in\mathbb{Z}^d\setminus\{\boldsymbol{0},\boldsymbol{p}-\boldsymbol{h}\}\\ -\boldsymbol{\ell}\cdot\boldsymbol{g}\equiv 0 \pmod N\\ (\boldsymbol{p}-\boldsymbol{\ell}-\boldsymbol{h})\cdot\boldsymbol{g}\equiv 0 \pmod N}} \frac{1}{r_{2\alpha,\boldsymbol{\gamma}}(\boldsymbol{\ell}+\boldsymbol{h})}$$

$$= \sum_{\substack{\boldsymbol{\ell}\in\mathbb{Z}^d\setminus\{\boldsymbol{0},\boldsymbol{p}-\boldsymbol{h}\}\\ \boldsymbol{\ell}\cdot\boldsymbol{g}\equiv 0 \pmod N\\ (\boldsymbol{h}-\boldsymbol{p})\cdot\boldsymbol{g}\equiv 0 \pmod N}} \frac{1}{r_{2\alpha,\boldsymbol{\gamma}}(\boldsymbol{\ell}+\boldsymbol{h})}.$$

Since the condition $(\boldsymbol{h}-\boldsymbol{p})\cdot\boldsymbol{g}\equiv 0 \pmod N$ in the latter sum is independent of the summation index $\boldsymbol{\ell}$, we can write, for $\boldsymbol{h},\boldsymbol{p}\in\mathcal{A}_{d,M}$,

$$\langle \tau_{\boldsymbol{h}},\tau_{\boldsymbol{p}}\rangle_{\mathrm{kor},d,\alpha,\boldsymbol{\gamma}} = \begin{cases} \displaystyle\sum_{\substack{\boldsymbol{\ell}\in\mathbb{Z}^d\setminus\{\boldsymbol{0},\boldsymbol{p}-\boldsymbol{h}\}\\ \boldsymbol{\ell}\cdot\boldsymbol{g}\equiv 0 \pmod N}} \frac{1}{r_{2\alpha,\boldsymbol{\gamma}}(\boldsymbol{h}+\boldsymbol{\ell})} & \text{if } (\boldsymbol{h}-\boldsymbol{p})\cdot\boldsymbol{g}\equiv 0 \pmod N,\\ 0 & \text{otherwise.}\end{cases} \tag{14.2}$$

Furthermore, for the proof of Proposition 14.1 below, we extend the definition of $\tau_{\boldsymbol{h}}$ given in (14.1) such that for $\boldsymbol{h}\in\mathbb{Z}^d\setminus\mathcal{A}_{d,M}$ we put

$$\tau_{\boldsymbol{h}}(\boldsymbol{x}) := \frac{\mathrm{e}^{2\pi\mathrm{i}\boldsymbol{h}\cdot\boldsymbol{x}}}{r_{2\alpha,\boldsymbol{\gamma}}(\boldsymbol{h})}.$$

This yields $\langle f,\tau_{\boldsymbol{h}}\rangle_{\mathrm{kor},d,\alpha,\boldsymbol{\gamma}} = \widehat{f}(\boldsymbol{h})$ for $\boldsymbol{h}\in\mathbb{Z}^d\setminus\mathcal{A}_{d,M}$.

For a given generating vector $\boldsymbol{g}$ of a rank-1 lattice rule with N points (N prime) and given set $\mathcal{A}_{d,M}$ as above, we define a $|\mathcal{A}_{d,M}|\times|\mathcal{A}_{d,M}|$ matrix $T_{\boldsymbol{g},d,M}$ by

$$T_{\boldsymbol{g},d,M} := \left(\langle \tau_{\boldsymbol{h}},\tau_{\boldsymbol{p}}\rangle_{\mathrm{kor},d,\alpha,\boldsymbol{\gamma}}\right)_{\boldsymbol{h},\boldsymbol{p}\in\mathcal{A}_{d,M}}. \tag{14.3}$$

The fundamental error estimate

The following result due to Kuo, Wasilkowski, and Woźniakowski [166] is the basis of the error analysis in this section. We will use the upper bound (14.4) for the construction of good lattice rules yielding a small approximation error. The lower bound is given as a reference and shows that the upper bound is, up to an absolute constant, best possible.

Proposition 14.1 *Let $\alpha > 1/2$, let $\boldsymbol{\gamma} = (\gamma_j)_{j\geq 1}$ be product weights, let M be a positive real, and let N be a prime number. Let $\boldsymbol{g}$ be the generating vector of a rank-1 lattice rule with N points and define the algorithm $A_{N,M,d,\alpha,\boldsymbol{\gamma}}(\boldsymbol{g})$ as in* (13.2). *Let $T_{\boldsymbol{g},d,M}$ be given by* (14.3). *Then it is true that*

$$\operatorname{err}^{L_\infty\text{-app}}(\mathcal{H}_{\mathrm{kor},d,\alpha,\boldsymbol{\gamma}}, A_{N,M,d,\alpha,\boldsymbol{\gamma}}(\boldsymbol{g})) \le \left(\sum_{\boldsymbol{h}\in\mathbb{Z}^d\setminus\mathcal{A}_{d,M}} \frac{1}{r_{2\alpha,\boldsymbol{\gamma}}(\boldsymbol{h})} + 3\Sigma(T_{\boldsymbol{g},d,M}) \right)^{1/2} \tag{14.4}$$

and

$$\operatorname{err}^{L_\infty\text{-app}}(\mathcal{H}_{\mathrm{kor},d,\alpha,\boldsymbol{\gamma}}, A_{N,M,d,\alpha,\boldsymbol{\gamma}}(\boldsymbol{g})) \ge \left(\sum_{\boldsymbol{h}\in\mathbb{Z}^d\setminus\mathcal{A}_{d,M}} \frac{1}{r_{2\alpha,\boldsymbol{\gamma}}(\boldsymbol{h})} + \Sigma(T_{\boldsymbol{g},d,M}) \right)^{1/2},$$

where $\Sigma(T)$ denotes the sum of all elements of a matrix T.

Proof In analogy to (13.4), it is easy to see that, for any fixed $f \in \mathcal{H}_{\mathrm{kor},d,\alpha,\boldsymbol{\gamma}}$ and any fixed $\boldsymbol{x} \in [0,1)^d$, we have

$$\begin{aligned}(f - A_{N,M,d,\alpha,\boldsymbol{\gamma}}(\boldsymbol{g})(f))(\boldsymbol{x}) &= \sum_{\boldsymbol{h}\notin\mathcal{A}_{d,M}} \widehat{f}(\boldsymbol{h})\,\mathrm{e}^{2\pi\mathrm{i}\boldsymbol{h}\cdot\boldsymbol{x}} \\ &\quad + \sum_{\boldsymbol{h}\in\mathcal{A}_{d,M}} \left(\widehat{f}(\boldsymbol{h}) - \frac{1}{N}\sum_{k=0}^{N-1} f\left(\left\{\frac{k\boldsymbol{g}}{N}\right\}\right) \mathrm{e}^{-2\pi\mathrm{i}k\boldsymbol{h}\cdot\boldsymbol{g}/N} \right) \mathrm{e}^{2\pi\mathrm{i}\boldsymbol{h}\cdot\boldsymbol{x}}.\end{aligned}$$

By our observations above, we can thus write

$$\begin{aligned}&(f - A_{N,M,d,\alpha,\boldsymbol{\gamma}}(\boldsymbol{g})(f))(\boldsymbol{x}) \\ &= \sum_{\boldsymbol{h}\notin\mathcal{A}_{d,M}} \langle f(\cdot), \tau_{\boldsymbol{h}}(\cdot)\,\mathrm{e}^{-2\pi\mathrm{i}\boldsymbol{h}\cdot\boldsymbol{x}}\rangle_{\mathrm{kor},d,\alpha,\boldsymbol{\gamma}} + \sum_{\boldsymbol{h}\in\mathcal{A}_{d,M}} \langle f(\cdot), \tau_{\boldsymbol{h}}(\cdot)\,\mathrm{e}^{-2\pi\mathrm{i}\boldsymbol{h}\cdot\boldsymbol{x}}\rangle_{\mathrm{kor},d,\alpha,\boldsymbol{\gamma}} \\ &= \left\langle f(\cdot), \sum_{\boldsymbol{h}\in\mathbb{Z}^d} \tau_{\boldsymbol{h}}(\cdot)\,\mathrm{e}^{-2\pi\mathrm{i}\boldsymbol{h}\cdot\boldsymbol{x}} \right\rangle_{\mathrm{kor},d,\alpha,\boldsymbol{\gamma}},\end{aligned}$$

where the inner product is with respect to the hidden variable $(\cdot)$.

Having expressed the approximation error for a fixed $\boldsymbol{x}$ as an inner product, we can now proceed in the by now familiar manner. We apply the Cauchy–Schwarz inequality and obtain in this way,

$$|(f - A_{N,M,d,\alpha,\boldsymbol{\gamma}}(\boldsymbol{g})(f))(\boldsymbol{x})| \le \|f\|_{\mathrm{kor},d,\alpha,\boldsymbol{\gamma}} \left\| \sum_{\boldsymbol{h}\in\mathbb{Z}^d} \tau_{\boldsymbol{h}}\,\mathrm{e}^{-2\pi\mathrm{i}\boldsymbol{h}\cdot\boldsymbol{x}} \right\|_{\mathrm{kor},d,\alpha,\boldsymbol{\gamma}}.$$

However, since the Cauchy–Schwarz inequality is sharp for

$$f(\boldsymbol{y}) = \frac{\sum_{\boldsymbol{h}\in\mathbb{Z}^d} \tau_{\boldsymbol{h}}(\boldsymbol{y})\, \mathrm{e}^{-2\pi \mathrm{i}\boldsymbol{h}\cdot\boldsymbol{x}}}{\|\sum_{\boldsymbol{h}\in\mathbb{Z}^d} \tau_{\boldsymbol{h}}\, \mathrm{e}^{-2\pi \mathrm{i}\boldsymbol{h}\cdot\boldsymbol{x}}\|_{\mathrm{kor},d,\alpha,\boldsymbol{\gamma}}},$$

we can even obtain equality. Accordingly we obtain

$$\begin{aligned}
&\mathrm{err}^{L_\infty\text{-app}}(\mathcal{H}_{\mathrm{kor},d,\alpha,\boldsymbol{\gamma}}, A_{N,M,d,\alpha,\boldsymbol{\gamma}}(\boldsymbol{g}))\\
&= \sup_{\boldsymbol{x}\in[0,1]^d} \left\| \sum_{\boldsymbol{h}\in\mathbb{Z}^d} \tau_{\boldsymbol{h}}\, \mathrm{e}^{-2\pi \mathrm{i}\boldsymbol{h}\cdot\boldsymbol{x}} \right\|_{\mathrm{kor},d,\alpha,\boldsymbol{\gamma}}\\
&= \sup_{\boldsymbol{x}\in[0,1]^d} \left| \sum_{\boldsymbol{h}\in\mathbb{Z}^d} \sum_{\boldsymbol{p}\in\mathbb{Z}^d} \langle \tau_{\boldsymbol{h}}, \tau_{\boldsymbol{p}}\rangle_{\mathrm{kor},d,\alpha,\boldsymbol{\gamma}}\, \mathrm{e}^{2\pi \mathrm{i}(\boldsymbol{p}-\boldsymbol{h})\cdot\boldsymbol{x}} \right|^{1/2}.
\end{aligned} \tag{14.5}$$

From (14.2), we already know expressions for $\langle \tau_{\boldsymbol{h}}, \tau_{\boldsymbol{p}}\rangle_{\mathrm{kor},d,\alpha,\boldsymbol{\gamma}}$ if both $\boldsymbol{h}$ and $\boldsymbol{p}$ are elements of $\mathcal{A}_{d,M}$. However, we also need information regarding the inner product if that is not the case. To this end, let us first suppose that $\boldsymbol{h} \in \mathcal{A}_{d,M}$, but $\boldsymbol{p} \notin \mathcal{A}_{d,M}$. Then we obviously have

$$\langle \tau_{\boldsymbol{h}}, \tau_{\boldsymbol{p}}\rangle_{\mathrm{kor},d,\alpha,\boldsymbol{\gamma}} = \begin{cases} (r_{2\alpha,\boldsymbol{\gamma}}(\boldsymbol{p}))^{-1} & \text{if } (\boldsymbol{p}-\boldsymbol{h})\cdot\boldsymbol{g} \equiv 0 \pmod{N},\\ 0 & \text{otherwise,} \end{cases} \tag{14.6}$$

and by symmetry the analog holds if $\boldsymbol{h} \notin \mathcal{A}_{d,M}$ but $\boldsymbol{p} \in \mathcal{A}_{d,M}$. If $\boldsymbol{h}, \boldsymbol{p} \notin \mathcal{A}_{d,M}$, we obtain

$$\langle \tau_{\boldsymbol{h}}, \tau_{\boldsymbol{p}}\rangle_{\mathrm{kor},d,\alpha,\boldsymbol{\gamma}} = \begin{cases} (r_{2\alpha,\boldsymbol{\gamma}}(\boldsymbol{p}))^{-1} & \text{if } \boldsymbol{h} = \boldsymbol{p},\\ 0 & \text{otherwise.} \end{cases} \tag{14.7}$$

Combining (14.2), (14.6), and (14.7), we have covered all possible cases for $\boldsymbol{h}, \boldsymbol{p} \in \mathbb{Z}^d$. Since in all cases we have $\langle \tau_{\boldsymbol{h}}, \tau_{\boldsymbol{p}}\rangle_{\mathrm{kor},d,\alpha,\boldsymbol{\gamma}} \geq 0$, it follows that the supremum in (14.5) is attained for $\boldsymbol{x} = \boldsymbol{0}$, which yields

$$\begin{aligned}
&\mathrm{err}^{L_\infty\text{-app}}(\mathcal{H}_{\mathrm{kor},d,\alpha,\boldsymbol{\gamma}}, A_{N,M,d,\alpha,\boldsymbol{\gamma}}(\boldsymbol{g})) = \left(\sum_{\boldsymbol{h}\in\mathbb{Z}^d} \sum_{\boldsymbol{p}\in\mathbb{Z}^d} \langle \tau_{\boldsymbol{h}}, \tau_{\boldsymbol{p}}\rangle_{\mathrm{kor},d,\alpha,\boldsymbol{\gamma}} \right)^{1/2}\\
&= \left(\sum_{\boldsymbol{h}\in\mathbb{Z}^d\setminus\mathcal{A}_{d,M}} \frac{1}{r_{2\alpha,\boldsymbol{\gamma}}(\boldsymbol{h})} + 2 \sum_{\boldsymbol{h}\in\mathcal{A}_{d,M}} \sum_{\substack{\boldsymbol{p}\notin\mathcal{A}_{d,M}\\ (\boldsymbol{p}-\boldsymbol{h})\cdot\boldsymbol{g}\equiv 0 \ (\mathrm{mod}\ N)}} \frac{1}{r_{2\alpha,\boldsymbol{\gamma}}(\boldsymbol{p})} + \Sigma(T_{\boldsymbol{g},d,M}) \right)^{1/2}.
\end{aligned}$$

This implies the lower bound. In order to prove the upper bound, we estimate the double sum in the middle term of the right-hand side of the previous expression by

$$\sum_{\boldsymbol{h}\in\mathcal{A}_{d,M}} \sum_{\substack{\boldsymbol{p}\notin\mathcal{A}_{d,M}\\ (\boldsymbol{p}-\boldsymbol{h})\cdot\boldsymbol{g}\equiv 0 \ (\mathrm{mod}\ N)}} \frac{1}{r_{2\alpha,\boldsymbol{\gamma}}(\boldsymbol{p})} \leq \sum_{\boldsymbol{h}\in\mathcal{A}_{d,M}} \sum_{\substack{\boldsymbol{p}\in\mathbb{Z}^d\setminus\{\boldsymbol{h}\}\\ (\boldsymbol{p}-\boldsymbol{h})\cdot\boldsymbol{g}\equiv 0 \ (\mathrm{mod}\ N)}} \frac{1}{r_{2\alpha,\boldsymbol{\gamma}}(\boldsymbol{p})}$$

$$= \sum_{\boldsymbol{h} \in \mathcal{A}_{d,M}} \sum_{\substack{\boldsymbol{\ell} \in \mathbb{Z}^d \setminus \{\boldsymbol{0}\} \\ \boldsymbol{\ell} \cdot \boldsymbol{g} \equiv 0 \pmod{N}}} \frac{1}{r_{2\alpha,\boldsymbol{\gamma}}(\boldsymbol{h}+\boldsymbol{\ell})}$$
$$= \mathrm{trace}(T_{\boldsymbol{g},d,M})$$
$$\leq \Sigma(T_{\boldsymbol{g},d,M}),$$

where $\mathrm{trace}(T_{\boldsymbol{g},d,M})$, as usual, denotes the sum of the diagonal elements of the matrix $T_{\boldsymbol{g},d,M}$. Hence

$$\mathrm{err}^{L_\infty\text{-app}}(\mathcal{H}_{\mathrm{kor},d,\alpha,\boldsymbol{\gamma}}, A_{N,M,d,\alpha,\boldsymbol{\gamma}}(\boldsymbol{g}))$$
$$\leq \left(\sum_{\boldsymbol{h} \in \mathbb{Z}^d \setminus \mathcal{A}_{d,M}} \frac{1}{r_{2\alpha,\boldsymbol{\gamma}}(\boldsymbol{h})} + 3\Sigma(T_{\boldsymbol{g},d,M}) \right)^{1/2}. \quad \square$$

Next, we use Proposition 14.1 for the construction of good lattice rules yielding a small approximation error. Before we do so, we give an estimate on the first term on the right-hand side of (14.4). To this end, we use a method shown by Kuo, Sloan, and Woźniakowski in [163, Lemma 6].

Lemma 14.2 *Let $\alpha > 1/2$ and $M \geq 1$ be given. For any $q \in (1/(2\alpha), 1)$ we have*

$$\sum_{\boldsymbol{h} \in \mathbb{Z}^d \setminus \mathcal{A}_{d,M}} \frac{1}{r_{2\alpha,\boldsymbol{\gamma}}(\boldsymbol{h})} \leq \frac{1}{(\gamma_1 M)^{(1/q-1)/(2\alpha)}} \frac{q}{1-q} \prod_{j=1}^{d} \left(1 + 2\gamma_j^q \zeta(2\alpha q)\right)^{1/q}.$$

Proof Suppose that the vectors $\boldsymbol{h} \in \mathbb{Z}^d$ are indexed by $i \in \mathbb{N}$ such that they are ordered according to the values of $(r_{2\alpha,\boldsymbol{\gamma}}(\boldsymbol{h}))^{-1}$ in a nonincreasing fashion. Let us denote this indexing by $\boldsymbol{h}^{(i)} \in \mathbb{Z}^d$. The ordering of vectors $\boldsymbol{h}$ with the same value of $(r_{2\alpha,\boldsymbol{\gamma}}(\boldsymbol{h}))^{-1}$ is of no importance and can be arbitrary. Then it is obvious that $\mathcal{A}_{d,M}$ contains the $\boldsymbol{h}^{(i)}$ with $i \in \{1, 2, \ldots, |\mathcal{A}_{d,M}|\}$, and hence

$$\sum_{\boldsymbol{h} \in \mathbb{Z}^d \setminus \mathcal{A}_{d,M}} \frac{1}{r_{2\alpha,\boldsymbol{\gamma}}(\boldsymbol{h})} = \sum_{i=|\mathcal{A}_{d,M}|+1}^{\infty} \frac{1}{r_{2\alpha,\boldsymbol{\gamma}}(\boldsymbol{h}^{(i)})}.$$

Furthermore, for any $i \in \mathbb{N}$ and any $q \in (1/(2\alpha), 1)$, we have, due to the ordering of the $\boldsymbol{h}^{(i)}$,

$$\frac{i}{(r_{2\alpha,\boldsymbol{\gamma}}(\boldsymbol{h}^{(i)}))^q} \leq \sum_{t=1}^{i} \frac{1}{(r_{2\alpha,\boldsymbol{\gamma}}(\boldsymbol{h}^{(t)}))^q},$$

and hence

$$\frac{1}{(r_{2\alpha,\boldsymbol{\gamma}}(\boldsymbol{h}^{(i)}))^q} \leq \frac{1}{i} \sum_{t=1}^{i} \frac{1}{(r_{2\alpha,\boldsymbol{\gamma}}(\boldsymbol{h}^{(t)}))^q}$$

$$\le \frac{1}{i}\sum_{t=1}^{\infty}\frac{1}{(r_{2\alpha,\gamma}(\boldsymbol{h}^{(t)}))^q}$$
$$= \frac{1}{i}\prod_{j=1}^{d}\left(1+2\gamma_j^q\zeta(2\alpha q)\right).$$

This yields

$$\sum_{\boldsymbol{h}\in\mathbb{Z}^d\setminus\mathcal{A}_{d,M}}\frac{1}{r_{2\alpha,\gamma}(\boldsymbol{h})} \le \sum_{i=|\mathcal{A}_{d,M}|+1}^{\infty}\frac{1}{i^{1/q}}\prod_{j=1}^{d}\left(1+2\gamma_j^q\zeta(2\alpha q)\right)^{1/q}$$
$$\le \left(\int_{|\mathcal{A}_{d,M}|}^{\infty}\frac{1}{x^{1/q}}\,\mathrm{d}x\right)\prod_{j=1}^{d}\left(1+2\gamma_j^q\zeta(2\alpha q)\right)^{1/q}$$
$$= \frac{1}{|\mathcal{A}_{d,M}|^{1/q-1}}\,\frac{q}{1-q}\prod_{j=1}^{d}\left(1+2\gamma_j^q\zeta(2\alpha q)\right)^{1/q}.$$

Recall that we know from Item 1 of Lemma 13.1 that for $\boldsymbol{h}=(h_1,\ldots,h_d)\in\mathcal{A}_{d,M}$ we have $|h_j|\le(\gamma_j M)^{1/(2\alpha)}$ for $j\in[d]$ and thus $|\mathcal{A}_{d,M}|\ge|\mathcal{A}_{1,M}|\ge(\gamma_1 M)^{1/(2\alpha)}$, where we used that $M\ge 1$, since otherwise $\mathcal{A}_{1,M}$ would be empty and the second inequality would not hold. This then yields

$$\sum_{\boldsymbol{h}\in\mathbb{Z}^d\setminus\mathcal{A}_{d,M}}\frac{1}{r_{2\alpha,\gamma}(\boldsymbol{h})} \le \frac{1}{(\gamma_1 M)^{(1/q-1)/(2\alpha)}}\,\frac{q}{1-q}\prod_{j=1}^{d}\left(1+2\gamma_j^q\zeta(2\alpha q)\right)^{1/q}. \quad \square$$

It remains to bound the term $\Sigma(T_{\boldsymbol{g},d,M})$ in (14.4) in a suitable way. For this task we now present two possible approaches.

Estimating $\Sigma(T_{\boldsymbol{g},d,M})$ by means of $S_{N,d,\alpha,\gamma}(\boldsymbol{g})$

The first way to bound $\Sigma(T_{\boldsymbol{g},d,M})$ is in observing that

$$\Sigma(T_{\boldsymbol{g},d,M}) = \sum_{\boldsymbol{h}\in\mathcal{A}_{d,M}}\ \sum_{\substack{\boldsymbol{p}\in\mathcal{A}_{d,M}\\ (\boldsymbol{h}-\boldsymbol{p})\cdot\boldsymbol{g}\equiv 0 \pmod N}}\ \sum_{\substack{\boldsymbol{\ell}\in\mathbb{Z}^d\setminus\{\boldsymbol{0},\boldsymbol{p}-\boldsymbol{h}\}\\ \boldsymbol{\ell}\cdot\boldsymbol{g}\equiv 0 \pmod N}}\frac{1}{r_{2\alpha,\gamma}(\boldsymbol{h}+\boldsymbol{\ell})}$$
$$\le |\mathcal{A}_{d,M}|\sum_{\boldsymbol{h}\in\mathcal{A}_{d,M}}\ \sum_{\substack{\boldsymbol{\ell}\in\mathbb{Z}^d\setminus\{\boldsymbol{0}\}\\ \boldsymbol{\ell}\cdot\boldsymbol{g}\equiv 0 \pmod N}}\frac{1}{r_{2\alpha,\gamma}(\boldsymbol{h}+\boldsymbol{\ell})}$$
$$\le |\mathcal{A}_{d,M}|\sum_{\boldsymbol{h}\in\mathcal{A}_{d,M}}\frac{M}{r_{2\alpha,\gamma}(\boldsymbol{h})}\ \sum_{\substack{\boldsymbol{\ell}\in\mathbb{Z}^d\setminus\{\boldsymbol{0}\}\\ \boldsymbol{\ell}\cdot\boldsymbol{g}\equiv 0 \pmod N}}\frac{1}{r_{2\alpha,\gamma}(\boldsymbol{h}+\boldsymbol{\ell})}$$
$$\le |\mathcal{A}_{d,M}|M[S_{N,d,\alpha,\gamma}(\boldsymbol{g})]^2,$$

where we used the definition of the set $\mathcal{A}_{d,M}$ in the penultimate step, and where $S_{N,d,\alpha,\boldsymbol{\gamma}}(\boldsymbol{g})$ is defined as in (13.11). Combining this estimate with Proposition 14.1 and Lemma 14.2 we get

$$[\mathrm{err}^{L_\infty-\mathrm{app}}(\mathcal{H}_{\mathrm{kor},d,\alpha,\boldsymbol{\gamma}}, A_{N,M,d,\alpha,\boldsymbol{\gamma}}(\boldsymbol{g}))]^2$$

$$\leq \frac{1}{(\gamma_1 M)^{(1/q-1)/(2\alpha)}} \frac{q}{1-q} \prod_{j=1}^{d} \left(1 + 2\gamma_j^q \zeta(2\alpha q)\right)^{1/q} + 3\,|\mathcal{A}_{d,M}| M [S_{N,d,\alpha,\boldsymbol{\gamma}}(\boldsymbol{g})]^2 \tag{14.8}$$

for any $q \in ((2\alpha)^{-1}, 1)$ as long as $M \geq 1$.

Fortunately, we are already familiar with the quantity $S_{N,d,\alpha,\boldsymbol{\gamma}}(\boldsymbol{g})$ and we have an algorithm for constructing reasonably good generating vectors $\boldsymbol{g}$ together with an upper bound on the corresponding value of $S_{N,d,\alpha,\boldsymbol{\gamma}}(\boldsymbol{g})$ from Theorem 13.5.

Theorem 14.3 *Let $N \geq 3$ be a prime number, let $\boldsymbol{\gamma} = (\gamma_j)_{j\geq 1}$ be product weights with $\gamma_j \in (0, 1]$ for all $j \in \mathbb{N}$, and assume that $\boldsymbol{g}$ has been found by Algorithm 13.3. Then for any $\tau \in (1/2, \alpha)$ we have*

$$\mathrm{err}^{L_\infty-\mathrm{app}}(\mathcal{H}_{\mathrm{kor},d,\alpha,\boldsymbol{\gamma}}, A_{N,M,d,\alpha,\boldsymbol{\gamma}}(\boldsymbol{g})) \leq \frac{C_{d,\alpha,\boldsymbol{\gamma},\tau}}{N^{\tau^2(2\tau-1)/(\alpha(2\tau+1)+\tau(2\tau-1))}},$$

where $M = N^{4\alpha\tau^2/(\alpha(2\tau+1)+\tau(2\tau-1))}$, and where

$$C_{d,\alpha,\boldsymbol{\gamma},\tau} = \left(3 + \frac{1}{\gamma_1^{(2\tau-1)/(2\alpha)}(2\tau-1)}\right)^{1/2} \prod_{j=1}^{d} \left(1 + 2^{4\alpha+1}\gamma_j^{1/(2\tau)} \zeta\left(\frac{\alpha}{\tau}\right)\right)^{2\tau+1/2}.$$

Proof Using (14.8), the bound on $|\mathcal{A}_{d,M}|$ in Item 4 of Lemma 13.1 with $\lambda = 1/(2\tau)$, Theorem 13.5, and setting also $q = 1/(2\tau)$, we obtain for any $\tau \in (1/2, \alpha)$ that

$$[\mathrm{err}^{L_\infty-\mathrm{app}}(\mathcal{H}_{\mathrm{kor},d,\alpha,\boldsymbol{\gamma}}, A_{N,M,d,\alpha,\boldsymbol{\gamma}}(\boldsymbol{g}))]^2$$

$$\leq \frac{C^{(1)}_{d,\alpha,\boldsymbol{\gamma},\tau}}{M^{(2\tau-1)/(2\alpha)}} + \frac{M^{1+1/(2\tau)} C^{(2)}_{d,\alpha,\boldsymbol{\gamma},\tau}}{N^{2\tau}}, \tag{14.9}$$

where

$$C^{(1)}_{d,\alpha,\boldsymbol{\gamma},\tau} = \frac{1}{\gamma_1^{(2\tau-1)/(2\alpha)}(2\tau-1)} \prod_{j=1}^{d} \left(1 + 2\gamma_j^{1/(2\tau)} \zeta\left(\frac{\alpha}{\tau}\right)\right)^{2\tau}$$

and

$$C^{(2)}_{d,\alpha,\boldsymbol{\gamma},\tau} = 3 \prod_{j=1}^{d} \left(1 + 2^{4\alpha+1}\gamma_j^{1/(2\tau)} \zeta\left(\frac{\alpha}{\tau}\right)\right)^{4\tau+1}.$$

We now balance the two terms in (14.9) with respect to the dependence on N and M by choosing

$$M = N^{4\alpha\tau^2/(\alpha(2\tau+1)+\tau(2\tau-1))}.$$

Then we obtain

$$[\mathrm{err}^{L_\infty\text{-app}}(\mathcal{H}_{\mathrm{kor},d,\alpha,\boldsymbol{\gamma}}, A_{N,M,d,\alpha,\boldsymbol{\gamma}}(\boldsymbol{g}))]^2 \leq \frac{C^{(1)}_{d,\alpha,\boldsymbol{\gamma},\tau} + C^{(2)}_{d,\alpha,\boldsymbol{\gamma},\tau}}{N^{2\tau^2(2\tau-1)/(\alpha(2\tau+1)+\tau(2\tau-1))}}.$$

Obviously, $C^{(1)}_{d,\alpha,\boldsymbol{\gamma},\tau} + C^{(2)}_{d,\alpha,\boldsymbol{\gamma},\tau} \leq (C_{d,\alpha,\boldsymbol{\gamma},\tau})^2$. Now the result follows by taking the square root. □

Since τ can be chosen arbitrarily close to α we achieve, using the result in Theorem 14.3, an error rate of order $O(N^{-\alpha/2+1/4+\delta})$ for arbitrarily small but positive δ.

We will analyze this result with respect to tractability in Section 14.3 below.

An alternative way of bounding $\Sigma(T_{\boldsymbol{g},d,M})$

Let us also outline an alternative way to proceed from the bound (14.4). Indeed, we can estimate

$$\Sigma(T_{\boldsymbol{g},d,M}) = \sum_{\boldsymbol{h}\in\mathcal{A}_{d,M}} \sum_{\boldsymbol{p}\in\mathcal{A}_{d,M}} \langle \tau_{\boldsymbol{h}}, \tau_{\boldsymbol{p}}\rangle_{\mathrm{kor},d,\alpha,\boldsymbol{\gamma}} \leq [X_{N,d,\alpha,\boldsymbol{\gamma}}(\boldsymbol{g})]^2,$$

where we put

$$X_{N,d,\alpha,\boldsymbol{\gamma}}(\boldsymbol{g}) := \sum_{\boldsymbol{h}\in\mathcal{A}_{d,M}} \|\tau_{\boldsymbol{h}}\|_{\mathrm{kor},d,\alpha,\boldsymbol{\gamma}}.$$

Note that we can express $X_{N,d,\alpha,\boldsymbol{\gamma}}(\boldsymbol{g})$ as

$$X_{N,d,\alpha,\boldsymbol{\gamma}}(\boldsymbol{g}) = \sum_{\boldsymbol{h}\in\mathcal{A}_{d,M}} \langle \tau_{\boldsymbol{h}}, \tau_{\boldsymbol{h}}\rangle^{1/2}_{\mathrm{kor},d,\alpha,\boldsymbol{\gamma}} = \sum_{\boldsymbol{h}\in\mathcal{A}_{d,M}} \left(\sum_{\substack{\boldsymbol{\ell}\in\mathbb{Z}^d\setminus\{\boldsymbol{0}\}\\ \boldsymbol{\ell}\cdot\boldsymbol{g}\equiv 0 \ (\mathrm{mod}\ N)}} \frac{1}{r_{2\alpha,\boldsymbol{\gamma}}(\boldsymbol{h}+\boldsymbol{\ell})} \right)^{1/2}.$$

Accordingly, we get from (14.4) and Lemma 14.2 that

$$\begin{aligned} &[\mathrm{err}^{L_\infty\text{-app}}(\mathcal{H}_{\mathrm{kor},d,\alpha,\boldsymbol{\gamma}}, A_{N,M,d,\alpha,\boldsymbol{\gamma}}(\boldsymbol{g}))]^2 \\ &\leq \sum_{\boldsymbol{h}\in\mathbb{Z}^d\setminus\mathcal{A}_{d,M}} \frac{1}{r_{2\alpha,\boldsymbol{\gamma}}(\boldsymbol{h})} + 3\,[X_{N,d,\alpha,\boldsymbol{\gamma}}(\boldsymbol{g})]^2 \\ &\leq \frac{1}{(\gamma_1 M)^{(1/q-1)/(2\alpha)}} \frac{q}{1-q} \prod_{j=1}^{d} \left(1+2\gamma_j^q \zeta(2\alpha q)\right)^{1/q} + 3\,[X_{N,d,\alpha,\boldsymbol{\gamma}}(\boldsymbol{g})]^2, \end{aligned} \tag{14.10}$$

as long as $M \geq 1$. This naturally leads to the strategy to use the quantity $X_{N,d,\alpha,\boldsymbol{\gamma}}(\boldsymbol{g})$ as a search criterion for good generating vectors of rank-1 lattice rules for L_∞-approximation in the Korobov space. This strategy was first pursued in [166], and we shall outline some of the main findings of that paper below. As it turns out, a search based on $X_{N,d,\alpha,\gamma}(\boldsymbol{g})$ leads to a better error bound than the search based on $S_{N,d,\alpha,\gamma}(\boldsymbol{g})$ that we presented above. The drawback of this result is that we require the smoothness parameter α to be strictly bigger than 1. We shall make this assumption for the rest of this section.

If we use $X_{N,d,\alpha,\gamma}(\boldsymbol{g})$ as a quality criterion for generating vectors of rank-1 lattice rules for L_∞-approximation, it is an obvious question whether we can optimize this criterion by using a CBC construction, and the answer to this question is affirmative. Indeed, the following algorithm is similar to Algorithm 13.3 that we used for finding $\boldsymbol{g}$ with low values of $S_{N,d,\alpha,\gamma}(\boldsymbol{g})$.

Algorithm 14.4 (CBC construction for L_∞-approximation in the weighted Korobov space) Let prime N and $d \in \mathbb{N}$ be given. Let $M \geq 1$ and $\alpha > 1$, and let $\boldsymbol{\gamma} = (\gamma_j)_{j\geq 1}$ be product weights. Construct a generating vector $\boldsymbol{g} = (g_1, \ldots, g_d) \in G_d^\varphi(N)$ as follows.

(1) Choose $g_1 = 1$.
(2) For s from 1 to $d-1$:
Assume that $g_1, \ldots, g_s \in G_1^\varphi(N)$ have already been found. Choose $g_{s+1} \in G_1^\varphi(N)$ as

$$g_{s+1} := \underset{g \in G_1^\varphi(N)}{\operatorname{argmin}} \; X_{N,s+1,\alpha,\gamma}((g_1, \ldots, g_s, g)).$$

End for.
(3) Set $\boldsymbol{g} = (g_1, \ldots, g_d)$.

Remark 14.5 As pointed out in [166], Algorithm 14.4 can be efficiently implemented, using methods outlined in [163] combined with the usual methods for fast CBC constructions (see Chapter 3), to obtain a runtime of $O(|\mathcal{A}_{d,M}| d N \log N)$ operations. This runtime can be seen as a slight disadvantage in comparison to Algorithm 13.3 for finding generating vectors with small values of $S_{N,d,\alpha,\gamma}(\boldsymbol{g})$, as the latter has a runtime of $O(d N \log N)$ operations.

Next we show that the generating vectors constructed by Algorithm 14.4 indeed yield a low value of $X_{N,d,\alpha,\gamma}$. However, as already pointed out, we have to assume that α is strictly greater than one in order to show this result.

Theorem 14.6 *Let $\alpha > 1$ and $M \geq 1$ be given. Let $N \geq 2M^{1/(2\alpha)}$ be a prime number, and let $\boldsymbol{\gamma} = (\gamma_j)_{j\geq 1}$ be product weights with $\gamma_j \in (0, 1]$ for $j \in \mathbb{N}$. Assume that $\boldsymbol{g}$ has been found by Algorithm 14.4. Then for arbitrary $\tau \in [1, \alpha)$ and for any $s \in [d]$ we have*

$$X_{N,s,\alpha,\gamma}((g_1, \ldots, g_s)) \leq \frac{4^\tau |\mathcal{A}_{s,M}|^\tau}{N^\tau} \prod_{j=1}^{s} \left(1 + 2^{2+\alpha/\tau} \gamma_j^{1/(2\tau)} \zeta\left(\frac{\alpha}{\tau}\right)\right)^\tau. \quad (14.11)$$

Proof We prove the result by induction on s. For $s = 1$, we use that $g_1 = 1$, and Jensen's inequality (see Lemma 2.25) for $\lambda \in (1/\alpha, 1]$ to obtain

$$
\begin{aligned}
(X_{N,1,\alpha,\gamma_1}(1))^\lambda &= \left(\sum_{h \in \mathcal{A}_{1,M}} \left(\sum_{\substack{\ell \in \mathbb{Z}\setminus\{0\} \\ \ell \equiv 0 \pmod N}} \frac{1}{r_{2\alpha,\gamma_1}(h+\ell)} \right)^{1/2} \right)^\lambda \\
&\le \sum_{h \in \mathcal{A}_{1,M}} \sum_{\substack{\ell \in \mathbb{Z}\setminus\{0\} \\ \ell \equiv 0 \pmod N}} \frac{1}{r_{\alpha\lambda,\gamma_1^{\lambda/2}}(h+\ell)}.
\end{aligned}
$$

For $h \in \mathcal{A}_{1,M}$ we have by Item 1 of Lemma 13.1,

$$
|h| \le (\gamma_1 M)^{1/(2\alpha)} \le \frac{N}{2},
$$

according to our assumption on N. In particular, this implies that for every $h \in \mathcal{A}_{1,M}$ and every $\ell \in \mathbb{Z} \setminus \{0\}$ with $\ell \equiv 0 \pmod N$ we have $h + \ell \neq 0$.

Now, for $h \in \mathcal{A}_{1,M}$, $h \ge 0$, we obtain

$$
\begin{aligned}
&\sum_{\substack{\ell \in \mathbb{Z}\setminus\{0\} \\ \ell \equiv 0 \pmod N}} \frac{1}{r_{\alpha\lambda,\gamma_1^{\lambda/2}}(h+\ell)} \\
&= \gamma_1^{\lambda/2} \left(\sum_{\ell=1}^\infty \frac{1}{(h+\ell N)^{\alpha\lambda}} + \sum_{\ell=1}^\infty \frac{1}{(\ell N - h)^{\alpha\lambda}} \right) \\
&= \gamma_1^{\lambda/2} \left(\sum_{\ell=1}^\infty \frac{1}{(h+\ell N)^{\alpha\lambda}} + \frac{1}{(N-h)^{\alpha\lambda}} + \sum_{\ell=1}^\infty \frac{1}{(\ell N + N - h)^{\alpha\lambda}} \right) \\
&\le \gamma_1^{\lambda/2} \left(\frac{2}{N^{\alpha\lambda}} \sum_{\ell=1}^\infty \frac{1}{\ell^{\alpha\lambda}} + \frac{2^{\alpha\lambda}}{N^{\alpha\lambda}} \right) \\
&= \frac{\gamma_1^{\lambda/2}}{N^{\alpha\lambda}} (2\zeta(\alpha\lambda) + 2^{\alpha\lambda}) \\
&\le \frac{\gamma_1^{\lambda/2} 2^{1+\alpha\lambda} \zeta(\alpha\lambda)}{N^{\alpha\lambda}}.
\end{aligned}
$$

For $h \in \mathcal{A}_{1,M}$, $h < 0$ we have

$$
\begin{aligned}
\sum_{\substack{\ell \in \mathbb{Z}\setminus\{0\} \\ \ell \equiv 0 \pmod N}} \frac{1}{r_{\alpha\lambda,\gamma_1^{\lambda/2}}(h+\ell)} &= \sum_{\substack{\ell \in \mathbb{Z}\setminus\{0\} \\ \ell \equiv 0 \pmod N}} \frac{1}{r_{\alpha\lambda,\gamma_1^{\lambda/2}}(-h-\ell)} \\
&= \sum_{\substack{\ell \in \mathbb{Z}\setminus\{0\} \\ \ell \equiv 0 \pmod N}} \frac{1}{r_{\alpha\lambda,\gamma_1^{\lambda/2}}(-h+\ell)} \\
&\le \frac{\gamma_1^{\lambda/2} 2^{1+\alpha\lambda} \zeta(\alpha\lambda)}{N^{\alpha\lambda}}.
\end{aligned}
$$

This yields

$$(X_{N,1,\alpha,\gamma_1}(1))^\lambda \le |\mathcal{A}_{1,M}| \frac{\gamma_1^{\lambda/2} 2^{1+\alpha\lambda} \zeta(\alpha\lambda)}{N^{\alpha\lambda}}.$$

Putting $\tau = 1/\lambda$, which implies $\tau \in [1, \alpha)$, implies

$$(X_{N,1,\alpha,\gamma_1}(1))^{1/\tau} \le \frac{|\mathcal{A}_{1,M}|}{N^{\alpha/\tau}} \gamma_1^{1/(2\tau)} 2^{1+\alpha/\tau} \zeta\left(\frac{\alpha}{\tau}\right).$$

Taking both sides of the latter inequality to the τ-th power gives

$$X_{N,1,\alpha,\gamma_1}(1) \le \frac{|\mathcal{A}_{1,M}|^\tau}{N^\alpha} \left(\gamma_1^{1/(2\tau)} 2^{1+\alpha/\tau} \zeta\left(\frac{\alpha}{\tau}\right)\right)^\tau$$

and this implies the result for $s = 1$.

In the induction step, let $\boldsymbol{g}^{(s)} = (g_1, \ldots, g_s) \in G_s^\varphi(N)$ denote the s-dimensional generating vector that has been obtained in the first s steps of the algorithm, and suppose that the claimed error bound (14.11) holds for the lattice rule generated by $\boldsymbol{g}^{(s)}$ (with $\tau = 1/\lambda$). Once again, we write $(\boldsymbol{g}^{(s)}, g_{s+1})$ for the vector $(g_1, \ldots, g_s, g_{s+1}) \in G_{s+1}^\varphi(N)$, where g_{s+1} is the $(s+1)$-st component of the generating vector selected by Algorithm 14.4. For $X_{N,s+1,\alpha,\boldsymbol{\gamma}}((\boldsymbol{g}^{(s)}, g_{s+1}))$ we can distinguish two cases according to whether the $(s+1)$-st component of $\boldsymbol{\ell} \in \mathbb{Z}^{s+1} \setminus \{\boldsymbol{0}\}$ is zero or not, which yields

$$\begin{aligned}
&X_{N,s+1,\alpha,\boldsymbol{\gamma}}((\boldsymbol{g}^{(s)}, g_{s+1})) \\
&= \sum_{(\boldsymbol{h},h_{s+1}) \in \mathcal{A}_{s+1,M}} \left(\frac{1}{r_{2\alpha,\gamma_{s+1}}(h_{s+1})} \sum_{\substack{\boldsymbol{\ell} \in \mathbb{Z}^s \setminus \{\boldsymbol{0}\} \\ \boldsymbol{\ell} \cdot \boldsymbol{g}^{(s)} \equiv 0 \pmod N}} \frac{1}{r_{2\alpha,\boldsymbol{\gamma}}(\boldsymbol{h} + \boldsymbol{\ell})} \right. \\
&\quad \left. + \sum_{\ell_{s+1} \in \mathbb{Z} \setminus \{0\}} \sum_{\substack{\boldsymbol{\ell} \in \mathbb{Z}^s \\ \boldsymbol{\ell} \cdot \boldsymbol{g}^{(s)} \equiv -\ell_{s+1} g_{s+1} \pmod N}} \frac{1}{r_{2\alpha,\boldsymbol{\gamma}}((\boldsymbol{h}, h_{s+1}) + (\boldsymbol{\ell}, \ell_{s+1}))} \right)^{1/2} \\
&\le \sum_{(\boldsymbol{h},h_{s+1}) \in \mathcal{A}_{s+1,M}} \left(\frac{1}{r_{2\alpha,\gamma_{s+1}}(h_{s+1})} \sum_{\substack{\boldsymbol{\ell} \in \mathbb{Z}^s \setminus \{\boldsymbol{0}\} \\ \boldsymbol{\ell} \cdot \boldsymbol{g}^{(s)} \equiv 0 \pmod N}} \frac{1}{r_{2\alpha,\boldsymbol{\gamma}}(\boldsymbol{h} + \boldsymbol{\ell})} \right)^{1/2} \\
&\quad + \theta_{N,s,\alpha,\boldsymbol{\gamma}}(\boldsymbol{g}^{(s)}, g_{s+1})
\end{aligned}$$

$$\le \sum_{h_{s+1}\in\mathbb{Z}} \frac{1}{r_{\alpha,\gamma_{s+1}^{1/2}}(h_{s+1})} \sum_{\boldsymbol{h}\in\mathcal{A}_{s,M}} \left(\sum_{\substack{\boldsymbol{\ell}\in\mathbb{Z}^s\setminus\{\boldsymbol{0}\} \\ \boldsymbol{\ell}\cdot\boldsymbol{g}^{(s)}\equiv 0 \pmod N}} \frac{1}{r_{2\alpha,\boldsymbol{\gamma}}(\boldsymbol{h}+\boldsymbol{\ell})} \right)^{1/2}$$
$$+\theta_{N,s,\alpha,\boldsymbol{\gamma}}(\boldsymbol{g}^{(s)}, g_{s+1})$$
$$= (1+2\gamma_{s+1}^{1/2}\zeta(\alpha))X_{N,s,\alpha,\boldsymbol{\gamma}}(\boldsymbol{g}^{(s)}) + \theta_{N,s,\alpha,\boldsymbol{\gamma}}(\boldsymbol{g}^{(s)}, g_{s+1}),$$

where we used Item 2 of Lemma 13.1, and where we write

$$\theta_{N,s,\alpha,\boldsymbol{\gamma}}(\boldsymbol{g}^{(s)}, g_{s+1})$$
$$:= \sum_{(\boldsymbol{h},h_{s+1})\in\mathcal{A}_{s+1,M}} \left(\sum_{\ell_{s+1}\in\mathbb{Z}\setminus\{0\}} \sum_{\substack{\boldsymbol{\ell}\in\mathbb{Z}^s \\ \boldsymbol{\ell}\cdot\boldsymbol{g}^{(s)}\equiv -\ell_{s+1}g_{s+1} \pmod N}} \frac{1}{r_{2\alpha,\gamma_{s+1}}(h_{s+1}+\ell_{s+1})} \frac{1}{r_{2\alpha,\boldsymbol{\gamma}}(\boldsymbol{h}+\boldsymbol{\ell})} \right)^{1/2}.$$

Next, we use Jensen's inequality and the standard averaging argument (see Remark 2.14), such that we obtain, for any $\lambda \in (1/\alpha, 1]$,

$$(X_{N,s+1,\alpha,\boldsymbol{\gamma}}((\boldsymbol{g}^{(s)}, g_{s+1})))^\lambda$$
$$\le (1+2\gamma_{s+1}^{\lambda/2}\zeta(\alpha\lambda))(X_{N,s,\alpha,\boldsymbol{\gamma}}(\boldsymbol{g}^{(s)}))^\lambda + \frac{1}{N-1}\sum_{g=1}^{N-1}(\theta_{N,s,\alpha,\boldsymbol{\gamma}}(\boldsymbol{g}^{(s)}, g))^\lambda. \tag{14.12}$$

Using a procedure that is long and tedious, but very similar to what we did in the proof of Theorem 13.5, it is possible to obtain the following estimate (we refer to [162, 163, 166] for further details),

$$\frac{1}{N-1}\sum_{g=1}^{N-1}(\theta_{N,s,\alpha,\boldsymbol{\gamma}}(\boldsymbol{g}^{(s)}, g))^\lambda$$
$$\le \frac{2\gamma_{s+1}^{\lambda/2}\zeta(\alpha\lambda)}{N-1}\left(\prod_{j=1}^{s}\left(1+2\gamma_j^{\lambda/2}\zeta(\alpha\lambda)\right)\right)|\mathcal{A}_{s,M}|$$
$$+\frac{1+2\gamma_{s+1}^{\lambda/2}\zeta(\alpha\lambda)}{N-1}\left(\prod_{j=1}^{s}\left(1+2\gamma_j^{\lambda/2}\zeta(\alpha\lambda)\right)\right)\sum_{h_{s+1}\in\mathbb{Z}\setminus\{0\}}|\mathcal{A}_{s,\widetilde{M}(h_{s+1})}|$$
$$+\frac{\gamma_{s+1}^{\lambda/2}2^{1+\alpha\lambda}\zeta(\alpha\lambda)}{N-1}\left(\prod_{j=1}^{s}\left(1+2\gamma_j^{\lambda/2}\zeta(\alpha\lambda)\right)\right)\sum_{h_{s+1}\in\mathbb{Z}\setminus\{0\}}|\mathcal{A}_{s,\widetilde{M}(h_{s+1})}|$$

$$\begin{aligned}
&\le \frac{2\gamma_{s+1}^{\lambda/2}\zeta(\alpha\lambda)}{N-1}\left(\prod_{j=1}^{s}\left(1+2\gamma_j^{\lambda/2}\zeta(\alpha\lambda)\right)\right)|\mathcal{A}_{s,M}| \\
&\quad+\frac{1+\gamma_{s+1}^{\lambda/2}2^{2+\alpha\lambda}\zeta(\alpha\lambda)}{N-1}\left(\prod_{j=1}^{s}\left(1+2\gamma_j^{\lambda/2}\zeta(\alpha\lambda)\right)\right)\sum_{h_{s+1}\in\mathbb{Z}\setminus\{0\}}|\mathcal{A}_{s,\widetilde{M}(h_{s+1})}| \\
&\le \frac{4\gamma_{s+1}^{\lambda/2}\zeta(\alpha\lambda)}{N}\left(\prod_{j=1}^{s}\left(1+2\gamma_j^{\lambda/2}\zeta(\alpha\lambda)\right)\right)|\mathcal{A}_{s,M}| \\
&\quad+\frac{2(1+\gamma_{s+1}^{\lambda/2}2^{2+\alpha\lambda}\zeta(\alpha\lambda))}{N}\left(\prod_{j=1}^{s}\left(1+2\gamma_j^{\lambda/2}\zeta(\alpha\lambda)\right)\right)\sum_{h_{s+1}\in\mathbb{Z}\setminus\{0\}}|\mathcal{A}_{s,\widetilde{M}(h_{s+1})}|,
\end{aligned}$$

where, as in Lemma 13.1, we used the notation $\widetilde{M}(h_{s+1}) := (\gamma_{s+1}M)/|h_{s+1}|^{2\alpha}$, and where we made use of the fact that $1/(N-1) \le 2/N$ in the last inequality.

Thus, inserting into (14.12) and using the induction assumption with $\tau = 1/\lambda$, we obtain

$$\begin{aligned}
&(X_{N,s+1,\alpha,\boldsymbol{\gamma}}((\boldsymbol{g}^{(s)}, g_{s+1})))^{\lambda} \\
&\le (1+2\gamma_{s+1}^{\lambda/2}\zeta(\alpha\lambda))\frac{4|\mathcal{A}_{s,M}|}{N}\prod_{j=1}^{s}\left(1+2^{2+\alpha\lambda}\gamma_j^{\lambda/2}\zeta(\alpha\lambda)\right) \\
&\quad+\frac{4\gamma_{s+1}^{\lambda/2}\zeta(\alpha\lambda)}{N}\left(\prod_{j=1}^{s}\left(1+2\gamma_j^{\lambda/2}\zeta(\alpha\lambda)\right)\right)|\mathcal{A}_{s,M}| \\
&\quad+\frac{2(1+\gamma_{s+1}^{\lambda/2}2^{2+\alpha\lambda}\zeta(\alpha\lambda))}{N}\left(\prod_{j=1}^{s}\left(1+2\gamma_j^{\lambda/2}\zeta(\alpha\lambda)\right)\right)\sum_{h_{s+1}\in\mathbb{Z}\setminus\{0\}}|\mathcal{A}_{s,\widetilde{M}(h_{s+1})}| \\
&\le \frac{4|\mathcal{A}_{s,M}|}{N}\prod_{j=1}^{s+1}\left(1+2^{2+\alpha\lambda}\gamma_j^{\lambda/2}\zeta(\alpha\lambda)\right) \\
&\quad+\frac{4}{N}\left(\sum_{h_{s+1}\in\mathbb{Z}\setminus\{0\}}|\mathcal{A}_{s,\widetilde{M}(h_{s+1})}|\right)\prod_{j=1}^{s+1}\left(1+2^{2+\alpha\lambda}\gamma_j^{\lambda/2}\zeta(\alpha\lambda)\right) \\
&= \frac{4|\mathcal{A}_{s+1,M}|}{N}\prod_{j=1}^{s+1}\left(1+2^{2+\alpha\lambda}\gamma_j^{\lambda/2}\zeta(\alpha\lambda)\right),
\end{aligned}$$

where we used Item 3 of Lemma 13.1, which is

$$|\mathcal{A}_{s,M}| + \sum_{h_{s+1}\in\mathbb{Z}\setminus\{0\}}|\mathcal{A}_{s,\widetilde{M}(h_{s+1})}| = |\mathcal{A}_{s+1,M}|.$$

Finally, set $\tau = 1/\lambda$ to obtain the result in the theorem. □

We can now show the following result.

Theorem 14.7 *Let $\alpha > 1$ be given and let $\boldsymbol{\gamma} = (\gamma_j)_{j\ge 1}$ be product weights with $\gamma_j \in (0, 1]$ for $j \in \mathbb{N}$. Let $N \ge 2^{(4\alpha-1)/(2\alpha-1)}$ be a prime number. Assume that $\boldsymbol{g}$ has been found by Algorithm 14.4. Then for any $\tau \in (1, \alpha)$ it is true that*

$$\mathrm{err}^{L_\infty\text{-app}}(\mathcal{H}_{\mathrm{kor},d,\alpha,\boldsymbol{\gamma}}, A_{N,M,d,\alpha,\boldsymbol{\gamma}}(\boldsymbol{g})) \le \frac{\widetilde{C}_{d,\alpha,\boldsymbol{\gamma},\tau}}{N^{\tau(2\tau-1)/(2\alpha+2\tau-1)}},$$

where $M = N^{(4\tau\alpha)/(2\alpha+2\tau-1)}$, and where

$$\widetilde{C}_{d,\alpha,\boldsymbol{\gamma},\tau} = \left(3\cdot 4^{2\tau} + \frac{1}{\gamma_1^{(2\tau-1)/(2\alpha)}(2\tau-1)}\right)^{1/2} \prod_{j=1}^{d}\left(1 + 2^{2+\alpha/\tau}\gamma_j^{1/(2\tau)}\zeta\left(\frac{\alpha}{\tau}\right)\right)^{2\tau}.$$

Proof Combining Theorem 14.6 with (14.10) where we set $q = 1/(2\tau)$, and using Item 4 of Lemma 13.1 with $\lambda = 1/(2\tau)$, we obtain

$$[\mathrm{err}^{L_\infty\text{-app}}(\mathcal{H}_{\mathrm{kor},d,\alpha,\boldsymbol{\gamma}}, A_{N,M,d,\alpha,\boldsymbol{\gamma}}(\boldsymbol{g}))]^2 \le \frac{\widetilde{C}^{(1)}_{d,\alpha,\boldsymbol{\gamma},\tau}}{M^{(2\tau-1)/(2\alpha)}} + \frac{M\,\widetilde{C}^{(2)}_{d,\alpha,\boldsymbol{\gamma},\tau}}{N^{2\tau}}, \tag{14.13}$$

whenever $N \ge 2M^{1/(2\alpha)}$, where

$$\widetilde{C}^{(1)}_{d,\alpha,\boldsymbol{\gamma},\tau} := \frac{1}{\gamma_1^{(2\tau-1)/(2\alpha)}(2\tau-1)} \prod_{j=1}^{d}\left(1 + 2\gamma_j^{1/(2\tau)}\zeta\left(\frac{\alpha}{\tau}\right)\right)^{2\tau}$$

and

$$\widetilde{C}^{(2)}_{d,\alpha,\boldsymbol{\gamma},\tau} := 3\cdot 4^{2\tau}\prod_{j=1}^{d}\left(1 + 2^{2+\alpha/\tau}\gamma_j^{1/(2\tau)}\zeta\left(\frac{\alpha}{\tau}\right)\right)^{4\tau}.$$

Again we balance the terms in (14.13) with respect to the dependence on N and M. This time we put

$$M = N^{4\tau\alpha/(2\alpha+2\tau-1)}.$$

This means that in order to fulfill the condition $N \ge 2M^{1/(2\alpha)}$, we need $N \ge 2^{(2\alpha+2\tau-1)/(2\alpha-1)}$. This is certainly satisfied if $N \ge 2^{(4\alpha-1)/(2\alpha-1)}$. For such N we obtain

$$[\mathrm{err}^{L_\infty\text{-app}}(\mathcal{H}_{\mathrm{kor},d,\alpha,\boldsymbol{\gamma}}, A_{N,M,d,\alpha,\boldsymbol{\gamma}}(\boldsymbol{g}))]^2 \le \frac{\widetilde{C}^{(1)}_{d,\alpha,\boldsymbol{\gamma},\tau} + \widetilde{C}^{(2)}_{d,\alpha,\boldsymbol{\gamma},\tau}}{N^{2\tau(2\tau-1)/(2\alpha+2\tau-1)}}.$$

Obviously, $\widetilde{C}^{(1)}_{d,\alpha,\boldsymbol{\gamma},\tau} + \widetilde{C}^{(2)}_{d,\alpha,\boldsymbol{\gamma},\tau} \le (\widetilde{C}_{d,\alpha,\boldsymbol{\gamma},\tau})^2$. Now the result follows by taking the square root. □

Since τ can be chosen arbitrarily close to α we achieve in this way an error rate of order $O(N^{-\alpha(2\alpha-1)/(4\alpha-1)+\delta})$ for arbitrarily small but positive δ. Note that we always have, as $\alpha > 1$,

$$\frac{\alpha(2\alpha-1)}{4\alpha-1} = \frac{\alpha}{2}\frac{4\alpha-2}{4\alpha-1} > \frac{\alpha}{2} - \frac{1}{4},$$

the latter being the corresponding value from the first approach. This means that we have an improved convergence rate for the worst-case error with this second method.

We will analyze also this result with respect to tractability in Section 14.3 below.

14.2 L_∞-Approximation of Functions in Korobov Spaces Using Splines

Finally, as an alternative to the method outlined in Section 14.1, we would like to describe a particular approach to L_∞-approximation based on splines. As we shall see, this makes it possible to obtain a convergence rate that can be as good as the one obtained in Theorem 14.7, but without having to make the additional assumption that $\alpha > 1$ (see Theorem 14.11 below).

The basic spline algorithm

We start with a summary of some definitions and facts regarding splines. Suppose that we are considering the Korobov space $\mathcal{H}_{\mathrm{kor},d,\alpha,\boldsymbol{\gamma}}$ of smoothness α with product weights $\boldsymbol{\gamma} = (\gamma_j)_{j\geq 1}$, and that we are given an N-element point set $\mathcal{P}$ with points $\boldsymbol{x}_0, \boldsymbol{x}_1, \ldots, \boldsymbol{x}_{N-1} \in [0,1)^d$, along with function data $f(\boldsymbol{x}_k)$, $k \in \{0,1,\ldots,N-1\}$, for $f \in \mathcal{H}_{\mathrm{kor},d,\alpha,\boldsymbol{\gamma}}$. We then define a spline

$$s_{f,d,\mathcal{P}} := \operatorname{argmin}\{\|h\|_{\mathrm{kor},d,\alpha,\boldsymbol{\gamma}} : h \in \mathcal{H}_{\mathrm{kor},d,\alpha,\boldsymbol{\gamma}}, h(\boldsymbol{x}_k) = f(\boldsymbol{x}_k) \ \forall\, k \in \{0,1,\ldots,N-1\}\}.$$

It can be shown that

$$s_{f,d,\mathcal{P}}(\boldsymbol{x}) = \sum_{k=0}^{N-1} K_{\mathrm{kor},d,\alpha,\boldsymbol{\gamma}}(\boldsymbol{x},\boldsymbol{x}_k)\, c_k \quad \text{for} \quad \boldsymbol{x} \in [0,1]^d,$$

where we recall that $K_{\mathrm{kor},d,\alpha,\boldsymbol{\gamma}}$ denotes the reproducing kernel of the weighted Korobov space. The coefficients c_k in the previous equation are determined by the so-called interpolating conditions

$$f(\boldsymbol{x}_\ell) = \sum_{k=0}^{N-1} K_{\mathrm{kor},d,\alpha,\boldsymbol{\gamma}}(\boldsymbol{x}_\ell,\boldsymbol{x}_k)\, c_k.$$

Now, we may use the concept of a Lagrange or cardinal basis, which we denote by ϕ_k, $k \in \{0,1,\ldots,N-1\}$. Each of the cardinal basis functions ϕ_k is given in terms of the kernel,

$$\phi_k(\boldsymbol{x}) = \sum_{\ell=0}^{N-1} K_{\mathrm{kor},d,\alpha,\boldsymbol{\gamma}}(\boldsymbol{x},\boldsymbol{x}_\ell)\,\widetilde{c}_{\ell,k} \quad \text{for} \quad \boldsymbol{x} \in [0,1]^d, \tag{14.14}$$

with the $\widetilde{c}_{\ell,k}$ are given by the conditions

$$\delta_{i,k} := \sum_{\ell=0}^{N-1} K_{\mathrm{kor},d,\alpha,\boldsymbol{\gamma}}(\boldsymbol{x}_i,\boldsymbol{x}_\ell)\,\widetilde{c}_{\ell,k},$$

where δ denotes the Kronecker delta function. Since the Korobov kernel is symmetric, we have $K_{\mathrm{kor},d,\alpha,\boldsymbol{\gamma}}(\boldsymbol{x}_i,\boldsymbol{x}_\ell) = K_{\mathrm{kor},d,\alpha,\boldsymbol{\gamma}}(\boldsymbol{x}_\ell,\boldsymbol{x}_i)$ and from this it is easy to see that also the $\widetilde{c}_{\ell,k}$ are symmetric, i.e., $\widetilde{c}_{\ell,k} = \widetilde{c}_{k,\ell}$. Using the cardinal basis, we can then write the spline in the form

$$s_{f,d,\mathcal{P}}(\boldsymbol{x}) = \sum_{k=0}^{N-1} f(\boldsymbol{x}_k)\,\phi_k(\boldsymbol{x}) \quad \text{for} \quad \boldsymbol{x} \in [0,1]^d. \tag{14.15}$$

Let us briefly discuss whether choosing $s_{f,d,\mathcal{P}}$ as an algorithm for L_∞-approximation is a good choice, at least among linear algorithms of the form

$$S_{f,\mathcal{P}}(\boldsymbol{x}) := \sum_{k=0}^{N-1} f(\boldsymbol{x}_k)\,a_k(\boldsymbol{x}) \quad \text{for} \quad \boldsymbol{x} \in [0,1]^d$$

for some suitably chosen a_k, where $f \in \mathcal{H}_{\mathrm{kor},d,\alpha,\boldsymbol{\gamma}}$. Using the reproducing property of $K_{\mathrm{kor},d,\alpha,\boldsymbol{\gamma}}$ and the Cauchy–Schwarz inequality, we obtain, for any $\boldsymbol{x} \in [0,1]^d$,

$$\begin{aligned}
&|f(\boldsymbol{x}) - S_{f,\mathcal{P}}(\boldsymbol{x})| \\
&= \left| \left\langle f, K_{\mathrm{kor},d,\alpha,\boldsymbol{\gamma}}(\cdot,\boldsymbol{x}) \right\rangle_{\mathrm{kor},d,\alpha,\boldsymbol{\gamma}} - \sum_{k=0}^{N-1} \left\langle f, K_{\mathrm{kor},d,\alpha,\boldsymbol{\gamma}}(\cdot,\boldsymbol{x}_k) \right\rangle_{\mathrm{kor},d,\alpha,\boldsymbol{\gamma}} a_k(\boldsymbol{x}) \right| \\
&= \left| \left\langle f, K_{\mathrm{kor},d,\alpha,\boldsymbol{\gamma}}(\cdot,\boldsymbol{x}) - \sum_{k=0}^{N-1} K_{\mathrm{kor},d,\alpha,\boldsymbol{\gamma}}(\cdot,\boldsymbol{x}_k)\,a_k(\boldsymbol{x}) \right\rangle_{\mathrm{kor},d,\alpha,\boldsymbol{\gamma}} \right| \\
&\le \|f\|_{\mathrm{kor},d,\alpha,\boldsymbol{\gamma}} \left\| K_{\mathrm{kor},d,\alpha,\boldsymbol{\gamma}}(\cdot,\boldsymbol{x}) - \sum_{k=0}^{N-1} K_{\mathrm{kor},d,\alpha,\boldsymbol{\gamma}}(\cdot,\boldsymbol{x}_k)\,a_k(\boldsymbol{x}) \right\|_{\mathrm{kor},d,\alpha,\boldsymbol{\gamma}}.
\end{aligned}$$

For an arbitrarily chosen but fixed $\boldsymbol{x} \in [0,1]^d$, it is easy to see that the latter inequality is sharp for an appropriate choice of f. This yields

$$\begin{aligned}
&\sup_{\substack{f \in \mathcal{H}_{\mathrm{kor},d,\alpha,\boldsymbol{\gamma}} \\ \|f\|_{\mathrm{kor},d,\alpha,\boldsymbol{\gamma}} \le 1}} \sup_{\boldsymbol{x} \in [0,1]^d} |f(\boldsymbol{x}) - S_{f,\mathcal{P}}(\boldsymbol{x})| \\
&\le \sup_{\boldsymbol{x} \in [0,1]^d} \left\| K_{\mathrm{kor},d,\alpha,\boldsymbol{\gamma}}(\cdot,\boldsymbol{x}) - \sum_{k=0}^{N-1} K_{\mathrm{kor},d,\alpha,\boldsymbol{\gamma}}(\cdot,\boldsymbol{x}_k)\,a_k(\boldsymbol{x}) \right\|_{\mathrm{kor},d,\alpha,\boldsymbol{\gamma}}
\end{aligned}$$

$$= \sup_{\boldsymbol{x}\in[0,1]^d} \left(K_{\mathrm{kor},d,\alpha,\boldsymbol{\gamma}}(\boldsymbol{x},\boldsymbol{x}) - 2\sum_{k=0}^{N-1} K_{\mathrm{kor},d,\alpha,\boldsymbol{\gamma}}(\boldsymbol{x},\boldsymbol{x}_k)\, a_k(\boldsymbol{x}) \right.$$
$$\left. + \sum_{k,\ell=0}^{N-1} a_\ell(\boldsymbol{x}) K_{\mathrm{kor},d,\alpha,\boldsymbol{\gamma}}(\boldsymbol{x}_\ell,\boldsymbol{x}_k)\, a_k(\boldsymbol{x}) \right)^{1/2}.$$

It is known (see [73]) that the choice of the a_k minimizing this expression is exactly the cardinal functions ϕ_k. If we make this choice, it can be shown that

$$\sum_{k=0}^{N-1} K_{\mathrm{kor},d,\alpha,\boldsymbol{\gamma}}(\boldsymbol{x},\boldsymbol{x}_k)\,\phi_k(\boldsymbol{x}) = \sum_{k,\ell=0}^{N-1} \phi_\ell(\boldsymbol{x})\, K_{\mathrm{kor},d,\alpha,\boldsymbol{\gamma}}(\boldsymbol{x}_\ell,\boldsymbol{x}_k)\,\phi_k(\boldsymbol{x}).$$

Consequently, we get

$$\inf_{a_0,\ldots,a_{N-1}} \sup_{\substack{f\in\mathcal{H}_{\mathrm{kor},d,\alpha,\boldsymbol{\gamma}}\\ \|f\|_{\mathrm{kor},d,\alpha,\boldsymbol{\gamma}}\le 1}} \sup_{\boldsymbol{x}\in[0,1]^d} |f(\boldsymbol{x}) - S_{f,\mathcal{P}}(\boldsymbol{x})|$$
$$= \sup_{\substack{f\in\mathcal{H}_{\mathrm{kor},d,\alpha,\boldsymbol{\gamma}}\\ \|f\|_{\mathrm{kor},d,\alpha,\boldsymbol{\gamma}}\le 1}} \sup_{\boldsymbol{x}\in[0,1]^d} |f(\boldsymbol{x}) - s_{f,d,\mathcal{P}}(\boldsymbol{x})|$$
$$= \sup_{\boldsymbol{x}\in[0,1]^d} \left(K_{\mathrm{kor},d,\alpha,\boldsymbol{\gamma}}(\boldsymbol{x},\boldsymbol{x}) - \sum_{k=0}^{N-1} K_{\mathrm{kor},d,\alpha,\boldsymbol{\gamma}}(\boldsymbol{x},\boldsymbol{x}_k)\,\phi_k(\boldsymbol{x}) \right)^{1/2}. \quad (14.16)$$

This means that the best we can do in this context is to choose $S_{f,\mathcal{P}}$ as the spline algorithm $s_{f,d,\mathcal{P}}$.

Spline algorithms and lattice rules

We are now ready to study L_∞-approximation using rank-1 lattice rules and spline algorithms in greater detail. According to (2.19) we can express the kernel $K_{\mathrm{kor},d,\alpha,\boldsymbol{\gamma}}$ as

$$\begin{aligned} K_{\mathrm{kor},d,\alpha,\boldsymbol{\gamma}}(\boldsymbol{x},\boldsymbol{y}) &= \sum_{\boldsymbol{h}\in\mathbb{Z}^d} \frac{1}{r_{2\alpha,\boldsymbol{\gamma}}(\boldsymbol{h})} \mathrm{e}^{2\pi\mathrm{i}\boldsymbol{h}\cdot(\boldsymbol{x}-\boldsymbol{y})} \\ &=: \widetilde{K}_{\mathrm{kor},d,\alpha,\boldsymbol{\gamma}}(\boldsymbol{x}-\boldsymbol{y}) \\ &= \widetilde{K}_{\mathrm{kor},d,\alpha,\boldsymbol{\gamma}}(\{\boldsymbol{x}-\boldsymbol{y}\}). \end{aligned}$$

If we use a rank-1 lattice point set $\mathcal{P}(\boldsymbol{g},N)$ as the point set underlying the spline defined in (14.15), and observe its group structure, the properties of the Korobov kernel carry over to the cardinal functions ϕ_k defined in (14.14). Indeed, by (14.14),

$$\phi_k(\boldsymbol{x}) = \sum_{\ell=0}^{N-1} \widetilde{K}_{\mathrm{kor},d,\alpha,\boldsymbol{\gamma}}(\{\boldsymbol{x}-\boldsymbol{x}_\ell\})\,\widetilde{c}_{\ell,k}$$

with the $\widetilde{c}_{\ell,k}$ determined by the conditions

$$\begin{aligned}\delta_{i,k} &:= \sum_{\ell=0}^{N-1} \widetilde{K}_{\mathrm{kor},d,\alpha,\boldsymbol{\gamma}}(\{\boldsymbol{x}_i - \boldsymbol{x}_\ell\})\, \widetilde{c}_{\ell,k} \\ &= \sum_{\ell=0}^{N-1} \widetilde{K}_{\mathrm{kor},d,\alpha,\boldsymbol{\gamma}}(\boldsymbol{x}_{i-\ell \pmod N})\, \widetilde{c}_{\ell,k}\end{aligned}$$

for $i, k \in \{0, 1, \ldots, N-1\}$. We define the $N \times N$ matrices

$$\begin{aligned}K &:= (\widetilde{K}_{\mathrm{kor},d,\alpha,\boldsymbol{\gamma}}(\boldsymbol{x}_{i-\ell \pmod N}))_{i,\ell\in\{0,1,\ldots,N-1\}}, \\ C &:= (\widetilde{c}_{\ell,k})_{\ell,k\in\{0,1,\ldots,N-1\}},\end{aligned}$$

and denote, as before, the $N \times N$-identity matrix by U_N. Obviously, the matrix K is a circulant matrix. The conditions determining the $\widetilde{c}_{\ell,k}$ can then be formulated as

$$U_N = KC,$$

and since K is circulant it is easily checked that also C is circulant. The circulant structure of C then implies that we can simplify the expressions for ϕ_k for $k \in \{0, 1, \ldots, N-1\}$, to

$$\phi_k(\boldsymbol{x}) = \sum_{\ell=0}^{N-1} \widetilde{K}_{\mathrm{kor},d,\alpha,\boldsymbol{\gamma}}(\{\boldsymbol{x} - \boldsymbol{x}_\ell\})\, \widetilde{c}_{\ell-k \pmod N}$$

for some suitably chosen $\widetilde{c}_0, \widetilde{c}_1, \ldots, \widetilde{c}_{N-1}$, and this implies that

$$\begin{aligned}\phi_k(\boldsymbol{x}) &= \sum_{\ell=0}^{N-1} \widetilde{K}_{\mathrm{kor},d,\alpha,\boldsymbol{\gamma}}(\{\boldsymbol{x} - \boldsymbol{x}_\ell\})\, \widetilde{c}_{\ell-k \pmod N} \qquad (14.17)\\ &= \sum_{\ell=0}^{N-1} \widetilde{K}_{\mathrm{kor},d,\alpha,\boldsymbol{\gamma}}(\{\boldsymbol{x} - \boldsymbol{x}_k + \boldsymbol{x}_k - \boldsymbol{x}_\ell\})\, \widetilde{c}_{\ell-k \pmod N} \\ &= \sum_{\ell=0}^{N-1} \widetilde{K}_{\mathrm{kor},d,\alpha,\boldsymbol{\gamma}}(\{\boldsymbol{x} - \boldsymbol{x}_k - \boldsymbol{x}_{\ell-k \pmod N}\})\, \widetilde{c}_{\ell-k \pmod N} \\ &= \phi(\{\boldsymbol{x} - \boldsymbol{x}_k\}) \\ &= \phi(\boldsymbol{x} - \boldsymbol{x}_k),\end{aligned}$$

where we use the notation $\phi := \phi_0$. Then, by (14.15), the spline for $f \in \mathcal{H}_{\mathrm{kor},d,\alpha,\boldsymbol{\gamma}}$ is given as

$$s_{f,d,\mathcal{P}(\boldsymbol{g},N)}(\boldsymbol{x}) = \sum_{k=0}^{N-1} f(\boldsymbol{x}_k)\, \phi(\boldsymbol{x} - \boldsymbol{x}_k) \quad \text{for} \quad \boldsymbol{x} \in [0,1]^d.$$

Error analysis

We define the worst-case error of the spline L_∞-approximation via $s_{f,d,\mathcal{P}(\boldsymbol{g},N)}$ for $f \in \mathcal{H}_{\mathrm{kor},d,\alpha,\boldsymbol{\gamma}}$ as

$$\begin{aligned}
&\mathrm{err}^{L_\infty\text{-app}}(\mathcal{H}_{\mathrm{kor},d,\alpha,\boldsymbol{\gamma}}, s_{d,\mathcal{P}(\boldsymbol{g},N)}) \\
&:= \sup_{\substack{f\in\mathcal{H}_{\mathrm{kor},d,\alpha,\boldsymbol{\gamma}} \\ \|f\|_{\mathrm{kor},d,\alpha,\boldsymbol{\gamma}}\le 1}} \sup_{\boldsymbol{x}\in[0,1]^d} |f(\boldsymbol{x}) - s_{f,d,\mathcal{P}(\boldsymbol{g},N)}(\boldsymbol{x})|,
\end{aligned}$$

and analyze this error in the following. We start with a general error bound.

Let $\boldsymbol{x}_0, \boldsymbol{x}_1, \ldots, \boldsymbol{x}_{N-1}$ denote the elements of the lattice point set $\mathcal{P}(\boldsymbol{g}, N)$. By (14.16), we obtain that the worst-case error of the spline L_∞-approximation equals

$$\begin{aligned}
&\mathrm{err}^{L_\infty\text{-app}}(\mathcal{H}_{\mathrm{kor},d,\alpha,\boldsymbol{\gamma}}, s_{d,\mathcal{P}(\boldsymbol{g},N)}) \\
&= \sup_{\boldsymbol{x}\in[0,1]^d} \left(\widetilde{K}_{\mathrm{kor},d,\alpha,\boldsymbol{\gamma}}(\boldsymbol{0}) - \sum_{k=0}^{N-1} \widetilde{K}_{\mathrm{kor},d,\alpha,\boldsymbol{\gamma}}(\boldsymbol{x}-\boldsymbol{x}_k)\,\phi(\boldsymbol{x}-\boldsymbol{x}_k)\right)^{1/2}. \quad (14.18)
\end{aligned}$$

By (14.17) and by the Fourier expansion of the Korobov kernel, we obtain

$$\begin{aligned}
\phi(\boldsymbol{x}) &= \sum_{\ell=0}^{N-1} \widetilde{K}_{\mathrm{kor},d,\alpha,\boldsymbol{\gamma}}(\{\boldsymbol{x}-\boldsymbol{x}_\ell\})\,\widetilde{c}_\ell \\
&= \sum_{\ell=0}^{N-1} \widetilde{c}_\ell \sum_{\boldsymbol{h}\in\mathbb{Z}^d} \frac{1}{r_{2\alpha,\boldsymbol{\gamma}}(\boldsymbol{h})}\,\mathrm{e}^{2\pi\mathrm{i}\boldsymbol{h}\cdot(\boldsymbol{x}-\boldsymbol{x}_\ell)} \\
&= \sum_{\boldsymbol{h}\in\mathbb{Z}^d} \frac{1}{r_{2\alpha,\boldsymbol{\gamma}}(\boldsymbol{h})}\,\mathrm{e}^{2\pi\mathrm{i}\boldsymbol{h}\cdot\boldsymbol{x}} \sum_{\ell=0}^{N-1} \widetilde{c}_\ell\,\mathrm{e}^{-2\pi\mathrm{i}\boldsymbol{h}\cdot\boldsymbol{x}_\ell}.
\end{aligned}$$

In particular, it must then be true that

$$\delta_{k,0} = \phi(\boldsymbol{x}_k) = \sum_{\boldsymbol{h}\in\mathbb{Z}^d} \frac{1}{r_{2\alpha,\boldsymbol{\gamma}}(\boldsymbol{h})}\,\mathrm{e}^{2\pi\mathrm{i}\boldsymbol{h}\cdot\boldsymbol{x}_k} \sum_{\ell=0}^{N-1} \widetilde{c}_\ell\,\mathrm{e}^{-2\pi\mathrm{i}\boldsymbol{h}\cdot\boldsymbol{x}_\ell}$$

for every $k \in \{0, 1, \ldots, N-1\}$. Let now $\boldsymbol{v} \in \mathbb{Z}^d$ be arbitrarily chosen but fixed. We multiply the above equation by $\mathrm{e}^{-2\pi\mathrm{i}\boldsymbol{v}\cdot\boldsymbol{x}_k}$, sum both sides over all $k \in \{0, 1, \ldots, N-1\}$, and use the group structure and character property of rank-1 lattices to obtain

$$\begin{aligned}
1 &= \sum_{\boldsymbol{h}\in\mathbb{Z}^d} \frac{1}{r_{2\alpha,\boldsymbol{\gamma}}(\boldsymbol{h})} \sum_{k=0}^{N-1} \mathrm{e}^{2\pi\mathrm{i}(\boldsymbol{h}-\boldsymbol{v})\cdot\boldsymbol{x}_k} \sum_{\ell=0}^{N-1} \widetilde{c}_\ell\,\mathrm{e}^{-2\pi\mathrm{i}\boldsymbol{h}\cdot\boldsymbol{x}_\ell} \\
&= \sum_{\boldsymbol{h}\in\mathbb{Z}^d} \frac{1}{r_{2\alpha,\boldsymbol{\gamma}}(\boldsymbol{h}+\boldsymbol{v})} \sum_{k=0}^{N-1} \mathrm{e}^{2\pi\mathrm{i}\boldsymbol{h}\cdot\boldsymbol{x}_k} \sum_{\ell=0}^{N-1} \widetilde{c}_\ell\,\mathrm{e}^{-2\pi\mathrm{i}(\boldsymbol{h}+\boldsymbol{v})\cdot\boldsymbol{x}_\ell}
\end{aligned}$$

$$= \sum_{\boldsymbol{h}\in\mathbb{Z}^d} \frac{1}{r_{2\alpha,\boldsymbol{\gamma}}(\boldsymbol{h}+\boldsymbol{\nu})} \sum_{\ell=0}^{N-1} \widetilde{c}_\ell\, \mathrm{e}^{-2\pi\mathrm{i}\boldsymbol{\nu}\cdot\boldsymbol{x}_\ell} \sum_{k=0}^{N-1} \mathrm{e}^{2\pi\mathrm{i}\boldsymbol{h}\cdot(\boldsymbol{x}_k-\boldsymbol{x}_\ell)}$$
$$= N \sum_{\boldsymbol{h}\in\mathcal{L}^\perp(\boldsymbol{g},N)} \frac{1}{r_{2\alpha,\boldsymbol{\gamma}}(\boldsymbol{h}+\boldsymbol{\nu})} \sum_{\ell=0}^{N-1} \widetilde{c}_\ell\, \mathrm{e}^{-2\pi\mathrm{i}\boldsymbol{\nu}\cdot\boldsymbol{x}_\ell},$$

where $\mathcal{L}^\perp(\boldsymbol{g},N)$ is the dual lattice corresponding to $\mathcal{P}(\boldsymbol{g},N)$, given by (1.7). Hence,

$$\sum_{\ell=0}^{N-1} \widetilde{c}_\ell\, \mathrm{e}^{-2\pi\mathrm{i}\boldsymbol{\nu}\cdot\boldsymbol{x}_\ell} = \frac{1}{N}\left(\sum_{\boldsymbol{h}\in\mathcal{L}^\perp(\boldsymbol{g},N)} \frac{1}{r_{2\alpha,\boldsymbol{\gamma}}(\boldsymbol{h}+\boldsymbol{\nu})}\right)^{-1}.$$

This yields a different expression for ϕ in terms of its Fourier coefficients, namely

$$\phi(\boldsymbol{x}) = \sum_{\boldsymbol{\nu}\in\mathbb{Z}^d} \widehat{\phi}(\boldsymbol{\nu})\, \mathrm{e}^{2\pi\mathrm{i}\boldsymbol{\nu}\cdot\boldsymbol{x}},$$

where

$$\widehat{\phi}(\boldsymbol{\nu}) = \frac{1}{N}\frac{1}{r_{2\alpha,\boldsymbol{\gamma}}(\boldsymbol{\nu})}\left(\sum_{\boldsymbol{h}\in\mathcal{L}^\perp(\boldsymbol{g},N)} \frac{1}{r_{2\alpha,\boldsymbol{\gamma}}(\boldsymbol{h}+\boldsymbol{\nu})}\right)^{-1}.$$

Using the Fourier expansions of $\widetilde{K}_{\mathrm{kor},d,\alpha,\boldsymbol{\gamma}}$ and ϕ, respectively, we get, for $\boldsymbol{x}\in[0,1]^d$,

$$\widetilde{K}_{\mathrm{kor},d,\alpha,\boldsymbol{\gamma}}(\boldsymbol{0}) - \sum_{k=0}^{N-1} \widetilde{K}_{\mathrm{kor},d,\alpha,\boldsymbol{\gamma}}(\boldsymbol{x}-\boldsymbol{x}_k)\,\phi(\boldsymbol{x}-\boldsymbol{x}_k)$$
$$= \sum_{\boldsymbol{h}\in\mathbb{Z}^d} \frac{1}{r_{2\alpha,\boldsymbol{\gamma}}(\boldsymbol{h})} - \sum_{k=0}^{N-1}\sum_{\boldsymbol{h},\boldsymbol{\nu}\in\mathbb{Z}^d} \frac{1}{r_{2\alpha,\boldsymbol{\gamma}}(\boldsymbol{h})}\, \mathrm{e}^{2\pi\mathrm{i}\boldsymbol{h}\cdot(\boldsymbol{x}-\boldsymbol{x}_k)}\, \widehat{\phi}(\boldsymbol{\nu})\, \mathrm{e}^{2\pi\mathrm{i}\boldsymbol{\nu}\cdot(\boldsymbol{x}-\boldsymbol{x}_k)}$$
$$= \sum_{\boldsymbol{h}\in\mathbb{Z}^d} \frac{1}{r_{2\alpha,\boldsymbol{\gamma}}(\boldsymbol{h})} - \sum_{\boldsymbol{h},\boldsymbol{\nu}\in\mathbb{Z}^d} \frac{1}{N}\frac{1}{r_{2\alpha,\boldsymbol{\gamma}}(\boldsymbol{h})}\frac{1}{r_{2\alpha,\boldsymbol{\gamma}}(\boldsymbol{\nu})}\left(\sum_{\boldsymbol{\ell}\in\mathcal{L}^\perp(\boldsymbol{g},N)} \frac{1}{r_{2\alpha,\boldsymbol{\gamma}}(\boldsymbol{\ell}+\boldsymbol{\nu})}\right)^{-1}$$
$$\times\, \mathrm{e}^{2\pi\mathrm{i}(\boldsymbol{h}+\boldsymbol{\nu})\cdot\boldsymbol{x}} \sum_{k=0}^{N-1} \mathrm{e}^{-2\pi\mathrm{i}(\boldsymbol{h}+\boldsymbol{\nu})\cdot\boldsymbol{x}_k}$$
$$= \sum_{\boldsymbol{h}\in\mathbb{Z}^d} \frac{1}{r_{2\alpha,\boldsymbol{\gamma}}(\boldsymbol{h})} - \sum_{\boldsymbol{\nu}\in\mathbb{Z}^d} \frac{1}{N}\frac{1}{r_{2\alpha,\boldsymbol{\gamma}}(\boldsymbol{\nu})}\left(\sum_{\boldsymbol{\ell}\in\mathcal{L}^\perp(\boldsymbol{g},N)} \frac{1}{r_{2\alpha,\boldsymbol{\gamma}}(\boldsymbol{\ell}+\boldsymbol{\nu})}\right)^{-1}$$
$$\times \sum_{\boldsymbol{h}\in\mathbb{Z}^d} \frac{1}{r_{2\alpha,\boldsymbol{\gamma}}(\boldsymbol{h})}\, \mathrm{e}^{2\pi\mathrm{i}(\boldsymbol{h}+\boldsymbol{\nu})\cdot\boldsymbol{x}} \sum_{k=0}^{N-1} \mathrm{e}^{-2\pi\mathrm{i}(\boldsymbol{h}+\boldsymbol{\nu})\cdot\boldsymbol{x}_k}$$

$$
\begin{aligned}
&= \sum_{\boldsymbol{h}\in\mathbb{Z}^d} \frac{1}{r_{2\alpha,\gamma}(\boldsymbol{h})} - \sum_{\boldsymbol{v}\in\mathbb{Z}^d} \frac{1}{N} \frac{1}{r_{2\alpha,\gamma}(\boldsymbol{v})} \left(\sum_{\boldsymbol{\ell}\in\mathcal{L}^\perp(\boldsymbol{g},N)} \frac{1}{r_{2\alpha,\gamma}(\boldsymbol{\ell}+\boldsymbol{v})} \right)^{-1} \\
&\qquad \times \sum_{\boldsymbol{h}\in\mathbb{Z}^d} \frac{1}{r_{2\alpha,\gamma}(-\boldsymbol{h}-\boldsymbol{v})} \mathrm{e}^{-2\pi \mathrm{i} \boldsymbol{h}\cdot\boldsymbol{x}} \sum_{k=0}^{N-1} \mathrm{e}^{2\pi \mathrm{i} \boldsymbol{h}\cdot\boldsymbol{x}_k} \\
&= \sum_{\boldsymbol{h}\in\mathbb{Z}^d} \frac{1}{r_{2\alpha,\gamma}(\boldsymbol{h})} - \sum_{\boldsymbol{v}\in\mathbb{Z}^d} \frac{1}{r_{2\alpha,\gamma}(\boldsymbol{v})} \left(\sum_{\boldsymbol{\ell}\in\mathcal{L}^\perp(\boldsymbol{g},N)} \frac{1}{r_{2\alpha,\gamma}(\boldsymbol{\ell}+\boldsymbol{v})} \right)^{-1} \\
&\qquad \times \sum_{\boldsymbol{h}\in\mathcal{L}^\perp(\boldsymbol{g},N)} \frac{1}{r_{2\alpha,\gamma}(\boldsymbol{h}+\boldsymbol{v})} \mathrm{e}^{-2\pi \mathrm{i} \boldsymbol{h}\cdot\boldsymbol{x}} \\
&= \sum_{\boldsymbol{v}\in\mathbb{Z}^d} \frac{1}{r_{2\alpha,\gamma}(\boldsymbol{v})} \left(\sum_{\boldsymbol{\ell}\in\mathcal{L}^\perp(\boldsymbol{g},N)} \frac{1}{r_{2\alpha,\gamma}(\boldsymbol{\ell}+\boldsymbol{v})} \right)^{-1} \sum_{\boldsymbol{h}\in\mathcal{L}^\perp(\boldsymbol{g},N)\setminus\{\boldsymbol{0}\}} \frac{1-\mathrm{e}^{-2\pi \mathrm{i} \boldsymbol{h}\cdot\boldsymbol{x}}}{r_{2\alpha,\gamma}(\boldsymbol{h}+\boldsymbol{v})},
\end{aligned}
$$

where we made use of the character property of rank-1 lattice point sets, and the fact that $r_{2\alpha,\gamma}(\boldsymbol{h}) = r_{2\alpha,\gamma}(-\boldsymbol{h})$ for any $\boldsymbol{h} \in \mathbb{Z}^d$.

By inserting the latter equality into the error formula (14.18), and since all coefficients $r_{2\alpha,\gamma}(\boldsymbol{h})$ are nonnegative, we can bound the squared worst-case error of the spline L_∞-approximation based on the rank-1 lattice point set $\mathcal{P}(\boldsymbol{g}, N)$ by

$$
\begin{aligned}
&[\mathrm{err}^{L_\infty-\mathrm{app}}(\mathcal{H}_{\mathrm{kor},d,\alpha,\gamma}, S_{d,\mathcal{P}(\boldsymbol{g},N)})]^2 \\
&\le \sup_{\boldsymbol{x}\in[0,1]^d} \sum_{\boldsymbol{v}\in\mathbb{Z}^d} \frac{1}{r_{2\alpha,\gamma}(\boldsymbol{v})} \left(\sum_{\boldsymbol{\ell}\in\mathcal{L}^\perp(\boldsymbol{g},N)} \frac{1}{r_{2\alpha,\gamma}(\boldsymbol{\ell}+\boldsymbol{v})} \right)^{-1} \sum_{\boldsymbol{h}\in\mathcal{L}^\perp(\boldsymbol{g},N)\setminus\{\boldsymbol{0}\}} \frac{|1-\mathrm{e}^{-2\pi \mathrm{i} \boldsymbol{h}\cdot\boldsymbol{x}}|}{r_{2\alpha,\gamma}(\boldsymbol{h}+\boldsymbol{v})} \\
&\le 2 \sum_{\boldsymbol{v}\in\mathbb{Z}^d} \frac{1}{r_{2\alpha,\gamma}(\boldsymbol{v})} \left(\sum_{\boldsymbol{\ell}\in\mathcal{L}^\perp(\boldsymbol{g},N)} \frac{1}{r_{2\alpha,\gamma}(\boldsymbol{\ell}+\boldsymbol{v})} \right)^{-1} \sum_{\boldsymbol{h}\in\mathcal{L}^\perp(\boldsymbol{g},N)\setminus\{\boldsymbol{0}\}} \frac{1}{r_{2\alpha,\gamma}(\boldsymbol{h}+\boldsymbol{v})}.
\end{aligned} \tag{14.19}
$$

Next, we use a special decomposition of the set $\mathbb{Z}^d$.

Lemma 14.8 *Let $\mathcal{L}$ be a rank-1 lattice with N elements and generating vector $\boldsymbol{g} \in \mathbb{Z}^d$, and let $\mathcal{L}^\perp(\boldsymbol{g}, N)$ be its dual lattice. Then there exists a set $V_{d,N} \subseteq \mathbb{Z}^d$ of N integer vectors with the following properties.*

1. *$\mathbb{Z}^d$ is the direct sum of the dual lattice $\mathcal{L}^\perp(\boldsymbol{g}, N)$ and $V_{d,N}$, i.e., $\mathbb{Z}^d = \mathcal{L}^\perp(\boldsymbol{g}, N) \uplus V_{d,N}$, and*
2. *for any two distinct elements of $V_{d,N}$ their difference is not in $\mathcal{L}^\perp(\boldsymbol{g}, N)$, i.e.,*

$$\text{if } \boldsymbol{v}, \boldsymbol{w} \in V_{d,N} \text{ and } \boldsymbol{v} \neq \boldsymbol{w}, \text{ then } \boldsymbol{v} - \boldsymbol{w} \notin \mathcal{L}^\perp(\boldsymbol{g}, N).$$

Furthermore, we can choose $V_{d,N}$ such that $\boldsymbol{0} \in V_{d,N}$ and that

$$r_{2\alpha,\gamma}(\boldsymbol{v}) \le r_{2\alpha,\gamma}(\boldsymbol{v}+\boldsymbol{w}) \tag{14.20}$$

holds for any $\boldsymbol{v} \in V_{d,N}$ *and* $\boldsymbol{w} \in \mathcal{L}^\perp(\boldsymbol{g}, N)$.

Proof For $\ell \in \{0, 1, \ldots, N-1\}$ define the sets

$$Z_\ell := \{\boldsymbol{h} \in \mathbb{Z}^d \ : \ \boldsymbol{h} \cdot \boldsymbol{g} \equiv \ell \pmod N\}.$$

Then we have $Z_0 = \mathcal{L}^\perp(\boldsymbol{g}, N)$ and

$$\mathbb{Z} = \bigcup_{\ell=0}^{N-1} Z_\ell.$$

From every set Z_ℓ we can choose an arbitrary vector, which we call $\boldsymbol{y}_\ell$ for $\ell \in \{0, 1, \ldots, N-1\}$, and set $V_{d,N} := \{\boldsymbol{y}_0, \boldsymbol{y}_1, \ldots, \boldsymbol{y}_{N-1}\}$. Obviously, $|V_{d,N}| = N$. Let $\boldsymbol{x} \in \mathbb{Z}^d$, then there exists an $\ell_* \in \{0, 1, \ldots, N-1\}$ such that $\boldsymbol{x} \in Z_{\ell_*}$. Then $\boldsymbol{y} := \boldsymbol{x} - \boldsymbol{y}_{\ell_*}$ belongs to $\mathcal{L}^\perp(\boldsymbol{g}, N)$, because

$$\boldsymbol{y} \cdot \boldsymbol{g} = (\boldsymbol{x} - \boldsymbol{y}_{\ell_*}) \cdot \boldsymbol{g} \equiv \ell_* - \ell_* = 0 \pmod N,$$

and hence $\boldsymbol{x} = \boldsymbol{y} + \boldsymbol{y}_{\ell_*} \in \mathcal{L}^\perp(\boldsymbol{g}, N) \uplus V_{d,N}$. This implies $\mathbb{Z}^d = \mathcal{L}^\perp(\boldsymbol{g}, N) \uplus V_{d,N}$. Next, for different $r, s \in \{0, 1, \ldots, N-1\}$ we have $(\boldsymbol{y}_r - \boldsymbol{y}_s) \cdot \boldsymbol{g} \equiv r - s \not\equiv 0 \pmod N$ which implies that $\boldsymbol{y}_r - \boldsymbol{y}_s \notin \mathcal{L}^\perp(\boldsymbol{g}, N)$.

Since $\boldsymbol{0} \in Z_0$ we can choose $\boldsymbol{y}_0 = \boldsymbol{0}$. Furthermore, we can guarantee the property (14.20) for any $\boldsymbol{v} \in V_{d,N}$ and any $\boldsymbol{w} \in \mathcal{L}^\perp(\boldsymbol{g}, N)$ by ranking the $\boldsymbol{h} \in \mathbb{Z}^d$ according to the value of $r_{2\alpha,\boldsymbol{\gamma}}(\boldsymbol{h})$, and then choosing from every set Z_ℓ the element which appears first in this list. □

Using a set $V_{d,N}$ as in Lemma 14.8 that satisfies the monotonicity condition (14.20), we easily derive, from (14.19), the bound

$$\begin{aligned}
&[\mathrm{err}^{L_\infty\text{-app}}(\mathcal{H}_{\mathrm{kor},d,\alpha,\boldsymbol{\gamma}}, s_{d,\mathcal{P}(\boldsymbol{g},N)})]^2 \\
&\le 2 \sum_{\boldsymbol{v} \in V_{d,N}} \left(\frac{1}{r_{2\alpha,\boldsymbol{\gamma}}(\boldsymbol{v})} \left(\sum_{\boldsymbol{\ell} \in \mathcal{L}^\perp(\boldsymbol{g},N)} \frac{1}{r_{2\alpha,\boldsymbol{\gamma}}(\boldsymbol{\ell} + \boldsymbol{v})} \right)^{-1} \sum_{\boldsymbol{h} \in \mathcal{L}^\perp(\boldsymbol{g},N)\setminus\{\boldsymbol{0}\}} \frac{1}{r_{2\alpha,\boldsymbol{\gamma}}(\boldsymbol{h} + \boldsymbol{v})} \right. \\
&\quad + \sum_{\boldsymbol{w} \in \mathcal{L}^\perp(\boldsymbol{g},N)\setminus\{\boldsymbol{0}\}} \frac{1}{r_{2\alpha,\boldsymbol{\gamma}}(\boldsymbol{v} + \boldsymbol{w})} \left(\sum_{\boldsymbol{\ell} \in \mathcal{L}^\perp(\boldsymbol{g},N)} \frac{1}{r_{2\alpha,\boldsymbol{\gamma}}(\boldsymbol{\ell} + \boldsymbol{v} + \boldsymbol{w})} \right)^{-1} \\
&\quad\quad \left. \times \sum_{\boldsymbol{h} \in \mathcal{L}^\perp(\boldsymbol{g},N)\setminus\{\boldsymbol{0}\}} \frac{1}{r_{2\alpha,\boldsymbol{\gamma}}(\boldsymbol{h} + \boldsymbol{v} + \boldsymbol{w})} \right) \\
&\le 4 \sum_{\boldsymbol{v} \in V_{d,N}} \sum_{\boldsymbol{w} \in \mathcal{L}^\perp(\boldsymbol{g},N)\setminus\{\boldsymbol{0}\}} \frac{1}{r_{2\alpha,\boldsymbol{\gamma}}(\boldsymbol{v} + \boldsymbol{w})}.
\end{aligned}$$

Before we proceed, we summarize the previous findings in the following proposition.

Proposition 14.9 *For the worst-case error of L_∞-approximation of the spline algorithm $s_{d,\mathcal{P}(\boldsymbol{g},N)}$ based on a rank-1 lattice point set $\mathcal{P}(\boldsymbol{g},N)$ we have*

$$[\mathrm{err}^{L_\infty-\mathrm{app}}(\mathcal{H}_{\mathrm{kor},d,\alpha,\boldsymbol{\gamma}}, s_{d,\mathcal{P}(\boldsymbol{g},N)})]^2 \le 4 \sum_{\boldsymbol{v}\in V_{d,N}} \sum_{\boldsymbol{w}\in\mathcal{L}^\perp(\boldsymbol{g},N)\setminus\{\boldsymbol{0}\}} \frac{1}{r_{2\alpha,\boldsymbol{\gamma}}(\boldsymbol{v}+\boldsymbol{w})},$$

where the N-element set $V_{d,N}$ is chosen as in Lemma 14.8 and satisfies the monotonicity condition (14.20).

From Proposition 14.9, we can now again take two different ways, as it was also the case for L_2-approximation in Section 13.1. Indeed, we can either use rank-1 lattices constructed for integration, or again construct rank-1 lattices directly for the problem in the present section.

Using rank-1 lattices constructed for integration

Similarly to our analysis for L_2-approximation above, let us start by using results on lattice rules for numerical integration from previous chapters. To this end, we assume that $\alpha > 3/2$ in the following and proceed directly with Proposition 14.9. Using (13.7), we obtain for $\boldsymbol{v} \in V_{d,N}$ and $\boldsymbol{w} \in \mathcal{L}^\perp(\boldsymbol{g},N) \setminus \{\boldsymbol{0}\}$

$$\frac{1}{r_{2\alpha,\boldsymbol{\gamma}}(\boldsymbol{v}+\boldsymbol{w})} \le \Gamma_{2\alpha,\boldsymbol{\gamma}} \frac{r_{2\alpha,\boldsymbol{\gamma}}(\boldsymbol{v})}{r_{2\alpha,\boldsymbol{\gamma}}(\boldsymbol{w})},$$

where

$$\Gamma_{2\alpha,\boldsymbol{\gamma}} := 2^{2\alpha\min(d_0,d)} \prod_{j=1}^{\min(d_0,d)} \gamma_j$$

with $d_0(\alpha,\boldsymbol{\gamma}) = d_0 \in \mathbb{N}_0$ again being the minimal index j for which it is true that $\gamma_{j+1} \le 2^{-2\alpha} < \gamma_j$ (recall that $d_0 := 0$ if $\gamma_1 \le 2^{-2\alpha}$, and $d_0 := \infty$ if no such index j exists). Let $\tau \in [3/2, \alpha)$ and recall that we have $r_{2\alpha,\boldsymbol{\gamma}}(\boldsymbol{v}) \le r_{2\alpha,\boldsymbol{\gamma}}(\boldsymbol{v}+\boldsymbol{w})$ for $\boldsymbol{v} \in V_{d,N}$ and $\boldsymbol{w} \in \mathcal{L}^\perp(\boldsymbol{g},N)$. Then we obtain

$$\begin{aligned}\frac{1}{r_{2\alpha,\boldsymbol{\gamma}}(\boldsymbol{v}+\boldsymbol{w})} &\le \Gamma_{2\alpha,\boldsymbol{\gamma}} \frac{(r_{2\alpha,\boldsymbol{\gamma}}(\boldsymbol{v}))^{(2\tau+1)/(2\tau-1)}}{r_{2\alpha,\boldsymbol{\gamma}}(\boldsymbol{w})(r_{2\alpha,\boldsymbol{\gamma}}(\boldsymbol{v}))^{2/(2\tau-1)}}\\ &\le \Gamma_{2\alpha,\boldsymbol{\gamma}} \frac{(r_{2\alpha,\boldsymbol{\gamma}}(\boldsymbol{v}+\boldsymbol{w}))^{(2\tau+1)/(2\tau-1)}}{r_{2\alpha,\boldsymbol{\gamma}}(\boldsymbol{w})(r_{2\alpha,\boldsymbol{\gamma}}(\boldsymbol{v}))^{2/(2\tau-1)}},\end{aligned}$$

which implies

$$\frac{1}{r_{2\alpha,\boldsymbol{\gamma}}(\boldsymbol{v}+\boldsymbol{w})} \le \left(\frac{\Gamma_{2\alpha,\boldsymbol{\gamma}}}{r_{2\alpha,\boldsymbol{\gamma}}(\boldsymbol{w})}\right)^{(2\tau-1)/(4\tau)} \frac{1}{(r_{2\alpha,\boldsymbol{\gamma}}(\boldsymbol{v}))^{1/(2\tau)}}.$$

Inserting this into the result in Proposition 14.9 yields

$$
\begin{aligned}
&[\mathrm{err}^{L_\infty\text{-app}}(\mathcal{H}_{\mathrm{kor},d,\alpha,\boldsymbol{\gamma}}, S_{d,\mathcal{P}(\boldsymbol{g},N)})]^2\\
&\le 4\Gamma_{2\alpha,\boldsymbol{\gamma}}^{(2\tau-1)/(4\tau)} \sum_{\boldsymbol{w}\in\mathcal{L}^\perp(\boldsymbol{g},N)\setminus\{\boldsymbol{0}\}} \frac{1}{(r_{2\alpha,\boldsymbol{\gamma}}(\boldsymbol{w}))^{(2\tau-1)/(4\tau)}} \sum_{\boldsymbol{v}\in V_{d,N}} \frac{1}{(r_{2\alpha,\boldsymbol{\gamma}}(\boldsymbol{v}))^{1/(2\tau)}}\\
&= 4\Gamma_{2\alpha,\boldsymbol{\gamma}}^{(2\tau-1)/(4\tau)} \sum_{\boldsymbol{w}\in\mathcal{L}^\perp(\boldsymbol{g},N)\setminus\{\boldsymbol{0}\}} \frac{1}{r_{(\tau-1/2)(\alpha/\tau),\boldsymbol{\gamma}^{(\tau-1/2)/(2\tau)}}(\boldsymbol{w})} \sum_{\boldsymbol{v}\in V_{d,N}} \frac{1}{r_{\alpha/\tau,\boldsymbol{\gamma}^{1/(2\tau)}}(\boldsymbol{v})}\\
&\le 4\Gamma_{2\alpha,\boldsymbol{\gamma}}^{(2\tau-1)/(4\tau)}\\
&\quad\times \sum_{\boldsymbol{w}\in\mathcal{L}^\perp(\boldsymbol{g},N)\setminus\{\boldsymbol{0}\}} \frac{1}{r_{(\tau-1/2)(\alpha/\tau),\boldsymbol{\gamma}^{(\tau-1/2)/(2\tau)}}(\boldsymbol{w})} \prod_{j=1}^{d}\left(1+2\gamma_j^{1/(2\tau)}\zeta\left(\frac{\alpha}{\tau}\right)\right),
\end{aligned}
\tag{14.21}
$$

where we note that $(\tau-1/2)(\alpha/\tau) > 1$ since $\tau \in [3/2, \alpha)$. Note that

$$
\sum_{\boldsymbol{w}\in\mathcal{L}^\perp(\boldsymbol{g},N)\setminus\{\boldsymbol{0}\}} \frac{1}{r_{(\tau-1/2)(\alpha/\tau),\boldsymbol{\gamma}^{(\tau-1/2)/(2\tau)}}(\boldsymbol{w})}
$$

is exactly the squared worst-case integration error of a lattice rule in the weighted Korobov space of smoothness $(\tau-1/2)\alpha/(2\tau)$ and with weights $\boldsymbol{\gamma}^{(\tau-1/2)/(2\tau)} = (\gamma_j^{(\tau-1/2)/(2\tau)})_{j\ge1}$. This observation suggests constructing a lattice rule with small worst-case integration error for these parameters. This however, has already been done before. For example, we may directly apply Algorithm 3.6 to construct a good generating vector $\boldsymbol{g}$ for given prime N. The estimate (14.21) in combination with Theorem 3.7 then leads to the following result.

Theorem 14.10 *Let $\alpha > 3/2$, let $\tau \in [3/2, \alpha)$, let N be a prime number, and let $\boldsymbol{\gamma} = (\gamma_j)_{j\ge1}$ be product weights. Assume that $\boldsymbol{g}$ has been found by Algorithm 3.6 using the weight sequence $\boldsymbol{\gamma}^{(\tau-1/2)/(2\tau)} = (\gamma_j^{(\tau-1/2)/(2\tau)})_{j\ge1}$ and smoothness parameter $(\tau-1/2)\alpha/(2\tau)$. Then, for any $\lambda \in [1/2, (\tau-1/2)\alpha/(2\tau))$ the error of the spline algorithm based on the lattice point set $\mathcal{P}(\boldsymbol{g}, N)$ satisfies*

$$
\mathrm{err}^{L_\infty\text{-app}}(\mathcal{H}_{\mathrm{kor},d,\alpha,\boldsymbol{\gamma}}, S_{d,\mathcal{P}(\boldsymbol{g},N)}) \le \frac{C_{d,\tau,\lambda}}{N^\lambda},
$$

where

$$
\begin{aligned}
C_{d,\tau,\lambda} &:= 2^{\lambda+1}\Gamma_{2\alpha,\boldsymbol{\gamma}}^{(2\tau-1)/(8\tau)}\\
&\quad\times\prod_{j=1}^{d}\left(\left(1+2\gamma_j^{(\tau-1/2)/(4\tau\lambda)}\zeta\left(\frac{(\tau-1/2)\alpha/(2\tau)}{\lambda}\right)\right)^{\lambda}\left(1+2\gamma_j^{1/(2\tau)}\zeta\left(\frac{\alpha}{\tau}\right)\right)^{1/2}\right).
\end{aligned}
$$

Since τ can be chosen arbitrarily close to α and λ can be chosen arbitrarily close to $(\tau-1/2)\alpha/(2\tau)$ we get in this way an error rate of order $O(N^{-\alpha/2+1/4+\delta})$ for arbitrarily small but positive δ in Theorem 14.10, which is exactly the same rate as in the very first approach in this chapter.

Direct construction of rank-1 lattices for L_∞-approximation

We now demonstrate another approach to obtaining an error bound for the spline algorithm which works for $\alpha > 1/2$ instead of requiring higher values of α. Let us once more go back to Proposition 14.9. Then we choose $M > 1$ and split the summation in the error bound there into two parts,

$$\sum_{\boldsymbol{v}\in V_{d,N}} \sum_{\boldsymbol{w}\in\mathcal{L}^\perp(\boldsymbol{g},N)\setminus\{\boldsymbol{0}\}} \frac{1}{r_{2\alpha,\boldsymbol{\gamma}}(\boldsymbol{v}+\boldsymbol{w})} = \sum_{\substack{\boldsymbol{v}\in V_{d,N}\\ r_{2\alpha,\boldsymbol{\gamma}}(\boldsymbol{v})>M}} \sum_{\boldsymbol{w}\in\mathcal{L}^\perp(\boldsymbol{g},N)\setminus\{\boldsymbol{0}\}} \frac{1}{r_{2\alpha,\boldsymbol{\gamma}}(\boldsymbol{v}+\boldsymbol{w})}$$
$$+ \sum_{\substack{\boldsymbol{v}\in V_{d,N}\\ r_{2\alpha,\boldsymbol{\gamma}}(\boldsymbol{v})\le M}} \sum_{\boldsymbol{w}\in\mathcal{L}^\perp(\boldsymbol{g},N)\setminus\{\boldsymbol{0}\}} \frac{1}{r_{2\alpha,\boldsymbol{\gamma}}(\boldsymbol{v}+\boldsymbol{w})}. \tag{14.22}$$

Let us begin by analyzing the first sum in the right-hand side of (14.22). Since $r_{2\alpha,\boldsymbol{\gamma}}(\boldsymbol{v}) \le r_{2\alpha,\boldsymbol{\gamma}}(\boldsymbol{v}+\boldsymbol{w})$ for any $\boldsymbol{v} \in V_{d,N}$ and $\boldsymbol{w} \in \mathcal{L}^\perp(\boldsymbol{g},N)$, we have

$$\sum_{\substack{\boldsymbol{v}\in V_{d,N}\\ r_{2\alpha,\boldsymbol{\gamma}}(\boldsymbol{v})>M}} \sum_{\boldsymbol{w}\in\mathcal{L}^\perp(\boldsymbol{g},N)\setminus\{\boldsymbol{0}\}} \frac{1}{r_{2\alpha,\boldsymbol{\gamma}}(\boldsymbol{v}+\boldsymbol{w})} \le \sum_{\substack{\boldsymbol{v}\in\mathbb{Z}^d\\ r_{2\alpha,\boldsymbol{\gamma}}(\boldsymbol{v})>M}} \frac{1}{r_{2\alpha,\boldsymbol{\gamma}}(\boldsymbol{v})}.$$

We will now show by induction on d that we have, for any $p \in [1/2, \alpha)$,

$$\sum_{\substack{\boldsymbol{v}\in\mathbb{Z}^d\\ r_{2\alpha,\boldsymbol{\gamma}}(\boldsymbol{v})>M}} \frac{1}{r_{2\alpha,\boldsymbol{\gamma}}(\boldsymbol{v})} \le M^{1/(2p)-1} \prod_{j=1}^{d} \left(1 + 2\gamma_j^{1/(2p)} \zeta\left(\frac{\alpha}{p}\right)\right). \tag{14.23}$$

Indeed, let us first consider $d = 1$ and assume that $M\gamma_1 \ge 1$. We then write $Y := \lfloor (M\gamma_1)^{1/(2\alpha)} \rfloor + 1$, and obtain

$$\begin{aligned}
\sum_{\substack{v\in\mathbb{Z}\\ r_{2\alpha,\gamma_1}(v)>M}} \frac{1}{r_{2\alpha,\gamma_1}(v)} &= \gamma_1 \sum_{\substack{v\in\mathbb{Z}\\ |v|>(M\gamma_1)^{1/(2\alpha)}}} \frac{1}{|v|^{2\alpha}} \\
&= 2\gamma_1 \sum_{v=Y}^{\infty} \frac{1}{v^{2\alpha}} \\
&= 2\gamma_1 \sum_{v=1}^{\infty} \sum_{\ell=0}^{Y-1} \frac{1}{(vY+\ell)^{2\alpha}} \\
&\le 2\gamma_1 \sum_{v=1}^{\infty} \frac{Y}{(vY)^{2\alpha}} \\
&= 2\gamma_1 Y^{1-2\alpha} \zeta(2\alpha).
\end{aligned}$$

Since $2\alpha > 1$, we get

$$Y^{1-2\alpha} \le \left((M\gamma_1)^{1/(2\alpha)}\right)^{1-2\alpha} = (M\gamma_1)^{1/(2\alpha)-1}.$$

Thus,

$$\sum_{\substack{\nu\in\mathbb{Z}\\ r_{2\alpha,\gamma_1}(\nu)>M}} \frac{1}{r_{2\alpha,\gamma_1}(\nu)} \le 2\gamma_1(M\gamma_1)^{1/(2\alpha)-1}\zeta(2\alpha) \le 2M^{1/(2p)-1}\gamma_1^{1/(2p)}\zeta\left(\frac{\alpha}{p}\right)$$

for any $p \in [1/2, \alpha)$, since $M\gamma_1 \ge 1$. This shows the desired bound in this case.

If, on the other hand, $d = 1$ and $M\gamma_1 < 1$, then

$$\sum_{\substack{\nu\in\mathbb{Z}\\ r_{2\alpha,\gamma_1}(\nu)>M}} \frac{1}{r_{2\alpha,\gamma_1}(\nu)} \le 2\sum_{\nu=1}^{\infty} \frac{1}{r_{2\alpha,\gamma_1}(\nu)} = 2\gamma_1\zeta(2\alpha).$$

As $M\gamma_1 < 1$, we have $(M\gamma_1)^{1/(2p)-1} \ge 1$, and so

$$\sum_{\substack{\nu\in\mathbb{Z}\\ r_{2\alpha,\gamma_1}(\nu)>M}} \frac{1}{r_{2\alpha,\gamma_1}(\nu)} \le 2\gamma_1\zeta(2\alpha)(M\gamma_1)^{1/(2p)-1} \le 2M^{1/(2p)-1}\gamma_1^{1/(2p)}\zeta\left(\frac{\alpha}{p}\right),$$

and this again yields the bound for $d = 1$.

Let us now assume that (14.23) holds for dimension d. We write $(\boldsymbol{\nu}_d, \nu_{d+1})$ to denote a $(d+1)$-dimensional vector with the first d components being given by $\boldsymbol{\nu}_d$, and the last component by ν_{d+1}, and obtain

$$\begin{aligned}
&\sum_{\substack{(\boldsymbol{\nu}_d,\nu_{d+1})\in\mathbb{Z}^{d+1}\\ r_{2\alpha,\boldsymbol{\gamma}}((\boldsymbol{\nu}_d,\nu_{d+1}))>M}} \frac{1}{r_{2\alpha,\boldsymbol{\gamma}}((\boldsymbol{\nu}_d,\nu_{d+1}))}\\
&= \sum_{\nu_{d+1}\in\mathbb{Z}} \frac{1}{r_{2\alpha,\gamma_{d+1}}(\nu_{d+1})} \sum_{\substack{\boldsymbol{\nu}_d\in\mathbb{Z}^d\\ r_{2\alpha,\boldsymbol{\gamma}}(\boldsymbol{\nu}_d)>M(r_{2\alpha,\boldsymbol{\gamma}}(\nu_{d+1}))^{-1}}} \frac{1}{r_{2\alpha,\boldsymbol{\gamma}}(\boldsymbol{\nu}_d)}\\
&\le M^{1/(2p)-1}\left(\prod_{j=1}^{d}\left(1+2\gamma_j^{1/(2p)}\zeta\left(\frac{\alpha}{p}\right)\right)\right)\sum_{\nu_{d+1}\in\mathbb{Z}} \frac{1}{(r_{2\alpha,\gamma_{d+1}}(\nu_{d+1}))^{1/(2p)}}\\
&= M^{1/(2p)-1}\prod_{j=1}^{d+1}\left(1+2\gamma_j^{1/(2p)}\zeta\left(\frac{\alpha}{p}\right)\right)
\end{aligned}$$

for any $p \in [1/2, \alpha)$. This concludes the proof of (14.23).

Let us now turn to the second term in (14.22). However, it is easy to see that this term does not exceed the second term on the right-hand side of (13.10), which we already bounded in Section 13.1 by means of the quantity $S_{N,d,\alpha,\boldsymbol{\gamma}}(\boldsymbol{g})$. This then yields

$$
\begin{aligned}
&[\mathrm{err}^{L_\infty\text{-app}}(\mathcal{H}_{\mathrm{kor},d,\alpha,\boldsymbol{\gamma}}, S_{d,\mathcal{P}(\boldsymbol{g},N)})]^2 \\
&\le M^{1/(2p)-1} \prod_{j=1}^{d} \left(1 + 2\gamma_j^{1/(2p)} \zeta\left(\frac{\alpha}{p}\right)\right) + M[S_{N,d,\alpha,\boldsymbol{\gamma}}(\boldsymbol{g})]^2 \qquad (14.24)
\end{aligned}
$$

for any $p \in [1/2, \alpha)$, where $S_{N,d,\alpha,\gamma}(\boldsymbol{g})$ is defined as in (13.11). This estimate holds for any $\boldsymbol{g} \in G_d^{\phi}(N)$. In particular, we can use Algorithm 13.3 to make $S_{N,d,\alpha,\gamma}(\boldsymbol{g})$ small. The corresponding bound is shown in Theorem 13.5. We summarize our results in the following theorem, which is in a similar vein as Theorem 13.6 for L_2-approximation.

Theorem 14.11 *Let $N \ge 3$ be a prime number, let $\alpha > 1/2$, let $\boldsymbol{\gamma} = (\gamma_j)_{j\ge1}$ be product weights, and assume that $\boldsymbol{g}$ has been found by Algorithm 13.3. Then for arbitrary $\tau \in [1/2, \alpha)$ we have*

$$
\mathrm{err}^{L_\infty\text{-app}}(\mathcal{H}_{\mathrm{kor},d,\alpha,\boldsymbol{\gamma}}, S_{d,\mathcal{P}(\boldsymbol{g},N)}) \le \frac{\sqrt{2}}{N^{\tau(2\tau-1)/(4\tau-1)}} \prod_{j=1}^{d} \left(1 + 2^{4\alpha+1}\gamma_j^{1/(2\tau)} \zeta\left(\frac{\alpha}{\tau}\right)\right)^{2\tau}. \qquad (14.25)
$$

Proof From (14.24) with $p = \tau$ and Theorem 13.5 we obtain for arbitrary $\tau \in [1/2, \alpha)$,

$$
\begin{aligned}
&[\mathrm{err}^{L_\infty\text{-app}}(\mathcal{H}_{\mathrm{kor},d,\alpha,\boldsymbol{\gamma}}, S_{d,\mathcal{P}(\boldsymbol{g},N)})]^2 \\
&\le M^{1/(2\tau)-1} \prod_{j=1}^{d} \left(1 + 2\gamma_j^{1/(2\tau)} \zeta\left(\frac{\alpha}{\tau}\right)\right) + \frac{M}{N^{2\tau}} \prod_{j=1}^{d} \left(1 + 2^{4\alpha+1}\gamma_j^{1/(2\tau)} \zeta\left(\frac{\alpha}{\tau}\right)\right)^{4\tau} \\
&\le \left(M^{1/(2\tau)-1} + \frac{M}{N^{2\tau}}\right) \prod_{j=1}^{d} \left(1 + 2^{4\alpha+1}\gamma_j^{1/(2\tau)} \zeta\left(\frac{\alpha}{\tau}\right)\right)^{4\tau}.
\end{aligned}
$$

We now choose

$$
M = N^{4\tau^2/(4\tau-1)}.
$$

With this choice we obtain

$$
\begin{aligned}
&[\mathrm{err}^{L_\infty\text{-app}}(\mathcal{H}_{\mathrm{kor},d,\alpha,\boldsymbol{\gamma}}, S_{d,\mathcal{P}(\boldsymbol{g},N)})]^2 \\
&\qquad \le \frac{2}{N^{2\tau(2\tau-1)/(4\tau-1)}} \prod_{j=1}^{d} \left(1 + 2^{4\alpha+1}\gamma_j^{1/(2\tau)} \zeta\left(\frac{\alpha}{\tau}\right)\right)^{4\tau}.
\end{aligned}
$$

Now the result follows by taking the square root. □

Since τ can be chosen arbitrarily close to α we get in this way an error rate of order $O(N^{-\alpha(2\alpha-1)/(4\alpha-1)+\delta})$ for arbitrarily small but positive δ in Theorem 14.11. This is the same convergence rate as in Theorem 14.7, but here we do not need to assume that $\alpha > 1$.

Remark 14.12 We note that one could also use the spline method for L_2-approximation in the weighted Korobov space. Regarding error convergence rates, one would obtain the same results as in Section 13.1.

14.3 Tractability of L_∞-Approximation Using Lattice Rules and Splines

We briefly discuss tractability results for the L_∞-approximation problem on the basis of Theorems 14.3, 14.6, and 14.11, using similar notation as in Section 13.3.

As the error criterion we choose the so-called absolute criterion (note that the initial error strictly exceeds 1 for the present approximation problem; for further details on this issue see the "Notes and Remarks" section at the end of this chapter), and define the information complexity as

$$N^{L_\infty\text{-app}}(\varepsilon, d, \Lambda)$$
$$:= \min\{N \in \mathbb{N} \,:\, \exists A_{N,d} \in \Lambda \text{ such that } \mathrm{err}^{L_\infty\text{-app}}(\mathcal{H}_{\mathrm{kor},d,\alpha,\boldsymbol{\gamma}}, A_{N,d}) \leq \varepsilon\}.$$

In all three cases (i.e., in Theorems 14.3, 14.6, and 14.11) we have a similar typical situation and it suffices to bound the product

$$\prod_{j=1}^{d}\left(1 + 2^{4\alpha+2}\gamma_j^{1/(2\tau)}\zeta\left(\frac{\alpha}{\tau}\right)\right).$$

This, however, is an easy task, since this product is not larger than

$$\mathrm{e}^{2^{4\alpha+2}\zeta(\alpha/\tau)\sum_{j=1}^{d}\gamma_j^{1/(2\tau)}},$$

which in turn is bounded uniformly in d if

$$\sum_{j=1}^{\infty}\gamma_j^{1/(2\tau)} < \infty. \tag{14.26}$$

So, if (14.26) holds for a suitable parameter τ, then the error bounds in Theorems 14.3, 14.6, and 14.11 hold uniformly in d, and from this one can deduce that the L_∞-approximation problem is strongly polynomially tractable. Exemplary, we elaborate on this in greater detail for the case of Theorem 14.11, and also include the notion of polynomial tractability.

Theorem 14.13 *The L_∞-approximation problem for the Korobov space, with smoothness $\alpha > 1/2$ and product weights $\boldsymbol{\gamma} = (\gamma_j)_{j\geq 1}$, $\gamma_j \in (0, 1]$ for $j \in \mathbb{N}$, with respect to the absolute error criterion and with respect to information from the class Λ^{std} is strongly polynomially tractable if*

$$\sum_{j=1}^{\infty} \gamma_j^{1/(2\tau)} < \infty \tag{14.27}$$

holds for some $\tau \in (1/2, \alpha)$. *Then the* ε*-exponent is at most* $(2\alpha+2\tau-1)/(\tau(2\tau-1))$. *If* (14.27) *even holds for every* $\tau \in (1/2, \alpha)$, *then the* ε*-exponent is at most*

$$\frac{2}{\alpha} + \frac{1}{\alpha(2\alpha-1)}.$$

Furthermore, polynomial tractability holds if

$$\limsup_{d\to\infty} \frac{1}{\log d} \sum_{j=1}^{d} \gamma_j^{1/(2\tau)} < \infty \tag{14.28}$$

holds for some $\tau \in (1/2, \alpha)$.

Proof Assume that (14.27) holds for some $\tau \in (1/2, \alpha)$. Then for every odd prime number N one can, by Theorem 14.11, construct a spline algorithm $s_{d,\mathcal{P}(\boldsymbol{g},N)}$ such that

$$\mathrm{err}^{L_\infty-\mathrm{app}}(\mathcal{H}_{\mathrm{kor},d,\alpha,\boldsymbol{\gamma}}, s_{d,\mathcal{P}(\boldsymbol{g},N)}) \le \frac{C_{\alpha,\tau,\boldsymbol{\gamma}}}{N^{\tau(2\tau-1)/(4\tau-1)}}, \tag{14.29}$$

where

$$C_{\alpha,\tau,\boldsymbol{\gamma}} = \sqrt{2} \prod_{j=1}^{\infty} \left(1 + 2^{4\alpha+1} \gamma_j^{1/(2\tau)} \zeta\left(\frac{\alpha}{\tau}\right)\right)^{2\tau}$$

is independent of the dimension d. Let $Q := \lceil (C_{\alpha,\tau,\boldsymbol{\gamma}}\varepsilon^{-1})^{(4\tau-1)/(\tau(2\tau-1))} \rceil$ and let N be the smallest odd prime number greater than or equal to Q. Note that obviously $N \in [Q, 2Q)$. Now construct a generating vector $\boldsymbol{g}$ according to Theorem 14.11. For this $\boldsymbol{g}$ we then obtain

$$\mathrm{err}^{L_\infty-\mathrm{app}}(\mathcal{H}_{\mathrm{kor},d,\alpha,\boldsymbol{\gamma}}, s_{d,\mathcal{P}(\boldsymbol{g},N)}) \le \varepsilon,$$

and thus the information complexity satisfies

$$N^{L_\infty-\mathrm{app}}(\varepsilon, d, \Lambda^{\mathrm{std}}) \le N \le 2Q = 2\left\lceil \left(C_{\alpha,\tau,\boldsymbol{\gamma}}\varepsilon^{-1}\right)^{(4\tau-1)/(\tau(2\tau-1))} \right\rceil.$$

Hence the problem is strongly polynomially tractable with an ε-exponent of at most

$$\frac{4\tau-1}{\tau(2\tau-1)}.$$

If (14.27) even holds for all $\tau \in (1/2, \alpha)$, then the ε-exponent is at most

$$\frac{4\alpha-1}{\alpha(2\alpha-1)} = \frac{2}{\alpha} + \frac{1}{\alpha(2\alpha-1)}.$$

Finally, let us assume that (14.28) holds for some $\tau \in (1/2, \alpha)$. Then we can start again from the bound in Theorem 14.11 and see by standard arguments as in the proof of Corollary 2.28, that

$$\prod_{j=1}^{d} \left(1 + 2^{4\alpha+1} \gamma_j^{1/(2\tau)} \zeta\left(\frac{\alpha}{\tau}\right)\right)^{2\tau}$$

depends only polynomially on d. In this way, we obtain polynomial tractability. □

Notes and Remarks

The problem of L_∞-approximation is also considered in the three volume-book [210]–[212] by Novak and Woźniakowski. What we outlined in Section 14.1 has mostly been shown for the first time in [166] (see also [168]).

The presentation of the spline method for L_∞-approximation in Section 14.2 follows the more general paper [270], but in our outline we restrict ourselves to the theory immediately relevant to Korobov spaces. For more general instances we refer to [270]. Furthermore, for general information on spline methods, radial basis functions, and related results that are partly used in Section 14.2, we refer to the monographs [21, 73, 254, 263].

There is a close relation between L_∞-approximation for $\mathcal{H}_{\mathrm{kor},d,\alpha,\gamma}$ in the worst-case setting and L_2-approximation in the average-case setting over a suitable function space that is equipped with a Gaussian probability measure. This implies that in many cases results for one problem can be transferred to the other and vice versa. For further information see, e.g., [70, 163, 210, 212].

In Section 14.3 we have restricted ourselves to the information complexity with respect to the absolute error criterion. We remark that the initial error of L_∞-approximation in the Korobov space $\mathcal{H}_{\mathrm{kor},d,\alpha,\gamma}$ equals

$$e^{L_\infty\text{-app}}(0,d) = \prod_{j=1}^{d} (1 + 2\gamma_j \zeta(2\alpha))^{1/2},$$

which is strictly greater than 1. In this sense the L_∞-approximation problem is not normalized (like, e.g., the L_2-approximation problem). This means that it is appropriate to study also the information complexity with respect to the normalized criterion. See, for example, [70]. Furthermore, in [70] also an overview of known weight conditions for most of the current standard notions of tractability beyond (strong) polynomial tractability of the L_∞-approximation problem in weighted Korobov spaces can be found.

In this chapter we only consider Korobov spaces of finite smoothness α. L_∞-approximation for Korobov spaces of analytic functions as introduced in Chapter 9 is studied in the papers [150, 265]. These papers also discuss results for L_p-approximation with $p \in [2, \infty]$.

Chapter 15
Multiple Rank-1 Lattice Point Sets

In this chapter, we would like to give an overview of *multiple rank-1 lattice point sets*, as presented in [128] and [131], which can be used to obtain a better convergence rate for approximation than when using ordinary rank-1 lattice point sets (see Chapters 13 and 14). The basic idea of multiple lattice point sets is to consider the "union" of several rank-1 lattice point sets and to use them suitably in an approximation algorithm. In order to find good multiple lattice point sets, one employs a probabilistic search algorithm. We outline the ideas underlying these results for approximation in the weighted Korobov space $\mathcal{H}_{\mathrm{kor},d,\alpha,\boldsymbol{\gamma}}$ in this chapter. In order to make our presentation not too technical, we restrict ourselves to considering only product weights $\boldsymbol{\gamma} = (\gamma_j)_{j\geq 1}$ with $\gamma_j \in (0,1]$ for $j \in \mathbb{N}$.

As we have seen in Section 13.2, using N function values at lattice points can at best yield a convergence rate of the L_2-approximation error of order $N^{-\alpha/2}$. This is due to aliasing of Fourier coefficients whose frequencies differ only by an element of the dual lattice. To avoid the aliasing, one could use a rank-1 lattice rule with generating vector $\boldsymbol{g}_1$ and consisting of N_1 points to only approximate a set of non-aliased Fourier coefficients, that is, a set of the form $\{\boldsymbol{h}_{1,1},\ldots,\boldsymbol{h}_{1,N_1}\}$, where $\boldsymbol{h}_{1,i} - \boldsymbol{h}_{1,j} \notin \mathcal{L}^{\perp}(\boldsymbol{g}_1, N_1)$ for all $i \neq j$.

Assume that we were to use an algorithm of the form (13.2), where we approximate all Fourier coefficients with frequencies in the set $\mathcal{A}_{d,M}$, for some given value of M. In Section 13.1 we used a rank-1 lattice rule as the underlying quadrature rule to approximate the Fourier coefficients of a given function. In that approach we did not assume that $\mathcal{A}_{d,M} \subseteq \{\boldsymbol{h}_{1,1},\ldots,\boldsymbol{h}_{1,N_1}\}$, and so some Fourier coefficients in the approximation may be aliased. If we demand that $\mathcal{A}_{d,M} \subseteq \{\boldsymbol{h}_{1,1},\ldots,\boldsymbol{h}_{1,N_1}\}$ then we would have to make N_1 large, but in any case, the lower bound in Theorem 13.7 shows that we cannot improve upon the convergence rate of order $N_1^{-\alpha/2}$.

The approach taken in this section is slightly different and based on an idea of Kämmerer, who suggested using a union of rank-1 lattice rules. In the following we provide some motivation for using several rank-1 lattice rules at once. Indeed, we could proceed as follows. Given $\mathcal{A}_{d,M}$, we first use a rank-1 lattice rule with generating vector $\boldsymbol{g}_1$ and N_1 points to approximate the Fourier coefficients in $\mathcal{A}_{d,M} \cap \{\boldsymbol{h}_{1,1},\ldots,\boldsymbol{h}_{1,N_1}\}$. Now we are left with approximating the Fourier coefficients in

J. Dick et al., *Lattice Rules*, Springer Series in Computational Mathematics 58,
https://doi.org/10.1007/978-3-031-09951-9_15

$\mathcal{A}_{d,M} \setminus \{\boldsymbol{h}_{1,1}, \ldots, \boldsymbol{h}_{1,N_1}\}$. For these, we use a new rank-1 lattice rule with generating vector $\boldsymbol{g}_2$ and N_2 points. Ideally, this second lattice rule should cover as many frequencies in $\mathcal{A}_{d,M} \setminus \{\boldsymbol{h}_{1,1}, \ldots, \boldsymbol{h}_{1,N_1}\}$ as possible. This means that the second lattice rule should be, in some sense, "very different" from the first lattice rule. After having used the second lattice rule, we are again left with some frequencies which we have not yet approximated and so we construct a third lattice rule, which again should be "very different" from the first two lattice rules so that it can be used to approximate as many of the remaining frequencies as possible. We continue this procedure using new rank-1 lattice rules with generating vectors $\boldsymbol{g}_\ell$ and number of points N_ℓ until all the Fourier coefficients in $\mathcal{A}_{d,M}$ have been approximated (say, $\ell \in \{1, 2, \ldots, L\}$). The number of different lattice rules needed to cover the set $\mathcal{A}_{d,M}$ (this corresponds to the so-called reconstruction property defined below) can be shown to be relatively small. The proof is based on probabilistic arguments. This approach yields a convergence rate of the L_∞-approximation error of order $N^{-\alpha+1/2+\delta}$ for any $\delta > 0$, which is close to the best possible convergence rate. Since L_2-approximation is not harder than L_∞-approximation of functions, we shall also obtain new (and, for significant choices of the smoothness parameter, better) results regarding L_2-approximation, see Section 15.3.

The precise details of this approach are given in the following section.

15.1 Multiple Rank-1 Lattice Point Sets for Approximation in Korobov Spaces

Let us assume that we have L rank-1 lattice point sets $\mathcal{P}(g_1, N_1), \ldots, \mathcal{P}(g_L, N_L)$ with generating vectors $\boldsymbol{g}_\ell \in G_d(N_\ell)$ for $\ell \in [L]$. For the sake of simplicity, we shall assume that $N_1, \ldots, N_L$ are prime numbers throughout this chapter. We remind the reader that all points of a single rank-1 lattice point set with a prime number of elements are distinct. Based on the $\mathcal{P}(g_\ell, N_\ell)$ for $\ell \in [L]$, we define the *multiple rank-1 lattice point set*,

$$\widetilde{\mathcal{P}} = \widetilde{\mathcal{P}}(g_1, N_1, \ldots, g_L, N_L) := \bigcup_{\ell=1}^{L} \mathcal{P}(\boldsymbol{g}_\ell, N_\ell). \tag{15.1}$$

Note that we obviously have $|\widetilde{\mathcal{P}}| \le 1 - L + \sum_{\ell=1}^{L} N_\ell$, where the term $1 - L$ appears due to the fact that every single lattice point set $\mathcal{P}(\boldsymbol{g}_\ell, N_\ell)$ contains the origin.

The general idea of approximating functions by multiple rank-1 lattice point sets is again to consider truncated Fourier series for $f \in \mathcal{H}_{\mathrm{kor},d,\alpha,\boldsymbol{\gamma}}$, and to approximate the relevant Fourier coefficients by a QMC rule, but with some modifications as compared to the preceding chapters. To this end, we assume that we are given an index set $\mathcal{A}_d \subseteq \mathbb{Z}^d$, and then approximate $f \in \mathcal{H}_{\mathrm{kor},d,\alpha,\boldsymbol{\gamma}}$ by the algorithm given by

$$A^{\mathrm{mult}}_{\mathcal{A}_d}(f)(\boldsymbol{x}) := \sum_{\boldsymbol{h} \in \mathcal{A}_d} \widehat{f}_{\widetilde{P}}(\boldsymbol{h})\, \mathrm{e}^{2\pi \mathrm{i} \boldsymbol{h} \cdot \boldsymbol{x}} \tag{15.2}$$

for $\boldsymbol{x} \in [0,1)^d$, where $\widehat{f}_{\overline{P}}(\boldsymbol{h})$ is a suitable approximation of the Fourier coefficients $\widehat{f}(\boldsymbol{h})$ for $\boldsymbol{h} \in \mathcal{A}_d$ that is based on a multiple rank-1 lattice point set and will be explained in greater detail below (see (15.11)).

The reconstruction property

When trying to find good instances of multiple rank-1 lattice point sets for approximation, it is desirable that they fulfill the so-called *reconstruction property*, which is to be understood in the following way.

Let $\mathcal{A}_d \subseteq \mathbb{Z}^d$ be an index set, let $\mathcal{P}$ be a finite point set in $[0,1)^d$, and let

$$F = F_{\mathcal{P},\mathcal{A}_d} := (\mathrm{e}^{2\pi \mathrm{i} \boldsymbol{h}\cdot\boldsymbol{x}})_{\boldsymbol{x}\in\mathcal{P},\boldsymbol{h}\in\mathcal{A}_d}$$

be the corresponding Fourier matrix, where we assume an arbitrary but fixed ordering of the elements in $\mathcal{P}$ and in $\mathcal{A}_d$. Note that here we use the term "Fourier matrix" in a similar but slightly different way than previously in (3.18).

This matrix can be used to compute the evaluation of trigonometric polynomials

$$p : [0,1)^d \to \mathbb{C}, \quad p(\boldsymbol{x}) = \sum_{\boldsymbol{h}\in\mathcal{A}_d} \widehat{p}(\boldsymbol{h})\, \mathrm{e}^{2\pi \mathrm{i} \boldsymbol{h}\cdot\boldsymbol{x}}$$

with index set $\mathcal{A}_d$ at all points of $\mathcal{P}$, by means of the matrix-vector product

$$\boldsymbol{p} = F\,\widehat{\boldsymbol{p}},$$

where $\boldsymbol{p} := (p(\boldsymbol{x}))^\top_{\boldsymbol{x}\in\mathcal{P}}$ and $\widehat{\boldsymbol{p}} := (\widehat{p}(\boldsymbol{h}))^\top_{\boldsymbol{h}\in\mathcal{A}_d}$. On the other hand, also the Fourier coefficients $\widehat{p}(\boldsymbol{h})$ of p with frequencies $\boldsymbol{h} \in \mathcal{A}_d$ can be uniquely reconstructed from the sample values $p(\boldsymbol{x})$, $\boldsymbol{x} \in \mathcal{P}$, if and only if the Fourier matrix F has full column rank. Usually, this is achieved by means of the pseudo-inverse matrix $F^\dagger := (F^*F)^{-1}F^*$, where F^* is the adjoint matrix of F, i.e., the conjugate transpose of F, $F^* := (\overline{F})^\top$. Note that the full column rank of F guarantees that F^*F is invertible. Then we have

$$\widehat{\boldsymbol{p}} = F^\dagger \boldsymbol{p},$$

see also [128, p. 704] for further information. In this sense the reconstruction property of a point set $\mathcal{P}$ for an index set $\mathcal{A}_d$ means that for every trigonometric polynomial p with index set $\mathcal{A}_d$,

$$p \in \operatorname{span}\left\{\boldsymbol{x} \mapsto \mathrm{e}^{2\pi \mathrm{i} \boldsymbol{h}\cdot\boldsymbol{x}} : \boldsymbol{h} \in \mathcal{A}_d\right\},$$

all Fourier coefficients $\widehat{p}(\boldsymbol{h})$ with $\boldsymbol{h} \in \mathcal{A}_d$ can be reconstructed from the pointwise evaluations of p at the points of $\mathcal{P}$. As already mentioned, this property is equivalent to having full column rank of the Fourier matrix F corresponding to $\mathcal{P}$ and $\mathcal{A}_d$. Likewise, the point set $\mathcal{P}$ is said to be reconstructing (with respect to $\mathcal{A}_d$).

Let now $\mathcal{P} = \mathcal{P}(\boldsymbol{g}, N)$ be a rank-1 lattice point set and let $\mathcal{A}_d \subseteq \mathbb{Z}^d$ be a finite index set. Then any two distinct columns of the corresponding Fourier Matrix $F = F_{\mathcal{P}(\boldsymbol{g},N),\mathcal{A}_d}$ are either orthogonal or equal. This follows because for any $\boldsymbol{h}, \boldsymbol{h}' \in \mathcal{A}_d$ the entry $(F^*F)_{\boldsymbol{h}',\boldsymbol{h}}$ of the matrix F^*F at position $(\boldsymbol{h}', \boldsymbol{h}) \in \mathcal{A}_d^2$ satisfies

$$(F^*F)_{\boldsymbol{h}',\boldsymbol{h}} = \sum_{k=0}^{N-1} \mathrm{e}^{2\pi \mathrm{i} k((\boldsymbol{h}-\boldsymbol{h}')\cdot \boldsymbol{g})/N} = \begin{cases} N & \text{if } \boldsymbol{h}\cdot\boldsymbol{g} \equiv \boldsymbol{h}'\cdot\boldsymbol{g} \pmod N, \\ 0 & \text{if } \boldsymbol{h}\cdot\boldsymbol{g} \not\equiv \boldsymbol{h}'\cdot\boldsymbol{g} \pmod N. \end{cases}$$

This implies that the Fourier matrix F has full column rank if and only if

$$\boldsymbol{h}\cdot\boldsymbol{g} \not\equiv \boldsymbol{h}'\cdot\boldsymbol{g} \pmod N \quad \text{for all } \boldsymbol{h}, \boldsymbol{h}' \in \mathcal{A}_d,\ \boldsymbol{h} \neq \boldsymbol{h}'. \tag{15.3}$$

Therefore, for a rank-1 lattice point set $\mathcal{P}(\boldsymbol{g}, N)$, the reconstruction property with respect to a finite index set $\mathcal{A}_d$ is equivalent to Condition (15.3).

Note that Condition (15.3) is equivalent to

$$\boldsymbol{h}\cdot\boldsymbol{g} \not\equiv 0 \pmod N \quad \text{for all } \boldsymbol{h} \in \mathcal{D}(\mathcal{A}_d) \setminus \{\boldsymbol{0}\}, \tag{15.4}$$

where $\mathcal{D}(\mathcal{A}_d)$ denotes the difference set of $\mathcal{A}_d$ given by $\mathcal{D}(\mathcal{A}_d) := \{\boldsymbol{h} \in \mathbb{Z}^d \,:\, \boldsymbol{h} = \boldsymbol{h}_1 - \boldsymbol{h}_2,\ \boldsymbol{h}_1, \boldsymbol{h}_2 \in \mathcal{A}_d\}$. If (15.4) is satisfied, then for every trigonometric polynomial p with $p(\boldsymbol{x}) = \sum_{\boldsymbol{h}\in\mathcal{D}(\mathcal{A}_d)} a_{\boldsymbol{h}}\, \mathrm{e}^{2\pi \mathrm{i}\boldsymbol{h}\cdot\boldsymbol{x}}$, for $\boldsymbol{x} \in [0,1)^d$, with the index set $\mathcal{D}(\mathcal{A}_d)$ it holds that

$$\begin{aligned} \frac{1}{N}\sum_{k=0}^{N-1} p\left(\left\{\frac{k}{N}\boldsymbol{g}\right\}\right) &= \sum_{\boldsymbol{h}\in\mathcal{D}(\mathcal{A}_d)} a_{\boldsymbol{h}} \frac{1}{N}\sum_{k=0}^{N-1} \mathrm{e}^{2\pi \mathrm{i} k(\boldsymbol{h}\cdot\boldsymbol{g})/N} \\ &= \sum_{\substack{\boldsymbol{h}\in\mathcal{D}(\mathcal{A}_d) \\ \boldsymbol{h}\cdot\boldsymbol{g}\equiv 0 \pmod N}} a_{\boldsymbol{h}} = a_{\boldsymbol{0}} = \int_{[0,1]^d} p(\boldsymbol{x})\,\mathrm{d}\boldsymbol{x}. \end{aligned}$$

From this it follows directly that for a reconstructing rank-1 lattice point set $\mathcal{P}(\boldsymbol{g}, N)$ with respect to an index set $\mathcal{A}_d$ we have

$$\widehat{p}(\boldsymbol{h}) = \frac{1}{N}\sum_{k=0}^{N-1} p\left(\left\{\frac{k}{N}\boldsymbol{g}\right\}\right) \mathrm{e}^{-2\pi \mathrm{i} k(\boldsymbol{h}\cdot\boldsymbol{g})/N}$$

for all $\boldsymbol{h} \in \mathcal{A}_d$ and for all trigonometric polynomials with the index set $\mathcal{A}_d$.

For a multiple rank-1 lattice point set consisting of single rank-1 lattice point sets $\mathcal{P}(\boldsymbol{g}_\ell, N_\ell)$, $\ell \in [L]$, the strategy will be to select index sets $\mathcal{A}_{d,\ell}$, $\ell \in [L]$, such that $\mathcal{P}(\boldsymbol{g}_\ell, N_\ell)$ satisfies the reconstruction property for $\mathcal{A}_{d,\ell}$ for $\ell \in [L]$.

Constructing multiple rank-1 lattices algorithmically

Below, we will present a probabilistic algorithm (Algorithm 15.2) that will return a good multiple rank-1 lattice point set $\widetilde{\mathcal{P}} = \widetilde{\mathcal{P}}(g_1, N_1, \dots, g_L, N_L)$ with a certain probability. The algorithm is designed such that each $\mathcal{P}(\boldsymbol{g}_\ell, N_\ell)$, $\ell \in [L]$, has the reconstruction property for an index set $\mathcal{A}_{d,\ell}$ for $\ell \in [L]$. Given $\mathcal{A}_d \subseteq \mathbb{Z}^d$, the single index sets $\mathcal{A}_{d,\ell}$ will be determined by the algorithm such that

$$\mathcal{A}_{d,\ell} := \left\{\boldsymbol{h} \in \mathcal{A}_d \colon \boldsymbol{h} \cdot \boldsymbol{g}_\ell \not\equiv \boldsymbol{h}' \cdot \boldsymbol{g}_\ell \pmod{N_\ell} \quad \forall \boldsymbol{h}' \in \mathcal{A}_d \setminus \{\boldsymbol{h}\}\right\}, \tag{15.5}$$

i.e., $\mathcal{A}_{d,\ell}$ consists of all frequencies $\boldsymbol{h} \in \mathcal{A}_d$ for which the Fourier coefficients $\widehat{p}(\boldsymbol{h})$ of any trigonometric polynomial p with index set $\mathcal{A}_d$ can be reconstructed exactly using $\mathcal{P}(\boldsymbol{g}_\ell, N_\ell)$. The algorithm also chooses the cardinalities $N_1, \dots, N_L$ of the single rank-1 lattice point sets $\mathcal{P}(\boldsymbol{g}_\ell, N_\ell)$ as distinct prime numbers from the set

$$P_{\mathcal{A}_d} := \{p \in \mathbb{N} \ : \ p \text{ prime such that } |\{\boldsymbol{h} \pmod{p} \colon \boldsymbol{h} \in \mathcal{A}_d\}| = |\mathcal{A}_d|\}, \tag{15.6}$$

where $\boldsymbol{h} \pmod{p}$ is to be understood component-wise, and where, naturally, $\boldsymbol{h} \pmod{p} \in G_d(p)$. To be more precise, given two parameters $\eta, L_{\max} \in \mathbb{N}$, the cardinalities $N_1, \dots, N_L$ will be chosen from a subset of $P_{\mathcal{A}_d}$ defined as

$$P_{\mathcal{A}_d,\eta,L_{\max}} := \{p_1, \dots, p_{L_{\max}} \in P_{\mathcal{A}_d}\}, \tag{15.7}$$

where the p_j for $j \in [L_{\max}]$ are given by

$$p_j := \begin{cases} \min\{p \in P_{\mathcal{A}_d} \colon p > \eta\} & \text{if } j = 1, \\ \min\{p \in P_{\mathcal{A}_d} \colon p > p_{j-1}\} & \text{if } 2 \le j \le L_{\max}. \end{cases} \tag{15.8}$$

I.e., $P_{\mathcal{A}_d,\eta,L_{\max}}$ contains the $L_{\max}$ smallest primes in $P_{\mathcal{A}_d}$ which are greater than η.

The condition $|\{\boldsymbol{h} \pmod{p} \colon \boldsymbol{h} \in \mathcal{A}_d\}| = |\mathcal{A}_d|$ in (15.6) implies that for arbitrary $\boldsymbol{g} \in G_d(p)$ the Fourier matrices $F_{\mathcal{P}(\boldsymbol{g},p),\mathcal{A}_d}$ and $F_{\mathcal{P}(\boldsymbol{g},p),\mathcal{A}_d \pmod{p}}$ coincide and hence the lattice point set $\mathcal{P}(\boldsymbol{g}, p)$ has the recovery property for $\mathcal{A}_d$ if and only if it has this property for $\mathcal{A}_d \pmod{p} := \{\boldsymbol{h} \pmod{p} \ : \ \boldsymbol{h} \in \mathcal{A}_d\}$.

Remark 15.1 Let

$$N_{\mathcal{A}_d} := \max_{j \in [d]} \left(\max_{\boldsymbol{k} \in \mathcal{A}_d} k_j - \min_{\boldsymbol{l} \in \mathcal{A}_d} l_j \right).$$

The number $N_{\mathcal{A}_d}$ quantifies the expansion of the index set $\mathcal{A}_d$. The set $P_{\mathcal{A}_d}$ contains all prime numbers greater than $N_{\mathcal{A}_d}$, which can be seen as follows. Assume that p is a prime number satisfying $p > N_{\mathcal{A}_d}$ but that $|\{\boldsymbol{h} \pmod{p} \colon \boldsymbol{h} \in \mathcal{A}_d\}| < |\mathcal{A}_d|$. Then there exist distinct $\boldsymbol{k}, \boldsymbol{l} \in \mathcal{A}_d$ with $\boldsymbol{k} \equiv \boldsymbol{l} \pmod{p}$. Since $\boldsymbol{k} - \boldsymbol{l} \neq \boldsymbol{0}$ there exists at least one index $j_0 \in [d]$, and a $t \in \mathbb{N}$ such that $|k_{j_0} - l_{j_0}| = tp \ge p > N_{\mathcal{A}_d}$ and this contradicts the definition of $N_{\mathcal{A}_d}$.

We can now formulate a probabilistic algorithm to find reconstructing multiple rank-1 lattices with a certain probability. The basic idea is to choose the lattice sizes $N_1, \ldots, N_L$ deterministically and then to randomly select the corresponding generating vectors $\boldsymbol{g}_1, \ldots, \boldsymbol{g}_L$ of rank-1 lattices $\mathcal{P}(\boldsymbol{g}_\ell, N_\ell)$ for $\ell \in [L]$. The following algorithm was first proposed by Kämmerer, see [128, Algorithm 4].

Algorithm 15.2 (Probabilistic algorithm to find good multiple rank-1 lattices) Let $\mathcal{A}_d \subseteq \mathbb{Z}^d$ be a finite frequency index set, let $c \in (1, \infty)$, and let $\delta \in (0, 1)$.

Find prime numbers $N_1, \ldots, N_L$ and generating vectors $\boldsymbol{g}_\ell \in G_d(N_\ell)$ for $\ell \in [L]$ as follows.

(1) Put

$$L_{\max} := \left\lceil \left(\frac{c}{c-1}\right)^2 \frac{\log|\mathcal{A}_d| - \log\delta}{2} \right\rceil$$

and $\eta := (|\mathcal{A}_d| - 1)c$.

(2) Determine $P_{\mathcal{A}_d,\eta,L_{\max}}$ according to (15.7) and (15.8) and order its elements in an increasing fashion, put $\widetilde{\mathcal{A}}_d := \emptyset$, and put $L := 0$.

(3) Increase L by 1, put $N_L := p_L \in P_{\mathcal{A}_d,\eta,L_{\max}}$, and choose $\boldsymbol{g}_L \in G_d(N_L)$ uniformly at random.

(4) If

$$\left\{\boldsymbol{h} \in \mathcal{A}_d \;:\; \nexists \boldsymbol{h}' \in \mathcal{A}_d \setminus \{\boldsymbol{h}\} \text{ such that } \boldsymbol{h} \cdot \boldsymbol{g}_L \equiv \boldsymbol{h}' \cdot \boldsymbol{g}_L \pmod{N_L}\right\} \not\subseteq \widetilde{\mathcal{A}}_d,$$

then put

$$\widetilde{\mathcal{A}}_d := \widetilde{\mathcal{A}}_d \cup \left\{\boldsymbol{h} \in \mathcal{A}_d \;:\; \nexists \boldsymbol{h}' \in \mathcal{A}_d \setminus \{\boldsymbol{h}\} \text{ such that } \boldsymbol{h} \cdot \boldsymbol{g}_L \equiv \boldsymbol{h}' \cdot \boldsymbol{g}_L \pmod{N_L}\right\},$$

otherwise reduce L by 1.

(5) Repeat Steps (3) and (4) as long as $|\widetilde{\mathcal{A}}_d| < |\mathcal{A}_d|$ and $L < L_{\max}$.

Algorithm 15.2 returns the cardinalities $N_1, \ldots, N_L$ and the generating vectors $\boldsymbol{g}_1, \ldots, \boldsymbol{g}_L$ of rank-1 lattices $\mathcal{P}(\boldsymbol{g}_\ell, N_\ell)$, $\ell \in [L]$, which are then combined to a multiple rank-1 lattice $\widetilde{\mathcal{P}}(g_1, N_1, \ldots, g_L, N_L)$. The runtime of Algorithm 15.2 is of order $O(|\mathcal{A}_d|(d + \log|\mathcal{A}_d|)\log|\mathcal{A}_d|)$ (see [128, p. 716]).

Remark 15.3 The parameter c in Algorithm 15.2 is sometimes referred to as the "oversampling factor", as it is related to how much oversampling occurs in the approximation (15.2), based on the multiple rank-1 lattice determined by the algorithm.

The "If"-condition in Step (4) of the algorithm ensures that in each step the additional sampling nodes of the corresponding single rank-1 lattice point set do actually contribute to getting closer to the underlying goal (see [128]).

As discussed in [131], with probability $1 - \delta$, Algorithm 15.2 yields a multiple rank-1 lattice $\widetilde{\mathcal{P}}(g_1, N_1, \ldots, g_L, N_L)$ that has the reconstruction property for $\mathcal{A}_d = \bigcup_{\ell=1}^{L} \mathcal{A}_{d,\ell}$, with the $\mathcal{A}_{d,\ell}$ as in (15.5). We omit the proof of this result and refer the interested reader to [128, Theorem 3.4].

The following proposition gives an estimate on the number of points in the multiple rank-1 lattice point set returned by Algorithm 15.2.

Proposition 15.4 *Assume that* $\eta := (|\mathcal{A}_d| - 1)c \geq \max(N_{\mathcal{A}_d}, 4L_{\max} \log L_{\max})$, *where* $|\mathcal{A}_d| \geq 2$, *and where* $N_{\mathcal{A}_d}$ *is defined as in Remark 15.1. Then the cardinality of the multiple rank-1 lattice point set* $\widetilde{\mathcal{P}} := \widetilde{\mathcal{P}}(\boldsymbol{g}_1, N_1, \ldots, \boldsymbol{g}_L, N_L)$ *obtained by Algorithm 15.2 satisfies*

$$|\widetilde{\mathcal{P}}| \leq 2L_{\max}(|\mathcal{A}_d| - 1)c.$$

Proof We have

$$|\widetilde{\mathcal{P}}| \leq \sum_{\ell=1}^{L} N_\ell \leq L_{\max} \max_{\ell \in [L]} N_\ell \leq L_{\max} \max P_{\mathcal{A}_d, \eta, L_{\max}},$$

where we used that all N_ℓ are chosen from the set $P_{\mathcal{A}_d, \eta, L_{\max}}$.

Since $\eta \geq N_{\mathcal{A}_d}$ it follows from Remark 15.1 that $P_{\mathcal{A}_d}$ contains all prime numbers in the interval $(\eta, 2\eta]$. Thus it suffices to show that this interval contains at least $L_{\max}$ prime numbers. To this end we use a quantitative version of Bertrand's postulate which states that the number of primes in the interval $(\eta, 2\eta]$ is bounded from below by $3\eta/(5 \log \eta)$ when $\eta \geq 20.5$ (see [221, Corollary 3]). We distinguish four cases.

(1) If $L_{\max} \geq 4$ we have by assumption $\eta \geq 4L_{\max} \log L_{\max} > 22$, and hence the number of primes in the interval $(\eta, 2\eta]$ is at least

$$\frac{3\eta}{5 \log \eta} \geq \frac{3 \cdot 4L_{\max} \log L_{\max}}{5 \log(4L_{\max} \log L_{\max})} = L_{\max} \frac{\log L_{\max}^{12/5}}{\log(4L_{\max} \log L_{\max})} \geq L_{\max},$$

where for the first estimate we used that the function $\eta \mapsto 3\eta/(5 \log \eta)$ is monotonically increasing for $\eta > 22$, and for the last estimate we used that for $L_{\max} \geq 4$ we have $L_{\max}^{12/5} \geq 4L_{\max} \log L_{\max}$.

(2) If $L_{\max} = 3$ we have $\eta \geq 12 \log 3 > 13$. It follows from Case (1) that for $\eta > 22$ there are at least four prime numbers in $(\eta, 2\eta]$. For $13 < \eta \leq 22$ the interval $(\eta, 2\eta]$ still contains at least three primes, since

$$\{17, 19, 23\} \subseteq (\eta, 2\eta] \quad \text{for } \eta \in [11.5, 17)$$

and

$$\{23, 29, 31\} \subseteq (\eta, 2\eta] \quad \text{for } \eta \in [17, 22).$$

(3) If $L_{\max} = 2$, this implies $\eta \geq 8 \log 2 > 5.5$. For $\eta \in [5.5, 17)$ there are at least two prime numbers in $(\eta, 2\eta]$, namely

$$\{7, 11\} \subseteq (\eta, 2\eta] \quad \text{for } \eta \in [5.5, 7),$$

$$\{11, 13\} \subseteq (\eta, 2\eta] \quad \text{for } \eta \in [7, 11)$$

and

$$\{17, 19\} \subseteq (\eta, 2\eta] \quad \text{for } \eta \in [11, 17).$$

For $\eta \geq 17$ there are at least three prime numbers in $(\eta, 2\eta]$, due to Cases (1) and (2).

(4) If $L_{\max} = 1$ we have $\eta = (|\mathcal{A}_d| - 1)c > 1$, and thus it follows directly from Bertrand's postulate that the interval $(\eta, 2\eta]$ contains at least one prime number.

These considerations show that $\max P_{\mathcal{A}_d,\eta,L_{\max}} \leq 2\eta = 2(|\mathcal{A}_d| - 1)c$, and so

$$|\widetilde{\mathcal{P}}| \leq 2L_{\max}(|\mathcal{A}_d| - 1)c.$$

This finishes the proof. □

Reconstruction based on multiple rank-1 lattices

We now introduce an algorithm which guarantees an efficient reconstruction of trigonometric polynomials based on multiple rank-1 lattices. This algorithm is essentially [131, Algorithm 2].

Algorithm 15.5 (Computing reconstructions based on multiple rank-1 lattices)
Let $\mathcal{A}_d \subseteq \mathbb{Z}^d$ be a finite frequency index set, and let p be a trigonometric polynomial with index set $\mathcal{A}_d$. Let $\widetilde{\mathcal{P}} = \widetilde{\mathcal{P}}(\boldsymbol{g}_1, N_1, \ldots, \boldsymbol{g}_L, N_L)$ be a multiple rank-1 lattice point set obtained by Algorithm 15.2.

(1) For all $\boldsymbol{h} \in \mathcal{A}_d$ initialize $\#\boldsymbol{h} := 0$ and $\widehat{p}_{\widetilde{\mathcal{P}}}(\boldsymbol{h}) := 0$.
(2) For ℓ from 1 to L:

(2a) Determine (see (15.5)) the set

$$\mathcal{A}_{d,\ell} := \{\boldsymbol{h} \in \mathcal{A}_d \ : \ \boldsymbol{h} \cdot \boldsymbol{g}_\ell \not\equiv \boldsymbol{h}' \cdot \boldsymbol{g}_\ell \pmod{N_\ell} \ \forall \boldsymbol{h}' \in \mathcal{A}_d \setminus \{\boldsymbol{h}\}\}.$$

(2b) For every $\boldsymbol{h} \in \mathcal{A}_{d,\ell}$:
Compute

$$\widehat{p}_{\mathcal{P}(\boldsymbol{g}_\ell,N_\ell)}(\boldsymbol{h}) := \frac{1}{N_\ell} \sum_{k=0}^{N_\ell - 1} p\left(\left\{\frac{k}{N_\ell}\boldsymbol{g}_\ell\right\}\right) \mathrm{e}^{-2\pi \mathrm{i} k(\boldsymbol{h}\cdot\boldsymbol{g}_\ell)/N_\ell},$$

and set

$$\widehat{p}_{\widetilde{\mathcal{P}}}(\boldsymbol{h}) := \widehat{p}_{\widetilde{\mathcal{P}}}(\boldsymbol{h}) + \widehat{p}_{\mathcal{P}(\boldsymbol{g}_\ell,N_\ell)}(\boldsymbol{h})$$

and

$$\#\boldsymbol{h} := \#\boldsymbol{h} + 1.$$

End for.

(3) For every $\boldsymbol{h} \in \mathcal{A}_d$ set

$$\widehat{p}_{\widetilde{\mathcal{P}}}(\boldsymbol{h}) := \frac{\widehat{p}_{\widetilde{\mathcal{P}}}(\boldsymbol{h})}{\#\boldsymbol{h}},$$

and the algorithm terminates.

Note that in Step (2b) of Algorithm 15.5 we have

$$\widehat{p}_{\mathcal{P}(\boldsymbol{g}_\ell, N_\ell)}(\boldsymbol{h}) = \widehat{p}(\boldsymbol{h})$$

for trigonometric polynomials p with index set $\mathcal{A}_{d,\ell}$, for $\ell \in [L]$.

Introducing, for each $\boldsymbol{h} \in \mathcal{A}_d$, the set

$$\mathcal{E}_{\boldsymbol{h},\widetilde{\mathcal{P}}} := \left\{\ell \in [L] : \boldsymbol{h} \cdot \boldsymbol{g}_\ell \not\equiv \boldsymbol{h}' \cdot \boldsymbol{g}_\ell \pmod{N_\ell}\ \forall \boldsymbol{h}' \in \mathcal{A}_d \setminus \{\boldsymbol{h}\}\right\},$$

the outputs $\widehat{p}_{\widetilde{\mathcal{P}}}(\boldsymbol{h})$ of Algorithm 15.5 can be written in the form

$$\widehat{p}_{\widetilde{\mathcal{P}}}(\boldsymbol{h}) = \frac{1}{|\mathcal{E}_{\boldsymbol{h},\widetilde{\mathcal{P}}}|} \sum_{\ell \in \mathcal{E}_{\boldsymbol{h},\widetilde{\mathcal{P}}}} \widehat{p}_{\mathcal{P}(\boldsymbol{g}_\ell, N_\ell)}(\boldsymbol{h}), \tag{15.9}$$

where for $\ell \in [L]$ and $\boldsymbol{h} \in \mathcal{A}_d$, $\widehat{p}_{\mathcal{P}(\boldsymbol{g}_\ell, N_\ell)}(\boldsymbol{h})$ denotes the approximation to $\widehat{p}(\boldsymbol{h})$ by applying the rank-1 lattice rule based on $\mathcal{P}(\boldsymbol{g}_\ell, N_\ell)$. Note that $|\mathcal{E}_{\boldsymbol{h},\widetilde{\mathcal{P}}}|$ is exactly the final value of the counter $\#\boldsymbol{h}$ in Algorithm 15.5, for $\boldsymbol{h} \in \mathcal{A}_d$. Note, furthermore, that for a given $\boldsymbol{h} \in \mathcal{A}_d$ the value of $|\mathcal{E}_{\boldsymbol{h},\widetilde{\mathcal{P}}}|$ corresponds to the number of $\ell \in [L]$ such that $\boldsymbol{h} \in \mathcal{A}_{d,\ell}$. Consequently, Equation (15.9) precisely implies the reconstruction property of $\widetilde{\mathcal{P}}$ for $\mathcal{A}_d = \bigcup_{\ell=1}^{L} \mathcal{A}_{d,\ell}$.

The approximation algorithm based on reconstructing multiple rank-1 lattices

We can now make the approximation algorithm in Equation (15.2) more concrete, namely in the following way. Similarly to the notation introduced above for trigonometric polynomials, we write, for $f \in \mathcal{H}_{\mathrm{kor},d,\alpha,\boldsymbol{\gamma}}$,

$$\widehat{f}_{\mathcal{P}(\boldsymbol{g}_\ell, N_\ell)}(\boldsymbol{h}) := \frac{1}{N_\ell} \sum_{k=0}^{N_\ell - 1} f\left(\left\{\frac{k}{N_\ell}\boldsymbol{g}_\ell\right\}\right) \mathrm{e}^{-2\pi \mathrm{i} k (\boldsymbol{h}\cdot\boldsymbol{g}_\ell)/N_\ell}$$

and

$$\widehat{f}_{\widetilde{\mathcal{P}}}(\boldsymbol{h}) := \frac{1}{|\mathcal{E}_{\boldsymbol{h},\widetilde{\mathcal{P}}}|} \sum_{\ell \in \mathcal{E}_{\boldsymbol{h},\widetilde{\mathcal{P}}}} \widehat{f}_{\mathcal{P}(\boldsymbol{g}_\ell, N_\ell)}(\boldsymbol{h}). \tag{15.10}$$

In order to approximate $f \in \mathcal{H}_{\mathrm{kor},d,\alpha,\boldsymbol{\gamma}}$ we apply the algorithm

$$A^{\mathrm{mult}}_{\mathcal{A}_d}(f)(\boldsymbol{x}) := \sum_{\boldsymbol{h} \in \mathcal{A}_d} \left(\frac{1}{|\mathcal{E}_{\boldsymbol{h},\widetilde{\mathcal{P}}}|} \sum_{\ell \in \mathcal{E}_{\boldsymbol{h},\widetilde{\mathcal{P}}}} \widehat{f}_{\mathcal{P}(\boldsymbol{g}_\ell, N_\ell)}(\boldsymbol{h}) \right) \mathrm{e}^{2\pi \mathrm{i} \boldsymbol{h}\cdot\boldsymbol{x}} \tag{15.11}$$

for $\boldsymbol{x} \in [0,1)^d$.

15.2 Error Analysis

Let us now proceed to the error analysis of multiple rank-1 lattices, in particular those found by Algorithm 15.2. Assume that we are given a finite index set $\mathcal{A}_d \subseteq \mathbb{Z}^d$, a multiple rank-1 lattice point set $\widetilde{\mathcal{P}}$, and that a function $f \in \mathcal{H}_{\mathrm{kor},d,\alpha,\gamma}$ is approximated by the algorithm in (15.11). We would like to analyze the error

$$\|f - A^{\mathrm{mult}}_{\mathcal{A}_d}(f)\|_{L_\infty}.$$

Our analysis is based on writing

$$f - A^{\mathrm{mult}}_{\mathcal{A}_d}(f) = f - \Sigma_{\mathcal{A}_d}(f) + \Sigma_{\mathcal{A}_d}(f) - A^{\mathrm{mult}}_{\mathcal{A}_d}(f),$$

where

$$\Sigma_{\mathcal{A}_d}(f) := \sum_{\boldsymbol{h} \in \mathcal{A}_d} \widehat{f}(\boldsymbol{h})\, \mathrm{e}^{2\pi \mathrm{i} \boldsymbol{h}\cdot\boldsymbol{x}}.$$

We then split the error into

- the truncation error $\|f - \Sigma_{\mathcal{A}_d}(f)\|_{L_\infty}$, and
- the so-called *aliasing error* $\|\Sigma_{\mathcal{A}_d}(f) - A^{\mathrm{mult}}_{\mathcal{A}_d}(f)\|_{L_\infty}$,

and analyze these separately.

Analyzing the aliasing error

The following lemma gives an estimate on the aliasing error.

Lemma 15.6 *Let* $f : [0,1]^d \to \mathbb{R}$ *be such that* $\sum_{\boldsymbol{h}\in\mathbb{Z}^d} |\widehat{f}(\boldsymbol{h})| < \infty$. *Let* $\mathcal{A}_d \subseteq \mathbb{Z}^d$, $2 \le |\mathcal{A}_d| < \infty$. *Let* $A^{\mathrm{mult}}_{\mathcal{A}_d}$ *be the approximation algorithm in* (15.11). *Then we have*

$$\left\|\Sigma_{\mathcal{A}_d}(f) - A^{\mathrm{mult}}_{\mathcal{A}_d}(f)\right\|_{L_\infty} \le L \sum_{\boldsymbol{h}\in\mathbb{Z}^d\setminus\mathcal{A}_d} |\widehat{f}(\boldsymbol{h})|.$$

Proof Let $\ell \in [L]$. For $k \in \{0, 1, \ldots, N_\ell - 1\}$ we have

$$f\left(\left\{\frac{k}{N_\ell}\boldsymbol{g}_\ell\right\}\right) = \sum_{\boldsymbol{m}\in\mathbb{Z}^d} \widehat{f}(\boldsymbol{m})\, \mathrm{e}^{2\pi \mathrm{i} k \boldsymbol{m}\cdot\boldsymbol{g}_\ell/N_\ell}.$$

Using the notation introduced above, we obtain

$$\begin{aligned}
\widehat{f}_{\mathcal{P}(\boldsymbol{g}_\ell,N_\ell)}(\boldsymbol{h}) &= \frac{1}{N_\ell}\sum_{k=0}^{N_\ell-1} f\left(\left\{\frac{k}{N_\ell}\boldsymbol{g}_\ell\right\}\right) \mathrm{e}^{-2\pi \mathrm{i} k \boldsymbol{h}\cdot\boldsymbol{g}_\ell/N_\ell} \\
&= \frac{1}{N_\ell}\sum_{k=0}^{N_\ell-1} \sum_{\boldsymbol{m}\in\mathbb{Z}^d} \widehat{f}(\boldsymbol{m})\, \mathrm{e}^{2\pi \mathrm{i} k (\boldsymbol{m}-\boldsymbol{h})\cdot\boldsymbol{g}_\ell/N_\ell}
\end{aligned}$$

$$= \frac{1}{N_\ell} \sum_{k=0}^{N_\ell - 1} \sum_{\boldsymbol{m} \in \mathbb{Z}^d} \widehat{f}(\boldsymbol{m} + \boldsymbol{h})\, \mathrm{e}^{2\pi \mathrm{i} k \boldsymbol{m} \cdot \boldsymbol{g}_\ell / N_\ell}$$

$$= \sum_{\boldsymbol{m} \in \mathbb{Z}^d} \widehat{f}(\boldsymbol{m} + \boldsymbol{h}) \frac{1}{N_\ell} \sum_{k=0}^{N_\ell - 1} \mathrm{e}^{2\pi \mathrm{i} k \boldsymbol{m} \cdot \boldsymbol{g}_\ell / N_\ell}$$

$$= \sum_{\substack{\boldsymbol{m} \in \mathbb{Z}^d \\ \boldsymbol{m} \cdot \boldsymbol{g}_\ell \equiv 0 \ (\mathrm{mod}\ N_\ell)}} \widehat{f}(\boldsymbol{m} + \boldsymbol{h}), \tag{15.12}$$

where we used the character property of the $\mathcal{P}(\boldsymbol{g}_\ell, N_\ell)$ as stated in Remark 1.10. Due to the fact that the reconstruction property of $\widetilde{\mathcal{P}}$ holds for $\mathcal{A}_d = \bigcup_{\ell=1}^{L} \mathcal{A}_{d,\ell}$, we have $\mathcal{E}_{\boldsymbol{h},\widetilde{\mathcal{P}}} \neq \emptyset$. Using (15.12), $\widehat{f}_{\widetilde{\mathcal{P}}}(\boldsymbol{h})$ can be written in the form

$$\widehat{f}_{\widetilde{\mathcal{P}}}(\boldsymbol{h}) = \frac{1}{|\mathcal{E}_{\boldsymbol{h},\widetilde{\mathcal{P}}}|} \sum_{\ell \in \mathcal{E}_{\boldsymbol{h},\widetilde{\mathcal{P}}}} \sum_{\substack{\boldsymbol{m} \in \mathbb{Z}^d \\ \boldsymbol{m} \cdot \boldsymbol{g}_\ell \equiv 0 \ (\mathrm{mod}\ N_\ell)}} \widehat{f}(\boldsymbol{m} + \boldsymbol{h})$$

$$= \widehat{f}(\boldsymbol{h}) + \frac{1}{|\mathcal{E}_{\boldsymbol{h},\widetilde{\mathcal{P}}}|} \sum_{\ell \in \mathcal{E}_{\boldsymbol{h},\widetilde{\mathcal{P}}}} \sum_{\substack{\boldsymbol{m} \in \mathbb{Z}^d \setminus \{\boldsymbol{0}\} \\ \boldsymbol{m} \cdot \boldsymbol{g}_\ell \equiv 0 \ (\mathrm{mod}\ N_\ell)}} \widehat{f}(\boldsymbol{m} + \boldsymbol{h}).$$

This implies, for any $\boldsymbol{x} \in [0, 1)^d$,

$$\left(\Sigma_{\mathcal{A}_d}(f) - A^{\mathrm{mult}}_{\mathcal{A}_d}(f)\right)(\boldsymbol{x}) = \sum_{\boldsymbol{h} \in \mathcal{A}_d} \left(\widehat{f}(\boldsymbol{h}) - \widehat{f}_{\widetilde{\mathcal{P}}}(\boldsymbol{h})\right) \mathrm{e}^{2\pi \mathrm{i} \boldsymbol{h} \cdot \boldsymbol{x}}$$

$$= - \sum_{\boldsymbol{h} \in \mathcal{A}_d} \frac{1}{|\mathcal{E}_{\boldsymbol{h},\widetilde{\mathcal{P}}}|} \sum_{\ell \in \mathcal{E}_{\boldsymbol{h},\widetilde{\mathcal{P}}}} \sum_{\substack{\boldsymbol{m} \in \mathbb{Z}^d \setminus \{\boldsymbol{0}\} \\ \boldsymbol{m} \cdot \boldsymbol{g}_\ell \equiv 0 \ (\mathrm{mod}\ N_\ell)}} \widehat{f}(\boldsymbol{m} + \boldsymbol{h})\, \mathrm{e}^{2\pi \mathrm{i} \boldsymbol{h} \cdot \boldsymbol{x}}.$$

Now, taking the L_∞-norm, we obtain

$$\left\|\Sigma_{\mathcal{A}_d}(f) - A^{\mathrm{mult}}_{\mathcal{A}_d}(f)\right\|_{L_\infty} \le \sum_{\boldsymbol{h} \in \mathcal{A}_d} \frac{1}{|\mathcal{E}_{\boldsymbol{h},\widetilde{\mathcal{P}}}|} \sum_{\ell \in \mathcal{E}_{\boldsymbol{h},\widetilde{\mathcal{P}}}} \sum_{\substack{\boldsymbol{m} \in \mathbb{Z}^d \setminus \{\boldsymbol{0}\} \\ \boldsymbol{m} \cdot \boldsymbol{g}_\ell \equiv 0 \ (\mathrm{mod}\ N_\ell)}} |\widehat{f}(\boldsymbol{m} + \boldsymbol{h})|$$

$$\le \sum_{\ell=1}^{L} \sum_{\boldsymbol{h} \in \mathcal{A}_{d,\ell}} \frac{1}{|\mathcal{E}_{\boldsymbol{h},\widetilde{\mathcal{P}}}|} \sum_{\substack{\boldsymbol{m} \in \mathbb{Z}^d \setminus \{\boldsymbol{0}\} \\ \boldsymbol{m} \cdot \boldsymbol{g}_\ell \equiv 0 \ (\mathrm{mod}\ N_\ell)}} |\widehat{f}(\boldsymbol{m} + \boldsymbol{h})|$$

$$\le \sum_{\ell=1}^{L} \sum_{\boldsymbol{h} \in \mathcal{A}_{d,\ell}} \sum_{\substack{\boldsymbol{m} \in \mathbb{Z}^d \setminus \{\boldsymbol{0}\} \\ \boldsymbol{m} \cdot \boldsymbol{g}_\ell \equiv 0 \ (\mathrm{mod}\ N_\ell)}} |\widehat{f}(\boldsymbol{m} + \boldsymbol{h})|,$$

where, for the second inequality, we changed the order of summation over $\boldsymbol{h}$ and $\boldsymbol{\ell}$, exploiting the definition of the index sets $\mathcal{A}_d$, $\mathcal{A}_{d,\ell}$, and $\mathcal{E}_{\boldsymbol{h},\widetilde{\mathcal{P}}}$.

Consequently it suffices to show that, for every $\ell \in [L]$, we have

$$\sum_{\boldsymbol{h} \in \mathcal{A}_{d,\ell}} \sum_{\substack{\boldsymbol{m} \in \mathbb{Z}^d \setminus \{\boldsymbol{0}\} \\ \boldsymbol{m} \cdot \boldsymbol{g}_\ell \equiv 0 \pmod{N_\ell}}} |\widehat{f}(\boldsymbol{m} + \boldsymbol{h})| \leq \sum_{\boldsymbol{h} \in \mathbb{Z}^d \setminus \mathcal{A}_d} |\widehat{f}(\boldsymbol{h})| \tag{15.13}$$

to finish the proof.

Fix $\ell \in [L]$. We are now going to show by contradiction that

$$\boldsymbol{h} + \boldsymbol{m} \neq \boldsymbol{h}' + \boldsymbol{m}' \tag{15.14}$$

for all $\boldsymbol{h} \in \mathcal{A}_{d,\ell}$, all $\boldsymbol{h}' \in \mathcal{A}_d \setminus \{\boldsymbol{h}\}$, and all $\boldsymbol{m}, \boldsymbol{m}'$ in the dual of $\mathcal{P}(\boldsymbol{g}_\ell, N_\ell)$.

Let us assume to the contrary that there exist $\boldsymbol{h} \in \mathcal{A}_{d,\ell}$, $\boldsymbol{h}' \in \mathcal{A}_d \setminus \{\boldsymbol{h}\}$, and $\boldsymbol{m}, \boldsymbol{m}'$ in the dual of $\mathcal{P}(\boldsymbol{g}_\ell, N_\ell)$ such that $\boldsymbol{h}+\boldsymbol{m} = \boldsymbol{h}'+\boldsymbol{m}'$. This implies that $\boldsymbol{h}-\boldsymbol{h}' = \boldsymbol{m}'-\boldsymbol{m}$, and thus also

$$(\boldsymbol{h} - \boldsymbol{h}') \cdot \boldsymbol{g}_\ell \equiv (\boldsymbol{m}' - \boldsymbol{m}) \cdot \boldsymbol{g}_\ell \pmod{N_\ell}.$$

Note, however, that also $\boldsymbol{m}' - \boldsymbol{m}$ is in the dual of $\mathcal{P}(\boldsymbol{g}_\ell, N_\ell)$, and so

$$(\boldsymbol{h} - \boldsymbol{h}') \cdot \boldsymbol{g}_\ell \equiv 0 \pmod{N_\ell},$$

or, equivalently,

$$\boldsymbol{h} \cdot \boldsymbol{g}_\ell \equiv \boldsymbol{h}' \cdot \boldsymbol{g}_\ell \pmod{N_\ell}.$$

However, the latter equivalence contradicts the construction of $\mathcal{A}_{d,\ell}$, so (15.14) is shown.

From (15.14), by setting $\boldsymbol{m}' = \boldsymbol{0}$, we obtain that $\boldsymbol{h} + \boldsymbol{m} \notin \mathcal{A}_d$ for any $\boldsymbol{h} \in \mathcal{A}_{d,\ell}$ and any $\boldsymbol{m}$ in the dual of $\mathcal{P}(\boldsymbol{g}_\ell, N_\ell)$. Moreover, for $\boldsymbol{h}, \boldsymbol{h}' \in \mathcal{A}_{d,\ell}$ with $\boldsymbol{h} \neq \boldsymbol{h}'$ we have $\boldsymbol{h} \cdot \boldsymbol{g}_\ell \not\equiv \boldsymbol{h}' \cdot \boldsymbol{g}_\ell \pmod{N_\ell}$, as $\mathcal{A}_{d,\ell} \subseteq \mathcal{A}_d$ and the reconstruction property holds.

Consequently, the sets

$$\begin{aligned}\left\{\boldsymbol{k} \in \mathbb{Z}^d \setminus \{\boldsymbol{h}\} \;:\; \boldsymbol{k} \cdot \boldsymbol{g}_\ell \equiv \boldsymbol{h} \cdot \boldsymbol{g}_\ell \pmod{N_\ell}\right\} \\ = \left\{\boldsymbol{m} + \boldsymbol{h} \;:\; \boldsymbol{m} \in \mathbb{Z}^d \setminus \{\boldsymbol{0}\},\ \boldsymbol{m} \cdot \boldsymbol{g}_\ell \equiv 0 \pmod{N_\ell}\right\}\end{aligned}$$

for $\boldsymbol{h} \in \mathcal{A}_{d,\ell}$ are pairwise disjoint, and they are all subsets of $\mathbb{Z}^d \setminus \mathcal{A}_d$. These considerations imply (15.13), and we are done. □

Choosing the set $\mathcal{A}_d$ of frequency indices

Next, we use a particular choice of the index set $\mathcal{A}_d$. In analogy to (13.1) we define

$$\mathcal{A}_{d,M,1/2,\gamma} := \left\{\boldsymbol{h} \in \mathbb{Z}^d : r_{1,\gamma}(\boldsymbol{h}) \leq M\right\}, \tag{15.15}$$

for a real $M \geq 1$, i.e., we consider an extension of the definition in (13.1) to the case where $\alpha = 1/2$. It is no major restriction to assume that M is chosen large enough to guarantee $M\gamma_1^{1/4} \geq 1$, which we shall do in this chapter for technical reasons. We have the following lemma regarding the cardinality of $\mathcal{A}_{d,M,1/2,\boldsymbol{\gamma}}$.

Lemma 15.7 *Let $d \in \mathbb{N}$ and let $M \geq 1$ be a real number. For $\mathcal{A}_{d,M,1/2,\boldsymbol{\gamma}}$ as defined in* (15.15), *and for any choice of $\tau > 1$ it is true that*

$$|\mathcal{A}_{d,M,1/2,\boldsymbol{\gamma}}| \leq M^\tau \prod_{j=1}^{d} \left(1 + 2\gamma_j^\tau \zeta(\tau)\right).$$

Proof The proof works analogously to that of the fourth item in Lemma 13.1. □

The L_∞-approximation error

We can now show the following theorem bounding the error of the algorithm $A_{\mathcal{A}_d}^{\text{mult}}$ for L_∞-approximation.

Theorem 15.8 *Let $\alpha, \lambda > 1/2$, and let $\boldsymbol{\gamma} = (\gamma_j)_{j\geq 1}$ be product weights with $\gamma_j \in (0, 1]$ for $j \in \mathbb{N}$. Let $f \in \mathcal{H}_{\text{kor},d,\alpha+\lambda,\boldsymbol{\gamma}}$. Let $M \geq 1$ be such that $M\gamma_1^{1/4} \geq 1$, let $\mathcal{A}_d = \mathcal{A}_{d,M,1/2,\boldsymbol{\gamma}^{1/4}}$, and assume that $\widetilde{\mathcal{P}} = \widetilde{\mathcal{P}}(\boldsymbol{g}_1, N_1, \ldots, \boldsymbol{g}_L, N_L)$ has been constructed by Algorithm 15.2 for $\mathcal{A}_{d,M,1/2,\boldsymbol{\gamma}^{1/4}}$. Let the approximated Fourier coefficients $\widehat{f}_{\widetilde{\mathcal{P}}}(\boldsymbol{h})$ of f be computed according to* (15.10) *for $\boldsymbol{h} \in \mathcal{A}_{d,M,1/2,\boldsymbol{\gamma}^{1/4}}$. Then we have*

$$\left\|f - A_{\mathcal{A}_d}^{\text{mult}}(f)\right\|_{L_\infty} \leq \frac{2^\alpha \gamma_1^{(1-\alpha)/4}(L+1)}{M^\alpha} \|f\|_{\text{kor},d,\alpha+\lambda,\boldsymbol{\gamma}} \prod_{j=1}^{d} \left(1 + 2\gamma_j^{1/2}\zeta(2\lambda)\right)^{1/2}.$$

Proof Throughout the proof, let $\mathcal{A}_d = \mathcal{A}_{d,M,1/2,\boldsymbol{\gamma}^{1/4}}$. By the triangle inequality we obviously have

$$\left\|f - A_{\mathcal{A}_d}^{\text{mult}}(f)\right\|_{L_\infty} \leq \left\|f - \Sigma_{\mathcal{A}_d}(f)\right\|_{L_\infty} + \left\|\Sigma_{\mathcal{A}_d}(f) - A_{\mathcal{A}_d}^{\text{mult}}(f)\right\|_{L_\infty}. \tag{15.16}$$

Regarding the first term on the right-hand side of (15.16), which is the truncation error, we have

$$\left(f - \Sigma_{\mathcal{A}_d}(f)\right)(\boldsymbol{x}) = \sum_{\boldsymbol{h} \in \mathbb{Z}^d \setminus \mathcal{A}_d} \widehat{f}(\boldsymbol{h})\, \mathrm{e}^{2\pi \mathrm{i} \boldsymbol{h}\cdot\boldsymbol{x}},$$

so

$$\left\|f - \Sigma_{\mathcal{A}_d}(f)\right\|_{L_\infty} \leq \sum_{\boldsymbol{h} \in \mathbb{Z}^d \setminus \mathcal{A}_d} |\widehat{f}(\boldsymbol{h})|.$$

For the aliasing error, which is the second term on the right-hand side of (15.16), we can directly apply Lemma 15.6, which yields

$$\left\|\Sigma_{\mathcal{A}_d}(f) - A_{\mathcal{A}_d}^{\text{mult}}(f)\right\|_{L_\infty} \leq L \sum_{\boldsymbol{h} \in \mathbb{Z}^d \setminus \mathcal{A}_d} |\widehat{f}(\boldsymbol{h})|.$$

Therefore, using (15.16), we obtain

$$\left\|f - A^{\mathrm{mult}}_{\mathcal{A}_d}(f)\right\|_{L_\infty} \leq (L+1) \sum_{\boldsymbol{h}\in\mathbb{Z}^d\setminus\mathcal{A}_d} |\widehat{f}(\boldsymbol{h})|. \tag{15.17}$$

Thus it remains to show that the right-hand side of (15.17) satisfies the asserted upper bound in the theorem. We have

$$\begin{aligned}
\sum_{\boldsymbol{h}\in\mathbb{Z}^d\setminus\mathcal{A}_d} |\widehat{f}(\boldsymbol{h})| &= \sum_{\boldsymbol{h}\in\mathbb{Z}^d\setminus\mathcal{A}_d} \frac{r_{\alpha,\boldsymbol{\gamma}^{1/4}}(\boldsymbol{h})}{r_{\alpha,\boldsymbol{\gamma}^{1/4}}(\boldsymbol{h})} |\widehat{f}(\boldsymbol{h})| \\
&\leq \left(\sup_{\boldsymbol{h}\in\mathbb{Z}^d\setminus\mathcal{A}_d} \frac{1}{r_{\alpha,\boldsymbol{\gamma}^{1/4}}(\boldsymbol{h})}\right) \sum_{\boldsymbol{h}\in\mathbb{Z}^d\setminus\mathcal{A}_d} r_{\alpha,\boldsymbol{\gamma}^{1/4}}(\boldsymbol{h})\,|\widehat{f}(\boldsymbol{h})| \\
&\leq \left(\sup_{\boldsymbol{h}\in\mathbb{Z}^d\setminus\mathcal{A}_d} \frac{1}{r_{\alpha,\boldsymbol{\gamma}^{1/4}}(\boldsymbol{h})}\right) \sum_{\boldsymbol{h}\in\mathbb{Z}^d} r_{\alpha,\boldsymbol{\gamma}^{1/4}}(\boldsymbol{h})\,|\widehat{f}(\boldsymbol{h})|.
\end{aligned}$$

We study the two factors in the latter expression separately. Let us first consider

$$\sup_{\boldsymbol{h}\in\mathbb{Z}^d\setminus\mathcal{A}_d} \frac{1}{r_{\alpha,\boldsymbol{\gamma}^{1/4}}(\boldsymbol{h})}.$$

For $\boldsymbol{h} \in \mathbb{Z}^d \setminus \mathcal{A}_d$, we have, by definition, $r_{1,\boldsymbol{\gamma}^{1/4}}(\boldsymbol{h}) > M$, and for $\widetilde{\boldsymbol{h}} \in \mathcal{A}_d$ we have $r_{1,\boldsymbol{\gamma}^{1/4}}(\widetilde{\boldsymbol{h}}) \leq M$. Therefore, $r_{1,\boldsymbol{\gamma}^{1/4}}(\boldsymbol{h}) \geq r_{1,\boldsymbol{\gamma}^{1/4}}(\widetilde{\boldsymbol{h}})$ for $\boldsymbol{h} \in \mathbb{Z}^d \setminus \mathcal{A}_d$ and $\widetilde{\boldsymbol{h}} \in \mathcal{A}_d$, and thus also $r_{\alpha,\boldsymbol{\gamma}^{1/4}}(\boldsymbol{h}) \geq r_{\alpha,\boldsymbol{\gamma}^{1/4}}(\widetilde{\boldsymbol{h}})$, which implies

$$\sup_{\boldsymbol{h}\in\mathbb{Z}^d\setminus\mathcal{A}_d} \frac{1}{r_{\alpha,\boldsymbol{\gamma}^{1/4}}(\boldsymbol{h})} \leq \frac{1}{\sup_{\boldsymbol{h}\in\mathcal{A}_d} r_{\alpha,\boldsymbol{\gamma}^{1/4}}(\boldsymbol{h})}.$$

By definition of $r_{\alpha,\boldsymbol{\gamma}}$ (see (2.22)), we have

$$r_{1,\boldsymbol{\gamma}^{1/4}}((k_1, \underbrace{0,\ldots,0}_{d-1 \text{ times}})) = \frac{|k_1|}{\gamma_1^{1/4}} \leq M$$

if $k_1 = \pm\lfloor M\gamma_1^{1/4}\rfloor$, such that $(k_1, 0, \ldots, 0) \in \mathcal{A}_d$, and we have

$$r_{\alpha,\boldsymbol{\gamma}^{1/4}}((k_1, \underbrace{0,\ldots,0}_{d-1 \text{ times}})) = \frac{|k_1|^\alpha}{\gamma_1^{1/4}} = \frac{(\lfloor M\gamma_1^{1/4}\rfloor)^\alpha}{\gamma_1^{1/4}} \geq \frac{\gamma_1^{(\alpha-1)/4} M^\alpha}{2^\alpha},$$

where we used that $\lfloor x \rfloor \geq x/2$ for $x \geq 1$. This yields

$$\frac{1}{\sup_{\boldsymbol{h}\in\mathcal{A}_d} r_{\alpha,\boldsymbol{\gamma}^{1/4}}(\boldsymbol{h})} \leq \frac{2^\alpha \gamma_1^{(1-\alpha)/4}}{M^\alpha}.$$

So we obtain

$$\sum_{\boldsymbol{h}\in\mathbb{Z}^d\setminus\mathcal{A}_d} |\widehat{f}(\boldsymbol{h})| \leq \frac{2^\alpha \gamma_1^{(1-\alpha)/4}}{M^\alpha} \sum_{\boldsymbol{h}\in\mathbb{Z}^d} r_{\alpha,\boldsymbol{\gamma}^{1/4}}(\boldsymbol{h})\, |\widehat{f}(\boldsymbol{h})|. \tag{15.18}$$

Finally, let us deal with

$$\sum_{\boldsymbol{h}\in\mathbb{Z}^d} r_{\alpha,\boldsymbol{\gamma}^{1/4}}(\boldsymbol{h})\, |\widehat{f}(\boldsymbol{h})|.$$

For $\lambda > 1/2$ we have, using the Cauchy–Schwarz inequality,

$$\begin{aligned}
&\sum_{\boldsymbol{h}\in\mathbb{Z}^d} r_{\alpha,\boldsymbol{\gamma}^{1/4}}(\boldsymbol{h})\, |\widehat{f}(\boldsymbol{h})| \\
&= \sum_{\boldsymbol{h}\in\mathbb{Z}^d} \frac{r_{\lambda,\boldsymbol{\gamma}^{1/4}}(\boldsymbol{h})}{r_{\lambda,\boldsymbol{\gamma}^{1/4}}(\boldsymbol{h})}\, r_{\alpha,\boldsymbol{\gamma}^{1/4}}(\boldsymbol{h})\, |\widehat{f}(\boldsymbol{h})| \\
&\leq \left(\sum_{\boldsymbol{h}\in\mathbb{Z}^d} \frac{1}{(r_{\lambda,\boldsymbol{\gamma}^{1/4}}(\boldsymbol{h}))^2}\right)^{1/2} \left(\sum_{\boldsymbol{h}\in\mathbb{Z}^d} (r_{\alpha,\boldsymbol{\gamma}^{1/4}}(\boldsymbol{h}) r_{\lambda,\boldsymbol{\gamma}^{1/4}}(\boldsymbol{h}))^2\, |\widehat{f}(\boldsymbol{h})|^2\right)^{1/2} \\
&= \left(\sum_{\boldsymbol{h}\in\mathbb{Z}^d} \frac{1}{r_{2\lambda,\boldsymbol{\gamma}^{1/2}}(\boldsymbol{h})}\right)^{1/2} \left(\sum_{\boldsymbol{h}\in\mathbb{Z}^d} r_{2(\alpha+\lambda),\boldsymbol{\gamma}}(\boldsymbol{h})\, |\widehat{f}(\boldsymbol{h})|^2\right)^{1/2} \\
&= \|f\|_{\mathrm{kor},d,\alpha+\lambda,\boldsymbol{\gamma}} \prod_{j=1}^{d} \left(1+\gamma_j^{1/2} 2\zeta(2\lambda)\right)^{1/2}.
\end{aligned}$$

Employing this estimate in (15.18) yields

$$\sum_{\boldsymbol{h}\in\mathbb{Z}^d\setminus\mathcal{A}_d} |\widehat{f}(\boldsymbol{h})| \leq \frac{2^\alpha \gamma_1^{(1-\alpha)/4}}{M^\alpha} \|f\|_{\mathrm{kor},d,\alpha+\lambda,\boldsymbol{\gamma}} \prod_{j=1}^{d} \left(1+\gamma_j^{1/2} 2\zeta(2\lambda)\right)^{1/2},$$

and inserting into (15.17) then gives

$$\left\|f - A^{\mathrm{mult}}_{\mathcal{A}_d}(f)\right\|_{L_\infty} \leq \frac{2^\alpha \gamma_1^{(1-\alpha)/4}(L+1)}{M^\alpha} \|f\|_{\mathrm{kor},d,\alpha+\lambda,\boldsymbol{\gamma}} \prod_{j=1}^{d} \left(1+\gamma_j^{1/2} 2\zeta(2\lambda)\right)^{1/2}.$$

This concludes the proof of the theorem. □

Let us now recall that L is the number of single rank-1 lattices that are used in a multiple rank-1 lattice $\widetilde{\mathcal{P}} = \widetilde{\mathcal{P}}(\boldsymbol{g}_1, N_1, \ldots, \boldsymbol{g}_L, N_L)$, and that $L \leq L_{\max}$ in Algorithm 15.2. If we assume that $\widetilde{\mathcal{P}}$ has been found by Algorithm 15.2, and that we choose $\mathcal{A}_d = \mathcal{A}_{d,M,1/2,\boldsymbol{\gamma}^{1/4}}$ in that algorithm, we get

$$L_{\max} \leq C_{c,\delta} \log |\mathcal{A}_{d,M,1/2,\boldsymbol{\gamma}^{1/4}}|,$$

where $C_{c,\delta}$ is a positive quantity depending only on the parameters c and δ in Algorithm 15.2. Due to Lemma 15.7, this implies that

$$L_{\max} \leq C_{c,\delta}\left(\tau \log M + \sum_{j=1}^{d} \log\left(1 + 2\gamma_j^{\tau/4}\zeta(\tau)\right)\right) \tag{15.19}$$

for any $\tau > 1$. Using Proposition 15.4 we know that under the technical condition $\eta \geq \max(N_{\mathcal{A}_d}, 4L_{\max} \log L_{\max})$ the multiple rank-1 lattice point set found by Algorithm 15.2 satisfies

$$\begin{aligned} N := |\widetilde{\mathcal{P}}| &\leq 2L_{\max}(|\mathcal{A}_{d,M,1/2,\gamma^{1/4}}| - 1)c \\ &\leq 2\,c\,C_{c,\delta}\left(\tau \log M + \sum_{j=1}^{d} \log\left(1 + 2\gamma_j^{\tau/4}\zeta(\tau)\right)\right) M^{\tau} \prod_{j=1}^{d}\left(1 + 2\gamma_j^{\tau/4}\zeta(\tau)\right), \end{aligned}$$

where we again used Lemma 15.7, and (15.19) for the second inequality. This implies that for any choice of a small $\nu > 0$, there is a real number $C_{c,\delta,\tau,\nu}$, depending only on c, δ, τ, and ν such that

$$N \leq C_{c,\delta,\tau,\nu}\, M^{\tau+\nu} \prod_{j=1}^{d}\left(1 + 2\gamma_j^{\tau/4}\zeta(\tau)\right)^{1+\nu},$$

or equivalently,

$$\frac{1}{M} \leq \frac{C_{c,\delta,\tau,\nu}^{1/(\tau+\nu)}}{N^{1/(\tau+\nu)}} \prod_{j=1}^{d}\left(1 + 2\gamma_j^{\tau/4}\zeta(\tau)\right)^{(1+\nu)/(\tau+\nu)}. \tag{15.20}$$

Let us return to (15.19) in order to obtain an upper bound on $L_{\max}$ and to get rid of the $\log M$-term there. It follows from the proof of Lemma 13.1 that

$$|\mathcal{A}_{d,M,1/2,\gamma^{1/4}}| \geq |\mathcal{A}_{1,M,1/2,\gamma^{1/4}}| = 1 + 2\lfloor \gamma_1^{1/4} M \rfloor,$$

which implies, again under the assumption that $\gamma_1^{1/4} M \geq 1$ (which implies $2\lfloor \gamma_1^{1/4} M \rfloor \geq \gamma_1^{1/4} M$) that

$$M \leq \frac{|\mathcal{A}_{d,M,1/2,\gamma^{1/4}}| - 1}{\gamma_1^{1/4}}.$$

Furthermore, if $\mathcal{A}_d$ is chosen equal to $\mathcal{A}_{d,M,1/2,\gamma^{1/4}}$ in Algorithm 15.2, then the algorithm chooses the cardinalities N_ℓ of the single rank-1 lattices $\mathcal{P}(\boldsymbol{g}_\ell, N_\ell)$, $\ell \in [L]$, such that they are at least $c(|\mathcal{A}_{d,M,1/2,\gamma^{1/4}}| - 1)$, where $c \in (1, \infty)$ is chosen as in Algorithm 15.2. This obviously implies $N \geq |\mathcal{A}_{d,M,1/2,\gamma^{1/4}}| - 1$, and consequently

$$\frac{N}{\gamma_1^{1/4}} \geq \frac{|\mathcal{A}_{d,M,1/2,\boldsymbol{\gamma}^{1/4}}| - 1}{\gamma_1^{1/4}} \geq M.$$

Plugging this inequality into (15.19) it follows that there is a positive real C_{c,δ,γ_1} such that

$$L \leq C_{c,\delta,\gamma_1} \tau \log N + C_{c,\delta} \sum_{j=1}^{d} \log\left(1 + 2\gamma_j^{\tau/4} \zeta(\tau)\right), \tag{15.21}$$

for any $\tau > 1$.

Combining Equations (15.20) and (15.21) with the bound on the worst-case approximation error in Theorem 15.8 leads to the estimate

$$\begin{aligned}
&\mathrm{err}^{L_\infty\text{-app}}(\mathcal{H}_{\mathrm{kor},d,\alpha+\lambda,\boldsymbol{\gamma}}, A^{\mathrm{mult}}_{\mathcal{A}_{d,M,1/2,\boldsymbol{\gamma}^{1/4}}}) \\
&\leq \frac{2^\alpha \gamma_1^{(1-\alpha)/4} C_{c,\delta,\tau,\nu}^{\alpha/(\tau+\nu)}}{N^{\alpha/(\tau+\nu)}} \prod_{j=1}^{d} \left(1 + 2\gamma_j^{\tau/4}\zeta(\tau)\right)^{\alpha(1+\nu)/(\tau+\nu)} \\
&\quad \times \left(1 + C_{c,\delta,\gamma_1} \tau \log N + C_{c,\delta} \sum_{j=1}^{d} \log\left(1 + 2\gamma_j^{\tau/4}\zeta(\tau)\right)\right) \\
&\quad \times \prod_{j=1}^{d} \left(1 + 2\gamma_j^{1/2}\zeta(2\lambda)\right)^{1/2}.
\end{aligned}$$

Setting

$$\widehat{C}_{c,\delta,\tau,\nu,\gamma_1} := 2^\alpha \gamma_1^{(1-\alpha)/4} C_{c,\delta,\tau,\nu}^{\alpha/(\tau+\nu)} \max(1, C_{c,\delta,\gamma_1}\tau, C_{c,\delta}), \tag{15.22}$$

the main result of this chapter can be summarized as follows.

Corollary 15.9 *Let $\alpha, \lambda > 1/2$, $\tau > 1$, $\nu > 0$, and let $\boldsymbol{\gamma} = (\gamma_j)_{j\geq 1}$ be product weights with $\gamma_j \in (0,1]$ for $j \in \mathbb{N}$. Let $M \geq 1$ be such that $M\gamma_1^{1/4} \geq 1$, let $\mathcal{A}_d = \mathcal{A}_{d,M,1/2,\boldsymbol{\gamma}^{1/4}}$, and assume that $\widetilde{\mathcal{P}} = \widetilde{\mathcal{P}}(\boldsymbol{g}_1, N_1, \ldots, \boldsymbol{g}_L, N_L)$ has been constructed by Algorithm 15.2 for $\mathcal{A}_{d,M,1/2,\boldsymbol{\gamma}^{1/4}}$. Let c and δ be as in Algorithm 15.2, and let $N := |\widetilde{\mathcal{P}}|$. Then we have for the worst-case error of the algorithm $A^{\mathrm{mult}}_{\mathcal{A}_{d,M,1/2,\boldsymbol{\gamma}^{1/4}}}$ in $\mathcal{H}_{\mathrm{kor},d,\alpha+\lambda,\boldsymbol{\gamma}}$ that*

$$\begin{aligned}
&\mathrm{err}^{L_\infty\text{-app}}(\mathcal{H}_{\mathrm{kor},d,\alpha+\lambda,\boldsymbol{\gamma}}, A^{\mathrm{mult}}_{\mathcal{A}_{d,M,1/2,\boldsymbol{\gamma}^{1/4}}}) \\
&\leq \frac{\widehat{C}_{c,\delta,\tau,\nu,\gamma_1}}{N^{\alpha/(\tau+\nu)}} \prod_{j=1}^{d} \left(1 + 2\gamma_j^{\tau/4}\zeta(\tau)\right)^{\alpha(1+\nu)/(\tau+\nu)} \\
&\quad \times \left(1 + \log N + \sum_{j=1}^{d} \log\left(1 + 2\gamma_j^{\tau/4}\zeta(\tau)\right)\right) \prod_{j=1}^{d} \left(1 + 2\gamma_j^{1/2}\zeta(2\lambda)\right)^{1/2},
\end{aligned}$$

where $\widehat{C}_{c,\delta,\tau,\nu,\gamma_1}$ *is as in* (15.22).

15.3 Comparison to Previous Results and Tractability

We recall that in Chapter 14 we obtained, by using an algorithm based on a single rank-1 lattice point set (see Theorem 14.7) or by using a spline algorithm (see Theorem 14.11), a convergence rate of order $O(N^{-\widetilde{\alpha}(2\widetilde{\alpha}-1)/(4\widetilde{\alpha}-1)+\eta})$. This estimate holds for arbitrarily small $\eta > 0$ in the Korobov space $\mathcal{H}_{\mathrm{kor},d,\widetilde{\alpha},\gamma}$, where we need to assume $\widetilde{\alpha} > 1$ in Theorem 14.7 and $\widetilde{\alpha} > 1/2$ in Theorem 14.11.

Considering Corollary 15.9 and the Korobov space $\mathcal{H}_{\mathrm{kor},d,\widetilde{\alpha},\gamma}$, where now $\widetilde{\alpha} = \alpha + \lambda$ with $\alpha, \lambda > 1/2$ (which implies $\widetilde{\alpha} > 1$), we obtain a convergence rate that is arbitrarily close to the order

$$O\left(N^{-(\widetilde{\alpha}-\lambda)/(\tau+\nu)}\right).$$

For $\eta > 0$ choose $\widetilde{\eta} > 0$ such that $\widetilde{\eta}(1 + \alpha/(1 + \widetilde{\eta})) = \eta$, i.e., choose

$$\widetilde{\eta} := \frac{-(1+\alpha-\eta) + \sqrt{(1+\alpha-\eta)^2 + 4\eta}}{2}.$$

Put then $\lambda = 1/2 + \widetilde{\eta}$, and choose $\tau > 1$ and $\nu > 0$ such that $\tau + \nu = 1 + \widetilde{\eta}$. With a few elementary computations, using the relations between the parameters involved, we see that

$$\frac{\widetilde{\alpha}-\lambda}{\tau+\nu} = \widetilde{\alpha} - \frac{1}{2} - \eta.$$

This means that for the error of L_∞-approximation in the Korobov space $\mathcal{H}_{\mathrm{kor},d,\widetilde{\alpha},\gamma}$ with $\widetilde{\alpha} > 1$ we can get a convergence rate of order

$$O\left(N^{-\widetilde{\alpha}+1/2+\eta}\right), \tag{15.23}$$

where $\eta > 0$ can be chosen arbitrarily small. This implies that Theorem 15.9, which is based on multiple rank-1 lattice point sets, yields an advantage over Theorem 14.7, which is based on single rank-1 lattice point sets, and Theorem 14.11, which is based on splines and single rank-1 lattice point sets, when the smoothness parameter exceeds 1/2, i.e., for all admissible choices of $\widetilde{\alpha}$.

Remark 15.10 Since the L_∞-norm dominates the L_2-norm it follows that also the L_2-approximation error satisfies the upper bound in Theorem 15.8. Hence, also the L_2-approximation error in $\mathcal{H}_{\mathrm{kor},d,\widetilde{\alpha},\gamma}$ decays at a rate of order

$$O\left(N^{-\widetilde{\alpha}+1/2+\eta}\right),$$

where $\eta > 0$ can be chosen arbitrarily small. If $\widetilde{\alpha} > 1$ this is an improvement over the rate

$$O\left(N^{-\widetilde{\alpha}/2+\eta}\right)$$

that we obtained in Theorem 13.6, based on single rank-1 lattice rules.

Tractability

Similarly to what we presented in the previous chapters, we can deduce an upper bound from Corollary 15.9 that is independent of the dimension, if the weights decay sufficiently fast.

Theorem 15.11 *Let $\alpha, \lambda > 1/2$, $\tau > 1$, $\nu > 0$, and let $\boldsymbol{\gamma} = (\gamma_j)_{j\geq 1}$ be product weights with $\gamma_j \in (0, 1]$ for $j \in \mathbb{N}$. Furthermore, assume that the weight sequence $\boldsymbol{\gamma}$ satisfies*

$$\sum_{j=1}^{\infty} \gamma_j^{\min(1/2,\tau/4)} < \infty. \tag{15.24}$$

Let $M \geq 1$ be such that $M\gamma_1^{1/4} \geq 1$ and assume that $\widetilde{\mathcal{P}} = \widetilde{\mathcal{P}}(\boldsymbol{g}_1, N_1, \ldots, \boldsymbol{g}_L, N_L)$ has been constructed by Algorithm 15.2 for $\mathcal{A}_{d,M,1/2,\boldsymbol{\gamma}^{1/4}}$. Let c and δ be as in Algorithm 15.2. Then there exists a quantity $C_{\alpha,\lambda,\tau,\nu,\boldsymbol{\gamma},c,\delta} > 0$, which is independent of $N := |\widetilde{\mathcal{P}}|$ and of the dimension d such that

$$\mathrm{err}^{L_\infty\text{-app}}(\mathcal{H}_{\mathrm{kor},d,\alpha+\lambda,\boldsymbol{\gamma}}, A^{\mathrm{mult}}_{\mathcal{A}_{d,M,1/2,\boldsymbol{\gamma}^{1/4}}}) \leq \frac{C_{\alpha,\lambda,\tau,\nu,\boldsymbol{\gamma},c,\delta}}{N^{\alpha/(\tau+\nu)}}. \tag{15.25}$$

In particular, this means that the approximation algorithm $A^{\mathrm{mult}}_{\mathcal{A}_{d,M,1/2,\boldsymbol{\gamma}^{1/4}}}$ yields strong polynomial tractability of L_∞-approximation in $\mathcal{H}_{\mathrm{kor},d,\alpha+\lambda,\boldsymbol{\gamma}}$ with respect to the absolute error criterion, and the same is true for L_2-approximation in $\mathcal{H}_{\mathrm{kor},d,\alpha+\lambda,\boldsymbol{\gamma}}$.

Proof The result follows by using Corollary 15.9 and by noting that, if the weight sequence $\boldsymbol{\gamma}$ satisfies (15.24), we have

$$\prod_{j=1}^{d}\left(1+2\gamma_j^{\tau/4}\zeta(\tau)\right) \leq \prod_{j=1}^{\infty}\left(1+2\gamma_j^{\tau/4}\zeta(\tau)\right) < \infty,$$

$$\sum_{j=1}^{d}\log\left(1+2\gamma_j^{\tau/4}\zeta(\tau)\right) \leq \sum_{j=1}^{\infty}\log\left(1+2\gamma_j^{\tau/4}\zeta(\tau)\right) < \infty,$$

and

$$\prod_{j=1}^{d}\left(1+2\gamma_j^{1/2}\zeta(2\lambda)\right) \leq \prod_{j=1}^{\infty}\left(1+2\gamma_j^{1/2}\zeta(2\lambda)\right) < \infty,$$

which can be shown using exactly the same methods that we have used in previous chapters. This proves (15.25).

Let now $\varepsilon \in (0,1)$. Then $N \geq (C_{\alpha,\lambda,\tau,\nu,\gamma,c,\delta}\,\varepsilon^{-1})^{(\tau+\nu)/\alpha}$, implies

$$\mathrm{err}^{L_\infty-\mathrm{app}}(\mathcal{H}_{\mathrm{kor},d,\alpha+\lambda,\gamma}, A^{\mathrm{mult}}_{\mathcal{A}_{d,M,1/2,\gamma^{1/4}}}) \leq \varepsilon.$$

This in turn shows that the information complexity $N^{L_\infty-\mathrm{app}}(\varepsilon, d, \Lambda^{\mathrm{std}})$ is bounded uniformly in d by $(C_{\alpha,\lambda,\tau,\nu,\gamma,c,\delta}\,\varepsilon^{-1})^{1/\alpha}$. Therefore we have strong polynomial tractability.

Since strong polynomial tractability for L_∞-approximation with respect to the absolute error criterion implies strong polynomial tractability for L_2-approximation, it is clear that also the statement on L_2-approximation in the theorem holds true. □

Remark 15.12 Further results on polynomial and weak tractability of L_∞- and L_2-approximation using multiple rank-1 lattice point sets can be derived in a similar manner, as it was done in previous chapters.

Notes and Remarks

In this chapter, we have mostly followed the paper [131] in outlining the results on multiple rank-1 lattice point sets for L_∞-approximation. These point sets were also dealt with in the earlier papers [127, 128]. In particular, in [128], the reader can find further technical details regarding Algorithm 15.2. We remark that the results in these papers are presented for a more general family of function spaces than the Korobov space. On the other hand, the setting in [131] does not include weights $\boldsymbol{\gamma}$, which we included here to be able to keep track of the influence of the number of variables, d, and to obtain tractability results.

The best possible convergence rate for L_∞-approximation in the Korobov space $\mathcal{H}_{\mathrm{kor},d,\widetilde{\alpha},\gamma}$ is $O(N^{-\widetilde{\alpha}+1/2}\,(\log N)^{(d-1)\widetilde{\alpha}})$, see [23, Theorem 6.10], and hence the rate in (15.23) that is obtained from Corollary 15.9 is almost best possible.

Chapter 16
Fast QMC Matrix-Vector Multiplication

In this chapter we discuss the approximation of integrals of the form

$$\int_{[0,1]^d} f(\boldsymbol{x}A)\,\mathrm{d}\boldsymbol{x}, \tag{16.1}$$

where A is a $d \times t$ matrix with entries from $\mathbb{R}$. Using a lattice rule with integration nodes $\boldsymbol{x}_0, \boldsymbol{x}_1, \ldots, \boldsymbol{x}_{N-1}$, we can approximate this type of integral by

$$\frac{1}{N}\sum_{k=0}^{N-1} f(\boldsymbol{x}_k A). \tag{16.2}$$

Integrals of the form (16.1) appear in practical applications, as for example in the study of partial differential equations (PDEs) with random coefficients (see Appendix A). There the main cost of approximating the integral (16.1) is computing the vector-matrix products $\boldsymbol{x}_0 A, \boldsymbol{x}_1 A, \ldots, \boldsymbol{x}_{N-1} A$.

Here we present an efficient way of computing such vector-matrix products, which is essentially based on ideas presented in Section 3.4 and that works in particular for node sets with a lattice structure. The presented approach is advantageous if N is significantly smaller than 2^d (say $N \approx d^\kappa$ for some $\kappa > 0$). In the context of PDEs with random coefficients this is often naturally satisfied. Or if d is large, say $d > 100$, then $2^d > 2^{100} \approx 10^{30}$, so it is not possible to use 10^{30} points and hence $N \ll 2^d$ by virtue of limitations of current computers.

16.1 The General Idea

In some examples where we encounter the problem described above, it is possible to modify the matrix A (and consequently the integrand f) to ensure that A has a certain structure (e.g., circulant, Toeplitz, tridiagonal, etc.), which can be used to

J. Dick et al., *Lattice Rules*, Springer Series in Computational Mathematics 58,
https://doi.org/10.1007/978-3-031-09951-9_16

speed up the computation of $\boldsymbol{x}_k A$ for $k \in \{0, 1, \ldots, N-1\}$. Our goal here is to design a method based on lattice point sets that makes a modification of the matrix A unnecessary.

The idea of this approach, called the fast QMC matrix-vector product, is the following. We introduce the $N \times d$ matrix

$$X := \begin{pmatrix} \boldsymbol{x}_0 \\ \boldsymbol{x}_1 \\ \vdots \\ \boldsymbol{x}_{N-1} \end{pmatrix} \tag{16.3}$$

and would like to have a fast method to compute, for the $d \times t$ matrix A,

$$XA =: B = \begin{pmatrix} \boldsymbol{b}_0 \\ \boldsymbol{b}_1 \\ \vdots \\ \boldsymbol{b}_{N-1} \end{pmatrix} \in \mathbb{R}^{N \times t},$$

where $\boldsymbol{b}_0, \boldsymbol{b}_1, \ldots, \boldsymbol{b}_{N-1} \in \mathbb{R}^t$ are the row vectors of the matrix B. In order to compute (16.2), we first compute the matrix-matrix product $XA = B$, store the matrix B, and then evaluate

$$\frac{1}{N} \sum_{k=0}^{N-1} f(\boldsymbol{b}_k). \tag{16.4}$$

Storing the matrix B requires $O(Nt)$ storage. In general, the computation of XA requires $O(d\,N\,t)$ operations, and the computation of the quadrature sum in (16.4) requires $O(N)$ operations. In the following we will construct quadrature points $\boldsymbol{x}_0, \boldsymbol{x}_1, \ldots, \boldsymbol{x}_{N-1}$ for which the matrix X permits a matrix-vector multiplication $X\boldsymbol{a}$ in $O(N \log N)$ operations, where $\boldsymbol{a}$ can be any column of the matrix A. The computation of XA then reduces to $O(t\,N \log N)$ operations, instead of $O(d\,N\,t)$ operations for the straightforward implementation. This leads to a significant speedup provided that N is much smaller than 2^d.

We may refer to the matrix X in (16.3) as the "QMC matrix". We are interested in a fast method for computing $X\boldsymbol{a}$, i.e., a fast method to compute the product of the QMC matrix X with a vector $\boldsymbol{a}$, which motivates why we refer to the method as "fast QMC matrix-vector multiplication".

The basic idea of the proposed approach is to find quadrature point sets $\{\boldsymbol{x}_0, \boldsymbol{x}_1, \ldots, \boldsymbol{x}_{N-1}\}$ in $\mathbb{R}^d$ with a specific ordering such that the truncated QMC matrix

$$X' := \begin{pmatrix} \boldsymbol{x}_1 \\ \vdots \\ \boldsymbol{x}_{N-1} \end{pmatrix} \in \mathbb{R}^{(N-1) \times d}$$

has a factorization of the form

$$X' = ZP,$$

where $Z \in \mathbb{R}^{(N-1)\times(N-1)}$ is a circulant matrix and $P \in \{0,1\}^{(N-1)\times d}$ is a matrix in which each column has at most one value which is 1, with the remaining entries being 0. The special structure of P means that, for a given column vector $\boldsymbol{a}$, the column vector $\boldsymbol{a}' = P\boldsymbol{a}$ can be obtained in at most $O(N)$ operations, and the matrix-vector multiplication $Z\boldsymbol{a}'$ can be computed in $O(N \log N)$ operations using FFT. The vector $\boldsymbol{x}_0$ is separated out because many QMC methods have a vector $\boldsymbol{x}_0$ for which all components are equal. Then, the computation of $\boldsymbol{x}_0\boldsymbol{a}$ requires at most $d-1$ additions and one multiplication. Indeed, in many instances of common QMC methods, as for example lattice rules, we even have $\boldsymbol{x}_0 = (0,\ldots,0)$, such that no extra computation is necessary.

In Sections 16.2 and 16.3, respectively, we will consider two important classes of QMC point sets whose structure facilitates the use of the presently proposed acceleration, namely

- (rank-1) lattice point sets, and
- the multi-set union of all Korobov lattice point sets (which is essentially one class of Korobov's p-sets as introduced in Definition 10.1).

16.2 Fast QMC Matrix-Vector Multiplication for Lattice Point Sets

Let us first explain the fast QMC matrix-vector multiplication for lattice point sets. We apply the main idea of the fast CBC construction of the generating vector $\boldsymbol{g}$ of (rank-1) lattice rules (see Section 3.4) to the matrix-vector multiplication $X\boldsymbol{a}$. For the sake of simplicity, we again confine the exposition to cases where the number of points N is prime. Based on the presently developed ideas, the general case can be handled analogously with the method of Nuyens and Cools [215].

Let q be a primitive root modulo N, i.e., we have

$$\{q^k \pmod{N} \ :\ k \in \{0,1,\ldots,N-2\}\} = \{1,2,\ldots,N-1\} = G_1^{\varphi}(N).$$

As is well known, $q^{N-1} \equiv q^0 \equiv 1 \pmod{N}$. Moreover, the multiplicative inverse of q, denoted by $q^{-1} \in G_1^{\varphi}(N)$, is also a primitive root modulo N. We write each component of the generating vector $\boldsymbol{g} = (g_1,\ldots,g_d) \in G_d^{\varphi}(N)$ as

$$g_j \equiv q^{v_j - 1} \pmod{N}, \quad \text{where } v_j \in \{1,2,\ldots,N-1\}.$$

Note that the fast CBC algorithm explained in Section 3.4 for constructing the generating vector computes the values v_j as a by-product, and thus no additional computation is needed to obtain the v_j if $\boldsymbol{g}$ has been found by this construction.

Since the ordering of the QMC points does not affect the quadrature sum, we can specify a particular (unconventional) ordering of the nodes of a rank-1 lattice rule, which permits a fast QMC matrix-vector multiplication. Note that with k also $q^{-(k-1)}$ runs through the whole set $G_1^{\varphi}(N)$. We put $\boldsymbol{x}_0 := (0,\ldots,0)$, and for

$k \in \{1, 2, \ldots, N-1\}$ we define

$$\begin{aligned}
\boldsymbol{x}_k &:= \left(\left\{ \frac{q^{-(k-1)} g_1}{N} \right\}, \left\{ \frac{q^{-(k-1)} g_2}{N} \right\}, \ldots, \left\{ \frac{q^{-(k-1)} g_d}{N} \right\} \right) \\
&= \left(\left\{ \frac{q^{-(k-1)} q^{\nu_1 - 1}}{N} \right\}, \left\{ \frac{q^{-(k-1)} q^{\nu_2 - 1}}{N} \right\}, \ldots, \left\{ \frac{q^{-(k-1)} q^{\nu_d - 1}}{N} \right\} \right) \\
&= \left(\left\{ \frac{q^{\nu_1 - k}}{N} \right\}, \left\{ \frac{q^{\nu_2 - k}}{N} \right\}, \ldots, \left\{ \frac{q^{\nu_d - k}}{N} \right\} \right).
\end{aligned}$$

In essence, we have changed the ordering by substituting the conventional index k with $q^{-(k-1)}$ and replacing each generating vector component g_j by $q^{\nu_j - 1}$.

This method not only works for lattice point sets, but also for transformed lattice point sets, where each component of each point has been transformed using the same mapping. In this case, the quadrature points are given by

$$\begin{aligned}
\boldsymbol{y}_k &= \psi(\boldsymbol{x}_k) \\
&= (\psi(x_{k,1}), \psi(x_{k,2}), \ldots, \psi(x_{k,d})) \quad \text{for } k \in \{0, 1, \ldots, N-1\},
\end{aligned}$$

where we apply the same univariate transformation $\psi : [0, 1] \to \mathbb{R}$ to every component of every point $\boldsymbol{x}_k = (x_{k,1}, x_{k,2}, \ldots, x_{k,d})$. One example of such a transformation is $\psi = \Phi^{-1}$, the inverse of the cumulative normal distribution function, which maps the points from $(0, 1)$ to $\mathbb{R}$. Another example is when ψ is given by $\psi(x) = \phi(x) = 1 - |2x - 1|$ for $x \in [0, 1]$, which is the tent transformation; results for lattice rules usually apply to periodic functions, and applying the tent transformation yields similar results for nonperiodic functions, see Chapter 7. The case where ψ is the identity mapping is included as a special case.

We define the $N \times d$ matrix

$$Y := \begin{pmatrix} \boldsymbol{y}_0 \\ \boldsymbol{y}_1 \\ \vdots \\ \boldsymbol{y}_{N-1} \end{pmatrix}$$

and discuss the multiplication of Y with a column vector $\boldsymbol{a} \in \mathbb{R}^d$. Since $\boldsymbol{y}_0 = (\psi(0), \psi(0), \ldots, \psi(0))$ we have

$$\boldsymbol{y}_0 \boldsymbol{a} = \psi(0) \sum_{j=1}^{d} a_j.$$

In particular, if ψ is the identity mapping, then $\boldsymbol{y}_0 \boldsymbol{a} = 0$. Thus the first component of $Y\boldsymbol{a}$ can be computed using at most $d - 1$ additions and one multiplication. We therefore disregard the first row of Y and consider the "remaining" $(N - 1) \times d$ matrix

$$Y' = \begin{pmatrix} \boldsymbol{y}_1 \\ \vdots \\ \boldsymbol{y}_{N-1} \end{pmatrix}.$$

In the following we show that Y' can be written as the product of a circulant matrix Z and a matrix P in which $N-1$ entries are 1 and the remaining entries are 0.

Recall that q is a primitive root modulo N. For $k \in \mathbb{Z}$ let

$$z_k := \psi\left(\left\{\frac{q^k}{N}\right\}\right).$$

Then we have $z_k = z_{k+\ell(N-1)}$ for all $\ell \in \mathbb{Z}$. Define the circulant $(N-1)\times(N-1)$ matrix

$$Z = \begin{pmatrix} z_0 & z_1 & z_2 & \cdots & z_{N-3} & z_{N-2} \\ z_{N-2} & z_0 & z_1 & \ddots & \ddots & z_{N-3} \\ z_{N-3} & z_{N-2} & z_0 & \ddots & \ddots & \vdots \\ \vdots & \ddots & \ddots & \ddots & \ddots & \vdots \\ z_2 & \ddots & \ddots & \ddots & z_0 & z_1 \\ z_1 & z_2 & \cdots & \cdots & z_{N-2} & z_0 \end{pmatrix},$$

and the $(N-1)\times d$ matrix $P = (p_{k,j})_{k\in\{1,2,\ldots,N-1\}, j\in[d]}$ over $\{0,1\}$ by

$$p_{k,j} = \begin{cases} 1 & \text{if } k = \nu_j, \\ 0 & \text{otherwise.} \end{cases}$$

Each column of the matrix P contains exactly one entry equal to 1, with the remaining elements being 0. It is now elementary to check that

$$Y' = ZP. \tag{16.5}$$

Note that the matrix Z directly corresponds to the matrix C_{N-1} in (3.21), used in the fast CBC algorithm in Section 3.4. In effect, the matrix P specifies which columns of Z to select (namely, the ν_1-th, the ν_2-th, …, and the ν_d-th) to recover Y'.

Let $\boldsymbol{a} \in \mathbb{R}^d$ be an arbitrary column vector. Then $\boldsymbol{a}' = P\boldsymbol{a}$ can be obtained in $O(N)$ operations due to the special structure of P, and the matrix-vector multiplication $Z\boldsymbol{a}'$ can be performed in $O(N \log N)$ operations using FFT (see Section 3.4) since Z is circulant. Thus $Y'\boldsymbol{a}$ can be computed in $O(N \log N)$ operations, and so the matrix-vector multiplication $X\boldsymbol{a}$ can be carried out using $O(N \log N)$ operations and at most $d-1$ additions.

Further, if one wants to store the transformed point set, i.e., the matrix Y, one can simply store the primitive root q and the d numbers $\nu_1, \ldots, \nu_d$.

We finish this section with a simple example to illustrate the idea presented above.

Example 16.1 Let $d = 3$, $N = 7$, and $(g_1, g_2, g_3) = (1, 3, 6)$. A primitive root modulo 7 is $q = 5$, with multiplicative inverse $q^{-1} = 3$, since $5 \cdot 3 = 15 \equiv 1 \pmod 7$. We have

$$\begin{aligned} g_1 &= 1 \equiv 5^{1-1} \pmod 7, &&\text{and hence} \quad v_1 = 1,\\ g_2 &= 3 \equiv 5^{6-1} \pmod 7, &&\text{and hence} \quad v_2 = 6,\\ g_3 &= 6 \equiv 5^{4-1} \pmod 7, &&\text{and hence} \quad v_3 = 4. \end{aligned}$$

The conventional ordering of the points and the new ordering are

$$\begin{cases} (0,0,0),\\ (\frac{1}{7},\frac{3}{7},\frac{6}{7}),\\ (\frac{2}{7},\frac{6}{7},\frac{5}{7}),\\ (\frac{3}{7},\frac{2}{7},\frac{4}{7}),\\ (\frac{4}{7},\frac{5}{7},\frac{3}{7}),\\ (\frac{5}{7},\frac{1}{7},\frac{2}{7}),\\ (\frac{6}{7},\frac{4}{7},\frac{1}{7}), \end{cases} \quad \text{versus} \quad \begin{cases} \boldsymbol{x}_0 = (0,0,0),\\ \boldsymbol{x}_1 = (\{\frac{5^{1-1}}{7}\},\{\frac{5^{6-1}}{7}\},\{\frac{5^{4-1}}{7}\}) = (\frac{1}{7},\frac{3}{7},\frac{6}{7}),\\ \boldsymbol{x}_2 = (\{\frac{5^{1-2}}{7}\},\{\frac{5^{6-2}}{7}\},\{\frac{5^{4-2}}{7}\}) = (\frac{3}{7},\frac{2}{7},\frac{4}{7}),\\ \boldsymbol{x}_3 = (\{\frac{5^{1-3}}{7}\},\{\frac{5^{6-3}}{7}\},\{\frac{5^{4-3}}{7}\}) = (\frac{2}{7},\frac{6}{7},\frac{5}{7}),\\ \boldsymbol{x}_4 = (\{\frac{5^{1-4}}{7}\},\{\frac{5^{6-4}}{7}\},\{\frac{5^{4-4}}{7}\}) = (\frac{6}{7},\frac{4}{7},\frac{1}{7}),\\ \boldsymbol{x}_5 = (\{\frac{5^{1-5}}{7}\},\{\frac{5^{6-5}}{7}\},\{\frac{5^{4-5}}{7}\}) = (\frac{4}{7},\frac{5}{7},\frac{3}{7}),\\ \boldsymbol{x}_6 = (\{\frac{5^{1-6}}{7}\},\{\frac{5^{6-6}}{7}\},\{\frac{5^{4-6}}{7}\}) = (\frac{5}{7},\frac{1}{7},\frac{2}{7}). \end{cases}$$

It is easy to see that indeed

$$\underbrace{\begin{pmatrix} \psi(\boldsymbol{x}_1)\\ \psi(\boldsymbol{x}_2)\\ \psi(\boldsymbol{x}_3)\\ \psi(\boldsymbol{x}_4)\\ \psi(\boldsymbol{x}_5)\\ \psi(\boldsymbol{x}_6) \end{pmatrix}}_{Y'} = \underbrace{\begin{pmatrix} \psi(\frac{1}{7}) & \psi(\frac{5}{7}) & \psi(\frac{4}{7}) & \psi(\frac{6}{7}) & \psi(\frac{2}{7}) & \psi(\frac{3}{7})\\ \psi(\frac{3}{7}) & \psi(\frac{1}{7}) & \psi(\frac{5}{7}) & \psi(\frac{4}{7}) & \psi(\frac{6}{7}) & \psi(\frac{2}{7})\\ \psi(\frac{2}{7}) & \psi(\frac{3}{7}) & \psi(\frac{1}{7}) & \psi(\frac{5}{7}) & \psi(\frac{4}{7}) & \psi(\frac{6}{7})\\ \psi(\frac{6}{7}) & \psi(\frac{2}{7}) & \psi(\frac{3}{7}) & \psi(\frac{1}{7}) & \psi(\frac{5}{7}) & \psi(\frac{4}{7})\\ \psi(\frac{4}{7}) & \psi(\frac{6}{7}) & \psi(\frac{2}{7}) & \psi(\frac{3}{7}) & \psi(\frac{1}{7}) & \psi(\frac{5}{7})\\ \psi(\frac{5}{7}) & \psi(\frac{4}{7}) & \psi(\frac{6}{7}) & \psi(\frac{2}{7}) & \psi(\frac{3}{7}) & \psi(\frac{1}{7}) \end{pmatrix}}_{Z} \underbrace{\begin{pmatrix} 1&0&0\\ 0&0&0\\ 0&0&0\\ 0&0&1\\ 0&0&0\\ 0&1&0 \end{pmatrix}}_{P}.$$

The matrix P specifies that we select the first, the sixth, and the fourth columns of Z, as indicated by the values of v_1, v_2, and v_3, to recover Y'.

Remark 16.2 The method presented here does *not* work when we apply general randomization techniques such as shifting of lattice rules (see Section 7.1). This is because the corresponding transformation ψ in the mapping $\boldsymbol{y}_k = \psi(\boldsymbol{x}_k)$ fails to be the same mapping in all coordinate directions. If we were to restrict all random shifts to be of the form $\boldsymbol{\Delta} = (\Delta, \ldots, \Delta) \in [0, 1]^d$, then the method would work.

16.3 Fast QMC Matrix-Vector Multiplication for a Special Case of Korobov's p-Sets

The ideas presented above can also be applied to the (in comparison to Chapter 10, slightly modified) p-set consisting of the elements

$$\boldsymbol{x}_{g,k} = \left(\left\{\frac{k}{p}\right\}, \left\{\frac{gk}{p}\right\}, \ldots, \left\{\frac{g^{d-1}k}{p}\right\}\right) \quad \text{for } g, k \in \{1, 2, \ldots, p-1\},$$

where p is a prime number. The number of points in this p-set is $N = (p-1)^2$, and it is essentially the union of all Korobov lattice point sets for given p (see Remark 10.2). Note that in the definition of the p-set $\mathcal{R}_{p^2,d}$ in Definition 10.1 we also included the cases $k = 0$ or $g = 0$, or both $k = g = 0$, but these only yield the zero vector, which is why we do not consider them here.

We now specify a particular ordering of the points $\boldsymbol{x}_{g,k}$ to allow for fast matrix-vector multiplications. Let again q be a primitive root modulo p. In a similar manner as above, we replace the index k in the conventional ordering by $q^{-(k-1)}$, and the index g by q^{g-1}. That is, for $g, k \in \{1, 2, \ldots, p-1\}$, we define

$$\begin{aligned}\boldsymbol{x}_{g,k} &:= \left(\left\{\frac{q^{-(k-1)}q^{0(g-1)}}{p}\right\}, \left\{\frac{q^{-(k-1)}q^{1(g-1)}}{p}\right\}, \ldots, \left\{\frac{q^{-(k-1)}q^{(d-1)(g-1)}}{p}\right\}\right) \\ &= \left(\left\{\frac{q^{\nu_{g,1}-k}}{p}\right\}, \left\{\frac{q^{\nu_{g,2}-k}}{p}\right\}, \ldots, \left\{\frac{q^{\nu_{g,d}-k}}{p}\right\}\right),\end{aligned}$$

with

$$\nu_{g,j} = (j-1)(g-1) + 1 \pmod{p-1} \qquad \text{for } j \in [d].$$

We also define

$$\boldsymbol{y}_{g,k} := \psi(\boldsymbol{x}_{g,k})$$

for a given transformation $\psi : [0,1] \to \mathbb{R}$ that is applied component-wise to the points $\boldsymbol{x}_{g,k}$.

Finally we define the $N \times d$ matrix

$$Y' := \begin{pmatrix} Y'_1 \\ Y'_2 \\ \vdots \\ Y'_{p-1} \end{pmatrix} \quad \text{with} \quad Y'_g := \begin{pmatrix} \boldsymbol{y}_{g,1} \\ \boldsymbol{y}_{g,2} \\ \vdots \\ \boldsymbol{y}_{g,p-1} \end{pmatrix} \quad \text{for } g \in \{1, 2, \ldots, p-1\}, \tag{16.6}$$

where we note that $N = (p-1)^2$.

For the matrices Y'_g we can apply the method discussed in Section 16.2 to write them as $Y'_g = ZP_g$, using the values of $\nu_{g,1}, \ldots, \nu_{g,d}$, so that a matrix-vector multiplication for a single instance $Y'_g\boldsymbol{a}$, for $\boldsymbol{a} \in \mathbb{R}^d$, can be computed in at most

$O(p \log p)$ operations. This has to be done for all $g \in \{1, 2, \ldots, p-1\}$, i.e., $p-1$ times. Thus one matrix-vector product for the matrix Y' can be evaluated in $O(N \log N)$ operations (recall again that $N = (p-1)^2$).

16.4 Applications

In this section, we present some concrete applications of the fast QMC matrix-vector product multiplication.

Generation of normally distributed points with general covariance matrix

In many applications one requires realizations of random variables in $\mathbb{R}^d$ which are normally distributed according to $\mathcal{N}(\boldsymbol{\mu}, \Sigma)$ with mean $\boldsymbol{\mu} = (\mu_1, \mu_2, \ldots, \mu_d)$ and covariance matrix $\Sigma \in \mathbb{R}^{d \times d}$. An algorithm to generate such random variables is described for instance in [114, Section 11.1.6] and works as follows.

Let $A \in \mathbb{R}^{d \times d}$ be such that $A^\top A = \Sigma$; for example, A can be the upper triangular matrix in the Cholesky decomposition of Σ. To generate a point $(Z_1, Z_2, \ldots, Z_d) \sim \mathcal{N}(\boldsymbol{\mu}, \Sigma)$, one generates i.i.d. standard normal random variables $Y_1, Y_2, \ldots, Y_d \sim \mathcal{N}(0, 1)$ with mean 0 and variance 1 and then computes $(Z_1, Z_2, \ldots, Z_d) = (Y_1, Y_2, \ldots, Y_d)A + (\mu_1, \mu_2, \ldots, \mu_d)$.

This procedure can for instance be implemented in the following way using deterministic QMC point sets. Let $\{\boldsymbol{x}_0, \boldsymbol{x}_1, \ldots, \boldsymbol{x}_{N-1}\}$ be a lattice point set in $[0, 1)^d$. Let Φ^{-1} be the inverse of the cumulative standard normal distribution function. Set

$$\boldsymbol{y}_k = \Phi^{-1}\left(\boldsymbol{x}_k + \frac{1}{2N}\mathbf{1}\right) \qquad \text{for } k \in \{0, 1, \ldots, N-1\},$$

where $\mathbf{1} = (1, \ldots, 1)$ is a vector of length d containing only ones, and where Φ^{-1} is applied component-wise, and let furthermore

$$\boldsymbol{z}_k = \boldsymbol{y}_k A + \boldsymbol{\mu} \quad \text{for } k \in \{0, 1, \ldots, N-1\}.$$

Note that we do not need to assume any structure in the matrix A, but we can directly apply the methods outlined above.

Partial differential equations with "uniform" random coefficients

The fast QMC matrix-vector multiplication can also be used in the study of PDEs with random coefficients. We first discuss the case of uniform random coefficients outlined in the Appendix in Section A.1. We use notation as in that section.

The three sources of errors, namely the truncation error, the finite element error, and the quadrature error, are independent of each other but need to be balanced to reduce the overall error. For instance, in the case that

$$\sum_{j=1}^{\infty} \|\psi_j\|_{L_\infty(D)}^{2/3} < \infty, \tag{16.7}$$

and that the representer g of the linear functional G is in $L_2(D)$, then for a continuous, piecewise linear finite element discretization of D on a quasi-uniform mesh we should choose

$$d \asymp M^{2/s} \asymp N, \tag{16.8}$$

where N is the number of samples, d is the truncation dimension, s is the dimension of the domain D, and M is the resolution of the employed piecewise linear finite elements, and where $\asymp$ indicates that the terms should be of the same order of magnitude. This can be deduced in the following way. For $\beta_j = \|\psi_j\|_{L_\infty(D)}/a_{\min}$ (see (A.18)), we have from (16.7) and (A.19) that $p = 2/3$. To reduce the overall error (A.22), we need to balance the truncation error (A.23) of order $d^{-2(1/p-1)}$, the finite element error (A.24) of order $M^{-2/s}$ and the integration error (A.25) of order $N^{-(1-\delta)}$ for any $\delta > 0$. (We slightly cheat here by replacing $N^{-(1-\delta)}$ for any $\delta > 0$ with N^{-1} for the sake of simplifying the discussion.) In general we have $N \asymp d^{\kappa}$ for some small $\kappa > 0$ (i.e., $N \ll 2^d$). Thus the fast QMC matrix-vector multiplication can be applied in this scenario.

In the standard approach, one defines, for $j \in [d]$, the symmetric matrices $A_j := (a_{j,\ell,m})_{\ell,m\in\{1,2,\ldots,M\}}$, and puts

$$B(\boldsymbol{y}_k) := A_0 + \sum_{j=1}^{d} y_{k,j} A_j \qquad \text{for } k \in \{0,1,\ldots,N-1\}.$$

Note that the A_j are usually sparse, with only $O(M)$ nonzero entries in the same position for each $j \in [d]$, depending only on the relative supports of the basis functions ϕ_ℓ, which are thus in particular independent of j. The cost of computing $B(\boldsymbol{y}_k)$ for all $k \in \{0,1,\ldots,N-1\}$ is therefore $O(d\,M\,N)$ operations.

The fast QMC matrix-vector approach is implemented as follows. Let

$$Y = \begin{pmatrix} \boldsymbol{y}_0 \\ \boldsymbol{y}_1 \\ \vdots \\ \boldsymbol{y}_{N-1} \end{pmatrix}$$

be the $N \times d$ matrix whose rows are the quadrature points of the QMC rule, and let $\boldsymbol{a}_{\ell,m} = (a_{1,\ell,m},\ldots,a_{d,\ell,m})^\top$. Then compute

$$\boldsymbol{b}_{\ell,m} = (a_{0,\ell,m},\ldots,a_{0,\ell,m})^\top + Y\boldsymbol{a}_{\ell,m} \quad \text{for all } \ell, m \in \{1,2,\ldots,M\}, \tag{16.9}$$

where $\boldsymbol{b}_{\ell,m} = (b_{0,\ell,m}, \ldots, b_{N-1,\ell,m})^\top$, and where each matrix-vector multiplication $Y\boldsymbol{a}_{\ell,m}$ is done using the fast QMC matrix-vector multiplication. Since only $O(M)$ vectors $\boldsymbol{a}_{\ell,m}$ are nonzero, this approach for obtaining $B(\boldsymbol{y}_k)$ for all $k \in \{0, 1, \ldots, N-1\}$ therefore only requires $O(M\,N \log N)$ operations.

The improvement in the computational cost is that we have replaced a factor of $O(d)$ by $O(\log N)$ (or $O(\log d)$ when $N \asymp d$, see (16.8)). However, this method requires us to store all the vectors $\boldsymbol{b}_{\ell,m}$. Using the sparsity of the stiffness matrices $B(\boldsymbol{y}_k)$, which is of order $O(M)$, we require $O(MN)$ storage.

Partial differential equations with "log-normal" random coefficients

The fast QMC matrix-vector multiplication can also be used in treating PDEs with random log-normal coefficients. We consider the PDE (A.16) in Appendix A, but now we assume that the random diffusion coefficient is log-normal (see (A.26)).

We need to estimate the coefficients $b_{k,\ell,m}$ defined in (A.28). The number of these nonzero inner products is $O(M)$ since, for a fixed ℓ, the number of m such that the intersection of the supports of ϕ_ℓ and ϕ_m is nonempty does not depend on M. The standard approach to obtaining $\widehat{B}(\boldsymbol{y}_k)$ for all $k \in \{0, 1, \ldots, N-1\}$ therefore requires $O(d\,N_I\,M\,N)$ operations, where N_I is the number of quadrature points (A.29) for the approximation of $b_{k,\ell,m}$.

We now describe the fast approach. Let

$$\Theta_{\ell,m} := (\theta_{i,\ell,m,k})_{i\in\{1,2,\ldots,N_I\},k\in\{0,1,\ldots,N-1\}} \in \mathbb{R}^{N\times N_I},$$

$$\widehat{\Psi}_{i,\ell,m} := \begin{pmatrix} \psi_0(x_{i,\ell,m}) \\ \psi_0(x_{i,\ell,m}) \\ \vdots \\ \psi_0(x_{i,\ell,m}) \end{pmatrix} \in \mathbb{R}^N, \quad \widehat{\Psi}_{\ell,m} := (\widehat{\Psi}_{1,\ell,m}, \ldots, \widehat{\Psi}_{N_I,\ell,m}) \in \mathbb{R}^{N\times N_I},$$

$$\Psi_{i,\ell,m} := \begin{pmatrix} \psi_1(x_{i,\ell,m}) \\ \psi_2(x_{i,\ell,m}) \\ \vdots \\ \psi_d(x_{i,\ell,m}) \end{pmatrix} \in \mathbb{R}^d, \quad \Psi_{\ell,m} := (\Psi_{1,\ell,m}, \ldots, \Psi_{N_I,\ell,m}) \in \mathbb{R}^{d\times N_I}.$$

Then (A.31) can be written in matrix form as

$$\Theta_{\ell,m} = \widehat{\Psi}_{\ell,m} + Y\Psi_{\ell,m}, \tag{16.10}$$

where the multiplication $Y\Psi_{\ell,m}$ is done as described in Section 16.2. Thus, (A.31) can be computed using the fast QMC matrix-vector multiplication and $\widehat{B}(\boldsymbol{y}_k)$ for all $k \in \{0, 1, \ldots, N-1\}$ can be computed in $O(N_I\,M\,N \log N)$ operations. Again, the saving is that we have replaced the factor of $O(d)$ by $O(\log N)$ in the computational cost.

16.5 Numerical Experiments

In this section we carry out numerical experiments for the three applications from the previous section. In all of these numerical experiments the times are averaged over five independent runs. Tables 16.1 and 16.2 have been calculated using Matlab R2013b on an Intel®Core^{TM} Xeon E5-2650v2 CPU @ 2.6GHz. These tables are taken from [47] (with the last digit rounded). Table 16.3 has been calculated using Matlab R2020a on an Intel®Core i7 CPU @ 2.2GHz.

Experiment 1: normally distributed points

We are interested in comparing the computation times using the standard approach of multiplying $\boldsymbol{y}_k A$ for $k \in \{0, 1, \ldots, N-1\}$, and the fast QMC matrix-vector multiplication using lattice point sets.

Table 16.1 shows the computation times in seconds for various values of N and d. In the single cells, the values on top show the standard approach, whereas the values below show the fast QMC matrix-vector approach. For our experiments we chose $\boldsymbol{\mu} = (0, 0, \ldots, 0)$, and A as a random upper triangular matrix with positive diagonal entries (such that A corresponds to the Cholesky factor of a random matrix Σ). The computation times do not include the fast component-by-component construction of the lattice generating vectors (see Sections 3.4 and 3.5), the computation of $v_1, \ldots, v_d$ (since this information can be obtained from the fast component-by-component construction), nor the computation of $\Phi^{-1}(k/N)$ for $k \in \{0, 1, \ldots, N-1\}$ (since this computation is the same for both methods).

The numerical experiments in Table 16.1 show that there is an advantage in using the fast QMC matrix-vector product if the dimension is large, and the advantage grows as the dimension increases. This is in agreement with the theory since the computational cost in the standard approach is of order $O(d^2 N)$ operations, whereas in the fast QMC matrix-vector approach it is of order $O(d\, N \log N)$ operations. Recall that the fast QMC matrix-vector method incurs a storage cost of order $O(d\, N)$. As we can also see from Table 16.1, the advantage of the fast QMC matrix-vector approach over the standard approach particularly occurs for cases where the computational cost is dominated by the terms depending on d.

Experiment 2: PDEs with "uniform" random coefficients

We use the example (A.1) with random field (A.2) and boundary conditions (A.8) in Appendix A in the uniform case with $\eta = 3/2$.

Table 16.2 shows the computation times comparing the standard approach with the fast QMC matrix-vector method based on lattice point sets. In this case the mapping in $\boldsymbol{y}_k = \psi(\boldsymbol{x}_k)$ is $\psi(x) = x - 1/2$, since the lattice points need to be translated from the standard unit cube $[0, 1)^d$ to $[-1/2, 1/2)^d$. Since the dimension d is large, the fast QMC matrix-vector method is very effective in reducing the computation

Method	N	$d = 200$	$d = 400$	$d = 600$	$d = 800$	$d = 1000$
std.	16001	0.31	0.74	1.30	1.62	2.15
fast		0.16	0.30	0.45	0.59	0.74
std.	32003	0.59	1.47	2.44	3.06	4.24
fast		0.60	1.20	1.79	2.40	2.99
std.	64007	1.17	2.97	4.92	6.00	8.35
fast		1.80	3.85	5.55	7.58	9.83
std.	127997	2.58	5.89	9.49	11.9	16.8
fast		2.33	4.66	7.32	9.98	12.3
std.	256019	4.28	11.1	17.6	23.1	33.5
fast		5.40	10.9	16.2	24.1	26.9
std.	512009	8.88	23.4	31.9	48.1	66.4
fast		10.9	22.1	35.5	45.2	56.2

Table 16.1: Times (in seconds) to generate normally distributed points with random covariance matrix. In each cell, the top row is the time required by using the standard approach, whereas the bottom row shows the time required using the fast QMC matrix-vector approach.

$M = d = 2N$									
N	67	127	257	509	1021	2053	4001	8009	16001
std.	1	5	31	190	1346	10610	74550	≈144h	≈1000h
fast	0.035	0.042	0.11	0.46	1.56	5.59	19.68	87.2	343
$M = d = \lceil\sqrt{N}\rceil$									
N	67	127	257	509	1021	2053	4001	8009	16001
std.	0.066	0.16	0.47	1.27	3.57	10.8	30.1	89.4	273
fast	0.012	0.015	0.028	0.059	0.13	0.27	0.52	1.11	2.44
$d = N$ and $M = N^2$									
N	67	127	257	509					
std.	6	82	1699	27935					
fast	0.24	1.39	11.3	107					

Table 16.2: Times (in seconds) to obtain the average value of the finite element coefficients of the approximation (A.21) for a uniform random field (top: $M = d = 2N$, middle: $M = d = \lceil\sqrt{N}\rceil$, bottom: $d = N$ and $M = N^2$).

times. In Table 16.2, for the case $M = d = 2N$, the times for the standard method for $N = 8009$ and $N = 16001$ are in hours and are estimated from extrapolating on previous values in the table. The experiments show that there is a clear advantage of the fast QMC matrix-vector approach, especially for large values of M, N, and d.

$M = d = 2N$									
N	67	127	257	509	1021	2053	4001	8009	16001
std.	0.0090	0.028	0.078	0.33	2.27	17.8	138	1016	7646
fast	0.0065	0.019	0.048	0.19	0.71	2.82	9.98	42.0	237

$M = d = \lceil\sqrt{N}\rceil$									
N	67	127	257	509	1021	2053	4001	8009	16001
std.	0.0059	0.011	0.0093	0.018	0.040	0.092	0.21	0.90	1.47
fast	0.0061	0.0099	0.0080	0.016	0.034	0.075	0.16	0.76	0.92

$d = N$ and $M = N^2$				
N	67	127	257	509
std.	0.090	0.69	6.39	74.8
fast	0.10	0.65	5.9	60.6

Table 16.3: Times (in seconds) to obtain the average value of the finite element coefficients of the approximation (A.21) for a log-normal random field (top: $M = d = 2N$, middle: $M = d = \lceil\sqrt{N}\rceil$, bottom: $d = N$ and $M = N^2$).

Experiment 3: PDEs with "log-normal" random coefficients

We use the example (A.16) with random field (A.27) in Appendix A in the log-normal case with $\eta = 3/2$. To compute (A.30), we use an equal-weight quadrature rule with M (i.e., $N_I = M$) points.

Table 16.3 shows the computation times for the log-normal case with different choices of the number of finite elements M, the number of QMC points N, and the truncated dimension d. As one would expect from the theory, the most significant advantage of the fast QMC matrix-vector multiplication occurs when 2^d is large in comparison to N, which is also reflected in the numerical results.

Notes and Remarks

The fast QMC matrix-vector multiplication has been introduced in [47]. Here we have presented the application for the case of lattice point sets and for a special kind of a p-set. A similar approach also works for other types of QMC point sets (e.g., polynomial lattice point sets).

The fast QMC matrix-vector multiplication can also be used in the randomized setting, namely in Algorithm 11.2 for integrands of the form $g(\boldsymbol{x}) = f(\boldsymbol{x}A)$ with $g \in \mathcal{H}_{\mathrm{kor},d,\alpha,\gamma}$ and also together with the tent transformation for integrands $g(\boldsymbol{x}) = f(\boldsymbol{x}A)$ with $g \in \mathcal{H}_{\cos,d,\alpha,\gamma}$, see Sections 11.1 and 11.2.

Appendix A
Partial Differential Equations With Random Coefficients

Partial differential equations (PDEs) with random coefficients have become an important application area of QMC methods, and in particular lattice rules. In this chapter we introduce basic background information regarding this field as it is relevant for lattice rules. The areas of PDEs with random coefficients and finite element methods are too vast to give a broad overview of or a detailed introduction to the topic. In particular, this chapter is meant for readers who are not familiar with numerical techniques for PDEs. Our goal is to provide some insight into how lattice rules are applied in this context, what the computational algorithm looks like, and how the error can be analyzed. We illustrate the ideas by some examples.

A.1 Uniform Random Coefficients

We first discuss the case where the random coefficients in a partial differential equation are uniformly distributed.

An illustrative example

The prototypical example of a PDE with random coefficients is a diffusion equation with random diffusion coefficients. We consider a simple concrete version of this example for illustration, where the domain D is just the interval $(0, 1)$. In this case the equation is given by

$$-\frac{\partial}{\partial z}\left(a(z, \boldsymbol{y})\frac{\partial}{\partial z}u(z, \boldsymbol{y})\right) = f(z) \quad \text{for } z \in (0, 1), \tag{A.1}$$

where the function f on the right-hand side of (A.1) is a given function in, say, $L_2([0, 1])$. As a concrete example, we first consider the random field

J. Dick et al., *Lattice Rules*, Springer Series in Computational Mathematics 58,
https://doi.org/10.1007/978-3-031-09951-9

$$a(z, \boldsymbol{y}) = c + \sum_{j=1}^{\infty} y_j \frac{\sin(2\pi j z)}{j^\eta}, \tag{A.2}$$

where $c \in \mathbb{R}$ is a constant and the y_j are independent and identically uniformly distributed random variables in $[-1/2, 1/2]$, hence the name PDE with random coefficients. We define the sequence $\boldsymbol{y} = (y_1, y_2, \ldots) \sim \mathcal{U}([-1/2, 1/2]^{\mathbb{N}})$, where $\mathcal{U}$ denotes the continuous uniform distribution in the sense of probability theory. We have

$$a(z, \boldsymbol{y}) \leq c + \frac{\zeta(\eta)}{2} \quad \text{for all } \boldsymbol{y} \in \left[-\frac{1}{2}, \frac{1}{2}\right]^{\mathbb{N}}, \tag{A.3}$$

where we remind the reader that $\zeta(\eta) = \sum_{j=1}^{\infty} j^{-\eta}$ is the Riemann zeta function. In order for a to be well-defined and differentiable we impose the assumption that $\eta > 1$, which ensures that $\zeta(\eta) < \infty$. Some instances of this random field for $\eta = 2$ and $\eta = 1.3$ are shown in the top row of Figure A.1, where we truncated the infinite sum (A.2) to 10^4 terms.

To illustrate the problem, let us first also impose boundary conditions which lead to an explicit solution of (A.1), namely

$$u(0, \boldsymbol{y}) = b, \quad \text{and} \quad a(1, \boldsymbol{y}) \left.\frac{\partial}{\partial z} u(z, \boldsymbol{y})\right|_{z=1} = q, \tag{A.4}$$

where for simplicity we choose b and q to be constants. Integrating (A.1) twice with respect to z and using the boundary conditions (A.4) leads to the explicit solution

$$u(z, \boldsymbol{y}) = b + \int_0^z \frac{1}{a(v, \boldsymbol{y})} \left(q + \int_z^1 f(w)\,\mathrm{d}w\right) \mathrm{d}v. \tag{A.5}$$

In order for the solution u to be well-defined, we also assume

$$a(x, \boldsymbol{y}) \geq c - \frac{\zeta(\eta)}{2} > 0 \quad \text{for all } \boldsymbol{y} \in \left[-\frac{1}{2}, \frac{1}{2}\right]^{\mathbb{N}}. \tag{A.6}$$

The conditions (A.3) and (A.6) are the so-called *uniform ellipticity assumptions*.

To numerically illustrate the problem, let $b = 0$, $c = 2$, $q = 1/2$, and let f be given by $f(z) = z$ for $z \in [0, 1]$. Then the explicit solution is given by

$$u(z, \boldsymbol{y}) = \int_0^z \frac{1 - v^2/2}{a(v, \boldsymbol{y})} \,\mathrm{d}v. \tag{A.7}$$

To compute approximate values of u for some given z and $\boldsymbol{y}$ we need to estimate the integral and compute approximate values of the random field a. To estimate the infinite sum (A.2) of the random field, we truncate this infinite sum after, say, d, terms. We call d the truncation dimension.

The random field a, the solution u for various instances of the random field, and a histogram of instances of the random variable $u(1/2, \cdot)$ are shown in Figure A.1.

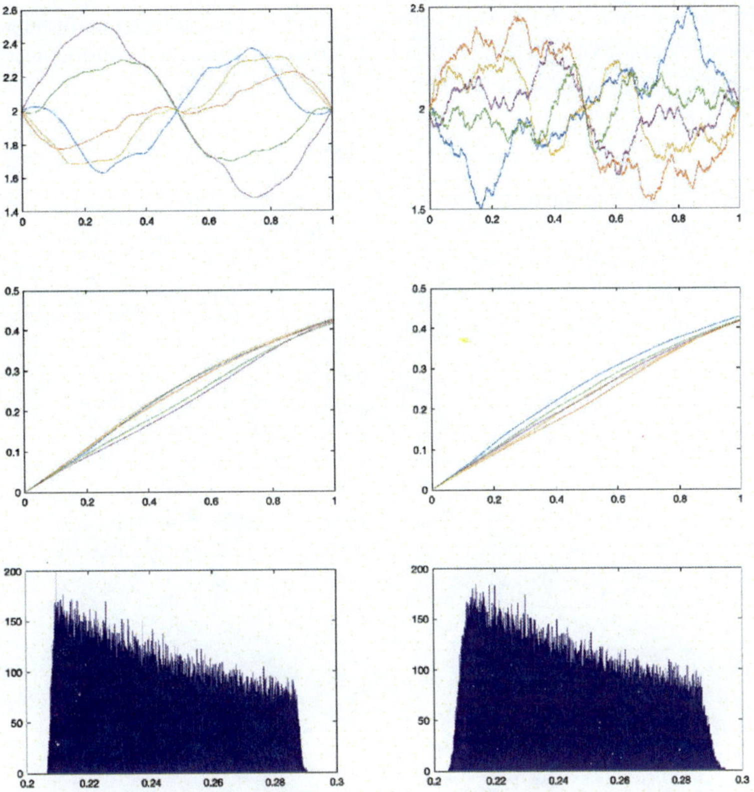

Fig. A.1: The left column uses $\eta = 2$, whereas the right column uses $\eta = 1.3$. In the top row, instances of the random field $a(\cdot, \boldsymbol{y})$ are shown, as well as various random selections of the random coefficients $\boldsymbol{y}$. In the middle row, the corresponding solutions $u(\cdot, \boldsymbol{y})$ of (A.7) are shown. The truncation dimension is 10^4. The bottom row shows a histogram of $u(1/2, \cdot)$ using $2 \cdot 10^5$ samples and 2000 bins, where we have restricted the truncation dimension to 100.

The case of no explicit solution

The boundary condition (A.4) is very artificial (as it is chosen such that we get an explicit solution). We continue with the previous example (A.1), but now change the boundary conditions to the more natural boundary conditions

$$u(0, y) = u(1, y) = 0. \tag{A.8}$$

In this case we do not have an explicit solution of the differential equation. We briefly describe how one can use piecewise linear finite elements to obtain an approximation of the solution u in this case.

Let $M \in \mathbb{N}$, let $z_\ell = \ell/M$ for $\ell \in \{0, 1, \dots, M\}$, and for $\ell \in \{1, 2, \dots, M-1\}$ define the hat function

$$\phi_\ell(z) := \begin{cases} (z - z_{\ell-1})M & \text{if } z_{\ell-1} \le z \le z_\ell, \\ (z_{\ell+1} - z)M & \text{if } z_\ell \le z \le z_{\ell+1}, \\ 0 & \text{otherwise.} \end{cases} \tag{A.9}$$

The graphs of several hat functions are shown in Figure A.2, and an approximation of the function f with $f(z) = z(1-z)$, for $z \in [0, 1]$, as a linear combination of such hat functions is shown in Figure A.3.

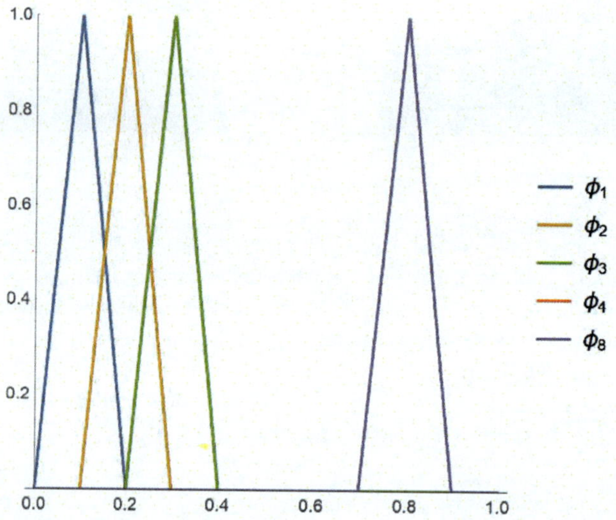

Fig. A.2: Finite element basis functions $\phi_1, \phi_2, \phi_3, \phi_4$, and ϕ_8 for $M = 10$.

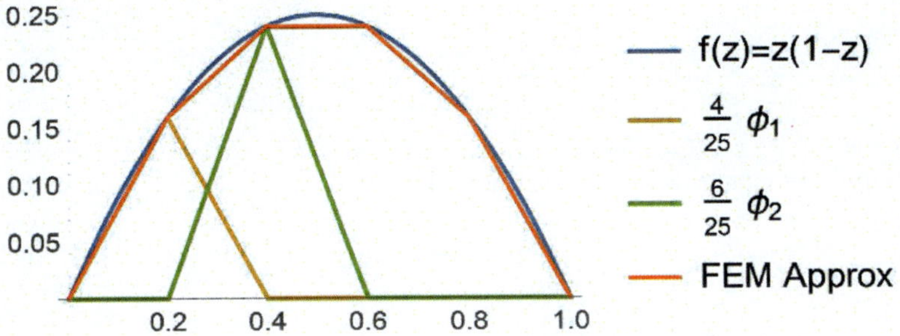

Fig. A.3: Approximation of the function f with $f(z) = z(1-z)$, for $z \in [0, 1]$, using piecewise linear hat functions as defined in Equation (A.9) with $M = 5$. The approximation is then $(4/25)\phi_1(z) + (6/25)\phi_2(z) + (6/25)\phi_3(z) + (4/25)\phi_4(z)$.

The goal is to obtain an approximation of the solution u of the form

$$u(z, \boldsymbol{y}) \approx \sum_{\ell=1}^{M-1} \widehat{u}_\ell(\boldsymbol{y})\, \phi_\ell(z). \tag{A.10}$$

To do so, we multiply (A.1) by ϕ_m and integrate on both sides with respect to z over the unit interval $[0, 1]$ to obtain

$$-\int_0^1 \frac{\partial}{\partial z}\left(a(z, \boldsymbol{y})\frac{\partial}{\partial z}u(z, \boldsymbol{y})\right)\phi_m(z)\,\mathrm{d}z = \int_0^1 f(z)\,\phi_m(z)\,\mathrm{d}z. \tag{A.11}$$

We use integration by parts on the left-hand side of (A.11) and the fact that $\phi_m(0) = \phi_m(1) = 0$ to obtain

$$\int_0^1 a(z, \boldsymbol{y})\frac{\partial}{\partial z}u(z, \boldsymbol{y})\frac{\partial}{\partial z}\phi_m(z)\,\mathrm{d}z = \int_0^1 f(z)\,\phi_m(z)\,\mathrm{d}z. \tag{A.12}$$

Set $\widehat{f}_m := \int_0^1 f(z)\,\phi_m(z)\,\mathrm{d}z$ for $m \in \{1, 2, \ldots, M-1\}$. Since f is a given function, the values $\widehat{f}_m$ can be computed, either explicitly, or approximately using a quadrature rule. Next, we substitute the approximation (A.10) into (A.12) and use (A.2) to obtain a system of linear equations in $\widehat{u}_\ell(\boldsymbol{y})$,

$$\sum_{\ell=1}^{M-1} \widehat{u}_\ell(\boldsymbol{y})\left(c\, a_{0,\ell,m} + \sum_{j=1}^{\infty} y_j\, a_{j,\ell,m}\right) = \widehat{f}_m, \tag{A.13}$$

where

$$a_{0,\ell,m} := \int_0^1 \phi'_\ell(z)\,\phi'_m(z)\,\mathrm{d}z,$$

and

$$a_{j,\ell,m} := \int_0^1 \frac{\sin(2\pi jz)}{j^\eta}\,\phi'_\ell(z)\,\phi'_m(z)\,\mathrm{d}z$$

for $j \in \mathbb{N}$ and $\ell, m \in \{1, 2, \ldots, M-1\}$.

In the present example we can calculate the values $a_{j,\ell,m}$ explicitly to obtain

$$a_{0,\ell,m} = \begin{cases} 2M & \text{if } \ell = m, \\ -M & \text{if } |\ell - m| = 1, \\ 0 & \text{otherwise,} \end{cases}$$

and, for $j \in \mathbb{N}$,

$$a_{j,\ell,m} = \int_0^1 \frac{\sin(2\pi jz)}{j^\eta}\,\phi'_\ell(z)\phi'_m(z)\,\mathrm{d}z$$

$$= \begin{cases} \frac{M^2}{2\pi j^{\eta+1}}\left(\cos\left(2\pi\frac{j(\ell-1)}{M}\right) - \cos\left(2\pi\frac{j(\ell+1)}{M}\right)\right) & \text{if } \ell = m, \\ \frac{M^2}{2\pi j^{\eta+1}}\left(\cos\left(2\pi\frac{j\min(\ell,m)}{M}\right) - \cos\left(2\pi\frac{j\max(\ell,m)}{M}\right)\right) & \text{if } |\ell - m| = 1, \\ 0 & \text{otherwise.} \end{cases}$$

In problems where these integrals cannot be computed explicitly one needs to resort to some quadrature rule to approximate them numerically.

Now set $A_j := (a_{j,\ell,m})_{\ell,m\in\{1,2,\ldots,M-1\}}$ for $j \in \mathbb{N}_0$ and, for $\boldsymbol{y} = (y_1, y_2, \ldots)$, set

$$B(\boldsymbol{y}) := cA_0 + \sum_{j=1}^{\infty} y_j A_j. \tag{A.14}$$

Since the matrices A_j, $j \geq 0$, are tridiagonal $(M-1) \times (M-1)$ matrices, so is $B(\boldsymbol{y})$. In the context of finite element methods for solving PDEs the matrix $B(\boldsymbol{y})$ is often called *stiffness matrix*. This matrix represents the linear system that has to be solved to find an approximation of a random solution of (A.1), with the boundary conditions given by (A.8).

We also define $\widehat{\boldsymbol{f}} := (\widehat{f}_m)_{m\in\{1,2,\ldots,M-1\}}$ and $\widehat{\boldsymbol{u}} := (\widehat{u}_\ell)_{\ell\in\{1,2,\ldots,M-1\}}$, where both vectors are considered as column vectors. The linear system (A.13) can now be written as

$$B(\boldsymbol{y})\,\widehat{\boldsymbol{u}} = \widehat{\boldsymbol{f}}.$$

In order to be able to obtain solutions to this linear system, we need to truncate the infinite sum (A.14) to

$$B(\boldsymbol{y}^{(d)}) = cA_0 + \sum_{j=1}^{d} y_j A_j, \tag{A.15}$$

where $d \in \mathbb{N}$ is called the *truncation dimension* in this context, and where $\boldsymbol{y}^{(d)} = (y_1, \ldots, y_d)$.

Consider now a numerical example where $f(z) = 1$ for $z \in [0, 1]$. Then

$$\widehat{f}_m = \int_0^1 \phi_m(z)\,\mathrm{d}z = \frac{1}{M} \quad \text{for all } m \in \{1, 2, \ldots, M-1\}.$$

Solving the linear system

$$B(\boldsymbol{y}^{(d)})\,\widehat{\boldsymbol{u}} = \widehat{\boldsymbol{f}}$$

for a random vector $\boldsymbol{y}^{(d)}$ then yields an approximation of a random solution to (A.1), with the boundary conditions given by (A.8). The random field and the solution u for various instances of the random field and a histogram of the random variable $u(1/2, \cdot)$ are shown in Figure A.4.

The general case

We consider the PDE with random coefficients

$$\begin{aligned} -\nabla_{\boldsymbol{z}} \cdot (a(\boldsymbol{z}, \boldsymbol{y})\nabla_{\boldsymbol{z}} u(\boldsymbol{z}, \boldsymbol{y})) &= f(\boldsymbol{z}) \quad \text{for } \boldsymbol{z} \in D, \\ u(\boldsymbol{z}, \boldsymbol{y}) &= 0 \quad \text{for } \boldsymbol{z} \in \partial D, \end{aligned} \tag{A.16}$$

where $f \in L_2(D)$, where $D \subseteq \mathbb{R}^s$ is a bounded Lipschitz domain with boundary ∂D, $\nabla_{\boldsymbol{z}} = (\partial/\partial z_1, \ldots, \partial/\partial z_s)^\top$ is a differential operator (the gradient), $\boldsymbol{y} \sim \mathcal{U}([-1/2, 1/2]^{\mathbb{N}})$, the random field is given by

$$a(\boldsymbol{z}, \boldsymbol{y}) = \psi_0(\boldsymbol{z}) + \sum_{j=1}^{\infty} y_j \psi_j(\boldsymbol{z}),$$

where $\psi_j : D \to \mathbb{R}$ and $\boldsymbol{y} = (y_1, y_2, \ldots)$, and where the uniform ellipticity assumptions hold,

$$0 < a_{\min} \leq a(\boldsymbol{z}, \boldsymbol{y}) \leq a_{\max} < \infty \quad \text{for } \boldsymbol{z} \in D \text{ and } \boldsymbol{y} \in \left[-\frac{1}{2}, \frac{1}{2}\right]^{\mathbb{N}}, \tag{A.17}$$

with $a_{\min}, a_{\max}$ being constants. We define the coefficients

$$\beta_j := \frac{\|\psi_j\|_{L_\infty(D)}}{a_{\min}} \quad \text{for } j \in \mathbb{N}, \tag{A.18}$$

and we assume that $\beta_1 \geq \beta_2 \geq \cdots$ satisfy the summability condition

$$\sum_{j=1}^{\infty} \beta_j^p < \infty \quad \text{for some } p \in (0, 1). \tag{A.19}$$

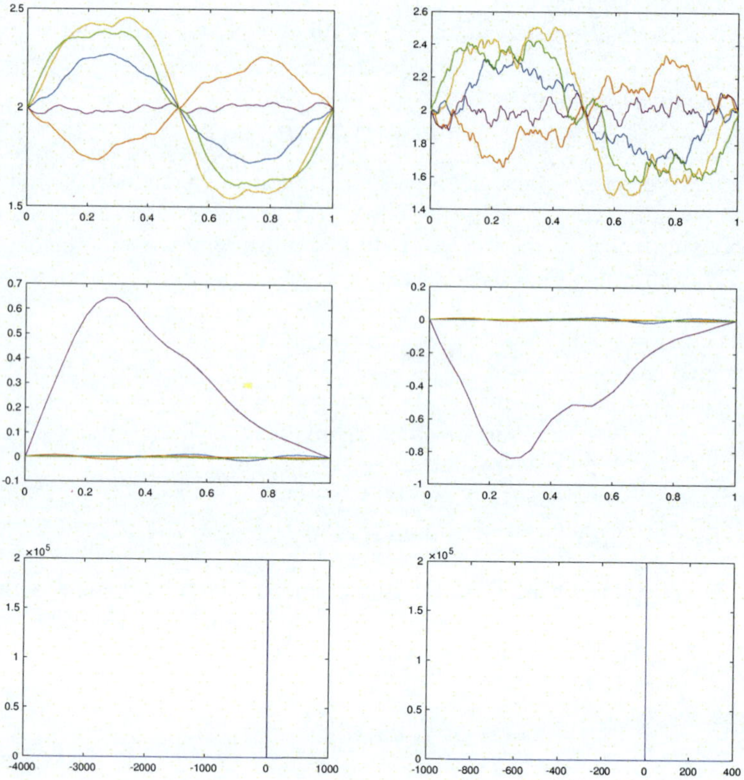

Fig. A.4: The left column uses $\eta = 2$, whereas the right column uses $\eta = 1.3$. In the top row, instances of the random field $a(\cdot, \boldsymbol{y})$ are shown, as well as various random selections of the random coefficients $\boldsymbol{y}$. In the middle row, the corresponding solutions $u(\cdot, \boldsymbol{y})$ of (A.1) with the boundary conditions (A.8) are shown. The bottom row shows a histogram of $u(1/2, \cdot)$ using $2 \cdot 10^5$ samples and 2000 bins. The truncation dimension is 100 in these examples.

Again, we do not have an explicit solution of the PDE and hence we have to resort to numerical methods, like a finite element method, to obtain an approximation of the PDE (A.16). We refer to [245] for more details on such methods. The goal then (under further assumptions) is to estimate $\mathbb{E}[G(u)]$, where G is a linear functional, and where the expectation is taken with respect to random $\boldsymbol{y}$. Under certain assumptions, we can represent G by $G(u(\cdot,\boldsymbol{y})) = \int_D u(z,\boldsymbol{y})g(z)\,\mathrm{d}z$ for some suitable function g. The approach in the general case has many similarities to the case shown in the previous section. We omit the technical details and refer to [160] for further information instead.

The computational approach

As we have seen above, the way to compute an approximation to $\mathbb{E}[G(u)]$ is as follows.

Algorithm A.1 Consider the boundary value problem (A.16) with uniform i.i.d. random coefficients. In order to find an approximation to $\mathbb{E}[G(u)]$, do the following.

(1) Generate samples $\boldsymbol{y}_0, \boldsymbol{y}_1, \ldots, \boldsymbol{y}_{N-1} \in [0,1]^d$ (using (Q)MC), where the points are given by $\boldsymbol{y}_k = (y_{k,1}, y_{k,2}, \ldots, y_{k,d})$, for $k \in \{0,1,\ldots,N-1\}$.
(2) Substitute these points into the truncated random field

$$a_d\left(z, \boldsymbol{y}_k - \tfrac{1}{2}\mathbf{1}\right) = \psi_0(z) + \sum_{j=1}^{d}\left(y_{k,j} - \tfrac{1}{2}\right)\psi_j(z) \quad \text{for } k \in \{0,1,\ldots,N-1\}. \tag{A.20}$$

(3) Use a PDE solver (for instance, the finite element method with mesh width h), in order to compute estimations of the solutions of the PDEs

$$-\nabla_z \cdot \left(a_d\left(z, \boldsymbol{y}_k - \tfrac{1}{2}\mathbf{1}\right)\nabla_z u\left(z, \boldsymbol{y}_k - \tfrac{1}{2}\mathbf{1}\right)\right) = f(z) \quad \text{for } z \in D,$$
$$u\left(z, \boldsymbol{y}_k - \tfrac{1}{2}\mathbf{1}\right) = 0 \quad \text{for } z \in \partial D,$$

to get approximations $u_d^h(\boldsymbol{y}_k - \frac{1}{2}\mathbf{1})$ for $k \in \{0,1,\ldots,N-1\}$.
In the case of a finite element method, this involves computing the stiffness matrices $B(\boldsymbol{y}_k)$ given in (A.15) and solving the linear systems $B(\boldsymbol{y}_k)\,\widehat{\boldsymbol{u}}_k = \widehat{\boldsymbol{f}}$ with unknown $\widehat{\boldsymbol{u}}_k$ for $k \in \{0,1,\ldots,N-1\}$.
(4) An estimation of the expected value $\mathbb{E}[G(u)]$ is then computed by

$$Q_{N,d}(G(u_d^h)) = \frac{1}{N}\sum_{k=0}^{N-1} G\left(u_d^h\left(\boldsymbol{y}_k - \tfrac{1}{2}\mathbf{1}\right)\right) \approx \mathbb{E}[G(u)]. \tag{A.21}$$

Error analysis

The error between the correct value $\mathbb{E}[G(u)]$ and the approximation $Q_{N,d}(G(u_d^h))$ can be split into three parts using add-and-subtract and the triangle inequality. One such approach leads to

$$\begin{aligned}
&|\mathbb{E}[G(u)] - Q_{N,d}(G(u_d^h))| \\
&\leq \underbrace{|\mathbb{E}[G(u)] - \mathbb{E}[G(u_d)]|}_{\text{truncation error}} + \underbrace{|\mathbb{E}[G(u_d)] - \mathbb{E}[G(u_d^h)]|}_{\text{finite element error}} \\
&\quad + \underbrace{|\mathbb{E}[G(u_d^h)] - Q_{N,d}(G(u_d^h))|}_{\text{integration error}},
\end{aligned} \tag{A.22}$$

where $\mathbb{E}[G(u)]$ is the expected value of $G(u)$, u_d is the solution of (A.16) using the truncated random field (A.20), and u_d^h is the approximation of u_d using a finite element method with mesh width h (in the example above $h = 2/M$).

In [160, Theorem 5.1] it was shown that the truncation error satisfies

$$\mathrm{err}^{\mathrm{trnc}}(d) := |\mathbb{E}[G(u)] - \mathbb{E}[G(u_d)]| \lesssim \left(\sum_{j=d+1}^{\infty} \beta_j\right)^2 \lesssim \frac{1}{d^{2(1/p-1)}}, \tag{A.23}$$

where the β_j are given by (A.18) with $a_{\min}$ according to (A.17), and $p \in (0,1)$ is given by (A.19). The multiplicative factors involved in the notation "$\lesssim$" are independent of the truncation dimension d, the finite element method, and the QMC method in use.

The convergence of the finite element error (as defined above) depends on the particular finite elements used, the properties of the function f, the domain D, the linear functional G and the random field a. Under certain assumptions the finite element error is of order

$$O(M^{-2/s}), \tag{A.24}$$

with an implied multiplicative factor independent of the truncation dimension d (see, for instance, [160, Theorem 7.2], where $t = t' = 1$, and $M \asymp h^{-s}$).

Now consider the integrand $F(\boldsymbol{y}) = G(u_d^h(\cdot, \boldsymbol{y}))$ in the integration problem. It was shown in [25] that the integrand is infinitely many times differentiable and satisfies the bound

$$\sup_{\boldsymbol{y} \in [0,1]^d} \left| \frac{\partial^{|\boldsymbol{\nu}|}}{\partial^{\nu_1} y_1 \cdots \partial^{\nu_d} y_d} F(\boldsymbol{y}) \right| \leq C |\boldsymbol{\nu}|! \prod_{j=1}^{d} \beta_j^{\nu_j} \quad \text{with } \boldsymbol{\nu} = (\nu_1, \ldots, \nu_d) \in \mathbb{N}_0^d,$$

where $|\boldsymbol{\nu}| = \nu_1 + \cdots + \nu_d$, and where $C > 0$ is an absolute constant. This estimate can be used to obtain bounds on the integration error using QMC rules. In particular, [160, Theorem 6.4] shows that the integration error can be bounded independently of the dimension, yielding a convergence of order

$$\begin{cases} O(N^{-(1-\delta)}) & \text{for any } \delta > 0 \text{ when } p \in (0, 2/3], \\ O(N^{-(1/p-1/2)}) & \text{when } p \in (2/3, 1), \\ O(N^{-1/2}) & \text{when } p = 1. \end{cases} \tag{A.25}$$

In order to reduce the overall error, the three sources of error need to be balanced. This implies that if we want to reduce the overall error, we need to increase the truncation dimension to reduce the truncation error. This in turn increases the dimension of the integration domain of the numerical integration problem and leads to a situation where we want to decrease the integration error while the dimension simultaneously increases. To be able to balance the three error terms independently, we require that the integration error can be bounded independently of the dimension, i.e., that the integration problem is strongly polynomially tractable and this property can be achieved by means of the employed integration rule. Otherwise, if the integration error depends on the dimension, then this would have to be compensated by increasing N, which reduces the convergence rate. Since the convergence rate of the truncation dimension is rather low, this approach does not yield good convergence rates in most cases.

A.2 Log-Normal Random Coefficients

We consider (A.1) again, but modify the definition of the random field (A.2) to

$$a(z, \boldsymbol{y}) = \psi_*(z) + \psi_0(z) \exp\left(\sum_{j=1}^{\infty} y_j \psi_j(z) \right), \tag{A.26}$$

where the y_j are i.i.d. standard normal random variables, and where $\psi_* \geq 0$ and $\psi_0 > 0$. If the functions $(\psi_j)_{j \geq 0}$ are the eigenfunctions of a correlated Gaussian random field, these functions are orthonormal in $L_2([0, 1])$.

Define

$$\underline{a}(\boldsymbol{y}) := \min_{z \in [0,1]} a(z, \boldsymbol{y}) \quad \text{and} \quad \overline{a}(\boldsymbol{y}) := \max_{z \in [0,1]} a(z, \boldsymbol{y}).$$

Under suitable assumptions (see [88]) the functions $\underline{a}$ and $\overline{a}$ are measurable and satisfy $\underline{a} > 0$ and $\overline{a} < \infty$ almost surely.

An illustrative example

We consider the illustrative example from the uniform case again, with the random field (A.2) replaced by

$$a(z, \boldsymbol{y}) = \exp\left(2 + \sum_{j=1}^{\infty} y_j \frac{\sin(2\pi j z)}{j^\eta}\right), \tag{A.27}$$

where the y_j are now i.i.d. standard normal random variables, and where we assume the boundary conditions given by (A.4). The explicit solution is again given by (A.5). A numerical illustration is shown in Figure A.5.

The case of no explicit solution

We now consider (A.1) using the zero boundary condition (A.8) and a log-normal random field (A.26). We use the same finite elements as in the uniform case. The weak form (A.12) is not influenced by the change to a log-normal random field and therefore still applies. However, the linear system (A.13) is now more difficult to compute as the random field is not linear anymore.

Define $\widehat{f}_m$ as in Section A.1. Then substituting (A.27) and the finite element expansion of the solution (A.10) into (A.12) we obtain for the sequence $\boldsymbol{y}_k = (y_{k,1}, y_{k,2}, \ldots) \in \mathbb{R}^{\mathbb{N}}$ that

$$\sum_{\ell=1}^{M-1} \widehat{u}_\ell(\boldsymbol{y}_k)\, b_{k,\ell,m} = \widehat{f}_m$$

for $m \in \{1, 2, \ldots, M-1\}$, where

$$b_{k,\ell,m} := \int_0^1 \exp\left(2 + \sum_{j=1}^{\infty} y_{k,j} \frac{\sin(2\pi j z)}{j^\eta}\right) \phi'_\ell(z)\, \phi'_m(z)\, \mathrm{d}z. \tag{A.28}$$

In contrast to the uniform case, we do not have an explicit solution for the integral in (A.28), and thus need to resort to a numerical method.

For $|\ell - m| > 1$ we have $b_{k,\ell,m} = 0$ since the supports of the functions ϕ_ℓ and ϕ_m intersect at most at one point, respectively. For the remaining cases, choose a set of N_I quadrature points

$$z_{0,\ell,m}, z_{1,\ell,m}, \ldots, z_{N_I-1,\ell,m} \in \operatorname{supp}\phi_\ell \cap \operatorname{supp}\phi_m \tag{A.29}$$

and a set of N_I integration weights $w_{0,\ell,m}, w_{1,\ell,m}, \ldots, w_{N_I-1,\ell,m} \in \mathbb{R}$. Here, N_I is usually chosen to be a small positive integer, which can be chosen as independent of M since the length of the support of ϕ_ℓ decreases as M increases. To approximate the integral in (A.28) we compute

$$b_{k,\ell,m} \approx \widehat{b}_{k,\ell,m} = \sum_{i=0}^{N_I-1} w_{i,\ell,m}\, a(z_{i,\ell,m}, \boldsymbol{y}_k)\, \phi'_\ell(z_{i,\ell,m})\, \phi'_m(z_{i,\ell,m}) \tag{A.30}$$

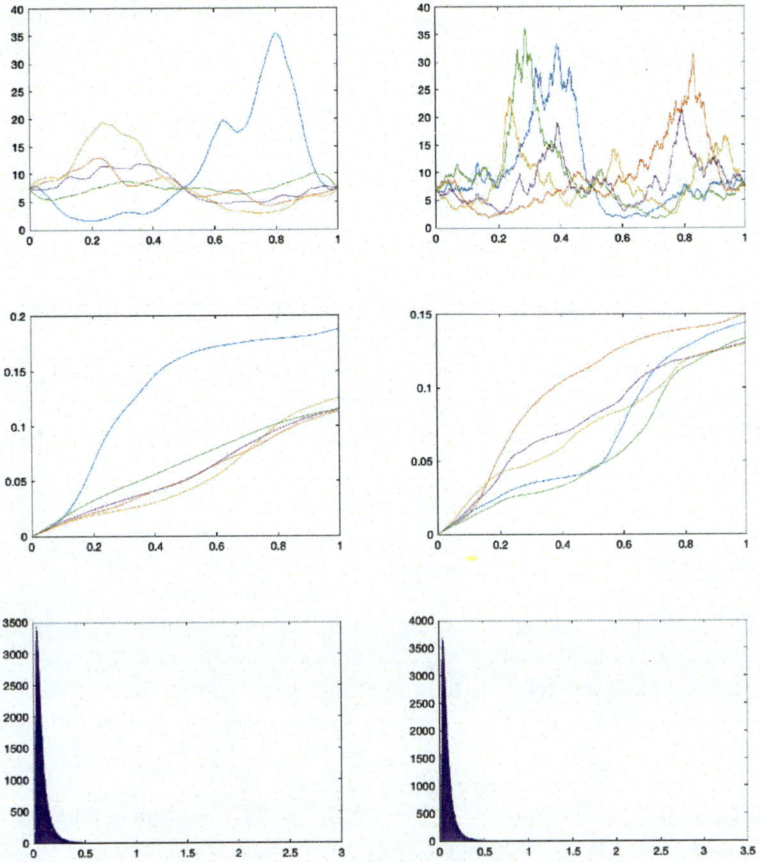

Fig. A.5: The left column uses $\eta = 2$, whereas the right column uses $\eta = 1.3$. In the top row, instances of the random field $a(\cdot, \boldsymbol{y})$ are shown, as well as various random selections of the random coefficients $\boldsymbol{y}$. In the middle row, the corresponding solutions $u(\cdot, \boldsymbol{y})$ of (A.7) are shown. Here the truncation dimension is 10^4. The bottom row shows a histogram of $u(1/2, \cdot)$ using $2 \cdot 10^5$ samples and 2000 bins, where we have restricted the truncation dimension to 100.

for $k \in \{0, 1, \ldots, N-1\}$ and $\ell, m \in \{1, 2, \ldots, M-1\}$ such that $|\ell - m| \le 1$. Let $\widehat{B}(\boldsymbol{y}_k) := (\widehat{b}_{k,\ell,m})_{\ell,m \in \{1,2,\ldots,M-1\}}$. To approximate a at the quadrature points, we use truncation, say after dimension d, and compute the truncated random field

$$a_d(z_{i,\ell,m}, \boldsymbol{y}_k) = \exp(\theta_{i,\ell,m,k}) \quad \text{with} \quad \theta_{i,\ell,m,k} = 2 + \sum_{j=1}^{d} y_{k,j} \frac{\sin(2\pi j z_{i,\ell,m})}{j^\eta},$$

for all $i \in \{0, 1, \ldots, N_I - 1\}$, $k \in \{0, 1, \ldots, N-1\}$, and $\ell, m \in \{1, 2, \ldots, M-1\}$ such that $|\ell - m| \le 1$. The number of these nonzero $b_{k,\ell,m}$ therefore is of order $O(N\,M)$. Consequently, the standard approach to obtaining the stiffness matrices $\widehat{B}(\boldsymbol{y}_k)$ for all $k \in \{0, 1, \ldots, N-1\}$ requires $O(d\,N_I\,M\,N)$ operations.

To get an approximation to the solution u we need to solve the linear system $\widehat{B}(\boldsymbol{y}_k)\,\widehat{\boldsymbol{u}}_k = \widehat{\boldsymbol{f}}$, as before. The approximation is then given by (A.10).

A numerical example is shown in Figure A.6. Note that if we set $a = 1$ and $f = 1$, then the solution of (A.1) and (A.8) is given by u with $u(z) = z(1-z)/2$ for $z \in [0, 1]$. The solutions for a random field are in some sense perturbations of this solution.

The general case

The PDE in the general case is again given by (A.16), but the random field is now log-normal like in (A.26), with a multidimensional domain $D \subseteq \mathbb{R}^s$.

We approximate the random field a at the quadrature points by the truncated random field a_d, where $d \in \mathbb{N}$ is the truncation dimension. To this end we need to compute

$$a(z_{i,\ell,m}, \boldsymbol{y}_k) \approx a_d(z_{i,\ell,m}, \boldsymbol{y}_k) = \psi_*(z_{i,\ell,m}) + \psi_0(z_{i,\ell,m}) \exp(\theta_{i,\ell,m,k}) \qquad \text{(A.31)}$$

with

$$\theta_{i,\ell,m,k} = \sum_{j=1}^{d} y_{k,j}\, \psi_j(z_{i,\ell,m}),$$

for all $i \in \{0, 1, \ldots, N_I - 1\}$, $k \in \{0, 1, \ldots, N-1\}$, and $\ell, m \in \{1, 2, \ldots, M-1\}$ such that $|\ell - m| \le 1$. The coefficients $\widehat{b}_{k,\ell,m}$ are then computed using (A.30). Setting $\widehat{B}(\boldsymbol{y}_k) := (\widehat{b}_{k,\ell,m})_{\ell,m \in \{1,2,\ldots,M-1\}}$, the coefficients $\widehat{\boldsymbol{u}}_k = (\widehat{u}_\ell(\boldsymbol{y}_k))_{\ell \in \{1,2,\ldots,M-1\}}$ are computed by solving the linear system $\widehat{B}(\boldsymbol{y}_k)\,\widehat{\boldsymbol{u}}_k = \widehat{\boldsymbol{f}}$. The approximation $u_d^h(\boldsymbol{y}_k)$ (where $h = 1/M$) to the solution u is then given by (A.10).

The computational approach

The algorithm proceeds in a similar way as for the uniform case, with the difference that we now need to sample from a log-normal random field.

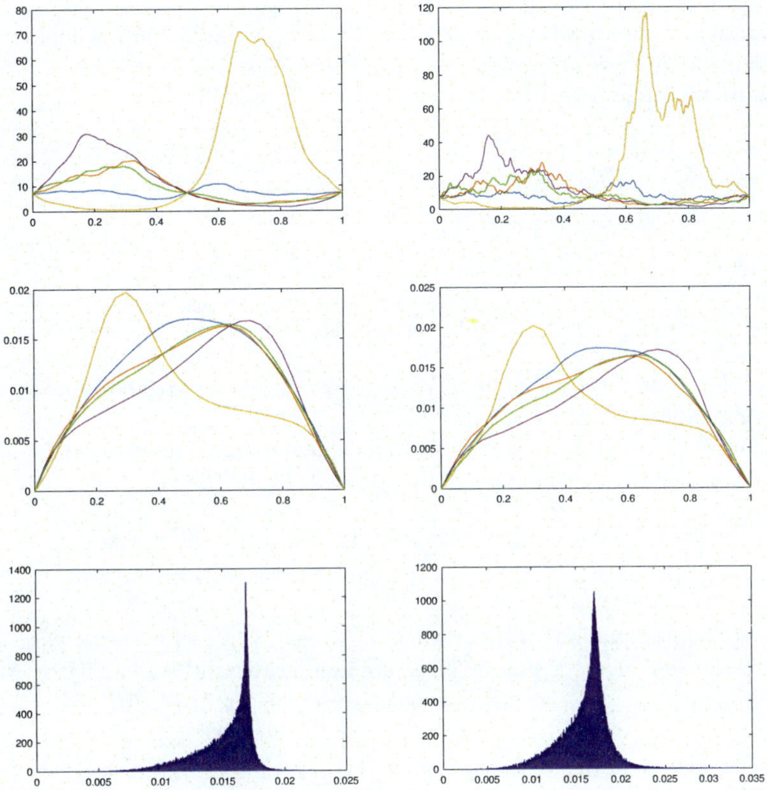

Fig. A.6: The left column uses $\eta = 2$, whereas the right column uses $\eta = 1.3$. In the top row, instances of the random field $a(\cdot, \boldsymbol{y})$ are shown, as well as various random selections of the random coefficients $\boldsymbol{y}$. In the middle row, the corresponding solutions $u(\cdot, \boldsymbol{y})$ of (A.7) and (A.8) are shown. The bottom row shows a histogram of $u(1/2, \cdot)$ using $2 \cdot 10^5$ samples and 2000 bins. The truncation dimension is 100 in these examples.

Algorithm A.2 Consider the boundary value problem (A.16) with log-normal i.i.d. random coefficients. In order to find an approximation to $\mathbb{E}[G(u)]$, do the following.

(1) Generate samples $\boldsymbol{x}_0, \boldsymbol{x}_1, \ldots, \boldsymbol{x}_{N-1} \in [0,1)^d$ (using (Q)MC), where the points are given by $\boldsymbol{x}_k = (x_{k,1}, x_{k,2}, \ldots, x_{k,d})$, for $k \in \{0, 1, \ldots, N-1\}$.
(2) Substitute these points into the truncated random field

$$a_d(z, \boldsymbol{y}_k) = \psi_*(z) + \psi_0(z) \exp\left(\sum_{j=1}^{d} y_{k,j}\, \psi_j(z)\right) \quad \text{for } k \in \{0, 1, \ldots, N-1\},$$

where

$$\boldsymbol{y}_k = (y_{k,1}, \ldots, y_{k,d}) = \Phi^{-1}(\boldsymbol{x}_k) = (\Phi^{-1}(x_{k,1}), \ldots, \Phi^{-1}(x_{k,d})) \tag{A.32}$$

and Φ^{-1} is the inverse cumulative distribution function of the standard normal distribution.
(3) Use a PDE solver (for instance the finite element method with mesh width h), in order to compute estimations of the solutions of the PDEs,

$$\begin{aligned} -\nabla_z \cdot (a_d(z, \boldsymbol{y}_k) \nabla_z u(z, \boldsymbol{y}_k)) &= f(z) \quad \text{for } z \in D,\\ u(z, \boldsymbol{y}_k) &= 0 \quad \text{for } z \in \partial D, \end{aligned}$$

to get approximations $u_d^h(\boldsymbol{y}_k)$ for $k \in \{0, 1, \ldots, N-1\}$. In the case of a finite element method, this involves computing the stiffness matrices $B(\boldsymbol{y}_k) = (b_{k,\ell,m})_{\ell,m \in \{1,2,\ldots,M-1\}}$ using a quadrature rule as in (A.30) and solving the linear systems $B(\boldsymbol{y}_k)\, \widehat{\boldsymbol{u}}_k = \widehat{\boldsymbol{f}}$ for $k \in \{0, 1, \ldots, N-1\}$.
(4) An estimation of the expected value $\mathbb{E}[G(u)]$ is then computed by

$$Q_{N,d}(G(u_d^h)) = \frac{1}{N} \sum_{k=0}^{N-1} G(u_d^h(\boldsymbol{y}_k)) \approx \mathbb{E}[G(u)].$$

Error analysis

The error can again be split into a truncation error, a finite element error, and an integration error. The numerical integration problem is now defined over $\mathbb{R}^d$, which we transformed to the unit cube $[0,1]^d$ using the inverse standard normal cumulative distribution function. This integration problem has different properties than the analogous problem defined over the unit cube, and QMC theory on this type of integration problem is in some sense not as advanced as integration over the unit cube (see also Chapter 8). In particular, to achieve integration error bounds independent of the dimension using QMC, by the current state of the art we need to use randomized QMC methods such as, e.g., randomly shifted lattice rules.

Consider the root mean square error of a randomly shifted lattice rule $Q_{N,d,\boldsymbol{\Delta}}$ with random shift $\boldsymbol{\Delta}$ (see Section 7.1),

$$E_{d,h,N} := \sqrt{\mathbb{E}_{\Delta}\left[\left(\mathbb{E}[G(u)] - Q_{N,d,\Delta}(G(u_d^h))\right)^2\right]}.$$

Since the random shift $\boldsymbol{\Delta}$ and the (Q)MC points $\boldsymbol{y}_k$ are independent, we have

$$E_{d,h,N}^2 = \left(\mathbb{E}[G(u) - G(u_d^h)]\right)^2 + \mathbb{E}_{\Delta}\left[\left(\mathbb{E}[G(u_d^h)] - Q_{N,d,\Delta}(G(u_d^h))\right)^2\right]. \quad \text{(A.33)}$$

The first term, $\mathbb{E}[G(u) - G(u_d^h)]$, can be further split into the truncation error and the finite element error, which can be bounded similarly to the uniform case, see [88] for more information.

The second term in (A.33) is the integration error using a randomly shifted lattice rule. As in the uniform case, the integrand $F(\boldsymbol{y}) = G(u_d^h(\cdot, \boldsymbol{y}))$ defined over $\mathbb{R}^d$ is infinitely many times differentiable and satisfies the bound

$$\left\| \frac{\partial^{|\boldsymbol{\nu}|}}{\partial^{\nu_1} y_1 \cdots \partial^{\nu_d} y_d} u(\cdot, \boldsymbol{y}) \right\|_V \leq \frac{|\boldsymbol{\nu}|!}{(\log 2)^{|\boldsymbol{\nu}|}} \left(\prod_{j=1}^{d} \beta_j^{\nu_j} \right) \frac{\|f\|_{V'}}{\overline{a}(\boldsymbol{y})},$$

where $\|\cdot\|_V$ is a certain norm for which $u(\cdot, \boldsymbol{y}) \in V$, and where V' is the corresponding dual space. For further details we refer to [88]. The bound on the norm of the partial derivatives of the solution u can again be used to obtain bounds on the integration error.

Appendix B
Numerical Experiments for Lattice Rule Construction Algorithms

By Adrian Ebert

In this chapter, we illustrate the computational cost of the various construction algorithms, which were introduced in Chapters 3 and 4, and the error convergence behavior of the resulting lattice rules, by means of numerical experiments. As before we consider the construction of generating vectors of rank-1 lattice rules for the weighted Korobov space $\mathcal{H}_{\mathrm{kor},d,\alpha,\gamma}$ of smoothness $\alpha > 1/2$. Additionally, we assume that the underlying weights $\boldsymbol{\gamma} = \{\gamma_{\mathfrak{u}}\}_{\mathfrak{u}\subseteq[d]}$ are either of product form, i.e., $\gamma_{\mathfrak{u}} = \prod_{j\in\mathfrak{u}} \gamma_j$ for $\mathfrak{u} \subseteq [d]$ (see Equation (2.21)), or of product and order dependent (POD) form, i.e., $\gamma_{\mathfrak{u}} = \Gamma_{|\mathfrak{u}|} \prod_{j\in\mathfrak{u}} \gamma_j$ for $\mathfrak{u} \subseteq [d]$ (see Equation (3.25)).

We recall that for product weights the squared worst-case error of an N-point rank-1 lattice rule with generating vector $\boldsymbol{g} \in \mathbb{Z}^d$ in the space $\mathcal{H}_{\mathrm{kor},d,\alpha,\gamma}$ is explicitly given by

$$[\mathrm{err}_{N,d,\alpha,\boldsymbol{\gamma}}(\boldsymbol{g})]^2 = -1 + \frac{1}{N}\sum_{k=0}^{N-1}\prod_{j=1}^{d}\left(1 + \gamma_j \sum_{h\in\mathbb{Z}\setminus\{0\}} \frac{\mathrm{e}^{2\pi \mathrm{i} hkg_j/N}}{|h|^{2\alpha}}\right), \tag{B.1}$$

while for POD weights it equals

$$[\mathrm{err}_{N,d,\alpha,\boldsymbol{\gamma}}(\boldsymbol{g})]^2 = \frac{1}{N}\sum_{k=0}^{N-1}\sum_{\ell=1}^{d}\Gamma_\ell \sum_{\substack{\mathfrak{u}\subseteq[d]\\|\mathfrak{u}|=\ell}}\left(\prod_{j\in\mathfrak{u}}\gamma_j \sum_{h\in\mathbb{Z}\setminus\{0\}} \frac{\mathrm{e}^{2\pi \mathrm{i} hkg_j/N}}{|h|^{2\alpha}}\right). \tag{B.2}$$

As we will use smoothness parameters $\alpha \in \mathbb{N}$ for the numerical experiments, we obtain by Remark 2.9 that the sum of exponentials occurring in (B.1) and (B.2) further simplifies to

$$\sum_{h\in\mathbb{Z}\setminus\{0\}} \frac{\mathrm{e}^{2\pi \mathrm{i} hkg_j/N}}{|h|^{2\alpha}} = \frac{(-1)^{\alpha+1}(2\pi)^{2\alpha}}{(2\alpha)!} B_{2\alpha}\left(\left\{\frac{kg_j}{N}\right\}\right),$$

where $B_{2\alpha}$ denotes the Bernoulli polynomial of degree 2α. For non-integer $\alpha > 1/2$, this sum can be numerically approximated.

J. Dick et al., *Lattice Rules*, Springer Series in Computational Mathematics 58,
https://doi.org/10.1007/978-3-031-09951-9

B.1 Numerical Results for the CBC Construction

Error convergence behavior: fast CBC algorithm with product weights

We examine the convergence behavior of $\mathrm{err}_{N,d,\alpha,\boldsymbol{\gamma}}(\boldsymbol{g})$ for generating vectors $\boldsymbol{g}$ constructed by the fast CBC algorithm (see Algorithm 3.6, and Section 3.4 (for prime N) and Section 4.2 (for prime-power N, where we set all reduction indices w_j equal to zero), respectively) for different sequences of product weights $\boldsymbol{\gamma} = (\gamma_j)_{j\geq 1}$ and different values of N. In particular, we consider prime N as well as prime powers $N = 2^m$ for our experiments and the product weight sequences $(\gamma_j)_{j\geq 1}$ given by $\gamma_j = j^{-2}, \gamma_j = j^{-3}, \gamma_j = (0.5)^j$, or $\gamma_j = (0.95)^j$. The dimension is fixed at $d = 100$.

We stress that the asymptotic error convergence rates stated in Theorems 3.7 and 3.9 may not always be visible for the ranges of N considered in the numerical experiments. Therefore, the graphs presented in Figures B.1 and B.2 are to be understood as a demonstration of the pre-asymptotic worst-case error behavior.

The convergence results presented are in accordance with Theorems 3.7 and 3.9, and the observed convergence rates in the considered examples are of order $O(N^{-\tau})$ with $\tau \in [1/2, \alpha)$. We also notice that the stronger the decay of the weight sequence $\boldsymbol{\gamma}$, the faster the worst-case error decays. Also this observation is in accordance with Theorems 3.7 and 3.9.

Computation time: fast CBC algorithm with product weights

In this section, we illustrate the computation time of the fast CBC construction for product weights (see Algorithm 3.6, and Section 3.4 (for prime N) and Section 4.2 (for prime-power N, where we set all reduction indices w_j equal to zero), respectively). To this end, let $m, d \in \mathbb{N}$ and $N = 2^m$, let $\alpha = 1$, and use the weight sequence $\boldsymbol{\gamma} = (\gamma_j)_{j\geq 1}$ with $\gamma_j = j^{-2}$. Note that the smoothness α and the chosen weight sequence do not influence the computation times. In Table B.1 and Figure B.3 we report and display the timings for the construction of generating vectors of rank-1 lattice rules via the fast CBC algorithm. In order to assure robustness of the timings, we display the average computation times over five independent executions of the numerical experiments. All timings here and in the following sections were performed on an Intel Core i5 CPU with 2.3 GHz using Matlab2019b, unless specified otherwise.

The obtained results in Table B.1 and Figure B.3 confirm the computational cost of $O(d\,N \log N)$ of the fast CBC construction. The linear dependence of the construction cost on the dimension d is well observable in Table B.1 while the dependence on N can be observed in Figure B.3.

Error convergence in the space $\mathcal{H}_{\mathrm{kor},d,\alpha,\gamma}$ with $d = 100$, $\alpha = 1$.

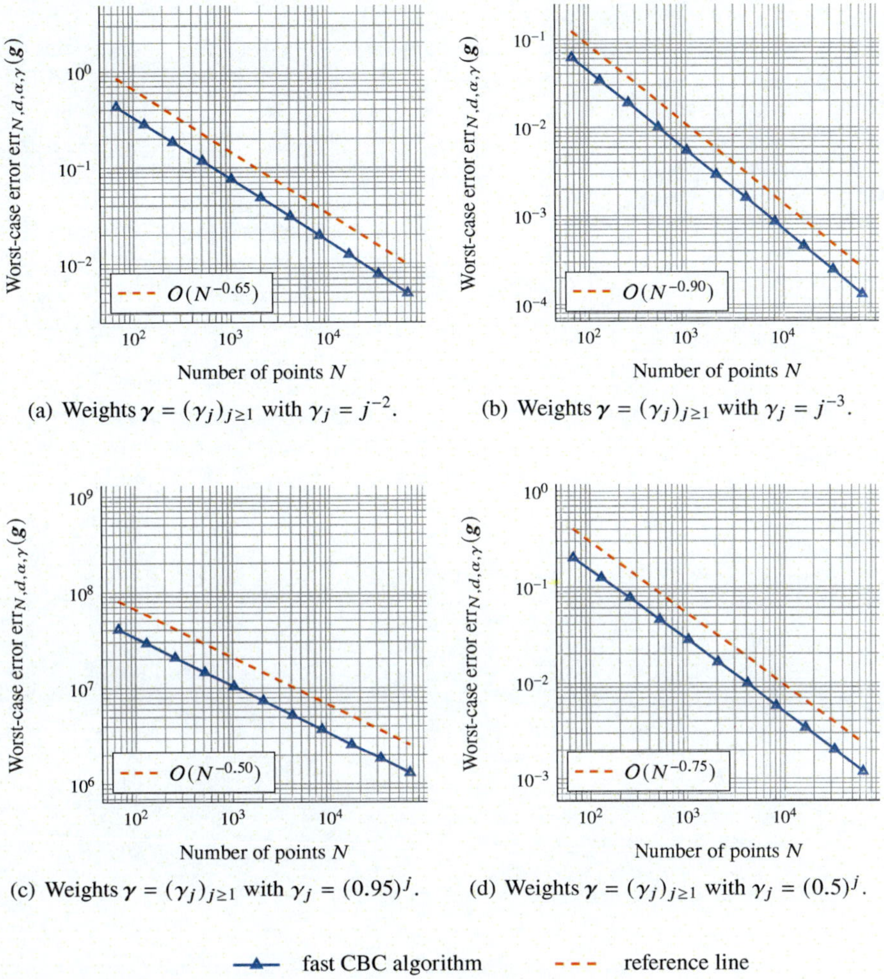

(a) Weights $\gamma = (\gamma_j)_{j\geq 1}$ with $\gamma_j = j^{-2}$.

(b) Weights $\gamma = (\gamma_j)_{j\geq 1}$ with $\gamma_j = j^{-3}$.

(c) Weights $\gamma = (\gamma_j)_{j\geq 1}$ with $\gamma_j = (0.95)^j$.

(d) Weights $\gamma = (\gamma_j)_{j\geq 1}$ with $\gamma_j = (0.5)^j$.

Fig. B.1: Convergence results for the worst-case error $\mathrm{err}_{N,d,\alpha,\gamma}(g)$ in the weighted Korobov space $\mathcal{H}_{\mathrm{kor},d,\alpha,\gamma}$ of smoothness $\alpha = 1$ with dimension $d = 100$. The generating vectors g have been constructed by the fast CBC algorithm for prime N using product weights.

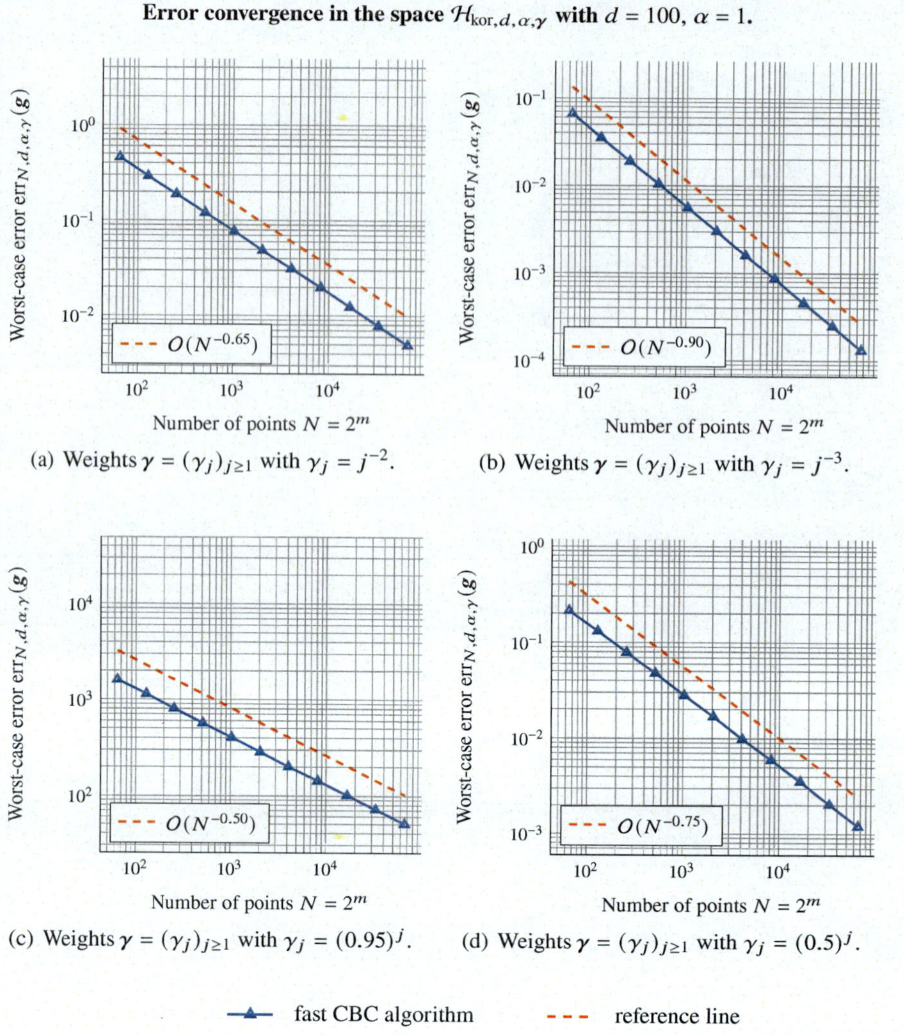

Fig. B.2: Convergence results for the worst-case error $\mathrm{err}_{N,d,\alpha,\gamma}(\boldsymbol{g})$ in the weighted Korobov space $\mathcal{H}_{\mathrm{kor},d,\alpha,\gamma}$ of smoothness $\alpha = 1$ with dimension $d = 100$. The generating vectors $\boldsymbol{g}$ have been constructed by the fast CBC algorithm for prime powers $N = 2^m$ using product weights.

Table B.1: Mean computation times (in seconds, mean over five runs) for constructing the generating vector $\boldsymbol{g}$ of a lattice rule with $N = 2^m$ points in d dimensions using the fast CBC algorithm.

	$d = 50$	$d = 100$	$d = 500$	$d = 1000$	$d = 2000$
$m = 10$	0.02972	0.03323	0.1644	0.3324	0.6546
$m = 12$	0.03367	0.0484	0.2393	0.4767	0.9555
$m = 14$	0.04972	0.08306	0.4121	0.8214	1.645
$m = 16$	0.114	0.2213	1.105	2.201	4.409
$m = 18$	0.3843	0.7457	3.702	7.363	14.71
$m = 20$	1.659	3.244	15.93	31.97	63.9

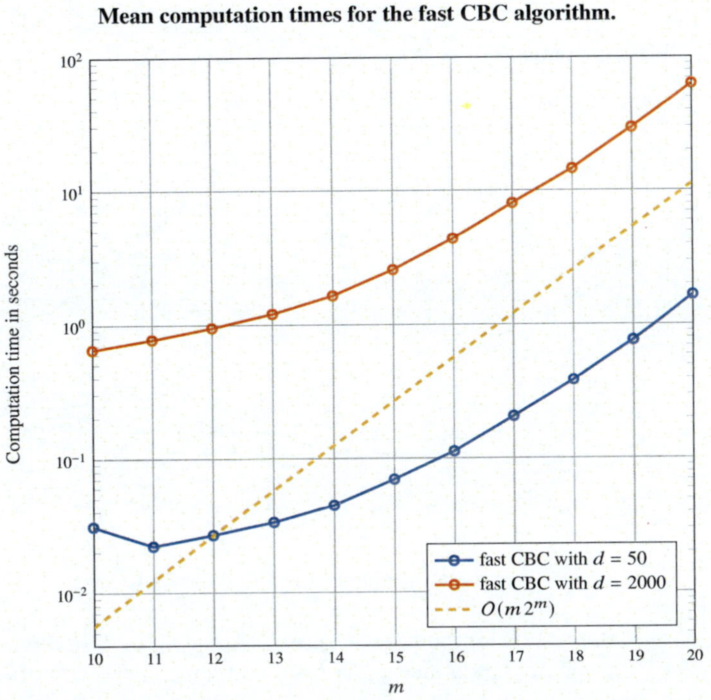

Fig. B.3: Mean computation times (in seconds, mean over five runs) for constructing the generating vector $\boldsymbol{g}$ of a lattice rule with $N = 2^m$ points in $d \in \{50, 2000\}$ dimensions using the fast CBC construction for product weights.

Error convergence behavior: fast CBC algorithm with POD weights

In this section, we illustrate the convergence behavior of $\mathrm{err}_{N,d,\alpha,\gamma}(\boldsymbol{g})$ for generating vectors $\boldsymbol{g}$ constructed by the fast CBC algorithm with POD weights (see Section 3.5) (for prime N) and Section 4.2 (for prime-power N, where we set all reduction indices w_j equal to zero), respectively). We use POD weights $\boldsymbol{\gamma} = \{\gamma_{\mathfrak{u}}\}_{\mathfrak{u} \subseteq [d]}$ of the general form

$$\gamma_{\mathfrak{u}} = \Gamma_{|\mathfrak{u}|} \prod_{j \in \mathfrak{u}} \gamma_j = |\mathfrak{u}|! \prod_{j \in \mathfrak{u}} j^{-a}$$

with positive $a > 0$, and consider different values of $N = 2^m$ for our experiments. The dimension is fixed at $d = 100$.

In Figure B.4 we display the worst-case errors of lattice rules constructed by the fast CBC algorithm for POD weights using two different weight sequences.

Error convergence in the space $\mathcal{H}_{\mathrm{kor},d,\alpha,\gamma}$ with $d = 100$, $\alpha = 1$.

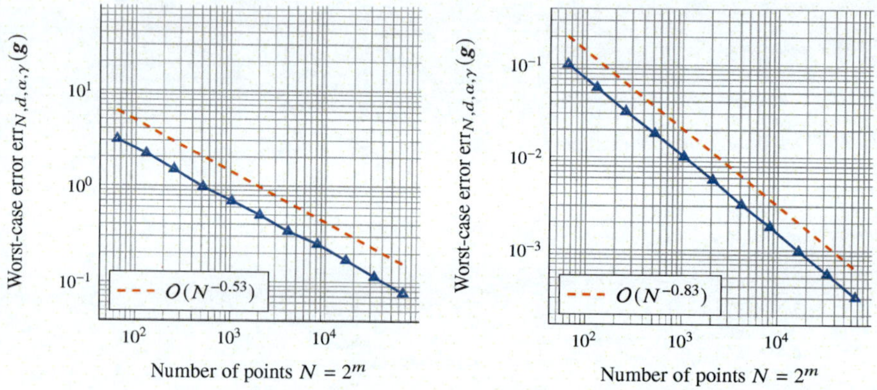

(a) Weights $\gamma_{\mathfrak{u}} = |\mathfrak{u}|! \prod_{j \in \mathfrak{u}} j^{-a}$ with $a = 2$. (b) Weights $\gamma_{\mathfrak{u}} = |\mathfrak{u}|! \prod_{j \in \mathfrak{u}} j^{-a}$ with $a = 4$.

fast CBC algorithm --- reference line

Fig. B.4: Convergence results for the worst-case error $\mathrm{err}_{N,d,\alpha,\gamma}(\boldsymbol{g})$ in the weighted Korobov space $\mathcal{H}_{\mathrm{kor},d,\alpha,\gamma}$ of smoothness $\alpha = 1$ with dimension $d = 100$ using POD weights. The generating vectors $\boldsymbol{g}$ have been constructed by the fast CBC algorithm for prime power $N = 2^m$.

The convergence results presented are again in accordance with Theorem 3.9, and the observed convergence rates in the considered examples are of order $O(N^{-\tau})$ with $\tau \in [1/2, \alpha)$. The stronger the decay of the POD weights (controlled by the decay parameter a in this case), the faster the worst-case error decays.

Computation time: fast CBC algorithm with POD weights

Here, we illustrate the computation time of the fast CBC algorithm for POD weights. In Sections 3.5 and 4.2, respectively, it was shown that the fast CBC algorithm for POD weights has an asymptotic computational cost of order $O(d\,N \log N + d^2 N)$.

Let $m, d \in \mathbb{N}$ and $N = 2^m$, let $\alpha = 1$, and use the weight sequence with weights given by $\gamma_{\mathfrak{u}} = |\mathfrak{u}|! \prod_{j \in \mathfrak{u}} j^{-2}$. Note again that the smoothness α and the chosen weight sequence do not influence the computation times. In Table B.2 and Figure B.5 we report and display the timings for the construction of generating vectors of rank-1 lattice rules via the fast CBC algorithm for POD weights. In order to assure robustness of the timings, we display the average computation times over five independent executions of the numerical experiments.

Table B.2: Mean computation times (in seconds, mean over five runs) for constructing the generating vector $\boldsymbol{g}$ of a lattice rule with $N = 2^m$ points in d dimensions using the fast CBC algorithm with POD weights.

	$d = 50$	$d = 100$	$d = 200$	$d = 500$
$m = 8$	0.2215	0.5515	2.116	12.4
$m = 10$	0.2128	0.7601	2.958	17.51
$m = 12$	0.293	1.115	4.287	25.58
$m = 14$	0.5664	1.946	7.571	46.05
$m = 16$	1.628	5.916	23.13	144.9

The results displayed in Table B.2 and Figure B.5 confirm the computational cost of $O(d\,N \log N + d^2 N)$ for the fast CBC construction for POD weights (see Section 3.5 (for prime N) and Section 4.2 (for prime-power N, where we set all reduction indices w_j equal to zero), respectively).

Mean computation times for the fast CBC algorithm with POD weights.

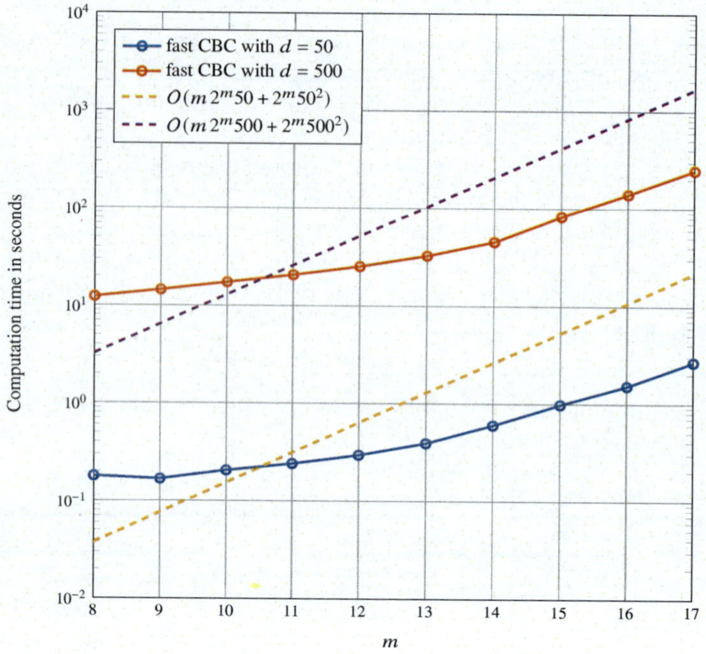

Fig. B.5: Mean computation times (in seconds, mean over five runs) for constructing the generating vector $\boldsymbol{g}$ of a lattice rule with $N = 2^m$ points in $d \in \{50, 500\}$ dimensions using the fast CBC construction for POD weights.

Error convergence behavior: reduced fast CBC algorithm

In this section, we illustrate the error convergence behavior of rank-1 lattice rules constructed by the reduced fast CBC algorithm, which was outlined in Sections 4.1 and 4.2. To this end, we assume that $N = b^m$ is a prime power with prime base b and $m \in \mathbb{N}$, and again consider the product weight sequences $\boldsymbol{\gamma} = (\gamma_j)_{j \geq 1}$ with $\gamma_j = j^{-2}, \gamma_j = j^{-3}, \gamma_j = (0.5)^j$, or $\gamma_j = (0.95)^j$. Furthermore, we choose a fixed dimension of $d = 100$ and reduction indices $w_1 \leq w_2 \leq \cdots \leq w_d$ of the form $w_j = \lfloor c \log_b(j) \rfloor$ with $c \in \{2, 3\}$. As a reference, we also display the worst-case errors of the corresponding lattice rules constructed by the fast CBC algorithm. The results of our experiments are displayed in Figures B.6 and B.7 below.

The results in Figures B.6 and B.7 reveal that the interplay between the weight sequence $\boldsymbol{\gamma}$ and the reduction indices $w_1 \leq w_2 \leq \cdots \leq w_d$ greatly influences the convergence behavior. If the weights decay sufficiently fast, the lattice rules constructed by the reduced fast CBC algorithm yield a similar convergence order as those constructed by the fast CBC algorithm. However, if the decay of the weights is too slow, or, in other words, the increase of the reduction indices is too high, then the

Error convergence in the space $\mathcal{H}_{\mathrm{kor},d,\alpha,\boldsymbol{\gamma}}$ with $d = 100$, $\alpha = 1$, $b = 2$, $w_j = \lfloor 2\log_b(j) \rfloor$.

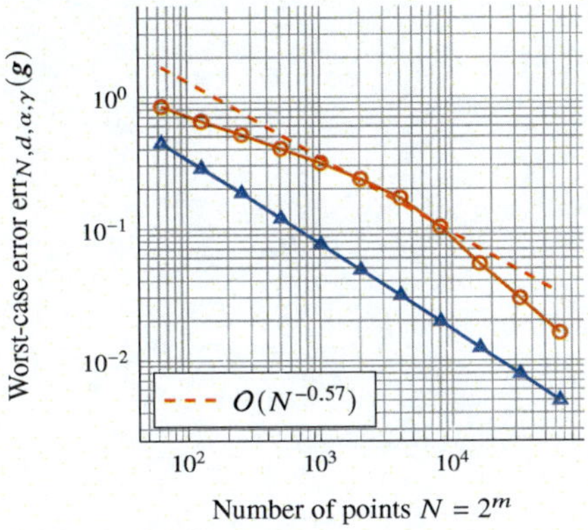

(a) Weights $\boldsymbol{\gamma} = (\gamma_j)_{j\geq 1}$ with $\gamma_j = j^{-2}$.

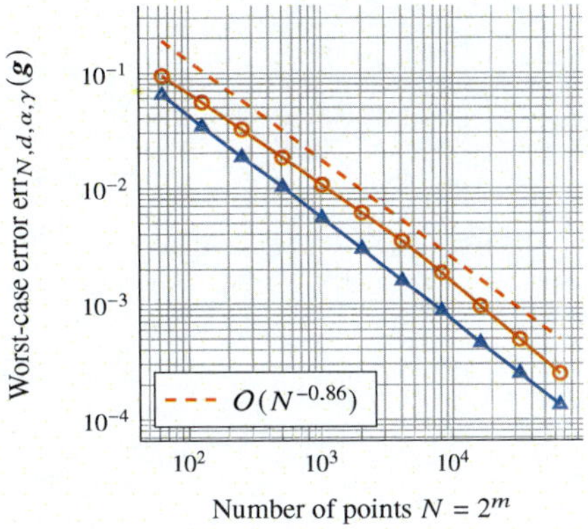

(b) Weights $\boldsymbol{\gamma} = (\gamma_j)_{j\geq 1}$ with $\gamma_j = j^{-3}$.

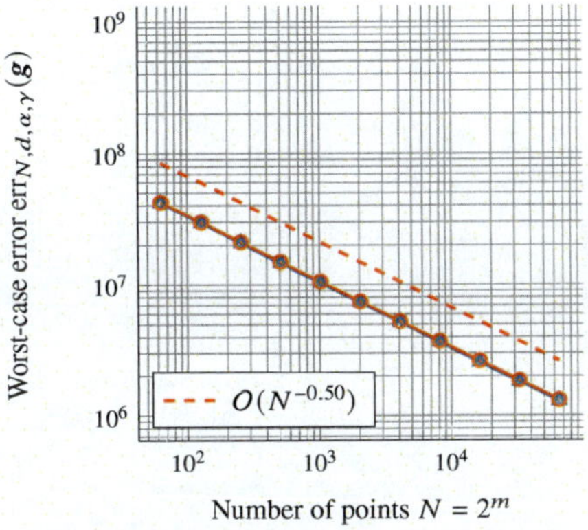

(c) Weights $\boldsymbol{\gamma} = (\gamma_j)_{j\geq 1}$ with $\gamma_j = (0.95)^j$.

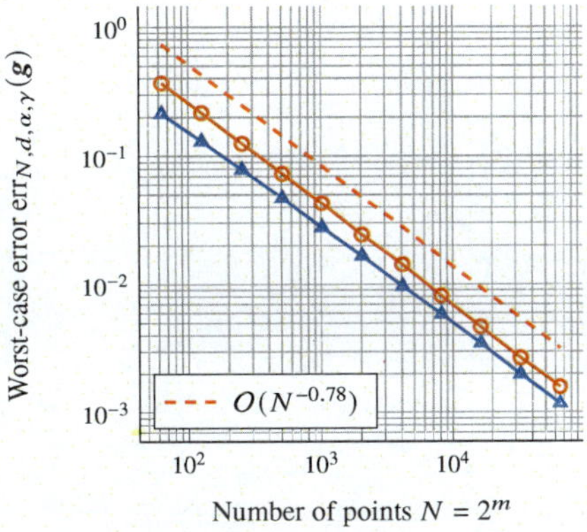

(d) Weights $\boldsymbol{\gamma} = (\gamma_j)_{j\geq 1}$ with $\gamma_j = (0.5)^j$.

fast CBC algorithm — reduced CBC algorithm — reference line

Fig. B.6: Convergence results for the worst-case error $\mathrm{err}_{N,d,\alpha,\boldsymbol{\gamma}}(\boldsymbol{g})$ in the weighted Korobov space $\mathcal{H}_{\mathrm{kor},d,\alpha,\boldsymbol{\gamma}}$ of smoothness $\alpha = 1$ with dimension $d = 100$. The generating vectors $\boldsymbol{g}$ have been constructed by the reduced fast CBC algorithm or the fast CBC algorithm, respectively, with indicated weights $\boldsymbol{\gamma} = (\gamma_j)_{j\geq 1}$ and reduction indices $w_j = \lfloor 2\log_b(j) \rfloor$ for $j \in [d]$.

Error convergence in the space $\mathcal{H}_{\mathrm{kor},d,\alpha,\boldsymbol{\gamma}}$ with $d = 100$, $\alpha = 1$, $b = 2$, $w_j = \lfloor 3\log_b(j) \rfloor$.

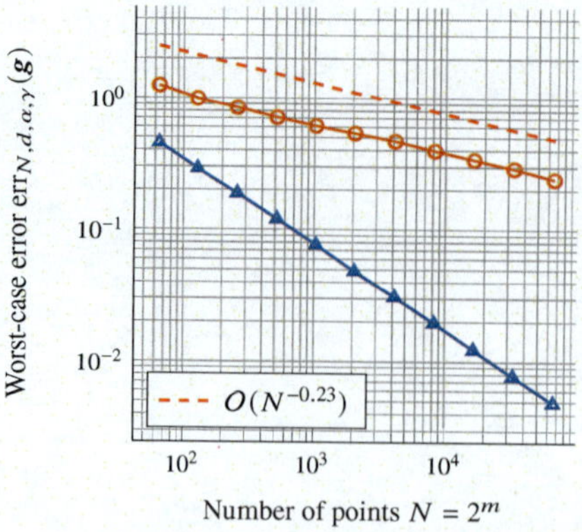

(a) Weights $\boldsymbol{\gamma} = (\gamma_j)_{j\geq 1}$ with $\gamma_j = j^{-2}$.

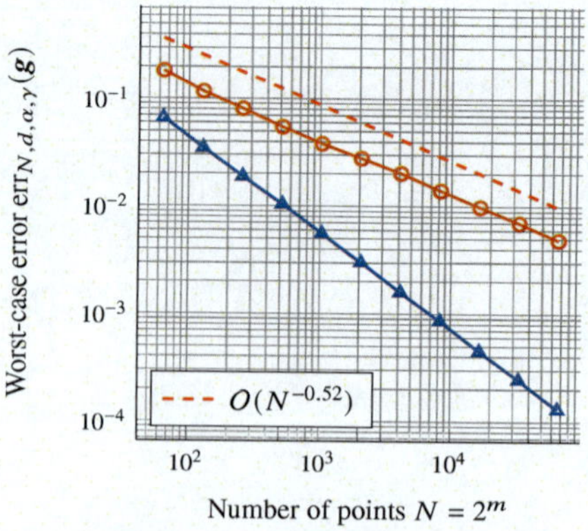

(b) Weights $\boldsymbol{\gamma} = (\gamma_j)_{j\geq 1}$ with $\gamma_j = j^{-3}$.

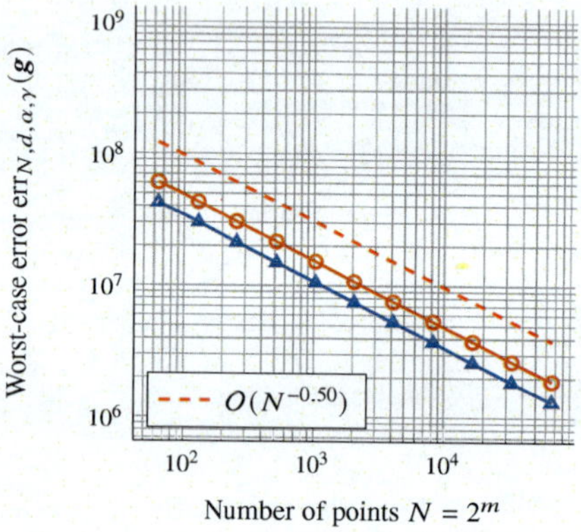

(c) Weights $\boldsymbol{\gamma} = (\gamma_j)_{j\geq 1}$ with $\gamma_j = (0.95)^j$.

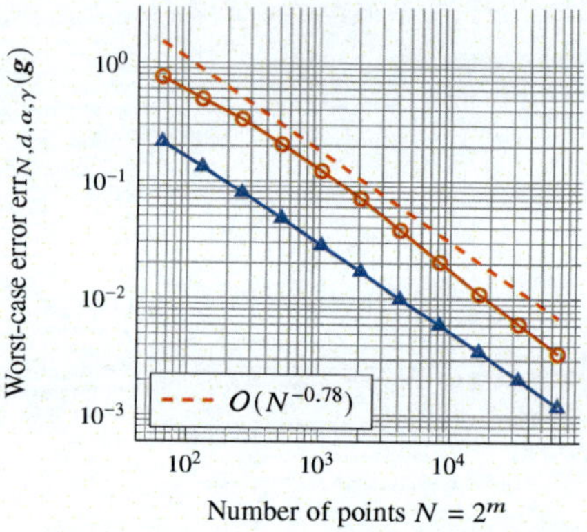

(d) Weights $\boldsymbol{\gamma} = (\gamma_j)_{j\geq 1}$ with $\gamma_j = (0.5)^j$.

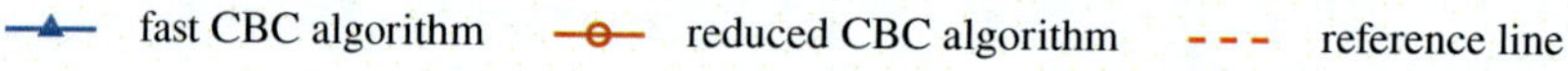

Fig. B.7: Convergence results for the worst-case error $\mathrm{err}_{N,d,\alpha,\boldsymbol{\gamma}}(\boldsymbol{g})$ in the weighted Korobov space $\mathcal{H}_{\mathrm{kor},d,\alpha,\boldsymbol{\gamma}}$ of smoothness $\alpha = 1$ with dimension $d = 100$. The generating vectors $\boldsymbol{g}$ have been constructed by the reduced fast CBC algorithm or the fast CBC algorithm, respectively, with indicated weights $\boldsymbol{\gamma} = (\gamma_j)_{j\geq 1}$ and reduction indices $w_j = \lfloor 3\log_b(j) \rfloor$ for $j \in [d]$.

experimental convergence rates of the reduced fast and the fast CBC constructions may differ. We also observe that using a larger increase parameter c for the reduction indices in general leads to a higher worst-case error of the lattice rules constructed by the reduced fast CBC algorithm.

Computation time: reduced fast CBC algorithm

In this section, we illustrate the computation time of the reduced fast CBC construction (see Sections 4.1 and 4.2). To this end, let $m, d \in \mathbb{N}$ and $N = 2^m$, let $\alpha = 1$, and use the weight sequence $\boldsymbol{\gamma} = (\gamma_j)_{j\geq 1}$ with $\gamma_j = j^{-2}$. In Table B.3 and Figure B.8 we illustrate the timings for the construction of generating vectors of rank-1 lattice rules via the fast CBC algorithm and the reduced fast CBC algorithm. For the latter we use the reduction indices given by $w_j = \lfloor 2 \log_b(j) \rfloor$. Again, we performed five independent executions of the numerical experiments in order to assure robustness of the timings.

Table B.3: Mean computation times (in seconds, mean over five runs) for constructing the generating vector $\boldsymbol{g}$ of a rank-1 lattice rule with $N = 2^m$ points in d dimensions using the fast CBC algorithm (normal font) or the reduced CBC algorithm (**bold font**), respectively.

	$d = 50$	$d = 100$	$d = 500$	$d = 1000$	$d = 2000$
$m = 10$	0.02972 **0.02559**	0.03323 **0.004809**	0.1644 **0.00549**	0.3324 **0.006638**	0.6546 **0.008283**
$m = 12$	0.03367 **0.01115**	0.0484 **0.01105**	0.2393 **0.01255**	0.4767 **0.01466**	0.9555 **0.01851**
$m = 14$	0.04972 **0.02233**	0.08306 **0.02625**	0.4121 **0.03607**	0.8214 **0.0493**	1.645 **0.07358**
$m = 16$	0.114 **0.0417**	0.2213 **0.06807**	1.105 **0.148**	2.201 **0.2007**	4.409 **0.3**
$m = 18$	0.3843 **0.1236**	0.7457 **0.176**	3.702 **0.635**	7.363 **0.7876**	14.71 **1.053**
$m = 20$	1.659 **0.573**	3.244 **0.9255**	15.93 **4.175**	31.97 **7.174**	63.9 **9.915**

The results in Table B.3 and Figure B.8 confirm that the times for constructing generating vectors of rank-1 lattice rules can be drastically decreased when using the reduced fast CBC algorithm (as compared to the fast CBC algorithm). The faster the reduction indices $w_1 \leq w_2 \leq \cdots \leq w_d$ decay, the larger are the savings in computational effort by using the reduced fast CBC construction.

Mean computation times for the fast and the reduced CBC algorithm.

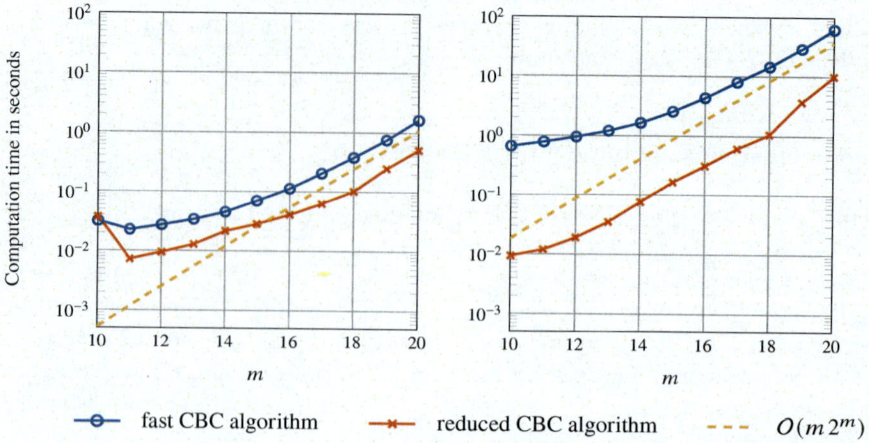

Fig. B.8: Mean computation times (in seconds, mean over five runs) for constructing the generating vector $\boldsymbol{g}$ of a lattice rule with $N = 2^m$ points in $d \in \{50, 2000\}$ dimensions using the fast CBC construction and the reduced fast CBC algorithm, respectively, with reduction indices $w_j = \lfloor 2 \log_b(j) \rfloor$ for $j \in [d]$.

Remark B.1 We obtain very similar results in terms of error convergence and computation time when using the reduced fast SCS algorithm (described in Section 4.4) instead of the reduced fast CBC construction. We therefore omit these results here as they do not give additional insights.

B.2 Numerical Results for Alternative Constructions

In this section we examine the error convergence behavior and the computation times of some alternative construction algorithms for the generating vectors of rank-1 lattice rules, such as the successive coordinate search (SCS) algorithm or the component-by-component digit-by-digit (CBC-DBD) construction, which were discussed in Sections 4.3 and 4.6, respectively.

Numerical results for the successive coordinate search algorithm

We begin with the SCS algorithm. Since we require an initial vector $\boldsymbol{g}^{(0)} = (g_1^{(0)}, \ldots, g_d^{(0)}) \in G_d^{\varphi}(N)$ as additional input for this algorithm, we consider the following two types of initial vectors.

- **Fixed initial vector:** We use the fixed vector $\boldsymbol{g}^{(0)} = (1, \dots, 1) \in G_d^{\varphi}(N)$ as the initial vector.
- **Random initial vector:** The initial vector $\boldsymbol{g}^{(0)} = (g_1^{(0)}, \dots, g_d^{(0)})$ is randomly and uniformly selected from the set $G_d^{\varphi}(N)$.

Error convergence behavior: fast SCS algorithm with product weights

We examine the convergence behavior of $\mathrm{err}_{N,d,\alpha,\boldsymbol{\gamma}}(\boldsymbol{g})$ for generating vectors $\boldsymbol{g}$ constructed by the fast SCS algorithm (see Section 4.3) for different sequences of product weights $\boldsymbol{\gamma} = (\gamma_j)_{j\geq 1}$ and different values of N. In particular, we consider prime N, a dimension of $d = 100$ and product weight sequences given by $\gamma_j = j^{-2}, \gamma_j = j^{-3}, \gamma_j = (0.5)^j$, or $\gamma_j = (0.95)^j$. Furthermore, we use fixed initial vectors as well as random initial vectors as inputs for the SCS algorithm. For comparison, we compute and display the worst-case errors of lattice rules computed by the fast CBC construction. The corresponding results are displayed in Figure B.9.

The convergence results displayed are in accordance with Theorem 4.10, and the observed convergence rates in the considered examples are of order $O(N^{-\tau})$ with $\tau \in [1/2, \alpha)$. We notice that the worst-case errors of lattice rules constructed by the fast SCS algorithm are virtually identical to those of lattice rules constructed by the fast CBC algorithm. Additionally, we see that both random and fixed initial vectors yield good lattice rules when used as inputs for the fast SCS algorithm.

Computation time: fast SCS algorithm with product weights

In this section, we illustrate the computation time of the successive coordinate search algorithm. We set $\alpha = 1$ and consider different values of prime N, different dimensions d, and use the weight sequence $\boldsymbol{\gamma} = (\gamma_j)_{j\geq 1}$ with $\gamma_j = j^{-2}$. For each of the displayed scenarios, we performed five independent timing runs and then calculated the mean computation time.

The results of the experiments are given in Table B.4 and Figure B.10. For comparison, we show the corresponding computation times that were obtained for the fast CBC algorithm.

Error convergence in the space $\mathcal{H}_{\text{kor},d,\alpha,\boldsymbol{\gamma}}$ with $d = 100$, $\alpha = 1$.

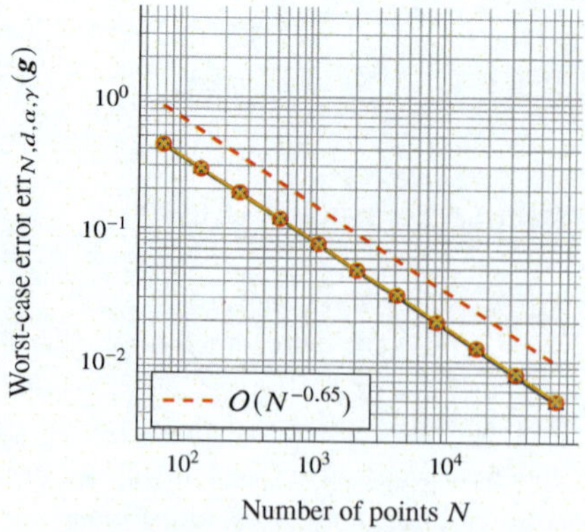

(a) Weights $\boldsymbol{\gamma} = (\gamma_j)_{j\geq 1}$ with $\gamma_j = j^{-2}$.

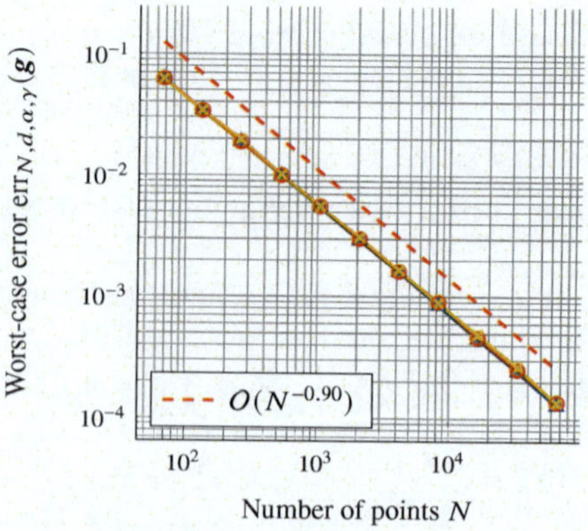

(b) Weighs $\boldsymbol{\gamma} = (\gamma_j)_{j\geq 1}$ with $\gamma_j = j^{-3}$.

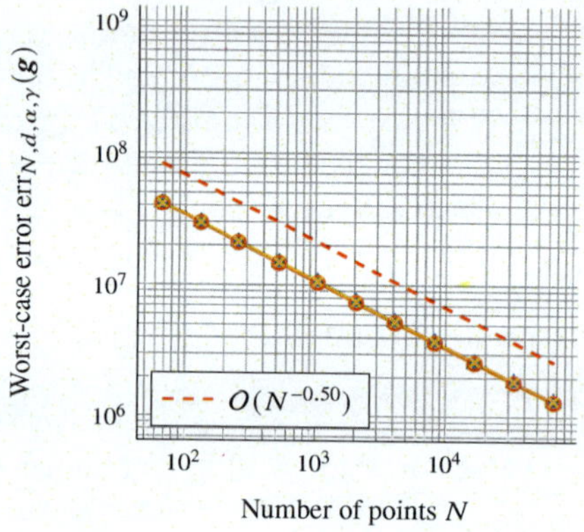

(c) Weights $\boldsymbol{\gamma} = (\gamma_j)_{j\geq 1}$ with $\gamma_j = (0.95)^j$.

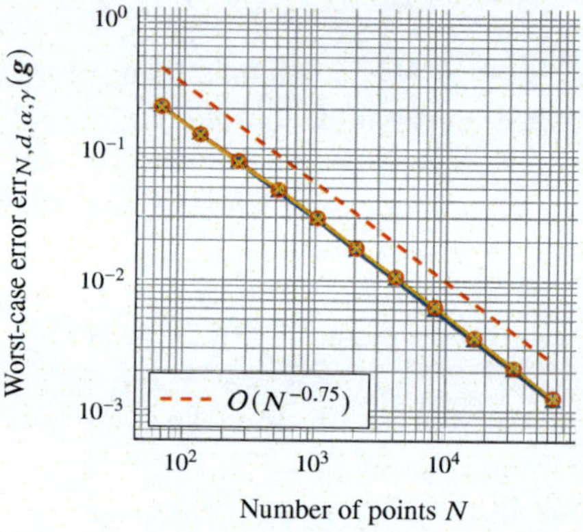

(d) Weighs $\boldsymbol{\gamma} = (\gamma_j)_{j\geq 1}$ with $\gamma_j = (0.5)^j$.

Fig. B.9: Convergence results for the worst-case error $\text{err}_{N,d,\alpha,\boldsymbol{\gamma}}(\boldsymbol{g})$ in the weighted Korobov space $\mathcal{H}_{\text{kor},d,\alpha,\boldsymbol{\gamma}}$ of smoothness $\alpha = 1$ with dimension $d = 100$. The generating vectors $\boldsymbol{g}$ have been constructed for prime N by the fast CBC algorithm and the fast SCS algorithm with fixed or random initial vectors, respectively.

Table B.4: Mean computation times (in seconds, mean over five runs) for constructing the generating vector $\boldsymbol{g}$ of a lattice rule with prime N points in d dimensions using the fast CBC construction (normal font) and the fast SCS algorithm (**bold font**), respectively.

	$d = 50$	$d = 100$	$d = 500$	$d = 1000$	$d = 2000$
$N = 257$	0.01563 **0.007508**	0.00321 **0.005391**	0.01274 **0.02105**	0.02334 **0.04033**	0.04384 **0.07741**
$N = 1031$	0.01084 **0.01205**	0.01413 **0.01723**	0.0486 **0.06401**	0.09233 **0.1222**	0.1797 **0.2375**
$N = 4099$	0.02519 **0.02904**	0.04551 **0.05216**	0.2211 **0.2558**	0.4404 **0.5095**	0.8807 **1.019**
$N = 16411$	0.07088 **0.08509**	0.129 **0.1546**	0.6324 **0.7566**	1.263 **1.51**	2.524 **3.012**
$N = 65537$	0.07908 **0.1283**	0.1304 **0.2147**	0.6284 **1.061**	1.28 **2.113**	2.548 **4.234**
$N = 262147$	0.6905 **0.8547**	1.314 **1.685**	6.811 **8.268**	13.66 **16.54**	27.13 **32.96**

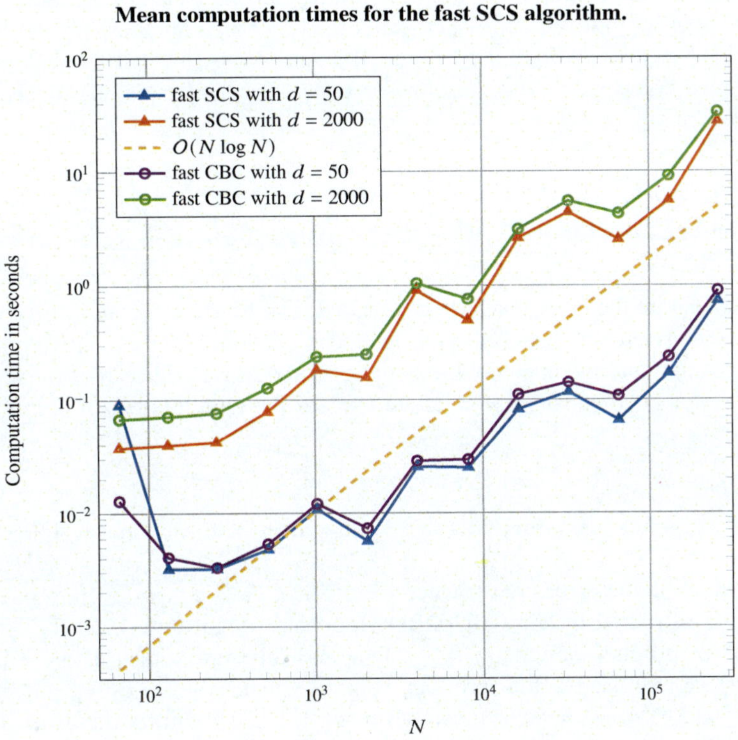

Fig. B.10: Mean computation times (in seconds, mean over five runs) for constructing the generating vector $\boldsymbol{g}$ of a lattice rule with prime N points in $d \in \{50, 2000\}$ dimensions using the fast CBC construction or the fast SCS algorithm, respectively.

The performed timings confirm the theoretical cost of $O(d\,N \log N)$ of the fast SCS algorithm (see Section 4.3). Moreover, we see that the computation times for the fast SCS algorithm are comparable to those for the fast CBC algorithm. However, we notice that the fast SCS algorithm always requires slightly more time. This can be explained by the fact that in the fast SCS algorithm we keep the final dimension d in each step, while in the CBC construction the dimension is increased gradually in each step of the algorithm. Furthermore, in the SCS construction an additional update based on the initial vector $\boldsymbol{g}^{(0)}$ is performed in each step which is not required in the CBC algorithm.

Remark B.2 We observe that the construction times displayed in Figure B.10 fluctuate stronger with respect to N than for the previously depicted timing results in Figure B.3. The reason for this difference is the fact that the fast Fourier transform (FFT) is faster when applied to a vector whose length can be factored into small primes. In our implementation of the fast CBC algorithm for $N = 2^m$ the fast Fourier transforms are always applied to vectors of length 2^k with $k \in \{1, 2, \ldots, m-2\}$. Therefore the timings are more robust with respect to different $N = 2^m$ and no fluctuations can be observed. In contrast, when using the fast CBC algorithm or the fast SCS algorithm for prime N, the fast Fourier transforms are applied to vectors of length $(N-1)/2$ (when exploiting the symmetry of the function φ_α (see Equation (3.7)). The measured computation times do then strongly depend on the prime factorization of $N-1$. In order to illustrate this, consider the two prime numbers $N_1 = 1031$ and $N_2 = 2053$. Since $(N_1-1)/2 = 515 = 5 \cdot 103$ and $(N_2-1)/2 = 1026 = 2 \cdot 3^3 \cdot 19$, the measured computation time for N_1 is actually higher than the one for the larger number N_2.

Numerical results for the CBC-DBD construction

Here we illustrate the error convergence behavior and the computation time for the CBC-DBD construction, which was introduced in Section 4.6. Due to the formulation of the algorithm and the considerations in Section 4.6, we will restrict ourselves to product weights and consider $N = 2^m, m \in \mathbb{N}$, as the total number of points of the lattice rules.

Error convergence behavior: CBC-DBD algorithm with product weights

We will examine the error convergence behavior of $\mathrm{err}_{N,d,\alpha,\boldsymbol{\gamma}^{2\alpha}}(\boldsymbol{g})$ for generating vectors $\boldsymbol{g}$ constructed by the CBC-DBD algorithm (Algorithm 4.20) for different sequences of product weights $\boldsymbol{\gamma} = (\gamma_j)_{j\geq 1}$ and different values of $N = 2^m$. In particular, we will use the four product weight sequences given by $\gamma_j = j^{-2}, \gamma_j = j^{-3}, \gamma_j = (0.5)^j$, or $\gamma_j = (0.95)^j$, we let $\alpha \in \{1, 1.5, 2\}$, and fix the dimension at $d = 100$. For comparison, we compute the worst-case error $\mathrm{err}_{N,d,\alpha,\boldsymbol{\gamma}^{2\alpha}}$ of lattice rules obtained by the fast CBC construction for the function space $\mathcal{H}_{\mathrm{kor},d,\alpha,\boldsymbol{\gamma}^{2\alpha}}$.

Error convergence in the space $\mathcal{H}_{\mathrm{kor},d,\alpha,\gamma^{2\alpha}}$ with $d = 100$, $\alpha = 1, 1.5, 2$.

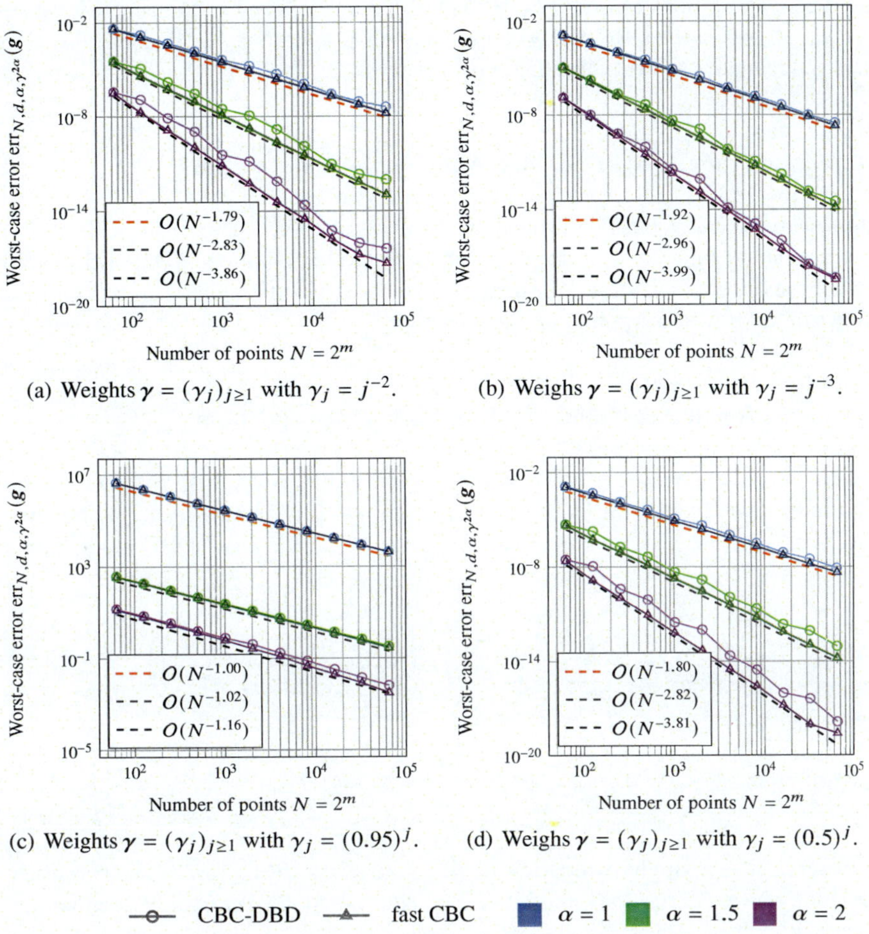

Fig. B.11: Convergence results for the worst-case error $\mathrm{err}_{N,d,\alpha,\gamma^{2\alpha}}(\boldsymbol{g})$ in the weighted Korobov space $\mathcal{H}_{\mathrm{kor},d,\alpha,\gamma^{2\alpha}}$ for smoothness parameters $\alpha \in \{1, 1.5, 2\}$ and dimension $d = 100$. The generating vectors $\boldsymbol{g}$ have been constructed via the CBC-DBD algorithm and the fast CBC algorithm (for each α considered) for $N = 2^m$, respectively.

The results shown in Figure B.11 confirm that the CBC-DBD algorithm reliably constructs good lattice rules with worst-case errors that are comparable to those of the corresponding lattice rules obtained by the fast CBC construction. We notice that the observed convergence rates are of similar order for both constructions, however, the worst-case errors of lattice rules constructed by the CBC-DBD algorithm are always slightly higher than the corresponding values for lattices constructed by the CBC algorithm. This observation is not surprising since the CBC algorithm constructs generating vectors of lattice rules that are tailored to the function space $\mathcal{H}_{\mathrm{kor},d,\alpha,\gamma^{2\alpha}}$, while the CBC-DBD construction uses a more general quality function that is independent of the smoothness parameter α.

Computation time: CBC-DBD algorithm with product weights

In this section, we discuss and illustrate the computation times for the CBC-DBD construction. At first, we will derive an efficient implementation of Algorithm 4.20 that only requires $O(d\,N\log N)$ operations to construct the generating vector of a lattice rule with $N = 2^m$ points in d dimensions.

Implementation and cost analysis of the CBC-DBD algorithm

Let $m, d \in \mathbb{N}$ and $N = 2^m$, let $\boldsymbol{\gamma} = (\gamma_j)_{j\geq 1}$ be product weights, and let x be an odd integer. We recall from (4.30) that for given $v \in [m]$ and $s \in [d]$, and positive integers $a_{1,m}, \ldots, a_{s-1,m}$, the quality function $h_{s,v,\boldsymbol{\gamma}}$ in the CBC-DBD algorithm is given by

$$h_{s,v,\boldsymbol{\gamma}}(x) = \sum_{t=v}^{m} \frac{1}{2^{t-v}} \sum_{\substack{k=1\\ k\equiv 1 \pmod 2}}^{2^t} \left(-1 + \left(1 + \gamma_s \log\left(\frac{1}{\sin^2(\pi k x/2^v)}\right)\right) \right.$$
$$\left. \times \prod_{j=1}^{s-1} \left(1 + \gamma_j \log\left(\frac{1}{\sin^2(\pi k a_{j,m}/2^t)}\right)\right)\right).$$

We observe that a single evaluation of $h_{s,v,\boldsymbol{\gamma}}$ requires $O(s\sum_{t=v}^{m} 2^{t-1})$ operations. Therefore, in a naive implementation of Algorithm 4.20, the number of calculations for each inner loop over $v = 1, 2, \ldots, m-1$ is

$$O\left(s \sum_{v=1}^{m-1} 2 \sum_{t=v+1}^{m} 2^{t-1}\right) = O\left(s \sum_{v=2}^{m}\sum_{t=v}^{m} 2^t\right)$$
$$= O\left(s\left(2^m m - 2(2^m - 1)\right)\right) = O(s\,N\log N).$$

Since there is an inner loop for each $s = 1, 2, \ldots, d-1$, the execution of a naive implementation of the CBC-DBD construction (Algorithm 4.20) requires $O(d^2 N \log N)$ operations. For large d this cost is prohibitive such that we aim for a more efficient implementation.

For $s \in [d-1]$, assume that the numbers $a_{1,m}, \ldots, a_{s,m}$ have been constructed by Algorithm 4.20. For integers $t \in \{2, 3, \ldots, m\}$ and odd $k \in \{1, 2, \ldots, 2^t - 1\}$, we introduce the term $q(s, t, k)$ as

$$q(s,t,k) = \prod_{j=1}^{s}\left(1 + \gamma_j \log\left(\frac{1}{\sin^2(\pi k a_{j,m}/2^t)}\right)\right)$$

and note that for the evaluation of $h_{s+1,v+1,\boldsymbol{\gamma}}(x)$ in Algorithm 4.20 we can compute and store $q(s,t,k)$ as it is independent of $v+1$ and x. In this way $h_{s+1,v+1,\boldsymbol{\gamma}}(x)$ can be rewritten as

$$\begin{aligned} &h_{s+1,v+1,\boldsymbol{\gamma}}(x) \\ &= \sum_{t=v+1}^{m} \frac{1}{2^{t-v-1}} \sum_{\substack{k=1 \\ k\equiv 1 \pmod 2}}^{2^t} \left(-1 + \left(1 + \gamma_{s+1} \log\left(\frac{1}{\sin^2(\pi k x/2^{v+1})}\right)\right) q(s,t,k)\right), \end{aligned} \tag{B.3}$$

where, after having determined $a_{s+1,m}$, the value of $q(s+1,t,k)$ is computed via the recurrence relation

$$q(s+1,t,k) = q(s,t,k)\left(1 + \gamma_{s+1} \log\left(\frac{1}{\sin^2(\pi k a_{s+1,m}/2^t)}\right)\right). \tag{B.4}$$

For an algorithmic realization of this finding, we introduce the auxiliary vector $\boldsymbol{p} = (p(1), \ldots, p(N-1))$ of length $N-1$ whose components, for the current $s \in [d-1]$, are given by

$$p(k\,2^{m-t}) = \prod_{j=1}^{s}\left(1 + \gamma_j \log\left(\frac{1}{\sin^2(\pi k a_{j,m}/2^t)}\right)\right)$$

for each $t \in [m]$ and each corresponding odd index $k \in \{1, 3, \ldots, 2^t - 1\}$. Note that $p(k\,2^{m-t})$ coincides with $q(s,t,k)$ for $t \in \{2, 3, \ldots, m\}$. Furthermore, we note that for the evaluation of $h_{s+1,v+1,\boldsymbol{\gamma}}$ we do not require the values of $q(s,t,k)$ (or $\boldsymbol{p}$) for $t \in \{2, 3, \ldots, v\}$. Additionally, due to the way the $a_{s,v}$ are constructed in Algorithm 4.20, we have that $a_{s,m} \equiv a_{s,v} \pmod{2^v}$ for $v \in [m]$ and thus, by the periodicity of $\sin^2(\pi x)$,

$$\sin^2\left(\pi \frac{k a_{s,m}}{2^v}\right) = \sin^2\left(\pi \frac{k a_{s,m} \pmod{2^v}}{2^v}\right) = \sin^2\left(\pi \frac{k a_{s,v}}{2^v}\right).$$

Hence, we can perform the update as in (B.4) for $k \in \{1, 3, \ldots, 2^v - 1\}$ with $a_{s,m}$ replaced by $a_{s,v}$ immediately after each $a_{s,v}$ has been determined.

These observations lead to the following fast implementation of Algorithm 4.20.

Algorithm B.3 (Fast component-by-component digit-by-digit algorithm)

Input: Integer $m \in \mathbb{N}$, dimension d, and positive weights $\boldsymbol{\gamma} = (\gamma_j)_{j\geq 1}$.

for $t = 2$ **to** m **do**

  for $k = 1$ **to** $2^t - 1$ **in steps of 2 do**

    $p(k\,2^{m-t}) = \left(1 + \gamma_1 \log\left(1/\sin^2(\pi k/2^t)\right)\right)$

  end for

end for

Set $a_{1,m} = 1$ and $a_{2,1} = \cdots = a_{d,1} = 1$.

for $s = 1$ **to** $d - 1$ **do**

  for $v = 1$ **to** $m - 1$ **do**

    $g^* = \underset{g\in\{0,1\}}{\operatorname{argmin}}\ h_{s+1,v+1,\boldsymbol{\gamma}}(a_{s+1,v} + 2^v g)$, where $h_{s+1,v+1,\boldsymbol{\gamma}}$ is evaluated using (B.3).

    $a_{s+1,v+1} = a_{s+1,v} + 2^v g^*$

    for $k = 1$ **to** $2^{v+1} - 1$ **in steps of 2 do**

      $p(k\,2^{m-1-v}) = p(k\,2^{m-1-v})\left(1 + \gamma_{s+1} \log\left(1/\sin^2(\pi k a_{s+1,v+1}/2^{v+1})\right)\right)$

    end for

  end for

end for

Set $\boldsymbol{g} = (g_1, \ldots, g_d)$ with $g_s := a_{s,m}$ for $s = 1, \ldots, d$.

Return: Generating vector $\boldsymbol{g} = (g_1, \ldots, g_d)$ for $N = 2^m$.

The computational cost of Algorithm B.3 is summarized in the following proposition.

Proposition B.4 *Let $m, d \in \mathbb{N}$ and $N = 2^m$. For a given positive sequence $\boldsymbol{\gamma} = (\gamma_j)_{j\geq 1}$ of product weights, Algorithm B.3 computes a generating vector $\boldsymbol{g} = (g_1, \ldots, g_d)$ of a rank-1 lattice rule using $O(d\,N \log N)$ operations and requiring $O(N)$ memory.*

Proof Due to the relation in (B.3), the cost of evaluating the quality function $h_{s,v,\boldsymbol{\gamma}}$ is reduced to $O(\sum_{t=v}^m 2^{t-1})$ operations. Thus, the number of calculations in the inner loop over $v = 1, 2, \ldots, m-1$ of Algorithm B.3 equals

$$O\left(\sum_{v=1}^{m-1} 2 \sum_{t=v+1}^{m} 2^{t-1}\right) = O\left(\sum_{v=2}^{m}\sum_{t=v}^{m} 2^t\right)$$
$$= O\left(2^m m - 2(2^m - 1)\right) = O\left(2^m m\right) = O\left(N \log N\right).$$

Thus, the outer loop over $s = 1, 2, \ldots, d-1$ can be carried out in $O(d\,N \log N)$ operations. Furthermore, we observe that initializing and updating the vector $\boldsymbol{p} \in \mathbb{R}^{N-1}$ can both be executed in $O(N)$ operations. To store the vector $\boldsymbol{p}$ itself, we require $O(N)$ memory. □

We note that the runtime of Algorithm B.3 can be further reduced by precomputing and storing the $N-1$ values

$$\log\left(\frac{1}{\sin^2(\pi k/N)}\right) \quad \text{for} \quad k \in [N-1].$$

Proposition B.4 reveals that the fast implementation of the CBC-DBD construction has the same order of computational complexity as the state of the art fast CBC construction (see Section 3.4). In the fast CBC construction, the speedup of the algorithm is achieved by exploiting the special (block-) circulant structure of the involved matrices and by employing a fast matrix-vector product which uses fast Fourier transforms (FFTs). We refer to Section 3.4 and the references there for details. In contrast, the fast CBC-DBD algorithm does not rely on the use of FFTs and its low complexity is the result of the smaller search space for the components g_j of $\boldsymbol{g}$ as well as the efficient implementation.

Timings for the CBC-DBD algorithm

In order to illustrate the theoretically proven cost of order $O(d\,N \log N)$, we measured the time required to construct the generating vector of a lattice rule via the fast CBC-DBD construction. As a comparison, we also measured the corresponding construction time required by the fast CBC algorithm. The timings were performed on an Intel Core i5 CPU with 2.3 GHz using Python 3.6.3.

We set $\alpha = 1$ and consider different prime powers $N = 2^m$, different dimensions d, and use the weight sequence $\boldsymbol{\gamma} = (\gamma_j)_{j\geq 1}$ with $\gamma_j = j^{-2}$. For each of the considered scenarios, we performed five independent timing runs and then calculated the mean computation times. The results of the numerical experiments are given in Table B.5 and Figure B.12.

The timings in Table B.5 confirm that both algorithms considered have a similar dependence on N and d. The linear dependence on the dimension d is well observable and in accordance with the theoretical complexity $O(d\,N \log N)$ of both algorithms. The computation times for both algorithms roughly differ by a factor between 2 and 4. We remark that the CBC algorithm, in particular the fast Fourier transforms, are based on compiled and optimized code via Python's Discrete Fourier Transform (numpy.fft) library. It is therefore noteworthy that the CBC-DBD algorithm, which is not based on any compiled libraries, is competitive nonetheless.

Table B.5: Mean computation times (in seconds, mean over five runs) for constructing the generating vector $\boldsymbol{g}$ of a lattice rule with $N = 2^m$ points in d dimensions using the fast CBC construction (normal font) and the fast CBC-DBD algorithm (**bold font**), respectively.

	$s = 50$	$s = 100$	$s = 500$	$s = 1000$	$s = 2000$
$m = 10$	0.019 **0.071**	0.037 **0.139**	0.179 **0.691**	0.36 **1.372**	0.716 **2.76**
$m = 12$	0.026 **0.107**	0.048 **0.21**	0.237 **1.059**	0.473 **2.097**	0.964 **4.233**
$m = 14$	0.043 **0.174**	0.079 **0.345**	0.37 **1.713**	0.736 **3.463**	1.478 **6.986**
$m = 16$	0.113 **0.363**	0.199 **0.719**	0.878 **3.594**	1.716 **7.167**	3.413 **14.501**
$m = 18$	0.548 **1.289**	0.966 **2.543**	4.255 **12.489**	8.414 **25.084**	16.624 **50.135**
$m = 20$	3.129 **7.803**	5.696 **15.78**	26.267 **79.511**	51.637 **158.982**	103.476 **318.403**

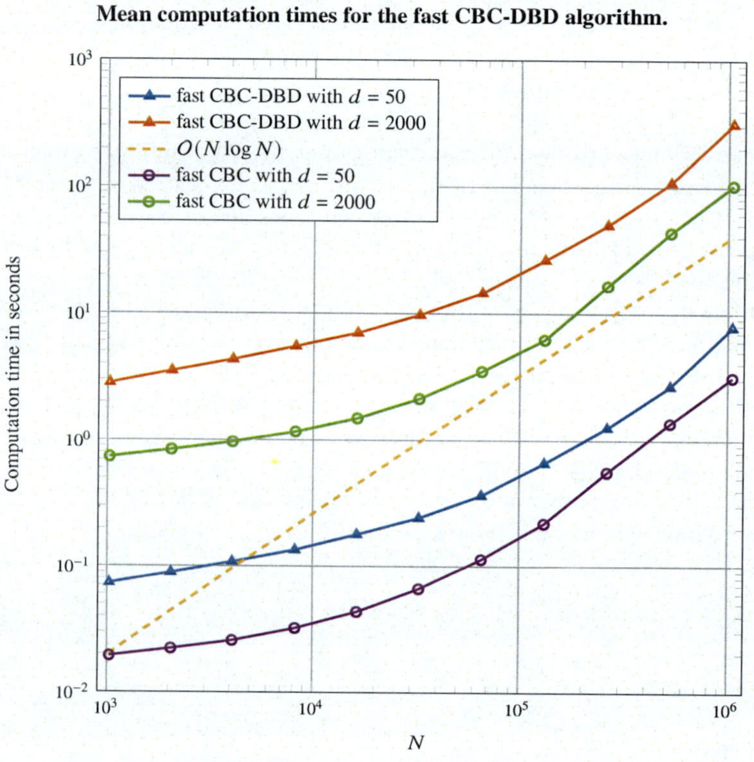

Fig. B.12: Mean computation times (in seconds, mean over five runs) for constructing the generating vector $\boldsymbol{g}$ of a lattice rule with $N = 2^m$ points in $d \in \{50, 2000\}$ dimensions using the fast CBC construction or the fast CBC-DBD algorithm, respectively.

References

1. Ch. Aistleitner. Tractability results for the weighted star-discrepancy. *J. Complexity*, 30(4):381–391, 2014. [219]
2. Ch. Aistleitner, J. S. Brauchart, and J. Dick. Point sets on the sphere $\mathbb{S}^2$ with small spherical cap discrepancy. *Discrete Comput. Geom.*, 48(4):990–1024, 2012. [211, 213, 218, 219]
3. Ch. Aistleitner, G. Larcher, F. Pillichshammer, S. Saad Eddin, and R. F. Tichy. On Weyl products and uniform distribution modulo one. *Monatsh. Math.*, 185(3):365–395, 2018. [185]
4. N. Aronszajn. Theory of reproducing kernels. *Trans. Amer. Math. Soc.*, 68:337–404, 1950. [53]
5. E. I. Atanassov. On the discrepancy of the Halton sequences. *Math. Balkanica (N.S.)*, 18(1–2):15–32, 2004. [52]
6. N. S. Bakhvalov. On the approximate computation of multiple integrals. *Vestnik Moskov. Univ. Ser. Mat. Meh. Astr. Fiz.*, 4:3–18, 1959. (In Russian). [52, 65, 138, 393]
7. N. S. Bakhvalov. Estimates in the mean of the remainder term of quadratic formulas. *Ž. Vyčisl. Mat i Mat. Fiz.*, 1:64–77, 1961. (In Russian). [377, 378, 393]
8. N. S. Bakhvalov. The optimality of linear operator approximation methods on convex function classes. *Ž. Vyčisl. Mat i Mat. Fiz.*, 11:1014–1018, 1971. (In Russian). [36]
9. N. S. Bakhvalov. On the approximate computation of multiple integrals. *J. Complexity*, 31(4):502–516, 2015. English translation of Russian original. [65, 138, 393]
10. J. Beck. Sums of distances between points on a sphere—an application of the theory of irregularities of distribution to discrete geometry. *Mathematika*, 31(1):33–41, 1984. [219]
11. J. Beck. On the discrepancy of convex plane sets. *Monatsh. Math.*, 105(2):91–106, 1988. [213]
12. J. Beck and W. W. L. Chen. *Irregularities of Distribution*. Cambridge University Press, Cambridge, 1987. [53]
13. R. Bellman. *Dynamic Programming*. Princeton University Press, Princeton, NJ, 1957. [38]
14. D. Bilyk, M. T. Lacey, and A. Vagharshakyan. On the small ball inequality in all dimensions. *J. Funct. Anal.*, 254(9):2470–2502, 2008. [29]
15. D. Bilyk, V. N. Temlyakov, and R. Yu. Fibonacci sets and symmetrization in discrepancy theory. *J. Complexity*, 28(1):18–36, 2012. [54, 317]
16. D. Bilyk, V. N. Temlyakov, and R. Yu. The L_2 discrepancy of two-dimensional lattices. In D. Bilyk, L. De Carli, A. Petukhov, A. M. Stokolos, and B. D. Wick, editors, *Recent Advances in Harmonic Analysis and Applications*, volume 25 of *Springer Proc. Math. Stat.*, pages 63–77. Springer, New York, NY, 2013. [54]
17. J. Bourgain and A. Kontorovich. On Zaremba's conjecture. *Ann. of Math. (2)*, 180(1):137–196, 2014. [45]
18. H. Brass and K. Petras. *Quadrature Theory*, volume 178 of *Mathematical Surveys and Monographs*. American Mathematical Society, Providence, RI, 2011. [349]

J. Dick et al., *Lattice Rules*, Springer Series in Computational Mathematics 58, https://doi.org/10.1007/978-3-031-09951-9

19. S. Breneis and A. Hinrichs. Fibonacci lattices have minimal dispersion on the two-dimensional torus. In D. Bilyk, J. Dick, and F. Pillichshammer, editors, *Discrepancy Theory*, volume 26 of *Radon Ser. Comput. Appl. Math.*, pages 117–132. De Gruyter, Berlin, 2020. [54]
20. Y. Bugeaud. *Approximation by Algebraic Numbers*, volume 160 of *Cambridge Tracts in Mathematics*. Cambridge University Press, Cambridge, 2004. [48]
21. M. D. Buhmann. *Radial Basis Functions: Theory and Implementations*. Cambridge University Press, Cambridge, 2003. [487]
22. V. A. Bykovskii. The discrepancy of the Korobov lattice points. *Izv. Math.*, 76(3):446–465, 2012. [201]
23. G. Byrenheid, D. Dũng, W. Sickel, and T. Ullrich. Sampling on energy-norm based sparse grids for the optimal recovery of Sobolev type functions in H^γ. *J. Approx. Theory*, 207:207–231, 2016. [508]
24. G. Byrenheid, L. Kämmerer, T. Ullrich, and T. Volkmer. Tight error bounds for rank-1 lattice sampling in spaces of hybrid mixed smoothness. *Numer. Math.*, 136(4):993–1034, 2017. [427, 455]
25. A. Cohen, R. DeVore, and Ch. Schwab. Convergence rates of best N-term Galerkin approximations for a class of elliptic sPDEs. *Found. Comput. Math.*, 10(6), 2010. [532]
26. J. W. Cooley and J. W. Tukey. An algorithm for the machine calculation of complex Fourier series. *Math. Comp.*, 19:297–301, 1965. [122]
27. R. Cools, F. Y. Kuo, and D. Nuyens. Constructing embedded lattice rules for multivariate integration. *SIAM J. Sci. Comput.*, 28(6):2162–2188, 2006. [52, 238, 239, 241, 243, 245, 263, 264]
28. R. Cools, F. Y. Kuo, D. Nuyens, and I. H. Sloan. Lattice algorithms for multivariate approximation in periodic spaces with general weight parameters. *Contemp. Math.*, 754:93–113, 2020. [437, 443, 456]
29. R. Cools, F. Y. Kuo, D. Nuyens, and I. H. Sloan. Fast component-by-component construction of lattice algorithms for multivariate approximation with POD and SPOD weights. *Math. Comp.*, 90(328):787–812, 2021. [456]
30. R. Cools, F. Y. Kuo, D. Nuyens, and G. I. Suryanarayana. Tent-transformed lattice rules for integration and approximation of multivariate non-periodic functions. *J. Complexity*, 36:166–181, 2016. [304, 317]
31. R. Cools and D. Nuyens. The role of structured matrices for the construction of integration lattices. *JNAIAM J. Numer. Anal. Ind. Appl. Math.*, 1(3):257–272, 2006. [138]
32. R. Cools and D. Nuyens. A Belgian view on lattice rules. In A. Keller, S. Heinrich, and H. Niederreiter, editors, *Monte Carlo and Quasi-Monte Carlo Methods 2006*, pages 3–21. Springer, Berlin, 2008. [54, 138]
33. D. Dũng, V. Temlyakov, and T. Ullrich. *Hyperbolic Cross Approximation*. Advanced Courses in Mathematics—CRM Barcelona. Birkhäuser/Springer, Cham, 2018. [16]
34. H. Davenport. Note on irregularities of distribution. *Mathematika*, 3:131–135, 1956. [317]
35. J. Dick. On the convergence rate of the component-by-component construction of good lattice rules. *J. Complexity*, 20(4):493–522, 2004. [138]
36. J. Dick. Random weights, robust lattice rules and the geometry of the cbc$r$c algorithm. *Numer. Math.*, 122(3):443–467, 2012. [408]
37. J. Dick. Numerical integration of Hölder continuous, absolutely convergent Fourier, Fourier cosine, and Walsh series. *J. Approx. Theory*, 183:14–30, 2014. [363, 366, 372, 373, 375]
38. J. Dick and T. Goda. Stability of lattice rules and polynomial lattice rules constructed by the component-by-component algorithm. *J. Comput. Appl. Math.*, 382:Paper No. 113062, 16 pp., 2021. [407]
39. J. Dick, T. Goda, and K. Suzuki. Component-by-component construction of randomized rank-1 lattice rules achieving almost the optimal randomized error rate. Submitted for publication, arXiv:2109.11694, 2021. [378, 383, 393]
40. J. Dick, D. Gomez-Perez, F. Pillichshammer, and A. Winterhof. Digital inversive vectors can achieve polynomial tractability for the weighted star discrepancy and for multivariate integration. *Proc. Amer. Math. Soc.*, 145(8):3297–3310, 2017. [219, 375]

41. J. Dick, A. Hinrichs, and F. Pillichshammer. Proof techniques in quasi–Monte Carlo theory. *J. Complexity*, 31(3):327–371, 2015. [53]
42. J. Dick, Ch. Irrgeher, G. Leobacher, and F. Pillichshammer. On the optimal order of integration in Hermite spaces with finite smoothness. *SIAM J. Numer. Anal.*, 56(2):684–707, 2018. [337]
43. J. Dick and P. Kritzer. On a projection-corrected component-by-component construction. *J. Complexity*, 32(1):74–80, 2016. [170, 193]
44. J. Dick, P. Kritzer, F. Y. Kuo, and I. H. Sloan. Lattice-Nyström method for Fredholm integral equations of the second kind with convolution type kernels. *J. Complexity*, 23(4–6):752–772, 2007. [417, 418]
45. J. Dick, P. Kritzer, G. Leobacher, and F. Pillichshammer. A reduced fast component-by-component construction of lattice points for integration in weighted spaces with fast decreasing weights. *J. Comput. Appl. Math.*, 276:1–15, 2015. [143, 192]
46. J. Dick, P. Kritzer, F. Pillichshammer, and H. Woźniakowski. Approximation of analytic functions in Korobov spaces. *J. Complexity*, 30(2):2–28, 2014. [361, 456]
47. J. Dick, F. Y. Kuo, Q. T. Le Gia, and Ch. Schwab. Fast QMC matrix-vector multiplication. *SIAM J. Sci. Comput.*, 37(3):A1436–A1450, 2015. [519, 521]
48. J. Dick, F. Y. Kuo, and I. H. Sloan. High-dimensional integration: the quasi-Monte Carlo way. *Acta Numer.*, 22:133–288, 2013. [138, 139, 316]
49. J. Dick, G. Larcher, F. Pillichshammer, and H. Woźniakowski. Exponential convergence and tractability of multivariate integration for Korobov spaces. *Math. Comp.*, 80(274):905–930, 2011. [47, 54, 345, 361]
50. J. Dick, D. Nuyens, and F. Pillichshammer. Lattice rules for nonperiodic smooth integrands. *Numer. Math.*, 126(2):259–291, 2014. [304, 307, 315, 317]
51. J. Dick and F. Pillichshammer. A note on the figure of merit of 2-dimensional rank 2 lattice rules. *Integers*, 5(3):A5, 10, 2005. [53]
52. J. Dick and F. Pillichshammer. *Digital Nets and Sequences: Discrepancy Theory and Quasi-Monte Carlo Integration*. Cambridge University Press, Cambridge, 2010. [52, 53, 219, 222, 283, 317]
53. J. Dick and F. Pillichshammer. Discrepancy theory and quasi-Monte Carlo integration. In W. W. L. Chen, A. Srivastav, and G. Travaglini, editors, *A Panorama of Discrepancy Theory*, volume 2107 of *Lecture Notes in Math.*, pages 539–619. Springer, Cham, 2014. [53, 317]
54. J. Dick and F. Pillichshammer. The inverse of the star-discrepancy problem and the generation of pseudo-random numbers. In K.-U. Schmidt and A. Winterhof, editors, *Sequences and Their Applications—SETA 2014*, volume 8865 of *Lecture Notes in Comput. Sci.*, pages 173–184. Springer, Cham, 2014. [375]
55. J. Dick and F. Pillichshammer. The weighted star discrepancy of Korobov's p-sets. *Proc. Amer. Math. Soc.*, 143(12):5043–5057, 2015. [219, 365]
56. J. Dick and F. Pillichshammer. Weighted integration over a hyperrectangle based on digital nets and sequences. *J. Comput. Appl. Math.*, 393:Paper No. 113509, 25 pp., 2021. [337]
57. J. Dick, F. Pillichshammer, K. Suzuki, M. Ullrich, and T. Yoshiki. Lattice-based integration algorithms: Kronecker sequences and rank-1 lattices. *Ann. Mat. Pura Appl. (4)*, 197(1):109–126, 2018. [54]
58. J. Dick, F. Pillichshammer, and B. J. Waterhouse. The construction of good extensible rank-1 lattices. *Math. Comp.*, 77(264):2345–2373, 2008. [52, 245, 251, 252, 253, 263, 264]
59. J. Dick, I. H. Sloan, X. Wang, and H. Woźniakowski. Good lattice rules in weighted Korobov spaces with general weights. *Numer. Math.*, 103(1):63–97, 2006. [81]
60. S. Disney and I. H. Sloan. Error bounds for the method of good lattice points. *Math. Comp.*, 56(193):257–266, 1991. [18]
61. S. Disney and I. H. Sloan. Lattice integration rules of maximal rank formed by copying rank 1 rules. *SIAM J. Numer. Anal.*, 29(2):566–577, 1992. [52]
62. B. Doerr and M. Gnewuch. Construction of low-discrepancy point sets of small size by bracketing covers and dependent randomized rounding. In A. Keller, S. Heinrich, and H. Niederreiter, editors, *Monte Carlo and Quasi-Monte Carlo Methods 2006*, pages 299–312. Springer, Berlin, 2007. [53]

63. B. Doerr, M. Gnewuch, P. Kritzer, and F. Pillichshammer. Component-by-component construction of low-discrepancy point sets of small size. *Monte Carlo Methods Appl.*, 14(2):129–149, 2008. [53]
64. B. Doerr, M. Gnewuch, and A. Srivastav. Bounds and constructions for the star discrepancy via δ-covers. *J. Complexity*, 21(5):691–709, 2005. [53]
65. B. Doerr, M. Gnewuch, and M. Wahlström. Algorithmic construction of low-discrepancy point sets via dependent randomized rounding. *J. Complexity*, 26(5):490–507, 2010. [53]
66. M. Drmota and R. F. Tichy. *Sequences, Discrepancies and Applications*. Springer, Berlin, 1997. [53, 195, 211]
67. A. Ebert and P. Kritzer. Constructing lattice points for numerical integration by a reduced fast successive coordinate search algorithm. *J. Comput. Appl. Math.*, 351:77–100, 2019. [160]
68. A. Ebert, P. Kritzer, and D. Nuyens. Constructing QMC finite element methods for elliptic PDEs with random coefficients by a reduced CBC construction. In B. Tuffin and P. L'Ecuyer, editors, *Monte Carlo and Quasi-Monte Carlo Methods*, pages 183–205. Springer, Cham, 2020. [158, 192]
69. A. Ebert, P. Kritzer, D. Nuyens, and O. Osisiogu. Digit-by-digit and component-by-component constructions of lattice rules for periodic functions with unknown smoothness. *J. Complexity*, 66:Paper No. 101555, 37 pp., 2021. [135, 139, 174, 193]
70. A. Ebert, P. Kritzer, and F. Pillichshammer. Tractability of approximation in the weighted Korobov space in the worst-case setting. arXiv:22201.09940. To appear in Z. Botev, A. Keller, Ch. Lemieux, and B. Tuffin, editors, *Advances in Modeling and Simulation*, Springer Nature., 2022. [487]
71. A. Ebert, H. Leövey, and D. Nuyens. Successive coordinate search and component-by-component construction of rank-1 lattice rules. In A. B. Owen and P. W. Glynn, editors, *Monte Carlo and Quasi-Monte Carlo Methods 2016*, pages 197–215. Springer, Cham, 2018. [158, 159, 160, 192]
72. A. Ebert and F. Pillichshammer. Tractability of approximation in the weighted Korobov space in the worst-case setting—a complete picture. *J. Complexity*, 67:Paper No. 101571, 15 pp., 2021. [93, 435, 456]
73. G. E. Fasshauer. *Meshfree Approximation Methods with MATLAB*. World Scientific, Singapore, 2007. [474, 487]
74. H. Faure, P. Kritzer, and F. Pillichshammer. From van der Corput to modern constructions of sequences for quasi-Monte Carlo rules. *Indag. Math. (N.S.)*, 26(5):760–822, 2015. [52, 222]
75. K. K. Frolov. Upper bounds for the errors of quadrature formulae on classes of functions. *Dokl. Akad. Nauk SSSR*, 231(4):818–821, 1976. (In Russian). [393]
76. S. D. Galbraith. *Mathematics of Public Key Cryptography*. Cambridge University Press, Cambridge, 2012. [215]
77. R. Gantner and Ch. Schwab. Computational higher order quasi-Monte Carlo integration. In R. Cools and D. Nuyens, editors, *Monte Carlo and Quasi-Monte Carlo Methods 2014*, pages 271–288. Springer, Cham, 2016. [169, 173, 193]
78. P. Glasserman. *Monte Carlo Methods in Financial Engineering*, volume 53 of *Stochastic Modelling and Applied Probability*. Springer-Verlag, New York, NY, 2003. [3, 52, 337]
79. M. Gnewuch and N. Hebbinghaus. Discrepancy bounds for a class of negatively dependent random points including Latin hypercube samples. *Ann. Appl. Probab.*, 31(4):1944–1965, 2021. [39]
80. M. Gnewuch, M. Hefter, A. Hinrichs, and K. Ritter. Embeddings of weighted Hilbert spaces and applications to multivariate and infinite-dimensional integration. *J. Approx. Theory*, 222:8–39, 2017. [317]
81. M. Gnewuch, M. Hefter, A. Hinrichs, K. Ritter, and G. W. Wasilkowski. Equivalence of weighted anchored and ANOVA spaces of functions with mixed smoothness of order one in L_p. *J. Complexity*, 40:78–99, 2017. [317]
82. M. Gnewuch, M. Hefter, A. Hinrichs, K. Ritter, and G. W. Wasilkowski. Embeddings for infinite-dimensional integration and L_2-approximation with increasing smoothness. *J. Complexity*, 54:Paper No. 101406, 32 pp., 2019. [317]

83. M. Gnewuch, M. Wahlström, and C. Winzen. A new randomized algorithm to approximate the star discrepancy based on threshold accepting. *SIAM J. Numer. Anal.*, 50(2):781–807, 2012. [53]
84. T. Goda. The b-adic symmetrization of digital nets for quasi-Monte Carlo integration. *Unif. Distrib. Theory*, 12(1):1–25, 2017. [317]
85. T. Goda and P. L'Ecuyer. Construction-free median quasi-Monte Carlo rules for function spaces with unspecified smoothness and general weights. Submitted for publication, arXiv:2201.09413, 2022. [394]
86. T. Goda, K. Suzuki, and T. Yoshiki. Lattice rules in non-periodic subspaces of Sobolev spaces. *Numer. Math.*, 141(2):399–427, 2019. [307, 317]
87. P. J. Grabner, P. Hellekalek, and P. Liardet. The dynamical point of view of low-discrepancy sequences. *Unif. Distrib. Theory*, 7(1):11–70, 2012. [224]
88. I. G. Graham, F. Y. Kuo, J. A. Nichols, R. Scheichl, Ch. Schwab, and I. H. Sloan. Quasi-Monte Carlo finite element methods for elliptic PDEs with lognormal random coefficients. *Numer. Math.*, 131(2):329–368, 2015. [533, 539]
89. S. Haber. Experiments on optimal coefficients. In S. K. Zaremba, editor, *Applications of Number Theory to Numerical Analysis (Proc. Sympos., Univ. Montréal, Montreal, Que., 1971)*, pages 11–37. Academic Press, New York, NY, 1972. [138]
90. S. Haber. Parameters for integrating periodic functions of several variables. *Math. Comp.*, 41(163):115–129, 1983. [138]
91. G. Halász. On Roth's method in the theory of irregularities of point distributions. In H. Halberstam and C. Hooley, editors, *Recent Progress in Analytic Number Theory, Vol. 2 (Durham, 1979)*, pages 79–94. Academic Press, London-New York, 1981. [29]
92. J. H. Halton. On the efficiency of certain quasi-random sequences of points in evaluating multi-dimensional integrals. *Numer. Math.*, 2:84–90, 1960. [52]
93. G. H. Hardy and E. M. Wright. *An Introduction to the Theory of Numbers*. The Clarendon Press, Oxford University Press, New York, NY, fifth edition, 1979. [47, 96]
94. M. Hefter and K. Ritter. On embeddings of weighted tensor product Hilbert spaces. *J. Complexity*, 31(3):405–423, 2015. [317]
95. M. Hefter, K. Ritter, and G. W. Wasilkowski. On equivalence of weighted anchored and ANOVA spaces of functions with mixed smoothness of order one in L_1 or L_∞. *J. Complexity*, 32(1):1–19, 2016. [296, 317]
96. S. Heinrich, E. Novak, G. W. Wasilikowski, and H. Woźniakowski. The inverse of the star-discrepancy depends linearly on the dimension. *Acta Arith.*, 96(3):279–302, 2001. [39]
97. P. Hellekalek. On the assessment of random and quasi-random point sets. In P. Hellekalek and G. Larcher, editors, *Random and Quasi-Random Point Sets*, volume 138 of *Lecture Notes in Statist.*, pages 49–108. Springer, New York, NY, 1998. [54]
98. F. J. Hickernell. Quadrature error bounds with applications to lattice rules. *SIAM J. Numer. Anal.*, 33(5):1995–2016, 1996. [53]
99. F. J. Hickernell. A generalized discrepancy and quadrature error bound. *Math. Comp.*, 67(221):299–322, 1998. [53]
100. F. J. Hickernell. Lattice rules: how well do they measure up? In P. Hellekalek and G. Larcher, editors, *Random and Quasi-Random Point Sets*, volume 138 of *Lecture Notes in Statist.*, pages 109–166. Springer, New York, NY, 1998. [268]
101. F. J. Hickernell. Obtaining $O(N^{-2+\epsilon})$ convergence for lattice quadrature rules. In K.-T. Fang, H. Niederreiter, and F. J. Hickernell, editors, *Monte Carlo and Quasi-Monte Carlo Methods 2000*, pages 274–289. Springer, Berlin, 2002. [301, 302, 304, 314]
102. F. J. Hickernell and H. S. Hong. Computing multivariate normal probabilities using rank-1 lattice sequences. In G. H. Golub, L. Shui-Hong, T. L. Franklin, and R. J. Plemmons, editors, *Scientific Computing (Proceedings of the Workshop, 10–12 March 1997, Hong Kong)*, pages 209–215. Springer, Singapore, 1997. [221, 223]
103. F. J. Hickernell, H. S. Hong, P. L'Ecuyer, and Ch. Lemieux. Extensible lattice sequences for quasi-Monte Carlo quadrature. *SIAM J. Sci. Comput.*, 22(3):1117–1138, 2000. [52, 222, 223, 238, 245]

104. F. J. Hickernell, P. Kritzer, F. Y. Kuo, and D. Nuyens. Weighted compound integration rules with higher order convergence for all N. *Numer. Algorithms*, 59(2):161–183, 2012. [52, 235, 237, 263]
105. F. J. Hickernell and H. Niederreiter. The existence of good extensible rank-1 lattices. *J. Complexity*, 19(3):286–300, 2003. [52, 135, 139, 223, 225, 226, 230, 232, 245, 263]
106. F. J. Hickernell and H. Woźniakowski. Tractability of multivariate integration for periodic functions. *J. Complexity*, 17(4):660–682, 2001. [93]
107. A. Hinrichs. Covering numbers, Vapnik-Červonenkis classes and bounds for the star-discrepancy. *J. Complexity*, 20(4):477–483, 2004. [39]
108. A. Hinrichs. Optimal importance sampling for the approximation of integrals. *J. Complexity*, 26(2):125–134, 2010. [22]
109. A. Hinrichs, P. Kritzer, F. Pillichshammer, and G. W. Wasilkowski. Truncation dimension for linear problems on multivariate function spaces. *Numer. Algorithms*, 80(2):661–685, 2019. [317]
110. A. Hinrichs and J. Oettershagen. Optimal point sets for quasi-Monte Carlo integration of bivariate periodic functions with bounded mixed derivatives. In R. Cools and D. Nuyens, editors, *Monte Carlo and Quasi-Monte Carlo Methods*, pages 385–405. Springer, Cham, 2016. [54]
111. A. Hinrichs, F. Pillichshammer, and W. Ch. Schmid. Tractability properties of the weighted star discrepancy. *J. Complexity*, 24(2):134–143, 2008. [219]
112. E. Hlawka. Funktionen von beschränkter Variation in der Theorie der Gleichverteilung. *Ann. Mat. Pura Appl. (4)*, 54:325–333, 1961. (In German). [29]
113. E. Hlawka. Zur angenäherten Berechnung mehrfacher Integrale. *Monatsh. Math.*, 66:140–151, 1962. (In German). [18, 52]
114. W. Hörmann, J. Leydold, and G. Derflinger. *Automatic Nonuniform Random Variate Generation*. Statistics and Computing. Springer-Verlag, Berlin, 2004. [516]
115. L. K. Hua and Y. Wang. *Applications of Number Theory to Numerical Analysis*. Springer, Berlin, 1981. [52, 316, 364, 365, 366, 375, 455]
116. Ch. Irrgeher, P. Kritzer, G. Leobacher, and F. Pillichshammer. Integration in Hermite spaces of analytic functions. *J. Complexity*, 31(3):380–404, 2015. [337, 361]
117. Ch. Irrgeher, P. Kritzer, and F. Pillichshammer. Integration and approximation in cosine spaces of smooth functions. *Math. Comput. Simulation*, 143:35–45, 2018. [361]
118. Ch. Irrgeher, P. Kritzer, F. Pillichshammer, and H. Woźniakowski. Approximation in Hermite spaces of smooth functions. *J. Approx. Theory*, 207:98–126, 2016. [337, 361]
119. Ch. Irrgeher, P. Kritzer, F. Pillichshammer, and H. Woźniakowski. Tractability of multivariate approximation defined over Hilbert spaces with exponential weights. *J. Approx. Theory*, 207:301–338, 2016. [337, 353, 361, 456]
120. S. Joe. Component by component construction of rank-1 lattice rules having $O(n^{-1}(\ln(n))^d)$ star discrepancy. In H. Niederreiter, editor, *Monte Carlo and Quasi-Monte Carlo Methods 2002*, pages 293–298. Springer, Berlin, 2004. [203, 219]
121. S. Joe. Construction of good rank-1 lattice rules based on the weighted star discrepancy. In H. Niederreiter and D. Talay, editors, *Monte Carlo and Quasi-Monte Carlo Methods 2004*, pages 181–196. Springer, Berlin, 2006. [219]
122. S. Joe and I. H. Sloan. Imbedded lattice rules for multidimensional integration. *SIAM J. Numer. Anal.*, 29(4):1119–1135, 1992. [12, 52]
123. S. Joe and I. H. Sloan. On computing the lattice rule criterion R. *Math. Comp.*, 59(200):557–568, 1992. [54]
124. S. Joe and I. H. Sloan. Implementation of a lattice method for numerical multiple integration. *ACM Trans. Math. Softw.*, 19:523–545, 1993. [52]
125. L. Kämmerer. Reconstructing hyperbolic cross trigonometric polynomials by sampling along multiple rank-1 lattices. *SIAM J. Numer. Anal.*, 51(5):2773–2796, 2013. [455]
126. L. Kämmerer. Reconstructing multivariate trigonometric polynomials from samples along multiple rank-1 lattices. In G. E. Fasshauer and L. L. Schumaker, editors, *Approximation Theory XIV: San Antonio 2013*, pages 255–271, Berlin, 2014. Springer. [455]

127. L. Kämmerer. Multiple rank-1 lattices as sampling schemes for multivariate trigonometric polynomials. *J. Fourier Anal. Appl.*, 24(1):17–44, 2018. [508]
128. L. Kämmerer. Constructing spatial discretizations for sparse multivariate trigonometric polynomials that allow for a fast discrete Fourier transform. *Appl. Comput. Harmon. Anal.*, 47(3):702–729, 2019. [489, 491, 494, 508]
129. L. Kämmerer, D. Potts, and T. Volkmer. Approximation of multivariate periodic functions by trigonometric polynomials based on rank-1 lattice sampling. *J. Complexity*, 31(4):543–576, 2015. [455]
130. L. Kämmerer, D. Potts, and T. Volkmer. Approximation of multivariate periodic functions by trigonometric polynomials based on sampling along rank-1 lattice with generating vector of Korobov form. *J. Complexity*, 31(3):424–456, 2015. [455]
131. L. Kämmerer and T. Volkmer. Approximation of multivariate periodic functions based on sampling along multiple rank-1 lattices. *J. Approx. Theory*, 246:1–27, 2019. [489, 494, 496, 508]
132. A. Keller. Myths of computer graphics. In H. Niederreiter and D. Talay, editors, *Monte Carlo and Quasi-Monte Carlo Methods 2004*, pages 217–243. Springer, Berlin, 2006. [52]
133. A. Keller. Quasi-Monte Carlo image synthesis in a nutshell. In J. Dick, F. Y Kuo, G. W. Peters, and I. H. Sloan, editors, *Monte Carlo and Quasi-Monte Carlo Methods 2012*, pages 213–249. Springer, Berlin, Heidelberg, 2013. [52]
134. J. F. Koksma. A general theorem from the theory of uniform distribution modulo 1. *Mathematica, Zutphen. B.*, 11:7–11, 1942. (In Dutch). [29]
135. J. F. Koksma. Some integrals in the theory of uniform distribution modulo 1. *Mathematica, Zutphen. B.*, 11:49–52, 1942. (In Dutch). [316]
136. H. König and M. Rudelson. On the volume of non-central sections of a cube. *Adv. Math.*, 360:Paper No. 106929, 30 pp., 2020. [215]
137. N. M. Korobov. Approximate calculation of repeated integrals by number-theoretical methods. *Dokl. Akad. Nauk SSSR (N.S.)*, 115:1062–1065, 1957. (In Russian). [364, 375]
138. N. M. Korobov. Approximate evaluation of repeated integrals. *Dokl. Akad. Nauk SSSR*, 124:1207–1210, 1959. (In Russian). [52]
139. N. M. Korobov. Properties and calculation of optimal coefficients. *Dokl. Akad. Nauk SSSR*, 132:1009–1012, 1960. (In Russian). [45, 96, 138]
140. N. M. Korobov. *Number-Theoretic Methods in Approximate Analysis*. Gosudarstv. Izdat. Fiz.-Mat. Lit., Moscow, 1963. (In Russian). [15, 52, 100, 174, 185, 364, 375, 455]
141. N. M. Korobov. On the calculation of optimal coefficients. *Dokl. Akad. Nauk SSSR*, 267(2):289–292, 1982. (In Russian). [174, 264]
142. D. Krieg and E. Novak. A universal algorithm for multivariate integration. *Found. Comput. Math.*, 17(4):895–916, 2017. [393]
143. P. Kritzer, F. Y. Kuo, D. Nuyens, and M. Ullrich. Lattice rules with random n achieve nearly the optimal $O(n^{-\alpha-1/2})$ error independently of the dimension. *J. Approx. Theory*, 240:96–113, 2019. [92, 378, 393]
144. P. Kritzer, F. Pillichshammer, L. Plaskota, and G. W. Wasilkowski. On efficient weighted integration via a change of variables. *Numer. Math.*, 146(3):545–570, 2020. [337]
145. P. Kritzer, F. Pillichshammer, and G. W. Wasilkowski. Very low truncation dimension for high dimensional integration under modest error demand. *J. Complexity*, 35:63–85, 2016. [297, 317]
146. P. Kritzer, F. Pillichshammer, and G. W. Wasilkowski. A note on equivalence of anchored and ANOVA spaces; lower bounds. *J. Complexity*, 38:31–38, 2017. [317]
147. P. Kritzer, F. Pillichshammer, and G. W. Wasilkowski. On quasi-Monte Carlo methods in weighted ANOVA spaces. *Math. Comp.*, 90(329):1381–1406, 2021. [299]
148. P. Kritzer, F. Pillichshammer, and H. Woźniakowski. Multivariate integration of infinitely many times differentiable functions in weighted Korobov spaces. *Math. Comp.*, 83(287):1189–1206, 2014. [345, 361]
149. P. Kritzer, F. Pillichshammer, and H. Woźniakowski. Tractability of multivariate analytic problems. In P. Kritzer, H. Niederreiter, F. Pillichshammer, and A. Winterhof, editors, *Uniform Distribution and Quasi-Monte Carlo Methods*, volume 15 of *Radon Ser. Comput. Appl. Math.*, pages 147–170. De Gruyter, Berlin, 2014. [361, 456]

150. P. Kritzer, F. Pillichshammer, and H. Woźniakowski. $\mathbb{L}_\infty$-approximation in Korobov spaces with exponential weights. *J. Complexity*, 41:102–125, 2017. [361, 488]
151. R. Kritzinger. L_p- and $S_{p,q}^r B$-discrepancy of the symmetrized van der Corput sequence and modified Hammersley point sets in arbitrary bases. *J. Complexity*, 33:145–168, 2016. [317]
152. R. Kritzinger and L. M. Kritzinger. L_2 discrepancy of symmetrized generalized Hammersley point sets in base b. *J. Number Theory*, 166:250–275, 2016. [317]
153. R. Kritzinger and F. Pillichshammer. L_p-discrepancy of the symmetrized van der Corput sequence. *Arch. Math. (Basel)*, 104(5):407–418, 2015. [317]
154. R. Kritzinger and J. Wiart. Improved dispersion bounds for modified Fibonacci lattices. *J. Complexity*, 63:Paper No. 101522, 14 pp., 2021. [54]
155. L. Kuipers and H. Niederreiter. *Uniform Distribution of Sequences*. John Wiley, New York, NY, 1974. Reprint, Dover Publications, Mineola, NY, 2006. [29, 33, 53, 195, 211]
156. F. Y. Kuo. Component-by-component constructions achieve the optimal rate of convergence for multivariate integration in weighted Korobov and Sobolev spaces. *J. Complexity*, 19(3):301–320, 2003. [103, 138]
157. F. Y. Kuo, W. T. M. Dunsmuir, I. H. Sloan, M. P. Wand, and R. S. Womersley. Quasi-Monte Carlo for highly structured generalised response models. *Methodol. Comput. Appl. Probab.*, 10(2):239–275, 2008. [337]
158. F. Y. Kuo and S. Joe. Component-by-component construction of good lattice rules with a composite number of points. *J. Complexity*, 18(4):943–976, 2002. [247]
159. F. Y. Kuo and S. Joe. Component-by-component construction of good intermediate-rank lattice rules. *SIAM J. Numer. Anal.*, 41(4):1465–1486, 2003. [52]
160. F. Y. Kuo, Ch. Schwab, and I. H. Sloan. Quasi-Monte Carlo finite element methods for a class of elliptic partial differential equations with random coefficients. *SIAM J. Numer. Anal.*, 50(6):3351–3374, 2012. [139, 174, 531, 532]
161. F. Y. Kuo, I. H. Sloan, G. W. Wasilkowski, and B. J. Waterhouse. Randomly shifted lattice rules with the optimal rate of convergence for unbounded integrands. *J. Complexity*, 26(2):135–160, 2010. [337]
162. F. Y. Kuo, I. H. Sloan, and H. Woźniakowski. Lattice rules for multivariate approximation in the worst case setting. In H. Niederreiter and D. Talay, editors, *Monte Carlo and Quasi-Monte Carlo Methods 2004*, pages 289–330. Springer, Berlin, 2006. [415, 435, 455, 469]
163. F. Y. Kuo, I. H. Sloan, and H. Woźniakowski. Lattice rule algorithms for multivariate approximation in the average case setting. *J. Complexity*, 24(2):283–323, 2008. [456, 462, 466, 469, 487]
164. F. Y. Kuo, G. W. Wasilkowski, and B. J. Waterhouse. Randomly shifted lattice rules for unbounded integrands. *J. Complexity*, 22(5):630–651, 2006. [337]
165. F. Y. Kuo, G. W. Wasilkowski, and H. Woźniakowski. Multivariate L_∞-approximation in the worst case setting over reproducing kernel Hilbert spaces. *J. Approx. Theory*, 152(2):135–160, 2008. [457]
166. F. Y. Kuo, G. W. Wasilkowski, and H. Woźniakowski. Lattice algorithms for multivariate l_∞ approximation in the worst-case setting. *Constr. Approx.*, 30(3):475–493, 2009. [459, 465, 466, 469, 487]
167. F. Y. Kuo, G. W. Wasilkowski, and H. Woźniakowski. On the power of standard information for multivariate approximation in the worst case setting. *J. Approx. Theory*, 158(1):97–125, 2009. [457]
168. F. Y. Kuo, G. W. Wasilkowski, and H. Woźniakowski. Correction to: Lattice algorithms for multivariate L_∞ approximation in the worst-case setting. *Constr. Approx.*, 52(1):177–179, 2020. [487]
169. H. Laimer. On combined component-by-component constructions of lattice point sets. *J. Complexity*, 38:22–30, 2017. [193]
170. G. Larcher. On the distribution of sequences connected with good lattice points. *Monatsh. Math.*, 101(2):135–150, 1986. [201]
171. G. Larcher. A best lower bound for good lattice points. *Monatsh. Math.*, 104(1):45–51, 1987. [201]

172. G. Larcher. On the distribution of the multiples of an s-tuple of real numbers. *J. Number Theory*, 31(3):367–372, 1989. [219]
173. G. Larcher and G. Leobacher. Quasi-Monte Carlo and Monte Carlo methods and their application in finance. *Surv. Math. Ind.*, 11(1–4):95–130, 2005. [52, 337]
174. G. Larcher and H. Niederreiter. Optimal coefficients modulo prime powers in the three-dimensional case. *Ann. Mat. Pura Appl. (4)*, 155:299–315, 1989. [46]
175. G. Larcher and F. Pillichshammer. Walsh series analysis of the L_2-discrepancy of symmetrisized point sets. *Monatsh. Math.*, 132(1):1–18, 2001. [317]
176. G. Larcher and F. Pillichshammer. On the L_2-discrepancy of the Sobol-Hammersley net in dimension 3. *J. Complexity*, 18(2):415–448, 2002. [317]
177. P. L'Ecuyer. Quasi-Monte Carlo methods with applications in finance. *Finance Stoch.*, 13(3):307–349, 2009. [52]
178. Ch. Lemieux. *Monte Carlo and Quasi-Monte Carlo Sampling*. Springer Series in Statistics. Springer, New York, NY, 2008. [3, 52]
179. G. Leobacher and F. Pillichshammer. Bounds for the weighted L^p discrepancy and tractability of integration. *J. Complexity*, 19(4):529–547, 2003. [219, 283, 317]
180. G. Leobacher and F. Pillichshammer. *Introduction to Quasi-Monte Carlo Integration and Applications*. Birkhäuser, Cham, 2014. [52, 53, 138]
181. W. J. LeVeque. *Fundamentals of Number Theory*. Dover Publications, Inc., Mineola, NY, 1996. Reprint of the 1977 original. [241]
182. D. Li and F. J. Hickernell. Trigonometric spectral collocation methods on lattices. In S. Y. Cheng, C.-W. Shu, and T. Tang, editors, *Recent Advances in Scientific Computing and Partial Differential Equations*, pages 121–132. American Mathematical Society, Providence, RI, 2003. [455]
183. R. Lidl and H. Niederreiter. *Finite Fields*, volume 20 of *Encyclopedia of Mathematics and its Applications*. Cambridge University Press, Cambridge, second edition, 1997. [366]
184. Y. Liu and G. Xu. A note on tractability of multivariate analytic problems. *J. Complexity*, 34:42–49, 2016. [361]
185. J. N. Lyness. Notes on lattice rules. *J. Complexity*, 19(3):321–331, 2003. [54]
186. J. N. Lyness and T. Sørevik. A search program for finding optimal integration lattices. *Computing*, 47(2):103–120, 1991. [138]
187. D. Maisonneuve. Recherche et utilisation des "bons treillis". Programmation et résultats numériques. In S. K. Zaremba, editor, *Applications of Number Theory to Numerical Analysis (Proc. Sympos., Univ. Montréal, Montreal, Que., 1971)*, pages 121–201. 1972. (In French). [138]
188. J. Matoušek. *Geometric Discrepancy*. Springer, Berlin, 1999. [53]
189. Th. Müller-Gronbach, E. Novak, and K. Ritter. *Monte Carlo-Algorithmen*. Springer-Lehrbuch. [Springer Textbook]. Springer, Heidelberg, 2012. (In German). [52]
190. R. C. Mullin. Classroom Notes: Some Trigonometric Products. *Amer. Math. Monthly*, 69(3):217–218, 1962. [185]
191. J. A. Nichols and F. Y. Kuo. Fast CBC construction of randomly shifted lattice rules achieving $O(n^{-1+\delta})$ convergence for unbounded integrands over $\mathbb{R}^s$ in weighted spaces with POD weights. *J. Complexity*, 30(4):444–468, 2014. [324, 337]
192. H. Niederreiter. Application of Diophantine approximations to numerical integration. In C. F. Osgood, editor, *Diophantine Approximation and its Applications (Proc. Conf., Washington, D.C., 1972)*, pages 129–199, 1973. [317]
193. H. Niederreiter. Pseudo-random numbers and optimal coefficients. *Adv. in Math.*, 26(2):99–181, 1977. [43, 195]
194. H. Niederreiter. Existence of good lattice points in the sense of Hlawka. *Monatsh. Math.*, 86(3):203–219, 1978. [200]
195. H. Niederreiter. Quasi-Monte Carlo methods and pseudo-random numbers. *Bull. Amer. Math. Soc.*, 84(6):957–1041, 1978. [52, 53, 219, 316]
196. H. Niederreiter. Dyadic fractions with small partial quotients. *Monatsh. Math.*, 101(4):309–315, 1986. [45]

197. H. Niederreiter. Point sets and sequences with small discrepancy. *Monatsh. Math.*, 104(4):273–337, 1987. [52]
198. H. Niederreiter. Quasi-Monte Carlo methods for multidimensional numerical integration. In H. Braß and G. Hämmerlin, editors, *Numerical integration III (Proceedings of the Conference held at the Mathematisches Forschungsinstitut, Oberwolfach, Nov. 8–14, 1987)*, volume 85 of *Internat. Schriftenreihe Numer. Math.*, pages 157–171. Birkhäuser, Basel, 1988. [52]
199. H. Niederreiter. *Random Number Generation and Quasi-Monte Carlo Methods*, volume 63 of *CBMS-NSF Series in Applied Mathematics*. SIAM, Philadelphia, PA, 1992. [3, 10, 15, 18, 32, 33, 42, 43, 44, 52, 53, 138, 201, 202, 218, 219, 316, 400]
200. H. Niederreiter. Improved error bounds for lattice rules. *J. Complexity*, 9(1):60–75, 1993. [42]
201. H. Niederreiter and F. Pillichshammer. Construction algorithms for good extensible lattice rules. *Constr. Approx.*, 30(3):361–393, 2009. [52, 264]
202. H. Niederreiter and J. M. Wills. Diskrepanz und Distanz von Maßen bezüglich konvexer und Jordanscher Mengen. *Math. Z.*, 144(2):125–134, 1975. (In German). [219]
203. H. Niederreiter and J. M. Wills. Berichtigung zu der Arbeit "Diskrepanz und Distanz von Massen bezüglich konvexer und Jordanscher Mengen" (Math. Z. **144** (1975), no. 2, 125–134). *Math. Z.*, 148(1):99, 1976. (In German). [219]
204. H. Niederreiter and A. Winterhof. *Applied Number Theory*. Springer, Cham, 2015. [10, 138, 218]
205. E. Novak. *Deterministic and Stochastic Error Bounds in Numerical Aanalysis*, volume 1349 of *Lecture Notes in Mathematics*. Springer Verlag, Berlin, 1988. [392]
206. E. Novak. Numerische Verfahren für hochdimensionale Probleme und der Fluch der Dimension. *Jahresber. Deutsch. Math.-Verein.*, 101:151–177, 1999. (In German). [53]
207. E. Novak. Some results on the complexity of numerical integration. In R. Cools and D. Nuyens, editors, *Monte Carlo and Quasi-Monte Carlo Methods*, pages 161–183. Springer, Cham, 2016. [53, 433]
208. E. Novak, I. H. Sloan, J. F. Traub, and H. Woźniakowski. *Essays on the Complexity of Continuous Problems*. European Mathematical Society Publishing House, Zürich, 2009. [53]
209. E. Novak, I. H. Sloan, and H. Woźniakowski. Tractability of approximation for weighted Korobov spaces on classical and quantum computers. *Found. Comput. Math.*, 4(2):121–156, 2004. [435, 455]
210. E. Novak and H. Woźniakowski. *Tractability of Multivariate Problems. Volume I: Linear Information*. European Mathematical Society Publishing House, Zürich, 2008. [36, 53, 60, 92, 316, 393, 433, 434, 455, 487]
211. E. Novak and H. Woźniakowski. *Tractability of Multivariate Problems. Volume II: Standard Information for Functionals*. European Mathematical Society Publishing House, Zürich, 2010. [53, 73, 84, 85, 93, 205, 219, 283, 316, 393]
212. E. Novak and H. Woźniakowski. *Tractability of Multivariate Problems. Volume III: Standard Information for Operators*. European Mathematical Society Publishing House, Zürich, 2012. [23, 53, 455, 487]
213. D. Nuyens and R. Cools. Fast algorithms for component-by-component construction of rank-1 lattice rules in shift-invariant reproducing kernel Hilbert spaces. *Math. Comp.*, 75(254):903–920, 2006. [115, 138]
214. D. Nuyens and R. Cools. Fast component-by-component construction, a reprise for different kernels. In H. Niederreiter and D. Talay, editors, *Monte Carlo and Quasi-Monte Carlo Methods 2004*, pages 373–387. Springer, Berlin, 2006. [115, 138]
215. D. Nuyens and R. Cools. Fast component-by-component construction of rank-1 lattice rules with a non-prime number of points. *J. Complexity*, 22(1):4–28, 2006. [115, 138, 139, 241, 243, 263, 511]
216. A. B. Owen. A constraint on extensible quadrature rules. *Numer. Math.*, 132(3):511–518, 2016. [235]
217. F. Pillichshammer. A lower bound for rank 2 lattice rules. *Math. Comp.*, 73(246):853–860, 2004. [219]

218. F. Pillichshammer. Tractability properties of the weighted star discrepancy of regular grids. *J. Complexity*, 46:103–112, 2018. [204]
219. F. Pillichshammer. A note on Korobov lattice rules for integration of analytic functions. *J. Complexity*, 63:Paper No. 101524, 8 pp., 2021. [361]
220. F. Pillichshammer and M. Sonnleitner. A note on isotropic discrepancy and spectral test of lattice point sets. *J. Complexity*, 58:Paper No. 101441, 7 pp., 2020. [219]
221. J. B. Rosser and L. Schoenfeld. Approximate formulas for some functions of prime numbers. *Illinois J. Math.*, 6:64–94, 1962. [385, 389, 495]
222. K. F. Roth. On irregularities of distribution. *Mathematika*, 1:73–79, 1954. [29, 53]
223. A. I. Saltykov. Tables for evaluating multiple integrals by the method of optimal coefficients. *Ž. Vyčisl. Mat i Mat. Fiz.*, 3:181–186, 1963. (In Russian). [138]
224. W. M. Schmidt. Irregularities of distribution VII. *Acta Arith.*, 21:45–50, 1972. [29]
225. W. M. Schmidt. Irregularities of distribution. IX. *Acta Arith.*, 27:385–396, 1975. [213]
226. W. M. Schmidt. Irregularities of distribution. X. In *Number Theory and Algebra*, pages 311–329. Academic Press, New York, 1977. [29]
227. V. Sinescu. Shifted lattice rules based on a general weighted discrepancy for integrals over Euclidean space. *J. Comput. Appl. Math.*, 232(2):240–251, 2009. [219]
228. V. Sinescu and S. Joe. Good lattice rules based on the general weighted star discrepancy. *Math. Comp.*, 76(258):989–1004, 2007. [131, 219]
229. V. Sinescu and S. Joe. Good lattice rules with a composite number of points based on the product weighted star discrepancy. In A. Keller, S. Heinrich, and H. Niederreiter, editors, *Monte Carlo and Quasi-Monte Carlo Methods 2006*, pages 645–658. Springer, Berlin, 2008. [203, 219]
230. I. H. Sloan and S. Joe. *Lattice Methods for Multiple Integration*. Oxford Univ. Press, New York and Oxford, 1994. [3, 10, 11, 12, 18, 43, 52, 53, 96, 138, 316]
231. I. H. Sloan and P. J. Kachoyan. Lattice methods for multiple integration: theory, error analysis and examples. *SIAM J. Numer. Anal.*, 24(1):116–128, 1987. [7, 16]
232. I. H. Sloan, F. Y. Kuo, and S. Joe. On the step-by-step construction of quasi-Monte Carlo integration rules that achieve strong tractability error bounds in weighted Sobolev spaces. *Math. Comp.*, 71(240):1609–1640, 2002. [276]
233. I. H. Sloan and J. N. Lyness. The representation of lattice quadrature rules as multiple sums. *Math. Comp.*, 52(185):81–94, 1989. [9, 10, 11]
234. I. H. Sloan and J. N. Lyness. Lattice rules: projection regularity and unique representations. *Math. Comp.*, 54(190):649–660, 1990. [10, 11]
235. I. H. Sloan and A. V. Reztsov. Component-by-component construction of good lattice rules. *Math. Comp.*, 71(237):263–273, 2002. [100]
236. I. H. Sloan and L. Walsh. Lattice rules—classification and searches. In H. Braß and G. Hämmerlin, editors, *Numerical integration III (Proceedings of the Conference held at the Mathematisches Forschungsinstitut, Oberwolfach, Nov. 8–14, 1987)*, volume 85 of *Internat. Schriftenreihe Numer. Math.*, pages 251–260. Birkhäuser, Basel, 1988. [96]
237. I. H. Sloan and L. Walsh. A computer search of rank-2 lattice rules for multidimensional quadrature. *Math. Comp.*, 54(189):281–302, 1990. [96]
238. I. H. Sloan and H. Woźniakowski. An intractability result for multiple integration. *Math. Comp.*, 66(219):1119–1124, 1997. [40, 345]
239. I. H. Sloan and H. Woźniakowski. When are quasi-Monte Carlo algorithms efficient for high-dimensional integrals? *J. Complexity*, 14(1):1–33, 1998. [53, 75, 78, 203, 205, 219, 283, 317]
240. I. H. Sloan and H. Woźniakowski. Tractability of multivariate integration for weighted Korobov classes. *J. Complexity*, 17(4):697–721, 2001. [92, 393]
241. I. M. Sobol'. On quasi-Monte Carlo integrations. *Math. Comput. Simulation*, 47(2–5):103–112, 1998. [235]
242. M. Sonnleitner. *The power of random information for numerical approximation and integration*. PhD thesis, University of Passau, 2022. [214, 219]

243. M. Sonnleitner and F. Pillichshammer. On the relation of the spectral test to isotropic discrepancy and L_q-approximation in Sobolev spaces. *J. Complexity*, 67:Paper No. 101576, 9 pp., 2021. [217, 219]
244. E. M. Stein and G. Weiss. *Introduction to Fourier analysis on Euclidean spaces*. Princeton University Press, Princeton, N.J., 1971. Princeton Mathematical Series, No. 32. [373]
245. G. Strang and G. J. Fix. *An Analysis of the Finite Element Method*. Prentice-Hall, Inc., Englewood Cliffs, NJ, 1973. Prentice-Hall Series in Automatic Computation. [531]
246. W. Stute. Convergence rates for the isotrope discrepancy. *Ann. Probability*, 5(5):707–723, 1977. [213]
247. V. N. Temlyakov. Approximate recovery of periodic functions of several variables. *Mat. Sbornik*, 128:256–268, 1985. (In Russian; English translation in Math. USSR Sbornik 56, 249–261, 1987). [455]
248. V. N. Temlyakov. Error estimates for quadrature formulas for classes of functions with a bounded mixed derivative. *Mat. Zametki*, 46(2):128–134, 160, 1989. (In Russian; English translation in Math. Notes 46, no. 1-2, 663–668, 1990). [54]
249. V. N. Temlyakov. On approximate recovery of functions with bounded mixed derivative. *J. Complexity*, 9(1):41–59, 1993. [455]
250. V. N. Temlyakov. Cubature formulas, discrepancy, and nonlinear approximation. *J. Complexity*, 19(3):352–391, 2003. [93]
251. J. F. Traub, G. W. Wasilkowski, and H. Woźniakowski. *Information-Based Complexity*. Computer Science and Scientific Computing. Academic Press, Inc., Boston, MA, 1988. [53]
252. M. Ullrich. On "Upper error bounds for quadrature formulas on function classes" by K. K. Frolov. In R. Cools and D. Nuyens, editors, *Monte Carlo and Quasi-Monte Carlo Methods*, pages 571–582. Springer, Cham, 2016. [393]
253. M. Ullrich. A Monte Carlo method for integration of multivariate smooth functions. *SIAM J. Numer. Anal.*, 55(3):1188–1200, 2017. [389, 393]
254. G. Wahba. *Spline Models for Observational Data*, volume 59 of *CBMS-NSF Regional Conference Series in Applied Mathematics*. SIAM, Philadelphia, PA, 1990. [487]
255. J. Waldvogel. Towards a general error theory of the trapezoidal rule. In W. Gautschi, G. Mastroianni, and T. M. Rassias, editors, *Approximation and Computation*, volume 42 of *Springer Optim. Appl.*, pages 267–282. Springer, New York, NY, 2011. [349]
256. H. Wang. A note about EC-(s,t)-weak tractability of multivariate approximation with analytic Korobov kernels. *J. Complexity*, 55:101412, 19, 2019. [361]
257. X. Wang and I. H. Sloan. Why are high-dimensional finance problems often of low effective dimension? *SIAM J. Sci. Comput.*, 27(1):159–183, 2005. [337]
258. X. Wang and I. H. Sloan. Quasi-Monte Carlo methods in financial engineering: an equivalence principle and dimension reduction. *Oper. Res.*, 59(1):80–95, 2011. [337]
259. X. Wang, I. H. Sloan, and J. Dick. On Korobov lattice rules in weighted spaces. *SIAM J. Numer. Anal.*, 42(4):1760–1779, 2004. [97, 99]
260. T. T. Warnock. Computational investigations of low discrepancy point sets. In S. K. Zaremba, editor, *Applications of Number Theory to Numerical Analysis (Proc. Sympos., Univ. Montréal, Montreal, Que., 1971)*, pages 319–343. Academic Press, 1972. [316]
261. G. W. Wasilkowski and H. Woźniakowski. Weigthed tensor product algorithms for linear multivariate problems. *J. Complexity.*, 15(3):402–447, 1999. [435]
262. A. Weil. On some exponential sums. *Proc. Nat. Acad. Sci. U. S. A.*, 34:204–207, 1948. [366]
263. H. Wendland. *Scattered Data Approximation*. Cambridge University Press, Cambridge, 2004. [487]
264. H. Weyl. Über die Gleichverteilung von Zahlen mod. Eins. *Math. Ann.*, 77(3):313–352, 1916. (In German). [26]
265. G. Xu. EC-tractability of $\mathbb{L}_p$-approximation in Korobov spaces with exponential weights. *J. Approx. Theory*, 249:Paper No. 105309, 20 pp., 2020. [361, 488]
266. S. K. Zaremba. Good lattice points, discrepancy, and numerical integration. *Ann. Mat. Pura Appl. (4)*, 73:293–317, 1966. [44]
267. S. K. Zaremba. La discrépance isotrope et l'intégration numérique. *Ann. Mat. Pura Appl. (4)*, 87:125–136, 1970. (In French). [219]

268. S. K. Zaremba. La méthode des “bons treillis” pour le calcul des intégrales multiples. In S. K. Zaremba, editor, *Applications of Number Theory to Numerical Analysis (Proc. Sympos., Univ. Montreal, Montreal, Que., 1971)*, pages 39–119, 1972. (In French). [43, 44, 52]
269. S. K. Zaremba. Good lattice points modulo composite numbers. *Monatsh. Math.*, 78:446–460, 1974. [44]
270. X. Y. Zeng, P. Kritzer, and F. J. Hickernell. Spline methods using integration lattices and digital nets. *Constr. Approx.*, 30(3):529–555, 2009. [457, 487]
271. X. Y. Zeng, K. T. Leung, and F. J. Hickernell. Error analysis of splines for periodic problems using lattice designs. In H. Niederreiter and D. Talay, editors, *Monte Carlo and Quasi-Monte Carlo Methods 2004*, pages 501–514. Springer, Berlin, 2006. [455]

Index

J. Dick et al., *Lattice Rules*, Springer Series in Computational Mathematics 58, https://doi.org/10.1007/978-3-031-09951-9